Geophysical Monograph Series

Including

IUGG Volumes
Maurice Ewing Volumes
Mineral Physics Volumes

Geophysical Monograph Series

86 Space Plasmas: Coupling Between Small and Medium Scale Processes *Maha Ashour-Abdalla, Tom Chang, and Paul Dusenbery (Eds.)*

87 The Upper Mesosphere and Lower Thermosphere: A Review of Experiment and Theory *R. M. Johnson and T. L. Killeen (Eds.)*

88 Active Margins and Marginal Basins of the Western Pacific *Brian Taylor and James Natland (Eds.)*

89 Natural and Anthropogenic Influences in Fluvial Geomorphology *John E. Costa, Andrew J. Miller, Kenneth W. Potter, and Peter R. Wilcock (Eds.)*

90 Physics of the Magnetopause *Paul Song, B.U.Ö. Sonnerup, and M.F. Thomsen (Eds.)*

91 Seafloor Hydrothermal Systems: Physical, Chemical, Biological, and Geological Interactions *Susan E. Humphris, Robert A. Zierenberg, Lauren S. Mullineaux, and Richard E. Thomson (Eds.)*

92 Mauna Loa Revealed: Structure, Composition, History, and Hazards *J. M. Rhodes and John P. Lockwood (Eds.)*

93 Cross-Scale Coupling in Space Plasmas *James L. Horowitz, Nagendra Singh, and James L. Burch (Eds.)*

94 Double-Diffusive Convection *Alan Brandt and H. J. S. Fernando (Eds.)*

95 Earth Processes: Reading the Isotopic Code *Asish Basu and Stan Hart (Eds.)*

96 Subduction Top to Bottom *Gray E. Bebout, David Scholl, Stephen Kirby, and John Platt (Eds.)*

97 Radiation Belts: Models and Standards *J. F. Lemaire, D. Heynderickx, and D. N. Baker (Eds.)*

98 Magnetic Storms Bruce *T. Tsurutani, Walter D. Gonzalez, Yohsuke Kamide, and John K. Arballo (Eds.)*

99 Coronal Mass Ejections *Nancy Crooker, Jo Ann Joselyn, and Joan Feynman (Eds.)*

100 Large Igneous Provinces *John J. Mahoney and Millard F. Coffin (Eds.)*

101 Properties of Earth and Planetary Materials at High Pressure and Temperature *Murli Manghnani and Takehiki Yagi (Eds.)*

102 Measurement Techniques in Space Plasmas: Particles *Robert F. Pfaff, Joseph E. Borovsky, and David T. Young (Eds.)*

103 Measurement Techniques in Space Plasmas: Fields *Robert F. Pfaff, Joseph E. Borovsky, and David T. Young (Eds.)*

104 Geospace Mass and Energy Flow: Results From the International Solar-Terrestrial Physics Program *James L. Horwitz, Dennis L. Gallagher, and William K. Peterson (Eds.)*

105 New Perspectives on the Earth's Magnetotail *A. Nishida, D. N. Baker, and S. W. H. Cowley (Eds.)*

106 Faulting and Magmatism at Mid-Ocean Ridges *W. Roger Buck, Paul T. Delaney, Jeffrey A. Karson, and Yves Lagabrielle (Eds.)*

107 Rivers Over Rock: Fluvial Processes in Bedrock Channels *Keith J. Tinkler and Ellen E. Wohl (Eds.)*

108 Assessment of Non-Point Source Pollution in the Vadose Zone *Dennis L. Corwin, Keith Loague, and Timothy R. Ellsworth (Eds.)*

109 Sun-Earth Plasma Interactions *J. L. Burch, R. L. Carovillano, and S. K. Antiochos (Eds.)*

110 The Controlled Flood in Grand Canyon *Robert H. Webb, John C. Schmidt, G. Richard Marzolf, and Richard A. Valdez (Eds.)*

111 Magnetic Helicity in Space and Laboratory Plasmas *Michael R. Brown, Richard C. Canfield, and Alexei A. Pevtsov (Eds.)*

112 Mechanisms of Global Climate Change at Millennial Time Scales *Peter U. Clark, Robert S. Webb, and Lloyd D. Keigwin (Eds.)*

113 Faults and Subsurface Fluid Flow in the Shallow Crust *William C. Haneberg, Peter S. Mozley, J. Casey Moore, and Laurel B. Goodwin (Eds.)*

114 Inverse Methods in Global Biogeochemical Cycles *Prasad Kasibhatla, Martin Heimann, Peter Rayner, Natalie Mahowald, Ronald G. Prinn, and Dana E. Hartley (Eds.)*

115 Atlantic Rifts and Continental Margins *Webster Mohriak and Manik Talwani (Eds.)*

116 Remote Sensing of Active Volcanism *Peter J. Mouginis-Mark, Joy A. Crisp, and Jonathan H. Fink (Eds.)*

117 Earth's Deep Interior: Mineral Physics and Tomography From the Atomic to the Global Scale *Shun-ichiro Karato, Alessandro Forte, Robert Liebermann, Guy Masters, Lars Stixrude (Eds.)*

118 Magnetospheric Current Systems *Shin-ichi Ohtani, Ryoichi Fujii, Michael Hesse, and Robert L. Lysak (Eds.)*

119 Radio Astronomy at Long Wavelengths *Robert G. Stone, Kurt W. Weiler, Melvyn L. Goldstein, and Jean-Louis Bougeret (Eds.)*

120 GeoComplexity and the Physics of Earthquakes *John B. Rundle, Donald L. Turcotte, and William Klein (Eds.)*

121 The History and Dynamics of Global Plate Motions *Mark A. Richards, Richard G. Gordon, Rob D. van der Hilst (Eds.)*

Geophysical Monograph 122

Dynamics of Fluids in Fractured Rock

Boris Faybishenko
Paul A. Witherspoon
Sally M. Benson
Editors

American Geophysical Union
Washington, DC

Published under the aegis of the AGU Books Board

Library of Congress Cataloging-in-Publication Data

Dynamics of fluids in fractured rock / Boris Faybishenko, Paul A. Witherspoon, Sally M. Benson, editors

p.cm.--(Geophysical monograph; 122)

Papers selected from a symposium held at Ernest Orlando Lawrence Berkeley National Laboratory on February 10-12, 1999.

Includes bibliographical references.

ISBN 0-87590-980-9

1. Groundwater flow--simulation methods--Congresses. 2. Rocks--Permeability--Simulation methods--Congresses. I. Faybishenko, Boris. II. Witherspoon, Paul Adams, 1919-. III. Benson, Sally. IV. Series

GB1197.7.D92 2000
551.49--dc21 00-045106

ISBN 0-87590-980-9
ISSN 0065-8448

2000 Florida Avenue, N.W.
Washington, DC 20009

Cover photograph by Larry R. Myer of LBNL at Mesa Verde National Park, Colorado

Printed in the United States of America.

CONTENTS

CONTENTS

CONTENTS

Geothermal Reservoirs

Remediation in Fractured Systems

Paul A. Witherspoon

PREFACE

Among the current problems that hydrogeologists face, perhaps there is none as challenging as the characterization of fractured rock. Within hydrogeological systems, general issues concerning groundwater flow and environmental remediation cannot be resolved in any practical manner prior to investigating the nature and vagaries of the fracture networks themselves. Comparable difficulties arise when developing economic programs for the exploitation of oil, gas, and geothermal reservoirs in fractured rock. Equal, if not greater, difficulties have commanded our attention relatively recently in regard to the storing of spent fuel generated by nuclear power plants. For example, if we are to isolate spent nuclear fuel in underground rock systems, we must construct a repository to protect the biosphere from contamination by radioactivity while subjecting the total rock system to a significant thermal field for many thousands of years. Predicting the behavior of a waste repository under such conditions, especially in fractured rock, is a formidable task.

Of course, fractured rock exhibits a unique feature: flow and transport processes within discontinuities are drastically different from those in the porous matrix. In fact, contrasts in hydraulic conductivity between fracture and matrix can be extreme and localized. In one case, the permeability of a fracture may be many orders of magnitude higher than that of the matrix; in another, the fracture may be completely sealed with different kinds of plugging material. Thus, when examining the interconnected network of several sets of fractures, we must be able to determine the percentage of fractures in each set that is conductive, the degree of conductivity, and the effect such conductivity has on the anisotropy of the rock mass.

Another major problem in characterizing a fractured rock system concerns the interaction between the fracture and the matrix. In working with unsaturated systems, determining the specific data needed to ascertain the division of fluid flow between matrix and fracture is a key factor. Because of the difficulty in obtaining direct measurements in situ of the fracture and matrix components of flow, determining what data are specific to the phenomenon presents its own set of problems. Another complication concerns the heterogeneity of the flow paths that affect both saturated and unsaturated systems. There is an increasing amount of field and laboratory data concerning the complicated flow paths that can exist.

Faced with the complexity of such issues, we organized an international symposium on "Dynamics of Fluids in Fractured Rocks: Concepts and Recent Advances." The symposium was held at Ernest Orlando Lawrence Berkeley National Laboratory on February 10-12, 1999. For the event, we brought together an international group of scientists and engineers from the different Earth science fields to discuss the various problems in these fields and to exchange ideas on new approaches for research on fractured rock. This monograph contains 26 papers selected from over 100 papers submitted by the more than 200 symposium participants.

The symposium was organized to celebrate the 80th birthday of Paul A. Witherspoon, who initiated early investigations on flow and transport in fractured rock at the University of California, Berkeley, and at Lawrence Berkeley National Laboratory. Dr. Witherspoon has played a key role at these institutions in developing research on basic concepts, modeling, and investigations in the laboratory and field on fluid flow and contaminant transport in fractured rock systems. In fact, the first paper in this monograph chronicles the evolution of ideas on the dynamics of flow in fractured rock over a 35-year period by Paul Witherspoon, his students, and his scientific colleagues. Succeeding papers cover a wide variety of topics that are organized in sections under the following headings: Field Testing, Laboratory Testing, Fracture Network Models, Modeling Flow and Transport, Isotopic Methods, Geothermal Reservoirs, and Remediation in Fractured Systems.

The editorial board would like to express its gratitude to the scientists who contributed to the monograph. We are grateful for the considerable time and effort these authors took in preparing their contributions and for their generosity in sharing their expertise with other researchers in the Earth sciences.

We thank the U.S. Department of Energy (Oakland Operations Office, Office of Environmental Management, Office of Science and Technology, Subsurface Contaminants Focus Area, Office of Civilian Radioactive Waste Management), Lawrence Berkeley National Laboratory, Idaho National Engineering and Environmental Laboratory, the U.S. Nuclear Regulatory Commission, the U.S. Geological Survey, and the American Institute of Hydrology for their support for the symposium and for the preparation of this monograph. We also thank Julie McCullough and Daniel Hawkes for their careful editing of the monograph and preparation of the figures, and Roy Kaltschmidt for the photograph of Paul Witherspoon.

Finally, we would like to thank Paul A. Witherspoon for the many decades of insight, enthusiasm, friendship, and encouragement he has given us. We also thank Paul for giving us this wonderful opportunity to grapple with the challenge of understanding the dynamics of flow in fractured rocks in the company of so many fine scientists.

Boris Faybishenko
Sally M. Benson
Lawrence Berkeley National Laboratory

Editorial Board

Investigations at Berkeley on Fracture Flow in Rocks: From the Parallel Plate Model to Chaotic Systems

Paul A. Witherspoon

Earth Sciences Division, Ernest Orlando Lawrence Berkeley National Laboratory and Department of Material Sciences and Mineral Engineering, University of California, Berkeley

This is a review of research at Berkeley over the past 35 years on characterization of fractured rocks and their hydrologic behavior when subjected to perturbations of various kinds. The parallel plate concept was useful as a first approach, but researchers have found that it has limitations when used to examine rough fractures and understand effects of aperture distributions on heterogeneous flow paths, especially when the fracture is deformed under stress. Results of investigations have been applied to fractured and faulted geothermal systems, where the inherent, nonisothermal conditions produce a different kind of perturbation. In 1977, the Stripa project in Sweden provided an unusual underground laboratory excavated in granite where new methods of investigating fractured rock were developed. New theoretical studies have been carried out on the fundamental role of heterogeneous flow paths in controlling fluid migration in fractured rocks. A major field study is now underway at the Yucca Mountain Project in Nevada, where a site for a radioactive waste repository may be constructed. The main effort has been to characterize the rock mass (fractured tuff) in sufficient detail so that a site scale model can be constructed and used to simulate operation of the repository. A new and entirely different problem has been identified through infiltration tests in the fractured basalt layers of the Eastern Snake River Plane in Idaho. Water flow through the unusual heterogeneities of these layers is so erratic that a model based on a hierarchy of scales is being investigated.

1. INTRODUCTION

This work reviews the research activities that have been carried out on fractured rocks over the past 35 years by a large number of investigators at Ernest Orlando Lawrence Berkeley National Laboratory (Berkeley Lab) and the University of California, Berkeley (collectively referred to as "Berkeley"). The principal directions of this work have been characterizing fractured rocks and investigating their behavior when subjected to perturbations of various kinds. The research has been focused primarily on the hydrologic factors that affect the flow of fluids through such rocks, but other aspects such as the location of fractures and their thermal and mechanical behaviors have also been investigated. In a review of this kind, where the range of investigation is quite large and spans such a long period of time, only the highlights of the results can be presented here.

Dynamics of Fluids in Fractured Rock
Geophysical Monograph 122

Published 2000 by the American Geophysical Union

2. EARLY INVESTIGATIONS

Early investigations were based on the idea that the parallel-plate concept could be used to develop models for flow of fluids in single fractures or networks of fractures. *Snow* [1965] was one of the early workers at Berkeley who used this concept to simulate real fractures. With this idealization, he treated flow along intersecting fractures as being proportional to the cube of the apertures and to the projection of a field gradient generally parallel to neither fracture. The discharge of one planar conductor or any set of conductors can be represented by a second-rank tensor. A tensor therefore describes the permeability of a continuous rock medium with the same discharge as a fractured medium under the same hydraulic gradient and laminar flow conditions. *Snow* [1969] proposed a method of determining principal permeabilities using an arrangement of three orthogonal drill holes from which information could be obtained using water-injection tests.

With this model, *Snow* [1965, 1969] was able to examine the effect of random variations in orientation and aperture of extensive fractures on the permeability of the rock mass. As each fracture was randomly generated, the total permeability tensor progressively accumulated. *Snow* found that an increase in sample size increased the geometric mean permeability. Most of his aperture distributions were for sample sizes of about 200 fractures.

Wilson [1970] extended the use of the parallel-plate concept to conducting channels that form networks of arbitrary geometry. The fractures may have any spacing and orientation, any aperture, and may intersect at any angle. He used two types of fracture elements in his modeling: (a) the triangular element, which allows simultaneous computation of porous media flow within the matrix of the rock blocks; and (b) the line element, which is used only when matrix permeability is sufficiently low to be ignored. *Wilson* [1970] demonstrated the applicability of his mathematical model in several problems involving seepage beneath dams and the effect of tunnel diameter in a given network of fractures on seepage into the tunnel. He also carried out a laboratory investigation of the magnitude of intersection resistance for flow through pipes [*Wilson and Witherspoon*, 1976]. Under conditions of laminar flow, which is generally the case in fracture networks, interference effects at intersections become negligible.

A method of analyzing pump test data from fractured aquifers was developed by *Gringarten and Witherspoon* [1972]. When fractures are intersected by the pumping well, the nonsteady behavior of the aquifer differs in significant ways from that predicted by the traditional Theis solution. This has been demonstrated by analytical solutions for flow to a well that intersects a single horizontal fracture [*Gringarten and Ramey*, 1974] or a single vertical fracture [*Gringarten et al.*, 1974] while the well produces at a constant rate. An important finding from this work was that for both solutions, a log-log plot of fluid-level drawdowns versus time yields a characteristic straight line with a slope of 0.5 at early times, indicating linear flow. At later times, the drawdown behavior is the same as that given by the Theis solution shifted by a constant that depends on the point where measurements are being made.

Type curves were developed from the analytical solutions of *Gringarten and Ramey* [1974] and *Gringarten et al.* [1974]. With the aid of these type curves, it is possible to distinguish between aquifers with fractures of either orientation and to analyze the system as an "equivalent" anisotropic homogeneous porous medium with a single fracture of much higher permeability. It can also be shown that the use of the Theis solution in analyzing such systems can lead to significant errors.

3. FLOW IN DEFORMABLE FRACTURES

Noorishad [1971] made the first numerical investigation at Berkeley of fluid flow in fractured rock masses that considered the coupled effects of fluid forces, body forces, and boundary loads. He used an iterative procedure to solve two equations governing the force displacement and flow behavior aspects for fractured rock systems. These equations were obtained by combining two existing finite element formulations: (a) one for stress analysis [*Goodman et al.*, 1968] and (b) the other for fracture flow [*Wilson*, 1970]. Through an iterative process, the stress analysis was carried out to determine the deformations taking place in the fracture network. This was followed by a flow analysis in the deformed rock mass that produced a pressure distribution compatible with the state of stress within the deformable rock body [*Noorishad et al.*, 1971].

In the late 1960s, field evidence from observations of fractured rock systems suggested that, because of an increase in fluid pressure, the reduction of effective stress across discontinuities had contributed to the failure of several important engineering structures [*Lane*, 1969]. Also, a great deal of interest had been generated by a report [*Evans*, 1966] that the injection of water into a deep well at the Rocky Mountain Arsenal near Denver, Colorado, seemed to be related to earthquake activity in the region. *Evans* [1966] reported that drawdown in the Arsenal well was rapid during fluid withdrawal (increase in effective stress) and that the fractured metamorphic rock accepted large flow rates during injection (decrease in effective stress). *Lane* [1969] suggested that injection in the Arsenal well produced fluid pressure increases in the fractured reservoir that changed the state of stress in the rock mass and triggered the release of tectonic stress, thereby causing the earthquakes.

It was decided to approach these fluid pressure problems in a different manner. Another code was developed using the more direct Galerkin method rather than the variational approach of *Noorishad et al.* [1971]. A method involving the coupled action of flow forces, body forces, and boundary loads was designed as a finite element program [*Witherspoon et al.*, 1973]. This program was used in investigations of fluid injection in different kinds of fault systems and tectonically stressed regions with simultaneous injection and withdrawal of fluids [*Witherspoon et al.*, 1973, 1974]. There have been further modifications to this coupled thermohydromechanical approach to fractured rocks [*Noorishad et al.*, 1982, 1984], and *Noorishad and C.-F. Tsang* [1996] have published a comprehensive treatment of the methodology, with solutions obtained from a code (ROCMAS) that was developed in this work.

Gale [1975] investigated the behavior of deformable fractures when they open and close as a result of changes in fluid pressure. His research consisted of numerical, field, and laboratory studies. He developed a numerical model that was essentially the same as that of *Noorishad* [1971]. Gale showed that fractures open during injection and close during withdrawal, but the injection rates are much greater than the withdrawal rates for the same pressure gradients. He also found that, because of differences in fluid pressures, fractures with nonuniform apertures in models cannot be replaced with a system of fractures having equivalent uniform apertures.

In the third part of his research program, *Gale* [1975] carried out a laboratory study of injection and withdrawal in artificial fractures using large-diameter rock samples. A 4 ton core of granite, 38 in. (0.97 m) in diameter and 6 ft (1.8 m) in length, was cut at the midpoint with a wire saw to simulate a horizontal fracture perpendicular to the axis of the core. A 3.2 in. (8.1 cm) hole was drilled from one end along the axis so that water could be injected into, or withdrawn from, the saw-cut fracture under radial flow conditions. The core was then prepared for testing in a large triaxial cell as shown in Figure 1. Linear Variable Differential Transducers (LVDTs) were mounted across the fracture, and strain gauges were attached to the rock surface to measure vertical and horizontal strains. This assembly was then enclosed in a test chamber on a moveable steel base that could be placed under a loading frame.

The results for fracture and rock deformation with the saw-cut fracture surface (Figure 2) show how nonlinear the changes in aperture are at low levels of axial stress. With an axial stress of 600 psi (4.1 MPa), the fracture closure was ~4 × 10^{-4} ft (~120 μm), or about 80% of the total closure finally achieved at 2,400 psi (16.6 MPa). The anchor points for the LVDTs are 4.75 in. (12.1 cm) apart, and the strains in the rock over this interval have been accounted for. The pressure drop across this fracture was fixed at 40 psi (0.28 MPa), and the changes in flow rate with changes in aperture were measured during injection and withdrawal in the central borehole. Figure 3 shows the results, and the nonlinear behavior of the fracture is quite evident. Note that when the effective stress had reached 200 psi (1.4 MPa), flow rates were approaching a minimum value of about 0.005 cfm (71 cc/m). This indicates that a fracture subjected to increasing normal stress may not close and can have a residual flow rate as a lower limit.

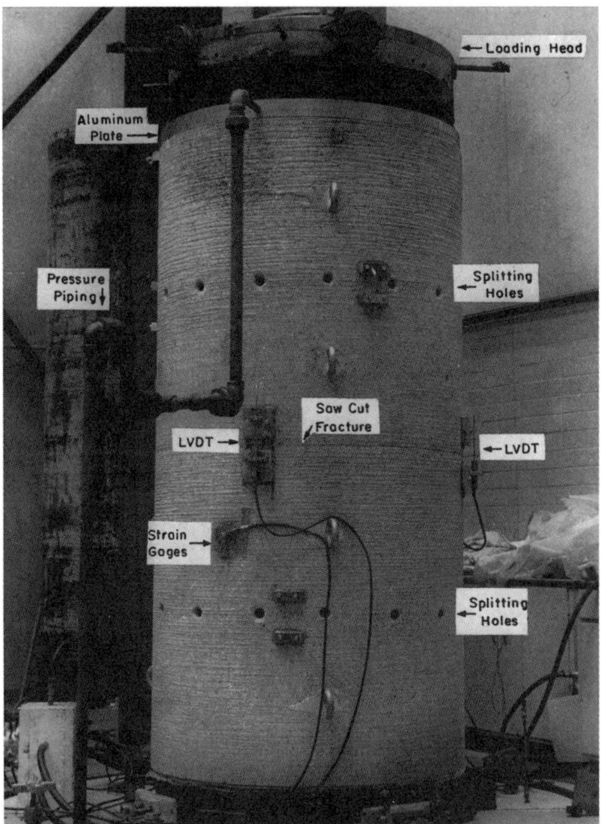

Figure 1. Large diameter core prepared for testing [after *Gale*, 1975].

Gale [1975] also carried out investigations on a tension fracture in this core. By driving wedges into the upper set of splitting holes (Figure 1), the core was split perpendicular to its axis into a fracture with a rough

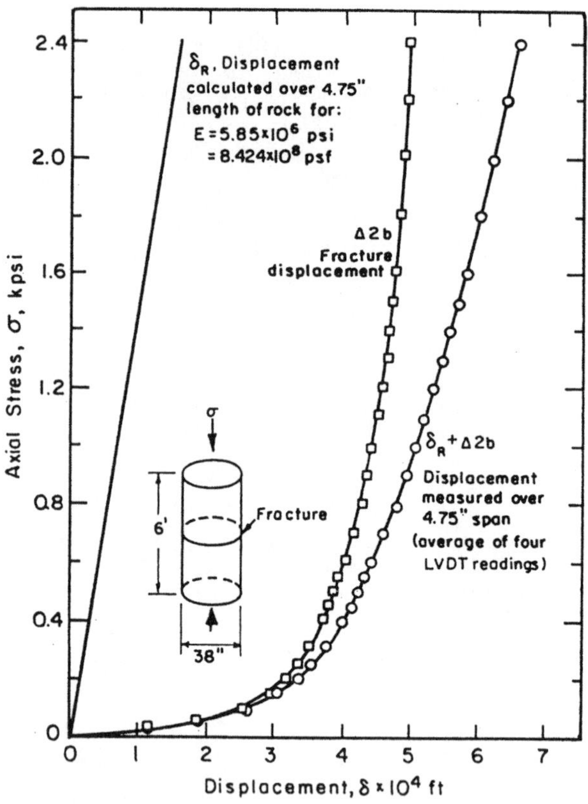

Figure 2. Fracture and rock deformation for saw-cut fracture [after *Gale*, 1975]

surface. This tension fracture was investigated using the same procedure as described above for the saw-cut fracture. Flow rates in this nonuniform fracture were one to two orders of magnitude greater, but the overall stress-flow behavior was essentially the same as that for the saw-cut fracture.

The validity of the cubic law for laminar flow of fluids through fractures with parallel planar plates has been of interest to many [*Boussinesq*, 1868; *Lomize*, 1951; *Polubarinova-Kochina*, 1962; *Snow*, 1965; *Romm*, 1966; *Louis*, 1969; and *Bear*, 1972]. For this idealized case, these researchers have shown that the hydraulic conductivity (K) of a fracture with an aperture 2b is given as:

$$K = (2b)^2 \frac{\rho g}{12\mu} , \tag{1}$$

where ρ is fluid density, g is acceleration of gravity, and μ is fluid viscosity. If flow is steady and isothermal, the flux per unit drop in head, $Q/\Delta h$, can be developed from Darcy's law and may be written in simplified form as:

$$\frac{Q}{\Delta h} = C(2b)^3 , \tag{2}$$

where C is a constant that depends on the geometry of the flow field. Equation (2) is the basis for what is often called the "cubic law." Note that it applies to an "open" fracture, i.e., the planar surfaces remain parallel and are not in contact.

The cubic law can be generalized in terms of the Reynolds number, Re, and a friction factor, Ψ. If one introduces D (hydraulic diameter) equal to 4 times the hydraulic radius, it can be shown that:

$$Re = \frac{Dv\rho}{\mu} , \tag{3}$$

$$\Psi = \frac{D}{v^2/2g} \Delta h , \tag{4}$$

where v is flow velocity. The cubic law then reduces to the simple relationship:

$$\Psi = \frac{96}{Re} . \tag{5}$$

The first comprehensive work on flow through open fractures was by *Lomize* [1951], who used parallel glass

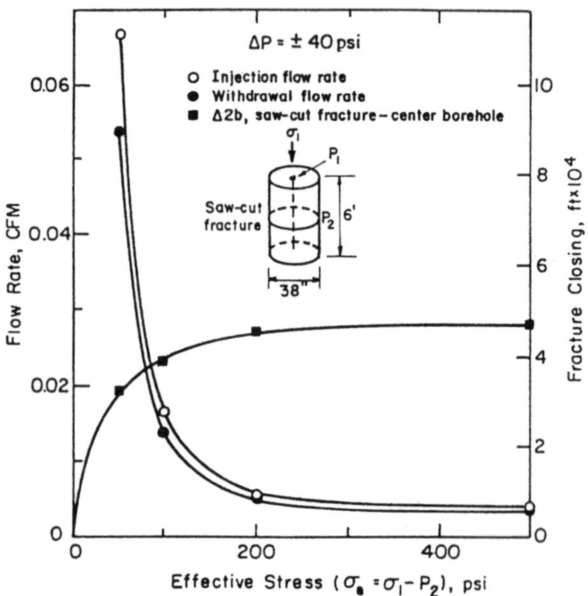

Figure 3. Change in flow rate and fracture aperture as a function of effective stress for saw-cut fracture [after *Gale*, 1975]

plates and demonstrated the validity of the cubic law. He also investigated effects of roughness and fractures with different shapes. He introduced the concept of roughness ε in terms of absolute height of the asperities and developed the empirical equation:

$$\Psi = \frac{96}{\text{Re}}\left[1.0 + 6.0\left(\frac{\varepsilon}{2b}\right)^{1.5}\right] \quad . \tag{6}$$

Equation (6) is simplified by introducing the factor f to account for deviations from the ideal conditions assumed in deriving (5). The equation is then rewritten as:

$$\Psi = \frac{96}{\text{Re}} f \quad . \tag{7}$$

In the case of roughness, $f > 1$, the cubic law in (2) becomes:

$$\frac{Q}{\Delta h} = \frac{C(2b)^3}{f} \quad . \tag{8}$$

In his comprehensive treatise on flow in fractured rocks, *Romm* [1966] presented the results of very careful laboratory studies on flow phenomena in fine (10–100 μm) and superfine (0.25–4.30 μm) fractures. His superfine fractures used optically smooth glass and were carefully constructed to provide open fractures. He demonstrated the validity of the cubic law for laminar flow in both fine and superfine fractures and concluded that laminar flow in a fracture obeys the cubic law down to apertures of 0.2 μm.

The transition from laminar to turbulent flow has been a matter of considerable interest. *Lomize* [1951] and *Romm* [1966] both showed how this transition depends on roughness. For a relatively smooth fracture surface ($\varepsilon/2b < 0.1$], the transition to turbulent flow is at a Reynolds number of about Re = 2400. They found that as the roughness factor increased, in the range of 0.5 to 0.8, Re for the transition to turbulent flow was significantly less than 2400. *Louis* [1969] reported essentially the same results.

In a different approach to this problem, *Iwai* [1976] carried out an investigation of flow in natural fractures where the sides are not parallel or planar. Usually, fracture surfaces have some degree of contact and the effective aperture will depend on the normal stress acting on the discontinuity. Prior to Iwai's work, the validity of the cubic law under these conditions had not been investigated.

Iwai [1976] selected samples of basalt, granite, and marble from which cylinders with a diameter of 0.15 m were used. A horizontal tension fracture was created near the middle of each sample; and a center hole, 0.022 m in diameter, provided an outward radial flow of water. A Riehle testing machine provided axial loads up to 20 MPa. The loading piston of the testing machine could also be attached to the upper half of the sample so that when the piston was raised sufficiently, an open fracture (no points of contact) of known aperture could be obtained. Three LVDTs placed 120° apart and across the fracture were used to measure apertures. This setup was capable of measuring changes in aperture as small as 0.4 μm.

Iwai's [1976] results for radial flow through the fracture in granite compared very favorably with the cubic law (Figure 4a). Theoretically, the value of the exponent in Equation (2) should be 3, and a least-squares fit of the results shown in Figure 4a differs by no more than 3% from this value. This was the case both during loading up to 20 MPa (open symbols) and during unloading (closed symbols). If one adopts a value of 3 in Equation (8), a least-squares fit of the results shown in Figure 4b indicates that the deviations from the cubic law can be accounted for using a value of $f = 1.49$. Results for the tension fracture in basalt were essentially the same as those in granite. In testing fractures at effective pressures well above those used by Iwai, *Engelder and Scholz* [1981] also found that the use of fracture aperture as the only variable in the cubic law is not enough to calculate permeability.

The results for radial flow through the tension fracture in marble (Figure 5) revealed another interesting feature. For apertures between 10 μm and 300 μm, results were in good agreement with the cubic law. This was the case during runs 2 and 3, when the system was unloading and fracture surfaces were no longer in contact. However, when the aperture was below 10 μm, deviations from the cubic law indicated that other effects needed to be considered. This was also revealed when *Gale* [1982] examined the behavior of fractures in gneissic granite and found that deviations from the cubic law were an artifact of the choice of reference point used in calculating fracture apertures. The reference point used was the residual fracture aperture calculated at the maximum normal stress (30 MPa). It was concluded from this work [*Gale*, 1982], and another series of investigations [*Raven and Gale*, 1985], that such a procedure to determine aperture is not valid for rough fractures because the size of the actual flow channels that still persist at maximum stress is underestimated.

4. GEOTHERMAL SYSTEMS

Geothermal systems are usually found in fractured and faulted rocks and, because of their nonisothermal

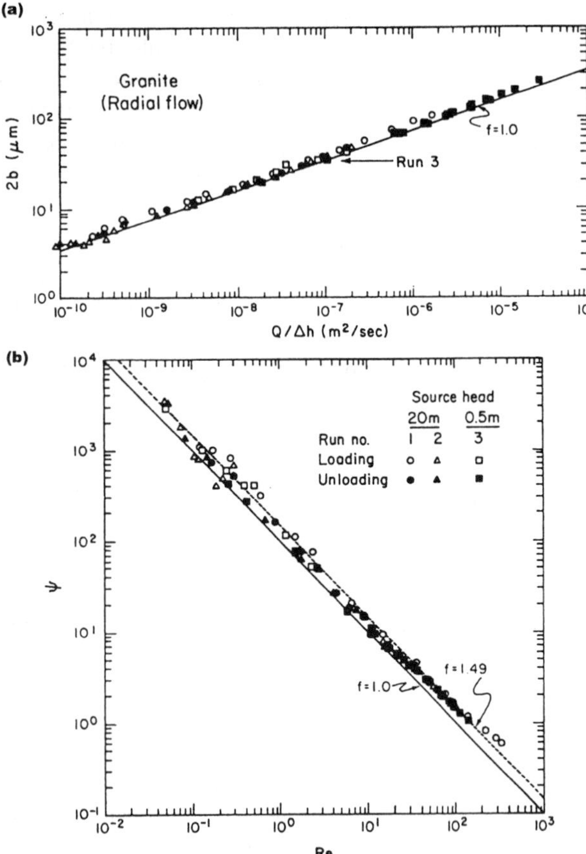

Figure 4. Comparison of experimental results for radial flow through tension fracture in granite with cubic law. In run 3, fracture surfaces were no longer in contact during unloading when aperture exceeded value indicated by arrow [after *Witherspoon, et al.*, 1980a].

conditions, it is more difficult to understand their behavior than is the case for the normal groundwater aquifer. Basically, a geothermal system includes a heat source, a permeable aquifer with a relatively impermeable caprock, and an adequate supply of water. Fractures and faults are usually the main paths of fluid flow in these systems, and their exploitation depends upon the success one has in locating boreholes that intersect these paths.

Geothermal systems are classified as being either vapor dominated or liquid dominated. In vapor-dominated systems, the pressure gradient is close to being vapor static, depending on the amount of the reservoir that is filled with saturated steam. The fluid temperature in these systems is approximately 240°C and the pressure is around 35 bars (3.43 MPa), corresponding to the maximum enthalpy for saturated steam. In liquid-dominated systems, the pressure profile is near hydrostatic, with reservoir temperatures generally ranging from 90°C up to 300°C, or more. Thus, the problems of characterizing geothermal systems are complicated by the coupling between the hydraulic and thermal behavior of the rock systems that exist in the earth's crust at elevated temperatures.

In the early 1970s, the Atomic Energy Commission (AEC), now part of the U.S. Department of Energy (DOE), was interested in supporting programs designed to exploit geothermal systems as alternative sources of energy. In 1973, the Geosciences Group within Berkeley Lab's Energy and Environment Division initiated a geothermal exploration program in northern Nevada that later was expanded to include geothermal reservoir engineering and geothermal energy conversion. By 1977, the research program had expanded further to include nuclear waste isolation, enhanced oil recovery, thermal energy storage, and fundamental geosciences. The Geosciences Group was reorganized into the present Earth Sciences Division of Berkeley Lab. The following subjects have been selected from a large number of published papers to illustrate the

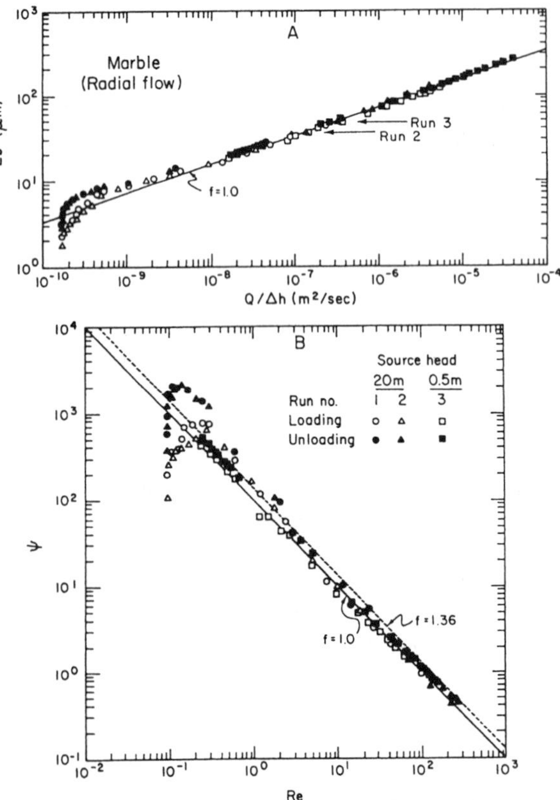

Figure 5. Comparison of experimental results for radial flow through tension fracture in marble with cubic law. In runs 2 and 3, fracture surfaces were no longer in contact during unloading when aperture exceeded value indicated by arrow [after *Witherspoon et al.*, 1980a].

approaches used in geophysics and reservoir modeling to investigate the importance of fractured rocks in geothermal systems.

4.1. Locating Fractures and Faults through Seismic Imaging

The primary focus of Berkeley's seismic research has been the use of a controlled seismic source to image the subsurface. The scale of resolution for this work has varied from submeter to hundreds of meters. Initial work in geothermal exploration involved surface vibrators and explosions to derive 2-D reflection sections over geothermal regions. The first work to delineate fractures and faults was in the Basin and Range geology of Nevada [*Majer*, 1978]. Similar work was done in the Imperial Valley of California [*Majer et al.*, 1980] and in The Geysers geothermal field in northern California [*Majer and McEvilly*, 1979; *Denlinger and Kovach*, 1980]. Although good images were obtained in many areas adjacent to the productive geothermal regions, "washed out" images were more the rule within the area of interest. This was thought to be because of the complex nature of faulting and fracturing common to geothermal reservoirs. Highly altered rocks and/or severe lateral variation in rock properties all resulted in scattered, attenuated, or dispersed energy, which made coherent imaging difficult.

A good example of this problem was obtained in a reflection profile over the Leach Hot Springs in Nevada. The image was degraded over the hot springs, but away from them the image was quite good, showing the normal faulting by offsets in the sedimentary layers. At that time, 2-D reflection imaging was the state-of-the-art (versus 3-D today), and reflection-imaging technology was focused on imaging the subsurface with P-wave energy by obtaining the best estimate of the velocity field and aligning the coherent reflections to estimate the spatial location of the layers and structure. It was obvious from the initial geothermal work that this approach would not be effective in a heterogeneous system that was as chaotic as a fractured medium with multiple length scales.

After trying surface methods, it was decided to pursue techniques that would yield greater resolution and wider bandwidth. To do so required higher frequencies, which implied borehole access. In the early 1980s, work started on a program utilizing vertical seismic profiling (VSP) to image fractures and faults. The advantage of this method was a closer proximity to the targets, validation of results, and higher resolution. VSP also could record multiple component data, utilizing both shear wave and compressional wave data. The disadvantage was that there was often a limited number of boreholes in the area of interest.

In the late 1970s and early 1980s, both equivalent and discrete media theories were being advanced to describe wave propagation in fractured media. On one side *Crampin* [1978, 1981, 1984a,b, 1985] advanced the theory that fractured media could be represented by an equivalent system of aligned cracks to explain cases of observed seismic anisotropy. A related but separate theory held that the media could be represented by discrete fractures in a matrix, with each fracture having a separate effect, related to its stiffness, on the seismic wave [*Myer et al.*, 1985; *Schoenberg*, 1980, 1983]. This theory differs from Crampin's in that at a fracture, the displacement across the surface is not required to be continuous as a seismic wave passes; only the stress must remain continuous. This displacement discontinuity is taken to be linearly related to the stress through the stiffness of the discontinuity. The implication is that very thin discontinuities (fractures) can have a significant effect upon the propagation of a wave.

This discrete-media theory is attractive from several points of view. Schoenberg shows that the ratio of the velocity of a seismic wave perpendicular and parallel to a set of stiffness discontinuities is a function of the spacing of the discontinuities as well as the stiffness. The subtle but important difference between equivalent media and discrete fractures is that if the stiffness theory were correct, one could in principle derive the actual properties of the fracture (location, density, aperture, filling) with seismic measurements.

Several experiments were carried out to test the fracture stiffness theory. The first experiment used VSP at The Geysers geothermal field in northern California [*Majer et al.*, 1988]. Several groups had pointed out the phenomenon of shear wave splitting and the anisotropy effects of SH (horizontal component) versus SV (vertical component) waves, in addition to P versus S wave velocity anisotropy [*Leary and Henyey*, 1985]. To test the applicability of multicomponent VSP surveys for fracture detection, our group at Berkeley had the opportunity to carry out a multi-offset VSP using compressional and shear-wave sources with a three-component geophone in a geothermal well at The Geysers field. We wanted to know what, if any, VSP techniques could be used to map the fracture content, dominant fracture orientation, or fracture spacing. This work resulted in uncovering the dual effect of shear-wave splitting and shear-wave polarization. Analyses of the rotated data clearly showed nonlinear polarization in the two different shear wave planes (H1 and H2) (see Figure 6). This was one of the first observations of shear-wave splitting that indicated shear-wave anisotropy in a fractured

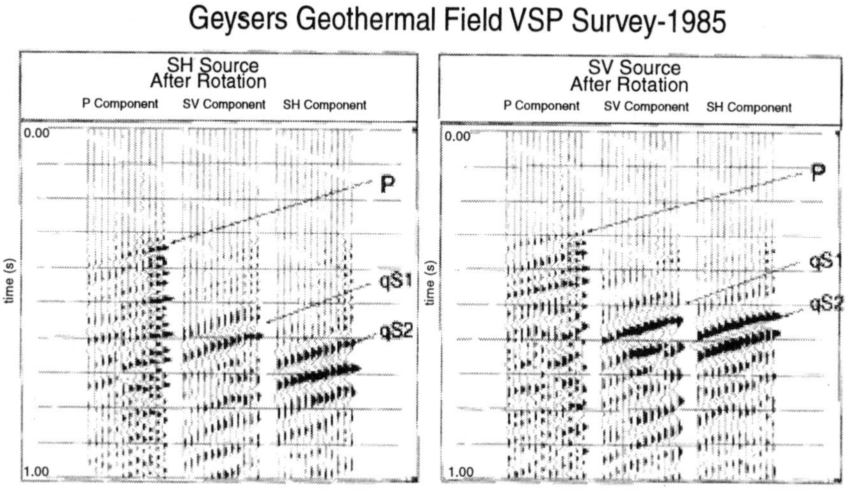

Geysers Geothermal Field VSP Survey-1985

Figure 6. Results of a 3-component VSP survey at The Geysers geothermal field demonstrating shear wave anisotropy [after *Majer et al.*, 1988].

medium. Although the data collected at The Geyser's were insufficient to determine fracture density and orientation, the results do indicate, as others have pointed out, the utility of using three-component data for determining fracture content and orientation. Additional information on fracture spacing may also be obtained by applying the theory developed by *Schoenberg* [1980, 1983] that relates SV and SH velocity differences to fracture stiffness. The data presented clearly showed that fractures have a significant effect on the propagation of shear waves.

4.2. *Characterizing Geothermal Systems through Reservoir Modeling*

Early work at Berkeley on geothermal systems was primarily concerned with the exploitation of liquid-dominated reservoirs. This research involved the development of analytical methods for solving ordinary or partial differential equations constrained by initial and boundary conditions. These methods included two different approaches: (1) the so-called "lumped-parameter" model, and (2) the "distributed-parameter" model. In the lumped-parameter model, the reservoir is characterized by homogeneous regions. Since the mathematical equations that govern single- or two-phase flow in geothermal reservoirs are highly nonlinear, many simplifying assumptions must be made to obtain a closed-form analytical solution to lumped-parameter problems. These assumptions and the difficulties of evaluating the potential of low-permeability reservoirs [*Bodvarsson et al.*, 1982a] have limited the usefulness of this approach. Distributed-parameter models, on the other hand, allow a much more detailed description of a reservoir system and the different flow regimes that can develop during exploitation. Investigations of geothermal reservoirs at Berkeley have primarily relied on distributed-parameter models.

Sorey [1975] developed one of the first numerical models for the investigation of geothermal systems under exploitation. He designed a 3-D single-phase simulator for mass and heat transport in geothermal reservoirs called SCHAFF (Slightly Compressible Heat and Fluid Flow). A numerical simulator, SHAFT (Simultaneous Heat and Fluid Transfer), was developed by *Lasseter et al.* [1975] and is an extension of an earlier code developed by *Lasseter* [1974]. The approach was based on an integrated finite difference method that *Edwards* [1972] originated to handle multidimensional, multiphase flow problems in porous media. The SHAFT code was useful in analyzing various aspects of geothermal systems and was extensively redeveloped, incorporating improved mathematical and numerical techniques. As a result, SHAFT78 [*Pruess et al.*, 1979] and SHAFT79 [*Pruess and Schroeder*, 1980] were developed.

Bodvarsson [1982] carried out an extensive analysis on the behavior of geothermal systems under exploitation. He developed a new 3-D numerical code named PT (Pressure and Temperature) that uses the integrated finite difference method and is quite general in its formulation. The new PT code and SHAFT79 were used to address several theoretical problems on geothermal systems.

An important development in modeling fractured geothermal reservoirs has been the application of the multiple interacting continuum (MINC) technique described by *Pruess and Narasimhan* [1982a]. Application of this method to reservoir conditions representative of The Geysers geothermal field [*Pruess and Narasimhan*, 1982b] produced a much better understanding of the potential reserves of this field. A new method for analyzing well test data from naturally fractured reservoirs was developed by *Lai et al.* [1983] using the MINC technique. The method applies primarily to petroleum reservoirs, but it can also be used for geothermal reservoirs, where the test conditions remain isothermal.

4.3. Some Results of Numerical Studies of Geothermal Reservoirs

Numerical techniques have been of great value in simulating the complexities of natural geothermal systems and in providing a basis for designing effective methods of exploitation. The first step is to understand the geology as well as the hydrogeology of these hydrothermal systems, which can be quite complex. The Cerro Prieto geothermal field is a good example of the kind of problems that can occur. This geothermal field is located in the Mexicali Valley of Baja, California. A program to investigate this geothermal system began in 1977 with the signing of a cooperative agreement between the Comisión Federal de Electricidad of Mexico and the DOE [*Witherspoon, et al.*, 1978].

An effort to understand the natural state of the Cerro Prieto field as well as its behavior under exploitation was carried out by *Lippmann and Bodvarsson* [1983]. The field is a complex geologic and hydrologic system. Natural flows through the reservoir are controlled by layered sedimentary units (sands and shales), major fault zones, buoyancy effects, and the regional hydraulic gradient. All of these factors had to be considered in modeling the reservoir. Figure 7 shows a layout of the field with the geothermal wells that had been drilled as of 1983. The initial development of the field was on the southwest side of the railroad line (bisecting the figure), where the reservoir depths ranged from 800 m to 1500 m. Subsequent drilling on the northeast side reached reservoir depths as much as 2500 m.

Several authors have presented generalized models describing the natural (initial) conditions of the Cerro Prieto field and its evolution under production [*Mercado*, 1976; *Elders et al.*, 1984; *Grant et al.*, 1984]. A model for the natural flow pattern in the Cerro Prieto system has been a matter of considerable interest. *Elders et al.* [1984] used mineralogical and isotopic data from well cuttings and cores to develop flow patterns for the field in its natural state. The heat source for the hydrothermal system is

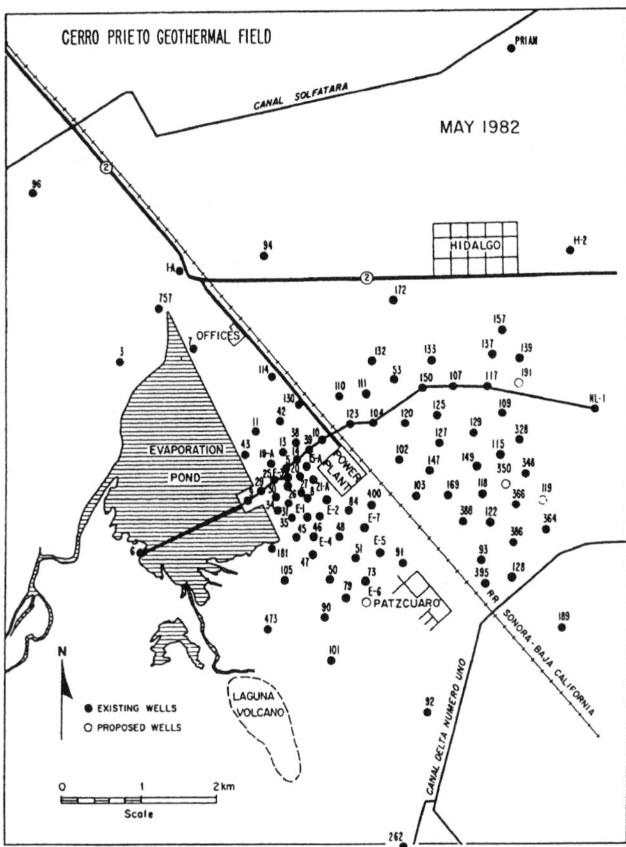

Figure 7. Layout of geothermal wells in Cerro Prieto geothermal field and location of SW-NE cross section shown on Figure 8 [after *Lippmann and Bodvarsson*, 1983].

located to the east, near well NL-1 (Figure 7), in an area where wells have been drilled through mafic and silicic dikes. A plume of hot water ascends upwards and horizontally to the west, forming hot springs and fumaroles at the surface.

Halfman et al. [1984] at Berkeley developed a detailed model of the Cerro Prieto system, based on well-log and reservoir-engineering data. They identified permeable and less permeable zones and postulated the fluid flow patterns, as illustrated in Figure 8. This figure depicts a cross section that goes through the line of wells shown on Figure 7. The cross section shows the distribution of sandstones, sandy shales, and shales; the temperature logs of various wells (thicker lines indicate where temperature is equal to or higher than 300°C) and production intervals; and postulated flow paths. The hot fluids move through a thick sandstone unit (Reservoir β) from the northeast, ascend through a gap in the shale layers, and flow towards the west through the more permeable units (Reservoir α at about 1200 m and Reservoir β at about 1700 m depth). The A/B contact on

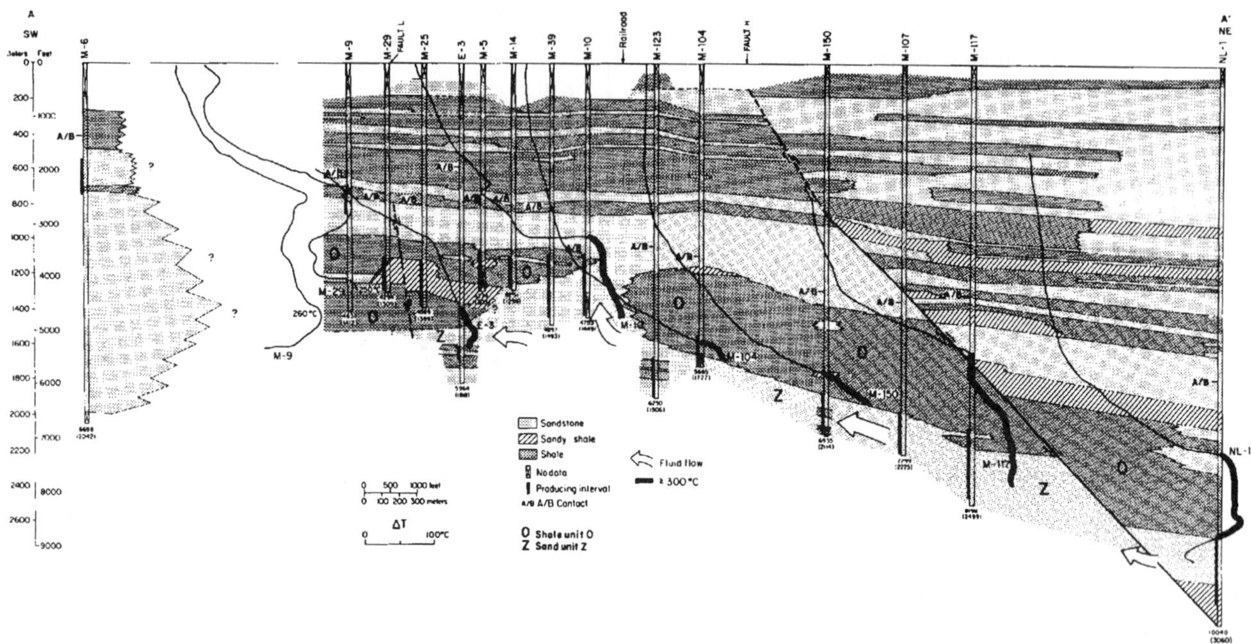

Figure 8. Southwest-northeast cross section of Cerro Prieto field showing well locations, lithofacies groups, faults, temperature profiles, producing intervals, A/B contacts, shale unit O, sand unit Z, and arrows indicating direction of fluid flow. Parts of temperature profiles shown by heavy lines indicate temperatures of 300°C or higher [after *Halfman et al.*, 1984].

Figure 8 is associated with a sharp increase in temperature gradient. Observations indicate that the heat of the geothermal system has greatly modified the sediments in Unit B, whereas the sediments in the overlying Unit A have remained relatively unaffected.

On the basis of this detailed picture of Cerro Prieto, *Lippmann and Bodvarsson* [1983] set up a 2-D east-west model of the field and simulated the natural state, using a 3-D, multiphase, multicomponent code called MULKOM [*Pruess*, 1983]. The steady-state pressure and temperature distributions were computed and compared against observed preproduction data with good results. A natural hot-water recharge rate of about 0.01 kg/s per meter of field width (measured in a north-south direction) was also obtained.

The model was then used to simulate the behavior of the field during the 1973–1978 production period. Results from the model agreed with that observed in the field or postulated by other investigators. There was a decrease in temperature and pressure in the producing region, but because of the strong fluid recharge, no extensive two-phase zone had developed in the reservoir. Most of the fluid recharging of the system comes from colder regions located above and west of the producing reservoir.

Another example of the complex problems in modeling geothermal systems is the comprehensive study of the Krafla field in Iceland, conducted by Berkeley in association with the State Electric Power Works of Iceland (SEPW) and the Icelandic National Energy Authority (NEA). The study covered four basic areas: (1) an analysis of Krafla well test data [*Bodvarsson et al.*, 1984a]; (2) a modeling study of the natural state of the reservoir system (*Bodvarsson et al.*, 1984b]; (3) the determination of the generating capacity of the reservoir [*Bodvarsson et al.*, 1984c); and (4) the development of a reservoir model that can be used to predict future performance of wells and reservoir depletion [*Pruess et al.*, 1984].

The Krafla field is in one of the most complex geothermal systems described in the literature [*Stefansson*, 1981]. The heat source is a magma chamber located at depths of 3 to 7 km below the ground surface [*Einarsson*, 1978]. The geothermal system is in a neovolcanic zone in northeastern Iceland that is characterized by fissure swarms and central volcanoes [*Saemundsson*, 1974, 1978]. The producing wells are located in the caldera (8 × 10 km) of the Krafla central volcano (Figure 9).

Stefansson [1981] has presented a detailed description of the reservoir system at Krafla. In the "old wellfield"

(Figure 9), pressure and temperature data in wells 1–13 and 15 indicate the presence of two reservoirs. The upper reservoir contains single-phase liquid water at a mean temperature of 205°C. This reservoir extends from a depth of 200 m to about 1100 m. The deeper reservoir is two phase, and extends down to depths greater than 2200 m with temperatures and pressures following the boiling curve. The two reservoirs seem to be connected near the Hveragil fracture. In the "new wellfield" (south of Mt. Krafla; wells 14, 16–18), the upper reservoir has not been identified, and only the two-phase liquid-dominated reservoir seems to be present

The Krafla reservoirs are primarily two phase, and heavily fractured conditions in all wells are a dominant feature controlling fluid flow [*Bodvarsson et al.,* 1984a]. Reservoir temperatures are high (>300°C), but composition of the fluids and effects of noncondensable gases cause frequent scaling problems in many wells. Volcanic activity at the site has altered the chemical composition of produced fluids and increased the scaling problems. Far fewer chemical and production problems have been encountered in the new wellfield (Figure 9), and these wells are generally more productive than those in the old wellfield.

A detailed analysis of injection test data from the Krafla wells was carried out [*Bodvarsson et al.,* 1984a]. This analysis was complicated by the various factors mentioned above. Theoretical studies were conducted to assess the impact of the different complications, and a simple model for the analysis of the injection test data was developed. The model results indicate that average transmissivity of the Krafla reservoirs is 2.0 Dm (Darcy meters), with values

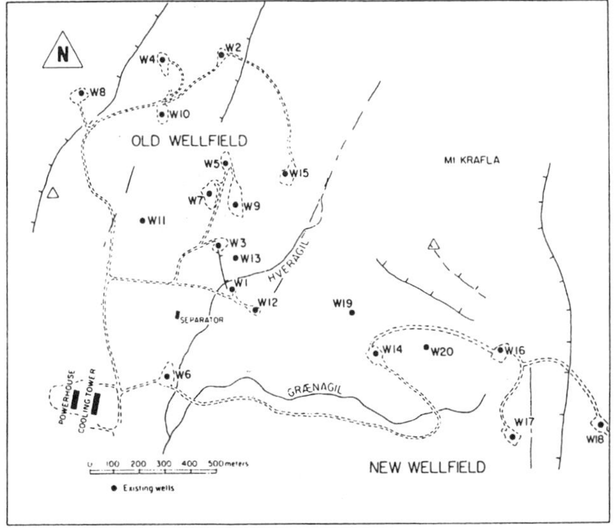

Figure 9. The Krafla wellfields [modified from *Stefansson*, 1981].

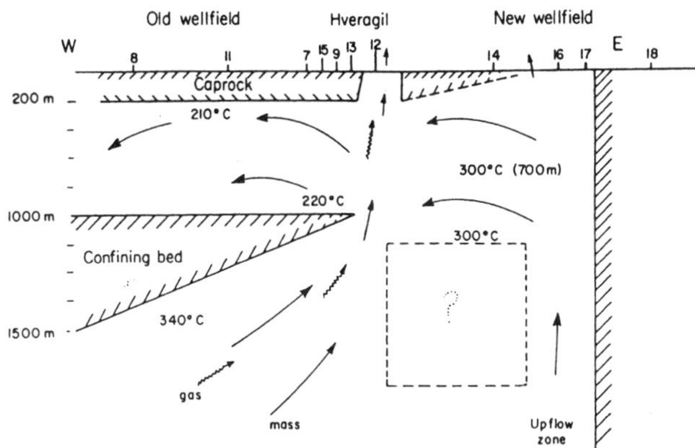

Figure 10. Conceptual model of the Krafla geothermal field [after *Bodvarsson et al.*, 1984b].

for most of the wells in the range of 1.5–2.5 Dm. These results are an order of magnitude lower than those reported for most commercially successful geothermal fields.

A model of the natural state of the Krafla reservoir system was also developed by *Bodvarsson et al.* [1984b]. As shown in Figure 10, the conceptual model of the field is a vertical cross section, which includes reservoirs in both the old and new wellfields. Thermodynamic conditions in both fields are controlled by upflow zones in the Hveragil fracture and the new wellfield. The lower reservoir in the old wellfield and the reservoir in the new wellfield are two phase, with average vapor saturations of 10–20% in the fracture system. An upflow zone recharges the reservoir in the new wellfield at a rate of 0.010 kg/s m. The fluids flow laterally along a high-permeability fracture zone at a depth of about 1 km and mix with fluids from the lower reservoir. The natural fluid flows are highest in the Hveragil fault zone, where about 0.008 kg/s m of high-enthalpy steam is discharged to surface springs; the remainder (about 0.013 kg/s m) recharges the upper reservoir in the old wellfield. In the Hveragil fault zone, extensive boiling takes place so that lower temperature waters feed the upper reservoir. The temperatures in other parts of the reservoir system are high—about 300°C at a depth of 1000 m and 345°C at a depth of 2000 m.

Although convection dominates heat transfer in the system, conductive heat loss through the caprock is substantial (about 1 W/m^2 or 26 heat flow units). The heat flux from the bottom of the reservoir system is estimated at 2.0 W/m^2. With this background and reservoir engineering data from the field, *Pruess et al.* [1984] designed a model used to develop predictions of future field performance. They were able to demonstrate the value of modeling in providing a range of solutions from which an optimum course of action can be selected.

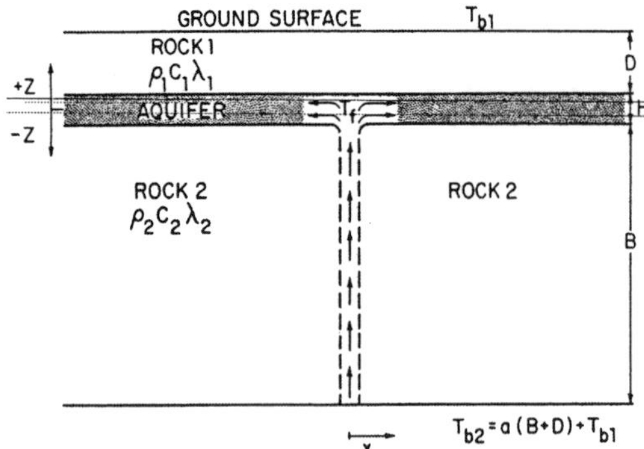

Figure 11. Mathematical model used for fault-charged aquifer problem [after *Bodvarsson*, 1982].

4.4. Development of Geothermal Systems Charged by Vertical Faults

Fault-charged geothermal systems are found in most geothermally active areas in the world. Berkeley has investigated a number of these systems in the western United States, including East Mesa geothermal field in Imperial Valley, California [*Narasimhan et al.*, 1977]; Raft River geothermal project in Idaho [*Narasimhan and Witherspoon*, 1977]; Susanville hydrothermal system in California [*Benson et al.*, 1981]; and Klamath Falls KGRA in Oregon [*Prucha*, 1987].

Fault-charged reservoirs commonly display atypical temperature profiles that are characterized by a reversal indicative of lateral hot water flow and conductive heat losses to confining beds. The distinct characteristics of these geothermal systems suggested the need for a theoretical analysis of such behavior [*Bodvarsson*, 1982]. Figure 11 shows the type of system for which a mathematical model was developed. Initially, temperatures increase linearly with depth in accord with the geothermal gradient a. At time t = 0, hot water starts to flow up the vertical fault and is recharged into a relatively thin horizontal aquifer under forced convection. The behavior of the system is then controlled by the following assumed conditions:

1. Temperature at the ground surface, T_{b1}, remains constant.
2. No heat losses occur as fluid moves up the fault and enters the aquifer at constant temperature T_f.
3. Within the aquifer: (a) mass flow rate is constant; (b) horizontal conduction is neglected; (c) temperatures in the vertical direction across the relatively thin aquifer are uniform; and (d) thermal equilibrium between fluid and solid is instantaneous.
4. Within the confining beds (caprock and bedrock): (a) permeability is so low that movement of heat is controlled only by conduction; (b) horizontal conduction is neglected; and (c) there is resistance to heat transfer at the interfaces within the aquifer.
5. At some depth, B, below the aquifer, temperature in the bedrock, T_{b2}, is constant.
6. Thermal properties of formations above and below the aquifer may be different, but all thermal parameters (thermal conductivity, κ, and volumetric heat capacity, ρc) for liquid and rock are constant.

On the basis of this model and the above assumptions, *Bodvarsson* [1982] obtained a solution for the temperature in the aquifer at any time t. An example, in dimensionless terms, is shown in Figure 12. This is for the particular case where dimensionless distance, $\varepsilon_1 = (\lambda_1 x)/(\rho_w c_w q D) = 0.1$. It can be seen how dimensionless temperature T_D increases rapidly at early values of dimensionless time τ_1 and then reaches a steady state as $\tau_1 \rightarrow \infty$.

The first attempt to validate this model was its application to data from the geothermal system at Susanville, California. The more than 20 exploration wells in Susanville had located a low temperature (<80°C) shallow geothermal aquifer of limited areal extent [*Benson et al.,* 1981]. In analyzing the field data using the model, calculated temperature contours compared very well with the observed ones in the hottest region of the field. Further analysis indicated that the size of the geothermal source and the rate of recharge could be significantly larger than an early evaluation and that this source of thermal energy needed further investigation [*Bodvarsson et al.*, 1982b].

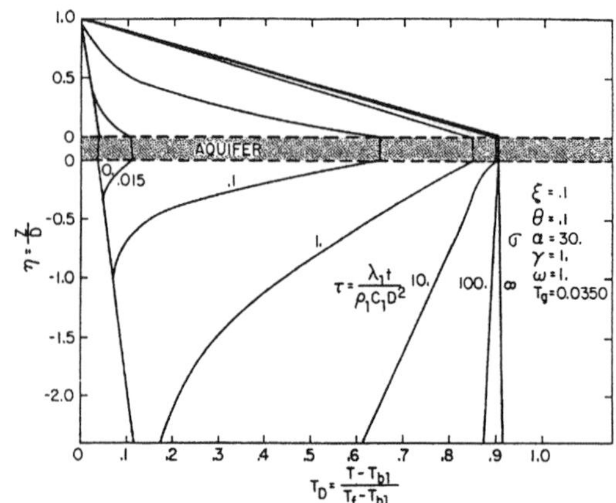

Figure 12. Evolution of a fault-charged hydrothermal system for $\varepsilon_1 = 0.1$ [after *Bodvarsson*, 1982].

4.5. Problems in the Exploitation of Geothermal Systems

An important problem in the exploitation of geothermal systems is the effect of cold-water injection on the overall thermal system. Reinjection of geothermal wastewater into some part of the thermal system is a preferred means of disposal, but it is imperative that the reinjected fluids not have undesired effects. During injection, the cold water will advance along fractures, extract heat from the adjacent rock matrix, and eventually arrive at the production wells. If the injected water has not become fully reheated, detrimental effects on energy production from decreasing fluid enthalpies may result.

Bodvarsson [1982] developed a model for this problem that consists of an injection well drilled completely through a fractured geothermal reservoir containing equally spaced horizontal fractures. Water at a constant temperature is injected into the fractured rock, and by assuming that the rock matrix is impermeable, flow is restricted to the fractures. The effect of heat conduction on the advancement of the cold water along the fractures was then investigated.

Key dimensionless parameters that describe the physical system were identified, an analytical solution was developed, and type curves were generated to illustrate the behavior of the model [*Bodvarsson*, 1982]. These curves can be used in designing injection/production systems, mainly to determine the appropriate locations and flow rates for injection wells [*Bodvarsson and C.-F. Tsang*, 1982].

Another problem that can occur during the exploitation of geothermal systems is subsidence of the land surface. The production of fluids may result in significant ground surface displacements. Such displacements are caused by lowering of pressures in the reservoir and surrounding rock systems. It is important to investigate the potential of adverse effects from such displacements as soon as adequate data on the geologic structure, stratigraphy, rock properties, and proposed reservoir development become available. It is also important to realize that the subsurface stratigraphy may create unexpected lateral migrations of the effects of subsidence. *Stillwell et al.* [1975] reported an unusual case for the Wairakei geothermal field in New Zealand, where the surface of the land had subsided about a kilometer east of the area where producing wells were located.

To investigate the difficulties caused by land subsidence, *Lippmann et al.* [1976, 1977] developed a mathematical model called CCC (Conduction, Convection, Consolidation) to simulate the transport effects and vertical deformations produced by changes in effective stress that are caused by fluid withdrawals. The PT code created by *Bodvarsson* [1982], mentioned above, was developed from CCC and operates much more efficiently through the use of more powerful mathematical and numerical techniques.

5. THE STRIPA PROJECT

The Stripa project was the first project in which Berkeley could carry out large-scale underground investigations on fractured rock masses. The site was an abandoned iron ore mine in central Sweden that had a long history of mining activity, spanning back to the 15th century. The project started July 1, 1977, when the Energy Research and Development Administration (ERDA) (successor to AEC and now part of DOE) executed a bilateral agreement with the Swedish Nuclear Fuel Supply Company (SKBF) to carry out a program of investigations at Stripa [*Witherspoon and Degerman*, 1978]. Berkeley was designated as the lead participant for the United States.

ERDA had originally set up an Office of Waste Isolation (OWI) in the Nuclear Division at Oak Ridge National Laboratory, which had been conducting investigations on radioactive waste isolation in the salt beds of Kansas. OWI wanted to broaden its investigations to include hard rock and was interested in working underground in the granitic rocks surrounding the ore body in the Stripa mine.

A plan view of the Stripa site is shown in Figure 13. The greater part of the underground works for the iron ore mine is south of the area shown on this figure. The surface outcrops are mainly glacial deposits, which are not shaded on Figure 13, with one granite and a number of leptite outcrops. To reach the granite rock mass for the underground laboratory, the Swedes mined a new drift at a depth of about 330 m. This drift is shown by the dashed lines that start at local mine coordinates 250 North and 975 East, and extend due north about 150 m.

5.1. Hydrogeological Investigations

To investigate the fractured granitic rock, *Gale and Witherspoon* [1979] set up a program to drill three slant boreholes (SBH), whose locations are shown on Figure 13. These boreholes were drilled with different orientations at angles of 45°, 50°, and 52° from the vertical and collected oriented core starting from the surface. Pressure measurements were taken at regular intervals during the drilling, and results are included in Figure 14 [*Witherspoon et al.*, 1980b]. Below a slant depth of about 120 m, the fractured rock system is saturated, but mining operations caused a drawdown in hydraulic head below normal values.

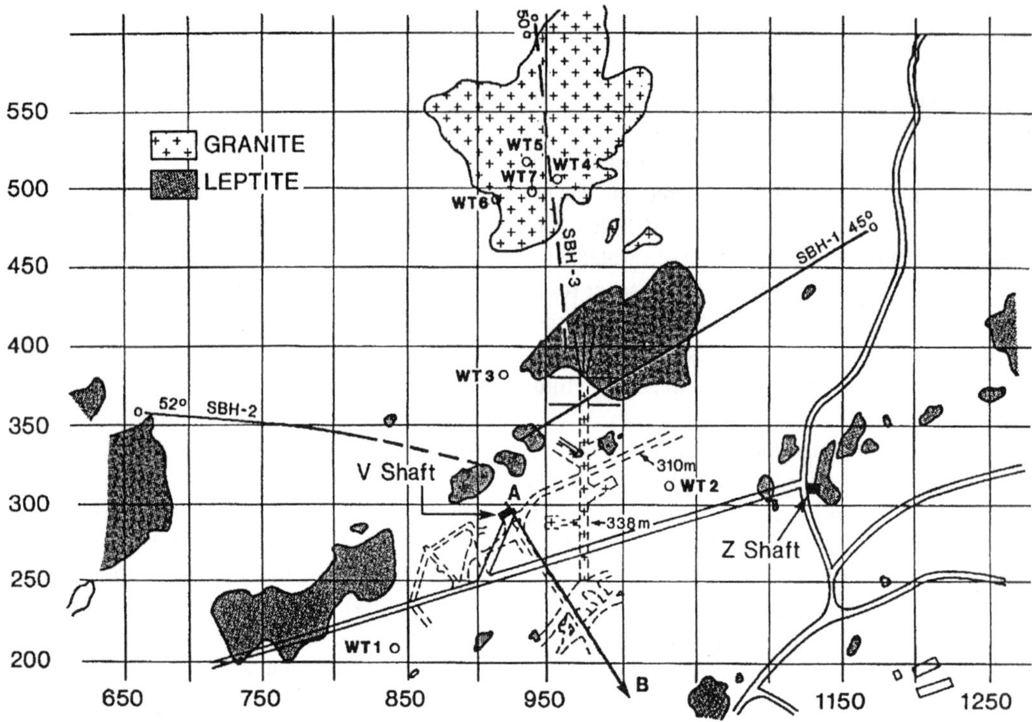

Figure 13. Areal geologic map of Stripa project on the north edge of iron-ore mine showing locations of exploratory boreholes and underground experimental area. Glacial deposits prevail except where leptite and granite outcrop. Coordinate system refers to mine coordinates in meters [after *Gale and Witherspoon*, 1979].

Mapping the fractures in the oriented core provided an excellent record of the fracture systems (Figure 15). The results at the top of this figure are for borehole SBH-1 at 45° [*Rouleau and Gale*, 1985], with one stereonet that shows data obtained over the total length (4130 poles) and a second stereonet that shows data from the lower part of the hole (1796 poles). The results from SBH-1 can be compared with the stereonets obtained at four other locations in the underground laboratory. These latter results [*Thorpe*, 1979; *Paulsson*, 1983; *Rouleau and Gale*, 1985] are included in the lower half of Figure 15. Each of the stereonets shows evidence of several sets of fractures in the granitic rock mass.

The orientations of these fracture sets and their effects on rock permeability were of considerable interest. During drilling of the three SBH boreholes (Figure 13), over 1000 packer tests were made, using a 2 m test interval, from which equivalent porous media permeabilities were computed. Figure 16 shows the results of a linear regression analysis of these data [*Gale et al.*, 1987] and the trends for each borehole of decreasing permeability with depth. The 95% confidence bands within which data from any given test interval will fall are indicated. The spread of the bands is an indication of scatter in the data at different depths.

In a subsequent analysis of fracture geometry, *Olsson and Gale* [1995] identified two scales of discontinuities. They found that orientations of large-scale fracture zones duplicated orientations of small-scale fracture systems, which were the dominant feature in the fractured rock mass at Stripa. They also found that the large-scale features, when extrapolated to ground surface, coincided with regional lineaments on topographic maps.

The principal stress in the rock mass at Stripa is about 20 MPa [*Doe et al.*, 1983; *Gale et al.*, 1987] and is oriented approximately N70°W with a shallow plunge southeast. The intermediate stress is subhorizontal, and the minor stress, of about 5 MPa, is vertical. Hence, *Gale et al.* [1987] suggested that fractures oriented subperpendicular to SBH-1 and SBH-2 are most likely to be intersected by these boreholes and also be oriented subperpendicular to the principal stress. Similarly, fractures subperpendicular to SBH-3, and hence preferentially intersected by that borehole, are oriented parallel to the principal stress. Thus, *Gale et al.* [1987] concluded that if fracture permeability is a function of stress acting normal in relation to fracture

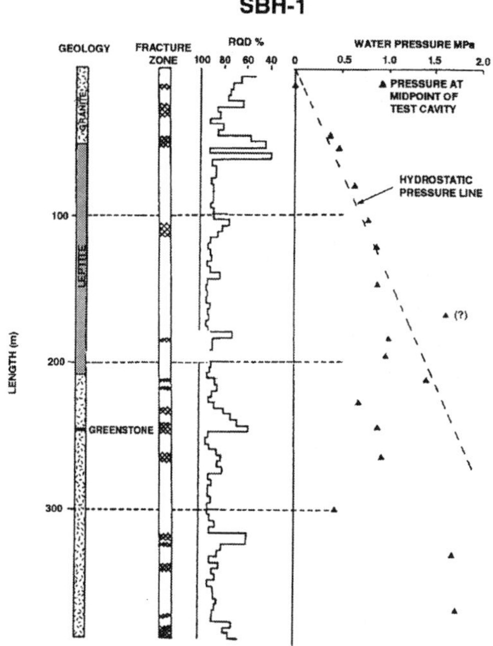

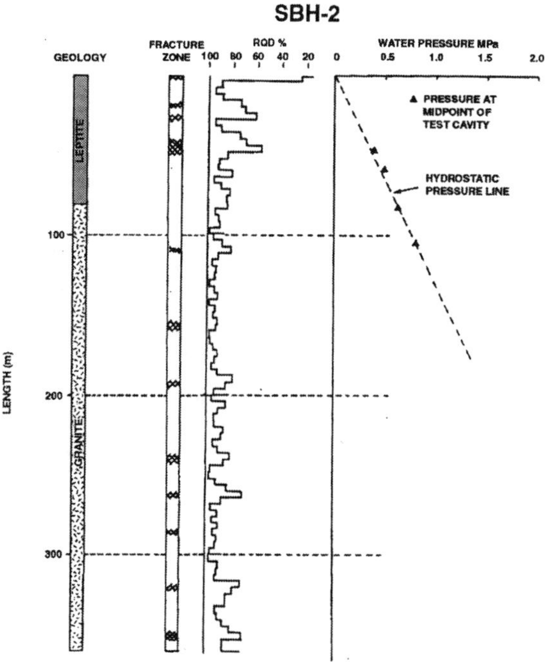

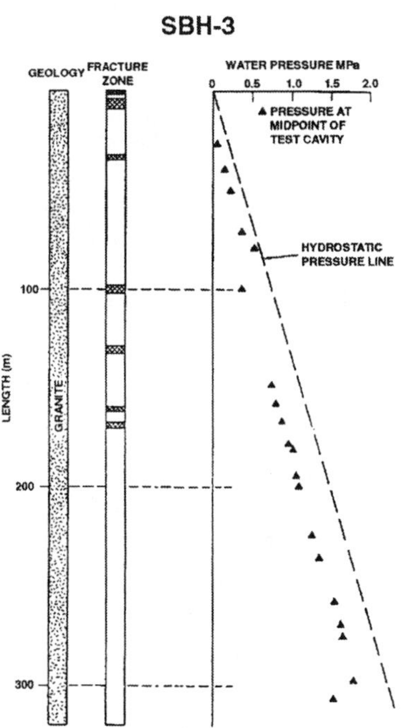

Figure 14. Fracture hydrology results from SBH-1, SBH-2, and SBH-3 showing general geology, fracture zones, RQD values, and bottom-hole hydrostatic pressures measured during drilling [after *Witherspoon et al.*, 1980b].

planes, then SBH-1 and SBH-2 should have lower average permeabilities than SBH-3. The regression (solid) lines on Figure 16 support this interpretation.

5.2. Geochemistry and Isotope Hydrology

Another component of investigations at Stripa was the geochemistry and isotope hydrology of the groundwaters [*Fritz et al.*, 1979]. This work provided an independent approach to the problem of the overall permeability of the fractured rock system. If the deep waters entered the groundwater system many thousands of years ago and percolated downward at very low velocities because of inherently low hydraulic conductivities in the rocks, waters at various depths should differ significantly. This approach must, of course, take into account the geochemistry of these systems because changes in the environment of groundwaters can also produce significant effects.

Water samples were collected from the surface, shallow private wells, and boreholes drilled at the 330 m level. In addition, a deep borehole drilled by the Swedish Geological Survey from 410 m (the deepest operating level in the mine) to about 840 m below surface provided the deepest sample point. The analytical results provide important information on the geochemical evolution, origin, and age of Stripa groundwaters [*Fritz et al.*, 1979].

The results showed an increase in total dissolved solids with depth due to only a few elements, notably calcium,

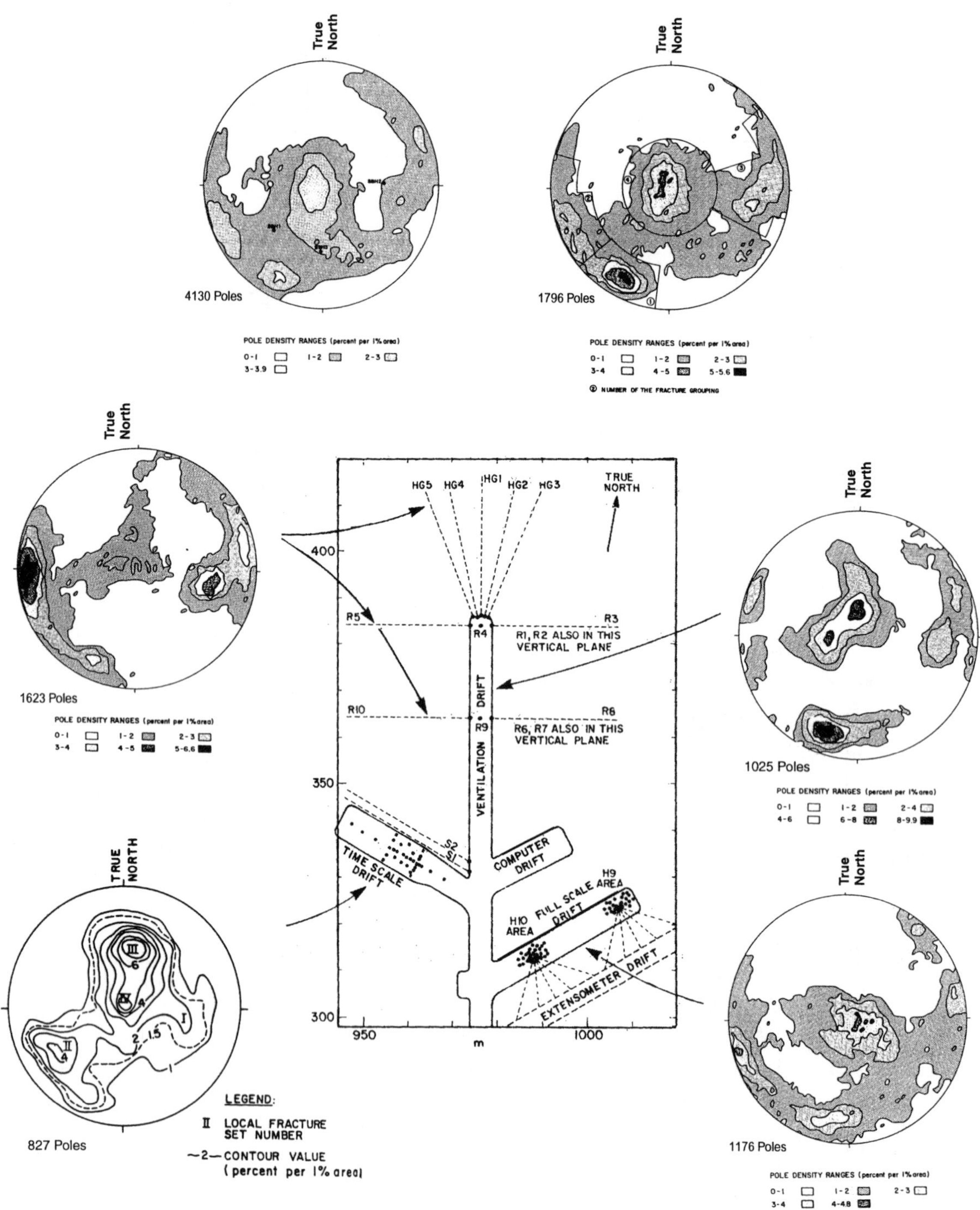

Figure 15. Contoured stereonets of poles to fracture planes measured in different areas of the Stripa site. The contoured values are in percent of points per 1% surface area [after *Rouleau and Gale*, 1985].

sodium, and chloride. Bicarbonate (or total inorganic carbon) decreases dramatically below the 100 m depth, and both magnesium and potassium contents drop from higher levels (2–10 ppm) in shallow groundwaters (<l00 m) to below 1 ppm in the mine waters. It was surprising to find a rise in pH, from approximately 7.0 in the shallow waters to as high as 9.8 in the deepest groundwaters (801–838 m). This rise in pH is probably linked to the dissolution of primary silicates such as feldspars and to the formation of clay minerals.

Simple mixing of freshwater and fossil seawater cannot explain the observed chemistry. The geochemical history of the deep groundwaters at Stripa is more complex than that of groundwaters from other localities in similar rocks [*Jacks*, 1973]. It was tentatively concluded that the deep groundwaters at Stripa have a different origin and are not related to fossil seawater, which would have infi ltrated less than 10,000 years ago.

The abundances of stable isotopes ^{18}O, ^{2}H, and ^{13}C were determined in the hope that this information would explain the origin of these waters. The results of the ^{18}O and ^{2}H analyses are shown in Figure 17, which illustrates that all groundwaters sampled except the surface waters fall close to the global meteoric waterline. They are thus "normal"

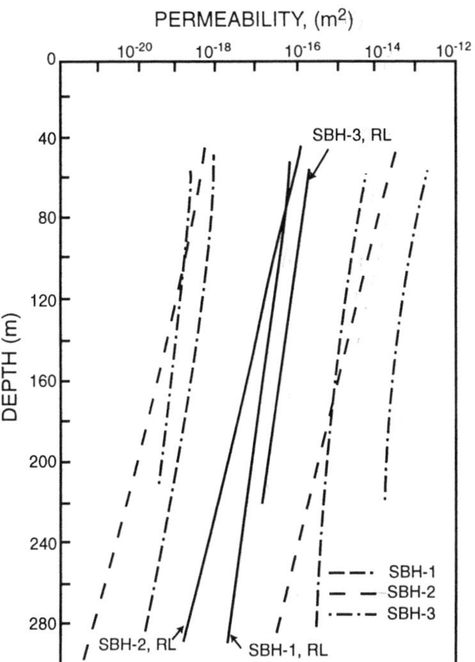

Figure 16. Results of linear regression analysis of permeability data from the three SBH boreholes showing variation of equivalent porous media permeabilities with depth [after *Gale et al.*, 1987].

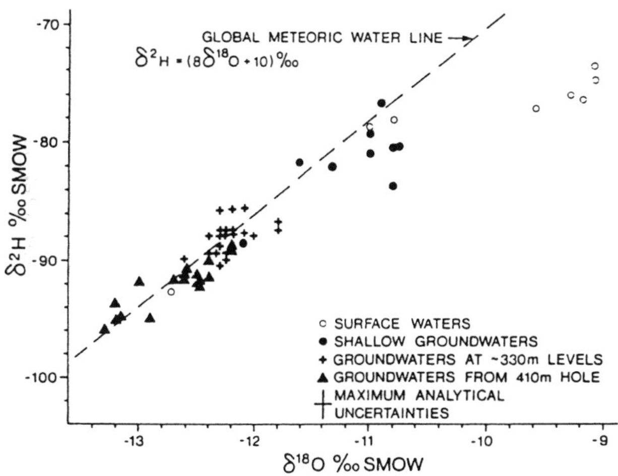

Figure 17. Comparison of $\delta^{18}O$ and $\delta^{2}H$ values for Stripa groundwaters. The analyses are reported as $\delta^{0}/_{00}$ values with reference to SMOW. A $\delta^{18}O$ of $-100^{0}/_{00}$ signifies that the sample has $10^{0}/_{00}$ (per mil) less ^{18}O than the reference standard, which closely reflects average seawater [after *Fritz et al.*, 1979].

groundwaters for which ^{18}O and ^{2}H contents reflect climatic conditions in the original recharge area.

As a general rule, lower concentrations of heavy isotopes signify lower average annual temperatures at the recharge area. Therefore, the deep "saline" groundwaters, which have the lowest ^{18}O and ^{2}H contents, must have recharged at lower average annual temperatures than the shallower groundwaters. This has been confirmed by rare-gas analyses performed on all samples [*Fritz et al.*, 1979]. One must therefore conclude that the deep groundwaters have an origin different from that of the shallower groundwaters.

This conclusion was further substantiated by comparing ^{18}O with chloride concentrations for groundwaters at different depths [*Fritz et al.*, 1979]. The deep groundwaters, especially those at the bottom of the 410 m hole, are distinctly different from the shallow groundwaters. In other words, the different fracture systems in the granite at Stripa carry different types of water because they are isolated from each other.

5.3. Electric Heater Tests

Two major heater experiments were conducted in the underground laboratory (Figure 18), at a depth of about 330 m. These experiments included a full-scale heater test and a time-scaled heater test [*Cook and Witherspoon*, 1981; *Chan et al.*, 1978]. The full-scale test utilized two steel canisters equipped with electric heaters [*Burleigh et al.*, 1979] and

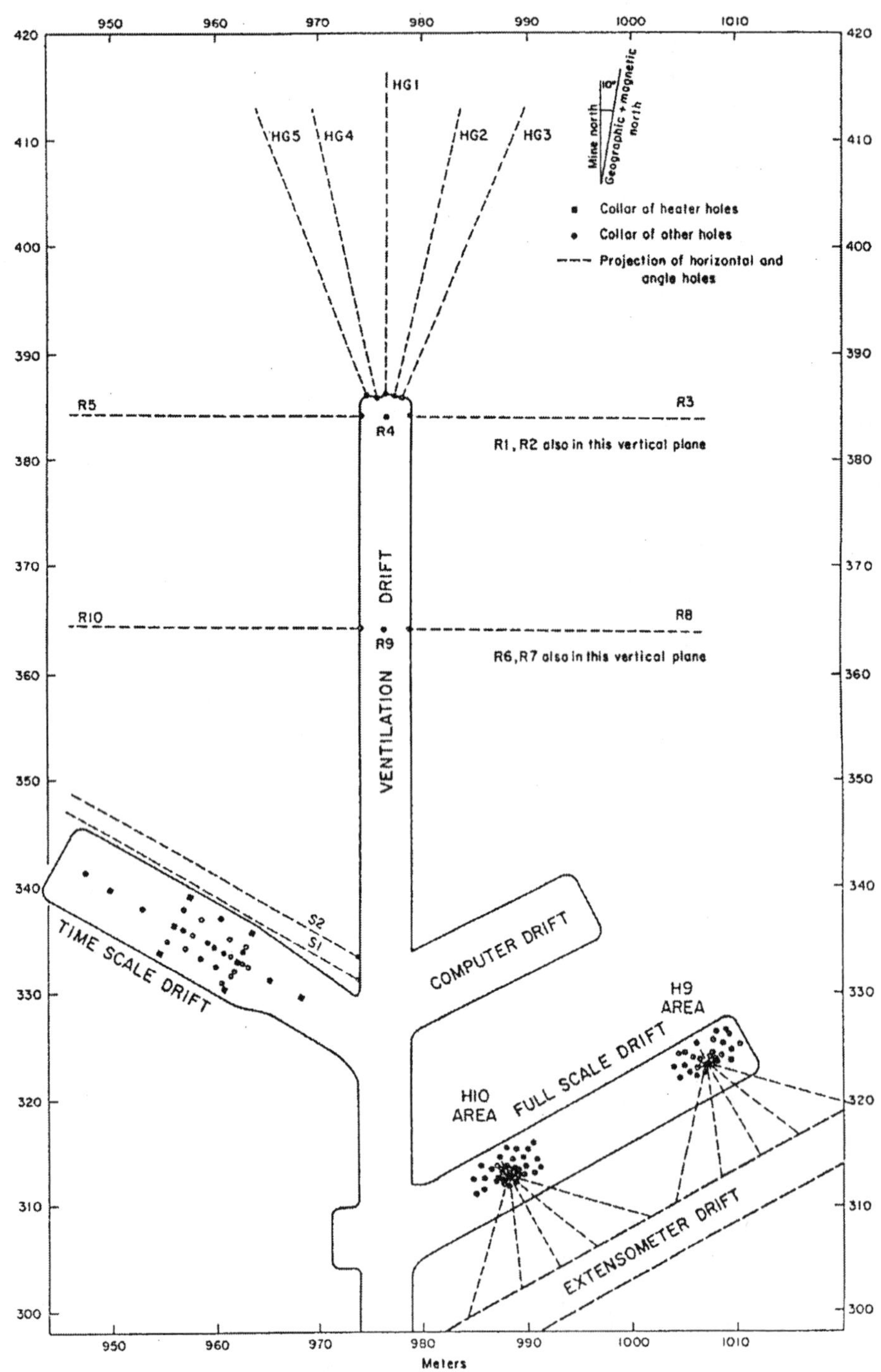

Figure 18. Plan view of experimental area about 330 m underground at Stripa. Coordinate system refers to mine coordinates in meters [after *Gale*, 1981].

placed in vertical holes drilled about 20 m apart in the full-scale drift (Figure 18). Each canister was cylindrical in shape, with a length of 2.6 m and a diameter of 0.324 m. The holes were drilled deep enough so that the midpoint of each canister could be placed 4.25 m below the floor of the drift. One canister was equipped with electric heaters that could generate a 5 kW output, and the other canister was equipped to generate a 3.6 kW output.

To measure displacements that would be produced by the thermal fields, another drift at a second elevation, slightly lower than the full-scale heater drift, was excavated. A series of horizontal extensometer holes were drilled from this second drift (Figure 18), and vertical extensometer holes was also drilled from the floor of the full-scale drift. Thermocouple holes were installed around each canister to measure the temperature fields.

The computer drift on Figure 18 housed a special computer for the collection and storage of data [*McEvoy*, 1979]. This system consisted of about 750 terminals connected to various instruments needed in the heater tests. Investigators were thus able to observe the results of the tests in real time. A database with the predicted thermal behavior of the tests [*Chan et al.*, 1978] was also stored on the computer.

Figure 19 is an example of a comparison of predicted and measured temperatures (°C) on the midplane of the 5 kW full-scale heater test [*Hood*, 1980]. Figure 19a shows an areal distribution on a horizontal plane, with predicted temperatures indicated by dashed lines and measured results shown by the values next to the plotted squares. Figure 19b shows a comparison, at a point 0.5 m from the center of the heater, of results as a function of time. Difficulties with instrumentation were the cause of fluctuations in the measured results. In making predictions, the effects of fractures were ignored, and good agreement between measured and predicted results was still obtained. As other researchers have found, the transfer of heat is controlled by conduction, and this process is not significantly affected by such discontinuities.

A more difficult problem was the prediction of displacements that would develop in the rock mass as a result of thermal expansion of the fractured granite. Because the mechanical properties of fractured granite were not known, predictions of displacements for unfractured granite served as a first approximation [*Chan and Cook*, 1979]. The agreement between predicted and measured displacements over a period of 180 days was not very good. The measured results for either orientation of the extensometers were only 30% of what was expected [*Hood*, 1980]. Predicting displacements in fractured rocks subjected to a thermal field is not an easy task because of

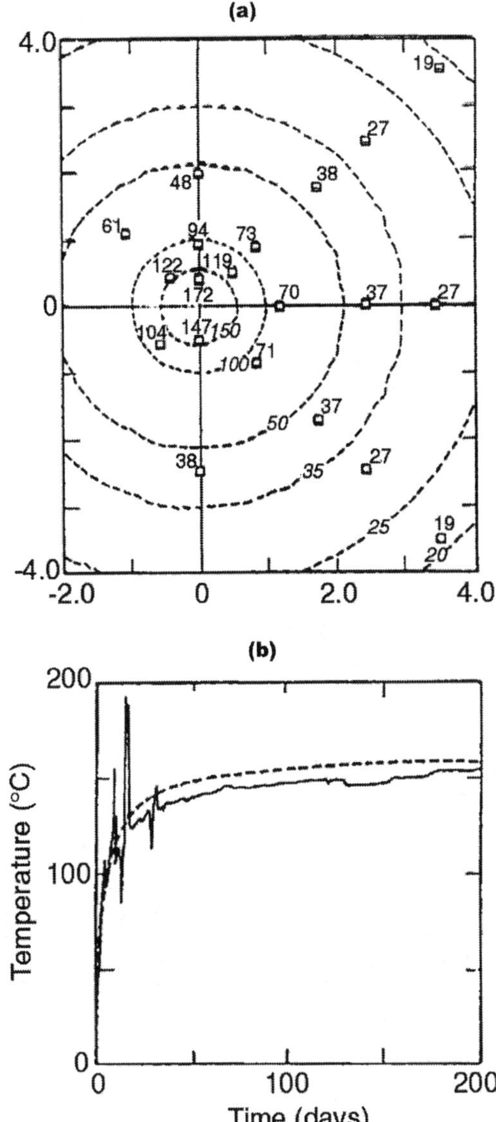

Figure 19. Predicted isotherms and measured temperatures on a horizontal plane through the center of the 5 kW full-scale heater: (a) 190 days after heating had started, and (b) as a function of time at a radius of 0.5 m from heater. Dimensions in meters [after *Hood*, 1980].

the nonlinear deformations that fractures undergo as the state of stress changes.

The initial state of the heater holes was dry. Once drill water had been removed, water levels did not change significantly. However, when electric heaters were turned on, a significant amount of water flowed into the heater holes, requiring redesign of equipment to avoid flooding. This increase in flow was caused by either fracture closure or a shear movement that caused dilation. Flow rates

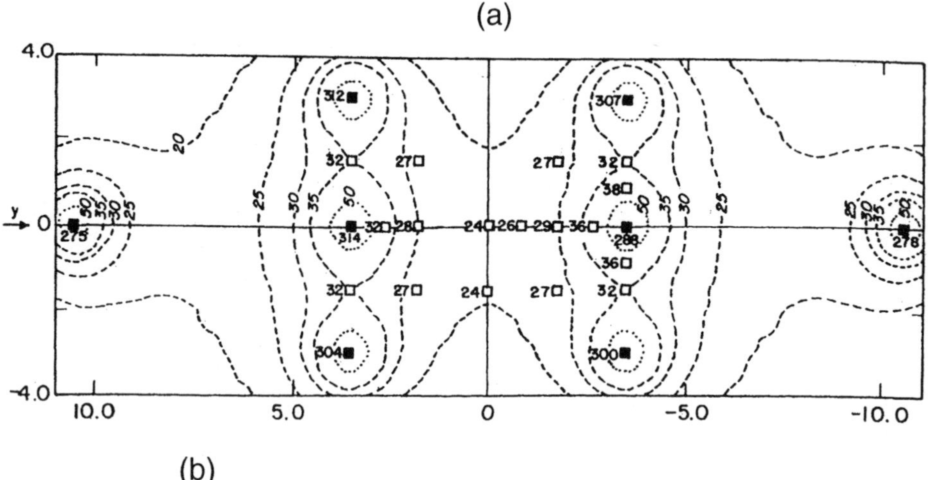

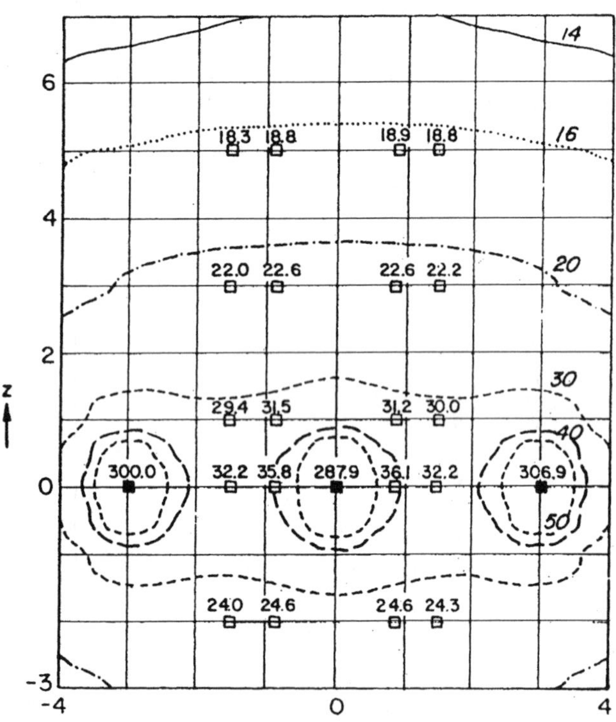

Figure 20. Predicted isotherms and measured temperatures in time-scaled heaters after 190 days: (a) for horizontal plane through centers of heaters, and (b) for vertical plane through centers of heaters at y = –3.5 m. Heater locations are marked in black. Dimensions in meters [after *Hood et al.*, 1980].

decreased with time, except when heat output was increased, which generated a similar surge in water inflow. These inflows to heater holes were a clear indication of the effects of fractures as they accommodated the thermal rock expansion.

The full-scale heater tests were designed to investigate local effects of temperature changes on granite. The time-scaled heater test (location on Figure 18) was the result of an idea that Neville Cook, Professor of Mining at the University of California, Berkeley, had for extending the results of the full-scale heater test over a longer period of time. The idea was to investigate the effects of an array of heaters as the thermal fields merge with time to appear as a single larger source.

As in all diffusion processes, the effects of a thermal perturbation propagate over distance as a function of the square root of thermal diffusivity and time. Thus, 2 years of a properly designed time-scaled experiment would yield the equivalent of 20 years of data from the full-scale heater test. To accomplish this, the linear scale must be reduced to $1/\sqrt{10} \approx 0.32$ of the full scale. An array of 8 heaters, spaced 7 m apart along the axis of the time-scaled drift and 3 m apart in the other direction, was used (Figure 20). Each heater was installed in a canister 1 m in length, which was placed in a borehole such that midplane of the canister was 10.5 m below the floor. Scaling of the power output for these heaters showed that 1 kW per heater would be representative of an initial power output for the base case. This power level was varied to simulate the decay in power output of radioactive waste over 20 years.

An example of the thermal field that developed after 190 days in the time-scaled test is shown in Figure 20 [*Hood et al.*, 1980]. Figure 20a shows temperatures (°C) in a horizontal plane through the center of the time-scaled heaters, and Figure 20b shows temperatures in a vertical plane through 3 of the heaters located at y = –3.5 m. Good agreement exists between predicted and measured temperatures, and it is of interest to observe the symmetry of the induced temperature field on both horizontal and vertical planes.

5.4. Large-Scale Permeability Test

As part of the hydrogeological investigations, an attempt was made to measure the in situ permeability of the fractured granite. This was done on a large scale using the ventilation drift (Figure 18). Figure 21 shows the arrangement of equipment and boreholes used [*Wilson et al.*, 1981]. A central question was how to quantify groundwater flowing through the fractured granite. Visual inspection of the drift showed the walls were almost totally dry. Minor amounts of water were seeping out of the rock mass from a few wet cracks, but when the walls were scanned with an infrared sensor, there was evidence of additional leakage at a few locations. A slight drop in temperature indicated water inside a crack that was not visible on the wall.

As shown in Figure 14, the depth of the ventilation drift at 330 m is well below the water table, and since air pressure in the drift is atmospheric, there had to be a significant hydraulic gradient in the surrounding rock to force groundwater into the drift. To measure this gradient, a series of 15 boreholes were drilled in all directions out from the drift (Figure 21). The boreholes were 30 m to 40 m in length, and were equipped with packers to seal off 6 sections so that pressures in the fractured rock could be measured.

Figure 22 shows the results of pressure measurements in 5 radial boreholes (R01–R05) drilled near the end of the drift (Figure 18). Note that pressures generally indicate a gradient for flow toward the drift. In drilling the 15 boreholes, it was observed that R01 was the only one that collected any significant amount of water. Therefore, it was decided that this monitor hole would be the last one shut off. Pressure increases at the other 4 boreholes on Figure 22, although not uniform in magnitude, indicated a satisfactory degree of connectedness within the fracture network.

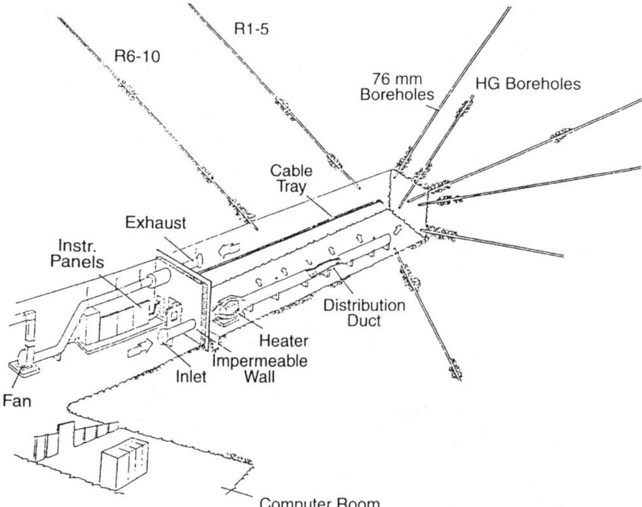

Figure 21. Schematic view and layout of equipment used in large-scale permeability test [after *Wilson et al.*, 1981].

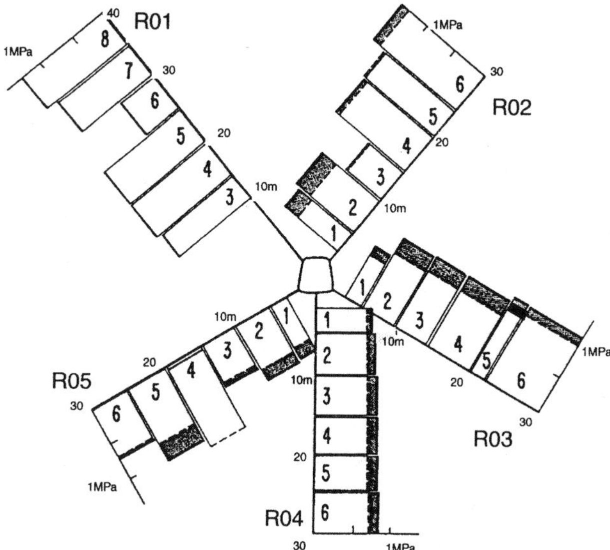

Figure 22. Pressure measurements in radial boreholes of large-scale permeability test. Stippled area shows pressure increase 8 days after packing off R01 [after *Wilson et al.*, 1981].

The remaining problem was to determine the flow rate of water into the drift. This was accomplished with the equipment shown in Figure 21 [*Wilson et al.*, 1981; *Galbraith et al.*, 1981]. A 30 m length of drift was walled off, and a fan used to withdraw air from the room. A duct on the inside of the room distributed air over all exposed surfaces. A heater was used to maintain a fixed air temperature, and the circulating air picked up whatever water vapor was present. Liquid water on the floor of the drift was insignificant, and as soon as ventilating conditions reached a steady state, water flux from the rock mass was computed from the change in humidity of the inlet and outlet air streams.

Radial cylindrical flow was assumed for this system because drawdown varies with the log of distance from a linear sink or source. This relationship is illustrated in Figure 23, which is a distance-drawdown plot for observed pressures when air temperature was fixed at 30°C and water flow rate, calculated from humidity data, was 42 mL/min. Note that a significant hydraulic gradient is evident in pressures measured at radial distances between 6 m and 30 m [*Wilson et al.*, 1983]. The weighted average gradient is shown by the solid line.

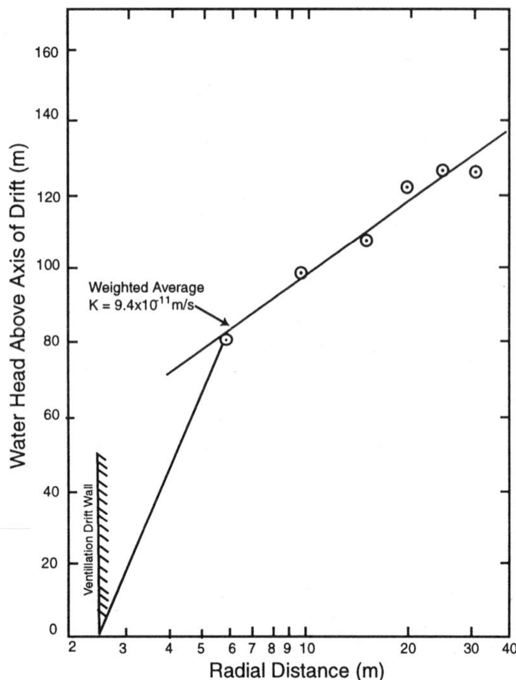

Figure 23. Distance-drawdown plot for observed pressures with air temperature of 30°C [after *Wilson et al.*, 1983].

Note also that the indicated gradient will not extrapolate to zero at the drift wall, although it should because air pressure in the sealed-off room is atmospheric. This is probably due to a combination of factors: (1) removal of rock produced a low-permeability "skin" that reflects the effects of fracture closure imposed by tangential stress concentrations around the opening; (2) effects of the blast itself could also have produced additional fracture closure; and (3) later work at Stripa showed that there was a release of gas as pressures on the groundwater fell, and this could have reduced the relative permeability of water due to interference effects of two-phase flow conditions in the walls of the tunnel.

The flow rate of 42 mL/min and the weighted average gradient were used to calculate a hydraulic conductivity of 9.4×10^{-11} m/s, which is equivalent to a permeability of about 10^{-17} m^2 [*Wilson et al.*, 1983]. From the dimensions of the system involved, this result for permeability applies to a rock volume of approximately a million cubic meters.

6. FURTHER INVESTIGATIONS OF FRACTURE NETWORKS

6.1. Equivalent Porous Media Concept

By the 1980s, the problem of how to determine equivalent porous media permeability of a network of fractures had not progressed very far beyond the work of *Snow* [1965, 1969]. He had assumed fractures to be of infinite extent, whereas it can be seen in the field that fractures have finite dimensions. Another unsolved problem was how to obtain data on effective aperture distributions.

Long [1983] adopted a different approach by investigating the possibility of determining an appropriate permeability tensor for a fracture system when it behaves like a porous medium. Fractured rock can be said to behave like an equivalent porous medium when: (1) there is an insignificant change in the value of permeability with either a small addition or subtraction to the test volume; and (2) apermeability tensor exists that can be used to predict the correct flux.

To investigate this approach, *Long* [1983] generated arbitrary fracture patterns designed to reproduce the geometry of real systems. She assumed the following conditions: (1) a two-dimensional system has fracture centers randomly located in the plane; (2) fractures are generally elliptical in three dimensions; (3) orientations are distributed normally; (4) trace lengths are distributed either exponentially or lognormally; and (5) apertures are distributed lognormally. A set of field data could then be used to generate a 2-D fracture pattern [*Long*, 1983], as shown in Figure 24. Simulated flow tests through fracture models generated in this fashion provide a system on which directional permeability, K_g, can be measured. A polar coordinate plot of $1/\sqrt{K_g}$ will be an ellipse if the fracture model behaves like an equivalent anisotropic, homogeneous medium [*Long*, 1983]. When the measured

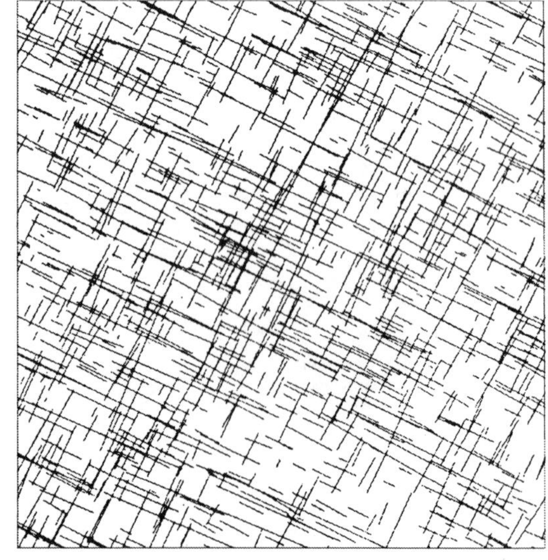

Figure 24. Example of two-dimensional network of fractures [after *Long*, 1983].

values of $1/\sqrt{K_g}$ did not all plot on a single, unique ellipse, *Long* [1983] developed a procedure for computing the best-fit ellipse for such results, no matter what the shape of the plot.

Using *Long's* [1983] method, *Long and Witherspoon* [1985] investigated the effect of the degree of interconnection by numerically simulating flow in networks where fracture density and extent varied inversely, while the product of these two parameters was held fixed in systems where matrix permeability could be ignored. The results illustrate that as fracture length increases, the degree of interconnection increases. Thus, for a given fracture frequency as measured in a borehole, permeability of the system increases as fracture length increases and density is proportionately decreased. Also, fracture systems with shorter but denser fractures behave less like porous media than do systems with longer but less dense fractures. Therefore, knowledge of fracture frequency and orientation alone is inadequate when determining permeability or deciding whether a given system behaves like a porous medium.

Long et al. [1985] extended *Long's* [1983] 2-D approach to 3-D, random networks of fractures where the fractures are disc-shaped discontinuities in an impermeable matrix. Figure 25 shows a cubic region within which three orthogonal sets of fractures have been generated. The discs can be arbitrarily located within the rock volume and can have any desired distribution of aperture, radius, orientation, and density.

A 2-D analysis of flow through some plane in such a network would only include traces of fractures in the specified plane. Such traces cannot interconnect to the same degree as a 3-D array of fractures, and a 2-D analysis would underestimate permeability of such a network. Thus, where the 3-D disk model is appropriate, one can analyze fracture networks that are statistically similar to those that occur in rock systems. After boundary conditions and fractures are identified, steady flow through the network is calculated using a mixed analytical-numerical technique [*Long et al.*, 1985; *Gilmour et al.*, 1986].

Long and Billaux [1987] applied the concept of an equivalent porous medium to field data collected on a fractured rock mass at Fanay-Augeres, a uranium mine in France. They adopted a 2-D approach and developed a technique that accounts for observed spatial variability. They focused their effort on a section of a drift where fractures had been mapped and permeability tests performed in 10 boreholes. To account for spatial variability, they generated a 100 m × 100 m fracture network with a total of 65,740 fractures distributed among

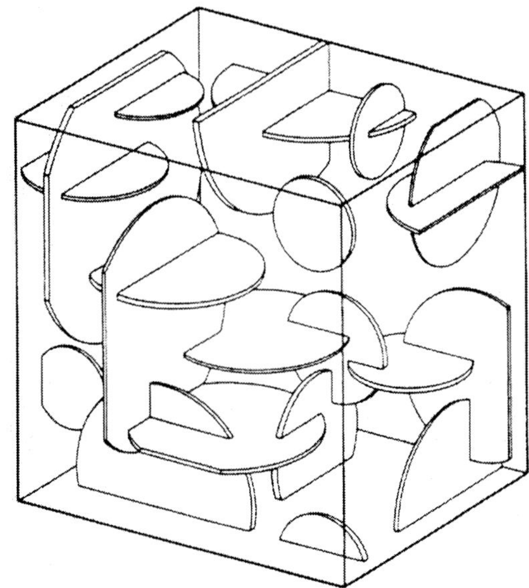

Figure 25. Cubic region with three-dimensional fracture network of disc-shaped orthogonal fractures [after *Long et al.*, 1985].

100 statistically homogeneous subregions. When directional permeabilities were tested, the results showed that the system was barely connected; about 0.1% of the fractures essentially controlled permeability.

Long and Billaux [1987] concluded that a 3-D analysis of the rock mass at Fanay-Augeres was needed. They suspected that the assumed parallel-plate concept was invalid because the effects of stress and filling materials produced conditions where flow in fractures only occurred in sinuous channels. These results indicated the need to investigate how fracture connectedness can be determined.

Hestir and Long [1990] investigated the possibility of combining percolation theory and equivalent media theory to characterize permeability of random 2-D Poisson fracture networks. Such theories are usually applied on regular lattices where the lattice elements are present with probability p. To apply these theories to random systems, it is necessary to define: (1) the equivalent to the case where $p = 1$; (2) p in terms of the statistical parameters of the random network; and (3) the equivalent of the coordination number z.

6.2. *Inverse Methods for Equivalent Media Investigations*

One of the characteristics of flow and transport in fractured rock is that fluid migration may be largely confined to a poorly connected network of fractures. To investigate this problem, *Mauldon et al.* [1993] developed a

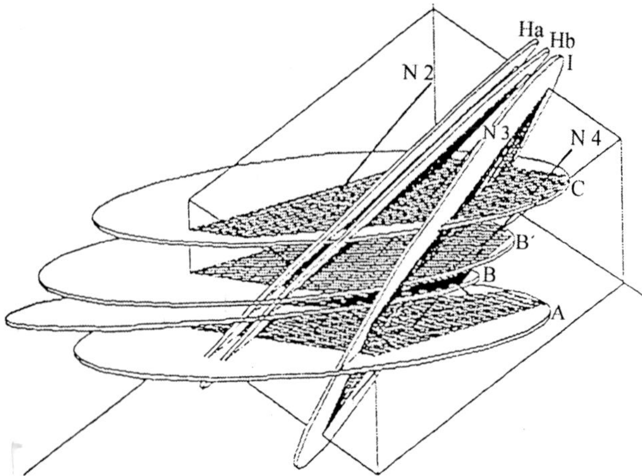

Figure 26. Three-dimensional conceptual model of principal fracture zones at the SCV site at Stripa [after *Long et al.*, 1990].

new type of model that they called an "equivalent discontinuum" model. They represented the discontinuous nature of the flow system by using a lattice that is only partially filled with fractures. This was done through a statistical inverse technique called "simulated annealing." The model is "annealed" by continually modifying an initial model, or "template," so that with each modification the model behaves more and more like the observed system. The template is constructed using geological and geophysical data to identify regions that may be permeable and the orientation of specific channels that are probable conductors.

This method was used in a field investigation during the international project at Stripa [*Gnirk*, 1993], in what is known as the Site Characterization and Validation (SCV) site. Using the methods described above, *Long et al.* [1990] developed a 3-D conceptual model of the principal fracture zones at the SCV site, as shown in Figure 26. A 2-D mesh of the H zone in this model was constructed that includes seven boreholes (C1, C2, C3, C4, C5, W1, W2). Each of these boreholes was fitted with packers that isolated a test interval in the H zone. An interference test was run by withdrawing water from the C1 borehole at a constant rate while responses were measured in the packed-off intervals in the other holes.

The 2-D model was annealed against the observed pressure effects in the H zone, and a comparison was made of the measured drawdown curves with those obtained from simulated annealing (Figure 27). The process of annealing is described by *Mauldon et al.* [1993]. This procedure allows the algorithm to bypass local minima and seek a more global solution. The process is repeated through successive iterations until the energy level has decreased to an acceptable minimum. The process does not, however, guarantee a unique solution for the flow system that is finally developed because other configurations could yield identical energy levels.

Doughty [1995] pointed out that the iterative procedure described above can be improved by restricting the search to parameters that represent self-similar (fractal) hydrologic property distributions. Far fewer parameters are needed, improving the efficiency and robustness of the inversion. Additionally, each parameter set produces a hydrologic property distribution with a hierarchical structure, which mimics the multiple scales of heterogeneity often seen in natural geologic media. The parameters varied during the inversion create fractal sets known as attractors, using an iterated function system (IFS). Application of the IFS inverse method to synthetic data shows that the method works well for simple heterogeneities, and application to field data from a sand/clay sedimentary sequence produces reasonable results. *Doughty et al.* [1994] investigated the same data, described above, that were obtained at the SCV site in the Stripa mine [*Long et al.*, 1992]. They obtained results that reveal several complications in the flow behavior of the H zone and indicated the need for a heterogeneous description of this zone.

In the process of applying inverse methods to fractured rocks, it is usually desirable to apply forward methods of analysis to obtain results for comparison. A finite element code TRINET was developed by *Karasaki* [1987] that can be used to calculate fluid flow and solute transport on a lattice of one-dimensional elements (or pipes) of porous media.

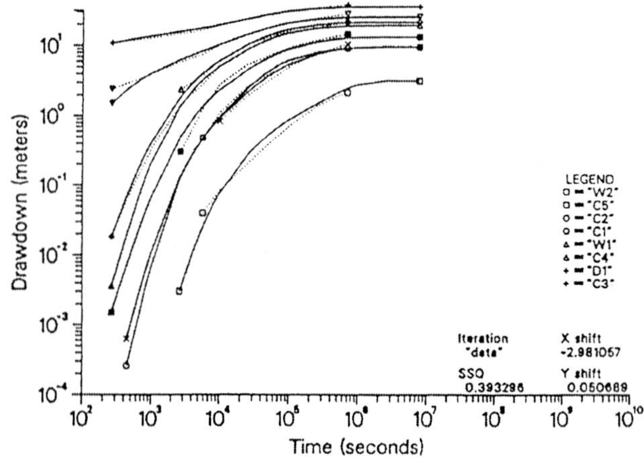

Figure 27. Drawdown curves (dotted lines) obtained at observation points during withdrawal from borehole C1, compared to results (solid lines) of simulated annealing analysis. [after *Long et al.*, 1992].

6.3. Analytical Methods of Analyzing Well Test Data

As mentioned above in Section 4, *Lai et al.* [1983] developed a method for analyzing constant-rate well test data from naturally fractured reservoirs. In addition, *Karasaki et al.* [1988a] developed an analytical model for analyzing well test data from fracture-dominated reservoirs. This is a concentric composite model with a finite-radius well located in the center. Flow is assumed to be linear in the inner region of the model, and radial in the outer region. Solutions were obtained analytically, and type curves for ranges of dimensionless parameters were developed. Solutions to various models of slug tests for this concentric composite model also were developed by *Karasaki et al.* [1988b].

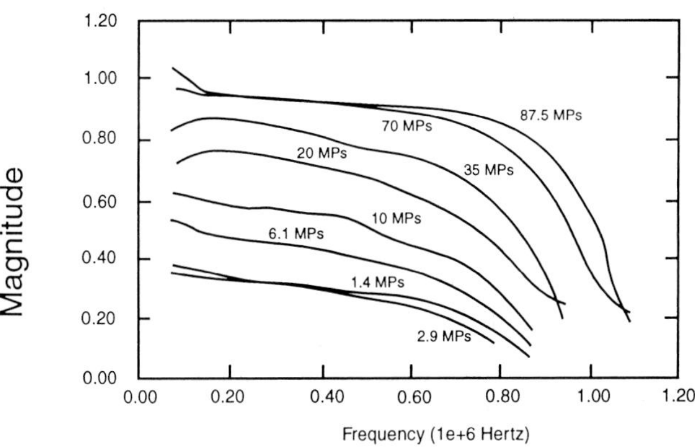

Figure 28. Attenuation of shear waves in propagating across a dry fracture in granite [after *Pyrak-Nolte et al.*, 1987].

6.4. Seismic Detection of Fractures

Various investigations have been carried out to determine how seismic data can be used to improve the characterization of fractured rocks. *Pyrak-Nolte et al.* [1990] examined how the effects of fractures on seismic wave propagation can be modeled as a boundary condition in the seismic wave equation. Seismic stress is continuous across such a boundary, but seismic particle displacement and seismic particle velocity are not. The complete solutions for seismic wave reflection, conversion, and transmission across a displacement and velocity discontinuity between two half spaces with different densities and elastic properties have been derived for all angles of the incident wave [*Pyrak-Nolte*, 1988; *Pyrak-Nolte et al.*, 1990].

Laboratory investigations have revealed the significant effect that a fracture can have on wave propagation. *Pyrak-Nolte et al.* [1986] reported the results of measurements on the attenuation of seismic waves passing across a single fracture in a sample of dry granite (Figure 28). The results of attenuation are presented in terms of a ratio of the shear wave spectral amplitude in the fractured sample relative to the spectral amplitude in the intact granite. Fracture stiffnesses are computed from measured fracture displacements, and over the stress range of 5 to 50 MPa, the specific stiffness increased by a factor of 10, from 2 to 22×10^{-6} MPa/m. These results are in good agreement with Schoenberg's theory [*Schoenberg*, 1980, 1983].

A major borehole experiment was carried out in a controlled environment as part of a program of detecting fractures for nuclear waste isolation. This experiment was over a period of 3 years [1987–1989] in the Grimsel underground research facility in Switzerland [*Majer et al.*, 1990]. Two boreholes were drilled through a fracture zone that had been located in the underground works. Crosshole imaging was then carried out as the fracture was subjected to pressurization (inflation). The purpose was to evaluate and develop high-resolution seismic imaging of fractured rock masses, as well as to validate the fracture stiffness theories. This work was highly successful and demonstrated the usefulness of seismic methods at high frequency (kilohertz) for mapping fractures in rock [*Majer et al.*, 1990].

The next major effort was to scale these methods up from tens of meters used in the work at Grimsel to hundreds of meters needed for applications in such areas as nuclear waste, geothermal, and fossil energy. Since the early 1990s, the focus of this effort has been to develop and apply borehole seismic methods for fracture definition and characterization. The primary thrust has been to define fractures that control flow and transport. This work is being carried out in a number of environments to gain fundamental knowledge on the scaling of seismic wave propagation in fractured media.

As part of the research effort on fossil energy, investigations have been performed to determine how borehole seismic methods could be used to characterize naturally fractured gas reservoirs. Work was carried out in conjunction with Conoco, Inc., at its Newkirk borehole test facility in Oklahoma [*Majer et al.*, 1997]. Investigations used an arrangement of 5 GW wells (Plate 1) that were completed in a shallow (15 to 35 m), water-saturated, fractured limestone (Fort Riley) sequence. Previous crosswell seismic and hydrologic interpretations indicated strong evidence [*Majer et al.*, 1997] for conductive fractures trending N70°E (i.e., between GW-2 and GW-5).

To increase visibility of the fracture zone to seismic imaging, air was injected in GW-5 at the same time as water was withdrawn from GW-2. Crosswell seismic

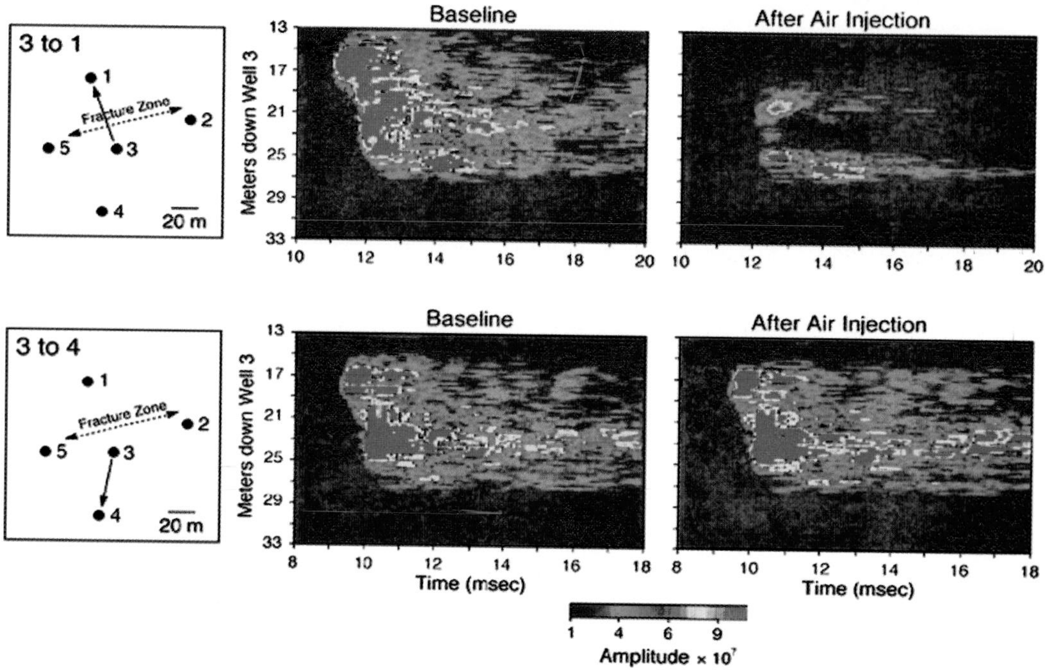

Plate 1. Crosswell amplitude data as a function of depth and time between well pairs GW-3/GW-1 (a, top row) and GW-3/GW-4 (b, bottom row), before (left tomograms) and after (right tomograms) air injection [after *Majer et al.*, 1997].

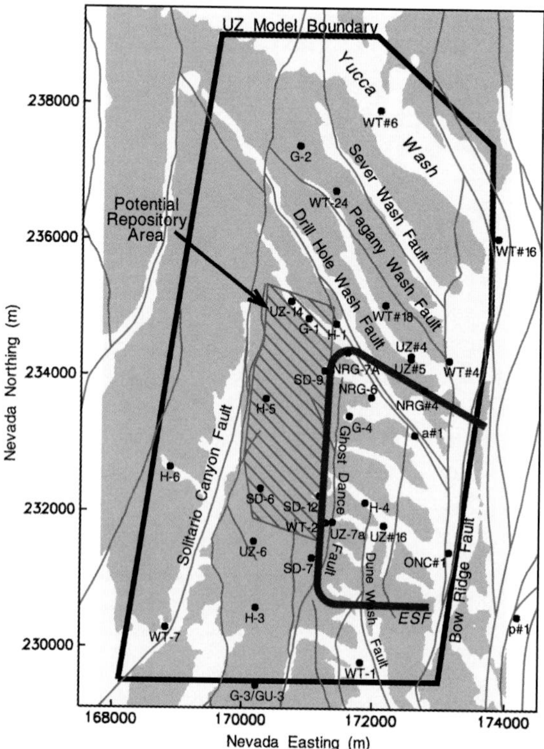

Plate 2. Map view of the site-scale model domain with GFM3.0 faults [*Clayton* 1998], selected borehole locations, the ESF, and the outline of the potential repository. Shaded areas represent exposed bedrock and white areas indicate alluvium [after *Hinds, et al.*, 1998].

surveys were performed before and after air injection using well pairs GW-3/GW-1 and GW-3/GW-4. Crosswell amplitude data as a function of depth are included on Plate 1. It is apparent that air followed the preferred pathway predicted by hydrologic modeling. In addition, single-well seismic imaging also detected the fracture zone in a location consistent with crosswell and hydrologic inversion results. It was concluded [*Majer et al.,* 1997] that very small aperture fractures can have a large effect on seismic wave propagation.

An example of fracture characterization using tomographic methods on a much larger scale is currently being carried out at Yucca Mountain, Nevada, at a DOE site that is a potential radioactive waste repository. (For a more detailed discussion of the Yucca Mountain facility, see Section 8 of this paper.) In previous surface geophysical work carried out from 1994 through 1996 [*Majer et al.,* 1997], it was concluded that the mountain as a whole was difficult to characterize from the surface because of topographic variations, surface noise, near-surface weathering, and lithology, as well as a variety of other access issues. Therefore, by using borehole-to-borehole seismic imaging, or at least by placing sensors in an underground drift, it was thought that one could increase resolution and ground truth of results by directly observing such properties as lithology and fracture and fault patterns, and relate these properties to the seismic data.

To meet a subset of the above needs, it was decided in December 1997 to use the combined methodologies of crosshole, crossdrift, drift-to-borehole, and surface-to-drift tomographic seismic imaging. The general concept was to place receivers at regular intervals in subsurface tunnels of the Exploratory Studies Facility (ESF) at Yucca Mountain to record seismic energy from the surface and other drifts. The concept was to first image from the surface into the ESF, predict geology in the crossdrift area, and use mapping information from the crossdrift data to calibrate seismic results. Then the data would be refined and improved by performing tunnel-to-tunnel imaging between the crossdrift and ESF. Vibroseis sources were used on the surface, and two component (vertical and horizontal) geophones were cemented in place in the ESF at 15 m intervals. Source intervals at the surface were 30 m along a line 5 km in length. The receiver line was approximately 3 km in length and offset 3 km parallel to the source line. The tomography experiment proved to be a successful test. This work, when combined with numerical modeling and prototype experiments, has demonstrated the potential of large-scale, surface-to-tunnel, seismic tomography at Yucca Mountain.

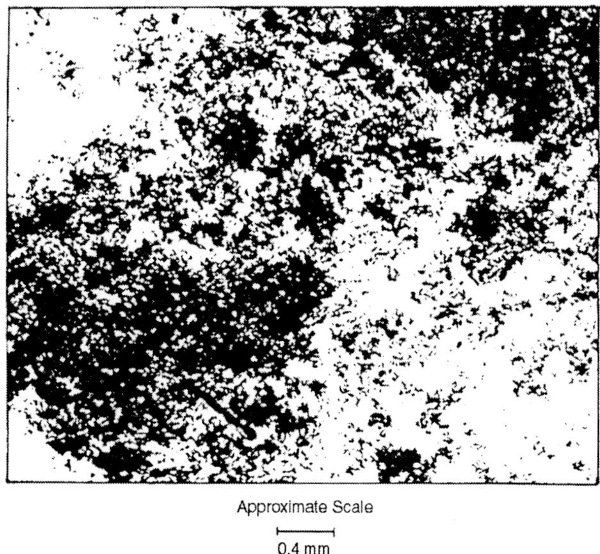

Figure 29. Composite from micrographs of fracture surfaces in a sample of Stripa granite at an effective stress of 85MPa. Regions penetrated by metal appear white [after *Pyrak-Nolte et al.,* 1987].

7. HETEROGENEOUS FLOW IN SATURATED FRACTURED ROCK

7.1. Evidence for Heterogeneous Flow in Fractures

As more experience was obtained from investigations on fractured rocks, it became clear that characterizing the heterogeneity of flow paths is a fundamental problem. *Pyrak-Nolte et al.* [1987] conducted one of the first laboratory studies at Berkeley that demonstrated the nature of these heterogeneities. They are well illustrated by the micrograph in Figure 29, which shows the surface of a natural fracture in Stripa granite (quartz monzonite). The picture gives the locations where Wood's metal (white) was found in open pore spaces of this fracture. The sample was prepared by enclosing it in a special cell, subjecting it to an effective stress of 85 MPa, and holding it at a temperature just above the melting point of Wood's metal (~80°C). After molten metal was injected, both fluid pressure and axial load were maintained while temperature was decreased until the metal solidified. The micrograph is a composite picture of images from each of the two fracture surfaces.

The complex nature of flow paths that Wood's metal had to follow is evident from the distribution of metal over the fracture surface. *Pyrak-Nolte et al.* [1987] also measured aperture distributions at two lower levels of effective stress

(3 MPa and 33 MPa), and as might be expected, there was significantly more open space in the fracture aperture. In other words, the degree of heterogeneity in flow paths through a fracture can change as a function of the stress field acting on the discontinuity.

Earlier, other researchers at Stripa [*Abelin et al.*, 1985; *Neretnieks*, 1987] carried out field experiments on solute migration in single fractures. Results showed that flow was very unevenly distributed along fracture planes that were being investigated, and large areas did not carry any water. The flow paths, or channels, made up only 5–20% of the fracture area. Furthermore, *Abelin et al.* [1985] found that if one assumes volumetric flow is proportional to the cube of fracture aperture, the equivalent value for this parameter, as derived from constant head permeability measurements, was much smaller than that derived from tracer migration measurements of the residence volume for the tracer solution. Similar observations of flow occurring in only a few channels in fractured rock were found on a much larger scale in migration experiments in granite in Cornwall [*Heath*, 1985; *Bourke*, 1987].

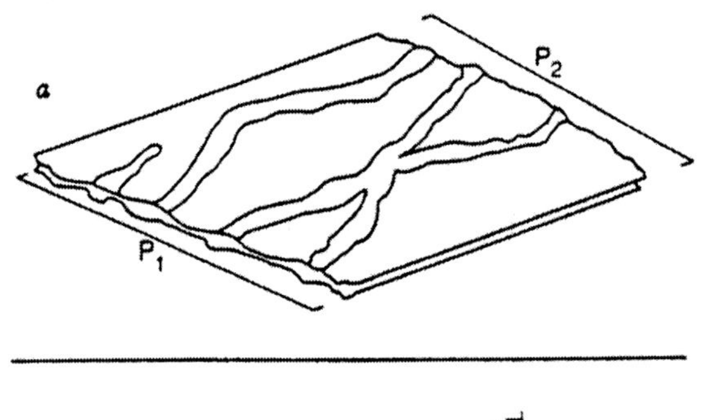

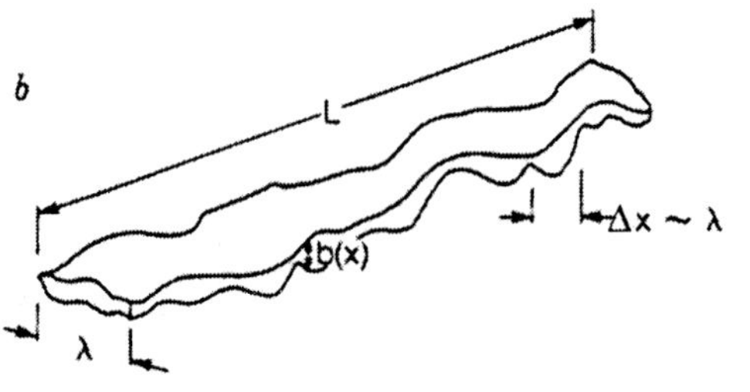

Figure 30. (a) Schematic diagram of the channel representation of fluid flow in a single fracture, and (b) Schematic sketch showing parameters for one channel [after *Y. Tsang and C.-F. Tsang*, 1987].

7.2. The Variable-Aperture Channel Model

These field observations on the complicated nature of fluid flow in saturated fractured rocks have led to various attempts to develop appropriate models for this behavior. *Y. Tsang and C.-F. Tsang* [1987] studied fluid flow and solute transport in a fractured medium in terms of flow through one-dimensional channels of variable aperture. They characterized the channels by: (1) aperture-density distribution, or the relative probability of occurrence of a given aperture value; (2) effective channel length and width; and (3) aperture spatial correlation length. They visualized channels in a single fracture in Figure 30a, and they identified the parameters of a channel in Figure 30b. This 1-D channel is defined by aperture density distribution n(b) along its length. Channel width is assumed to be a constant of same order as correlation length, λ. By definition, correlation length is the spatial range within which apertures have similar values.

With this geometric picture in mind, *Y. Tsang and C.-F. Tsang* [1987] statistically generated aperture profiles along channels and compared results with laboratory measurements of fracture surfaces. Tracer transport between two points in fractured medium is by way of a number of such channels. Tracer breakthrough curves display features that correspond well with data reported by *Moreno et al.* [1985], which lends support to the validity of this model.

Moreno et al. [1988] extended this work by investigating 2-D flow in a single fracture that was discretized into a square mesh. They used geostatistical methods based on a given aperture probability density distribution and a specified spatial correlation length. Fluid potential at each node of the mesh was computed, and steady-state flow rates between all nodes were obtained. Results showed that fluid flow occurs predominantly in a few preferred paths. Hence, a large range of apertures gives rise to flow channeling.

Solute transport was investigated using a particle-tracking method, and both spatial and time variations of tracer breakthrough were obtained. Spatial variation of tracer transport between a line of injection points and a line of observation points was displayed in contour plots. Results indicated that such plots can provide information on spatial correlation length of heterogeneity in the fracture. The tracer breakthrough curve obtained from a line of point measurements was found to be controlled by aperture results and is insensitive to statistical realization and spatial correlation length.

7.3. Evidence of Channel Flow from the Stripa 3-D Experiment

C.-F. Tsang et al. [1991] used the variable-aperture channel model to analyze results from the Stripa 3-D experiment. This experiment was one of the most

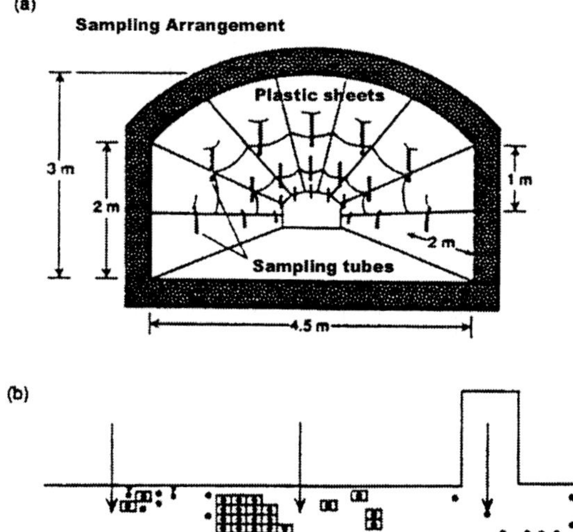

Figure 31. (a) Sampling arrangement in the Stripa 3-D experiment showing placement of plastic sheets for tracer collection, and (b) tracer distribution in the test site. Arrows indicate positions of injection holes, solid circles indicate sheets with significant water flow, and rectangles indicate sheets where tracers were collected [adapted from *Abelin et al.*, 1987].

comprehensive, tracer investigations in fractured rocks carried out at Stripa by Neretnieks and coworkers [*Neretnieks*, 1987; *Abelin et al.*, 1987] during the period 1984–1987. The investigations used two drifts arranged in the form of a cross, as shown in Figure 31b. Each drift is 4.5 m wide and 3 m high; the longer drift is 75 m long and the shorter intersecting drift is 25 m long. Three vertical boreholes were drilled into the ceilings of the drift (Figure 31b), and packed-off intervals were used to inject nine different tracers into the rock mass. The roof and sides of the drifts were covered with over 300 plastic sheets (1 m × 2 m) to collect tracers in the groundwater dripping from the walls (Figure 31a). Details of the experiment are given by *Neretnieks* [1987] and *Abelin et al.* [1987].

Flow rates at the collection sheets were approximately constant in time, indicating that steady-state flow conditions prevailed. However, the majority of these collection sheets did not receive significant flow, and as shown in Figure 31b, tracer mass tended to concentrate at discrete regions in the main drift. Only five of the nine tracers were found at significant concentrations in the waters that collected in different plastic sheets during the first 30 months of the experiment. These were Eosin B, Eosin Y, Uranine, Elbenyl, and iodide. It is apparent that

Table 1. Major conclusions from application of variable-aperture channel model to Stripa-3-D data.

Parameter	Value
Number of injection locations	5
Transport distance	10—43 m
Mean travel time	2200—16000 hr
Peclet number (all channels)	3.0—400
Dispersivity (first channels)	0.6—2.9 m
Mean aperture (first channels)	0.3—340 μm
Aperture variance	0.35—0.75
Number of channels	2—4

nearly all the tracer mass was collected in a limited section (Figure 31b) of the main drift.

In analyzing data from this experiment, *C.-F. Tsang et al.* [1991] applied the variable-aperture channel model [*Y. Tsang and C.-F. Tsang*, 1987, 1989], which assumes that flow is dominated by migration in just a few channels. Two methods were developed to address the strong time variation in injection flow rate. The first approximates the early part of injection history by an exponential decay function. The second is a deconvolution method using Toeplitz matrices, applicable over the complete period of variable injection of the tracers. The results led to a confirmation of the concept of multiple variable-aperture channels and to the determination of a number of parameter values for the fractures. As shown in Table 1, these parameters include not only estimates of mean aperture and its standard deviation, but also number of channels involved in tracer transport, Peclet numbers, and dispersivities. *C.-F. Tsang et al.* [1991] obtained a surprising result—the data indicate the Peclet number increases with mean travel time. This is interpreted in terms of the strong heterogeneity of the flow system.

7.4. A Three-Dimensional Variable-Aperture Model

The effects of high variance in fracture transmissivity on transport and sorption at different scales were investigated by *Nordqvist et al.* [1992, 1996]. They used a 3-D variable-aperture fracture network model for flow and transport to study the effects of this variability on sorbing and nonsorbing tracer transport, as well as scale effects on transport distance. With this model, *Nordqvist et al.* [1992] found they could produce multipeak breakthrough curves, even for relatively moderate values of fracture-transmissivity variance. Breakthrough curves displayed dispersion on two different scales, as was observed in field tracer experiments [e.g., *C.-F. Tsang et al.*, 1991]. The multipeak structure was interpreted as evidence of channeling.

When linear sorption was included, *Nordqvist et al.* [1996] found that dispersion characteristics of the breakthrough curves also changed. The degree of change depends strongly on fracture-transmissivity variance, as does the translation. In particular, when there is a high fracture-transmissivity variance, the translation in mean residence time due to sorption is significantly smaller compared to those cases with low variance. In view of the high variability in the model output data, Nordqvist and coworkers concluded that extrapolation of results from a particular tracer experiment would be highly uncertain.

7.5. The Stochastic Continuum Model of Fractured Rock

In a new approach to characterizing heterogeneity, *Y. Tsang et al.* [1996] developed a stochastic continuum model of fractured rock and were able to condition the model on a specific set of hydrological field data from the Underground Hard Rock Laboratory at Äspö, Sweden. The model is a single-continuum model, yet the fracture zones are distinguished from the matrix by the assignment of long-range correlation structures along preferred planes of fractures with high hydraulic conductivity. Since the conceptual model is a continuum, hydraulic continuity is not an issue. However, the inclusion of hydraulic conductivities from Äspö that range over eight orders of magnitude implies that their generated stochastic field represents a strongly heterogeneous medium.

Y. Tsang et al. [1996] presented the results of flow and transport simulations in 3-D to illustrate the large spatial variability of point measurements in the data from Äspö. Since predictions based on spatially integrated properties provide less uncertainty and decreased sensitivity to different levels of heterogeneity, *Y. Tsang et al.* [1996] suggested that spatially integrated quantities may provide a better choice for predicting flow and transport. They pointed out that the strong heterogeneity of fractured systems manifests itself in a large spatial variability or large uncertainty in the point measurements. Therefore, such results may provide a poor representation of system behavior. On the other hand, spatially integrated quantities average out the variability, providing more stable results and a better representation of the average system behavior.

In this stochastic model, *Y. Tsang et al.* [1996] accounted for the fractures by assigning a long-range correlation structure, in the preferred orientation of the fracture zones, to a small fraction (approximately 11%) of the highest hydraulic conductivities. Their results show that, in the presence of the long-range correlation structure, the transport parameters derived from short-distance transport do not represent those for transport over a longer distance. This is an issue of concern that is being investigated further.

7.6. Heterogeneous Flow in Fractures from the Standpoint of Fluid Mechanics

Another approach to characterizing heterogeneous flow has been to consider the effects of asperities on the tortuous flow path developed around these features. The permeability of the fracture is controlled primarily by the geometry of these tortuous flow paths. *Chen* [1990] used the boundary element method to investigate the effects of tortuosity on permeability of an idealized fracture, consisting of two parallel plates propped open by arrangements of isolated asperities.

Zimmerman et al. [1992] used the boundary element method and a code, FLOW, described by *Chen* [1990], to investigate permeabilities of regions containing contact areas of various shapes. Since fluid flow can be described by Laplace's equation, with contact areas serving as impermeable boundaries, this problem is analogous to flow of electrical current in a thin sheet with holes. Experiments were carried out by *Zimmerman et al.* [1992] on such sheets to measure overall electrical conductivity (the analogue of fracture permeability) to validate the numerical code. Holes with the desired shapes, sizes, and locations were cut in the sheet, and overall conductance was measured.

This problem is typical of the application of effective-medium theory to heterogeneous systems. Unobstructed areas between obstacles are regions with some known permeability, k_o, whereas the obstacles are regions of zero permeability. The objective of effective-medium theory is to determine an effective macroscopic permeability, k^*, that can be used in Darcy's law, to model flow through fractures on length scales large enough to encompass many asperities. *Walsh* [1981] applied this method to a fracture with "randomly" located circular obstructions; *Zimmerman et al.* [1992] extended the method to cases in which obstacles are elliptical in shape, with random orientations. They found that permeability depends not only on the amount of contact area, but also on the shape of asperities. For circular or elliptical asperities, very accurate estimates were obtained using this approach.

Zimmerman and Bodvarsson [1996] reviewed the basic problem of single-phase fluid flow through a rough-walled fracture within the context of fluid mechanics. They examined the various geometric and kinematic conditions

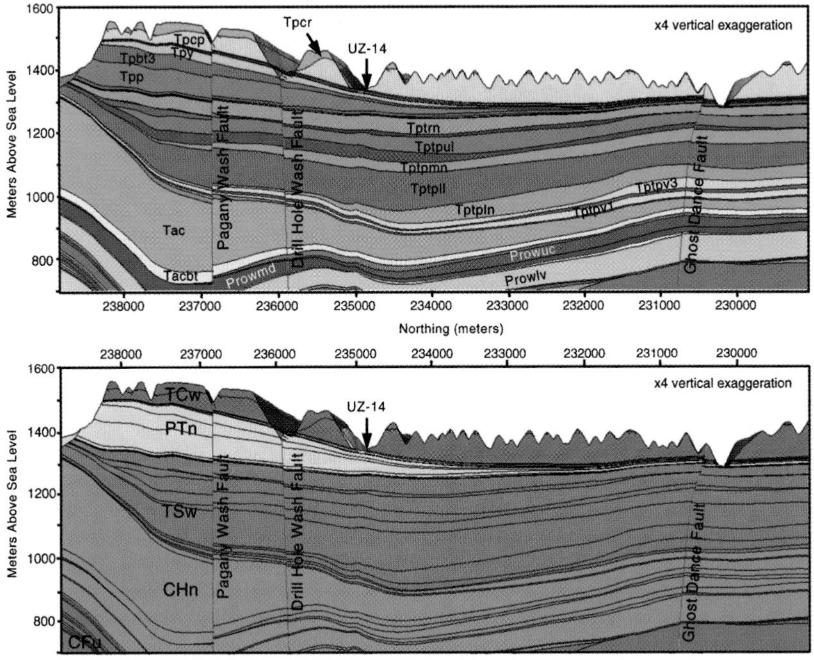

Plate 3. North-south cross section of Yucca Mountain near borehole UZ-14 showing (top) lithostratigraphic sublayers in GFM3.0 [*Clayton*, 1998] and (bottom) grouping of sublayers into hydrogeologic units of *Montazer and Wilson* [1984].

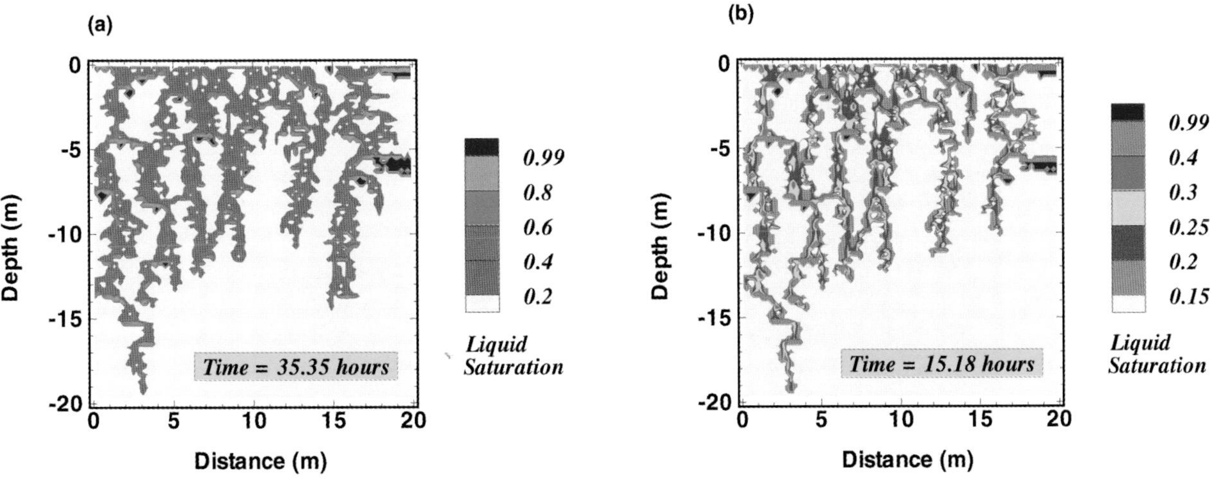

Plate 4. Liquid seepage at the time of breakthrough at a depth of –19.5 m for water injection at a constant rate of 10^{-3} kg/s over the entire top of the vertical fracture. Reference permeabilities are (a) 10^{-9} m^2 for the top result and (b) 10^{-7} m^2 for the bottom result [after *Pruess*, 1999].

that are necessary in order for the Navier-Stokes equations to be replaced by more tractable lubrication [*Zimmerman et al.*, 1991] or Hele-Shaw equations. In general, this requires a sufficiently low flow rate and some restrictions on spatial rate of change in aperture profile. Various analytical and numerical results were reviewed. These studies all led to the conclusion that effective hydraulic aperture is less than mean aperture by a factor that depends on the ratio of the aperture's mean value to its standard deviation.

7.7. Heterogeneous Flow in Fractures from Standpoint of Rock Mechanics

Changes in void space within a fracture that result from changes in stress also must be considered when evaluating heterogeneous flow in fractured rocks. *Pyrak-Nolte et al.* [1987] were able to demonstrate how heterogeneities in fracture surfaces (Figure 29) could be changed significantly as effective stress increases from 3 MPa to 85 MPa. Such variations in fracture geometry can have significant effects on global hydraulic properties.

Hopkins [1991] developed a model in which she could account for realistic mechanical deformations of fracture surfaces with changes in normal stress. The fracture was represented in the model by asperities between elastic half spaces. These asperities were modeled as cylinders that deform elastically, and a replica of the fracture surface was constructed from which fracture surfaces could be calculated from the theory of elasticity. *Hopkins* [1991] was then able to account for the mechanical interaction among all contacting asperities. Parametric studies using this model have clearly shown the role not only of the contact area in global mechanical behavior of a fracture, but also the role of spatial distribution of the contact area [*Hopkins*, 1999].

Gentier and Hopkins [1996] used the *Hopkins* [1991] model to investigate an actual granite fracture. A geometric model of surface roughness was calculated based on the void space map obtained from a cast of the fracture. It was thus possible to map aperture and contact area across the fracture for any specified applied normal stress and determine normal stresses acting at each contact point. Evaluating fracture deformations in this fashion led to estimates of contact area much smaller than are obtained assuming uniform closure across fractures, which has important implications for flow.

In a number of laboratory experiments, *Hopkins* [1999] has observed that, at high stresses, measured flow rate through a fracture can become independent of stress and an irreducible flow persists. Mechanically, the jointed specimen behaves like intact rock, meaning the cubic law for fracture flow no longer holds. This has been observed by *Gale* [1975] (Figure 3), *Iwai* [1976] (Figure 5), and others [*Engelder and Scholz,* 1981; *Raven and Gale*, 1985; *Pyrak-Nolte et al.*, 1987; *Gentier*, 1987, 1990; *Cook et al.*, 1990]. This experimental evidence for irreducible flow is consistent with the results presented by *Hopkins* [1999], showing that joints can achieve normal stiffnesses comparable to intact rock, with relatively small amounts of total contact between joint surfaces. Even at very high stresses, the tallest asperities have the effect of propping the joint open, while pockets of connected void space will remain for flow around the contacting asperities. *Hopkins* [1999] concludes that geometry-dependent deformation of the joint means that average aperture is a relatively unimportant parameter in determining hydraulic properties of joints.

8. THE YUCCA MOUNTAIN PROJECT

The Yucca Mountain Project has been set up in the state of Nevada by the Department of Energy (DOE) to investigate a potential site for a radioactive waste repository. A rock layer known as the Topopah Spring tuff has been selected as the layer in which the waste will be placed. This is a welded and fractured tuff in the vadose zone some 300 m below the surface. The basic plan and strategies were set forth in the Site Characterization Plan (SCP) issued by DOE in 1988 [*DOE*, 1988]. Initial investigations involved detailed geological field studies carried out by the U.S. Geological Survey (USGS). In recent years, these studies have been expanded to include a wider collection of geophysical, hydrological, and geochemical data. These data are being integrated into a detailed conceptual model and into a 3-D numerical model known collectively as the "site-scale model."

Yucca Mountain is a fault-bounded volcanic plateau in the arid, south-central part of the Great Basin. Within the 600–700 m thick unsaturated zone that would be affected by a repository, the volcanic rocks are dominated by Miocene ash flow and bedded tuffs. These tuffs vary considerably in their degree of welding, porosity, permeability, saturation, fracturing, and alteration. As a result, some of the volcanic units have the potential to focus fluid movement, while others may retard it.

Plate 2 shows a plan view of the model domain, including selected borehole locations, mapped faults, and the potential repository outline. The upper boundary of the site-scale model coincides with the bedrock surface, defined as topography minus alluvium, while the lower boundary is the water table at an elevation of 730 m. The

Table 2. Description of hydrogeologic units at Yucca Mountain.

<table>
<tr><th colspan="2">Rock-Stratigraphic Unit</th><th>Hydrogeologic Unit[a]</th><th>Approximate Range of Thickness (m)</th><th>Lithology[b]</th></tr>
<tr><td colspan="2">Alluvium</td><td>QAL</td><td>0-30</td><td>Irregularly distributed surficial deposits of alluvium and colluvium</td></tr>
<tr><td rowspan="4">Paintbrush Group</td><td>Tiva Canyon Tuff</td><td>TCw</td><td>0-150</td><td>Moderately to densely welded, devitrified ash-flow tuff</td></tr>
<tr><td>Yucca Mountain Tuff</td><td rowspan="2">PTn</td><td rowspan="2">20-100</td><td rowspan="2">Partially welded to nonwelded, vitric and occasionally divitrified tuffs</td></tr>
<tr><td>Pah Canyon Tuff</td></tr>
<tr><td>Topopah Spring Tuff</td><td>TSw</td><td>290-360</td><td>Moderately to densely welded, devitrified ash-flow tuffs that are locally lithophysae-rich in the upper part, includes basal vitrophyre</td></tr>
<tr><td colspan="2">Calico Hills Formation</td><td rowspan="2">CHnv / CHnz</td><td rowspan="2">100-400</td><td rowspan="2">Nonwelded to partially welded ash-flow tuff; Vitric / Zeolitized</td></tr>
<tr><td rowspan="2">Crater Flat Group</td><td>Prow Pass Tuff</td></tr>
<tr><td>Bullfrog Tuff</td><td>Cfu</td><td>0-200</td><td>Undifferentiated, welded and nonwelded, vitric, devitrified, and zeolitized ash-flow and air-fall tuffs</td></tr>
</table>

Sources: *Montazer and Wilson* [1984], except as noted.
[a]QAL = Quaternary alluvium and collumium; TCw = Tiva Canyon welded unit; PTn = Paintbrush nonwelded unit; TSw = Topopah Spring welded unit; CHn = Calico Hills nonwelded unit; CHnv = Calico Hills nonwelded vitric unit; CHnz = Calico Hills nonwelded zeolitzed unit; Cfu = Crater Flat undifferentiated unit.
[b]Lithology summarized from *Ortiz et al.* [1985].

vertical boundaries enclose a vadose zone of up to 750 m in thickness.

The potential repository would be constructed 300 to 325 m below the surface. As shown on Plate 2, the area of the facility is approximately 1 km wide and 3 km long,. just east of the Solitario Canyon Fault in the midst of a number of other faults. The solid black line shows the location of an Exploratory Studies Facility (ESF), which is a 6.8 m tunnel constructed to provide access along the eastern edge of the potential repository area. The ESF was placed close to the Ghost Dance Fault so that various investigations could be made of this particular fault zone.

The various volcanic rocks within the unsaturated zone of the repository block were assigned to three formations. From youngest to oldest, these include the Paintbrush Group, the Calico Hills Formation, and the Crater Flat Group [*Scott and Bonk*, 1984]. These formations include several beds of tuff that have been organized into stratigraphic units, as shown in Table 2. For the purposes of modeling, these stratigraphic units have been reorganized

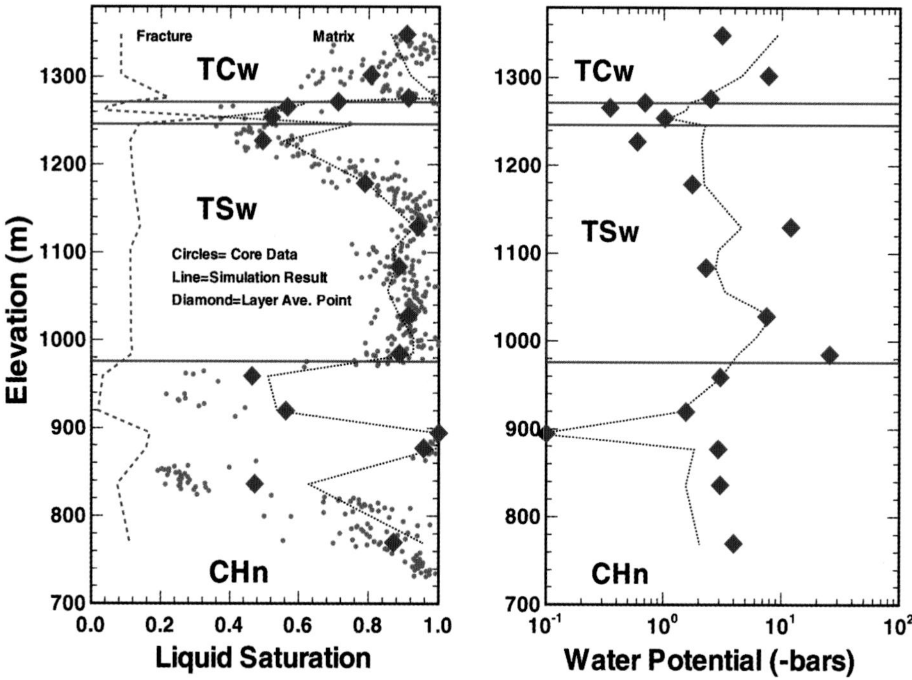

Plate 5. Matches at borehole SD-7 for observed core sample saturation and water potential data using calibrated parameter set. Fracture saturation is also shown. Core data from USGS. [after *Bandurraga and Bodvarsson*, 1997].

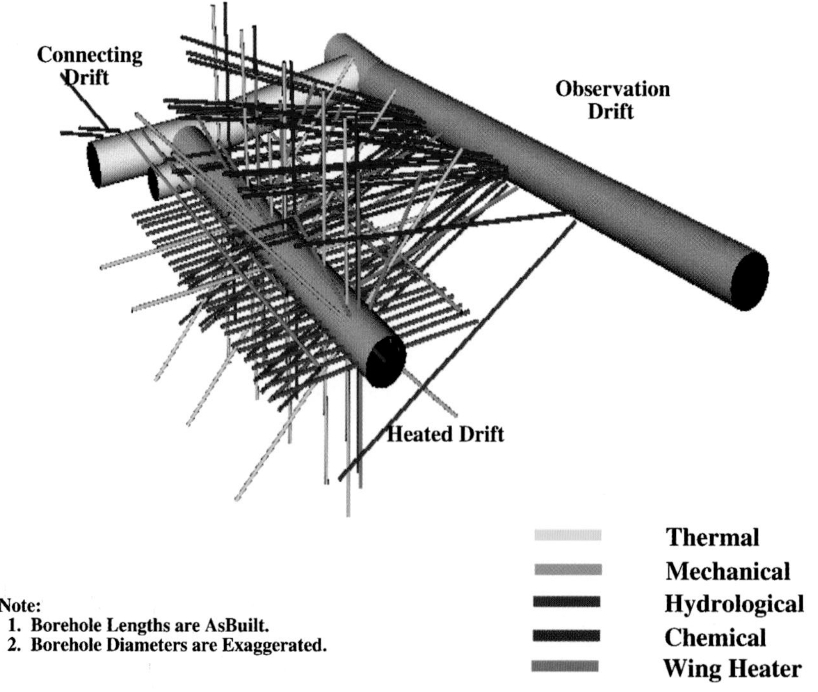

Plate 6. A 3-D perspective of the DST with wing heaters and various observation boreholes [after *Y. Tsang et al.*, 1999b].

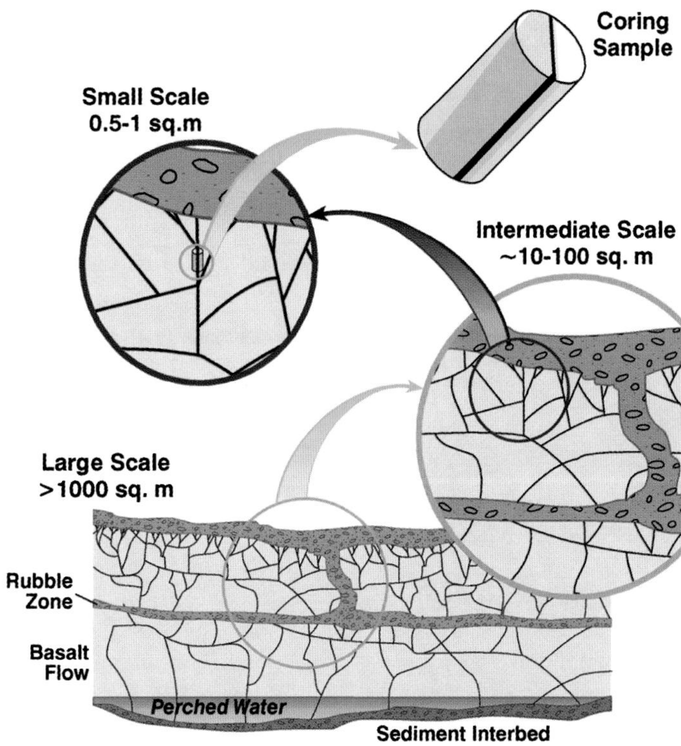

Coring Sample

Small Scale 0.5-1 sq.m

Intermediate Scale ~10-100 sq. m

Large Scale >1000 sq. m

Rubble Zone

Basalt Flow

Perched Water

Sediment Interbed

Plate 7. A 4-level hierarchy of scales for the hydrogeological components in fractured basalt [after *Faybishenko et al.*, 1999a].

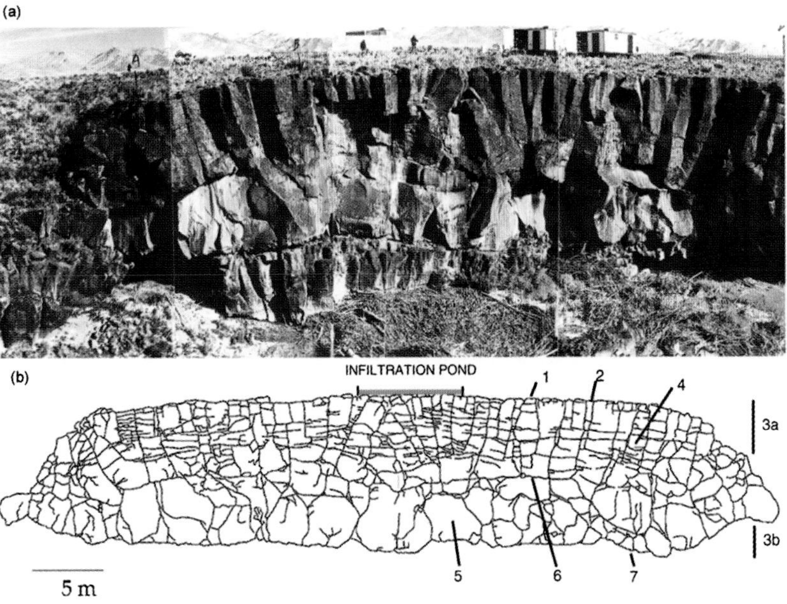

Plate 8. (a) Photograph of cliff face along Box Canyon that is about 30 m from location of infiltration pond. (b) Map of fracture pattern in upper basalt flow exposed on cliff face showing principal flow components: (1) soils, flow-top breccia, and boulders; (2) soil infilled near-surface fractures; (3) upper and lower vesicular zones; (4) isolated vesicular layers; (5) massive basalt; (6) central fracture zone; and (7) bottom rubble zone [after *Faybishenko et al.*, 1999b].

Table 3. Parameters of hydrogeologic units at Yucca Mountain.

Name	Symbol	Status	Sub-layers	Porosity[a]	Permeability (m^2) Matrix[b]	Fracture[b]
Tiva Canyon	TCw	welded	3	0.082—0.253	5.5E-18—3.9E-15	4.6E-12—3.0E-11
Paintbrush Tuff	PTn	nonwelded	7	0.254—0.499	3.2E-15—3.3E-13	3.0E-13—3.0E-12
Topopah Spring	TSw	welded	9	0.036—0.173	4.1E-18—1.7E-16	1.6E-13—1.3E-12
Calico Hills	CHn	nonwelded	10	0.266—0.345	4.6E-18—1.6E-12	2.5E-14—2.2E-13
Crater Flat	CFu	undifferentiated	4	0.115—0.325	7.7E-18—1.9E-15	2.5E-14—2.2E-13

[a]From Table 7.3-1 of *Cushey* [1998].
[b]From Table 7.2-1 of *Cushey* [1998].

into six hydrogeologic units, and a brief description of each hydrogeologic unit is given in Table 2 [*DOE*, 1988].

The process of characterizing the unsaturated hydrogeologic system at Yucca Mountain has required development of a site-scale model that could be used in investigations, first under ambient conditions, and then under the impact of perturbations that will be created by radioactive waste. Model development, using the integral finite difference approach, essentially started with the work of *Wittwer et al.* [1995]. Succeeding versions of updated site-scale models have been generated in an effort to describe the parameters of this system of volcanic layers as accurately as possible. As of this review, three comprehensive Berkeley reports have been made of the unsaturated zone at Yucca Mountain [*Bodvarsson and Bandurraga*, 1996; *Bodvarsson et al.*, 1997a; 1998].

The model architecture represents known geological information about Yucca Mountain as of 1998, as interpreted in the Geological Framework Model, GFM3.0, of *Clayton* [1998]. In addition to lithostratigraphic and structural data, relevant hydrogeological information pertaining to distribution of mineral alteration below the repository horizon is essential to any comprehensive flow and transport model. The potential of clays and zeolites to greatly reduce permeability has important implications for groundwater flow paths and travel times, and for perched water development. Sorption potential of zeolites is needed to assess potential retardation of radionuclides migrating away from the repository horizon. These issues stress importance of including mineralogic data in the model.

A brief summary of the parameters of the five hydrogeologic units [*Montazer and Wilson*, 1984] used in constructing the site-scale model is given in Table 3. In examining an enormous amount of field data, it was necessary to divide each unit into sublayers, (Table 3). This provided a means of assigning important variations in parameters within a given unit to individual sublayers that could then be incorporated into the model to simulate variations in geologic conditions. Plate 3 is a north-south cross section through the unsaturated zone near borehole UZ14 showing: (top) sublayers from GFM3.0 [*Clayton*, 1998], and (bottom) the arrangement of sublayers within the five hydrogeologic units.

8.1. Fracture and Fault Properties in the Site-Scale Model

The usefulness of the site scale model will depend on the accuracy with which data on fractures and faults can be gathered that are representative of the potential site. It is evident that the network of faults (Plate 2) that surround the repository area is an important factor in determining overall fluid flow behavior of Yucca Mountain. A second factor is the degree to which fractures have formed in the welded and nonwelded layers of the volcanic tuffs (Table 2). There are two main types of fractures: cooling joints, which are developed shortly after emplacement; and fractures of tectonic origin, which result from the imposed stress field [*Sweetkind et al.*, 1997].

Measurements of fracture permeabilities by air-injection testing [*LeCain and Patterson*, 1997] were carried out in the various sublayers of the five main hydrogeologic units (Table 3), with a large majority of the work in the TSw. Figure 32 illustrates the uncertainty and variability of these permeability data [*Cushey*, 1998]. Only sublayers in which more than 10 intervals were sampled are shown. The 95% confidence intervals represent ranges of one to almost four orders of magnitude. *Sonnenthal et al.* [1997] developed a consistent set of fracture properties from such measurements and an analysis of other fracture data collected during the construction of the ESF.

Because near-vertical fractures dominate permeability at Yucca Mountain [*Ahlers, et al.*, 1996], the vertical

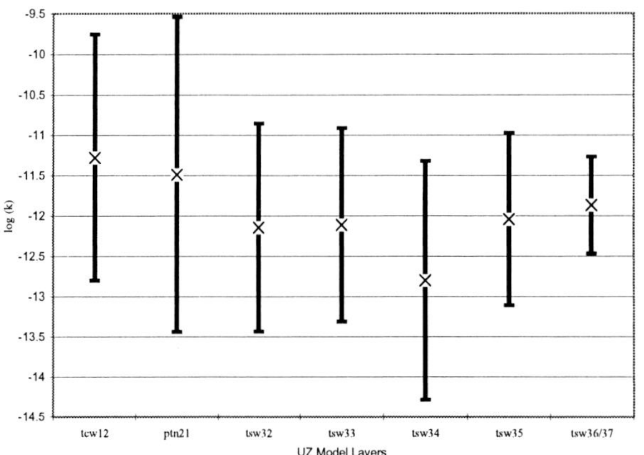

Figure 32. The geometric mean (X) and 95% confidence intervals for fracture permeabilities of sublayers in site-scale model [after *Cushey*, 1998].

permeability is expected to be larger than results obtained by air-injection testing in vertical boreholes, referenced above. Permeabilities from pneumatic monitoring in the TSw are an order of magnitude higher than the values obtained from air-injection testing, indicating a vertical permeability 10 times higher than the horizontal [*Sonnenthal et al.*, 1997].

Fracture orientations for the TSw have been mapped in the ESF [*Day and Beason*, 1996], and in analyzing these data, *Sweetkind et al.* [1997] show that it is possible to discriminate different orientations for cooling and tectonic joints (Table 4). Similar results for the TSw from oriented cores and by borehole television have been reported by *Lin et al.* [1993]. Most units show strong north-south orientations except for the TSw middle nonlithophysal unit (C3), which has a strong east-west component owing to numerous subhorizontal cooling joints.

Sonnenthal et al. [1997] have made an interesting analysis of the heterogeneity in the TSw that provides more insight on this important parameter than the average properties given above. From an analysis of data collected in the ESF, they obtained a picture of the variation in fracture frequency for three adjacent sublayers in the TSw, as shown in Figure 33. There is a much higher frequency of fractures in TSw34 (crystal-poor nonlithophysal) than in TSw32 (crystal-rich nonlithophysal) and TSw33 (crystal-rich lithophysal).

Sonnenthal et al. [1997] also examined the ESF fracture mapping results to determine how the preferred orientations of different size classes of fractures influence the permeability tensor. They determined the frequency of fractures ranging in size from 0.3 m to 34.0 m in the TSw34 sublayer. As would be expected, the number of fractures increases substantially as fracture size decreases, such that over half of the fractures are represented in the 0.3 to 1.0+ m size class, and only 1% of the fractures are represented in the largest class (10.0 to 34.0 m).

In addition, *Sonnenthal et al.* [1997] drew some important conclusions from the relationship of spacing to fracture size. In the 1.0 to 3.0 m fracture-size class, results suggest the network is well above its percolation threshold, and should form a well-connected network. For larger fractures, the network is probably also above its percolation threshold. Of course, geometric connectivity does not necessarily imply a hydraulically well-connected network. Aperture distributions in the fractures and intersections determine hydraulic connectivity.

Because of the lack of data on faults, *Sonnenthal et al.* [1997] had to develop estimates of fracture permeabilities that could be assigned to the different fault types in the TCw, PTn, and TSw hydrogeologic units. Ambient pneumatic pressure data were used to estimate pneumatic diffusivity of the fracture continuum in faults. With the addition of fracture continuum permeability results from air-injection tests and fracture intensity data, permeability and porosity of the fracture continuum could be estimated. In some units, porosity of the matrix contributes to pneumatic diffusivity of the fault. *Ahlers et al.* [1997] discuss factors contributing to pneumatic diffusivity. After grouping faults in each hydrogeologic unit according to fault type, *Sonnenthal et al.* [1997] assembled a set of fracture permeabilities for each unit, as shown in Table 5.

8.2. Hydraulic Characterization of Faults

Shan [1990] investigated the problem of characterizing vertical leaky faults for either saturated or unsaturated conditions. If a vertical fault cuts through two aquifers in a saturated system, characterization can be achieved by pumping water from one of the aquifers (assumed to be horizontal) and observing effects in the pumped and/or unpumped aquifer. Analytical solutions were obtained for pressure effects in both aquifers from which permeability of the fault can be obtained. In the case of an unsaturated rock-fault system, characterization can be achieved by observing effects of airflow through the system caused by changes in atmospheric pressure. Assuming any water in the system is immobile, analytical solutions for effects of transient air pressures in the vertical fault on pressures in adjacent rock have been obtained [*Shan*, 1990]. Once flow parameters of surrounding rocks are known, these solutions can be used to determine permeability of the leaky fault.

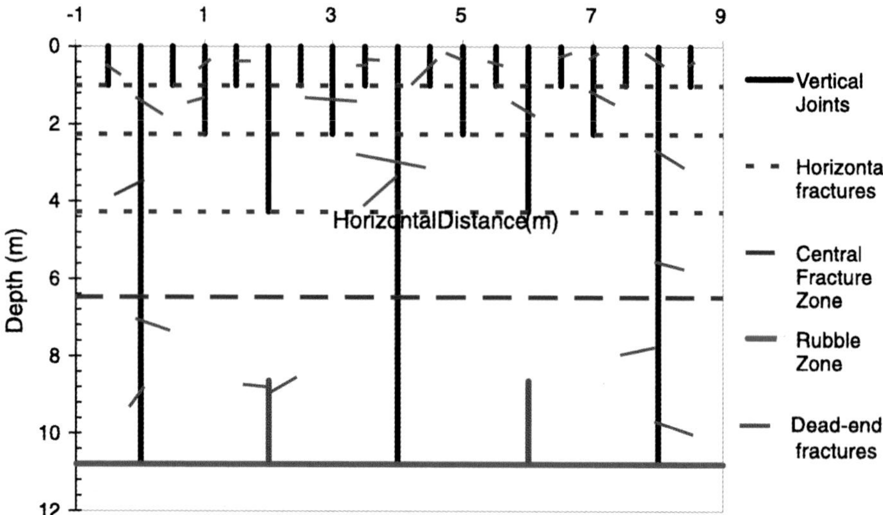

Plate 9. A conceptual tree-type model for the fracture pattern and fracture connectivity in a basalt flow. Locations for nonconductive fractures are arbitrary [after *Faybishenko et al.*, 1999a].

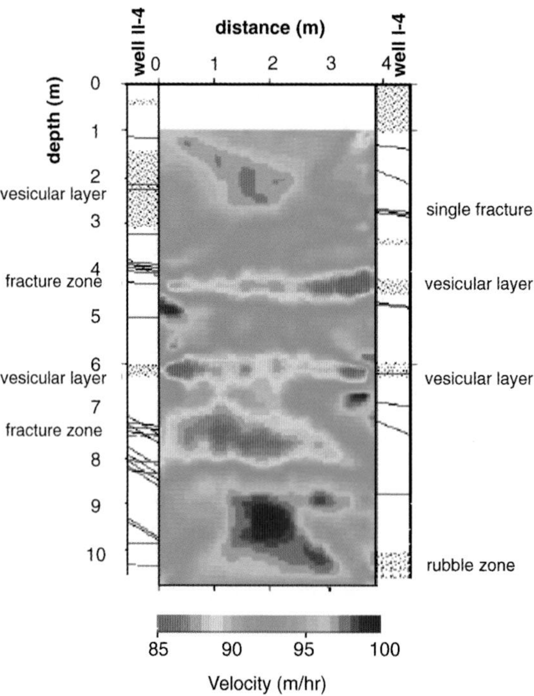

Plate 10. An example of radar velocities between two boreholes located 4 m apart just outside the pond. Red color indicates low radar velocity zones associated with higher moisture content, and blue color indicates higher velocity zones associated with lower moisture content [after *Faybishenko et al.*, 1999a].

Table 4. Median orientations (strike/dip) of fractures in the Topopah Spring tuff.[a]

Cooling Joints			Tectonic Joints		
C1	C2	C3	T1	T2	T3
N38W/77SW	N75E/86SE	N84E/21S	N06W/86W	N31W/84W	N55E/87SE

[a]From *Sweetkind et al.* [1997].

8.3. Current and Predicted Infiltration Rates

The site-scale model provides a method of numerically simulating behavior of the unsaturated system; therefore, a number of complicated processes are involved in its construction. The model must effectively simulate these processes so that critical features in the operation of the repository can be revealed and understood.

One of these processes is net infiltration rate, which varies spatially and temporally and is the ultimate source of the percolation flux at the repository horizon [*Hedegaard et al.*, 1998]. This flux provides water for flow and transport mechanisms that move radionuclides from the potential repository to the water table. Infiltration is spatially and temporally variable because of the nature of storm events that supply precipitation, seasonal changes in evapotranspiration, and variations in topography, soil cover, surface storage content, and runoff. The current conceptual model of infiltration is based on periodic measurements of water-content profiles in shallow boreholes at Yucca Mountain.

Near-surface infiltration data from Flint and coworkers at the USGS suggest that significant infiltration into the mountain occurs only once every few years [*Flint et al.*, 1996]. As water travels through the nonwelded PTn, with relatively high matrix permeability and porosity and low fracture densities, these infiltration pulses are attenuated. Spatial and temporal attenuation of infiltration pulses by the PTn leads to near steady-state conditions in underlying TSw and CHn [*Wang and Narasimhan,* 1985]. Therefore, in most cases ambient conditions below the PTn can be modeled and described using steady-state infiltration rates.

Recent present-day net infiltration estimates, based on measurements of average annual precipitation, provide average infiltration rates over the site-scale model domain ranging from 0.35 mm/yr to greater than 10 mm/yr [*Flint and Flint* 1994; *Hevesi and Flint,* 1995; *Flint et al.*, 1996, *Hudson and Flint,* 1996].

Future climatic conditions and infiltration may also impact waste isolation. USGS researchers estimate long-term average and super-pluvial infiltration based on expected changes in precipitation from present day. The long-term average is 32.5 mm/yr, and ranges from approximately 0 to 120 mm/yr. The super-pluvial infiltration average is much higher, 118 mm/yr over the model domain, and ranges from 11 to 570 mm/yr [*Wu et al.*, 1998a].

8.4. Gas and Heat Flow Processes

Characterizing Yucca Mountain also entails determining implications of ambient processes involving gas and heat flow. In some cases, there is an interaction between these processes that must be recognized so that the effects of each can be determined in the analysis of the field data.

Gas-flow processes are caused by barometric pumping, wind, and density-driven flow [*Hedegaard et al.*, 1998]. Barometric pumping is the response of subsurface pneumatic pressure to changes in atmospheric pressure [*Ahlers et al.*, 1997]. In welded tuff units, this translates into little change in the pneumatic pressure signal with depth. However, *Ahlers et al.* [1996] found that equal fracture and matrix permeability in the PTn, as well as high gas-filled porosity, serve to attenuate and lag the response to barometric pumping between top and bottom of this unit. They showed that degree of attenuation and lag observed between top and bottom of the PTn could be correlated to thickness of the bed. Two observations of ambient pneumatic pressures below the TSw in borehole SD-12 (see location on Plate 2) show significant attenuation and lag of the signal, suggesting extremely low gas permeability [*Rousseau et al.*, 1997].

Steady-state gas flow occurs in the ambient system because of density and wind-driven processes [*Weeks,* 1987, 1991]. Density-driven flow occurs where topography is steep because of differences between density of dry air in the atmosphere and wet air in the mountain. This causes flow from the base of the mountain toward the crest. Wind-driven flow occurs because of higher pressure exerted on the windward side of the mountain and lower pressure in the lee of the crest. Thus, wind-driven flow also promotes flow toward the crest. Measurements of air flow in an open borehole near the crest of Yucca Mountain (at UZ-6s) show that density-driven and wind-driven flow occurs mainly in

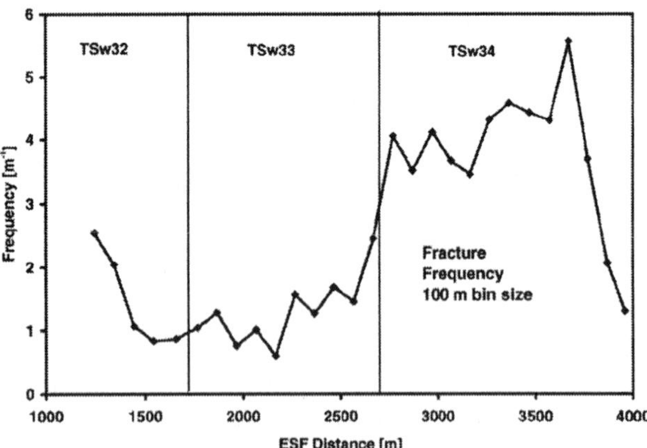

Figure 33. Variation in fracture frequency within the TSw as a function of distance in the ESF for 100 m intervals [after *Sonnenthal et al.*, 1997].

the TCw [*Thorstenson et al.*, 1989]. *Ahlers et al.* [1996] performed simulations showing that the PTn acts as a barrier to development of significant density-driven and wind-driven flow below it.

The flow of heat due to the geothermal gradient is another ambient process that must be investigated to determine its importance. This area is near the southern boundary of a regional heat flow anomaly, the Eureka Low. According to *Sass et al.* [1988], the average heat flux in the Eureka Low is about half that for adjacent regions. *Fridrich et al.* [1994] suggested that the anomaly results from cool underflow in the deep carbonate aquifer.

Yucca Mountain heat flow is characterized through borehole temperature measurements and thermal conductivity. The temperature measurements have been obtained from borehole logging [*Sass et al.*, 1988] over short time periods or from long-term monitoring of instrumented boreholes [*Rousseau et al.*, 1998]. *Bodvarsson et al.* [1997b] applied both conductive and convective heat-transfer methods to these temperature data to infer estimates for percolation flux through TSw and CHn. Percolation flux estimates ranged from 5 to 10 mm/yr in the TSw, consistent with modeled infiltration from *Flint et al.* [1996]. Somewhat lower values estimated for the CHn were attributed to lateral flow above low-permeability zeolitic tuffs. *Bodvarsson et al.* [1997b] concluded that estimated percolation fluxes were affected by proximity to known fault zones. With few exceptions, boreholes yielding highest flux estimates were located near faults. It was concluded there was an interaction between gas and heat flow such that flow of air and vapor through highly fractured fault zones affected the thermal gradient, biasing percolation estimates.

8.5. Fracture-Matrix Interactions and Heterogeneities

As discussed above, there are very significant differences between welded and nonwelded tuffs. The frequency of fracturing is much higher in welded tuffs; consequently, the fracture permeability of welded tuffs is significantly higher than that of the matrix (Table 3). In the nonwelded tuffs, there is much less fracturing and there are smaller differences between fracture and matrix permeabilities. Fracture flow is expected to dominate where percolation flux exceeds hydraulic conductivity of the matrix, such as within densely welded units of TSw and TCw. On the other hand, matrix-to-matrix flow may be the main flow mechanism for water crossing PTn and vitric portions of CHn.

The percentage of fracture flow that is included in total percolation flux has important implications for flow into drifts at the repository and for bypassing zeolitic units in the CHn unit [*Hedegaard et al.*, 1998]. In the latter case, radionuclide transport through CHn may not effectively retard transport via adsorption in zeolitic matrix rocks if the majority of water flows through fractures instead of matrix blocks. Therefore, partitioning between fracture and matrix flow may directly impact long-term performance of the unsaturated zone. In addition, continuous fracture flow may provide preferential fast flow paths from ground surface to repository level and from repository level to water table.

Another very perplexing aspect of the interactions between fracture and matrix is the effect of spatial structures within the matrix as well as the fractures. These structures result from evolution of soil formations and layering of stratigraphic units. Likewise, spatial variability of fracture and matrix flow properties can result in rapid movement of fluids and radionuclides through preferential flowpaths.

Considerable effort has been made at Berkeley to investigate the factors involved in fingering and meandering of flow paths within individual fractures. In addition to the early laboratory work mentioned above [*Pyrak-Nolte et al.*, 1987], there has been a series of mechanistic and numerical studies of heterogeneous flow in fractures [*Y. Tsang and C.-F. Tsang*, 1989; *Pruess*, 1999]. There also have been a number of laboratory investigations [*Kneafsey and Pruess*, 1995; *Persoff and Pruess*, 1998; *Geller et al.*, 1996; Su et al., 1999] revealing complicated flow patterns that develop due to fracture heterogeneity.

A numerical result for fingering that *Pruess* [1999] obtained is shown in Plate 4 and illustrates the complexity of dendritic flow patterns that can develop with significant ponding and bypassing. In this particular case, water was

Table 5. Fracture permeabilities for faults at Yucca Mountain.

FAULT TYPE	Normal Large Displacement	Normal Small Displacement	Strike-Slip
Hydrogeologic Unit			
TCw	2.0×10^{-11} m^2	3.0×10^{-13} m^2	2.0×10^{-11} m^2
PTn	2.2×10^{-12} m^2	4.0×10^{-12} m^2	1.0×10^{-12} m^2
TSw	1.0×10^{-11} m^2	6.0×10^{-10} m^2	6.0×10^{-10} m^2

provided across the top boundary of a single vertical fracture and allowed to migrate downward until breakthrough occurred at the time shown. The fracture plane was designed to be strongly heterogeneous, with permeability that is variable throughout the plane. The idea was to capture what are believed to be the essential features of fracture aperture distributions in hard rocks, such as tuffs, basalts, granites, or graywackes [*Wang and Narasimhan*, 1985; *Pruess*, 1998]. Pervasive time-varying and intermittent flows along preferential pathways have also been reported in experimental work of *Geller et al.* [1996] and *Su et al.* [1999].

The current understanding of fracture and matrix flow is primarily based on indirect evidence from field studies and numerical modeling investigations. The problem is further complicated by considerable uncertainties that still exist associated with fracture and matrix properties, their spatial distributions in each of the geological units, and mechanisms governing fracture-matrix interactions. In discussing these difficulties, *Pruess et al.* [1997] observed that modeling investigations have employed a number of approximations for flow in fractured rocks. Regardless of the approximation used, these models generally require adjustment of parameters to sustain fracture flow in the presence of a partially saturated matrix, and to obtain fast groundwater travel as dictated by observations of environmental isotopes (^{36}Cl) in the ESF [*Fabryka-Martin et al.*, 1997, 1998]. *Pruess et al.* [1997a] indicated that typical adjustments made to enhance fracture flow to achieve a desired result are not very satisfactory and leave serious doubts about the realism and predictive power of the models being used. Similar criticisms have been made by others [*Doughty and Bodvarsson*, 1997; *CRWMS M&O*, 1999].

To overcome the limitations mentioned above, *Liu et al.* [1998] proposed a new formulation for the fracture-matrix interaction based on an active fracture model, which is briefly described in Chapter 8 of *Bodvarsson et al.* [1998]. While this new approach is more physically rigorous than other schemes used in the current site-scale model, some of its major hypotheses/assumptions still need to be validated.

8.6. Effects of Major Faults

Ahlers and Ritcey [1998] examined how faults may play a major role in shaping the ambient hydrogeologic system at Yucca Mountain. Plate 2 shows locations of these faults with respect to the areal geology of Yucca Mountain and locations of the ESF and selected boreholes.

According to *Day et al.* [1998, in review], the main area of the proposed repository is an undeformed section bounded by long, north-trending, normal faults, such as the Solitario Canyon and Bow Ridge faults, that have displacements of tens to hundreds of meters. Intrablock faults such as the Ghost Dance, Abandoned Wash and Dune Wash faults are also north-trending, but tend to be shorter in length with variable offsets ranging from meters up to a hundred meters. Northwest-trending faults, such as the Pagany Wash, Drill Hole Wash, Sundance, and Sever Wash faults, generally have less offset and may correspond to northwest-striking drainages.

Ahlers and Ritcey [1998] reviewed the available information and data relevant to faults, including results of direct testing of the Ghost Dance Fault. They also reviewed the passive monitoring and characterization of ambient conditions in and around faults. In addition, they examined other data, mainly pneumatic pressure monitoring, which have shown the influence of nearby faults. From this analysis and geologic characterization of the major faults, *Ahlers and Ritcey* [1998] proposed a conceptual model of fault behavior in the unsaturated zone.

This model is based on knowledge that faulting has induced an intensified fracturing in rocks surrounding the offset plane. Near the surface, the fault zone is likely to widen or "horse tail" because of reduced confining stress. Near the water table, ancient hydrothermal alteration is likely to have reduced permeability of fault zones. Gas will preferentially flow in faults because of higher permeability than the surrounding rock. Liquid flow in the TSw and above is likely to be impeded by the capillary barrier presented by wider fractures in the fault zone. At the bottom of TSw, structural traps formed by stratigraphic offset combined with reduced fault permeability will induce formation of perched water. Though relatively impermeable, fault zones below TSw will continue to be preferential flow paths because they will still be more permeable than surrounding hydrothermally altered units. *Ahlers and Ritcey* [1998] realized that the limited data available restricted their ability to provide reliable conclusions about hydraulic behavior of the major faults.

8.7. Vitric and Zeolitic Components

In addition to understanding the role of fractures and faults, it is important to recognize the role that secondary alteration plays in flow and transport within the unsaturated zone [*Hedegaard et al.*, 1998]. From the repository horizon down through the Calico Hills (CHn) hydrogeologic unit, zeolites have been found at a number of locations [*Bodvarsson et al.*, 1996]. Waste isolation performance may therefore be very dependent on sorptive properties of these zeolitic layers and the degree to which they are contacted by water.

Zeolites are associated with significant bodies of perched water located in the unsaturated zone beneath the proposed repository at an elevation of ~1080 m. These perched zones have been identified at a number of boreholes in zones that are either in or overlying geologic units extensively altered to zeolites [*Bodvarsson et al.,* 1996]. This condition is often encountered in material underlying the densely welded and fractured TSw. Figure 34 shows elevations of perched water reported for these boreholes projected to a north-south cross section. The figure also shows the top elevation of the nearest underlying zeolitic layer in the borehole. Note the close association of the vitric/zeolitic boundary to elevations of perched water zones. The low permeability of the altered material (1×10^{-18} m^2 or less) is one reason why perched water bodies form above zeolite layers when lateral movement is prevented [*Bodvarsson et al.*, 1996].

The close association of perched water with low-permeability sorptive zeolites implies that vertical percolation flux locally exceeds the saturated hydraulic conductivity of the perching layer. Thus, *Wu et al.* [1998c] believe that predominant flow paths through CHn may not be vertical. Instead, water may be diverted laterally to a fault zone, fracture network, or relatively high-permeability vitric zone that focuses flow to the water table. If so, a large amount of water may bypass underlying zeolitic units; consequently, radionuclides may not be retarded by highly sorptive zeolites. Furthermore, low-permeability values associated with zeolitic rocks may prevent a large proportion of matrix flow through sorptive rocks. Thus,

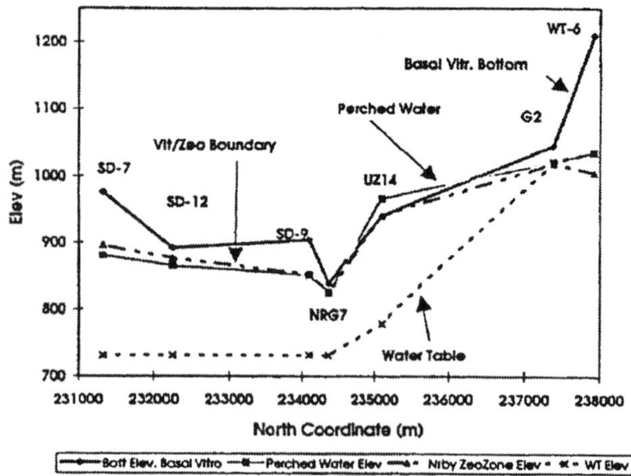

Figure 34. Perched water elevation data and the association with zeolites [after *Bodvarsson et al.*, 1996].

understanding the sorptive potential and distribution of the more vitric tuffs (containing <10% zeolite) is needed to assess performance of the geologic repository.

8.8. Geochemistry in the Ambient System

Dam [1998] developed a conceptual model of the Yucca Mountain hydrogeologic system, and from an evaluation of geochemical and isotopic data for minerals, various types of water, and unsaturated zone gases, he concluded the following:

- Most $^{36}Cl/Cl$ ratio samples from the ESF suggest recharge occurred within the past 30 ka. Some bomb-pulse samples near faults and fractures indicate water had migrated from the surface down to TSw within the past 50 years [*Fabryka-Martin et al.*, 1998].
- A comparison of stable isotopes were collected at the Nevada Test Site (NTS) [*Tyler et al.,* 1996] to Yucca Mountain data suggests that major isotopic shifts occurred in the top 50 m, for which no Yucca Mountain data were available. Therefore, NTS samples for that top 50 m were correlated to Pleistocene ages for δD. This correlation suggests that NTS pore water, and therefore Yucca Mountain pore water, was primarily recharged during the last super-pluvial ice age.
- In comparing stratigraphic units, it was found that PTn contained the highest concentrations of Ca^{2+}, Mg^{2+}, SO_4^{2-}, Cl^-, and P_{CO2}, which generally declined with depth down into the CHn for these parameters. The CHn also contained the highest concentrations of Na^+ and HCO_3^-, and had the most alkaline pH values.

Perched water was dilute but similar in composition to pore water, groundwater, and Rainier Mesa tunnel seeps. SiO_2 was consistently high in all water samples. Nitrate was persistent in all unsaturated zone and perched water samples, suggesting microbial activity was low during recharge. Nitrate occurrence would be more likely if recharge took place under colder climates rather than warm climatic conditions.

- Secondary minerals, including calcite, were dated to Pleistocene and older, based on $\delta^{13}C$. Calcite and opal deposits are abundant in the TSw, which correlates more with fault and shear zones than with infiltration [*Paces* 1998].
- Smectite in the PTn was formed by syngenetic processes when Miocene tuffs were cooling. Diagenetic alteration has not yet been determined. There is evidence for lateral diversion in PTn [*Levy and Chipera*, 1997].
- Zeolites in the CHn are related to a past SZ water level, suggesting a rise of more than 100 m in the water table [*Chipera et al.*, 1996].

Sonnenthal [1998] is investigating another approach to understanding the geochemistry of the ambient system, which involves reaction-transport modeling. His preliminary examination looks at distribution of mineral alteration in the ambient system, using estimated water and gas compositions and kinetics and thermodynamics of mineral precipitation and dissolution for a multicomponent multimineral system. Preliminary results show that calcite distribution in the TSw unit may be inversely correlated with infiltration and at the same time positively correlated within the TCw unit [*Sonnenthal*, 1998]. Smectite precipitation, however, is greater under regions of higher infiltration. Reactions involving feldspar and calcite with water to form zeolites are consistent with the mechanism proposed by *Vaniman and Chipera* [1996] for the absence of calcite in the "barren zone" and possibly its irregular distribution in the unsaturated zone. *Sonnenthal* [1998] concludes that simulation of ambient system water-gas-rock interaction in the unsaturated zone can therefore be invaluable in assessing conceptual models of hydrologic behavior.

8.9. Calibrating Hydrogeologic Parameters for Site-Scale Model

An important part of characterizing the unsaturated zone is calibration of parameters for the site-scale model. This process can be rather complex, and a good example is the investigation by *Bandurraga and Bodvarsson* [1997] on the problem of calibrating matrix and fracture properties in the site-scale model using inverse modeling. The investigation involved core-sample saturation, porosity, and density measurements for a number of boreholes penetrating a significant portion of TSw, including boreholes SD-7, SD-9, SD-12, NRG-6, NRG-7a, UZ-14, and UZ-16 (see Plate 2). (Core-sample water-potential data are also available for boreholes SD-7, SD-9, and UZ-14.) Details of the calibration procedure are described by *Bandurraga and Bodvarsson* [1997].

Developing the site-scale model also requires knowing extent of fracture-matrix (F/M) interaction; i.e., knowing what fraction of total flow is in fractures and what fraction is in matrix. Based on field observations and some preliminary work, it was determined that matrix flow is dominant in nonwelded layers and fracture flow in welded layers. Therefore, as an initial estimate for model inversions (see directly below for more information on model inversion), the F/M interaction factor for nonwelded layers was assumed to be 0.5, to provide for significant flow in the matrix. For the welded layers, the initial factor was assumed to be 0.0005, resulting primarily in fracture flow. To achieve a more consistent result during inverse modeling, these factors were then allowed to vary in each model layer. However, this method of handling the F/M interaction area has been found to be unsatisfactory [*Doughty and Bodvarsson*, 1997; *CRWMS M&O*, 1999].

Moisture-flow calibration results also help to determine what fraction of the total flow is in the fractures and what fraction is in the matrix. Examples of such results at SD-7 are illustrated in Plate 5. The plate shows matches of average matrix saturation with water-potential data used in model calibration from ITOUGH2 inverse modeling [*Finsterle*, 1997a,b, 1999]. The plate also shows simulated fracture saturations in the model. In general, the plots indicate good agreement with layer-average values, shown by diamonds, which were calculated from borehole data. Layer-averaged values include high and low saturations and potentials reported for geologic layers represented by a model layer. Inversions are attempts to match data from 5 boreholes simultaneously to provide a layer-averaged parameter set. Thus, the results may not match the data for an individual borehole as well as individual inversions with refined columns.

Matches to pneumatic sensor data for borehole SD-12 obtained from pneumatic inversion are illustrated in Figure 35. The plots show relatively good matches to sensor data. In this case, the F/M problem does not arise. Note how pressures increase with depth within the TSw unit.

This calibration work has utilized the TOUGH2 code [*Pruess*, 1991] because of its flexibility and robustness in handling multidimensional, multiphase, and multicomponent fluid flow and heat transfer in porous and

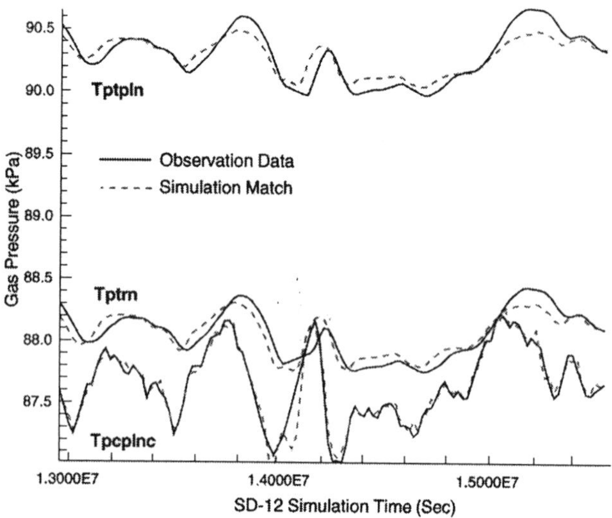

Figure 35. Gas pressure calibration results at SD-12. Tpcplnc is a sublayer in the TCw unit. Tptrn and Tptpln are both sublayers in the TSw unit [after *Bandurraga and Bodvarsson*, 1997].

fractured rocks. The code has proven itself to be a reliable, accurate, and efficient predictive tool, and various extensions and modifications have led to a TOUGH2 family of codes [*Moridis and Pruess*, 1995; *Pruess et al.*, 1996, 1997b; *Moridis and Pruess*, 1998]. Inverse modeling capabilities for TOUGH2 were developed by *Finsterle* [1997a, 1997b, 1999]. The main purpose of ITOUGH2 ("Inverse of TOUGH2") is to estimate hydrogeologic parameters for the TOUGH2 code by automatically calibrating the model against observed laboratory and field data. In addition, ITOUGH2 provides a detailed residual and error analysis of the estimated parameter set. It also allows an examination of uncertainty in model predictions by either first-order/second-moment error propagation analysis or Monte Carlo simulations.

8.10. Impact of Repository Heat

Haukwa et al. [1996] made a detailed analysis of the impact of repository heat by examining effects of thermal loading at the repository horizon using a 2-D vertical north-south cross section. This section crosses the entire repository block through its center and extends about 500 m below the water table. The purpose was to investigate fluid and heat transfer in the unsaturated zone and the long-term response of the unsaturated zone and saturated zone SZ to thermal loading in the proposed repository. At the time of the investigation by *Haukwa et al.*, the design for thermal loading had been set at 83 kW/acre, with emplacement drifts on a spacing of 25 m. As will be shown below, this leads to boiling conditions that coalesce between drifts and migrate for significant distances above and below the repository horizon.

This design has recently been changed considerably by increasing the drift spacing to 81 m and reducing thermal loading to 60 kW/acre [*Bodvarsson*, personal communication, 1999]. Since details of the effects of this lower thermal loading were not available at the time of this review, results from *Haukwa et al.* [1996] are presented to provide a frame of reference for some important differences in temperature fields.

One of the critical differences in the new thermal loading of 60 kW/acre is that temperature fields above and below the level of emplacement drifts will be significantly cooler. With this loading, boiling will not extend more than 5 m from the walls of the drift [*Bodvarsson*, personal communication, 1999]. Since drift diameter is 5 m, a drift spacing of 81 m means that boiling conditions will essentially be isolated in a relatively small region around each drift.

In previous investigations with the higher thermal loading, *Haukwa et al.* [1996] developed a set of temperature profiles (Figure 36) through the center of the repository block that show a much different picture in terms of boiling conditions. The infiltration rate was set at 4.4 mm/yr, and the fractured rock was analyzed using an equivalent continuum model (ECM). The ECM combines parameters of fracture and matrix into appropriate average values so that the rock mass can be modeled as an equivalent continuum.

It can be seen in Figure 36 that 10 years after waste emplacement, temperatures on the repository horizon have reached the boiling point, ~96°C at the elevation of Yucca Mountain. This condition spreads throughout the repository horizon and migrates vertically through action of heat-pipe activity. After 1,000 years, it can be seen that a region of boiling encloses the repository with a vertical thickness of almost 300 m. Note the temperature effects at the surface and water table.

In contrast to this, temperature profiles with a thermal load of 60 kW/acre would only show a peak temperature in the near vicinity of the repository horizon. Over the greater part of the repository horizon and its immediate vicinity, temperatures would remain below boiling. Since total heat to be released from waste remains the same, a much longer period of time would be needed for its dissipation throughout the mountain.

8.11. Single Heater Test

A Single Heater Test (SHT) was started in a special alcove of the ESF on August 26, 1996. The heating phase

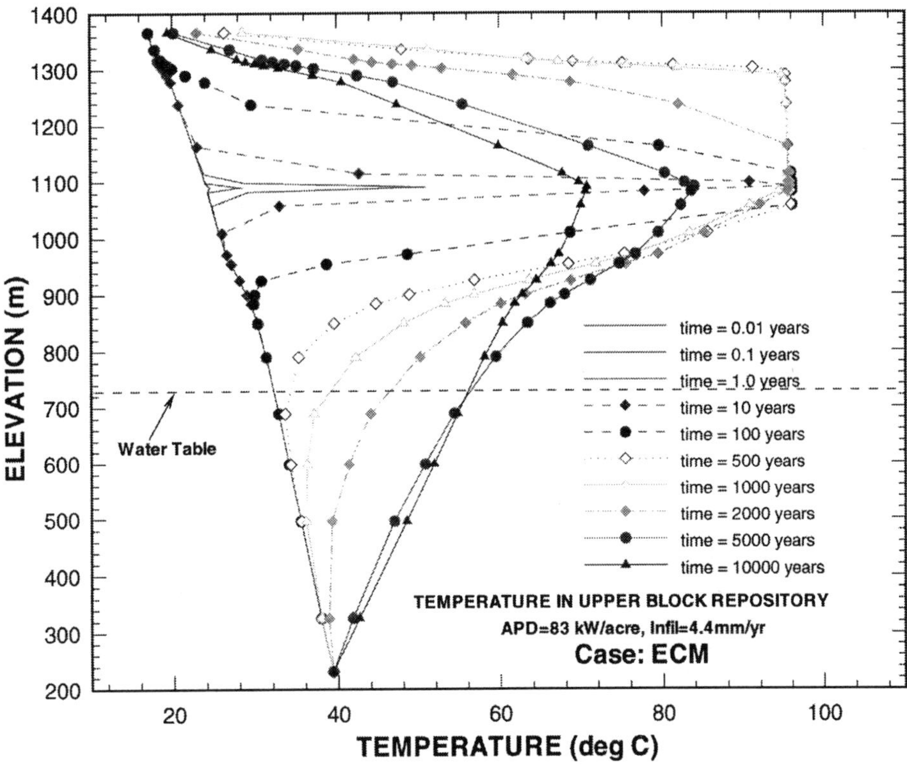

Figure 36. Vertical temperature profiles through the center of the repository block during 10,000 yr of thermal loading using ECM, 83 kW/acre, infiltration = 4.4 mm/yr [after *Haukwa et al.*, 1996].

lasted nine months, until May 28, 1997. The cooling phase continued until January 5, 1998, at which time postcooling field characterization activities began. A comprehensive report summarizing all the activities carried out during the SHT was published by *Y. Tsang et al.* [1999a]. The primary objective of this test was to acquire an understanding of the coupled thermal, mechanical, hydrological, and chemical processes likely to exist in the rock mass surrounding the proposed geologic repository at Yucca Mountain [*Y. Tsang*, 1999a].

Figure 37 is a layout of the alcove for SHT with the electric heater (borehole 1) surrounded by a 3-D arrangement of observation boreholes. Preheat characterization by air-injection tests was carried out in 22 of the 31 boreholes, as shown in Figure 37. Tests with isolated zone lengths ranging from 2 to 11 m illustrate permeability values ranging from 5.0×10^{-15} m^2 to 5.2×10^{-12} m^2. *Y. Tsang et al.* [1999a] attributed three-orders-of-magnitude difference in these values to flow through fractures of hierarchical scales, with microfractures accounting for the lower values and fracture zones (a few meters in extent) responsible for the highest values.

An interesting result revealing the effects of heat on the fractured tuff was obtained from two observation holes, boreholes 16 and 18 (Figure 37). Constant mass-flux air-injection tests, in addition to continuous monitoring of relative humidity, temperature, and pressure, were conducted periodically in these two boreholes. Air-injection data in packed-off zones at the end of each of these boreholes, which are only about 3 m from heater borehole 1, revealed an interesting effect on permeability. Temperatures in both boreholes only reached a modest level of 52°C, but air permeabilities decreased significantly during the heating period and then recovered in the postheating period to slightly higher values than original. *Y. Tsang et al.* [1999a] attributed the permeability reduction to thermal-hydrological coupling caused by water condensing in a large, continuous zone that decreased air flow to boreholes 16 and 18.

Radar velocity tomograms taken before and after heating showed significant differences. These tomograms and difference tomograms were quite effective in mapping changes in moisture content resulting from heating. *Y. Tsang et al.* [1999a] found that saturation changes

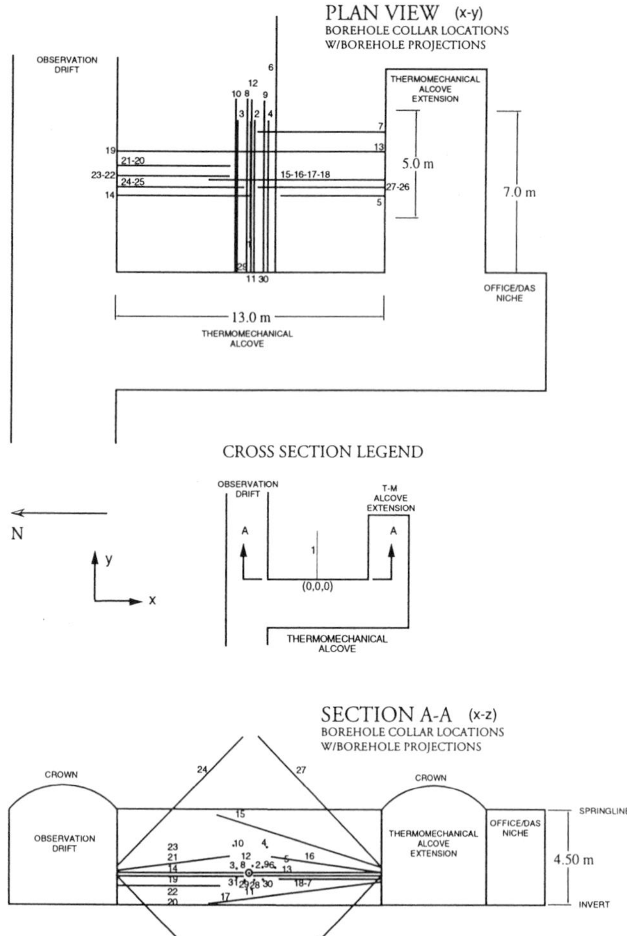

Figure 37. Plan view and cross-section of SHT showing layout of boreholes [after *Y. Tsang et al.*, 1999a].

indicate a region of extreme drying to about a radius of 1 m from the heater, where saturation decreases to between 20 and 40 %.

Birkholzer and Y. Tsang [1999a] conducted an interpretive analysis of the thermal-hydrological processes of the SHT during the heating and cooling phases by means of numerical simulations. A 3-D numerical model was developed to simulate transient coupled transport of water, vapor, gas, and heat in the rock mass. The dual-permeability method was applied to account for the combined effects of matrix and fracture flow. Good agreement was obtained between measured and simulated temperature results, showing that thermal-hydrological response in the SHT is well represented in the numerical model. Both measured and simulated temperatures suggest that while heat is mainly transported by conduction, the effects of thermal-hydrological coupling are also important. Overall, the signature of convective transport caused by heat-induced vapor and liquid fluxes is very subtle, indicating that hydrological properties of matrix and fractures in the SHT do not promote significant vapor-liquid counterflow during the heating phase.

Two sets of water samples were collected from borehole 16 during the heating period, and results of analyses were evaluated by *Glassley* [1997] and *Glassley and DeLoach* [1997]. These researchers concluded that the water consisted of steam condensate that underwent some minimal chemical interaction with surrounding rock. *Spycher et al.* [1999] carried out reaction-transport simulations to provide insight into thermal-hydrological coupling processes that were also useful for interpreting chemical compositions of waters collected in borehole 16. They found that borehole-16 water resulted from steam condensation in fractures, as reported above. They also found that the presence of three minerals (calcite, gypsum, and amorphous silica), detected in post-test analyses and attributed to reactions accompanying the SHT, appeared to have formed during the SHT by evaporation at low temperature, most likely sometime after the heater was turned off.

8.12. Drift Scale Test

The Drift Scale Test (DST) is a major underground thermal test at Yucca Mountain and is located near the SHT in the ESF. Its purpose is to carry out a long-term investigation of coupled thermal, mechanical, hydrological, and chemical processes likely to be encountered in the rock mass surrounding the proposed repository [*Birkholzer and Y. Tsang*, 1997]. The planned heating phase is 4 years, and the cooling phase following power shutoff will be of similar duration. Plate 6 shows a 3-D perspective of the DST, with heaters and different boreholes color-coded according to their functions for thermal, mechanical, hydrological, or chemical purposes. The DST consists of a 5 m drift, 47.5 m in length, with connecting and observation drifts of the same size. Heat is generated in 50 horizontal wing heaters and 9 horizontal floor heaters inside the 47.5 m drift. The heaters were activated on December 3, 1997.

Various processes in the DST are being monitored by a multitude of sensors to measure temperature, humidity, gas pressure, and mechanical displacement of the rock mass in response to temperature changes generated by electric heaters. In addition to collecting passive monitoring data, active hydrological and geophysical testing is also being carried out periodically. Active tests are intended to monitor changes in moisture redistribution in the rock mass, to collect water and gas samples for chemical and isotopic analysis, and to detect microfracturing due to heating. A report of the first 6 months of operations was published and includes analyses and predictions concerning: thermal-hydrological processes [*Birkholzer and Y. Tsang*, 1999b]; thermal-hydrological-chemical processes [*Sonnenthal et al.*, 1999]; and geophysical measurements [*Peterson and Williams*, 1999].

9. CHAOTIC FLOW IN VADOSE ZONE OF FRACTURED BASALT

Recent investigations in fractured basalt of the Eastern Snake River Plain of southeastern Idaho near Idaho National Engineering and Environmental Laboratory (INEEL) have revealed some unusual heterogeneities that exist in basalt flows. In an investigation that consisted of 3 ponded infiltration tests carried out in these basalt flows, *Faybishenko et al.* [1999a] found that flow of water in these fractured basalts is so erratic that it can only be described as the result of a chaotic flow system. To explain such behavior, they have proposed development of a conceptual model that is based on a hierarchy of scales, as illustrated in Plate 7.

An example of a chaotic flow system was revealed in one of the infiltration tests that was carried out at Box Canyon, which is near Arco, Idaho, and INEEL [*Long et al.*, 1995]. The surface consists of exposed weathered basalt and soils (clays and silts), which have infilled the near-surface fractures and columnar joints of basalt. The depth to the regional aquifer at this site is about 200 m, and a perched water zone is at a depth of about 20 m. The infiltration pond was centered over a basalt flow, and located about 30 m from a cliff face along Box Canyon (Plate 8). The photograph in Plate 8 shows 2 distinct basalt flows on this cliff face; the detailed fracture map was developed for the upper flow directly beneath the pond.

As *Faybishenko et al.* [1999b] pointed out, the fracture map of the basalt flow can be divided into upper and lower zones of distinctly different fracture geometries. Such geometry results from the different cooling rates and temperature gradients in the upper and lower portions of the flow. As can be seen in Plate 8, many fractures propagate from outer margins toward the interior of the basalt flow. Fractures of varying lengths that propagate from the top of the flow connect the surface with horizontal fractures and major vertical fractures. Fractures that propagate upward from the bottom of the flow are less interconnected, and therefore most may be considered dead-end, or nonconductive, fractures. On the basis of these observations, borehole logging, and gas-phase interference tests, *Faybishenko et al.* [1999b] suggested that the vadose zone in fractured basalt is made up of the following hydrogeological heterogeneities: (1) soils, flow-top breccia, and boulders; (2) soil infilled near-surface fractures; (3) upper and lower vesicular zones; (4) isolated vesicular layers; (5) massive basalt; (6) central fracture zone; and (7) a bottom rubble zone.

On the basis of these observations of such a heterogeneous system of fractures, *Faybishenko et al.* [1999a] proposed a four-component tree model, as shown in Plate 9. The model is based on the assumption that there are four generations of vertical column-bounding fractures. From the surface, the two smallest basalt columns of size n (1st generation) merge into a column of size 2n (2nd generation). Below this level, the 2n columns merge into a column of size 4n (3rd generation), and then the 4n columns merge into a final column of size 8n (4th generation). Visual inspection of several basalt outcrops showed that every set of vertical fractures is connected through horizontal vesicular layers and fractures. In addition, there are numerous dead-end, or nonconductive fractures. *Faybishenko et al.* [1999b] concluded that the minimal size for a fracture investigation should extend from top to bottom of the basalt flow, and its width should coincide with the maximum spacing between columnar joints. They found that at Box Canyon, average spacing between columnar joints is 0.46 m at the surface of the flow, 2.0 m near the center, and 1.3 m at the bottom.

During the infiltration test, results of an investigation of moisture distribution beneath the pond illustrated the effects of fracture heterogeneities on flow [*Faybishenko et al.*, 1999a]. Ground-penetrating radar (GPR) was used to measure radar velocities between two boreholes located 4 m apart just outside the pond. Plate 10 shows generalized lithological and fracture logs for these two boreholes and wide variations in radar velocities. The red color indicates a low radar velocity, interpreted as resulting from a higher moisture content; the blue color indicates a high velocity, interpreted as a lower moisture content. This plate is of three horizontal low-velocity zones, presumably wetted zones, located in vesicular basalt and the central fracture zone.

Field observations in heterogeneous systems, such as described above, often show random-looking fluctuations in pressure, flow rate, or temperature. *Faybishenko* [1999] pointed out that these fluctuations may not in fact be random, but rather a combination of deterministic-chaotic and random components. It is important to determine whether a system exhibits chaotic behavior because long-term predictability of a chaotic system is limited. One can determine only a range of possible results, rather than give a precise prediction. Some perspectives for using chaos theory to describe nonlinear flow processes in fractured media are discussed by *Faybishenko* [1999] and *Faybishenko et al.* [1999a].

10. SUMMARY

Over a period of 35 years, a large number of researchers at Berkeley have carried out an impressive number of investigations to develop a better understanding of factors controlling behavior of fractured rocks. The parallel-plate concept is useful as a first approach to many problems, but its usefulness is limited when one must examine the geometry of rough fractures to understand effects of aperture distributions on heterogeneous flow paths. It has been found experimentally that problems can occur when deformable fractures with rough surfaces are subjected to significant levels of stress (>20 MPa). To determine the effective aperture under such conditions, we tend to underestimate the size of actual flow channels that still exist at maximum stress.

Geothermal systems are found in fractured and faulted rocks. Because of nonisothermal conditions, their behavior is more complicated than that of a normal isothermal aquifer. The Cerro Prieto geothermal field in Mexico and Krafla field in Iceland are two examples where development and exploitation of geothermal resources were enhanced by application of numerical models. There is also the problem of locating major discontinuities in geothermal systems, although vertical seismic profiling has been used in field investigations on this problem with some success.

In 1977, Berkeley had the first opportunity to carry out large-scale investigations on fractured granitic rock. These investigations took place in an underground laboratory constructed in an abandoned iron ore mine in central Sweden. Electric heater tests provided the first opportunity for a field investigation on the effects of heat on hard rock. Effects of fractures on heat transfer were not significant, and transfer by conduction was the controlling process. A new method of measuring permeability of fractured granite on a large scale was developed. This involved using a 30 m drift that was sealed off and subjected to a ventilation scheme. Seepage of water into the drift was determined by change in humidity of a controlled flow of air within the drift. Rate of seepage and measurement of the hydraulic gradient in the walls of the drift provided data from which average permeability of a very large volume of rock was determined.

Characterizing the flow of groundwater through networks of fractures has not been easy. One approach has been to look for an equivalent porous medium that has the same flow behavior. Difficulties arise in measuring geometries of fractures and in determining which of those fractures control overall flow through a given network. Both forward and inverse numerical methods have been developed to address this problem, and various investigations carried out to determine how seismic methods can help characterize these complicated systems.

Through various investigations on fractured rocks, it has become clear that one of the fundamental problems of characterization is the heterogeneity of flow paths that control fluid migration. Both field and laboratory data have revealed the complicated flow paths that can exist. In a numerical approach to this problem, the concept of a 3-D variable-aperture model has been proposed that can be applied to both flow and transport. More recently, a stochastic continuum model for heterogeneous fractured rock has been proposed. This is a single-continuum model, yet the fracture zones are distinguished from the matrix by the assignment of long-range correlation structures along preferred planes of fracture zones for the high hydraulic conductivity part of the total system.

For several years, the Earth Sciences Division at Berkeley Lab has been carrying out a large number of investigations at Yucca Mountain, a potential site for a radioactive waste repository. The Topopah Spring tuff (TSw) has been selected from a series of volcanic sediments as the layer in which the repository would be excavated. This is a welded and heavily fractured tuff in the vadose zone some 300 m below the surface. Initial investigations involved detailed geological field studies carried out by the USGS. In recent years, these studies have been expanded to include the collection of geophysical, hydrological, and geochemical data. All of these data have been integrated into a detailed conceptual model from which a three-dimensional numerical model, known as the site-scale model, has evolved.

The process of characterizing the site at Yucca Mountain has required a considerable period of time and a large number of investigations. In addition to the geologic data collected by the USGS, properties of faults and fractures had to be assembled and organized into a site-scale model by which the 3-D spatial variations of these properties would be adequately represented in the rock mass. The process has also involved determining the implications of certain ambient processes, such as infiltration rates, gas and heat flow processes, and geochemical processes.

In addition to these ambient effects, the impact of heat released from the radioactive waste is a critical concern.

Repository heating will induce a variety of physical and chemical changes in the rock layers and their contained fluids at Yucca Mountain over very long periods of time. The site-scale model has been used in extensive studies to investigate fluid and heat transfer in the unsaturated zone, and the long-term response in both the unsaturated and saturated zones. Initially, the design for thermal loading had been set at 83 kW/acre, with emplacement drifts spaced 25 m apart. This loading leads to boiling conditions that coalesce between drifts and migrate for significant distances above and below the repository horizon. However, this design has recently been changed considerably by increasing the drift spacing to 81 m and reducing the thermal load to 60 kW/acre. Details of the effects of this lower thermal loading were not available for this review, but it is clear that the temperature rise above and below the level of the emplacement drifts will be significantly reduced. The thermal behavior of the repository is now being re-evaluated.

One of the critical problems in characterizing Yucca Mountain, independent of the thermal loading, is the question of fracture-matrix flow. In general, the study of this problem in the unsaturated zone is limited because of the difficulty in obtaining direct measurements of fracture and matrix components of flow in situ. Current understanding of fracture and matrix flow is primarily based on indirect evidence from field studies and numerical modeling investigations. The problem is further complicated by considerable uncertainties associated with fracture and matrix properties, their spatial distributions in each of the geological units, and mechanisms governing fracture-matrix interactions. The models generally require adjustments of parameters to sustain fracture flow in the presence of a partially saturated matrix and to obtain a component of relatively fast groundwater travel as dictated by field observations of ^{36}Cl in the ESF. These ad hoc adjustments of model parameters to achieve a desired result have raised questions about the realism and predictive power of the models.

Two important underground heater tests have been put in operation at Yucca Mountain. A Single Heater Test (SHT) was started in a special alcove of the ESF on August 26, 1996, with a heating phase that lasted 9 months and a cooling phase that lasted until January 8, 1998. The primary objective was to acquire an understanding of the coupled thermal, mechanical, hydrological, and chemical processes likely to exist in the rock mass surrounding the proposed repository. Significant decreases in air permeability of the heated rock mass were measured and are interpreted as an indication of thermal-hydrological effects. Radar velocity tomograms were quite effective in mapping changes in moisture content. A model using the dual-permeability method was used in an interpretive analysis of the thermal-hydrological processes in the SHT with excellent results.

The Drift Scale Test (DST) is a major underground thermal test at Yucca Mountain and is located near the SHT in the ESF. The purpose is to carry out a long-term investigation of the coupled thermal, mechanical, hydrological, and chemical processes likely to be encountered in the rock mass surrounding the proposed repository. The planned heating phase is 4 years, and the cooling phase following the power shutoff will be of similar duration. The DST consists of a 5 m drift 47.5 m in length, with heat generated in 50 horizontal wing heaters and 9 horizontal floor heaters inside the drift. The heaters were activated December 3, 1997. The various processes are being monitored by a multitude of sensors to measure the response of the rock mass. A report of the first 6 months of operations has been published that includes analyses and predictions concerning thermal-hydrological processes, thermal-hydrological-chemical processes, and geophysical measurements.

One of the most recent developments in investigations of heterogeneities in fractured rocks concerns the fractured basalt layers of the Eastern Snake River Plain in Idaho. An unusual level of heterogeneity exists in the basalt flows, and was revealed during an investigation of three ponded infiltration tests carried out in conjunction with the Idaho National Engineering and Environmental Laboratory (INEEL). The results show that flow of water in these fractured basalts is very erratic. To explain such behavior, it is proposed that one must adopt a conceptual model that is based on a hierarchy of scales.

An example of a chaotic flow system was revealed in one of the infiltration tests that was carried out at Box Canyon, near Arco, Idaho and INEEL. A nearby cliff face in the canyon provides an excellent exposure of the basalt flow that was investigated in this test. Careful mapping of the fractures and discontinuities in the cliff face has led to the adoption of a four-component tree model for the flow system. The model is based on the assumption that there are four generations of vertical column-bounding fractures. Visual inspection of several basalt outcrops in the field showed that every set of vertical fractures is connected through horizontal vesicular layers and fractures. In addition, there are numerous dead-end, or nonconductive fractures. It was concluded that the minimal size for a fracture investigation of this kind should extend from top to bottom of the basalt flow and its width should coincide with the maximum spacing between columnar joints.

Field observations in heterogeneous systems, such as described above, often show random-looking fluctuations in pressure, flow rate, or temperature. These fluctuations may not in fact be random, but rather a combination of deterministic-chaotic and random components. It is important to determine whether a system exhibits chaotic behavior, because the long-term predictability of such a system is limited. Under these circumstances, only a range of possible results can be determined, rather than a specific prediction.

This overview of the past four decades of research at Berkeley on the characterization of fractured rocks illustrates the wide range of projects that have been investigated, both in the United States and abroad. This paper also chronicles the various methods scientists have investigated in trying to understand the behavior of fractured rocks so as to predict the nature of flow in fractures—from the useful but limited parallel-plate concept to deterministic-chaotic modeling. Considerable progress has been made, but many problems remain.

Acknowledgments. The careful review of this report by John Gale, Gudmundur Bodvarsson, Karsten Pruess, and Yvonne Tsang is gratefully acknowledged. Their many suggestions have helped to clarify details and add important information on subjects that had been overlooked. The patient assistance of Katherine Wentworth and Maryann Villavert in helping to bring this lengthy manuscript together and the careful editing by Julie McCullough and Dan Hawkes to put the manuscript in final form are greatly appreciated.

REFERENCES

Abelin, H., I. Neretnieks, S. Tunbrant, and L. Moreno, *Final Report on the Migration in a Single Fracture; Experimental Results and Evaluations*, Svensk. Kärnbränsleförsorjning Tech. Rep. 85-03, Nucl. Fuel Safety Proj., Stockholm, Sweden, 1985.

Abelin, H., L. Birgersson, J. Gidlund, L. Moreno, I. Neretnieks, H. Widen, and J. Agren, *3D Migration Experiment—Report 3: Performed Experiments, Results and Evaluation*, Stripa Proj. Tech. Rep. 87-21, Swed. Nucl. Fuel and Waste Manage. Co. (SKB), Stockholm, Sweden, 1987.

Ahlers, C. F., C. Shan, C. Haukwa, A. J. B. Cohen, and G. S. Bodvarsson, *Calibration and Prediction of Pneumatic Response at Yucca Mountain, Nevada Using the LBNL/USGS Three-Dimensional, Site-Scale Model of the Unsaturated Zone*, Yucca Mountain Project Milestone OB12M, MOL. 19970206. 0285, Lawrence Berkeley National Laboratory, Berkeley, Calif., 1996.

Ahlers, C. F., T. M. Bandurraga, and G. S. Bodvarsson, Chapter 10: Pneumatic data analysis and the UZ model, in *The Site-Scale Unsaturated Zone Model of Yucca Mountain, for the Viability Assessment*, LBNL-40376, edited by G. S. Bodvarsson, T. M. Bandurraga, and Y. S., Wu, Lawrence Berkeley National Laboratory, Berkeley, Calif., 1997.

Ahlers, C. F., and A. C. Ritcey, Chapter 10, Effects of major faults, in UZ processes, in *Unsaturated Zone Flow and Transport Modeling of Yucca Mountain, Nevada—Fiscal Year 1998 Report*, Document ID# BAB000000-01717-2200-00016, edited by G. S. Bodvarsson, E. L. Sonnenthal, and Y. S. Wu, Lawrence Berkeley National Laboratory, Berkeley, Calif., 1998.

Bandurraga, T. M., S. Finsterle, and G. S. Bodvarsson, Chapter 3, Saturation and capillary pressure analysis, in *Development and Calibration of the 3D Site-Scale Unsaturated Zone Model of Yucca Mountain*, Report LBNL-39315, edited by G. S. Bodvarsson and T. M. Bandurraga, Lawrence Berkeley National Laboratory, Berkeley, Calif., 1996.

Bandurraga, T. M., and G. S, Bodvarsson, Chapter 6, Calibrating matrix and fracture properties using inverse modeling, in *The Site-Scale Unsaturated Zone Model of Yucca Mountain, for the Viability Assessment*, Report LBNL-40376, edited by G. S. Bodvarsson, T. M. Bandurraga, and Y. S. Wu, Lawrence Berkeley National Laboratory, Berkeley, Calif., 1997.

Bandurraga, T. M., and Y. S. Wu, Chapter 18, Recent development of numerical codes for Yucca Mountain Project, in *Unsaturated Zone Flow and Transport Modeling of Yucca Mountain, Nevada—Fiscal Year 1998 Report*, Document ID# BAB000000-01717-2200-00016, edited by Bodvarsson, G. S., E. L. Sonnenthal, and Y. S. Wu, Lawrence Berkeley National Laboratory, Berkeley, Calif., 1998.

Bear, J., *Dynamics of Fluids in Porous Media*, 764 pp., Elsevier, New York, 1972.

Benson, S. M., C. B. Goranson, J. Nobel, R. Schroeder, D. Corrigan, and H. Wollenberg, *Evaluation of the Susanville, California Geothermal Resource*, Report, LBL-11187, Lawrence Berkeley National Laboratory, Berkeley, Calif. 1981.

Birkholzer, J. T., and Y. W. Tsang, *Pretest Analysis of the Thermo-Hydrological Conditions of the ESF Drift Scale Test*, Yucca Mountain Project Level 4 Milestone SP9322M4, Lawrence Berkeley National Laboratory, Berkeley, Calif., 1997.

Birkholzer, J. T., and Y. W. Tsang, Chapter 3, Interpretive analysis of the thermo-hydrological aspects of the SHT, in *Yucca Mountain Single Heater Test Final Report, Yucca Mountain Site Characterization Project*, edited by Y. W. Tsang, J. Apps, J. T. Birkholzer, B. Freifeld, M. Q. Hu, J. Peterson, E. Sonnenthal, and N. Spycher, LBNL-42537, Lawrence Berkeley National Laboratory, Berkeley, Calif., 1999a.

Birkholzer, J. T., and Y. W. Tsang, Chapter 2, Interpretive analysis of the thermo-hydrological processes of the Drift Scale Test, in *Yucca Mountain Drift Scale Test Progress Report*, Lawrence Berkeley National Laboratory, Report LBNL-42538, Berkeley, Calif., 1999b.

Björnsson, A., G. V. Johnsen, S. Sigurdsson, G. Thorbergsson, and E. Tryggvason, Rifting of the plate boundary in North Iceland, 1975-1978, *J. Geophys. Res.*, *86*, 3024–3038, 1979.

Bodvarsson, G. S., *Mathematical Modeling of the Behavior of Geothermal Systems under Exploitation*, Ph.D. dissertation, 353 pp., Univ. of California, Berkeley, Calif., 1982.

Bodvarsson, G. S., and C.-F. Tsang, Injection and thermal breakthrough in fractured geothermal reservoirs, *J. Geophys. Res.*, *87*(B2), 1031–1048, 1982.

Bodvarsson, G. S., S. Vonder Haar, M. Wilt, and C.-F. Tsang, Preliminary studies of the reservoir capacity and the generating potential of the Baca geothermal field, New Mexico, *Water Resour. Res.*, *18*(6), 1713–1723, 1982a.

Bodvarsson, G. S., S. M. Benson, and P. A. Witherspoon, Theory of the development of geothermal systems charged by vertical faults, *J. Geophys. Res.*, *87*(B11), 9317–9328, 1982b.

Bodvarsson, G. S., S. M. Benson, O. Sigurdsson, V. Stefansson, and E. T. Eliasson, The Krafla geothermal field, Iceland 1. An analysis of well test data, *Water Resour. Res.*, *20*(11), 1515–1530, 1984a.

Bodvarsson, G. S., K. Pruess, V. Stefansson, and E. T. Eliasson, The Krafla geothermal field, Iceland 2. The natural state of the system, *Water Resour. Res.*, *20*(11), 1531–1544, 1984b.

Bodvarsson, G. S., K. Pruess, V. Stefansson, and E. T. Eliasson, The Krafla geothermal field, Iceland 3. The generating capacity of the field, *Water Resour. Res.*, *20*(11), 1545–1559, 1984c.

Bodvarsson, G. S., and T. M. Bandurraga (eds.), *Development and Calibration of the 3d Site-Scale Unsaturated Zone Model of Yucca Mountain*, Report LBNL-39315, Lawrence Berkeley National Laboratory, Berkeley, Calif., 1996.

Bodvarsson, G. S, T. M. Bandurraga, and Y. S. Wu, Chapter 1, Development of the unsaturated zone model in fiscal year 1996, in *Development and Calibration of the 3D Site-Scale Unsaturated Zone Model of Yucca Mountain*, edited by G. S. Bodvarsson and T. M. Bandurraga, Report LBNL-39315, Lawrence Berkeley National Laboratory, Berkeley, Calif., 1996.

Bodvarsson, G. S; T. M. Bandurraga, and Y. S. Wu (eds.), *The Site-Scale Unsaturated Zone Model of Yucca Mountain, for the Viability Assessment*, Report LBNL-40376, Lawrence Berkeley National Laboratory, Berkeley, Calif., 1997a.

Bodvarsson, G. S., C. Shan, A. Htay, A. Ritcey, and Y. S. Wu, Chapter 11, Estimation of percolation flux from temperature data, in *The Site-Scale Unsaturated Zone Model of Yucca Mountain, for the Viability Assessment*, edited by G. S. Bodvarsson; Bandurraga, T. M. ; and Wu, Y. S., Report LBNL-40376, Lawrence Berkeley National Laboratory, Berkeley, Calif., 1997b.

Bodvarsson, G. S., E. L. Sonnenthal, and Y. S. Wu, *Unsaturated Zone Flow and Transport Modeling of Yucca Mountain, Nevada—Fiscal Year 1998 Report*, Document ID# BAB000000-01717-2200-00016, Lawrence Berkeley National Laboratory, Berkeley, Calif., 1998.

Bourke, P. J., Channeling of flow through fractures in rock, *Proc. GEOVAL-87 International Symposium*, Swed. Nucl. Power Inspectorate, Stockholm, Sweden, 1987.

Boussinesq, J., Mémoire sur l'influence des frottements dan les movements réguliers des fluides, *J. Math. Pure. Appl., 2*, 13, 377–424, 1868.

Burleigh, R. H., E. P. Binnall, A. O. DuBois, D. U. Norgren, and A. R. Ortiz, *Electrical Heaters for Thermo-Mechanical Tests at the Stripa Mine*, Technical Project Report No. 13, LBL-7063, SAC-13, Lawrence Berkeley Laboratory, January 1979.

Chan, T., N. G. W. Cook, and C.-F. Tsang, *Theoretical Temperature Fields for the Stripa Heater Project,* Technical Project Report No. 09, LBL-7082, SAC-09, Lawrence Berkeley Laboratory, Berkeley, Calif., September 1978.

Chan, T., and N. G. W. Cook, *Calculated Thermally Induced Displacements and Stresses for Heater Experiments at Stripa, Sweden*, Technical Project Report No. 22, LBL-7061, SAC-22, Lawrence Berkeley Laboratory, December 1979.

Chen, D. W., Coupled Stiffness-Permeability Analysis of a Single Rough-Surfaced Fracture by the Three-Dimensional Boundary-Element Method, Ph.D. dissertation, Univ. of California, 75 pp., 1990.

Clayton, R. W., A 3-D Geologic Framework and Integrated Site Sub-Model (GFM3. 0) of Yucca Mountain, MO9804MWDGFM03. 001, 1998.

Cook, N. G. W., and P. A. Witherspoon, *Mechanical and Thermal Design Considerations for Radioactive Waste Repositories in Hard Rock*, Technical Project Report No. 10, LBL-7073, SAC-10, Lawrence Berkeley Laboratory, Berkeley, Calif., April 1981.

Cook, A. M., L. R. Myer, N. G. W. Cook, and F. M. Doyle, The effects of tortuosity on flow through a natural fracture, *Proc. 31st U.S. Symp. on Rock Mechanics*, edited by W. A. Hustrulid, and G. A. Johnson, Balkema, Rotterdam, The Netherlands, 1990.

Crampin, S., 1978, Seismic-wave propagation through a cracked solid: polarization as a possible dilatancy diagnostic, *Geophys. J. Roy. Astron. Soc.*, *53*, 467–496, 1978.

Crampin, S., A review of wave motion in anisotropic and cracked elastic-media, *Wave Motion*, *3*, 343–391, 1981.

Crampin, S., Effective anisotropic propagation through a cracked solid, *Proc. the First Internat. Workshop on Seismic Anisotropy, Geophys., Vol. 76,* edited by S. Crampin, R. G. Hipkin, and E. M. Chesnokov, pp. 135–145, J. Roy. Astron. Soc., 1984a.

Crampin, S., Anisotropy in exploration seismics, *First Break*, *2*, 19–21, 1984b.

Crampin, S., Evaluation of anisotropy by shear wave splitting, *Geophysics*, *50*(1), 142–152, 1985.

CRWMS M&O, *Total System Performance Assessment Peer Review Panel*, Final Report, Las Vegas, Nev., Jan. 11, 1999.

Cushey, M., Chapter 7, UZ fracture and matrix properties, in *Unsaturated Zone Flow and Transport Modeling of Yucca Mountain, Nevada—Fiscal Year 1998 Report*, edited by G. S. Bodvarsson, E. L. Sonnenthal, and Y. S. Wu, Document ID# BAB000000-01717-2200-00016, Lawrence Berkeley National Laboratory, Berkeley, Calif., 1998.

Dam, W. L., Chapter 15, Yucca Mountain geochemistry and development of conceptual models for incorporation into UZ flow modeling, in *Unsaturated zone flow and transport modeling of Yucca Mountain, Nevada—Fiscal Year 1998 Report*, Document ID# BAB000000-01717-2200-00016, edited by G. S. Bodvarsson, E. L. Sonnenthal, and Y. S. Wu, Lawrence Berkeley National Laboratory, Berkeley, Calif., 1998.

Day, W., and S. Beason, *Geological Structure at Yucca Mountain*, presentation given to U.S. Nuclear Waste Technical Review Board, Denver, Colo., 1996.

Day, W. C., C. J. Potter, D. S. Sweetkind, and R. P. Dickerson, *Mapping through 4/26/96, Preliminary Bedrock Geological Map of the Central Block Area, Yucca Mountain, Nevada*, U.S. Geological Survey Preliminary Map, U.S. Geological Survey, Denver, Colo., in review 1998.

Denlinger, R. P., and R. L. Kovach, *Seismic-Reflection Investigations at Castle Rock Springs in The Geysers Geothermal Area*, U.S. Geol. Survey Prof. Paper 1141, pp. 117–128, U.S. Geo. Survey, Denver, Colo., 1980.

Doe, T. W., K. Ingevald, L. Strindell, B. Leijon, W. Hustrulid, E. Majer, and H. Carlsson, *In situ Stress Measurements at the Stripa Mine, Sweden,* Technical Project Report No. 44, LBL-15009, SAC-44, Lawrence Berkeley Laboratory, Berkeley, Calif., Mar. 1983.

DOE (U.S. Department of Energy), *Site Characterization Plan: Yucca Mountain Site, Nevada Research and Development Area, Nevada*, DOE/RW-0199, Office of Civilian Radioactive Waste Management, Washington, D.C., 1988.

Doughty, C., Estimation of hydrologic properties of heterogeneous geologic media with an inverse method based on iterated function systems, Ph.D. dissertation, Univ. of California, 257 pp., Berkeley, Calif., 1995.

Doughty, C., J. C. S. Long, K. Hestir, and S. M. Benson, Hydrologic characterization of heterogeneous geologic media with an inverse method based on iterated function systems, *Water Resour. Res.*, *30*(6), 1721–1745, 1994.

Doughty, C., and G. S. Bodvarsson, Chapter 5, Investigation of conceptual and numerical approaches for evaluating moisture flow and chemical transport, in *The Site-Scale Unsaturated-Zone Model of Yucca Mountain, Nevada, for the Viability Assessment*, LBNL-40376, edited by G. S. Bodvarsson, T. M. Bandurraga, and Y. S. Wu, Lawrence Berkeley National Laboratory, Berkeley, Calif., 1997.

Edwards, A. L., TRUMP: A Computer Program for Transient and Steady State Temperature Distribution in Multidimensional Systems, UCRL-14754, Lawrence Livermore Laboratory, Livermore, California, 1972.

Einarsson, P., S-wave shadows in the Krafla caldera in NE Iceland: Evidence for a magma chamber in the crust, *Bull. Volcanol.*, *41*, 1–9, 1978.

Elders, W. A., D. K. Bird, A. E. Williams, and P. Schiffman, *Hydrothermal Flow Regime and Magmatic Heat Source of the Cerro Prieto Geothermal System*, Baja California, Mexico, *Geothermics*, *13*(1–2), 27–47, 1984.

Engelder, T., and C. H. Scholz, Fluid flow along very smooth joints at effective pressure up to 200 megapascals, in *Mechanical Behavior of Crustal Rocks*, Geophysical Monograph 24, edited by N. L. Carter et al., pp. 147–152, American Geophysical Union, Washington, D.C., 1981.

Evans, D. M., The Denver area earthquakes and the Rocky Mountain Arsenal well, *The Mountain Geologist*, *3*(1), 1966.

Fabryka-Martin, J. T. A. V. Wolfsberg, P. R. Dixon, S. Levy, J. Musgrave, and H. J. Turin, *Summary Report of Chlorine-36 Studies: Sampling, Analysis and Simulation of Chlorine-36 in the Exploratory Studies Facility*, LA-13352-MS, Los Alamos National Laboratory, Los Alamos, N.M., 1997.

Fabryka-Martin, J. T., A. V. Wolfsberg, J. L. Roach, S. S. Levy, S. T. Winters, L. E. Worfsberg, D. Elmore, and P. Sharma, Distribution of fast hydrologic paths in the unsaturated zone at Yucca Mountain, *Proc. the Eighth International Conference on High-Level Radioactive Waste Management, Las Vegas, Nev., May 11-14, 1998,* pp. 93–96, American Nuclear Society, La Grange Park, Ill., 1998.

Faybishenko, B., Evidence of chaotic behavior in flow through fractured rocks, and how we might use chaos theory in fractured rock hydrology, *Proc. Dynamics of Fluids in Fractured Rocks, Concepts and Recent Advances Feb. 10–12, 1999*, LBNL-42718, Lawrence Berkeley National Laboratory, Berkeley, Calif., 1999.

Faybishenko, B., P. A. Witherspoon, C. Doughty, T. R. Wood, R. K. Podgorney, and J. T. Geller, *Multi-Scale Investigations of Liquid Flow in a Fractured Basalt Vadose Zone*, LBNL-42910, Lawrence Berkeley National Laboratory, Berkeley, Calif., 1999a.

Faybishenko, B., C. Doughty, M. Steiger, J. C. S. Long, T. Wood, J. Jacobsen, J. Lore, and P. T. Zawislanski, *Conceptual Model of the Geometry and Physics of Water Flow in a Fracture Basalt Vadose Zone: Box Canyon Site, Idaho*, LBNL-42925, Lawrence Berkeley National Laboratory, 1999b.

Finsterle, S., *ITOUGH2 Command Reference, Version 3.1*, Report LBNL-40041, Lawrence Berkeley National Laboratory, Berkeley, Calif., 1997a.

Finsterle, S., *ITOUGH2 Sample Problems*, Report LBNL-40042, Lawrence Berkeley National Laboratory, Berkeley, Calif., 1997b.

Finsterle, S., *ITOUGH2 User's Guide*, Report LBNL-40040, Lawrence Berkeley National Laboratory, Berkeley, Calif., 1999.

Flint, L. E., and A. L. Flint, Shallow infiltration processes in arid watersheds at Yucca Mountain, Nevada, *Proc. the Fifth Annual High Level Radioactive Waste Management Conference*, *4*, pp. 2315–2322, Las Vegas, Nev., American Nuclear Society, La Grange Park, Ill., 1994.

Flint, A. L., J. A. Hevesi, and L. E. Flint, *Conceptual and Numerical Model of Infiltration for the Yucca Mountain Area, Nevada*, U.S. Geol. Surv. Water Res. Invest. Rep., U.S. Geological Survey, Denver, Colo., in review 1996.

Fridrich, C. J., W. W. Dudley, Jr., and J. S. Stuckless, Hydrogeologic Analysis of the Saturated-Zone Groundwater System, under Yucca Mountain, Nevada, *J. Hydrology*, *154*, 133–168, 1994.

Fritz, P., J. F. Barker, and J. E. Gale, *Geochemistry and Isotope Hydrology of Groundwaters in the Stripa Granite—Results And Preliminary Interpretation*, Technical Project Report No. 12, LBL-8285, SAC-12, Lawrence Berkeley Laboratory, Berkeley, Calif., April, 1979.

Galbraith, R. M., C. R. Wilson, A. O. DuBois, S. T. Lundgren, M. J. McPherson, and G. W. West, *Equipment Design, Installation, and Operation for the Macropermeability Experiment at Stripa, Sweden*, Technical Project Report No. 47, LBL-13392, SAC-47, Lawrence Berkeley Laboratory, Berkeley, Calif., Sept., 1981.

Gale, J. E., *A Numerical, Field and Laboratory Study of Flow in Rocks with Deformable Fractures*, Ph.D. dissertation, 255 pp., Univ. of California, Berkeley, Calif., 1975.

Gale, J. E., *Fracture and Hydrology Data from Field Studies at Stripa, Sweden*, Technical Project Report No. 46, LBL-13101, SAC-46, Lawrence Berkeley Laboratory, Berkeley, Calif., April, 1981.

Gale, J. E., The effects of fracture type (induced versus natural) on the stress-fracture closure-fracture permeability relationships, *Proc. 23rd U.S. Rock Mechanics Symposium, Berkeley, CA, August 1982*, edited by R. E. Goodman and F. E. Heuze, pp. 290–298, 1982.

Gale, J. E., and P. A. Witherspoon, *An Approach to the Fracture Hydrology at Stripa*, Technical Project Report No. 15, LBL-7079, SAC-15, Lawrence Berkeley Laboratory, Berkeley, Calif., May 1979.

Geller, J. T., G. Su, and K. Pruess, *Preliminary Studies of water Seepage through Partially Saturated Rough-Walled Fractures*, LBNL-38810, Lawrence Berkeley National Laboratory, Berkeley, Calif., 1996.

Gentier, S., *Morphologie et Comportement Hydromécanique d'une Fracture Naturelle dans un Granite Sous Contrainte Normale,* Document du BRGM, No. 154, Thèse de l'Université d'Orléans, France,1987.

Gentier, S., Hydromechanical behavior of a single natural fracture under normal stress, in *Selected Papers on Hydrogeology 28th Int. Geological Congress*, Vol. 1, edited by E. S. Simpson and J. M. Sharp, pp. 327–338, Verlag Heinz Heise, Hannover, Germany, 1990.

Gentier, S., and D. Hopkins, Comportement hydromécanique sous contrainte normale d'une fracture in-situ, in *Modélisation du Comportement Mécanique des Essais en Laboratoire*, Report ANDRA B RP 0ANT 96-119/A, 1996.

Gentier, S., and D. Hopkins, Mapping fracture aperture as a function of normal stress using a combination of casting, image analysis and modeling techniques, *Int. J. Rock Mech., and Min. Sci. Geomech. Abstr., 34*(3/4), 1997.

Gilmour, H. M. P., D. Billaux, and J. C. S. Long, Models for Calculating Fluid Flow in Randomly Generated Three-Dimensional Networks of Disc-Shaped Fractures, Theory and Design of FMG3D, DISCEL, and DIMES, LBL-19515, Lawrence Berkeley Laboratory, Berkeley, Calif., 1986.

Glassley, W. E., *Third quarter report, chemical analyses of waters collected from the Single Heater Test*, Yucca Mountain Project Level 4 Milestone SP9281M4, Lawrence Livermore National Laboratory, Livermore, Calif., 1997.

Glassley, W. E., and L. DeLoach, *Second quarter results of chemical measurements in the Single Heater Test*, Milestone Report for the CRWMS M&O, U.S. Dept. of Energy, SP9240M4, Lawrence Livermore National Laboratory, Livermore, Calif., 1997.

Gnirk, P., OECD/NEA International Stripa project, Overview Volume II Natural Barriers, SKB, Stockholm, Sweden, 1993.

Goodman, R. E., R. E. Taylor, and T. Brekke, A model for the mechanics of jointed rock, *J. Soil Mech., and Foundations Div., ASCE, 94*, SM3, 1968.

Grant, M. A., A. H. Trusdell, and A. Mañón M., Production induced boiling and cold water entry in the Cerro Prieto geothermal reservoir indicated by chemical and physical measurements, *Geothermics, 13*(1/2),117–140, 1984.

Gringarten, A. C., and P. A. Witherspoon, A method of analyzing pump test data from fractured aquifers, *Proc. Symp. on Percolation Through Fissured Rock of Int. Soc. for Rock Mech., Stuttgart, T2-C, Sept. 18–19, 1972*, 1972.

Gringarten, A. C., and H. J. Ramey, Unsteady-state pressure distributions created by a well with a single horizontal fracture, partial penetration, or restricted entry, *Soc. Petr. Engr. J.*, SPE 3819, 413–426, 1974.

Gringarten, A. C., H. J. Ramey, and R. Raghavan, Unsteady-state pressure distributions created by a well with a single infinite conductivity vertical fracture, *Soc. Petr. Engr. J.*, SPE 4051, 347–360, 1974.

Halfman, S. E., M. J. Lippmann, R. Zelwer, and J. H. Howard, A geological interpretation of the geothermal fluid movement in the Cerro Prieto Geothermal Field, Baja California, Mexico, *Assoc. Pet. Geol. Bull., 68*, 18–30, 1984.

Haukwa, C., Y. S. Wu, and G. S. Bodvarsson, Chapter 13, Thermal loading studies using the unsaturated zone model, in *Development and Calibration of the 3D Site-Scale Unsaturated Zone Model of Yucca Mountain*, edited by G. S. Bodvarsson and T. M. Bandurraga, Report LBNL-39315, Lawrence Berkeley National Laboratory, Berkeley, Calif., 1996.

Haukwa, C., and Y. S. Wu, Chapter 4, Grid generation and analysis, in *The Site-Scale Unsaturated Zone Model of Yucca Mountain, Nevada, for the Viability Assessment*, edited by G. S. Bodvarsson, T. M. Bandurraga, and Y. S. Wu, Report LBNL-40376, Lawrence Berkeley National Laboratory, Berkeley, Calif., 1997.

Heath, M. J., Solute migration experiments in fractured granite, South West England, in Design and instrumentation of in situ experiments in underground laboratories for radioactive waste disposal, in *Proc. of a Joint CEC-NEA Workshop*, pp. 191–200, edited by A. A. Balkema, Rotterdam, The Netherlands, 1985.

Hedegaard, R., C. F. Ahlers, T. M. Bandurraga, M. Cushey, C. Haukwa, J. Hinds, H. H. Lui, A. C. Ritcey, E. Sonnenthal, and Y. S. Wu, Chapter 6, UZ Processes in *Unsaturated Zone Flow and Transport Modeling of Yucca Mountain, Nevada—Fiscal Year 1998 Report*, edited by G. S. Bodvarsson, E. L. Sonnenthal, and Y. S. Wu, Document ID# BAB000000-01717-2200-00016, Lawrence Berkeley National Laboratory, Berkeley, Calif., 1998.

Hestir, K., and J. C. S. Long, Analytical expressions for the permeability of random two-dimensional Poisson fracture networks based on regular lattice percolation and equivalent media theories, *J. Geophys. Res., 95*(B13), 21,565–21,581, 1990.

Hevesi, J. A., and A. L. Flint, Geostatistical Model for Estimating Precipitation and Recharge in the Yucca Mountain Region, Nevada-California, Water Resour. Invest. Rep., U.S. Geological Survey, Denver, Colo., 1995.

Hinds, J., L. Pan, and R. Hedegaard, Chapter 4, Geological model and numerical grids, in *Unsaturated Zone Flow and Transport Modeling of Yucca Mountain, Nevada—Fiscal Year 1998 Report*, edited by G. S. Bodvarsson, E. L. Sonnenthal, and Y. S. Wu, Yucca Mountain Project Milestone SP3CKJM4; Document ID# BAB000000-01717-2200-00016, Lawrence Berkeley National Laboratory, Berkeley, Calif., 1998.

Hood, M., I. *Some Results from a Field Investigation of Thermo-Mechanical Loading of a Rock Mass When Heaters Are Emplaced in the Rock*, Technical Project Report No. 26, LBL-9392, SAC-26, Lawrence Berkeley Laboratory, Berkeley, Calif., March 1980.

Hood, M., H. Carlsson, and P. H. Nelson, II. *The Application of Field Data from Heater Experiments Conducted at Stripa,*

Sweden to Parameters for Repository Design, Technical Project Report No. 26, LBL-9392, SAC-26, Lawrence Berkeley Laboratory, Berkeley, Calif., March 1980.

Hopkins, D. L., The Effect of Surface Roughness on Joint Stiffness, Aperture, and Acoustic Wave Propagation, Ph.D. dissertation, 421 pp., Univ. of California, Berkeley, 1991.

Hopkins, D. L., The implications of joint deformation in analyzing the properties and behavior of fractured rock masses, underground excavations, and faults, *Int. J. Rock Mech. Min. Sci.,* submitted, 1999.

Hudson, D. B., and A. L. Flint, Estimation of shallow infiltration and presence of potential fast pathways for shallow infiltration in the Yucca Mountain area, Nevada, submitted for publication as U.S. Geol. Surv., Water Res. Invest. Rep., U.S. Geological Survey, Denver, Colo., in review 1996.

Iwai, K., *Fundamental Studies of Fluid Flow through a Single Fracture*, Ph.D. dissertation, 208 pp., Univ. of California, Berkeley, Calif., 1976.

Jacks, G., Chemistry of some groundwaters in igneous rocks, *Nordic Hydrology*, *4*(4) 207, 1973.

Karasaki, K., *A New Advection-Dispersion Code for Calculating Transport in Fracture Networks*, LBL-22090, Lawrence Berkeley Laboratory, Berkeley, Calif., 1987.

Karasaki, K., P. A. Witherspoon, and J. C. S. Long, A new analytical model for fracture-dominated reservoirs, *Soc. Petr. Engr. J., SPE 14171*, 242–250, 1988a.

Karasaki, K., J. C. S. Long, and P. A. Witherspoon, Analytical models of slug tests, *Water Resour. Res.*, *24*(1), 115–126, 1988b.

Kneafsey, T., and K. Pruess, Laboratory experiments in natural and artificial rock fractures, *Water Resour. Res.*, *34*(12), 3349–3367, 1998.

Lai, C. H., G. S. Bodvarsson, C.-F. Tsang, and P. A. Witherspoon, *A New Model for Well Test Data Analysis for Naturally Fractured Reservoirs*, *SPE 11688*, paper presented at California Regional Meeting of SPE, Ventura, Calif., 1983.

Lane, K. S., Engineering problems due to fluid pressure in rock, *Eleventh Symposium on Rock Mechanics*, Univ. California, Berkeley, Calif., 1969.

Lasseter, T. J., *Underground Storage of Liquefied Natural Gas in Cavities Created by Nuclear Explosives*, Ph.D. dissertation, 273 pp., Univ. of California, Berkeley, Calif., 1974.

Lasseter, T. J., P. A. Witherspoon, and M. J. Lippmann, Multiphase multidimensional simulation of geothermal reservoirs, *Proc. of Second U. N. Symposium on the Development and Use of Geothermal Resources, San Francisco, California*, *3*, pp. 1715–1723, 1975.

Leary, P. C., and T. L. Henyey, Anisotropy and fracture zones about a geothermal well from p-wave velocity profiles, *Geophysics*, *50*(1), 25–36, 1985.

LeCain, G. D., and G. Patterson, *Memo to Robert Craig to Satisfy Level 4 Milestone SPH35EM4, Technical Analysis/ Interpretation, Air-Permeability and Hydrochemistry Data through January 31, 1997,* GS970383122410. 004, U.S. Geological Survey, Denver, Colo., March 11, 1997.

Levy, S. S., and S. J. Chipera, *Mineralogy, Alterations and Hydrologic Properties of the Ptn Hydrogeologic Unit, Yucca Mountain, Nevada*, (Draft) Yucca Mountain Project Level 4 Milestone SP344DM4, Los Alamos National Laboratory, Los Alamos, N.M., 1997.

Lin, M., M. P. Hardy, and S. J. Bauer, Fracture Analysis and Rock Quality Designation Estimation for the Yucca Mountain Site Characterization Project, SAND92-0449, Sandia National Laboratories, Albuquerque, N.M., 1993.

Lippmann, M. J., T. N. Narasimhan, and P. A. Witherspoon, *Numerical Simulation of Reservoir Compaction in Liquid Dominated Systems*, LBL-4462, Lawrence Berkeley Laboratory, Berkeley, Calif., December 1976.

Lippmann, M. J., T. N. Narasimhan, and P. A. Witherspoon, *Modeling Subsidence Due to Geothermal Fluid Production*, LBL-7007, Lawrence Berkeley Laboratory Report, Berkeley, Calif., October 1977.

Lippmann, M. J., and G. S. Bodvarsson, Numerical studies of the heat and mass transport in the Cerro Prieto geothermal field, Mexico, *Water Resour. Res.*, *19*(3), 753–767, 1983.

Liu, H. H., and G. S. Bodvarsson, Chapter 8, A fracture-matrix interaction model, in *Unsaturated Zone Flow and Transport Modeling of Yucca Mountain, Nevada—Fiscal Year 1998 Report*, edited by G. S. Bodvarsson, E. L. Sonnenthal, and Y. S. Wu, Document ID# BAB000000-01717-2200-00016, Lawrence Berkeley National Laboratory, Berkeley, Calif., 1998.

Lomize, G. M., *Flow in Fractured Rocks* (in Russian), 127 pp., Gosenergoizdat, Moscow, 1951.

Long, J. C. S., Investigation of Equivalent Porous Medium Permeability in Networks of Discontinuous Fractures, Ph.D. dissertation, 277 pp., Univ. of California, Berkeley, Calif., 1983.

Long, J. C. S., and P. A. Witherspoon, The relationship of the degree of interconnection to permeability in fracture networks, *J. Geophys. Res.*, *90*(B4), 3087–3098, 1985.

Long, J. C. S., P. Gilmour, and P. A. Witherspoon, A model for steady fluid flow in random three-dimensional networks of disc-shaped fractures, *Water Resour. Res.*, *21*(8), 1105–1115, 1985.

Long, J. C. S., and D. M. Billaux, From field data to fracture network modeling: An example incorporating spatial structure, *Water Resour. Res.*, *23*(7), 1201–1216, 1987.

Long, J. C. S., K. Karasaki, A. Davey, J. Peterson, M. Landsfield, J. Kemeny, and S. Martel, *Preliminary Prediction of Inflow into the D-holes at the Stripa Mine*, Stripa Project TR 90-04, Swedish Nuclear Fuel Company, Stockholm, Sweden, 1990.

Long, J. C. S., A. Mauldon, K. Nelson, S. Martel, P. Fuller, and K. Karasaki, *Prediction of Flow and Drawdown for the Site Characterization and Validation Site in the Stripa Mine*, Stripa Project TR 92-05, Swedish Nuclear Fuel Company, Stockholm, Sweden, 1992.

Long, J. C. S., C. Doughty, B. Faybishenko, A. Aydin, B. Feifeld, K. Grossenbacher, P. Holland, J. Horsman, J. Jacobsen, T. Johnson, K.-H. Lee, J. Lore, K. Nihei, J. Peteron, R. Salve, J. Sisson, B. Thapa, D. Vasco, K. Williams, T. Wood, and P. Zawislanski, *Analog Site for Fractured Rock Characterization, Annual Report FY 1995*, LBNL-38095, Lawrence Berkeley National Laboratory, Berkeley, Calif., 1995.

Louis, C., *A Study of Groundwater Flow in Jointed Rock and its Influence on the Stability of Rock Masses, Rock Mech. Res. Rep. 10*, 90 pp., Imp. Coll., London, England, 1969.

Majer, E. L., *Seismological Investigation in Geothermal Areas*, Ph.D. dissertation, University of California, Berkeley, Calif., 1978.

Majer, E. L., and T. V. McEvilly, Seismological investigations at The Geysers geothermal field, *Geophysics*, *44*(2), 246–269, 1979.

Majer, E. L., T. V. McEvilly, A. Albores, and S. Diaz, Seismological studies at the Cerro Prieto geothermal field, *Geothermics*, *9*(1–2), 79–89, 1980.

Majer, E. L., T. V. McEvilly, F. Eastwood, and L. Myer, Fracture detection using P-wave and S-wave vertical seismic profiling at The Geysers, *Geophysics*, *53*, 76–84, 1988.

Majer, E. L., L. R. Myer, J. E. Peterson Jr., K. Karasaki, J. C. S. Long, S. J. Martel, P. Blumling, and S. Vomvoris, *Joint Seismic, Hydrogeological, and Geomechanical Investigations of a Fracture Zone in the Grimsel Rock Laboratory*, Switzerland, LBL-27913, Lawrence Berkeley Laboratory Berkeley, Calif., June 1990.

Majer, E. L., J. E. Peterson, T. M. Daley, B. Kaelin, J. Queen, P. D'Onfro, and W. Rizer, Fracture detection using crosswell and single well surveys, *Geophysics*, *62*(2), 495–504, 1997.

Mauldon, A. D., K. Karasaki, S. L. Martel, J. C. S. Long, M. Landsfeld, and A. Mensch, An inverse technique for developing models for fluid flow in fracture systems using simulated annealing, *Water Resour. Res.*, *29*(11), 3775–3789, 1993.

McEvoy, M. B., *Data Acquisition, Handling, and Display for the Heater Experiments at Stripa*, Technical Project Report No. 14, LBL-7062, SAC-14, Lawrence Berkeley Laboratory, Berkeley, Calif., February 1979.

Mercado, G., S., Migración de fluidos geotérmicos y distribución de temperaturas en el subsuelo del campo geotérmico de Cerro Prieto, Baja California, Mexico, *Proc. Second United Nations Symposium on the Development and Use of Geothermal Resources*, pp. 487–492, U.S. Gov. Printing Office, Washington, D.C., 1976.

Montazer, P., and W. E, Wilson, *Conceptual Hydraulic Model of Flow in the Unsaturated Zone, Yucca Mountain, Nevada*, Water Resources Investigations Report 84-4355, 55 pp., U.S. Geological Survey, Denver Colo., 1984.

Moreno, L., I. Neretnieks, and T. Eriksen, Analysis of some laboratory tracer runs in natural fractures, *Water Resour. Res.*, *21*(7), 951–958, 1985.

Moreno, L., Y. W. Tsang, C.-F. Tsang, F. V. Hall, and I. Neretnieks, Flow and tracer transport in a single fracture: A stochastic model and its relation to some field operations, *Water Resour. Res.*, *24*(12), 2033–2048, 1988.

Moridis, G. J., and K. Pruess, *T2CG1: A Package of Preconditioned Conjugate Gradient Solvers for the TOUGH2 Family of Codes*, LBL-36235, Lawrence Berkeley Laboratory, Berkeley, Calif., 1995.

Moridis, G. J., and K. Pruess, T2SOLV: An enhanced package of solvers for the TOUGH2 family of reservoir simulation codes, *Geothermics*, *27*(4), 415–444, 1998.

Myer, L. R., D. L. Hopkins, and N. G. W. Cook, Effects of contact area of an interface on acoustic wave transmission characteristics, *Proc. 25th U. S. Symposium on Rock Mechanics*, 565-572, 1985.

Narasimhan, T. N., and P. A. Witherspoon, *Reservoir Evaluation Tests on RRGE 1 and RRGE 2, Raft River Geothermal Project*, LBL-5958, Lawrence Berkeley Laboratory, Berkeley, Calif., May 1977.

Narasimhan, T. N., D. G. McEdwards, and P. A. Witherspoon, *Results of Reservoir Evaluation Tests, 1976 East Mesa Geothermal Field*, LBL-6369, Lawrence Berkeley Laboratory, Berkeley, Calif., July 1977.

Neretnieks, I., *Channeling Effects in Flow and Transport in Fractured Rocks—Some Recent Observations and Models*, (paper presented at GEOVAL-87 International Symposium, Swed. Nucl. Insp., Stockholm, Sweden, April 7–9, 1987), 1987.

Noorishad, J., *Seepage in Fractured Rock: Finite Element Analysis of Rock Mass Behavior under Coupled Action of Body Forces, Flow Forces, and External Loads*, Ph.D. dissertation, 128 pp., Univ. of California, Berkeley, Calif., 1971.

Noorishad, J., P. A. Witherspoon, and T. L. Brekke, *A Method for Coupled Stress and Flow Analysis of Fractured Rock Masses*, Geotechnical Engr. Pub. No. 71-6, Univ. of California, Berkeley, Calif., 1971.

Noorishad, J., M. S. Ayatollahi, and P. A. Witherspoon, A finite-element method for coupled stress and fluid flow analysis in fractured rock masses, *Int. J. Rock Mech., Min. Sci. & Geomech. Abstr.*, *19*, 185–193, 1982.

Noorishad, J., C.-F. Tsang, and P. A. Witherspoon, Coupled thermal-hydraulic-mechanical phenomena in saturated fractured porous rocks: numerical approach, *J. Geophysical Res.*, *89*(12), 10365–10373, 1984.

Noorishad, J., and C.-F. Tsang, Coupled thermohydroelasticity phenomena in variably saturated fractured porous rocks—Formulation and numerical solution, in *Coupled Thermo-Hydro-Mechanical Processes of Fractured Media*, edited by O. Stephansson, L. Jing, and C.-F. Tsang, pp. 93–134, Elsevier, New York, N.Y., 1996.

Nordqvist, A. W., Y. W. Tsang, C.-F. Tsang, B. Dverstorp, and J. Anderson, A variable aperture fracture network model for flow and transport in fractured rocks, *Water Resour. Res., 28*(6), 1703–1713, 1992.

Nordqvist, A. W., Y. W. Tsang, C.-F. Tsang, B. Dverstorp, and J. Anderson, Effects of high variance of fracture transmissivity on transport and sorption at different scales in a discrete model for fractured rocks, *J. Cont. Hydrol*, *22*, 39–66, 1996.

Olsson, O., and J. E. Gale, Site assessment and characterization for high-level nuclear waste disposal: Results from the Stripa Project, Sweden, *Quart. J. Engr. Geology*, *28*, s17–s30, 1995.

Ortiz, T. S., R. L. Williams, F. B. Nimick, B. C. Whittet, and D. L. South, *A Three-Dimensional Model of Reference Thermal/Mechanical and Hydrological Stratigraphy at Yucca Mountain, Southern Nevada*, Sandia Report SAND84-1076, Sandia National Laboratories, Albuquerque, N.M., 1985.

Paces, J. B. L. A. Newmark, B. D. Marshall, J. F. Whelan, and Z. E. Peterman, Inferences for Yucca Mountain unsaturated zone hydrology from secondary minerals, *Proc. Eighth International Conference on High-Level Radioactive Waste Management, Las Vegas, Nev., May 11–14, 1998*, pp. 36–39, American Nuclear Society, La Grange Park, Ill., 1998.

Paulsson, B. N. P., *Seismic Velocities and Attenuation in a Heated Granitic Repository*, Technical Project Report No. 51, LBL-

16346, SAC-51, Lawrence Berkeley Laboratory, Berkeley, Calif., January 1983.

Persoff, P., and K. Pruess, Two-phase flow visualization and relative permeability measurement in natural rough-walled rock fractures, *Water Resour. Res.*, *31*(5), 1175–1186, 1995.

Peterson, J. E., and K. H. Williams, Chapter 4, Interpretive analysis of the geophysical measurements: Ground penetrating radar and acoustic emission, in *Drift Scale Test Progress Report, Lawrence Berkeley National Laboratory*, Level 4 Milestone SP2930M4, Lawrence Berkeley National Laboratory, Berkeley, Calif., July 14, 1999.

Polubarinova-Kochina, P. Ya., *Theory of Groundwater Movement,* translated from Russian by J. M. R. DeWiest, 613 pp., Princeton University Press, Princeton, N.J., 1962.

Prucha, R. H., *Heat and Mass Transfer in the Klamath Falls, Oregon, Geothermal System*, M. S. thesis, 164 pp., Univ. of California, Berkeley, Calif., 1987.

Pruess, K., *Development of the General Purpose Simulator MULKOM*, LBL-15500, Lawrence Berkeley Laboratory, Berkeley, Calif., 1983.

Pruess, K., *TOUGH2—A General Purpose Numerical Simulator for Multiphase Fluid and Heat Flow*, LBL-29400, Lawrence Berkeley Laboratory Report, Berkeley, Calif., 1991.

Pruess, K., On water seepage and fast preferential flow in heterogeneous unsaturated rock fractures, *J. Contam. Hydrol.*, *30*(3–4), 333–362, 1998.

Pruess, K., A mechanistic model for water seepage through thick unsaturated zones in fractured rocks of low matrix permeability, *Water Resour. Res.*, *35*(4), 1039–1051, 1999.

Pruess, K., R. C. Schroeder, J. M. Zerzan, and P. A. Witherspoon, *SHAFT78, a Two-Phase Multidimensional Computer Program for Geothermal Reservoir Simulation*, LBL-8264, Lawrence Berkeley Laboratory, Berkeley, Calif., November 1979.

Pruess, K., and R. C. Schroeder, *SHAFT79 User's Manual*, LBL-10861, Lawrence Berkeley Laboratory, Berkeley, Calif., March 1980.

Pruess, K., and T. N. Narasimhan, *A Practical Method for Modeling Heat and Fluid Flow in Fractured Porous Media, Soc. Pet. Engr., SPE-10509* (presented at Sixth SPE Symposium on Reservoir Simulation New Orleans, La., Feb. 1982), 1982a.

Pruess, K., and T. N. Narasimhan, On fluid reserves and the production of superheated steam from fractured, vapor-dominated geothermal reservoirs, *J. Geophys. Res., 87*(B11), 9329–9339, 1982b.

Pruess, K., G. S. Bodvarsson, V. Stefansson, and E. T. Eliasson, The Krafla geothermal field, Iceland 4. History match and prediction of individual well performance, *Water Resour. Res.*, *20*(11), 1561–1584, 1984.

Pruess, K., and Y. W. Tsang, On two-phase relative permeability and capillary pressure of rough-walled rock fractures, *Water Resour. Res.*, *26*(9),1915–1926, 1990.

Pruess, K., A. Simmons, Y. S. Wu, and G. Moridis, *TOUGH2 Software Qualification,* LBL-38384, Lawrence Berkeley National Laboratory, Berkeley, Calif., 1996.

Pruess, K., B. Faybishenko, and G. S. Bodvarsson, Chapter 24, Alternative concepts and approaches for modeling unsaturated flow and transport in fractured rocks, in *The Site-Scale Unsaturated-Zone Model of Yucca Mountain, Nevada, for the Viability Assessment*, LBNL-40376, edited by G. S. Bodvarsson, T. M. Bandurraga, and Y. S. Wu, Lawrence Berkeley National Laboratory, Berkeley, Calif., 1997a.

Pruess, K., S. Finsterle, G. Moridis, C. Oldenburg, and Y. S. Wu, General-purpose reservoir simulators: the TOUGH2 family (originally published as Report LBL-40140, Lawrence Berkeley National Laboratory, Berkeley Calif., 1997b), *GRC Bull.*, *25*(2). 53–57, 1997.

Pyrak-Nolte, L. J., N. G. W. Cook, and L. R. Myer, *The Effect of Stress on the Hydraulic, Mechanical and Seismic Properties of a Natural Fracture* (presented at Spring 1986 Mtg., Amer. Geophys. Union, Baltimore, Md., Apr. 19–23, 1986), 1986.

Pyrak-Nolte, L. J., L. R. Myer, N. G. W. Cook, and P. A. Witherspoon, Hydraulic and mechanical properties of natural fractures in low permeability rock, *Proc. 6th International Congress on Rock Mechanics*, pp. 224–231, edited by G. Herget and S. Vongpaisal, Balkema, Rotterdam, The Netherlands, 1987.

Pyrak-Nolte, L. J., *Seismic Visibility of Fractures*, Ph.D. dissertation, Univ. of California, Berkeley, Calif., 1988.

Pyrak-Nolte, L. J., L. R. Myer, and N. G. W. Cook, Transmission of seismic waves across single natural fractures, *J. Geophys. Res.*, *95*(B6), 8617–8638, 1990.

Raven, K. G., and J. E. Gale, Water flow in natural fractures as a function of stress and sample size, *Int. J. Rock Mech. Min. Sci. Geomech. Abstr., 22*, 251–261, 1985.

Romm, E. S., *Flow Characteristics of Fractured Rocks* (in Russian) 283 pp., Nedra, Moscow, Russia, 1966.

Rouleau, A., and J. E. Gale, *Characterization of the Fracture System at Stripa with Emphasis on the Ventilation Drift*, Technical Project Report No. 52, LBL-14875, SAC-52, Lawrence Berkeley Laboratory, Berkeley, Calif., October 1985.

Rousseau, J. P., C. L. Loskot, F. Thamir, and N. Lu, *Results of Borehole Monitoring in the Unsaturated Zone within the Main Drift Area of the Exploratory Studies Facility, Yucca Mountain, Nevada,* TIC: 238150, U.S. Geological Survey, Denver, Colo., in review 1997.

Rousseau, J. P., E. M. Kwicklis, and D. C. Gillies, D. C. (eds.) *Hydrogeology of the Unsaturated Zone, North Ramp Area of the Exploratory Studies Facility, Yucca Mountain, Nevada*, USGS-WRIR-98-4050 Yucca Mountain Project Milestone 3GUP667M, U.S. Geological Survey. Denver, Colo., in press 1998.

Saemundsson, K., Evolution of the axial rifting zone in northern Iceland and the Tjörnes fracture zone, *Geol. Soc. Am. Bull.*, *85*, 495–504, 1974.

Saemundsson, K., Fissure swarms and central volcanoes of the neovolcanic zones of Iceland, *Geol. J.*, *Special Issue*, 415–432, 1978.

Sass, J. H., A. H. Lachenbruch, W. W. Dudley, Jr., S. S. Priest, and R. J. Munroe, *Temperature, Thermal Conductivity and Heat Flow Near Yucca Mountain, Nevada: Some Tectonic and Hydrologic Implications*, U.S. Geol. Surv. Open File Rept. 87-649, U.S. Geological Survey, Denver, Colo., 1988.

Schoenberg, M., Elastic wave behavior across linear slip interfaces, *J. Acoust. Soc. Am.*, *68*(5), 1516–1521, 1980.

Schoenberg, M., Reflection of elastic waves from periodically stratified media with interfacial slip, *Geophys. Prosp.*, *31*, 265–292, 1983.

Scott, R. B., and J. Bonk, Preliminary geologic map of Yucca Mountain with geologic sections, Nye County Nevada, scale 1:12,000, in *U.S.G.S. Open-File Report*, pp. 484–494, U.S. Geological Survey, Denver, Colo., 1984.

Shan, C., *Characterization of Leaky Faults*, Ph.D. dissertation, 201 pp., Univ. of California, Berkeley, Calif., 1990.

Snow, D. T., *A Parallel Plate Model of Fractured Permeable Media*, Ph.D. dissertation, 331 pp., Univ. of California, Berkeley, Calif., 1965.

Snow, D. T., Anisotropic permeability of fractured media, *Water Resour. Res.*, *5*(6), 1273–1289, 1969.

Sonnenthal, E. L., Chapter 16, Simulation of water-gas-rock interaction in the ambient system, in *Unsaturated Zone Flow and Transport Modeling of Yucca Mountain, Nevada—Fiscal Year 1998 Report*, Document ID# BAB000000-01717-2200-00016, edited by G. S. Bodvarsson, E. L. Sonnenthal, and Y. S. Wu, Lawrence Berkeley National Laboratory, Berkeley, Calif., 1998.

Sonnenthal, E. L., C. F. Ahlers, and G. S. Bodvarsson, Chapter 7, Fracture and fault properties for the UZ Site-Scale flow model, in *The Site-Scale Unsaturated Zone Model of Yucca Mountain, for the Viability Assessment*, LBNL-40376, edited by G. S. Bodvarsson, T. M. Bandurraga, and Y. S. Wu, Lawrence Berkeley National Laboratory, Berkeley, Calif., 1997.

Sonnenthal, E. L., N. Spycher, and J. Apps, Chapter 3, Interpretive analysis of the thermo-hydrological-chemical processes of the Drift-Scale Test, in *Drift Scale Test Progress Report, Lawrence Berkeley National Laboratory*, Level 4 Milestone SP2930M4, Lawrence Berkeley National Laboratory, Berkeley, Calif., July 14, 1999.

Sorey, M. L., *Numerical Modeling of Liquid Geothermal Systems*, Ph.D. dissertation, 66 pp., Univ. of California, Berkeley, Calif., 1975.

Spycher, N., E. Sonnenthal, and J. Apps, Chapter 4, Interpretive analysis of the thermo-hydrological-chemical aspects of the single heater test, in *Yucca Mountain Single Heater Test Final Report, Yucca Mountain Site Characterization Project,* LBNL-42537, edited by Y. W. Tsang, J. Apps, J. T. Birkholzer, B. Freifeld, M. Q. Hu, J. Peterson, E. Sonnenthal, and N. Spycher, Lawrence Berkeley National Laboratory, Berkeley, Calif., 1999.

Stefansson, V., *The Krafla Geothermal Field, Northeast Iceland, in Geothermal Systems*, edited by L. Ryback and L. J. P. Muffler, pp. 273–294, John Wiley, New York, N.Y., 1981.

Stillwell, W. B., W. K. Hall, and J. Tawhai, Ground movement in New Zealand geothermal fields, *Proceedings Second U. N. Symposium on the Development and Use of Geothermal Resources, San Francisco, California, Vol. 2,* pp. 1427–1434, 1975.

Su, G. W., J. T. Geller, K. Pruess, and F. Wen, Experimental studies of water seepage and intermittent flow in unsaturated, rough-walled fractures, *Water Resour. Res.*, *35*(4), 1019–1037, 1999.

Sweetkind, D. S., D. I. Barr, D. K. Polacseck, and L. O. Anna, *Integrated Fracture Data in Support of Process Models, Yucca Mountain, Nevada*, U.S.G.S. Administrative Report, U.S. Geological Survey, Denver, Colo., 1997.

Thorpe, R., *Characterization of Discontinuities in the Stripa Granite Time-Scale Heater Experiment*, Technical Project Report No. 20, LBL-7083, SAC-20, Lawrence Berkeley Laboratory, Berkeley, Calif., July 1979.

Thorstenson, D. C., H. Haas, E. P. Weeks, and J. C. Woodward, Physical and chemical characteristics of topographically affected airflow in an open borehole at Yucca Mountain, Nevada, *Proc. of the Topical Meeting on Nuclear Waste Isolation in the Unsaturated Zone FOCUS '89, Las Vegas, Nev., September 17–21, 1989*, pp. 256–270, American Nuclear Society, La Grange Park, Ill., 1989.

Tsang, C. F., Y. W. Tsang, and F. V. Hale, Tracer transport in fractures: Analysis of field data based on a variable-aperture channel model, *Water Resour. Res.*, *27*(12), 3095–3106, 1991.

Tsang, Y. W., Chapter 1, Introduction, in *Yucca Mountain Single Heater Test Final Report, Yucca Mountain Site Characterization Project*, LBNL-42537, edited by Y. W. Tsang, J. Apps, J. T. Birkholzer, B. Freifeld, M. Q. Hu, J. Peterson, E. Sonnenthal, and N. Spycher, Lawrence Berkeley National Laboratory, Berkeley, Calif., 1999a.

Tsang, Y. W., Chapter 1, Overview—Status of Drift Scale Test at six months of heating, in *Drift Scale Test Progress Report,* Level 4 Milestone SP2930M4, Lawrence Berkeley National Laboratory, Berkeley, Calif., July 14, 1999b.

Tsang Y. W., and C.-F. Tsang, Channel model of flow through fractured media, *Water Resour. Res.*, *23*(3), 467–479, 1987.

Tsang, Y. W., and C.-F. Tsang, Flow channeling in a single fracture as a two-dimensional strongly heterogeneous permeable medium, *Water Resour. Res.*, *25*(9), 2076–2080, 1989.

Tsang, Y. W., C.-F. Tsang, and F. V. Hale, Tracer transport in a stochastic continuum model of fractured media, *Water Resour. Res.*, *32*(10), 3077–3092, 1996.

Tsang, Y. W., and J. T. Birkholzer, Predictions and observations of the thermal-hydrological conditions in the Single Heater Test, *J. Contam. Hydrol.*, *38*, 385–425, 1999.

Tsang, Y. W., J. Apps, J. T. Birkholzer, B. Freifeld, M. Q. Hu, J. Peterson, E. Sonnenthal, and N. Spycher, *Yucca Mountain Single Heater Test Final Report, Yucca Mountain Site Characterization Project*, LBNL-42537, Lawrence Berkeley National Laboratory, Berkeley, Calif., 1999a.

Tsang Y. W., J. Apps, J. T. Birkholzer, B. Freifeld, M. Q. Hu, J. Peterson, E. Sonnenthal, and N. Spycher, *Yucca Mountain Drift Scale Test Progress Report,* LBNL-42538, Lawrence Berkeley National Laboratory, Berkeley, Calif., 1999b.

Tyler, S. W., J. B. Chapman, S. H. Conrad, D. P. Hammermeister, D. O. Blout, J. J. Miller, M. J. Sully, and J. M. Ginanni, Soil-water flux in the Southern Great Basin, United States: Temporal and spatial variations over the last 120,000 years, *Water Resour. Res.*, *32* (6), 1481–1499, 1996.

Vaniman, D. T., and S. J. Chipera, Paleotransport of lanthanides and strontium recorded in calcite compositions from tuffs at Yucca Mountain, Nevada, USA, *Geochemica et Cosmochemica Acta*, *60*(22), 4417–4433, 1996.

Walsh, J. B., The effect of pore pressure and confining pressure on fracture permeability, *Int. J. Rock Mech., 18*, 429–435, 1981.

Wang, J. S. Y., and T. N. Narasimhan, Hydrologic mechanisms governing fluid flow in a partially saturated, fractured, porous medium, *Water Resour. Res.*, *21*(12), 1861–1874, 1985.

Warren, J. E., and P. J. Root, The behavior of naturally fractured reservoirs, *Soc. Pet. Eng. J.*, *228*, 245–255, 1963.

Weeks, E. P., Effect of topography on gas flow in unsaturated fractured rock: Concepts and observations, in *Flow and Transport Through Unsaturated Fractured Rock, Geophysical Monograph 42,* edited by D. D. Evans and T. J. Nicholson, pp. 165–170, American Geophysical Union, Washington, D.C., 1987.

Weeks, E. P., Does the wind blow through Yucca Mountain, *Proc. Workshop V: Flow and Transport through Unsaturated Fractured Rock—Related to High-Level Radioactive Waste Disposal*, , pp. 43–53, Nuclear Regulatory Commission, Washington, D.C., 1991.

Wilson, C. R., *An Investigation of Laminar Flow in Fractured Porous Rocks*, Ph.D. dissertation, 178 pp., Univ. of California, Berkeley, Calif., 1970.

Wilson, C. R., and P. A. Witherspoon, Flow interference effects at fracture intersections, *Water Resour. Res., 12*(1), 102–104, 1976.

Wilson, C. R., J. C. S. Long, R. M. Galbraith, K. Karasaki, H. K. Endo, A. O. DuBois, M. J. McPherson, and G. Ramquist, *Geohydrological Data from the Macropermeability Experiment at Stripa, Sweden*, Technical Project Report No. 37, LBL-12520, SAC-37, Lawrence Berkeley Laboratory, Berkeley, Calif., March 1981.

Wilson, C. R., P. A. Witherspoon, J. C. S. Long, R. M. Galbraith, A. O. DuBois, and M. J. McPherson, Large-scale hydraulic conductivity measurements in fractured granite, *Int. J. Mech. Min. Sci. & Geomech. Abstr. 20*(6), 269–276, 1983.

Witherspoon, P. A., and O. Degerman, *Swedish-American Cooperative Program on Radioactive Waste Storage in Mined Caverns Program Summary*, Technical Project Report No. 01, LBL-7049, SAC-01, Lawrence Berkeley Laboratory, Berkeley, Calif., May 1978.

Witherspoon, P. A., J. E. Gale, R. L. Taylor and M. S. Ayatollahi, *Investigation of Fluid Injection in Fractured Rock and Effect on Stress Distribution*, Report No. TE-73-2, Geotechnical Engineering, Univ. of California, Berkeley, Calif., 1973.

Witherspoon, P. A., J. E. Gale, R. L. Taylor and M. S. Ayatollahi, *Investigation of Fluid Injection in Fractured Rock and Effect on Stress Distribution*, Geotechnical Engineering, Report No. TE-74-4, Univ. of California, Berkeley, Calif., 1974.

Witherspoon, P. A., E. H. Alonso, M. J. Lippmann, A. Mañón M., and H. A. Wollenberg, Mexican-*American Cooperative Program at the Cerro Prieto Geothermal Field*, LBL-7095, Lawrence Berkeley Laboratory, Berkeley, Calif., August 1978.

Witherspoon, P. A., J. S. Y. Wang, K. Iwai, and J. E. Gale, Validity of cubic law for fluid flow in a deformable rock fracture, *Water Resour. Res. 16*(6), 1016–1024, 1980a.

Witherspoon, P. A., N. G. W. Cook, and J. E. Gale, *Progress with Field Investigations at Stripa*, Technical Project Report No. 27, LBL-10559, SAC-27, Lawrence Berkeley Laboratory, Berkeley, Calif., February 1980b.

Wittwer, C., G. Chen, G. S. Bodvarsson, M. Chornack, A. Flint, L. Flint, E. Kwicklis, and R. Spengler, *Preliminary Development of the LBL/USGS Three-Dimensional Site-scale Model of Yucca Mountain, Nevada*, LBL-37356, UC-814, Lawrence Berkeley National Laboratory, Berkeley, Calif., 1995.

Wu, Y. S., A. C. Ritcey, C. F. Ahlers, J. J. Hinds, A. K. Mishra, C. Haukwa, H. H. Liu, E. L. Sonnenthal, and G. S. Bodvarsson, *3-D UZ Site-Scale Model for Abstraction in TSPA-VA*, Yucca Mountain Project Level 4 Milestone SLX01LB3.: Lawrence Berkeley National Laboratory, Berkeley, Calif., 1998a.

Wu, Y. S., W, Zhang, A. C. Ritcey, L. H. Pan, and G. S. Bodvarsson, Chapter 17, 3-D simulation using the UZ flow and transport model, in *Unsaturated Zone Flow and Transport Modeling of Yucca Mountain, Nevada—Fiscal Year 1998 Report*, edited by G. S. Bodvarsson, E. L. Sonnenthal, and Y. S. Wu, Document ID# BAB000000-01717-2200-00016, Lawrence Berkeley National Laboratory, Berkeley, Calif., 1998b.

Wu, Y. S., J. Hinds, W. Zhang, C. Haukwa, and G. S. Bodvarsson, Chapter 13, Calico Hills and perched-water models, in *Unsaturated Zone Flow and Transport Modeling of Yucca Mountain, Nevada—Fiscal Year 1998 Report*, Document ID# BAB000000-01717-2200-00016, edited by G. S. Bodvarsson, E. L. Sonnenthal, and Y. S. Wu, Lawrence Berkeley National Laboratory, Berkeley, Calif., 1998c.

Zimmerman, R. W., S. Kumar, and G. S. Bodvarsson, Lubrication theory analysis of the permeability of rough-walled fractures, *Int. J. Rock Mech., 28*(4), 325–331, 1991

Zimmerman, R. W., D. W. Chen, and N. G. W. Cook, The effect of contact area on the permeability of fractures, *J. Hydrology, 139*, 79–86, 1992.

Zimmerman, R. W., and G. S. Bodvarsson, Hydraulic conductivity of rock fractures, *Trans. Porous Media*, *23*, 1–30, 1996.

Paul A. Witherspoon, Earth Sciences Division, Ernest Orlando Lawrence Berkeley National Laboratory, and Department of Material Sciences and Mineral Engineering, University of California, Berkeley.

A Brief Survey of Hydraulic Tests in Fractured Rocks

Paul A. Hsieh

U.S. Geological Survey, Menlo Park, CA

During the past 50 years, significant developments have advanced the application of hydraulic tests for investigating fluid flow through fractured media. An extensive body of literature from the groundwater, petroleum, and geotechnical engineering fields has established a firm knowledge base for analyzing hydraulic tests in granular media. Application of hydraulic tests to fractured rocks is difficult because of the highly heterogeneous nature of fractured rocks. In addition, many fractured rocks of interest have low permeability, a factor that presents special problems for field testing. This paper presents a brief survey of methods to deal with these difficulties, with an example from the U.S. Geological Survey fractured-rock research site near Mirror Lake, New Hampshire. Areas of future advances in hydraulic test analysis include better integration with geophysical methods that can image the subsurface, and effective use of numerical models and inverse methods to infer subsurface heterogeneity.

1. INTRODUCTION

Hydraulic testing is the primary field method for determining the hydraulic properties of the subsurface. The traditional use of hydraulic tests in the groundwater and petroleum fields is to determine water or oil production by pumping from one well and monitoring pressure or hydraulic-head response in the pumped well and in nearby observation wells. In the geotechnical field, hydraulic tests are often conducted by injecting water into a packer-isolated portion of a borehole to determine the grouting requirements of foundations. Such a test is known as a Lugeon test—a Lugeon being a unit defined as a water injection rate of 1 liter per minute into 1 meter of borehole at an injection pressure of 10 MPa (10 bars).

In the past 50 years, a sustained development effort by numerous workers has established a firm knowledge base for analysis of hydraulic tests. Application of classical mathematical methods (Laplace, Hankel, and Fourier transforms) to solve the groundwater flow equation has yielded a large number of analytical solutions. These solutions guide the interpretation of hydraulic tests by using type curves (plots of dimensionless drawdown versus dimensionless time) to infer flow geometry (linear, radial, spherical), anisotropy, mechanism of fluid storage and release (compressibility versus water-table drainage, fractures versus porous rock blocks), borehole conditions (skin, wellbore storage, borehole intercepting a highly transmissive fracture), and the presence of semiconfining layers and boundaries. The introduction of the numerical Laplace inversion to the groundwater field [*Moench and Ogata*, 1981] further bolstered the ability to obtain solutions for more complicated subsurface settings. Compendiums of well-testing solutions in the petroleum field can be found in the publications of *Matthews and Russell* [1967], *Earlougher* [1977], and *Streltsova* [1988]. In the groundwater literature, texts on aquifer test analysis include the works of *Kruseman and de Ridder* [1983], *Dawson and Istok* [1991], *Walton* [1996], and *Batu* [1998].

Dynamics of Fluids in Fractured Rock
Geophysical Monograph 122

Published 2000 by the American Geophysical Union

The application of traditional hydraulic testing methods to fractured rocks, however, presents special challenges. Two major issues are notable. First, rocks of low permeability are increasingly the focus of investigation because of their role in subsurface disposal of hazardous wastes. Withdrawal or injection of fluid into low-permeability rocks is difficult to measure and control. Second, fractured rocks typically exhibit a high degree of heterogeneity. With a few exceptions, however, analytical solutions are developed with the assumption that the rock is homogeneous. This assumption imposes significant limitations on analyzing test data from heterogeneous systems. This paper presents a brief survey of methods for hydraulic testing of low-permeability rocks and the treatment of heterogeneity in hydraulic test analysis. The latter is illustrated by an example from the U.S. Geological Survey fractured-rock research site near Mirror Lake, New Hampshire.

2. HYDRAULIC TESTING OF LOW PERMEABILITY ROCKS

Application of hydraulic testing has traditionally focused on determining the hydraulic properties of productive strata or aquifers. The properties of semiconfining units (aquitards) are of interest primarily because they represent leakage of fluid into the pumped aquifer, leading to a reduction of drawdown compared to the case in which the aquifer is fully confined. During the 1960s, however, the semiconfining units themselves became the focus of interest in studies to evaluate the use of aquifers for underground storage of natural gas. Whether or not an aquifer is suitable for gas storage depends on the presence of a sufficiently tight caprock to prevent gas leakage out of the storage zone. In a series of landmark papers, *Neuman and Witherspoon* [1969a,b, 1972] presented the general theory of well hydraulics for a multiple aquifer-aquitard system. They extended previous work by accounting for fluid storage in aquitards and drawdown in unpumped aquifers. A key contribution of this work is the development of the "Ratio Method" for evaluating the hydraulic properties of aquitards. *Neuman and Witherspoon* [1972] showed that the hydraulic diffusivity of an aquitard can be determined from the ratio of drawdown in the aquitard to that in the aquifer at the same time and same radial distance from the pumped well. Without this method, type curve matching would be very difficult owing to the number of parameters involved. Even for a two-aquifer, one-aquitard system, the analytical solution involves six dimensionless parameters, which made it impractical to construct a sufficient number of type curves to cover the range of values needed for data analysis.

The 1977 Invitational Well Testing Symposium at Berkeley, California, saw the introduction of the pressure pulse test (also known as the pressurized slug test) to investigate transient flow in tight fractures [*Wang et al.*, 1978]. A similar test was also developed by *Bredehoeft and Papadopulos* [1980] to determine the hydraulic properties of tight formations. The pressure pulse test is a variant of the slug test. In a slug test, the hydraulic head in the test interval is abruptly changed by raising or lowering the water level in the open well or a standpipe connected to the test interval. The hydraulic head is monitored as the water level recovers to the equilibrium state. In a pressure pulse test, the test interval is shut-in (isolated by packers) after the initial change in hydraulic head. Because the test interval is isolated from the water level in the well, the subsequent recovery of hydraulic head is controlled by water compressibility and the compliance of the downhole packer equipment [*Neuzil*, 1982]. The response time of the pressure pulse test is several orders of magnitude faster than the response time of a conventional slug test and has significantly extended the lower limit of hydraulic conductivity determination in practical applications.

A limitation of the pressure pulse test is that the volume of rock tested is relatively small. In 1979, to determine the large-scale permeability of fractured crystalline rocks, investigators conducted a unique experiment at the Stripa research mine in Sweden [*Witherspoon et al.*, 1981]. To set up the "Ventilation Shaft Experiment" (Figure 1), a 33 m length mine shaft was sealed at its entrance. In effect, the shaft served as a (horizontal) pumped well. Withdrawal of fluid was effected by a ventilation system to evaporate water seeping into the drift. The water withdrawal rate was determined by measuring the air-flow rate and the difference in temperature and humidity of inflow and outflow air. Hydraulic-head response in the rock mass was monitored by packer-isolated intervals in 15 boreholes drilled radially away from the shaft in different directions, thus spanning a rock volume of about 10^5 to 10^6 m^3. Based on the observed hydraulic gradients and the measured withdrawal rate of about 50 milliliters per minute, the average hydraulic conductivity of the rock was estimated to be on the order of 10^{-11} m/s. Rigorous analysis of the experimental results is difficult because of the stress perturbation of the shaft opening on the surrounding rocks and the presence of multiphase (air/water) conditions at the shaft wall. Nonetheless, this landmark experiment set the basic design for subsequent experiments of fluid flow and solute transport conducted in mine shafts [for example, *Abelin et al.*, 1991].

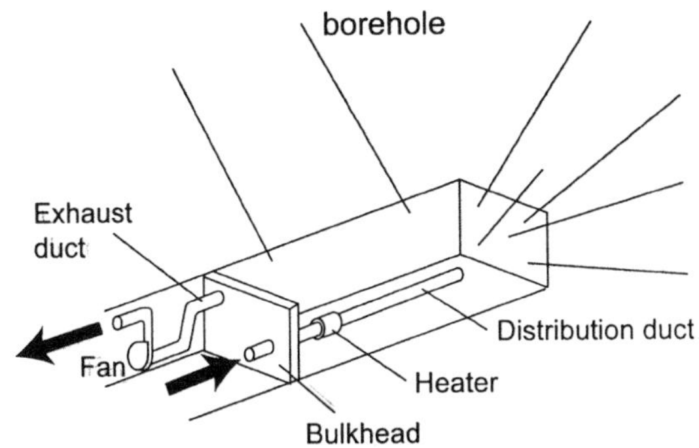

Figure 1. "Ventilation Shaft Experiment" at the Stripa Mine, Sweden [after *Witherspoon et al.*, 1981].

3. WRESTLING WITH HETEROGENEITY

A central problem in field investigations of fluid flow in fractured rocks is how to characterize a medium that is highly heterogeneous. For example, hydraulic conductivity of crystalline rocks can vary by several orders of magnitude over distances of several meters. Fractured rocks often contain zones of highly transmissive fractures embedded within a network of less permeable fractures. The presence of irregularly shaped high-transmissivity zones could lead to test responses that are difficult to analyze within a traditional well-testing context. For example, consider a field setting in which a pumped well, a nearby observation well, and a distant observation well are all located along a straight line. An irregularly shaped high-permeability zone connects the pumped well with the distant observation well, but not with the nearby observation well. During pumping, greater drawdown is observed at the distant observation well than at the nearby observation well. This type of test response cannot be readily interpreted by analytical solutions based on the homogeneous model assumption.

Although there does not currently exist a common practice for analyzing hydraulic tests in heterogeneous rocks, developments in this area will likely involve three ingredients: numerical modeling, geophysical "imaging" of the subsurface, and optimizing model calibration and evaluation of alternative conceptualizations. A numerical model is needed to treat all but the simplest kind of heterogeneity. Regardless of whether a continuum model or a discrete-fracture network model is employed, major features that control fluid flow will have to be delineated in the field and explicitly represented in the model. The delineation of these controlling features is not likely accomplished solely by inference from drawdown data (by model calibration). Rather, these features will have to be located by geophysical techniques that can "image" the subsurface. For example, under favorable conditions, crosswell tomography using seismic or electromagnetic waves can be used to map the distribution of wave velocity and attenuation in the vertical section between two boreholes separated by tens of meters. Because wave velocity and attenuation are affected by rock properties and the presence of fractures and fluids, tomography offers a possibility for delineating zones of contrasting hydraulic properties.

The counterpart to the traditional method of matching type curves to data is an optimization method for model calibration or inversion. From an application perspective, inverse methods for analyzing hydraulic tests in fractured rocks can be broadly classified into two categories: those involving few degrees of freedom and those involving many degrees of freedom. The classical approach to model inversion typically subdivides the model domain into several zones of contrasting hydraulic properties. Within each zone, the hydraulic properties are uniform or may vary with a known trend. By limiting the number of zones (and thus limiting the degrees of freedom), we can usually determine a unique set of parameters to "best fit" the observed drawdowns. By contrast, methods that involve many degrees of freedom are generally not concerned with uniqueness. For example, methods developed by *Long et al.* [1991], *Mauldon et al.* [1993], and *Hestir et al.* [1998] represent a fracture network by a lattice of conductor elements or links. During inversion, the lattice is successively modified by turning links "on" or "off" following a simulated annealing algorithm. Because each link can be modified, the inverse method allows many degrees of freedom. This approach yields multiple configurations of lattices—the simulated drawdowns in each lattice may match the observed drawdowns equally well. Each lattice can be considered a probable scenario in the field.

Related to the issue of uniqueness is the treatment of alternative conceptualization. Even in the classical inverse approach employing few degrees of freedom, the zonation pattern (or model structure) is often based on incomplete knowledge of the field and thus open to varying interpretations. By using inverse methods, the analyst can efficiently determine the optimal parameters for a variety of conceptualizations or model structures. These alternative conceptualizations can be assessed by the degree of fit, the reasonableness of the estimated parameters, and the consistency with other field data.

4. AN EXAMPLE FROM MIRROR LAKE, NEW HAMPSHIRE

Since 1990, the U. S. Geological Survey has conducted field studies at a fractured rock research site near Mirror

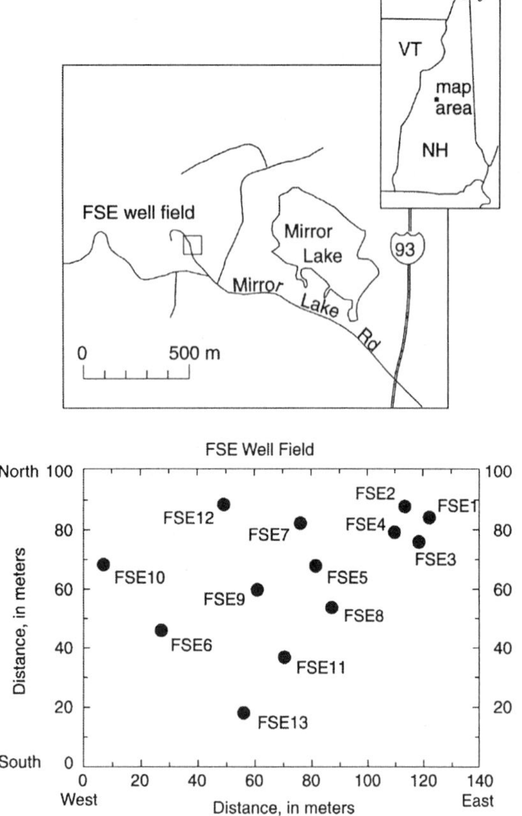

Figure 2. Location of wells in the FSE well field, Mirror Lake site, New Hampshire.

Lake, New Hampshire. Bedrock underlying the Mirror Lake Research Site consists of sillimanite-grade schist extensively intruded by granite, pegmatite, and lesser amounts of lamprophyre. At the FSE well field (Figure 2), 13 wells were drilled in a 120 × 80 m area to investigate the upper 60 m of bedrock, which is overlaid by approximately 20 m of glacial deposits. Single-well straddle-packer tests conducted at 4.5 m test intervals in the FSE wells suggest that the fracture transmissivities are highly variable. As an example, the borehole televiewer log of well FSE11 shows that the well intersects approximately 25 fractures in the 60 m of open hole below the casing. However, the results of straddle-packer tests (Figure 3) indicate that only a few fractures have high transmissivity (greater than 10^{-5} m^2/s), whereas most of the fractures have a much lower transmissivity (less than 2×10^{-8} m^2/s). The presence of a few highly transmissive fractures among a background of less transmissive fractures was observed in nearly all the FSE wells.

Based on borehole observations, geophysical measurements, and crosshole pressure responses, *Hsieh and Shapiro* [1996] conceived of the bedrock as containing four clusters of highly transmissive fractures embedded within a network of less transmissive fractures. The hypothesized distribution of fractures in the vertical section between wells FSE1 and FSE6 is shown in Figure 4. The four clusters of highly transmissive fractures are labeled I to IV. Each cluster occupies a near-horizontal, tabular-shaped volume that extends tens of meters in the horizontal direction and several meters in the vertical direction. Imaging of these fracture clusters was attempted using crosswell seismic tomography between selected well pairs [*Ellefsen et al.*, 1998]. Figure 5 shows the seismic velocity tomogram between wells FSE2 and FSE3, and the inferred location of fracture cluster II. The tomogram was ambiguous. *Ellefsen et al.* [1999] compared hydraulic conductivities and seismic velocities measured along wellbores and found that high seismic velocities (greater than approximately 5,100 m/s) are strongly correlated to low hydraulic conductivity, but low seismic velocities (less than approximately 5,100 m/s) are associated with a wide range of hydraulic conductivity. Thus, for the FSE well field, seismic tomography does not yield clear delineation of the high-conductivity zones. Instead, the tomograms are viewed as one among several lines of evidence suggesting the possible presence of these zones.

A multiple-well hydraulic test was conducted by pumping at a constant rate (10 L/min) from fracture cluster III,

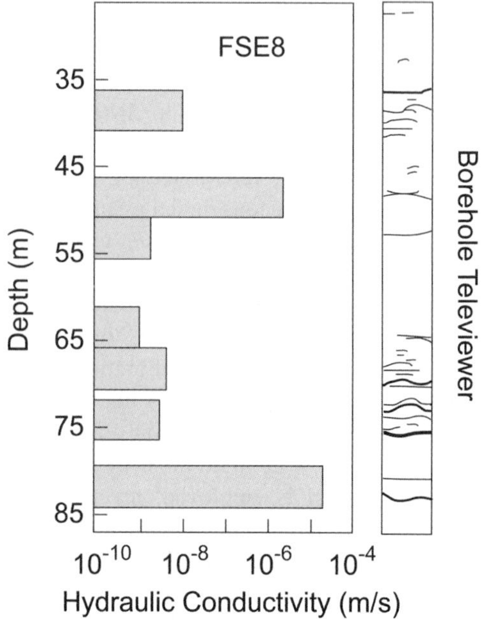

Figure 3. Distribution of hydraulic conductivity and location of fractures in well FSE8.

which was isolated with straddle packers in well FSE6. In observation wells that intersect two fracture clusters, packers were installed between the fracture clusters to minimize fluid flow and pressure transmission from one fracture cluster to another via the well. Observed drawdowns in the well intervals clearly show the effects of the highly transmissive fracture clusters, which tend to equalize the hydraulic head within the cluster. Figure 6 shows drawdowns in well intervals labeled according to the well name and the fracture cluster intersected by the interval. Note that well intervals intersecting fracture cluster II (FSE1-II, FSE4-II and FSE5-II) exhibit nearly identical drawdowns, even though well FSE5 is 58 m from the pumped well (FSE6) and well FSE1 is 101 m from the pumped well. Similar responses are observed in well intervals intersecting fracture cluster I (FSE1-I and FSE4-I), and fracture cluster IV (FSE6-IV and FSE9-IV). By contrast, well intervals that are not connected by a fracture cluster exhibit significantly different responses.

The drawdown response of the multiple-well hydraulic test cannot be analyzed by classical aquifer-testing methods that assume a homogeneous aquifer. Instead, *Hsieh et al.* [1999] employed a numerical model using MODFLOW-96, the U.S. Geological Survey modular finite-difference groundwater flow model [*Harbaugh and McDonald*, 1996]. The application of MODFLOW-96, a continuum model,

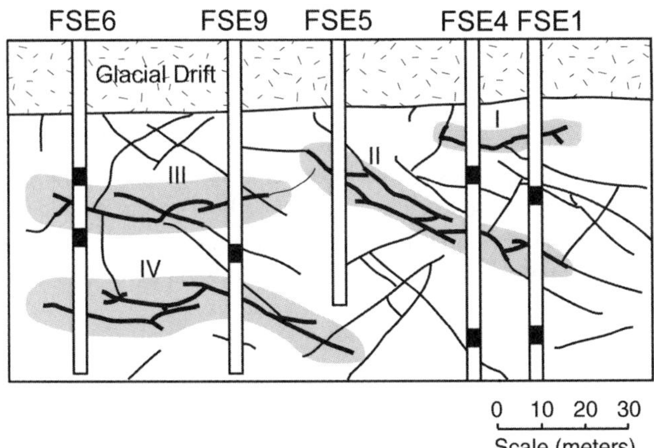

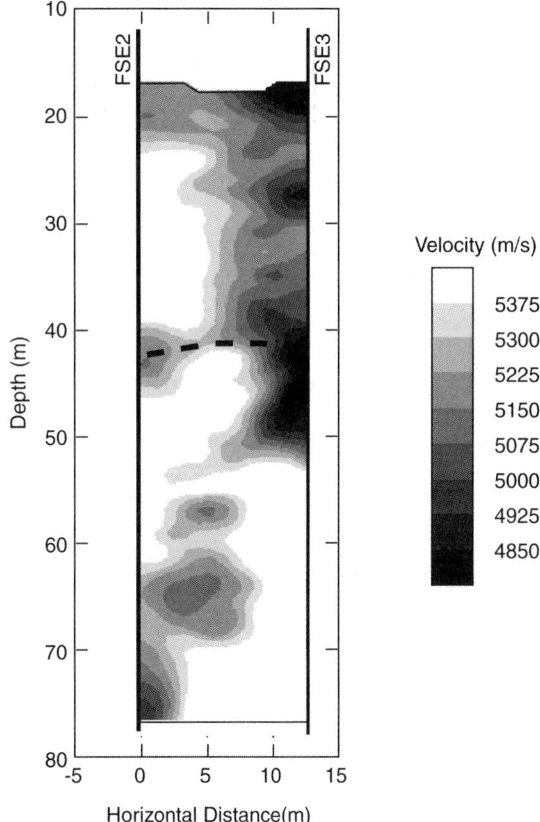

Figure 5. Seismic velocity tomogram between well FSE2 and FSE3. Dashed line indicates inferred location of fracture cluster II.

assumes that flow in the fractured rock underlying the FSE well field can be represented by flow in a heterogeneous continuum. The model grid (Figure 7) is composed of two types of cells. The highly transmissive fracture clusters (I to IV in Figure 4) are represented by cells of comparatively higher hydraulic conductivity. The surrounding bedrock with less transmissive fractures is represented by cells of comparatively lower hydraulic conductivity. The glacial deposits are treated as a constant-head (zero-drawdown) boundary at the bedrock surface. Because the water table lies in the glacial deposits, and storage in the glacial deposits is significantly larger than storage in the fractured bedrock, the glacial deposits are the primary source of pumped water. Although a small drawdown was observed in piezometers installed in the glacial deposits, this drawdown is negligible when compared to the much larger drawdown in the bedrock.

The model was calibrated using the nonlinear least-squares regression method of *Hill* [1992] to find parameter values that minimize the sum of squared differences be-

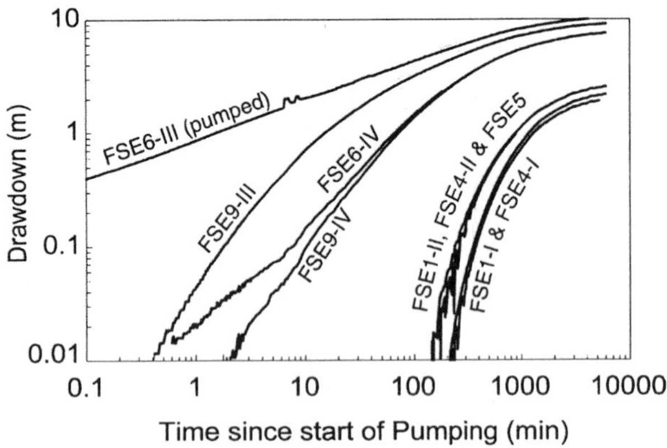

Figure 6. Observed drawdown in well intervals during multiple-well hydraulic test. A well interval is identified by the well name followed by the fracture cluster intersected by the interval.

tween model-simulated drawdowns and observed drawdowns. The model parameters are:

- K_{high}, isotropic hydraulic conductivity of cells representing highly transmissive fracture clusters
- $K_{h,low}$, horizontal hydraulic conductivity of cells representing the surrounding bedrock with less transmissive fractures
- $K_{v,low}$, vertical hydraulic conductivity of cells representing the surrounding network of less transmissive fractures
- C_V, vertical conductance of the interface between the glacial deposits and the bedrock, defined as the vertical hydraulic conductivity divided by the effective thickness of the interface
- S_s, specific storage of all cells in the bedrock

Because the highly transmissive fracture clusters are relatively thin (represented by 1.5 m thick cells in the model), their vertical hydraulic conductivity and specific storage have little effect on model-simulation results. Thus, cells representing highly transmissive fracture clusters are assumed to be isotropic and a single specific storage is used for all cells in the bedrock.

Table 1. Calibrated value of model parameters.

Model parameter	Calibrated value
K_{high}	2.0×10^{-4} m/s
$K_{h,low}$	1.8×10^{-8} m/s
$K_{v,low}$	3.7×10^{-7} m/s
C_V	1.8×10^{-9} s^{-1}
S_s	2.0×10^{-6} m^{-1}

Although the model consists of only 5 parameters, relatively good matches were achieved between the model-simulated drawdowns and the observed drawdowns. Although the matches are not perfect, the simulated and observed responses exhibit similar behavior and magnitudes. Representative examples are shown in Figure 8. The calibrated parameter values, given in Table 1, suggest that cells representing highly transmissive fracture clusters are three to four orders of magnitude more conductive than cells representing the surrounding bedrock with fewer transmissive fractures. The hydraulic conductivity of the surrounding bedrock is about 20 times higher in the vertical direction than in the horizontal direction, suggesting that near-vertical fractures in the surrounding bedrock might be denser, more transmissive, or better connected than near-horizontal fractures. If the interface between the glacial deposits and the bedrock is assumed to be 1 m thick, then C_V is numerically equal to the vertical hydraulic conductivity of the interface layer. The relatively low value of this con-

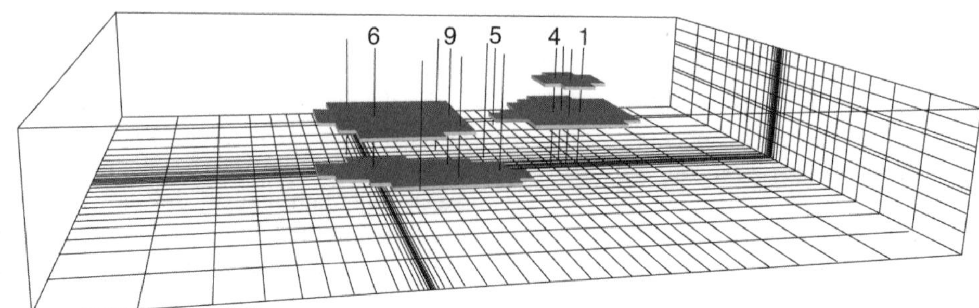

Figure 7. Numerical model of FSE well field. Cells representing four fracture clusters are shown. Cells representing background network of fractures are not shown.

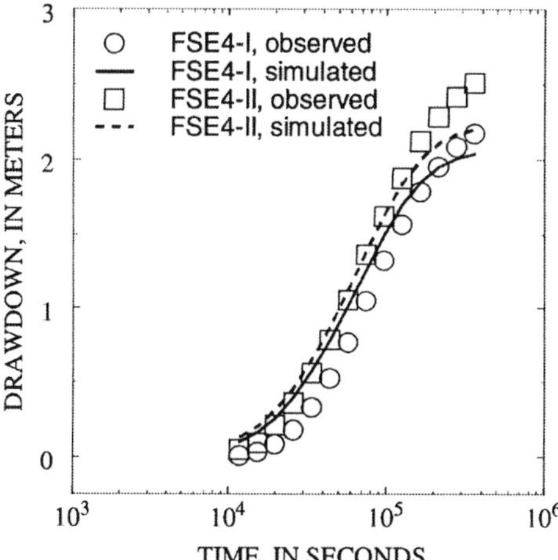

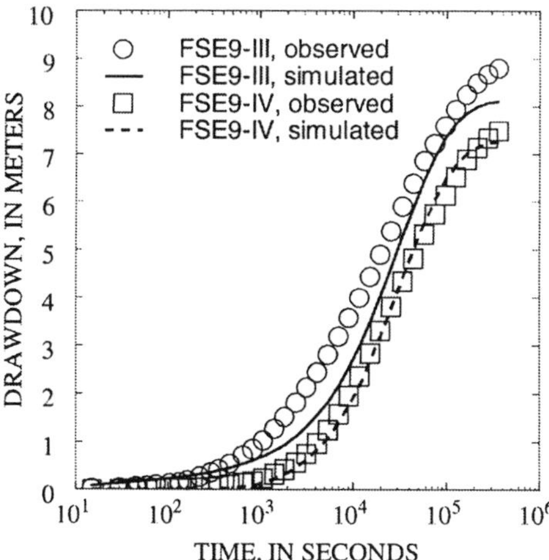

Figure 8. Comparison between observed and model-simulated drawdown in two well intervals of FSE4 and FSE9.

ductivity estimate (1.8×10^{-9} m/s) suggests that fractures at the bedrock surface might be partially clogged by fine-grained glacial deposits, causing a reduction of hydraulic conductivity.

As discussed earlier, the interpretation of hydraulic tests in heterogeneous systems is subject to nonuniqueness, even though the inverse problem is formulated with few parameters. In the Mirror Lake case, the spatial extent of the high-conductivity zones is not well known, especially in the regions away from the 13 drilled wells. Within certain constraints, the geometry of a high-conductivity zone can be modified in the model to yield different parameter estimates but similar fits to observed drawdowns. Improved methods of subsurface geophysical imaging can better constrain the zone geometry. If a stochastic model of fracture-zone distribution can be developed from geostatistical analysis of field measurements, variability in zone geometry may be assessed by geostatistical simulation.

4. CONCLUSIONS

This paper has focused on two sources of difficulties in the application of hydraulic tests to fractured rocks: low permeability and heterogeneity. Testing of low-permeability rocks is difficult to control, and test results characterize a small volume of rock around the borehole. Testing of heterogeneous rocks often yields field data that are difficult to interpret using standard well testing solutions. To overcome these difficulties, application of hydraulic testing on its own is insufficient. Rather, hydraulic tests should be considered as one component within a larger, multidisciplinary "tool box" that includes geophysical investigations, geologic studies, tracer tests, and geochemical methods (including environmental tracers and groundwater age dating). Effective integration of different investigation techniques is a key challenge for the future. Numerical modeling and inversion will play a central role in hydraulic test analysis. However, nonuniqueness in test interpretation cannot be entirely eliminated. In this regard, computer methods will likely remain part of a "computer-assisted" approach, requiring sound technical judgement by a knowledgeable analyst, rather than part of a "computer-automated" approach that replaces the analyst.

REFERENCES

Abelin, H., L. Birgersson, J. Gidlund, and I. Neretnieks, A large-scale flow and tracer experiment in granite 1. Experimental design and flow distributions, *Water Resour. Res.*, *27*(12), 3017–3117, 1991.

Batu, V., *Aquifer Hydraulics*, John Wiley and Sons, New York, N.Y. 1998.

Bredehoeft, J. D., and S. S. Papadopulos, A method for determining the hydraulic properties of tight formations, *Water Resour. Res.*, *16*(1), 233–238, 1980.

Dawson, K. J., and J. D. Istok, *Aquifer Testing: Design and Analysis of Pumping and Slug Tests*, Lewis Publishers, Chelsea, Mich., 1991.

Earlougher, R. C., *Advances in Well Test Analysis*, Society of Petroleum Engineers, Dallas, Tex., 1977.

Ellefsen, K. J., P. A. Hsieh, and A. M. Shapiro, Relation between seismic velocity and hydraulic conductivity at the USGS Fractured Rock Research Site, *U.S. Geological Survey Toxic Substances Hydrology Program—Proceedings of the Technical Meeting, Charleston, S.C., 1999*, edited by D. W. Morganwalp, and H. T. Buxton, Water Resources Investigations Report 99-4108C, pp. 735–745, U.S. Geological Survey, Denver, Colo., 1999.

Ellefsen, K. J., J. E. Kibler, P. A. Hsieh, and A. M. Shapiro, *Crosswell Seismic Tomography at the USGS Fractured Rock Research Site: Data Collection, Data Processing, and Tomograms*, Open-File Report 98-510, U. S. Geological Survey, Denver, Colo., 1998.

Hestir, K., S. J. Martel, S. Vail, J. Long, P. D'Onfro, and W. D. Rizer, Inverse hydrologic modeling using stochastic growth algorithms, *Water Resour. Res., 34*, 3335–3347, 1998.

Hsieh, P.A., and A.M. Shapiro, Hydraulic characteristics of fractured bedrock underlying the FSE well field at the Mirror Lake Site, Grafton County, New Hampshire, *U.S. Geological Survey Toxics Substances Hydrology Program—Proceedings of the Technical Meeting, Colorado Springs, Colorado, September 20–24, 1993*, edited by D. W. Morganwalp, and D. A. Aronson, Water Resources Investigations Report 94-4105, pp. 127–130, U.S. Geological Survey, Denver, Colo., 1996.

Hsieh, P. A., A. M. Shapiro, and C. E. Tiedeman, Computer Simulation of Fluid Flow in Fractured Rocks at the Mirror Lake FSE Well Field, in *U.S. Geological Survey Toxic Substances Hydrology Program—Proceedings of the Technical Meeting, Charleston, South Carolina, March 8–12, 1999*, edited by D. W. Morganwalp, and H. T. Buxton, Water Resources Investigations Report 99-4108C, pp. 777–781, U.S. Geological Survey, Denver, Colo., 1999.

Kruseman, G. P., and N. A. de Ridder, *Analysis and Evaluation of Pumping Test Data, 3rd Edition*, International Institute for Land Reclamation and Improvement, The Netherlands, 1983.

Long, J. C. S., K. Karasaki, A. Davey, J. Peterson, M. Landsfeld, J. Jemeny, and S. Martel, An inverse approach to the construction of fracture hydrology models conditioned by geophysical data, Abstract, *Int'l J. Rock Mech. Mining Sci. and Geomechan.*, *28*, 121–142, 1991.

Matthews, C. S., and D. G. Russell, *Pressure Build-Up and Flow Tests in Wells*, Society of Petroleum Engineers, Dallas, Tex., 1967.

Mauldon, A. D., K. Karasaki, S. J. Martel, J. C. S. Long, M. Landsfeld, and A. Mensch, An inverse technique for developing models for fluid flow in fractured systems using simulated annealing, *Water Resour. Res., 29*(11), 3775–3789, 1993.

Moench, A. F., and A. Ogata, A numerical inversion of the Laplace transform solution to radial dispersion in a porous medium, *Water Resour. Res., 17*(1), 250–252, 1981.

Neuman, S. P., and P. A. Witherspoon, Theory of flow in a confined two aquifer system, *Water Resour. Res.*, *5*(4), 803–816, 1969a.

Neuman, S. P., and P. A. Witherspoon, Applicability of current theories of flow in leaky aquifers, Water Resour. Res., *5*(4), 817–829, 1969b.

Neuman, S. P., and P. A. Witherspoon, Field determination of hydraulic properties of leady multiple aquifer system, *Water Resour. Res., 8*(5), 1284–1289, 1972.

Neuzil, C. E., On conducting the modified "slug" test in tight formations, *Water Resour. Res., 18*(2), 439–441, 1982.

Streltsova, T. D., *Well Testing in Heterogeneous Formations*, John Wiley and Sons, New York, N.Y., 1988.

Walton, W. C., *Aquifer Test Analysis with Windows Software*, CRC Press, Boca Raton, Fla., 1996.

Wang, J. S. Y., T. N. Narasimhan, C.-F. Tsang, and P. A. Witherspoon, Transient flow in tight fractures, in *Invitational Well Testing Symposium Proceedings*, edited by T. N. Narasimhan, pp. 103–113, University of California, Berkeley, Calif., 1978.

Witherspoon, P. A., N. G. W. Cook, and J. E. Gale, Geologic storage of radioactive waste: Field studies in Sweden, *Science, 211*, 894–900, 1981.

Witherspoon, P. A., I. Javandel, S. P. Neuman, and R. A. Freeze, *Interpretation of Aquifer Gas Storage Conditions from Water Pumping Tests*, 273 pp., American Gas Association, Inc., New York, N.Y., 1967.

Paul A. Hsieh, U.S. Geological Survey, Menlo Park, CA.

Unresolved Problems in Vadose Zone Hydrology and Contaminant Transport

William A. Jury and Zhi Wang

University of California, Riverside, CA 92521

Despite a great deal of research, many unsolved problems remain in vadose zone hydrology. Because of the difficulties in sampling spatially variable soil and in monitoring at great depths, the effect of heterogeneities on flow and transport is poorly understood. Observations from the few comprehensive field studies that have been conducted suggest that preferential flow, or rapid flow through a small part of the soil volume, is present under a variety of conditions and can be a significant or even dominant part of the total flow regime. Reasons for the occurrence of preferential flow are soil dependent, but include flow through structural voids, channeling caused by discrete obstacles within the soil matrix, and fluid instability (arising from a host of causes). Once believed to occur exclusively in structured soils, preferential flow is now recognized as prevalent under a wide range of conditions in permeable, structureless soils as well as those containing cracks and crevices. Monitoring of preferential flow has thus far only been possible using dyes, densely replicated soil coring, or analysis of tile-drain effluent. These tools either fail to characterize the speed of preferential flow, as is the case with dyes or coring, or provide no spatial resolution of the preferential flow event, as with tile-drain monitoring. To make any real progress in characterizing preferential flow, we need rapid-response tensiometers and solution samplers that offer minimum disturbance of the soil, and we need some means of monitoring a substantial portion of the soil volume at a high degree of spatial resolution.

1. INTRODUCTION

The vadose zone is an important region to a number of different research and applied disciplines. It is the crop-management zone where plants receive nutrients and water through their roots. It is both a waste repository site and an accidental recipient of waste that is spilled or dumped on the land surface. It is also the buffer zone protecting groundwater from contamination by toxic chemicals or pathogens. Thus, the vadose zone is of importance to agronomists, soil scientists, waste engineers, and a host of management professionals who are concerned with protection of groundwater. With its obvious significance to so many disciplines, we might expect that research activity in the vadose zone would have produced a thorough characterization of its important transport, retention, and reaction properties, but this is not the case. Although a number of experiments have been conducted in the vadose zone to understand water or chemical movement and fate, most studies have been monitored only in the top 1–2 m [*Flury*, 1996]. Moreover, few experiments have been conducted in

Dynamics of Fluids in Fractured Rock
Geophysical Monograph 122

clay-rich soils because of the difficulties in monitoring and sampling, and the effect of soil layers on transport in any soil type is largely unknown. Even though large areas are often involved in applications of vadose zone hydrology, most of the information we have has been obtained through experiments conducted at the scale of the soil plot or smaller. Moreover, although replicated plot experiments have provided some information on the spatial variability of transport processes, little knowledge exists about the variation of these processes over time [*Jury and Fluhler*, 1992].

In addition to experimental limitations, a number of theoretical gaps persist in our understanding of the vadose zone. Soil water hysteresis continues to provide dilemmas for modelers, particularly when secondary scanning loops are involved. Many coupled flow interactions (involving, for example, water and heat movement) are incompletely understood. Multiphase flow of nonaqueous phase liquids and water is an important area of research in the vadose zone, but attempts to model it have been limited by failure to represent unstable flow (among other problems). Upscaling, or the transformation of experimental or theoretical results from one space or time scale to another, is an active research area, but no consensus has been reached about the best approach to use in information transfer. Stochastic-continuum modeling, which has enjoyed success in groundwater studies, has limited application to vadose zone hydrology because its assumptions are not well met in unsaturated soil. In particular, flow in the vadose zone is normally perpendicular to the direction of stratification, so that the properties of the medium relevant to transport are often nonstationary.

Flow and transport processes in the vadose zone differ from those in groundwater in several important respects. The resistance offered by the matrix to water flow is a nonlinear function of the degree of saturation or the energy state of the water. Because of the transient processes occurring at the inlet boundary, flow and transport processes in the vadose zone rarely, if ever, reach a steady state. Since temperature and air-pressure changes can significantly influence the soil environment, representation of the flow and transport regime requires multiphase characterization. And whereas in groundwater the direction of flow is generally parallel to the natural stratification, in the vadose zone water and chemicals often move perpendicular to the natural layering pattern of soil. In addition, spatial and temporal variability of important flow and transport properties is substantial, leading to extreme data demands for both deterministic and stochastic modeling. As a result, theories for flow and transport processes in the vadose zone have not evolved as rapidly as those describing movement in groundwater. At present, the Richards' flow equation is used in virtually all simulations of water flow in the vadose zone, despite its limiting assumptions, and solute transport is represented by the advection dispersion equation, which suffers from similar drawbacks.

Field experiments of flow and transport have yielded mixed support for either of these equations. Such experiments have frequently revealed the existence of varying degrees of so-called preferential flow, wherein water moves at much higher than average rates through a small portion of the soil volume. The reasons why preferential flow occurs are soil dependent, but include flow through structural voids [*White*, 1985], channeling caused by discrete obstacles within the soil matrix [*Kung*, 1990], and fluid instability, arising from a host of causes [*Hillel and Baker*, 1988]. Once believed to occur exclusively in structured soils, preferential flow is now recognized as prevalent under a wide range of conditions in permeable, structureless soils as well as those containing cracks and crevices [*Flury*, 1996]. Since the theme of this conference is flow in fractured rock, I will tailor the rest of the discussion to preferential flow in the vadose zone, which is the transport process most relevant to the discipline of fractured rock hydrology.

2. FIELD STUDIES OF PREFERENTIAL FLOW

Preferential flow has been observed by researchers and field workers for many years. However, until very recently, experimental tests designed specifically to measure properties of preferential flow have been lacking. One of the difficulties with monitoring preferential flow is finding a tracer that can be detected when preferential flow is rapid and also confined to a small portion of the soil matrix. Two strategies have dominated field research thus far: use of visible dyes that are added to the surface and later exposed by digging a trench to reveal the pattern along a vertical face, and use of a strongly adsorbed compound that will migrate under preferential flow to a depth substantially greater than it would reach by complete interaction with the entire soil matrix.

Jury et al. [1986] conducted a comprehensive field study with a strongly adsorbed pesticide (napropamide) that was added to the surface of a 1.44 ha sandy loam field. The pesticide was leached by sprinkler irrigation for two weeks until 25 cm of water had been added. Then, 19 soil core samples were taken at random locations on the field and sampled in 10 cm increments for napropamide concentration. About 72% of the recovered pesticide was found in the top 20 cm, where it would be expected to be if the water flow rate were uniform, but the remaining 28% was distributed erratically between 30 and 190 cm.

A second pesticide transport study was conducted by *Ghodrati and Jury* [1990] on the same field. In this study, sixty-four 1 m × 1 m plots were divided into a set of treatments that used one of four different methods of water application (continuous and intermittent sprinkling or ponding), three different methods of pesticide application (wettable powder, emulsifiable concentrate, or dissolved), and two different surface preparation methods (sieving and replacking the top 20 cm or leaving it undisturbed). Each plot received exactly 12 cm of water and pulses of three pesticides (atrazine, napropamide, and prometryn) having different mobilities, together with a mobile tracer (bromide). At the conclusion of the water application, each plot was sampled in 10 cm increments to a 2 m depth by three soil cores that were combined prior to analysis. Preferential flow to varying degrees was observed in all plots regardless of treatment, with the sole exception that wettable powder could not penetrate the repacked soil layer. The amount of preferential flow averaged over all treatments was 22%, similar to what had been found earlier by *Jury et al.* [1986] on the same field. Although somewhat less preferential flow was observed for the most strongly adsorbed compound, the maximum depth of penetration into the soil was independent of the extent of adsorption potential. As part of the study, *Ghodrati and Jury* [1990] added a mobile red dye to the surface of the plots and excavated along a lateral trench face to observe the flow patterns. In virtually every case, the wetting front revealed leading-edge plumes that resembled instabilities, and the paths followed by the plumes appeared to lead through the same soil material as that bypassed.

Kung [1990] conducted a comprehensive dye-trace study on a field containing Plainfield sand, a highly permeable soil without pronounced layers or significant structural features. He added a pulse of the dye to the surface of a 4 m × 4 m plot, washed it in, and excavated the entire plot to a depth of 6 m. Although much of the plot area was covered with dye near the surface, the water channeled into smaller and smaller zones at greater depths, finally occupying less than 2% of the area at the bottom of the excavation. Detailed examination of the flow paths revealed that the water was flowing around discrete sand lenses imbedded in the matrix, joining with other isolated flow channels that were also meandering through the matrix. He termed the phenomenon "funnel flow."

Other studies of pesticide transport under field conditions have reported higher than expected mobility for compounds that adsorb to the soil. In a field study on a crop silt- loam soil, *Gish et al.* [1986] reported that the laboratory-measured adsorption coefficient for atrazine greatly underestimated its mobility relative to bromide. *Hornsby et al.* [1990] monitored bromide together with aldicarb and its degradation products for 200 days under a citrus grove and sandy soil in Florida. They observed that bromide and aldicarb had comparable mobility, even though the latter compound was predicted to have a retardation factor between 1.1 and 2.0, based on laboratory measurements. Because of the large amount of rainfall in this area, both bromide and aldicarb reached groundwater at 7.2 m during the 200 day monitoring period.

Bowman and Rice [1986] conducted a large study on a sandy loam field in which a water tracer and bromacil, a mildly adsorbed herbicide, were applied and leached under periodic ponding. Nearly all the individual soil cores showed the bromacil to be retarded with respect to the water tracer, but both were moving faster than predicted by piston flow. Tile drains have been used to monitor nutrients and pesticide-leaching losses below cropland for many years. Recently, they have been used to detect the extremely high mobility of dissolved chemicals. *Richard and Steenhuis* [1988] reported chloride arrival at 80 cm in the tile-drain effluent of a sandy loam, with the first outflow following application of the chemical to the surface. Subsequent rainfall events also triggered additional outflow pulses of chloride. A similar phenomenon was reported by *Kladivko et al.* [1991] on a tile-drain silt-loam soil. Following a single application to the surface, traces of the pesticides appeared in the tile effluent at 0.75 m after only 2 cm of net drainage. These pulses tapered off before the water quit flowing out of the tile, but returned with subsequent rainfall events through the season.

In the majority of these studies, preferential flow was a significant but not major part of the total flow regime. In more heterogeneous soil, however, it has been observed to dominate. *Roth et al.* [1991] performed a field study on a layered soil with a highly variable texture and structure and found that preferential flow was responsible for about 55% of the total mass transport through the top 2 m. *Flury et al.* [1994] conducted comprehensive dye-trace studies on the 14 major agricultural soil types in Switzerland and concluded that preferential flow was a significant or dominant feature of the flow regime in all but one of them. In a subsequent review article that surveyed all existing pesticide transport studies in unsaturated field soil, *Flury* [1996] concluded that preferential flow of even strongly adsorbed compounds was commonly observed in structureless soils, particularly loamy textures, without any apparent cause.

3. CAUSES OF PREFERENTIAL FLOW

Despite all of the evidence and effort at monitoring, no correlation has been observed between the local values of soil properties and the location where preferential flow occurs in structureless, vertically homogeneous soil. Various mechanisms have been postulated, including water repellency [*van Dam et al.*, 1990], air entrapment [*Peck*, 1965], and small-scale (i.e., 1 cm) variations in soil properties [*Roth*, 1995]. But definitive proof relating the pathways, intensity, and extent of preferential flow to measurable soil matrix properties has not been found. One reason why this might be true has been largely overlooked: the preferential flow may be the result of an instability in the flow field that is not caused exclusively by local permeability variations. Furthermore, the instability may be a consequence of one of the most common hydrologic events in soil: redistribution following the cessation of rainfall or irrigation. Early theoretical analyses by *Raats* [1973] and *Philip* [1975], using the Green and Ampt infiltration model, showed that redistribution was inherently unstable, even in homogeneous soil, and that any deviation from perfect one-dimensional advance of the wetting front would grow into a finger. The condition responsible for this instability is the matric potential gradient behind the wetting or draining front: whenever it opposes the flow, instabilities will develop out of perturbations in the uniformity of the wetting front as it advances into the soil.

This conclusion was tempered somewhat by *Diment et al.* [1982]. They conducted a simplified stability analysis of the Richards' equation in homogeneous soil and showed that instabilities would only develop if the initial water content of the soil ahead of the advancing front were extremely dry. They also demonstrated that the tendency toward instability increased as the wavelength of the disturbance increased, implying that the unrestricted width of the field environment may be an important factor in the development of instability. Their analysis was limited to a restricted set of soil conditions (perturbed one-dimensional infiltration and redistribution into homogeneous soil with uniform initial wetting) and contained some simplifications (no hysteresis, uniform velocity of wetting front).

Laboratory validation of unstable flow in homogeneous soil has been demonstrated repeatedly in Hele-Shaw cells, in which redistribution follows infiltration into oven- or air-dry soil [*Glass et al.*, 1989; *Selker et al.*, 1992; *Wang*, 1997]. However, *Diment and Watson* [1985] demonstrated that redistribution became stable in their Hele-Shaw cell when the initial water content of the soil was increased by only a few percent. The reason for this is that the wavelength of the fingers becomes large enough that the side walls of the column will not permit an instability to propagate. This observation also offered an explanation as to why preferential flow is so rarely observed in the laboratory. If the dimension of the column perpendicular to the direction of flow is not large compared to the wavelength of the unstable front, a finger will not propagate because it will not have sufficient lateral flow occurring to feed the advance of the instability.

Unstable flow is considerably more likely to occur when soil conditions deviate from ideality. Increasing air compression ahead of an advancing wetting front will induce instability at the time when the soil water potential (matric + air pressure) at the front exceeds that above it [*Peck*, 1965; *Parlange and Hill*, 1976]. Abrupt soil layering, with a finer-textured soil above a coarser textured one, will interrupt flow until the pressure reaches a critical point, at which time infiltration will advance into the lower zone at one particular location [*Hill and Parlange*, 1976]. Furthermore, discrete soil lenses of either coarser or finer soil than the matrix surrounding them can cause channeling. This channeling will induce a preferential flow in part of the soil volume as long as the matrix conductivity is large enough to support it [*Kung*, 1990]. If preferential flow in structureless soils is caused by fluid instability rather than simply being a consequence of variations in local soil properties, the implications for water flow and chemical transport research and management are profound. Simply by virtue of their widespread use in research and management, water flow models have gained an acceptance for simulation of flow under field conditions that arguably is not warranted by their record of achievement. They either ignore the possibility of preferential flow entirely by using the volume-averaged Richards' equation to model water movement, or introduce parameter-intensive methods of inducing preferential flow entirely as a result of soil property variations.

If preferential flow is initiated and reinforced by fluid instability rather than property variability, models that use the Richards' equation will not predict its occurrence with their current approach. Despite the wide variation in soil conditions and experimental design, field studies of preferential flow have all shared one common feature: the time of sampling was always scheduled 24 hr or more after cessation of water application, allowing substantial time for redistribution. According to theory, redistribution can favor the propagation of an instability should any perturbation occur in a sharp draining front. Since numerous local features in a heterogeneous field soil could create a perturbation in an advancing wetting front, efforts to characterize the exact location where the process starts, in terms of a measurable soil property, are unfeasible with current technology. In the interim, a more practical and fruitful approach might be to develop an understanding of the water application conditions that will turn a flow perturbation into an instability.

4. CONCLUDING REMARKS

Improved understanding of field-scale preferential flow will emerge from new developments in monitoring and more experimental testing under natural conditions. At the moment, we do not have answers to a number of basic questions that limit our ability to manage or model this phenomenon: Which measurable soil properties are important in predicting the onset of preferential flow? Which measurable soil properties are important in predicting the extent of preferential flow? How fast do water and dissolved chemicals move in a preferential flow channel? How can preferential flow be enhanced or reduced in a given environment? What is the role of a water application method or surface water management in producing preferential flow? To what extent do sorption reactions occur during preferential flow, and how can they be characterized as a function of measurable soil properties?

The standard tools for monitoring water flow in soil are inadequate for addressing these questions. Field monitoring of preferential flow has thus far been possible only using dyes, densely replicated soil coring, or analysis of tile-drain effluent. These tools either fail to characterize the speed of preferential flow, as is the case with dyes or coring, or provide no spatial resolution of the preferential flow event, as with tile-drain monitoring. To make real progress in characterizing preferential flow, we need rapid- response tensiometers and solution samplers that offer minimum disturbance of the soil, and we need some means of monitoring a substantial portion of the soil volume at a high degree of spatial resolution.

REFERENCES

Bowman, R. C., and R. C. Rice, Accelerated herbicide leaching resulting from preferential flow phenomena and its implications for groundwater contamination, in *Proceedings of the Conference on Southwestern Groundwater Issues, Phoenix, Arizona*, pp. 122–133, National Water Well Association, Dublin, Ohio, 1986.

Diment, G. A., K. K. Watson, and P. J. Blennerhassett, Stability analysis of water movement in unsaturated porous materials, I, Theoretical considerations, *Water Resour. Res., 18*, 1248–1254, 1982.

Diment, G. A., and K. K. Watson, Stability analysis of water movement in unsaturated porous materials. I. Experimental Studies, *Water Resour. Res., 21*, 979–984, 1985.

Flury, M., Experimental evidence of transport of pesticides through field soils—A review, *J. Environ. Qual., 25*, 25–45, 1996.

Flury, M., H. Fluhler, W. A. Jury, and J. Leuenberger, Susceptibility of soils to preferential flow: A field study, *Water Resour. Res., 30*, 1945–1954, 1994.

Ghodrati, M., and W. A. Jury, A field study using dyes to characterize preferential flow of water, *Soil Sci. Soc. Amer. J., 54*, 1558–1563, 1990.

Ghodrati, M., and W. A. Jury, A field study of the effects of water application method and surface preparation method on preferential flow of pesticides in unsaturated soil, *J. Contaminant Hydrol., 11*, 101–125, 1992.

Gish, T. J., C. S. Helling, and P. C. Kearney, Chemical transport under no-till field conditions, *Geoderma, 38*, 251–259, 1986.

Glass, R. J., T. Steenhuis, and J. Y. Parlange, Wetting front instability, 2, Experimental determination of relationships between system parameters and two-dimensional unstable flow field behavior in initially dry porous media, *Water Resour. Res., 25*, 1195–1207, 1989.

Hill, D. E., and J. Y. Parlange, Wetting front instability in layered soils, *Soil Sci. Soc. Amer. J., 36*, 697–702, 1976.

Hillel, D., and R. Baker, A descriptive theory of fingering during infiltration into layered soils, *Soil Sci., 146*, 51–56, 1988.

Hornsby, A. G., P. S. C. Rao, and R. L. Jones, Fate of aldicarb in the unsaturated zone beneath a citrus grove, *Water Resour. Res., 26*, 2287–2302, 1990.

Isensee, A. R., R. G. Nash, and C. Helling, Effect of conventional versus no-tillage on pesticide leaching to shallow ground water, *J. Environ. Qual., 19*, 434–440, 1990.

Jury, W. A., H. Elabd, and M. Resketo, 1986. Field study of napropamide movement through unsaturated soil, *Water Resour. Res., 22*, 749–755, 1990.

Kladivko, E., G. Van Scoyoc, E. Monte, and K. Oates, Pesticide and nutrient movement into subsurface tile drains on a silt loam in Indiana, *J. Environ. Qual, 2*, 264–270, 1991.

Kung, K. J. S., Preferential flow in a sandy vadose zone, 1, Field observation, *Geoderma, 46*, 51–58, 1990.

Parlange, J. Y., and D. E. Hill, Theoretical analysis of wetting front instability in soils, *Soil Sci., 122*, 236–239, 1976.

Peck, A. J., Moisture profile development and air compressions during water uptake by air-confining porous bodies, 3, Vertical columns, *Soil Sci., 100*, 44–51, 1965.

Philip, J. R., Stability analysis of infiltration, *Soil Sci. Soc. Amer. J., 39*, 1042–1053, 1975.

Raats, P. A. C., Unstable wetting fronts in uniform and nonuniform soils, *Soil Sci. Soc. Amer. J., 37*, 681–685, 1973.

Richard, T. L. and T. Steenhuis, Tile-drain sampling of preferential flow on a field, *J. Contaminant Hydrol., 3*, 307–325, 1988.

Roth, K., Steady-state flow in an unsaturated, two-dimensional macroscopically homogeneous Miller-similar medium, *Water Resour. Res., 31*, 2127–2140, 1995.

Roth, K., W. A. Jury, H. Fluhler, and W. Attinger, Transport of chloride through an unsaturated field soil, *Water Resour. Res., 27*, 2533–2541, 1991.

Roth, K., and K. Hammel, Transport of conservative chemical through an unsaturated, two-dimensional Miller-similar medium with steady-state flow, *Water Resour. Res.*, *32*, 1653–1663, 1996.

Selker, J., P. Leclerq, J. Y. Parlange, and T. Steenhuis, Fingered flow in two dimensions, 1, Measurement of matric potential, *Water Resour. Res.*, *28*, 2513–2522, 1992.

van Dam, J. C., J. Hendrick, H. van Ommen, M. Bannink, M. Th. van Genuchten, and L. W. Dekker, Water and solute movement in a coarse-textured water-repellent field soil, *J. of Hydrol.*, *120*, 359–379, 1990.

Wang, Z., *Dynamic Simulation of Liquid Air Displacement and Preferential Flow in Porous Media*, Ph.D. dissertation, Kathoielke Universiteit Leuvin, 1997.

White, R. E., The influence of macropores on the transport of dissolved and suspended matter through soil, *Advan. Soil Sci.*, *3*, 95–120, 1985.

William A. Jury and Zhi Wang, University of California, Riverside, CA 92521.

Geostatistical, Type-Curve, and Inverse Analyses of Pneumatic Injection Tests in Unsaturated Fractured Tuffs at the Apache Leap Research Site Near Superior, Arizona

Guoliang Chen,[1] Walter A. Illman,[2] Dick L. Thompson,[3]
Velimir V. Vesselinov,[4] and Shlomo P. Neuman[4]

Over 270 single-hole [*Guzman et al.*, 1996] and 44 cross-hole pneumatic injection tests [*Illman et al.*, 1998; *Illman*, 1999] have been conducted at the Apache Leap Research Site (ALRS) near Superior, Arizona. In this paper, we describe a geostatistical analysis of the single-hole data and type-curve as well as numerical inverse interpretations of one cross-hole test, PP4. Our geostatistical analysis yields information about the spatial structure of air permeabilities measured on a nominal scale of 1 m, as well as of other variables such as fracture density. Our type-curve and inverse interpretations of cross-hole test PP4 yield information about pneumatic connections, directional air permeabilities, and air-filled porosities on scales ranging from a few meters to a few tens of meters. The numerical model can be applied simultaneously to pressure data from multiple borehole intervals (and multiple cross-hole tests), which amounts to "pneumatic tomography" of the rock. Our analyses suggest that (a) pneumatic pressure behavior of unsaturated fractured tuffs at the ALRS can be interpreted by treating the rock as a continuum on scales ranging from meters to tens of meters; (b) this continuum is representative primarily of interconnected fractures; (c) its pneumatic properties nevertheless correlate poorly with fracture density; and (d) air permeability exhibits multiscale random variations in space.

1. INTRODUCTION

Issues associated with the site characterization of fractured rock terrains, the analysis of fluid flow and contaminant transport in such terrains, and the efficient handling of contaminated sites are typically very difficult to resolve. A major source of this difficulty is the complex nature of the subsurface "plumbing systems" of pores and fractures through which flow and transport in rocks take place. There is at present no well-established field methodology to characterize the fluid flow and contaminant transport properties of unsaturated fractured rocks. In order to characterize the ability of such rocks to conduct water, and to transport dissolved or suspended contaminants, one would ideally want to observe these phenomena directly by conducting controlled field hydraulic injection and tracer experiments within the rock. In order to characterize the ability of unsaturated fractured rocks to conduct nonaqueous phase liquids such as chlorinated solvents, one would ideally want to observe the movement of such liquids under controlled conditions in the field. In practice, there are severe logisti-

[1]Broken Hill Proprietary, Florence, Arizona
[2]CNWRA, Southwest Research Institute, San Antonio, Texas
[3]City of Tucson Water Department, Tucson, Arizona
[4]University of Tucson, Tucson, Arizona

Dynamics of Fluids in Fractured Rock
Geophysical Monograph 122

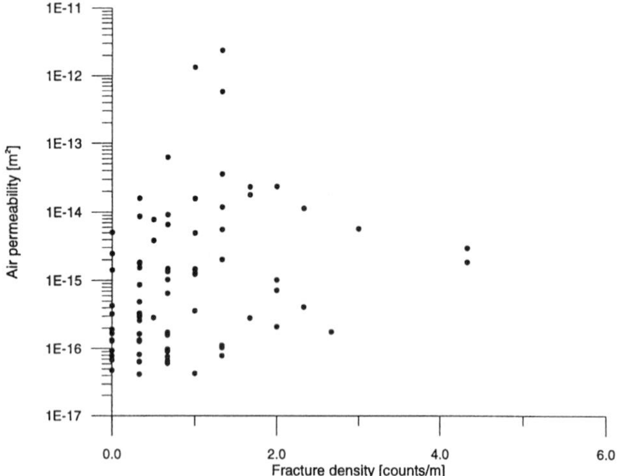

Figure 1. Air permeability versus fracture density data. From *Rasmussen et al.* [1990].

cal obstacles to the injection of water into unsaturated geologic media, and logistical as well as regulatory obstacles to the injection of nonaqueous liquids. There also are important technical reasons why the injection of liquids, and dissolved or suspended tracers, into fractured rocks may not be the most feasible approach to site characterization when the rock is partially saturated with water. Injecting liquids and dissolved or suspended tracers into an unsaturated rock would cause them to move predominantly downward under the influence of gravity, and would therefore yield at best limited information about the ability of the rock to conduct liquids and chemical constituents in directions other than the vertical. It would further make it difficult to conduct more than a single test at any location because the injection of liquid modifies the ambient saturation of the rock, and the time required to recover ambient conditions may be exceedingly long.

Many of these limitations can be overcome by conducting field tests with gases rather than with liquids, and with gaseous tracers instead of chemicals dissolved in water. Experience with pneumatic injection and gaseous tracer experiments in fractured rocks is limited. Much of this experience has been gained recently during pneumatic injection tests in tuffs at Yucca Mountain, Nevada [*LeCain*, 1996a,b; *Wang et al.*, 1998], and at the Apache Leap Research Site (ALRS) near Superior, Arizona. Earlier pneumatic injection tests were conducted by *Kirkham* [1947], *Boardman and Skrove* [1966], *Montazer* [1982], and *Mishra et al.* [1987].

This paper focuses on single- and cross-hole pneumatic injection tests conducted by the University of Arizona at the ALRS. The site is situated near Superior in central Arizona. It consists of a cluster of 22 vertical and inclined (at 45°) boreholes that have been completed to a maximum vertical depth of 30 m within a layer of slightly welded unsaturated tuff. The boreholes span a surface area of 55 m by 35 m and a volume of rock on the order of 60,000 m^3. The upper 1.8 m of each borehole was cased, and a surface area of 1500 m^2 was covered with a plastic sheet to minimize infiltration and evaporation. Core data and borehole television images are available for many of the boreholes.

Early work related to our area of study at the ALRS is described by *Evans* [1983], *Schrauf and Evans* [1984], *Huang and Evans* [1984], *Green and Evans* [1987], *Rasmussen and Evans* [1987, 1989, 1992], *Tidwell et al.* [1988], *Yeh et al.* [1988], *Weber and Evans* [1988], *Chuang et al.* [1990], *Rasmussen et al.* [1990, 1996], *Evans and Rasmussen* [1991], and *Bassett et al.* [1994]. This early work included drilling 16 boreholes and conducting numerous field and laboratory investigations. Information about the location and geometry of fractures in the study area has been obtained from surface observations, the examination of oriented cores, and borehole televiewer records. Fracture density, defined by *Rasmussen et al.* [1990] as the number of fractures per meter in a 3 m borehole interval, ranges from zero to a maximum of 4.3 per meter. Though the fractures exhibit a wide range of inclinations and trends, most of them are near vertical, strike north-south, and dip steeply to the east. Surface fracture traces reveal a steeply dipping east-west set. An experimental study of aperture distributions in a large natural fracture at the ALRS has been published by *Vickers et al.* [1992].

Single-hole pneumatic injection tests were conducted in 87 intervals of 3 m length in 9 boreholes by *Rasmussen et al.* [1990, 1993]. Those tests were conducted by injecting air at a constant mass rate between two inflated packers while monitoring pressure within the injection interval. Pressure was said to have reached stable values within minutes in most test intervals, allowing the calculation of air permeability by means of steady-state formulas. Figure 5b of *Rasmussen et al.* [1993] suggests a good correlation (r = 0.876) between pneumatic and hydraulic permeabilities at the ALRS. Figure 1 herein shows a scatter plot of pneumatic permeability versus fracture density for 3 m borehole intervals based on the data of *Rasmussen et al.* [1990]. It suggests a lack of correlation between fracture density and air permeability.

The single-hole tests of *Rasmussen et al.* [1990, 1993] were of relatively short duration and involved relatively long test intervals. *Guzman et al.* [1994, 1996] and *Guzman and Neuman* [1996] conducted a much larger number of single-hole pneumatic injection tests of considerably longer duration over shorter intervals in six boreholes. A total of

184 borehole segments were tested by setting the packers 1 m apart, as shown in Figure 2. Additional tests were conducted in segments of lengths 0.5, 2.0, and 3.0 m, bringing the total number of tests to over 270. The tests were conducted by maintaining a constant injection rate until air pressure became relatively stable and remained so for some time. The injection rate was then incremented by a constant value and the procedure repeated. In most tests, three or more such incremental steps were conducted in each borehole segment while the air injection rate, pressure, temperature and relative humidity were recorded. For each relatively stable period of injection rate and pressure, air permeability was estimated by treating the rock around each test interval as a uniform, isotropic continuum within which air flows as a single phase under steady state, in a pressure field exhibiting prolate spheroidal symmetry.

The results of these steady state interpretations of single-hole air injection tests are listed in *Guzman et al.* [1996]. They reveal that:

1. Air permeabilities determined *in situ* from steady state single-hole test data are much higher than those determined on core samples of rock matrix in the laboratory, suggesting that the in situ permeabilities represent the properties of fractures at the site.
2. It is generally not possible to distinguish between the permeabilities of individual fractures and the bulk permeability of the fractured rock in the immediate vicinity of a test interval by means of steady-state single-hole test data.
3. The time required for pressure in the injection interval to stabilize typically ranges from 30 to 60 min, increases with flow rate, and may at times exceed 24 hr, suggesting that steady-state permeability values published in the literature for this and other sites, based on much shorter air injection tests, may not be entirely valid.
4. Steady-state interpretation of single-hole injection tests, based on the assumption of radial flow, which correspond closely to prolate spheroidal flow, is acceptable for intervals of length equal to or greater than 0.5 m in boreholes having a radius of 5 cm, as is the case at the ALRS.
5. Pressure in the injection interval typically rises to a peak prior to stabilizing at a constant value, possibly due to a two-phase flow effect whereby water in the rock is displaced by air during injection.
6. In most test intervals, pneumatic permeabilities show a systematic increase with applied pressure as air apparently displaces water under two-phase flow.
7. In a few test intervals, intersected by widely open fractures, air permeabilities decrease with applied pressure due to apparent inertial effects.
8. Air permeabilities exhibit a hysteretic variation with applied pressure.
9. The pressure-dependence of air permeability suggests that it is advisable to conduct single-hole air injection tests at several applied flow rates and/or pressures;
10. Enhanced permeability due to slip flow—the Klinkenberg effect [*Klinkenberg* 1941]—appears to be of little relevance to the interpretation of single-hole air-injection tests at the ALRS.
11. Local-scale air permeabilities vary by orders of magnitude between test intervals across the site.
12. Spatial variability is much greater than that due to applied pressure and lends itself to meaningful statistical and geostatistical analysis.
13. Air permeabilities are poorly correlated with fracture densities, as is known to be the case for hydraulic conductivities at many water-saturated fractured rock sites worldwide [*Neuman*, 1987], providing further support for Neuman's conclusion that the permeability of fractured rocks cannot be reliably predicted from information about fracture geometry (density, trace lengths, orientations, apertures, and their roughness),

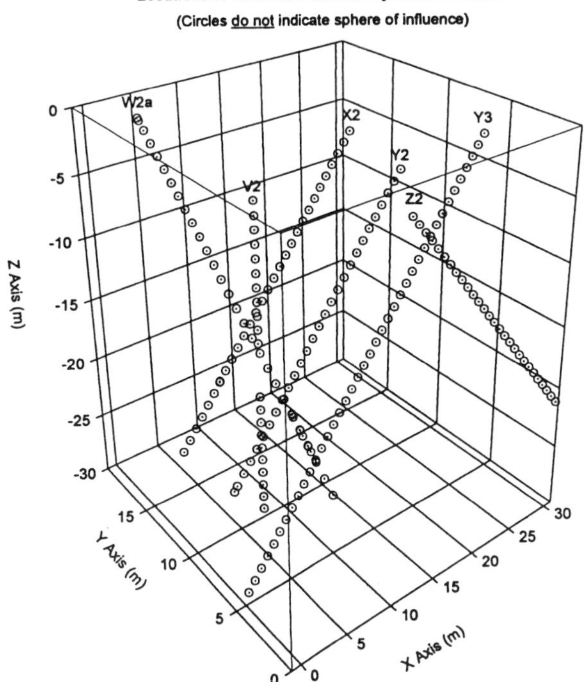

Figure 2. Perspective toward the Northeast showing center locations of 1-m single-hole pneumatic test intervals; overlapping circles indicate re-tested locations. After *Guzman et al.* [1996].

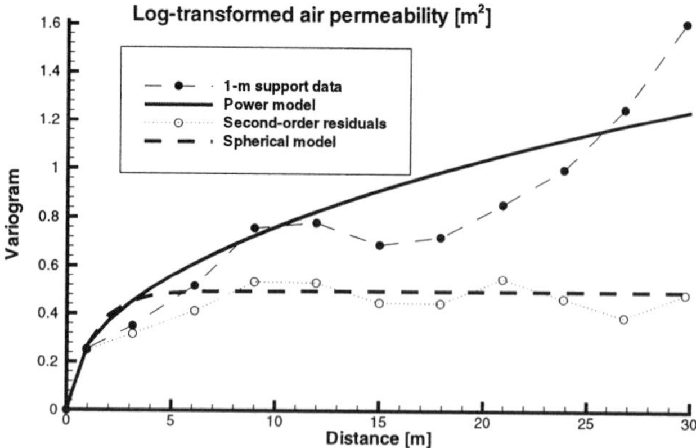

Figure 3. Omnidirectional sample and model variograms of air permeability and fracture density at a minimum separation distance of 3 m.

but must instead be determined directly by means of hydraulic and/or pneumatic tests.

14. Air permeabilities vary systematically with the scale of measurement as represented nominally by the distance between packers in an injection interval.

The work of *Guzman et al.* [1994, 1996] and *Guzman and Neuman* [1996] strongly suggests that air-injection tests yield properties of the fracture system, which are relevant to both unsaturated and saturated conditions. In particular, numerical simulations by these authors show that, whereas the intrinsic permeability one determines from such tests is generally lower than the intrinsic permeability to water of fractures which surround the test interval, it nevertheless approaches the latter as the applied pressure goes up. This is because, under ambient conditions, capillary forces tend to draw water from fractures into the porous (matrix) blocks of rock between the fractures, thereby leaving the fractures saturated primarily with air. Water saturation in the matrix blocks is therefore typically much higher than that within the fractures, making it relatively difficult for air to flow through such blocks. It follows that, during a pneumatic injection test, the air moves primarily through fractures (most of which contain relatively little water) and the test therefore yields flow and transport parameters that reflect the intrinsic properties of these largely air-filled fractures.

Core and single-hole measurements, conducted over short segments of a borehole, provide information only about a small volume of rock in the immediate vicinity of each measurement interval. Available data from the ALRS indicate that rock properties measured on such small scales vary erratically in space in a manner that renders the rock randomly heterogeneous and pneumatically anisotropic. Our analyses to date [*Chen et al.*, 1997] suggest that it is possible to interpolate some of the core and single-hole measurements at the ALRS between boreholes by means of geostatistical methods, which view the corresponding variables as correlated random fields. This is especially true about air permeability, porosity, fracture density, water content, and the van Genuchten water retention parameter α; for each of these we possess enough measurements to constitute a workable geostatistical sample. We describe in this paper our geostatistical analyses of air-permeability and fracture-density data from the ALRS.

Illman et al. [1998] and *Illman* [1999] discuss pressure and pressure-derivative type-curve analyses of transient data corresponding to the first among multiple steps of single-hole pneumatic tests conducted by *Guzman et al.* [1994, 1996] and *Guzman and Neuman* [1996]. These analyses provide additional information about the dimensionality of the corresponding flow regime, skin factors and compressible air storage effects. At the ALRS, airflow around the vast majority of the 1 m single-hole test intervals appears to be three-dimensional; borehole storage due to air compressibility is pronounced; and skin effects are minimal. Air permeabilities obtained from steady state and transient type-curve interpretations of these tests agree closely with each other but correlate poorly with fracture density data. Borehole storage makes it difficult to obtain reliable air-filled porosity values from type-curve analyses of data from the first step of a single-hole test. Inclusion of multiple steps, and especially recovery data (on which borehole storage has a relatively small effect), in our numerical inverse analysis of single-hole tests [*Vesselinov and Neuman*, 1999] makes it possible to obtain reasonable estimates of air-filled porosity (as well as air permeability) from these tests.

A total of 44 cross-hole pneumatic interference tests of various types (constant injection rate, multiple step injection rates, instantaneous injection) have been conducted during the years 1995–1997 using various configurations of injection and monitoring intervals [*Illman et al.*, 1998; *Illman*, 1999]. In this paper we discuss pressure and pressure-derivative type curve as well as numerical inverse interpretations of one of these tests.

2. GEOSTATISTICAL ANALYSIS OF SPATIAL VARIABILITY

A geostatistical analysis of several variables at the ALRS has been conducted by *Chen et al.* [1997]. We describe below a slightly modified version of their analysis of single-hole air-permeability and fracture-density data. Figure 3 shows an omnidirectional sample (semi) variogram of log (to base ten) air-permeability data, where k is given in

Table 1. Discrimination Among Log Permeability Geostatistical Models.

Drift Model NLL	No Drift 665.801	1st Order Polynomial 665.080	2nd Order Polynomial 655.849
Variogram model of Residuals	Power	Exponential	Exponential
Number of Parameters	2	6	12
Variance (Scaling Coefficient for Power model)	0.2715	0.5807	0.495
Integral scale (Power for power model)	0.4475	1.665	1.2602
AIC (Rank)	669.801 (1)	677.08 (2)	679.849 (3)
MAIC (Rank)	677.231 (1)	696.37 (2)	718.428 (3)
HIC (Rank)	672.407 (1)	684.898 (2)	695.486 (3)
KIC (Rank)	680.016 (1)	690.088 (2)	700.907 (3)

[m^2], obtained from steady-state analyses of 1 m scale single-hole pneumatic injection tests at the site. The variogram exhibits statistical nonhomogeneity and was analyzed by *Chen et al.* using two structural models, one that views the data as belonging to a random fractal field characterized by a power variogram, and another that views the data as belonging to a statistically homogeneous random field about a linear or quadratic spatial drift. To select the best among these models, they used formal model discrimination criteria based on the Maximum Likelihood Cross Validation (MLCV) approach of *Samper and Neuman* [1986a,b], coupled with the generalized least squares drift removal approach of *Neuman and Jacobson* [1984].

MLCV estimates variogram parameters by maximizing the likelihood of kriging (geostatistical interpolation) cross validation errors. As MLCV assumes that the variogram model is known, it leads to optimum parameters for a given model structure without regard to the question of how well this model represents the real system. Fortunately, a number of model identification criteria have been developed in the context of maximum likelihood estimation. Earlier work by *Carrera and Neuman* [1986a, b] and *Samper and Neuman* [1986a,b] have compared four such criteria—AIC [*Akaike*, 1974], MAIC [*Akaike*, 1977], HIC [*Hannan*, 1980], KIC [*Kashyap*, 1982]—and concluded that the one that comes closest to satisfying these requirements is that of *Kashyap* [1982]. Kashyap's criterion favors the model that, among all alternatives considered, is least probable (in an average sense) of being incorrect. Stated otherwise, the criterion minimizes the average probability of selecting the wrong model among a set of alternatives. It supports the principle of parsimony in that, everything else being equal, the model with the smallest number of parameters is favored. While this means favoring the simpler model, the criterion nevertheless allows considering models of growing complexity as the database improves in quantity and quality. In other words, the criterion recognizes that when the database is limited and/or of poor quality, one has little justification for selecting an elaborate model with numerous parameters. Instead, one should then prefer a simpler model with fewer parameters, which nevertheless reflects adequately the underlying structure of the rock, and the corresponding flow and transport regime. The cited works by *Carrera and Neuman* [1986a,b] and *Samper and Neuman* [1989a,b] clearly indicate that an inadequate model structure is far more detrimental to its predictive ability than is noise in data used to calibrate the model.

Chen et al. [1997] extended MLCV so as to render it applicable to statistically nonhomogeneous fields of the kind represented by the power variogram in Figure 3. The latter field possesses neither a finite variance nor a finite spatial correlation scale, although it has statistically homogeneous spatial increments. The exponential model represents a homogeneous field with finite variance and spatial correlation scale. Which of these models represents the data more accurately? Table 1 shows that, whereas the exponential variogram model with a quadratic drift fits the data best (as measured and implied by the smallest negative log likelihood model fit criterion, NLL), all four model discrimination criteria (AIC, MAIC, HIC, KIC) consistently rank the power model as most acceptable and the former model as least acceptable. The reason is that, whereas all three models fit the data almost equally well, the power model is most parsimonious with only two parameters and the exponential variogram model with second-order drift is least parsimonious with twelve parameters. As shown in Figure 4, both the power model (left) and exponential model with second order de-trending (right) yield very similar kri-

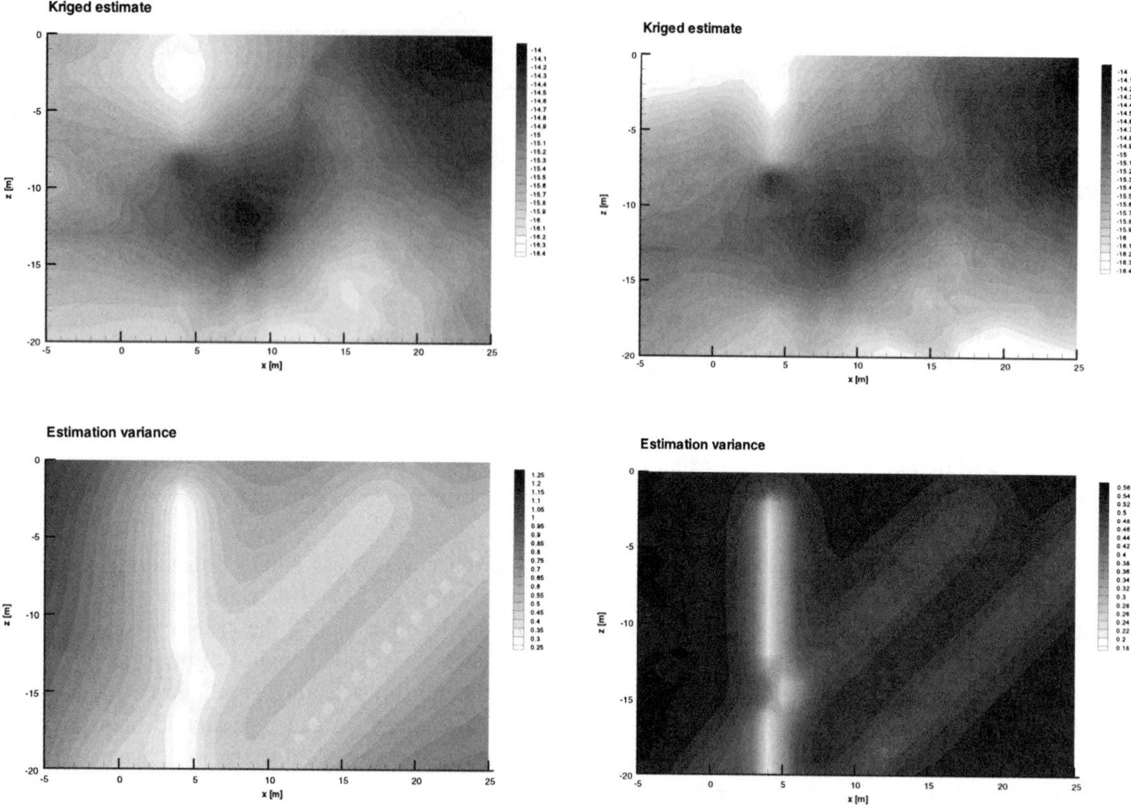

Figure 4. Kriged estimates and variance of log k at y = 7 m using power model (left) and exponential model with second order detrending (right).

ged (estimated) images of log k, but rather different measures of the associated estimation uncertainty.

The geostatistical analysis of 1 m scale air permeabilities, conducted by *Chen et al.* [1997], is based on data from only six boreholes (X2, Y2, Y3, Z2, V2, W2). As these boreholes span only a part of the domain we model for purposes of interpreting our cross-hole tests, it would be good if we could validly augment the 1 m data with air permeabilities obtained from 3 m test intervals in other boreholes (X1, X3, Y1, Z1, Z3). To check whether this is justified, we compare in Figure 5 the omnidirectional sample variograms of the available 1 m, 3 m, and combined 1 m and 3 m $\log_{10} k$ data. Although the sample variograms differ somewhat from each other at large separation distances, they are otherwise quite similar. Attempts to represent the 3 m data by a variogram model that views them as a sample from a statistically homogeneous random field with a linear or quadratic spatial drift were not successful. We therefore krige the combined set of 227 1 m and 3 m air-permeability data jointly, using the same power variogram model as that surmised on the basis of 1 m data by *Chen et al.* [1997]. The combined set of 1 m and 3 m scale $\log_{10}$-transformed air permeability ranges from −17.13 (7.41×10^{-18} m^2) to −11.62 (2.40×10^{-12} m^2) and is characterized by mean, variance, and coefficient of variation equal to −15.22 (6.3×10^{-16} m^2), 0.87, and −0.061, respectively.

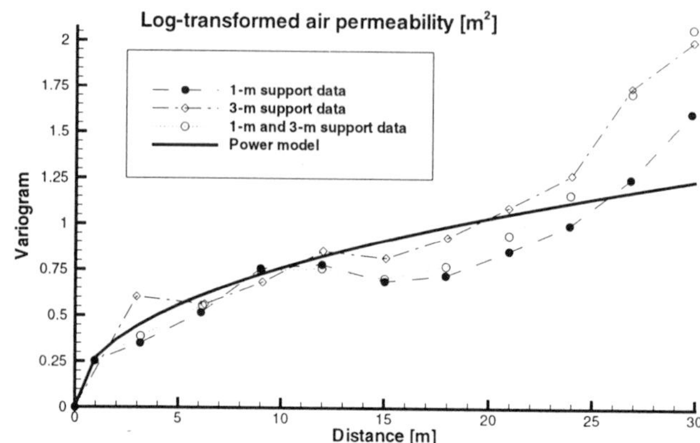

Figure 5. Omnidirectional sample and power model variograms of log k data with various supports. Power model fitted to 1 m data.

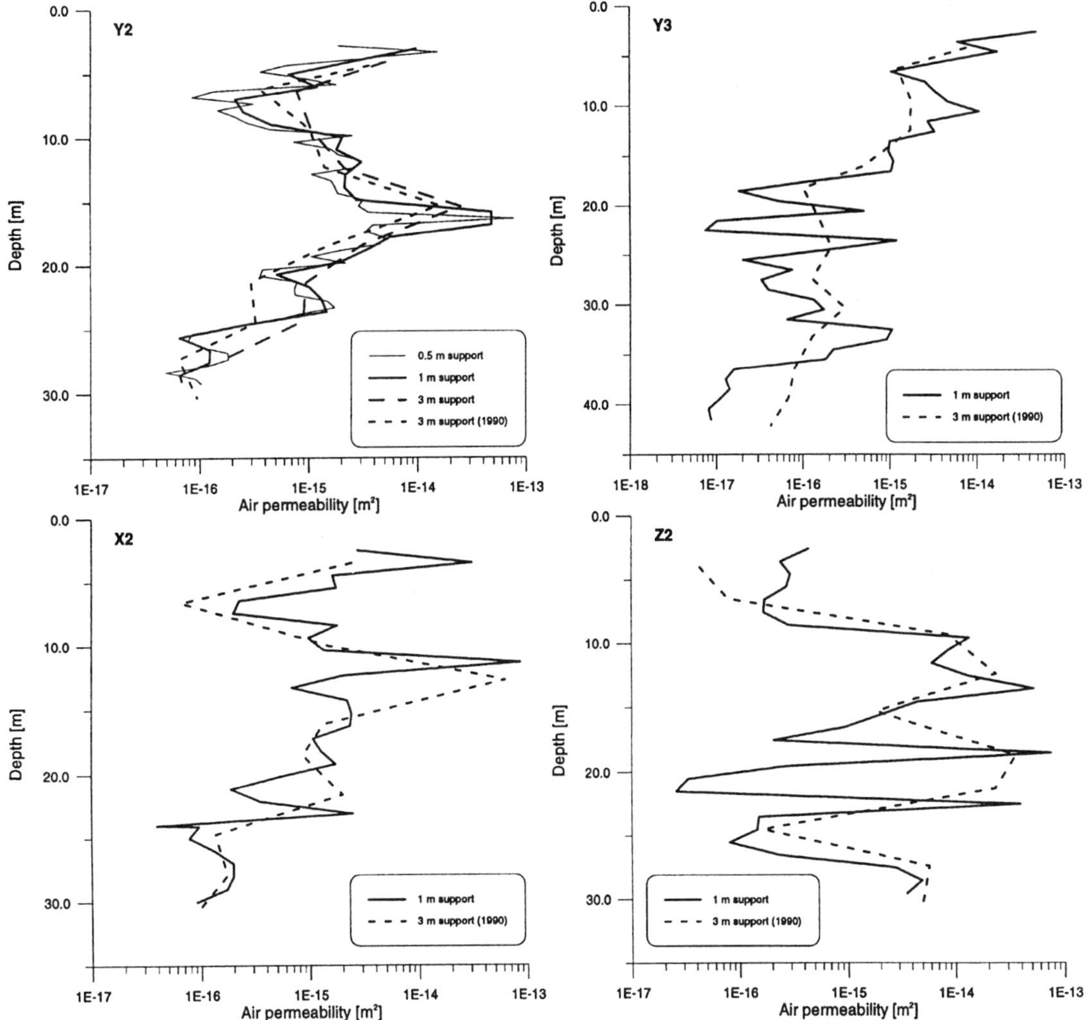

Figure 6. Variation of air permeability with depth and support scale. Data from *Guzman et al.* [1996] and *Rasmussen et al.* [1990].

One-meter and three-meter scale single-hole air-permeability data are available for boreholes X2, Y2, Y3, and Z2. Figure 6 shows how these and 0.5 m data from borehole Y2 vary with depth in each borehole. We see that 3 m scale permeabilities obtained for Y2 by *Rasmussen et al.* [1990] are consistently lower than those obtained by *Guzman et al.* [1996]; we attribute this systematic difference to the relatively short duration of tests conducted by the former authors. In general, as support scale increases, the amplitude and frequency of spatial variations in permeability decrease. Figure 7 compares kriged images of log air permeability we have generated along four vertical sections at y = 0, 5, 7, and 10 m using 1 m data (left column) and the combined set of 1 m and 3 m data, the latter from boreholes X1, X3, Y1, Z1, and Z3 (right column). The figure shows boreholes intersected by, or located very close to, each cross section. The two sets of kriged images are considerably different from each other. This is most pronounced at y = 0 m, which passes through the Z-series of boreholes. Here the inclusion of data from boreholes Z1 and Z3 has caused estimated permeability in the upper-right corner of the section to be much higher than it is without these data. The effect extends to all four cross sections, which exhibit elevated permeabilities near the upper-right corner. Along sections at y = 5 m and 7 m, which pass close to the Y and V series, respectively, the addition of data from Y1 affects the shape and size of a prominent high-permeability zone, which extends through Y2 (see corresponding peak in Fig-

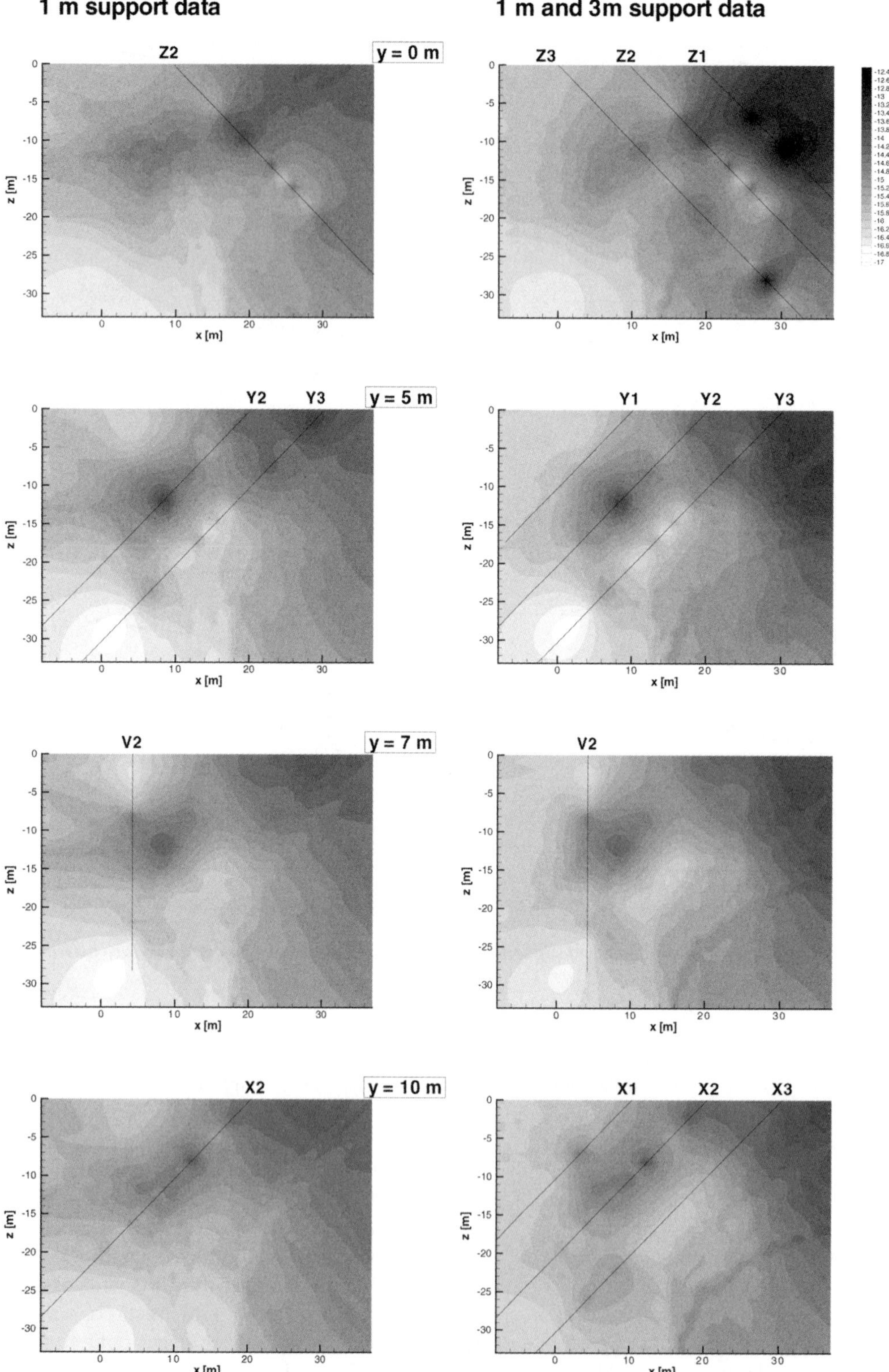

Figure 7. Kriged log k estimates obtained using 1 m scale data from boreholes X2, Y2, Y3, Z2, V2, and W2A (left) as well as the combined set of 1 m and 3 m data, the latter from boreholes X1, X3, Y1, Z1, and Z3 (right).

ure 6). The addition of data from X3 to the set reveals corresponding high- and low-permeability zones in section y = 10 m, which correlate well with similar zones intersected by Y3 in section y = 5 m. A three-dimensional representation of kriged log permeability, based on the combined set of data, is shown in Figure 8. The kriged estimates in Figure 8 have mean, variance, and coefficient of variation equal to –15.22 (6.3×10^{-16} m^2), 0.51 and –0.047, respectively. The mean of the estimates is similar to that of the input data, but the estimates' variance and coefficient of variation are lower due to the smoothing effect of kriging. Figures 9 through 11 show corresponding sections in the y-z, z-x, and x-y planes, respectively.

Another omnidirectional sample variogram in Figure 3 concerns fracture density in counts per meter. The corresponding best-fit exponential model (without drift) has a variance of 0.69 and an integral scale of 2.5 m [*Chen et al.*, 1997]. Figure 12 shows kriged images of log-transformed air permeability and fracture density in a vertical plane corresponding to y = 7.0 m, around which many of the data are clustered. There clearly is no correlation between log permeability and fracture density along this plane (nor anywhere else in our domain of investigation).

Kriged estimates of hydrogeologic variables are smooth relative to their random counterparts. To generate less smooth and more realistic images that honor the available data, we have used GCOSIM3D, a sequential Gaussian conditional simulation code developed for three-dimensional data by *Gómez-Hernández and Cassiraga* [1994]. The code is applied to log permeability (conditioned on the combined set of 1 m and 3 m data) and fracture density data on the assumption that both of these variables are Gaussian. Indeed, both have passed the Kolmogorof-Smirnoff test of Gaussianity at the 95% confidence level. Figure 13 shows conditionally simulated images of log air permeability and fracture density in a vertical plane corresponding to $y = 7\ m$. These images are clearly much less smooth than are their kriged counterparts in Figure 12.

Our analysis shows that air-permeability data (as well as fracture density, porosity, saturation, and van Genuchten's α) are amenable to continuum geostatistical analysis and exhibit distinct spatial correlation structures. This suggests that the data can be viewed as samples from a random field, or stochastic continuum, as proposed over a decade ago by *Neuman* [1987] and affirmed more recently by *Tsang et al.* [1996]. This is so despite the fact that the rock is fractured and therefore discontinuous. Our finding supports the application of continuum flow theories and models to fractured porous tuffs on scales of a meter or more.

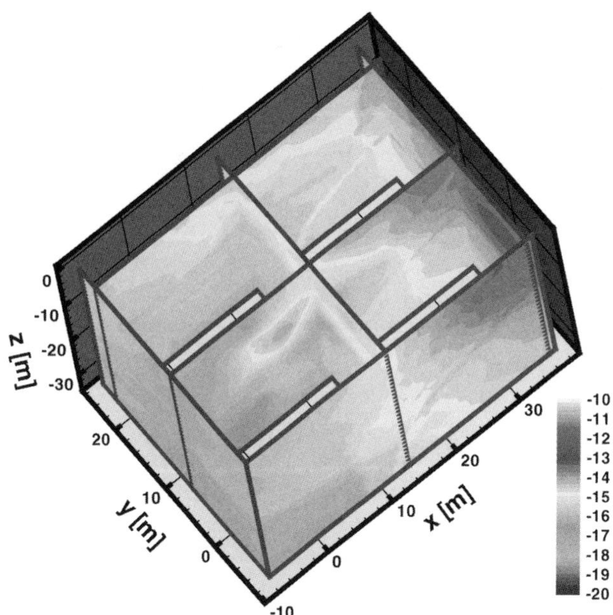

Figure 8. Three-dimensional representation of kriged log k.

3. TYPE-CURVE INTERPRETATION OF CROSS-HOLE TEST PP4

Cross-hole pneumatic interference tests were conducted at the ALRS using modular straddle packer systems that were easily adapted to various test configurations and allowed rapid replacement of failed components, modification of the number of packers, and adjustment of distances between them in both the injection and monitoring boreholes [*Illman et al.*, 1998; *Illman*, 1999]. A typical cross-hole test consisted of packer inflation, a period of pressure recovery, air injection, and another period of pressure recovery. Once packer inflation pressure had dissipated in all (monitoring and injection) intervals, air injection at a constant mass flow rate began. It generally continued for several days, until pressure in most monitoring intervals appeared to have stabilized. In some tests, injection pressure was allowed to dissipate until ambient conditions had been recovered. In other tests, air injection continued at incremental flow rates, each lasting until the corresponding pressure had stabilized, before the system was allowed to recover.

Cross-hole test PP4 is of particular interest because it involved injection into a high-permeability zone, which helped pressure to propagate rapidly across much of the site; a relatively high injection rate, which led to unambiguous pressure responses in a relatively large number of monitoring intervals; the largest number of pressure and

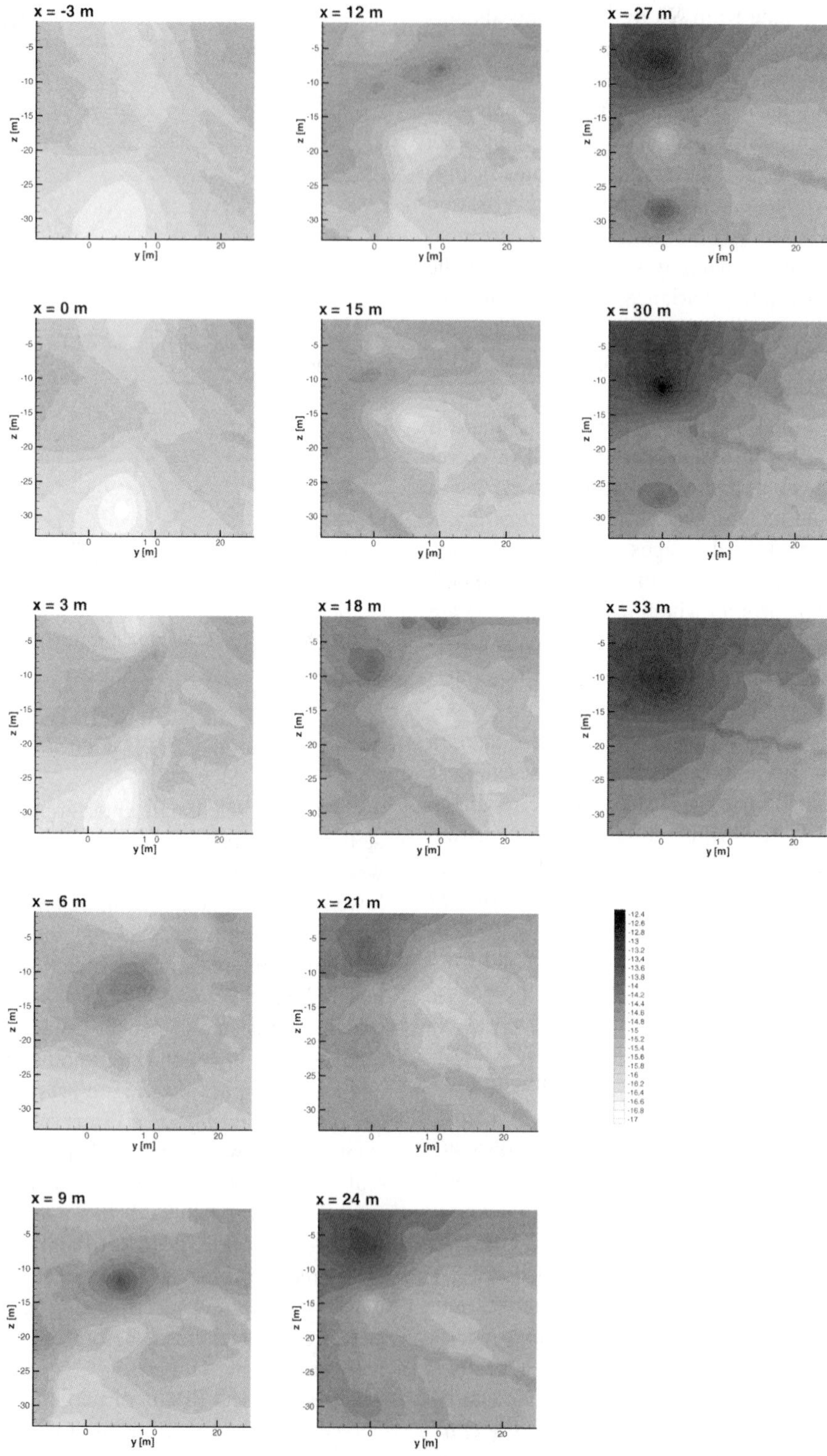

Figure 9. Kriged log k along various y-z planes.

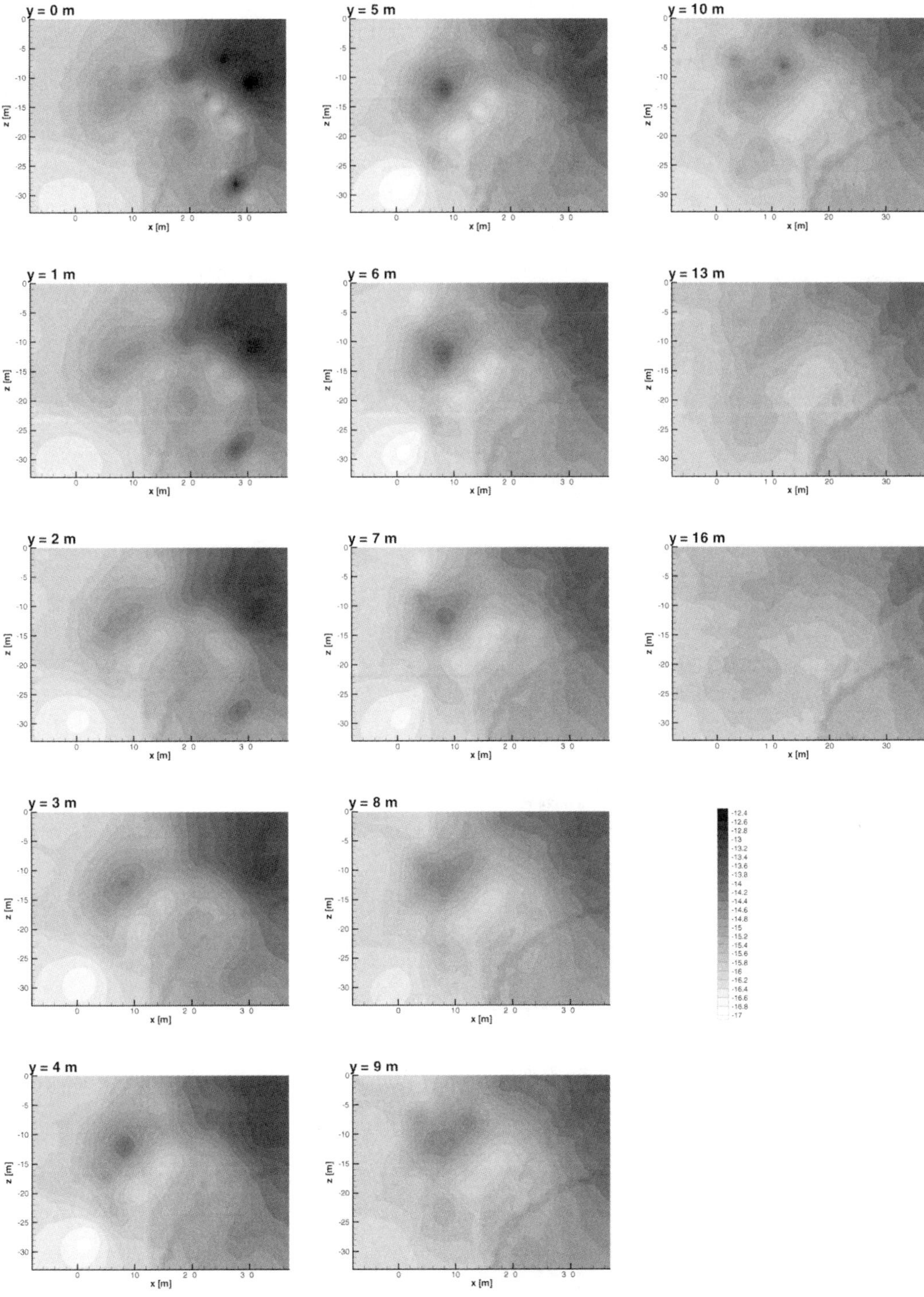

Figure 10. Kriged log *k* along various *x-z* planes.

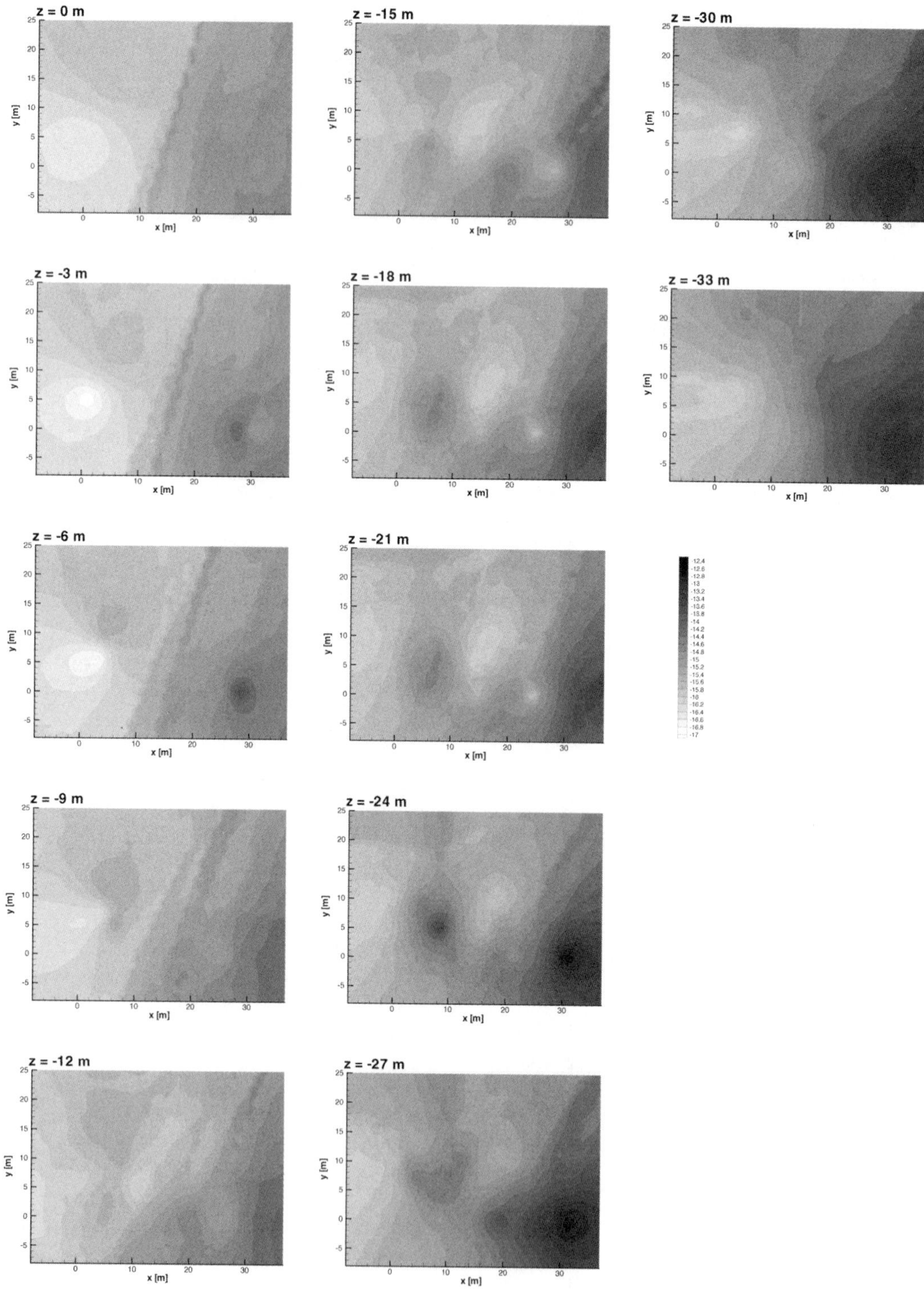

Figure 11. Kriged log k along various x-y planes.

temperature monitoring intervals among all tests; a complete record of relative humidity, battery voltage, atmospheric pressure, packer pressure, and injection pressure; the least number of equipment failures among all tests; flow conditions (such as injection rate, fluctuations in barometric pressure, battery voltage, and relative humidity) that were better controlled and more stable than in all other tests; minimum boundary effects due to injection into the central part of the tested rock mass; a relatively long injection period; rapid recovery; and a test configuration that allowed direct comparison of test results with those obtained from two line-injection/line-monitoring tests, and a point-injection/line-monitoring test at the same location. Stable flow rate and barometric pressure made type-curve analysis of test PP4 results relatively straightforward.

The test was conducted by injecting air at a rate of 50 slpm into a 2 m interval located 15–17 m below the lower lip of casing in borehole Y2, as indicated by a large solid circle in Figure 14. The figure also shows a system of

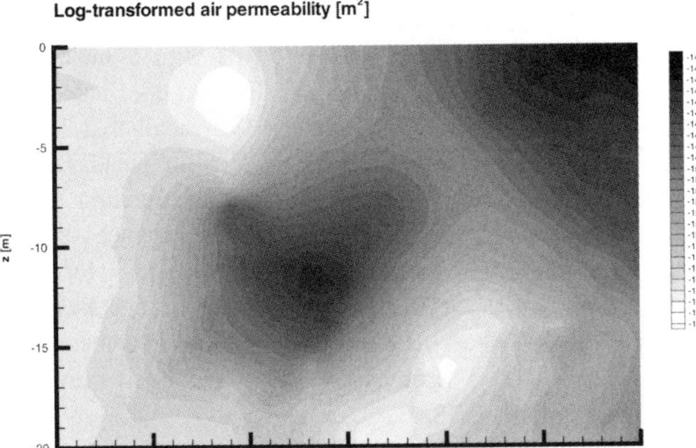

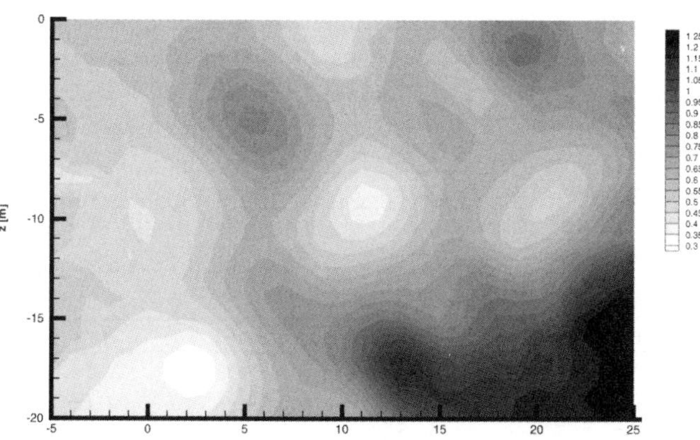

Figure 12. Kriged estimates of air permeability and fracture density at $y = 7$ m.

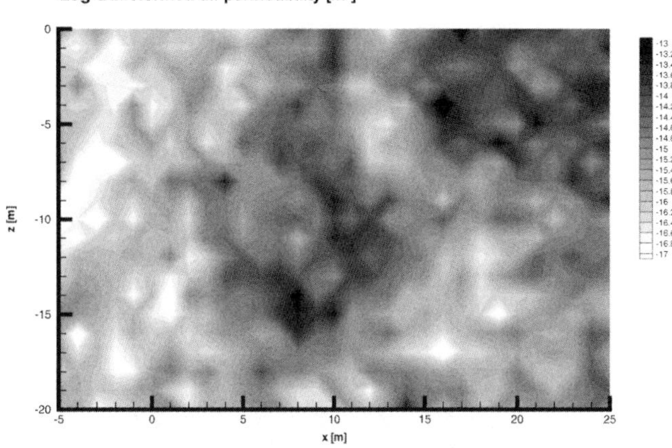

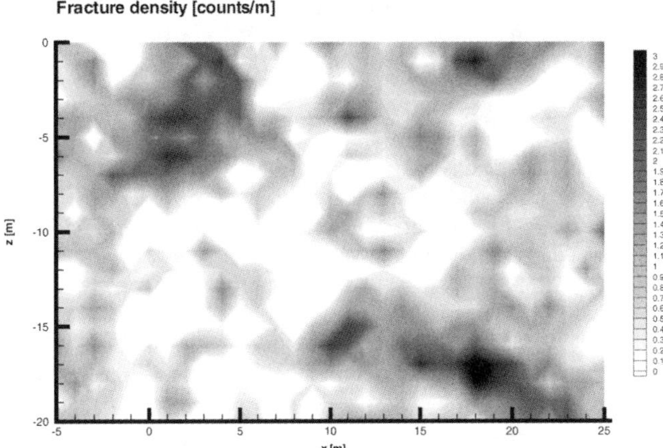

Figure 13. Conditional sequential Gaussian simulations of air permeability and fracture density at $y = 7$ m.

Cartesian coordinates x, y, z with origin at the center of the injection interval, which we use to identify the placement of monitoring intervals relative to this center. Responses were monitored in 13 relatively short intervals (0.5–2.0 m), whose centers are indicated in the figure by small white circles, and 24 relatively long intervals (4.0–42.6 m), whose centers are indicated by small solid circles, located in 16 boreholes. Several of the short monitoring intervals were designed to intersect a high-permeability region that extends across much of the site at a depth comparable to that of the injection interval.

For purposes of cross-hole test analysis by means of type curves, we represent the fractured rock by an infinite, three-dimensional, uniform, anisotropic continuum, as was done by *Hsieh and Neuman* [1985]; consider air to represent a single fluid phase; and linearize the corresponding airflow equations in terms of pressure [*Illman et al.*, 1998]. *Hsieh and Neuman* treat injection and observation intervals as

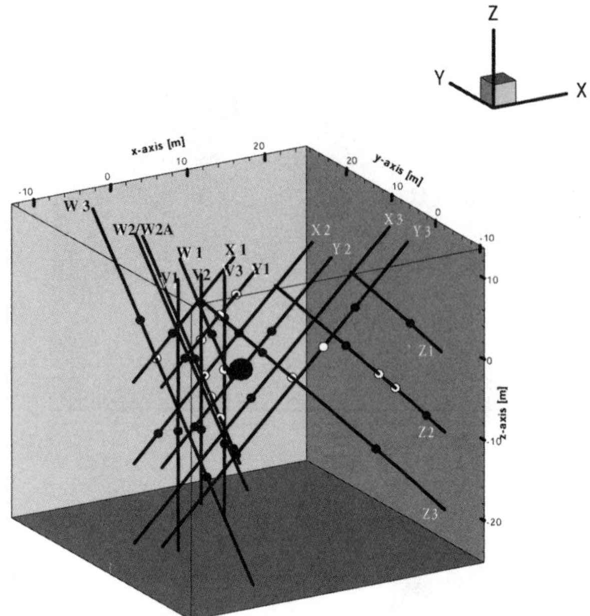

Figure 14. Locations of centers of injection and monitoring intervals. Large solid circle represents injection interval, small solid circles represent short monitoring intervals, and open circles represent long monitoring intervals.

points or lines; we consider the special case where injection takes place at a point and observation along a line. However, we modify their solution to account for the effects of storage and skin on pressure, and its derivative (not considered by these authors), in the observation interval. Type curves of pressure derivative versus the logarithm of time have become popular in recent years because they accentuate phenomena that might otherwise be missed, help diagnose the prevailing flow regime, and aid in constraining the calculation of corresponding flow parameters.

Type-curve interpretation of pressure data from cross-hole test PP4 included 32 intervals; pressure data from five intervals were not amenable to type-curve interpretation and have been excluded. A set of type curves was developed for each pressure monitoring interval by treating the medium as if it was pneumatically isotropic and uniform. However, since the analysis of pressure data from different monitoring intervals yields different values of pneumatic parameters (air permeability and air-filled porosity), our type-curve analysis ultimately yields information about the spatial and directional dependence of these parameters.

Figures 15–20 show how we matched typical records of pressure buildup and pressure derivatives to corresponding type curves. To eliminate spurious oscillations of pressure derivatives, we fitted cubic splines to pressure data over ten equal intervals along the logarithmic time axis, and took the derivatives of these splines. Most of the matches are excellent to good, but a few are poor. Some residual low-frequency fluctuations in late-time pressure derivative data may be caused by fluctuations in barometric pressure.

Our finding that most pressure buildup data are a very good match to the type curves is a clear indication that the majority of cross-hole test PP4 results are amenable to interpretation by means of a model that (a) treats the rock as a pneumatically uniform and isotropic continuum, and (b) describes airflow by means of linearized, pressure-based equations. The fact that some of our data do not fit this model shows that the model does not provide a complete description of pneumatic pressure behavior at the site. That the site is not pneumatically uniform or isotropic on the scale of cross-hole test PP4 is made evident by pneumatic parameters derived from our type-curve matches. Values of pneumatic permeability and air-filled porosity derived from these matches represent bulk properties of the rock between the corresponding monitoring interval and the injection interval. The permeabilities represent directional values along lines that connect the centers of these intervals. Their $\log_{10}$ transformed values range from –14.27 (5.4×10^{-15} m^2) to –12.33 (4.7×10^{-13} m^2) with a mean of –13.46 (3.5×10^{-14} m^2),variance of 0.34, and coefficient of variation equal to –0.043 (Table 2). We recall that $\log_{10}$ transformed air permeabilities derived from steady-state analyses of 1 m and 3 m scale single-hole injection tests had a wider range, a lower mean, a higher variance, and a larger coefficient of variation (Table 2). Log_{10} transformed air-filled porosities from the cross-hole test range from –4.47 (3.4×10^{-5}) to –1.08 (8.3×10^{-1}), with a mean of –2.11 (7.8×10^{-3}), variance of 0.65, and coefficient of variation equal to –0.38.

4. NUMERICAL INVERSE INTERPRETATION OF CROSS-HOLE TEST PP4

We are in the process of interpreting several cross-hole tests simultaneously by means of a three-dimensional finite-volume numerical simulator, FEHM [*Zyvoloski et al.*, 1988, 1996, 1997], coupled with a numerical inverse code, PEST [*Doherty et al.*, 1994], and a geostatistical package, GSTAT [*Pebesma and Wesseling*, 1998]. Such simultane-

Table 2. Summary statistics for $\log_{10} k$ [m^2] identified using different data and methods.

Source	Mean	Variance	CV
Single-hole tests			
-measurements	–15.22	0.87	–0.061
- kriging	–15.22	0.51	–0.047
Cross-hole test PP4			
- type curves	–13.46	0.34	–0.043
- uniform inverse	–13.53	0.69	–0.069
- nonuniform inverse	–15.49	2.51	–0.102

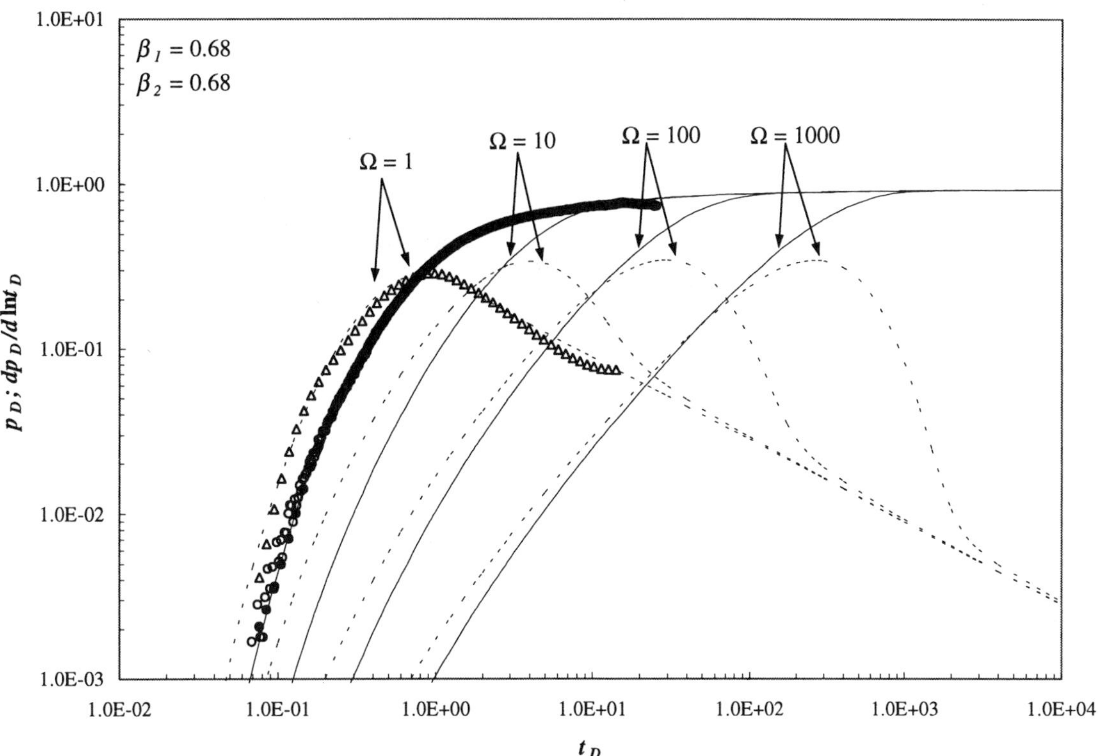

Figure 15. Type-curve match of pressure (open circles) and its derivatives (open triangles) from monitoring interval V1. Solid circles represent spline fitted to pressure data.

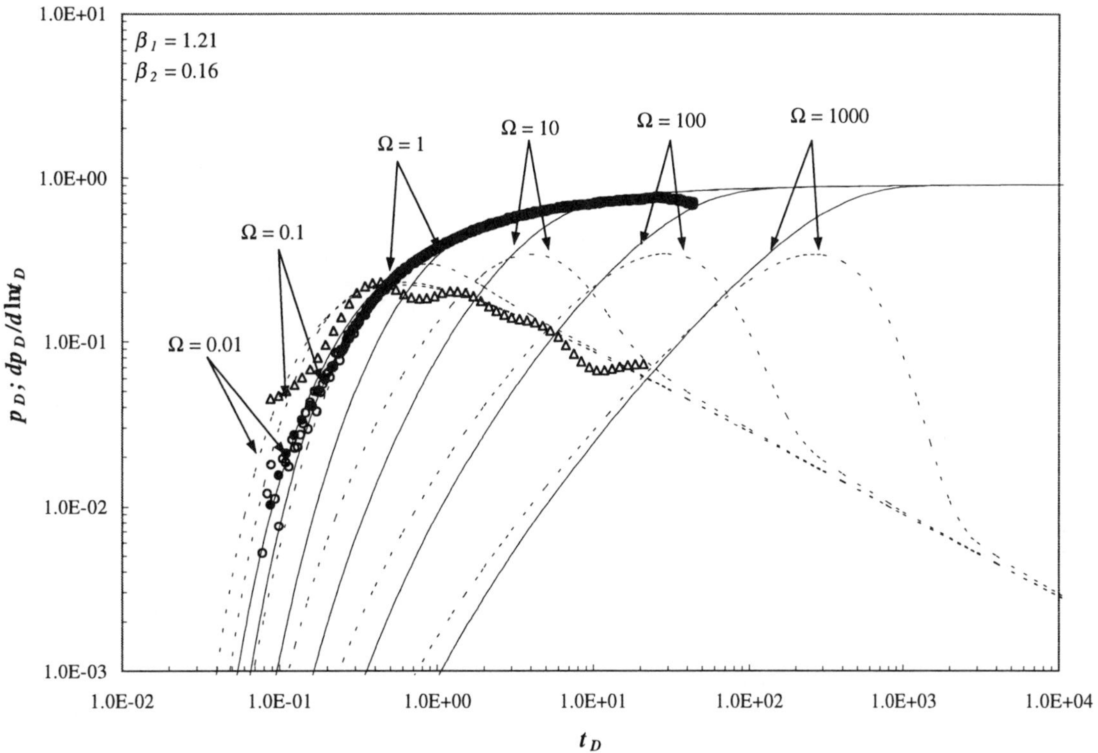

Figure 16. Type-curve match of pressure (open circles) and its derivatives (open triangles) from monitoring interval X1. Solid circles represent spline fitted to pressure data.

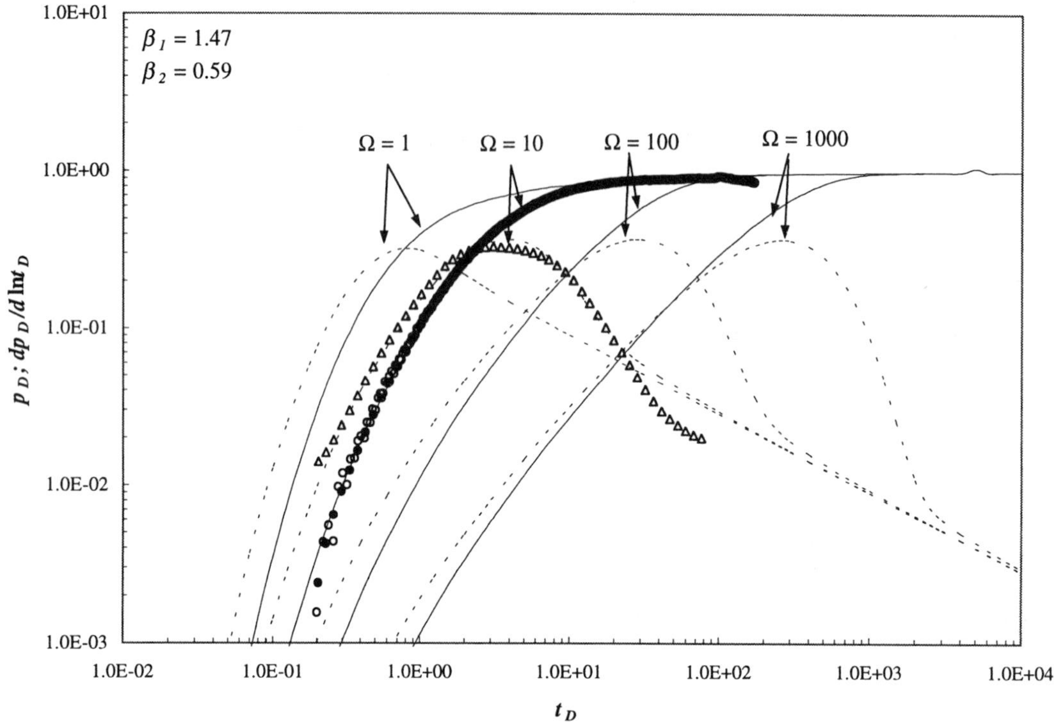

Figure 17. Type-curve match of pressure (open circles) and its derivatives (open triangles) from monitoring interval X2U. Solid circles represent spline fitted to pressure data.

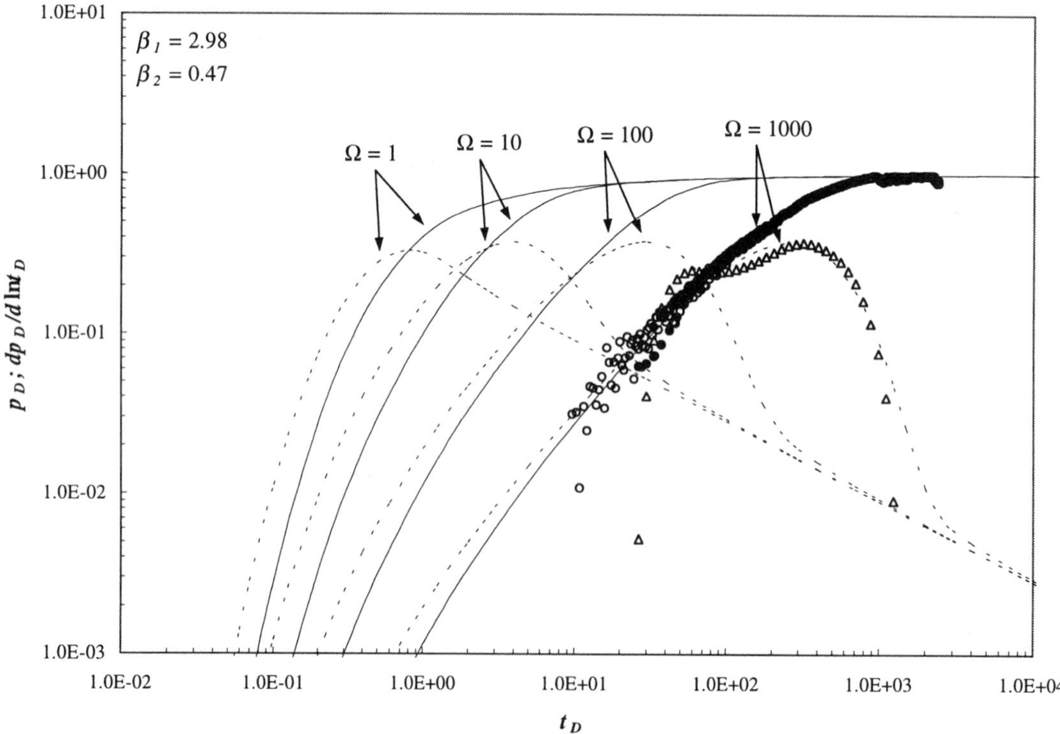

Figure 18. Type-curve match of pressure (open circles) and its derivatives (open triangles) from monitoring interval Z2U. Solid circles represent spline fitted to pressure data.

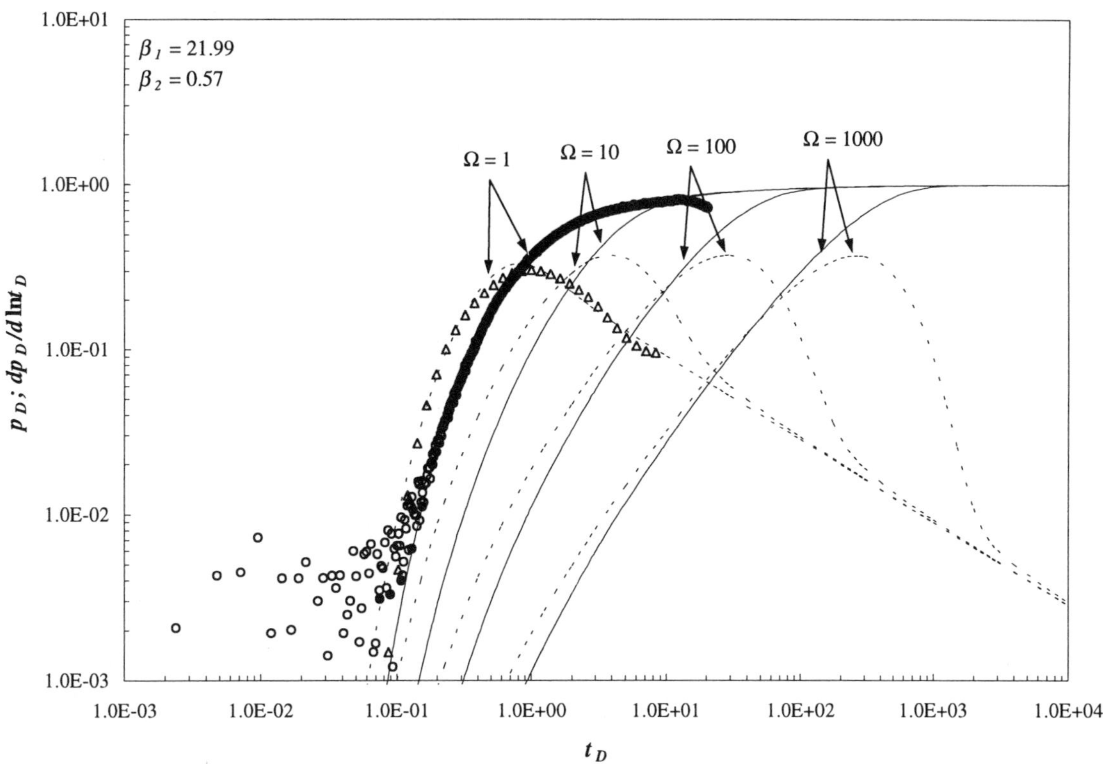

Figure 19. Type-curve match of pressure (open circles) and its derivatives (open triangles) from monitoring interval W3M. Solid circles represent spline fitted to pressure data.

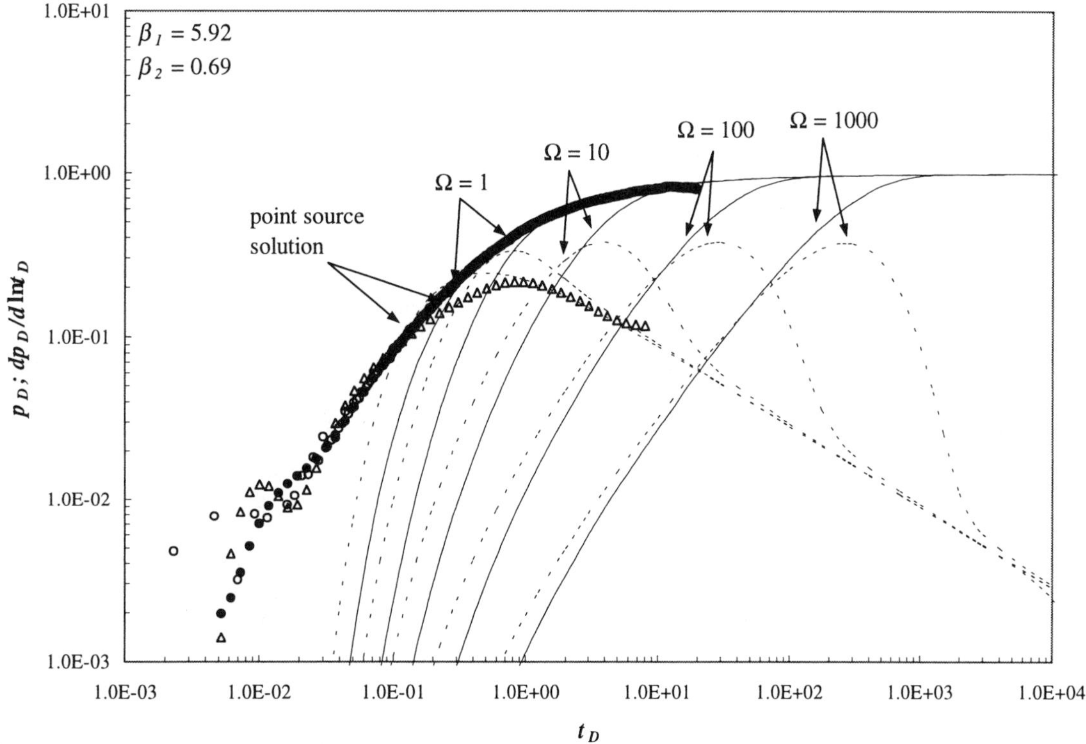

Figure 20. Type-curve match of pressure (open circles) and its derivatives (open triangles) from monitoring interval W2AL. Solid circles represent spline fitted to pressure data.

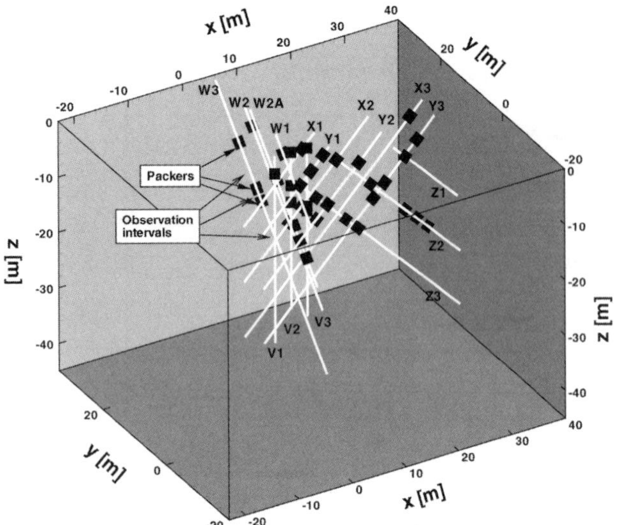

Figure 21. Three-dimensional perspective of site and computational region.

ous interpretation of multiple cross-hole tests amounts to "pneumatic tomography" of unsaturated fractured rock, a concept akin to that of "hydraulic tomography," which was originally proposed for saturated fractured rocks by *Neuman* [1987]. Here we discuss application of our numerical inverse procedure to a single cross-hole test, PP4. An earlier analysis, described by *Illman et al.* [1998] and *Vesselinov et al.* [1999], considers pressure data from each monitoring interval separately (while treating pneumatic permeability and air-filled porosity as if they were uniform across the site) and in a manner similar to that of our type-curve analysis. The results from these inverse and type-curve analyses were comparable.

Our numerical analysis is based on the assumption of single-phase airflow through a nonuniform, locally isotropic porous continuum. It accounts directly for the ability of all boreholes, and packed-off borehole intervals, to store and conduct air through the system. The model does so by treating these as high-permeability and high-porosity cylinders of finite length and radius. It solves the airflow equations in nonlinear form and is able to account for atmospheric pressure fluctuations at the soil surface. The model accounts geostatistically for spatial variations in air permeability and air-filled porosity of the fractured rock by treating them as random fields. It provides kriged estimates of these random fields by using an inverse method based on pilot points as proposed by de Marsily [*Zimmerman et al.*, 1998].

Our numerical model thus accounts more fully and accurately for nonlinear pressure propagation and storage through a nonuniform rock intersected by boreholes than do our type curves, which rely on linearized airflow equations, account for borehole storage only in one monitoring interval at a time, do not account for the effect of open boreholes on pressure distribution through the system, treat the rock as being uniform, and therefore cannot be fitted simultaneously to pressure data from multiple monitoring intervals when the rock is nonuniform.

There is little information in the literature about the effect that open borehole intervals may have on pressure propagation and response during interference tests. *Paillet* [1993] noted that the drilling of an additional observation borehole had an effect on drawdowns created by an aquifer test. We found through numerical simulations [*Illman et al.,* 1998] that the presence of open borehole intervals has a considerable impact on pressure propagation through the site, and on pressure responses within monitoring boreholes intervals, during cross-hole air-injection tests.

Our decision to analyze cross-hole tests at the ALRS by means of the FEHM code was based in part on its ability to simulate two-phase flow of air and water in dual porosity and/or dual permeability continua, and to account for discrete fractures, should the need to do so arise. To date, we found it unnecessary to activate these features of the code, as our data show neither dual continuum nor discrete fracture effects. We use FEHM in conjunction with a grid generator, X3D [*Trease et al.*, 1996], which automatically subdivides a three-dimensional domain into tetrahedral elements. We have supplemented these codes with a series of pre- and postprocessors to facilitate the handling, analysis, and visualization of massive input and output data files.

Our computational grid measures 63 m in the x direction, 54 m in the y direction, and 45 m in the z direction, encompassing a rock volume of 153,090 m^3 (Figure 21). Figures 22 and 23 are two-dimensional representations of the grid that illustrate cross-hole test PP4, during which injection takes place into a packed-off interval in borehole Y2. Figure 22 shows three views of the grid perpendicular to the x-y, x-z, and y-z planes. Because the grid in the vicinity of the boreholes is relatively fine, the corresponding areas appear dark in the figures. Figure 23 shows four cross-sectional views of the grid along vertical planes that contain selected boreholes. Since the grid is three-dimensional, its intersections with these planes do not necessarily occur along nodal points (i.e., what may appear as nodes in the figure need not be such).

The grid is divided into three parts: a regular grid at the center of the modeled area, which measures 30 × 20 × 25 (15,000) m^3 and has a node spacing of 1 m; a surrounding regular grid having a node spacing of 3 m; and a much finer and more complex unstructured grid surrounding each

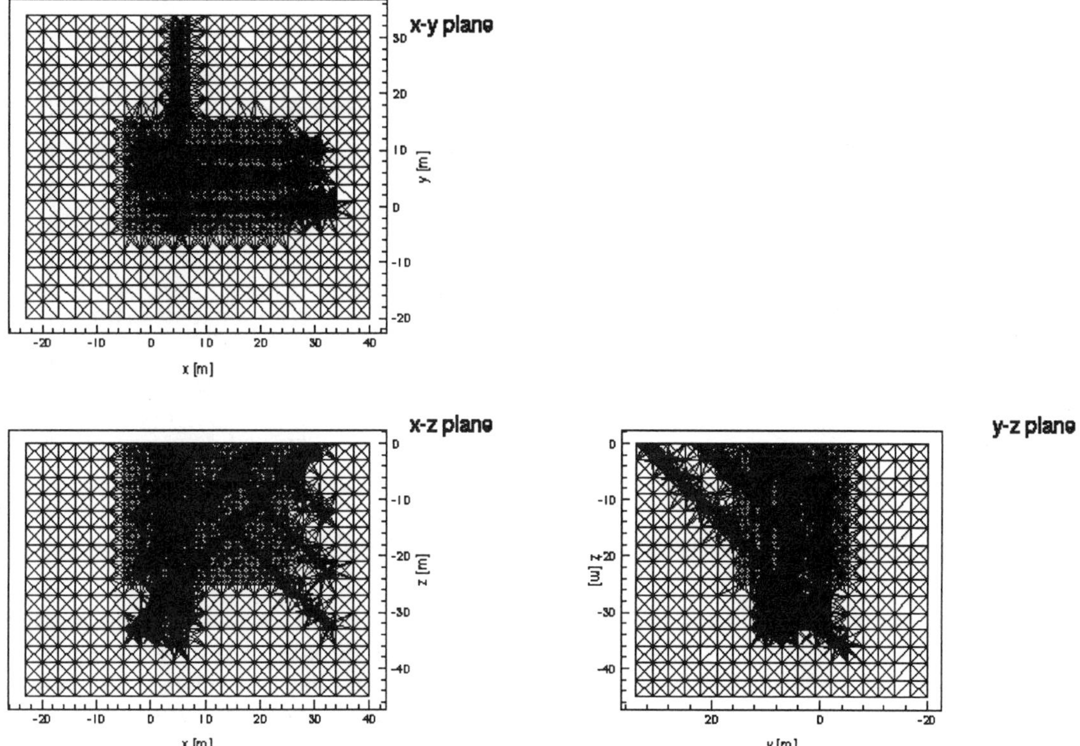

Figure 22. Side views of computational grid.

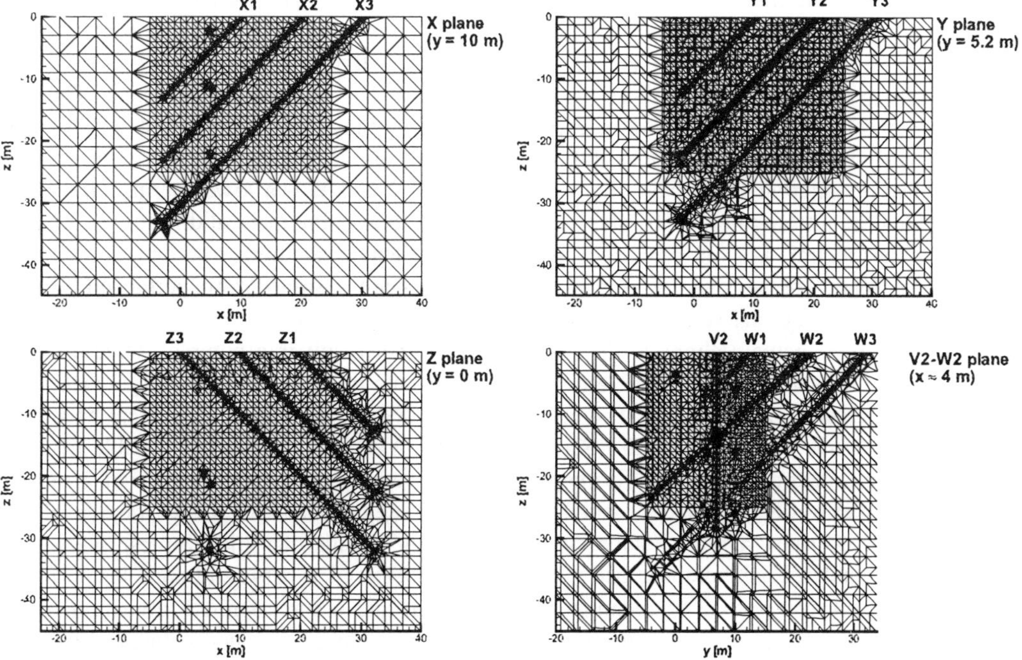

Figure 23. Vertical cross sections through computational grid.

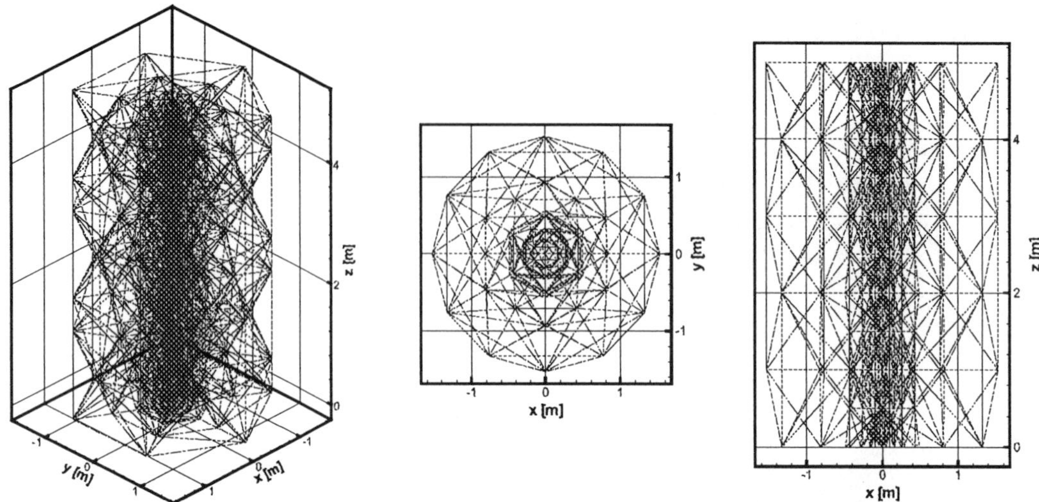

Figure 24. Three-dimensional representation of computational grid along the injection borehole.

borehole. The grid includes a total of 39,264 nodes and 228,035 tetrahedral elements.

FEHM is a node-based finite volume code in which parameters are defined at nodes, not at elements. A nodal parameter is viewed as an average over a control volume associated with the node. This volume is defined using three-dimensional Voronoï diagrams based on Delaunay tessellation [*Watson*, 1981]. Numerical calculations are based on finite difference expressions, which represent local mass balance over each such finite volume.

A 5 m long fragment of the three-dimensional computational grid associated with the injection borehole is shown in Figure 24. This grid is wider and finer than those associated with monitoring boreholes so as to allow accurate resolution of the relatively high pressure gradients that develop around the injection interval. However, the grids around both injection and monitoring boreholes have similar structures. Along the central axis of each borehole, nodes are spaced 0.5 m apart. The surrounding grid is designed so that the sum of computational volumes associated with these nodes is close to the actual volume of the borehole. Additional nodes are located along radii that are perpendicular to this axis within each borehole grid. The intervals between these nodes grow sequentially with distance from the borehole axis by a factor of 1.6, thus forming a geometric series. The number of rays and nodes associated with the injection borehole is larger than the number associated with monitoring boreholes. Each borehole grid is additionally refined near the soil surface so as to obtain an accurate resolution of conditions near this atmospheric boundary. Where boreholes are located close to each other (as in the cases of V2 and the W-series of boreholes; W2A and W2; W1 and Y1; W3 and Y3), the grid between them is made finer in order to resolve correctly processes that take place within this grid volume. The most complicated of the grid structures is that representing the region between boreholes V2 and W2, which intersect each other (Figure 23).

To simulate the effect of open borehole intervals on pressure propagation, we treat these intervals as porous media having much higher permeability and porosity than the surrounding rock. The permeability and porosity of nodes along open borehole intervals are set to 3.23×10^{-4} m^2 and 1.0, respectively. These correspond to a cylinder of radius equal to that of a typical borehole. The permeability and porosity of instrumented borehole intervals are set to 3.23×10^{-5} m^2 and 0.5, respectively. At the intersection of boreholes V2 and W2, a lower permeability of 10^{-10} m^2 is assigned to avoid numerical difficulties. This value is still orders of magnitude higher than that of the surrounding rock. Packers are assigned zero permeability and a porosity of 10^{-5}.

The net result is a complex three-dimensional grid that represents quite accurately the geometry, flow properties, and storage capabilities of vertical and inclined boreholes in the ALRS study area; is capable of resolving medium heterogeneity on a support scale of 1 m across the site; is able to represent, with a high degree of resolution, steep gradients around the injection-test interval, as well as pressure interference between boreholes no matter how closely spaced; and assures smooth transition between fine borehole grids having radial structures and surrounding coarser grids having regular structures.

As we consider only single-phase airflow the saturation

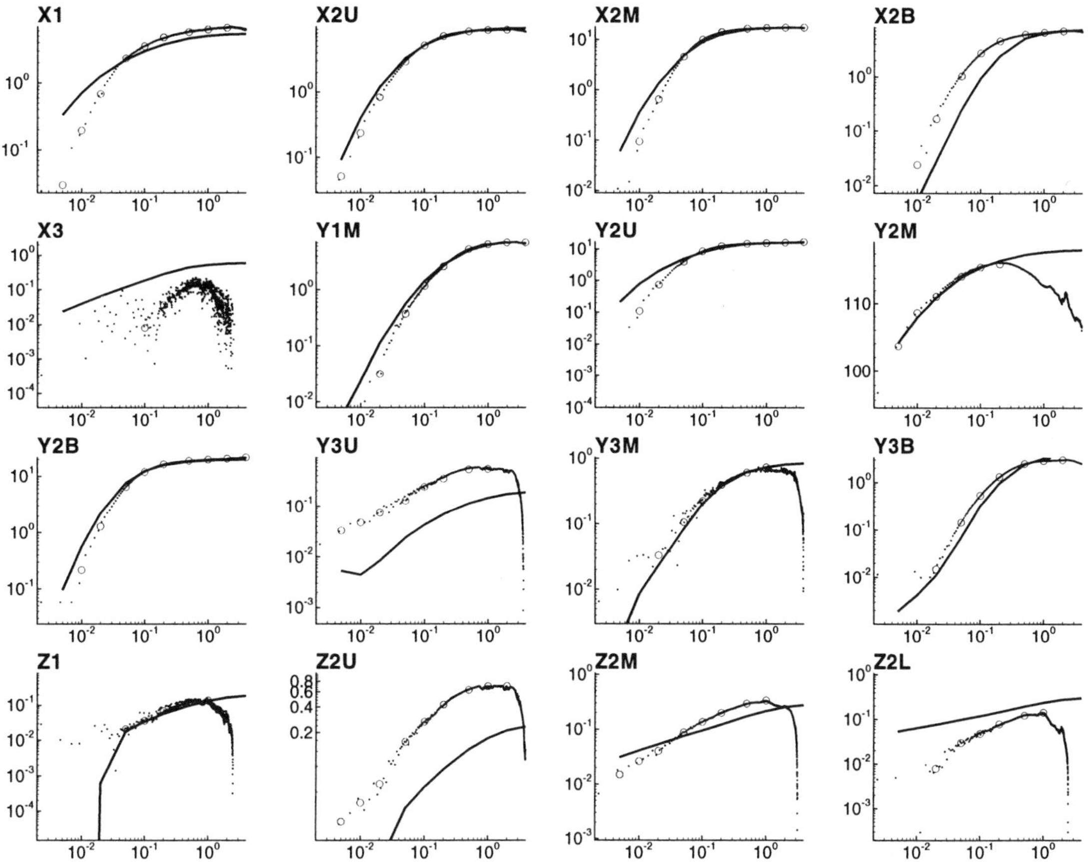

Figure 25a. Simultaneous numerical inverse matches between simulated (solid) and observed (dots) pressure data (*kPa*, vertical axes) versus time (*days*, horizontal axes) in all intervals during cross-hole test PP4. Match points are indicated by open circles. (X1, X2U, X2M, X2B, X3, Y1M, Y2U, Y2M, Y2B, Y3U, Y3M, Y3B, Z1, Z2U, Z2M, Z2L.)

of air, and associated pneumatic properties of the rock, remain constant during each simulation. The only initial condition we need to specify is air pressure, which we set equal to 0.1 MPa, the average barometric pressure at sea level. The mean barometric pressure at the ALRS, 4000 ft (1219.21 m) above mean sea level, is closer to 87 kPa. Our choice thus leads to a slight upward bias in calculated air-filled porosities. However, this is easy to correct *a posteriori*. The side and bottom boundaries of the flow model are impermeable to airflow. Our results suggest that these boundaries have been placed sufficiently far from injection intervals to have virtually no effect on simulated cross-hole tests. The top boundary coincides with the soil surface and is maintained at a constant and uniform pressure of 0.1 MPa. Although barometric pressure fluctuated during each cross-hole test, these fluctuations were small and are ignored in our analysis.

To interpret single-hole test data, we employed the optimization code PEST. The latter automatically estimates the pneumatic parameters of the rock and the open borehole intervals by minimizing a weighted square difference, $\Phi(\beta)$, between calculated and observed pressures,

$$\Phi(\beta) = \sum_{i=1}^{N} w_i \left[p_i^* - p_i(\beta) \right]^2, \qquad (1)$$

where β is a vector of parameters to be adjusted; N is the number of pressure match points in time; w_i is a relative weight ascribed to data point i; p_i^* is the corresponding observed pressure value; and $p_i(\beta)$ is its value calculated with parameters β. To analyze the tests, we set all weights w_i in (1) equal to unity and use match points that are more or less evenly distributed along a log-transformed time axis. The total number of match points, associated with one injection interval and 31 monitoring intervals, is $N = 256$.

PEST uses a variant of the Levenberg-Marquardt [*Marquardt*, 1963] algorithm to estimate the parameters β,

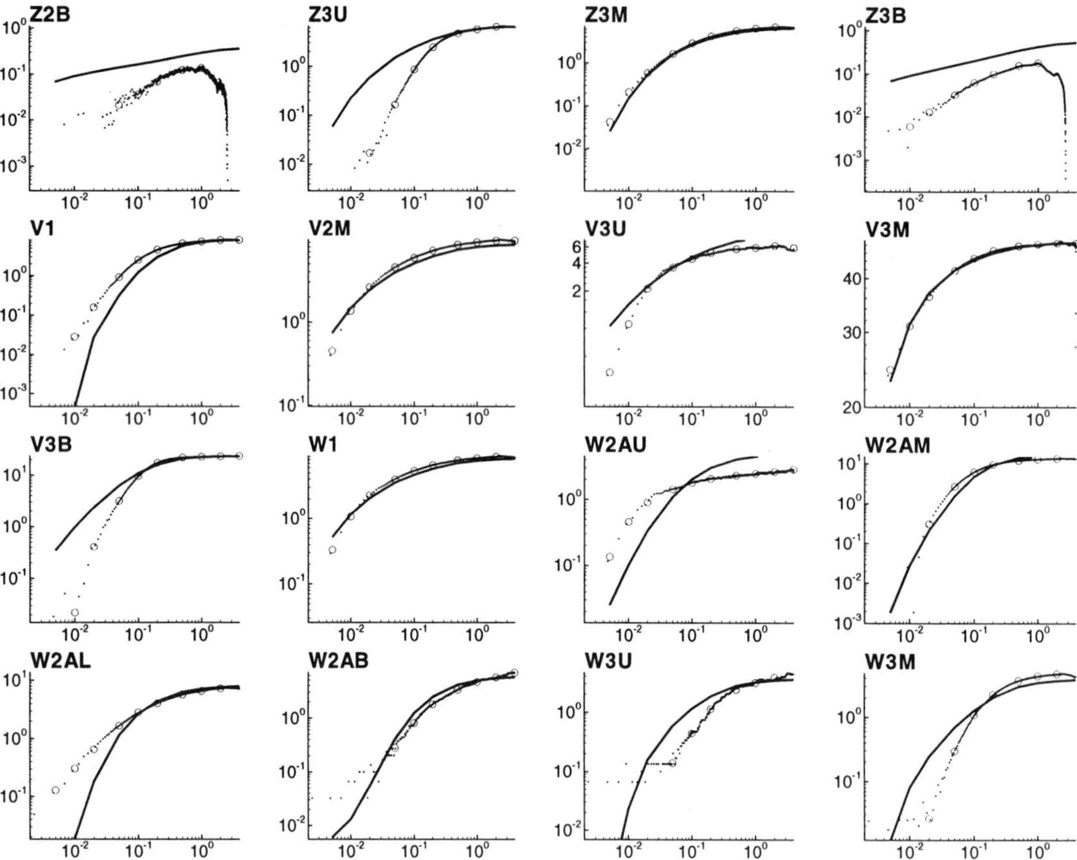

Figure 25b. Simultaneous numerical inverse matches between simulated (solid) and observed (dots) pressure data (*kPa*, vertical axes) versus time (*days*, horizontal axes) in all intervals during cross-hole test PP4. Match points are indicated by open circles. (Z2B, Z3U, Z3M, Z3B, V1, V2M, V3U, V3M, V3B, W1, W2AU, W2AM, W2AL, W2AB, W3U, W3M.)

which minimize $\Phi(\beta)$. PEST also calculates corresponding estimation covariance and correlation coefficient matrices, their normalized eigenvectors and eigenvalues, and 95% confidence limits for the optimized parameters. These are based on the assumptions that (a) the data are mutually uncorrelated, (b) their weights represent their inverse normalized variances, (c) the estimation errors are Gaussian, and (d) they can be assessed with the aid of a linear model. In our case, these assumptions are not expected to be fulfilled and we therefore consider confidence limits calculated by PEST merely as crude approximations.

In our inverse analysis, we associate one pilot point with each of 32 injection and monitoring test intervals. Most pilot points are located at the centers of the corresponding intervals. In the guard intervals Y2-1 and Y2-3, the pilot points are offset from the center toward the injection interval Y2-2, and in the long interval V1, toward the ground surface. The vector β consists of 66 unknown parameters: log-transformed air-permeability and air-filled porosity values at the 32 pilot points, and the exponents of corresponding power variograms. The values of air permeability and air-filled porosity at the pilot points are initially set close to the geometric average of each, as obtained from our earlier inverse interpretations of PP4 pressure data, corresponding to individual monitoring intervals [*Illman et al.*, 1998; *Vesselinov et al.*, 1999]. Changing these initial values has little effect on the final results. The exponent of each power variogram is initially set equal to 0.4475, the value obtained earlier through geostatistical analysis of single-hole air-permeability data.

The inverse analysis was conducted in parallel on all 32 processors of the University of Arizona SGI Origin 2000 supercomputer. We modified the parallel version of PEST [*Doherty*, 1997] so as to optimally utilize our UNIX multiprocessor system, and to allow restarting the numerical inverse computation (if and when it terminates prematurely) so as to minimize loss of computational time. The analysis required about 5000 forward model runs and took approximately 50 hours.

Figures 25a and b show simultaneous matches between computed and recorded pressures in one injection (Y2-2) and 31 monitoring borehole intervals during cross-hole test PP4. Some of these matches are poor (due in part to barometric pressure effects, which cause pressure in some intervals to drop during the late part of the test), some are of intermediate quality, and some are good to excellent. Corresponding kriged estimates of log (to base 10) air permeability, obtained by means of our pilot-point inverse approach, are shown by a three-dimensional fence diagram in Figure 26. The overall pattern of spatial variability in this diagram is quite similar to that obtained earlier from the combined kriging of 1 m and 3 m scale single-hole air-permeability data in Figure 7. However, the range of values in Figure 26 is wider by several orders of magnitude than is the range in Figure 7. In particular the mean, variance, and coefficient of variation of the cross-hole estimates in Figure 26 are –15.49 (3.2×10^{-16} m^2), 2.51, and –0.102, respectively. Table 2 shows that this mean value is very similar to that of the single-hole estimates. On the other hand, it is much lower than the mean values we obtain from type-curve and numerical inverse analyses of individual PP4 pressure monitoring records, in which we treat the rock us an equivalent uniform medium.

Kriged estimates of air-filled porosity, obtained from the

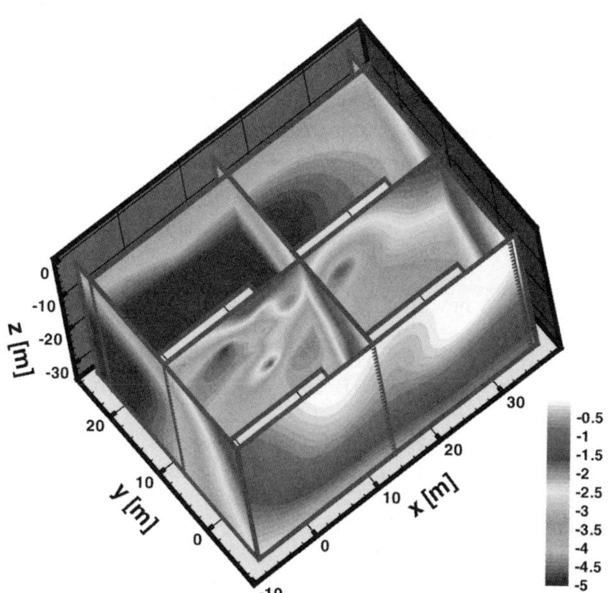

Figure 26. Three-dimensional representation of kriged log-transformed air permeability obtained by simultaneous numerical inversion of cross-hole test PP4 data.

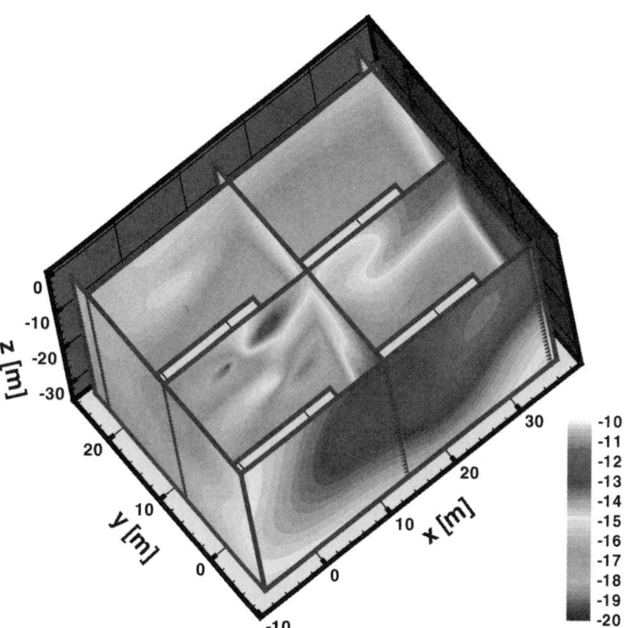

Figure 27. Three-dimensional representation of kriged log-transformed air-filled porosity obtained by simultaneous numerical inversion of cross-hole test PP4 data.

cross-hole test by means of our pilot-point inverse approach, are shown by a three-dimensional fence diagram in Figure 27. The overall pattern of spatial variability in this diagram is quite similar to that of air-permeability in Figure 26. The mean, variance, and coefficient of variation of $\log_{10}$ transformed air-filled porosity estimates are –3.14 (7.24×10^{-4}), 2.16, and –0.47, respectively. Log-transformed air-filled porosities obtained from type-curve and numerical inverse analyses of individual PP4 pressure monitoring records, in which we treat the rock us an equivalent uniform medium, have much higher mean values (Table 3).

It thus appears that treating the rock as uniform when it is nonuniform yields higher mean estimates of air permeability and air-filled porosity than are obtained when one treats the rock as being nonuniform.

5. CONCLUSIONS

The following major conclusions follow from this paper:

It is possible to interpret cross-hole pneumatic injection tests at the ALRS by means of analytically derived type curves and a numerical inverse model, which treat the rock as a continuum and air as a single mobile fluid phase. The rock continuum represents primarily interconnected air-filled fractures.

Pressure and pressure-derivative type-curve analyses of cross-hole test PP4 have provided a simple, fast, and reliable means of assessing pneumatic connections, directional

Table 3. Summary statistics for $\log_{10} \phi$ [m^3/m^3] identified for cross-hole test PP4 data using different methods.

Method	Mean	Variance	CV
Type curves	−2.05	0.65	−0.13
Uniform inverse	−1.57	0.13	−0.23
Nonuniform inverse	−3.14	2.16	−0.47

air permeabilities, and air-filled porosities on scales ranging from a few meters to over 20 m at the ALRS. Disadvantages of the type-curve approach include reliance on linearized airflow equations; inability to account for borehole storage in more than one monitoring interval at a time; inability to consider the effect of open boreholes on pressure distribution through the system; treatment of the rock as being uniform; and inability to simultaneously analyze pressure data from multiple monitoring intervals when the rock is nonuniform.

Our numerical inverse model is based on the assumption of single-phase airflow through a nonuniform, locally isotropic porous continuum. The model accounts directly for the ability of all boreholes and packed-off borehole intervals to store and conduct air through the system; solves the airflow equations in their original nonlinear form; can account for atmospheric pressure fluctuations at the soil surface; provides kriged estimates of spatial variations in air permeability and air-filled porosity throughout the tested fractured rock volume; and can be applied simultaneously to pressure data from multiple borehole intervals (as well as to multiple cross-hole tests, which amounts to "pneumatic tomography" of the rock). It has the disadvantage of being complex, difficult, and time consuming.

One-meter and three-meter scale air-permeability data from single-hole air-injection tests, as well as fracture density data, are amenable to geostatistical analysis, which considers them as samples from random fields defined over a continuum. Though this continuum is representative primarily of interconnected fractures, there is no correlation between air permeabilities and fracture densities at the site.

Air permeability at the ALRS is well characterized by a power variogram, which is representative of a random fractal field with multiple scales of spatial correlation.

Inverse estimates of air permeability from cross-hole test PP4 exhibit a spatial pattern that is very similar to that shown by kriged estimates of the same quantity from 1 m and 3 m scale single-hole tests. The two sets of estimates have similar geometric mean values, but the inverse estimates vary over a much broader range than do the single-hole test results. There is good correlation between the spatial patterns of air permeability and air-filled porosity as obtained by the inverse procedure. All of these patterns differ markedly from the spatial pattern of fracture densities.

Analyzing all pressure data from cross-hole test PP4 simultaneously, while allowing air permeability and air-filled porosity to vary in space, gave lower mean estimates of these quantities than had been obtained when analyzing separately the pressure records of individual injection and monitoring intervals, while treating the rock as being uniform.

Acknowledgements. This work was supported by the U.S. Nuclear Regulatory Commission under contracts NRC-04-95-038 and NRC-04-97-056. We wish to acknowledge with gratitude the support, advice, and encouragement of our NRC Project Manager, Thomas J. Nicholson. Walter Illman was supported in part by a National Science Foundation Graduate Traineeship during 1994–1995, a University of Arizona Graduate College Fellowship during 1997–1998, the Horton Doctoral Research Grant from the American Geophysical Union during 1997–1998, and the John and Margaret Harshbarger Doctoral Fellowship from the Department of Hydrology and Water Resources at the University of Arizona during 1998–1999. Velimir (Monty) Vesselinov conducted part of his simulation and inverse modeling work during a summer internship with the Geoanalysis Group at Los Alamos National Laboratory in 1997. We are grateful to Dr. George A. Zyvoloski for his help in the implementation of FEHM, and to Dr. Carl W. Gable for his assistance in the use of the X3D code to generate the corresponding computational grid. All pneumatic cross-hole tests at the ALRS were conducted by Walter Illman, with technical assistance by Dick Thompson. Type-curve development and analyses were performed by Walter Illman, geostatistical analyses by Guoliang Chen and Velimir Vesselinov, and the development and implementation of our inverse model by Velimir Vesselinov. We are grateful to Edwin P. Weeks and Gary D. LeCain of the U.S. Geological Survey for their constructive review of our manuscript.

REFERENCES

Akaike, H., A new look at statistical model identification, *IEEE Trans. Automat. Control*, *AC-19*, 716–722, 1974.

Akaike, H., On entropy maximization principle, in *Application of Statistics*, edited by P. R. Krishnaiah, North-Holland, Amsterdam, 1977.

Bassett, R. L., S. P. Neuman, T. C. Rasmussen, A. Guzman, G. R. Davidson, and C. L. Lohrstorfer, *Validation Studies for Assessing Unsaturated Flow and Transport Through Fractured Rock*, NUREG/CR-6203, U.S. Nuclear Regulatory Commission, Washington, D.C., 1994.

Boardman C. R., and J. Skrove, Distribution in fracture permeability of a granitic rock mass following a contained nuclear explosion, *J. Petrol. Tech.*, 619–623, 1966.

Carrera, J., and S. P. Neuman, Estimation of aquifer parameters under transient and steady state conditions: 1. Maximum likelihood method incorporating prior information, *Water Resour. Res.*, *22*(2), 199–210, 1986a.

Carrera, J., and S. P. Neuman, Estimation of aquifer parameters under transient and steady state conditions: 2. Uniqueness, stability, and solution algorithm, *Water Resour. Res.*, *22*(2), 212–227, 1986b.

Chen G, S. P. Neuman, and P. J. Wierenga, 4. Infiltration tests in fractured porous tuffs at the ALRS, in *Data Collection and Field Experiments at the Apache Leap Research Site, May 1995–1996*, edited by R. L. Bassett, S. P. Neuman, P. J. Wierenga, G. Chen, G. R. Davidson, E. L. Hardin, W. A. Illman, M. T. Murrell, D. M Stephens, M. J. Thomasson, D. L. Thompson, E. G. Woodhouse, NUREG/CR-6497, U.S. Nuclear Regulatory Commission, Washington, D.C., 1997.

Chuang, Y., W. R. Haldeman, T. C. Rasmussen, and D. D. Evans, *Laboratory Analysis of Fluid Flow and Solute Transport through a Variably Saturated Fracture Embedded in Porous Tuff*, NUREG/CR-5482, U.S. Nuclear Regulatory Commission, Washington, D.C., 1990.

Doherty, J., L. Brebber, and P. Whyte, *PEST: Model Independent Parameter Estimation*, Watermark Numerical Computing, Brisbane, Australia, 1994.

Doherty, J., *Parallel PEST*, Watermark Numerical Computing, Brisbane, Australia, 1997.

Evans, D. D., *Unsaturated Flow and Transport Through Fractured Rock—Related to High-Level Waste Repositories, NUREG/CR-3206*, U.S. Nuclear Regulatory Commission, Washington, D.C., 1983.

Evans, D. D., and T. C. Rasmussen, *Unsaturated Flow and Transport Through Fractured Rock Related to High-Level Waste Repositories, Final Report—Phase III*, NUREG/CR-5581, U.S. Nuclear Regulatory Commission, Washington, D.C., 1991.

Gomez-Hernandez, J. J., and E. F. Cassiraga, The theory of practice of sequential simulation, in *Geostatistical Simulations*, edited by M. Armstrong and P. A. Dowd, 111–124, Kluwer Academic, Norwell, Mass., 1994.

Green, R. T., and D. D. Evans, *Radionuclide Transport as Vapor through Unsaturated Fractured Rock*, NUREG/CR-4654, U.S. Nuclear Regulatory Commission, Washington, D.C., 1987.

Guzman, A. G., and S. P. Neuman, Field air injection experiments, in *Apache Leap Tuff INTRAVAL Experiments: Results and Lessons Learned*, edited by T. C. Rasmussen, S. C. Rhodes, A. Guzman, and S. P. Neuman, pp. 52–94 NUREG/CR-6096, U.S. Nuclear Regulatory Commission, Washington, D.C., 1996.

Guzman, A. G., S. P. Neuman, C. Lohrstorfer, and R. Bassett, Chapter 4, in *Validation Studies for Assessing Unsaturated Flow and Transport Through Fractured Rock*, edited by R. L. Bassett, S. P. Neuman, T. C. Rasmussen, A. G. Guzman, G. R. Davidson, and C. L. Lohrstorfer, pp. 4-1–4-58, NUREG/CR-6203, U.S. Nuclear Regulatory Commission, Washington, D.C., 1994.

Guzman, A. G., A. M. Geddis, M. J. Henrich, C. Lohrstorfer, and S. P. Neuman, *Summary of Air Permeability Data from Single-Hole Injection Tests in Unsaturated Fractured Tuffs at the Apache Leap Research Site: Results of Steady-State Test Interpretation*, NUREG/CR-6360, U.S. Nuclear Regulatory Commission, Washington, D.C., 1996.

Hannan, E. S., The estimation of the order of an ARMA process, *Ann. Stat.*, *8*, 197–181, 1980.

Hsieh, P. A., and S. P. Neuman, Field determination of the three-dimensional hydraulic conductivity tensor of anisotropic media, 1. Theory, *Water Resour. Res.*, *21*(11), 1655–1665, 1985.

Huang, C., and D. D. Evans, *A 3-Dimensional Computer model to Simulate Fluid Flow and Contaminant Transport Through a Rock Fracture System*, NUREG/CR-4042, U.S. Nuclear Regulatory Commission, Washington, D.C., 1985.

Illman W. A., Ph.D. Dissertation, the University of Arizona, Tucson, Ariz., 1999.

Illman W. A., D. L. Thompson, V. V. Vesselinov, G. Chen, and S. P. Neuman, *Single- and Cross-Hole Pneumatic Tests in Unsaturated Fractured Tuffs at the Apache Leap Research Site: Phenomenology, Spatial Variability, Connectivity, and Scale*, NUREG/CR-5559, U.S. Nuclear Regulatory Commission, Washington, D.C., 1998.

Kashyap, R. L., Optimal choice of AR and MA parts in autoregressive moving average models, *IEEE Trans. Pattern Anal. Mach. Intel. PAMI*, *4*, 99–104, 1982.

Kirkham, D., Field method for determination of air permeability of soil in its undisturbed state, *Soil Sci. Soc. Am. Proc.*, *11*, 93–99, 1947.

Klinkenberg, L. J., The permeability of porous media to liquids and gases, *Am. Petrol. Inst., Drilling and Production Practice*, 200–213, 1941.

LeCain, G. D., *Air-Injection Testing in Vertical Boreholes in Welded and Nonwelded Tuff, Yucca Mountain, Nevada*, USGS Milestone Report 3GUP610M, U.S. Geological Survey, Denver, Colo., 1996a.

LeCain, G. D., and J. N. Walker, Results of air-permeability testing in a vertical borehole at Yucca Mountain, Nevada, *Radioactive Waste Management*, 2782–2788, 1996b.

Marquardt, D. W., An algorithm for least-squares estimation of nonlinear parameters, *J. SIAM*, *11*, 431–441, 1963.

Mishra, S., G. S. Bodvarsson, and M. P. Attanayake, Injection and falloff test analysis to estimate properties of unsaturated fractured fractures, in *Flow and Transport Through Unsaturated Fractured Rock*, *42*, Geophysical Monograph, edited by D. D. Evans and T. J. Nicholson, American Geophysical Union, Washington, D.C., 1987.

Montazer, P. M., *Permeability of Unsaturated, Fractured metamorphic Rocks Near An Underground Opening*, Ph.D. dissertation, Colorado School of Mines, Golden, Colo., 1982.

Neuman, S. P., Stochastic continuum representation of fractured rock permeability as an alternative to the REV and fracture network concepts, *Proc. of the 28th U.S. Symposium, Rock Mechanics*, edited by I. W. Farmer, J. J. K. Dalmen, C. S. Desai, C. E. Glass and S. P. Neuman, Balkema, Rotterdam, pp. 533–561, 1987.

Neuman, S. P., and E. A. Jacobson, Analysis of nonintrinsic spatial variability by residual kriging with application to regional groundwater levels, *Math. Geology*, *16*, 491–521, 1984.

Paillet, F. L., Using borehole geophysics and cross-borehole flow testing to define connections between fracture zones in bedrock aquifers, *J. Appl. Geophys.*, *30*, 261–279, 1993.

Pebesma, E. J., and C. G. Wesseling, GSTAT: a program for

geostatistical modelling, prediction and simulation, *Computers & Geosciences*, *24*(1), 17–31, 1998.

Rasmussen, T. C., and D. D. Evans, *Unsaturated Flow and Transport Through Fractured Rock-Related to High-level Waste Repositories*, NUREG/CR-4655, U.S. Nuclear Regulatory Commission, Washington, D.C., 1987.

Rasmussen T. C., and D. D. Evans, *Fluid Flow and Solute Transport modeling Through Three-Dimensional Networks of Variably Saturated Discrete Fractures*, NUREG/CR-5239, U.S. Nuclear Regulatory Commission, Washington, D.C., 1989.

Rasmussen, T. C., and D. D. Evans, *Nonisothermal Hydrologic Transport Experimental Plan*, NUREG/CR-5880, U.S. Nuclear Regulatory Commission, Washington, D.C., 1992.

Rasmussen, T. C., D. D. Evans, P. J. Sheets, and J. H. Blanford, *Unsaturated Fractured Rock Characterization Methods and Data Sets at the Apache Leap Tuff Site*, NUREG/CR-5596, U.S. Nuclear Regulatory Commission, Washington, D.C., 1990.

Rasmussen, T. C., D. D. Evans, P. J. Sheets, and J. H. Blanford, Permeability of Apache Leap Tuff: borehole and core measurements using water and air, *Water Resour. Res.*, *29*(7), 1997–2006, 1993.

Rasmussen T. C., S. C. Rhodes, A. Guzman, and S. P. Neuman, *Apache Leap Tuff INTRAVAL Experiments: Results and Lessons Learned*, NUREG/CR-6096, U.S. Nuclear Regulatory Commission, Washington, D.C., 1996.

Samper, F. J., and S. P. Neuman, Estimation of spatial structures by adjoint state maximum likelihood cross validation: 1. Theory, *Water Resour. Res.*, *25*(3), 351–362, 1989a.

Samper, F. J., and S. P. Neuman, Estimation of spatial structures by adjoint state maximum likelihood cross validation: 2. Synthetic experiments, *Water Resour. Res.*, *25*(3), 363–372, 1989b.

Schrauf, T. W., and D. D. Evans, *Relationship Between the Gas Conductivity and Geometry of a Natural Fracture*, NUREG/CR-3680, U.S. Nuclear Regulatory Commission, Washington, D.C., 1984.

Tidwell, V. C., T. C. Rasmussen, and D. D. Evans, Saturated hydraulic conductivity estimates for fractured rocks in the unsaturated zone, in *Proc. of International Conference and Workshop on the Validation of Flow and Transport models for the Unsaturated Zone*, edited by P. J. Wierenga, New Mexico State University, Las Cruces, N.M., 1988.

Trease, H. E., D. George, ,C. W. Gable, J. Fowler, A. Kuprat, and A. Khamyaseh, The X3D grid generation system, in *Numerical Grid Generation in Computational Fluid Dynamics and Related Fields*, edited by B. K. Soni, J. F. Thompson, H. Hausser, and P. R. Eiseman, Engineering Research Center, Mississippi State Univ. Press, 1996.

Tsang, Y. W., C. F. Tsang, F. V. Hale, and B. Dverstorp, Tracer transport in a stochastic continuum model of fractured media, *Water Resour. Res.*, *32*(10), 3077–3092, 1996.

Vesselinov, V. V., and S. P. Neuman, Numerical inverse interpretation of multistep transient single-hole pneumatic tests in unsaturated fractured tuffs at the Apache Leap Research Site, in *Theory, Modeling and Field Investigation in Hydrogeology: A Special Volume in honor of Shlomo P. Neuman's 60th birthday*, Geological Society of America, Boulder, Colo., in press.

Vesselinov, V. V., Neuman, S. P., Illman W. A., Three-dimensional inverse modeling of pneumatic tests in unsaturated fractured rocks, in *ModelCare'99*, in press.

Vickers, B. C., S. P. Neuman, M. J. Sully, and D. D. Evans, Reconstruction geostatistical analysis of multiscale fracture apertures in a large block of welded tuff, *Geoph. Res. Let.*, *19*(10), 1029–1032, 1992.

Wang, J. S. Y., R. C. Trautz, P. J. Cook, and R. Salve, *Drift Seepage Test and Niche moisture Study: Phase 1 Report on Flux Threshold Determination, Air Permeability Distribution, and Water Potential Measurements*, Level 4 Milestone Report SPC315M4, Lawrence Berkeley National Laboratory, Berkeley, Calif., 1998.

Watson, D., Computing the n-dimensional Delaunay tessellation with application to Voronoï polytopes, *The Computer Journal*, 167–172, 1981.

Weber, D. S., and D. D. Evans, *Stable Isotopes of Authigenic minerals in Variably-Saturated Fractured Tuff*, NUREG/CR-5255, U.S. Nuclear Regulatory Commission, Washington, D.C., 1988.

Yeh, T. C., T. C. Rasmussen, and D. D. Evans, *Simulation of Liquid and Vapor Movement in Unsaturated Fractured Rock at the Apache Leap Tuff Site—Models and Strategies*, NUREG/CR-5097, 1988.

Zimmerman, D. M., G. de Marsily, C. A. Gotway, M. G. Marietta, C. L. Axness, R. L. Beauheim, R. L. Bras, J. Carrera, G. Dagan, P. B. Davies, D. P. Gallegos, A. Galli, J. Gómez-Hernández, P. Grindrod, A. L. Gutjahr, P. K. Kitanidis, A. M. Lavenue, D. McLaughlin, S. P. Neuman, B. S. RamaRao, C. Ravenne, and Y. Rubin, A comparison of seven geostatistically based inverse approaches to estimate transmissivities for modeling advective transport by groundwater flow, *Water Resour. Res.*, *34*(6), 1373–1413, 1998.

Zyvoloski, G. A., Z. V. Dash, and S. Kelkar, *FEHM: Finite Element Heat and Mass Transfer*, Tech. Rep. LA-11224-*MS*, Los Alamos National Laboratory, N.M., 1988.

Zyvoloski, G. A., B. A. Robinson, Z. V. Dash, and L. L. Trease, *Users Manual for the FEHMN Application*, Tech. Rep. LA-UR-94-3788, Los Alamos National Laboratory, N.M., 1996.

Zyvoloski, G. A., B. A. Robinson, Z. V. Dash, and L. L. Trease, *Summary of the Models and Methods for the FEHM Application—A Finite-Element Heat- and Mass-Transfer Code*, Tech. Rep. *LA-13307-MS*, Los Alamos National Laboratory, Los Alamos, N.M., 1997.

Guoliang Chen, Broken Hill Proprietary, Florence, AZ 85232; Walter A. Illman, CNWRA, Southwest Research Institute, San Antonio, TX 78238-5166; Dick L. Thompson, City of Tucson, Tucson Water Department, Tucson, AZ; Velimir V. Vesselinov, and Shlomo P. Neuman, University of Arizona, Tucson, AZ 85721.

Estimation of the Heterogeneity of Fracture Permeability by Simultaneous Modeling of Multiple Air-Injection Tests in Partially Saturated Fractured Tuff

Y. W. Tsang, K. Huang, and G. S. Bodvarsson

Earth Sciences Division, Ernest Orlando Lawrence Berkeley National Laboratory, Berkeley, California

Air-injection tests were used to investigate the flow characteristics of the fractured volcanic tuffs at Yucca Mountain, Nevada, the potential site for a high-level nuclear waste repository. Because the tuff matrix pores are saturated over 90% with water and the matrix permeability is on the order of microdarcies, the air component of flow is mainly in the fractures. Air-injection tests can therefore help to determine the flow characteristics and heterogeneity structure of the densely fractured welded tuff. The tests were carried out in the Exploratory Studies Facility, an 8 km long underground tunnel at the Yucca Mountain site, in twelve 40 m long boreholes, forming three clusters within a cubic rock volume of approximately 40 meters on each edge. Each borehole in the test block was packed off (or isolated) into four sections (or zones) by inflatable packers. The in situ field tests consisted of constant-rate air injection into one of the isolated borehole zones while the pressure response was monitored in all the isolated zones. The pressure data showed an almost universal response in all monitored zones to injection into any borehole -zone, indicating that the fractures are well connected for airflow. Air-injection tests were performed in succession for all isolated zones. A simultaneous inversion was performed for the pressure response of all the monitoring zones for all the injection tests in the test block. TOUGH2, a 3D numerical code for multiphase, multicomponent transport, was used for this purpose. Spatially variable fracture permeability was used as an adjustable parameter to fit the simulated pressure responses to those measured, assuming fixed fracture porosity. For most of the pneumatic experiments, the calculated pressure changes matched the data well, and the estimated permeability ranged over four orders of magnitude, from 10^{-15} m^2 to 10^{-11} m^2.

1. INTRODUCTION

The United States Office of Civilian Radioactive Waste Management/Department of Energy is investigating a site at Yucca Mountain, Nevada, to determine whether it is a suitable location for a deep-mined geological repository [*U.S. DOE*, 1988]. The potential repository would be located in fractured welded volcanic tuff, approximately 300 meters (m) below the surface and about the same distance above the water table. The strategy for waste isolation at Yucca Mountain consists of reliance on both a robust engineered barrier system and the natural system (natural barrier). Characterization of the natural system includes both surface-based and underground testing programs. As part of the underground investigations, there are two in situ ther-

Dynamics of Fluids in Fractured Rock
Geophysical Monograph 122

Published 2000 by the American Geophysical Union

mal tests: the Single Heater Test and the Drift Scale Test carried out in a side alcove off the main drift of the Exploratory Studies Facility (ESF), an underground 8 km long tunnel at the Yucca Mountain site. The objective of the thermal tests is to gain a more in-depth understanding of the coupled thermal, hydrological, mechanical, and chemical processes likely to occur in the repository rock mass in response to decay heat from the high-level nuclear waste. Figure 1 shows a plan view of the thermal test facility, with the Single Heater Test rock block to the south and the Drift Scale Test block to the north of the Observation Drift. The Single Heater Test, already completed, was located in a cubic rock block of approximately 13 meters on each edge surrounded by drifts on three sides. The Drift Scale Test, still in progress, centers on a 47.5 m long, 5 m diameter Heated Drift lying in the east-west direction parallel to the Observation Drift. The thermal test blocks reside in the middle nonlithophysal unit of the Topopah Spring welded tuff, the geological unit of the potential repository.

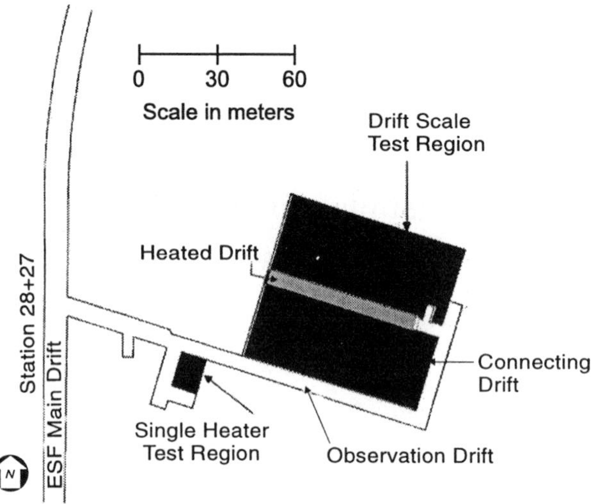

Figure 1. Thermal Test Facility layout, relative position of the Single Heater Test and Drift Scale Test rock blocks.

Prior to heat activation, crosshole air-injection tests were performed to characterize the hydrological characteristics of fractures within the thermal test blocks, since these characteristics will control the thermally driven moisture movement in the heater tests. Gas-phase rather than liquid-phase testing is the tool of choice for characterizing fluid flow in fractures. Gas-phase testing is preferred because the welded tuff at Yucca Mountain has very low matrix permeability, on the order of microdarcies, with matrix liquid saturation on the order of 90% [*Peters et al.,* 1984; *Flint*, 1998]. The fractures, on the other hand, are believed to be almost completely air filled and far more permeable. Fracture maps of ESF walls show that the fracture lengths range from less than a meter to several meters, and that on every scale, the fracture spacing is smaller than the fracture length. For instance, for fractures with length greater than one meter, the fracture spacing is less than 0.5 m [*Sonnenthal et al.,* 1997]. That the fracture length is larger than that of the average fracture spacing suggests good connectivity within the fracture system (e.g., *Long et al.,* 1982; *Dverstorp,* 1991). Given these conditions, air-injection and gas-tracer tests are appropriate field methods for characterizing fracture flow and transport characteristics. Gas-tracer test results in the thermal test blocks will be discussed in a separate manuscript under preparation. This paper focuses on the crosshole air-injection tests.

The Single Heater Test crosshole air-injection test analysis was reported in *Huang et al.* [1999]. Here, we focus on analysis of the preheating air-injection tests performed for the Drift Scale Test. The crosshole air-injection test data were analyzed by a simultaneous inversion of the pressure responses to multiple injection tests in the entire test block. This simultaneous-inversion approach contrasts with the more traditional type-curve fitting of single-hole or crosshole pressure data [e.g., *LeCain*, 1995; *Illman et al.*, 1998]. Instead, we performed simultaneous analysis of multiple injection tests. The densely spaced fractures were conceptualized as a heterogeneous continuum [*Neuman,* 1987; *Schwartz and Smith*, 1988; *Tsang et al.,* 1996]. The test block was divided into different zones to which different fracture-continuum permeability values were assigned. Air-permeability values of these zones were treated as fitting parameters in the simultaneous inversion of all crosshole pressure responses to multiple test-block injection tests. Numerical simulations were performed for these air-injection tests, using the multiphase, multicomponent code TOUGH2 [*Pruess*, 1991; *Pruess et al.*, 1996].

Rather than implementing the automatic inversion capability of TOUGH2 with its companion software ITOUGH2 [*Finsterle,* 1999], we chose a trial-and-error approach (in conjunction with forward modeling) to simultaneously match the pressure data for all the air-injection tests in each thermal-test rock block. This approach is an alternative to quantitative optimization schemes that rely on automatic adjustments of parameter values to reduce the sum of weighted variance of physical states to a given criterion [*Neuman*, 1973; *Yeh*, 1986; *Sun*, 1994; *Finsterle and Persoff*, 1997]. It relies on the subjective judgement of the researcher and is more flexible than an automatic optimization scheme in that it allows incorporation of soft information. Soft information such as fracture maps and borehole videos, when coupled with crosshole pressure data, can help to shed light on the flow connectivity and permeability characteristics of the test block. Such soft information is useful for the problem at hand: although we conceptualize the fractures as a continuum, flow is in fact channeled through preferred paths of least fluid resistance.

In a 12 m long borehole within the Single Heater Test block, air-permeability values measured in short testing intervals show that flow occurred in selective discrete features intersecting only a small fraction of the borehole zone into which air was injected [*Tsang and Birkholzer*, 1999; *Tsang et al.*, 1999a]. Sixteen 0.69 m long intervals were tested; permeability values ranged from less than 1×10^{-15} m^2 to 6.2×10^{-13} m^2, with only three zones having higher permeability. These results contrast with the permeability derived from air injection into the entire borehole length of 11 m, which yields a permeability of 5.1×10^{-14} m^2. The permeability variation clearly indicates that airflow occurs only in a fraction of the borehole length. This suggests that a trial-and-error approach requiring active input from the investigator affords more efficient incorporation of the soft information than an automatic optimization scheme. Granted, important statistical information routinely generated from automatic optimization schemes would be missed in our trial and error approach. Still, the solution to an inverse problem is intrinsically nonunique, and we are relying on the large amount of crosshole pressure data that must be simultaneously matched to constrain the inverted spatially variable fracture-continuum permeability.

2. AIR-INJECTION TESTS AND PRESSURE-RESPONSE DATA

Forty-six air-injection tests were carried out in twelve 40 m boreholes in the Drift Scale Test. All twelve boreholes originate from the Observation Drift, forming three vertical fans, each bracketing the Heated Drift. Configurations of these 12 boreholes are shown in 3D perspective in Figure 2. The first cluster of five boreholes 57 through 61 was located near the east end of the Heated Drift. The five boreholes 74 through 78 were constructed about 20 m west of the first cluster, intersecting the Heated Drift about mid-drift. The vertical fan of boreholes 185 and 186 lay near the west end of the Heated Drift, 15 m from the borehole cluster 74 through 78. Each vertical fan of boreholes spans about 40 m along the vertical direction. High-temperature inflatable packers were installed in each borehole to subdivide it into typically four (~ 8–10 m long) zones. Zone 1 is closest to the borehole collar at the Observation Drift, and zone 4 is closest to the bottom of the borehole. For boreholes 58 and 77, only three packers were installed because of intense fracturing and an abundance of lithophysal cavities along the borehole walls.

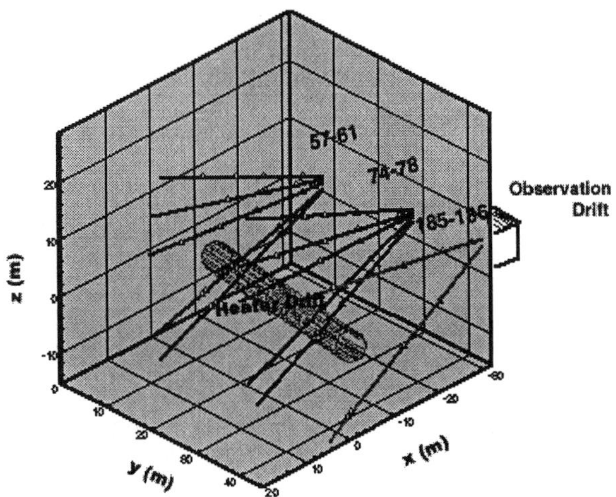

Figure 2. Three-dimensional view of borehole layouts and locations of air-injection and pressure measurements at the Drift Scale Test site, Yucca Mountain.

Constant flow-rate injection was conducted in these borehole zones isolated by the packers. A typical test consisted of injecting air at a constant flow rate into one chosen borehole zone, for the duration of about one hour, while the pressure responses in the same borehole zone and all other isolated zones were monitored. From November 4 through November 17, 1997, air-injection tests were conducted in all 46 isolated zones. After the air injection started, the pressure response in the injection hole typically reached steady state within minutes. Each injection test ran for three to five hours, consisting of initial background monitoring for a few minutes prior to air injection and ending with a period of pressure recovery following the termination of injection. Pressure responses were recorded every five seconds during the early part of injection and recovery because of rapid pressure changes. The data recording interval was increased as the rate of pressure changes decreased. Software for the data acquisition system was programmed to ensure that data recording was continuous and automatic during the entire test period. The designed automatic control monitors the maximum change in pressure during air injection and adjusts the flow rate accordingly to keep the total change in pressure in the injection zone below 50 kPa, ensuring that non-ideal-gas behavior associated with air compressibility is negligible. For each test, a starting injection rate of 100 standard liters per minute (SLPM) was initiated. If the pressure increase in the injection zone exceeded 25 kPa, the injection flow rate would be programmed to reduce to 20 SLPM. If the pressure buildup were still higher than 25 kPa, the flow rate would be cut by a factor of 10, to 2 SLPM. On the other hand, if the rock surrounding the injection zone were so permeable that the pressure increase was less than 4 kPa, then the flow rate would be adjusted to double the initial rate (200 SLPM).

Table 1 summarizes the borehole geometry and the injection rates for all the injection tests. Since pressure responses in all boreholes were monitored, 46 sets of pressure data were generated for each injection test; hence, a total of 46 × 46 sets of pressure response data were obtained and analyzed.

Figures 3, 4, and 5 show the flow rates of injected air for the 46 injection tests (top figure a) and the measured pressure changes in all boreholes zones (bottom figure b) in the Drift Scale Test for borehole clusters 57–61, 74–78, and 185–186, respectively. The number adjacent to the peak of the pressure response designates the borehole zone into which air was injected. The nomenclature is borehole number followed by the packer-isolated zone, where zone 1 is closest to the collar of the borehole at the Observation Drift, between packers 1 and 2, and zone 4 is the last zone between the fourth packer and the borehole bottom. Note that in some of the tests, the injection rate rises and falls sharply at the initiation of the injection as a result of the designed automatic control system. Corresponding to the sharp rise and fall of the injection rate, pressure in the injection zone responded and recovered rapidly with near-zero time lag, thus displaying little borehole storage effect.

Figures 3, 4, and 5 show that in response to the constant flow-rate air injection, most of the pressure changes in the injection zones occurred almost instantaneously and very quickly reached steady state. The steady-state pressure changes were then maintained as the constant flow continued. However, in a few instances, air pressure quickly increased as the injection started, but decreased slightly before arriving at a constant-pressure steady state. A typical example of this transient process occurred in the injection tests at Zone 77-1 (Figure 4b), where in response to a constant injection flow rate of 100 SLPM, pressure increased by about 25 kPa immediately after the injection started, but gradually decreased to 21.5 kPa when the injection stopped. This kind of pressure response to constant-rate air injection has been frequently observed in pneumatics tests performed in unsaturated fractured tuffs, at the Apache Leap Research Site [e.g., *Guzman*, 1995; *Guzman et al.*, 1996] and at other locations in the ESF [*LeCain*, 1998]. *Guzman* [1995] attributed the phenomenon to the displacement of residual water in fractures by air during the injection test.

As expected, Figures 3b, 4b, and 5b show that pressure changes are largest in the injection boreholes, while their magnitude in the observation borehole zones is much smaller and mostly not discernible on the linear scale plots.

Table 1. Injection-zone length, injection rate, steady-state pressure increase, and estimated permeabilities for borehole blocks from simultaneous inversion.

Borehole Zone ID	Zone length, m	Injection rate, L/min	Pressure Change, kPa	Borehole Block Permeability, m^2 ($\times 10^{-12}$)
57-1	8.84	20	4.58	0.165
57-2	6.10	100	18.5	0.4789
57-3	7.62	2	39.9	0.00288
57-4	10.55	200	12.6	0.38
58-1	6.1	20	5.12	0.347
58-2	8.54	20	3.18	0.33
58-3	17.98	171	3.74	7.22
59-1	10.06	20	4.25	1.70
59-2	7.62	100	8.95	0.88
59-3	8.54	100	10.8	0.44
59-4	7.19	200	7.8	2.59
60-1	5.49	20	2.75	0.646
60-2	10.67	100	5.8	1.72
60-3	5.49	2	7.2	0.038
60-4	11.19	20	45.5	0.115
61-1	7.01	100	14.6	0.305
61-2	8.54	100	3.85	9.57
61-3	6.10	20	16.3	0.0638
61-4	12.63	100	26.9	0.235
74-1	10.67	100	10.6	0.665
74-2	6.71	20	12.9	0.089
74-3	4.27	20	8.04	0.207
74-4	14.09	100	17.3	0.23
75-1	8.23	100	11.3	0.55
75-2	7.32	100	23.7	0.26
75-3	10.67	100	17.3	0.259
75-4	8.48	100	4.68	2.0
76-1	7.93	100	13.1	0.34
76-2	8.54	20	5.27	0.16
76-3	8.54	20	9.89	0.068
76-4	10.00	20	6.82	0.0945
77-1	8.84	100	21.5	0.225
77-2	5.49	20	21.2	0.0287
77-3	22.73	100	3.83	4.65
78-1	6.10	20	4.4	0.076
78-2	8.23	20	14.3	0.055
78-3	5.79	20	16	0.0829
78-4	14.49	20	4	0.19
185-1	5.79	20	2.75	0.543
185-2	8.54	100	15.6	0.165
185-3	15.24	100	20.9	0.0395
185-4	6.65	20	4.18	0.062
186-1	5.79	20	2.47	0.175
186-2	8.54	20	22.1	0.033
186-3	13.11	20	51.9	0.00164
186-4	5.09	2	11.4	0.00265

A careful study of the data indicates that the magnitude of pressure change tends to decrease as the separation between the injection and monitored packed borehole zone increases, typical of continuum behavior. Because of the

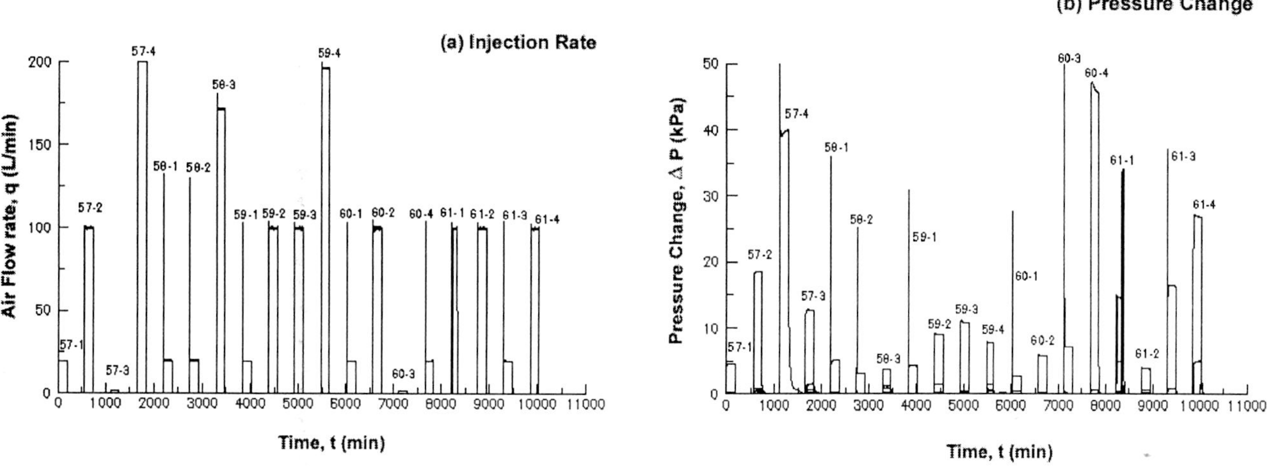

Figure 3. (a) Flow rates and (b) measured pressure changes for all air-injection tests at borehole cluster 57–61.

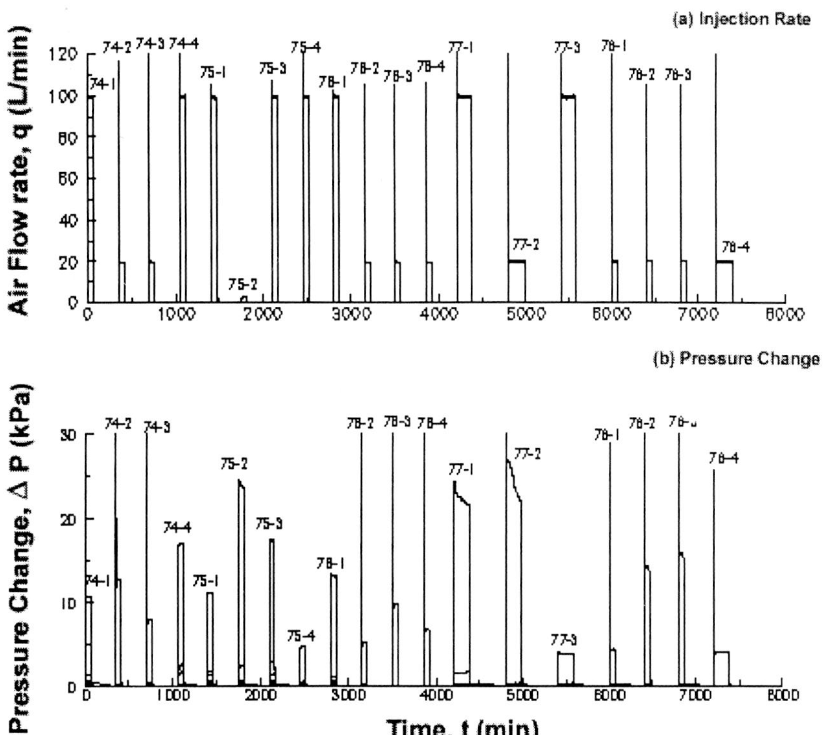

Figure 4. (a) Flow rates and (b) measured pressure changes for all air injection tests at borehole cluster 74–78.

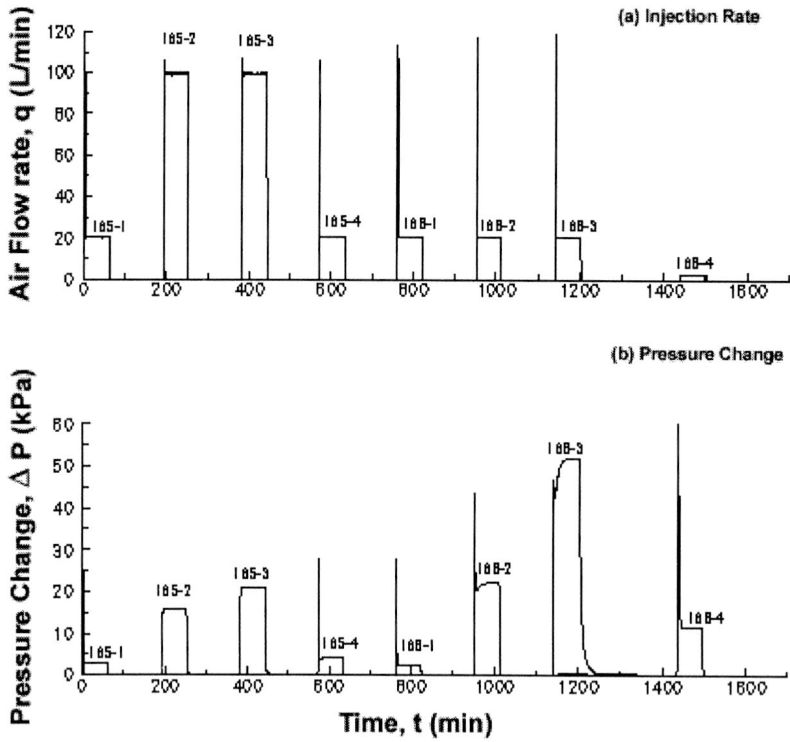

Figure 5. (a) Flow rates and (b) measured pressure changes for all air injection tests at borehole cluster 185–186.

large separation between neighboring clusters of boreholes in the Drift Scale Test (see Figure 2) pressure responses were extremely small, generally less than 0.2% of the injection pressure magnitude. Therefore, simultaneous inversions for the Drift Scale Test were performed independently for each of the three borehole clusters.

3. SIMULTANEOUS INVERSION OF THE PRESSURE DATA

3.1. Modeling Details—Discretization, Boundary and Initial Conditions

The model domain for simulating the pressure response to constant flow-rate air-injection tests for each vertical fan of boreholes is 55 m wide along the north-south (x), 70 m deep along the vertical (z), and 30 m thick east-west along the drift (y). Furthermore, symmetry allows the thickness of the block to be halved, thus reducing the model domain to $55 \times 15 \times 70$ m^3. Fitting between the simulated and observed pressure changes was made simultaneously for all borehole zones within each vertical fan. A constant ambient pressure and moisture boundary condition was imposed on the Observation Drift wall (about 5 m high), where the boreholes collared, and on the Heated Drift opening 30 m north of the Observation Drift within the simulated block. No-flow boundaries were applied everywhere else.

The 3D model domain was discretized into $47 \times 47 \times 68$ hexahedral elements. Nonuniform element spacing (generally ranging from 1.0 to 3.0 m) was used in each direction. A more refined discretization of 0.2 m was imposed in the vertical direction from $z = 0$ to 5 m where boreholes were clustered. Note that the origin (x = 0 and z = 0) of the domain was set at the center of the Heated Drift and y = 0 was located at the west end of the drift (see Figure 2). In addition, each 40 m long and 7.57 cm diameter borehole was explicitly discretized into several elements that were connected to the surrounding rock elements in which they were embedded. Near the collar of the boreholes at the Observation Drift, the boreholes are closely clustered, requiring finely discretized elements, thus giving rise to a relatively large ratio of borehole volume to that of the surrounding rock element. Our experience with simulations of the Single Heater Test block [*Huang et al.*, 1999] indicates that explicit borehole discretization is preferable when the ratio of borehole volume to the rock element volume in which it is embedded is not negligibly small. The borehole elements were assigned a large-enough permeability, 4.0×10^{-9} m^2, to ensure that the spatial pressure differential within each borehole injection zone is less than 0.01 kPa. Those borehole elements representing the impermeable packers were given a very low permeability value of 10^{-20} m^2.

To simulate the air-injection tests, we solved the isothermal problem by employing the version of the numerical code TOUGH2 that uses the EOS3 equation of state module (water, air and heat flow), as described in *Pruess et al.* [1996]. For the transient-gas-flow problem, we assumed that the matrix and fracture continua would behave as one effective continuum. that is, we anticipated that the gas flow between the matrix and fractures would occur quickly, so that the same-state variables such as gas pressure could be assumed to apply for both fracture and matrix continua. *Doughty* [1999] discussed in detail the importance of selecting a suitable conceptual representation of matrix and fracture continua when describing moisture, gas, and chemical transport in fractured-porous welded tuff. She indicated that an effective-continuum representation of the fractures and matrix is adequate for describing transient gas flow.

Prior to simulating the air-injection experiments, the initial equilibrium pressure and moisture distribution of the test block under gravity was obtained by running TOUGH2 until the system reached steady state for the assumed initial conditions. The temperature was set to 25°C, and air pressure at the center of the Heated Drift (z = 0 m) was set to 89 kPa, based on field measurements. Matrix liquid saturation, based on the laboratory data from grab and core samples of the experimental site [*Tsang et al.*, 1999a], was fixed at 0.92. The porosity of matrix cores for Yucca Mountain welded tuff (based on laboratory measurements) is about 0.11 [*Flint*, 1996]. Because no direct measurements for fracture porosity existed at the time of simulating the air-injection tests, we assumed an effective porosity of 0.01 for gas flow, which occurs predominantly in fractures. The effective porosity on the order of 0.01 was confirmed by gas-tracer data that we collected subsequently. This parameter value is not adjusted in the pressure-data analysis. Given the measured liquid saturation (0.92) for the matrix, the liquid saturation in the fractures (0.052) is dictated by the effective-continuum assumption of capillary pressure equilibrium between matrix and fracture [*Birkholzer and Tsang*, 1996].

3.2. Fracture Permeability Heterogeneity Structure

Assignment of the permeability heterogeneity structure is governed by the test configuration of the injection tests. Nineteen blocks, each representing the rock surrounding a borehole zone (isolated by two packers or by one packer and the borehole bottom) for the two boreholes clusters 57–61 and 74–78, were constructed for the simulation domain in the x-z plane (Figure 6). Each block was given different fracture-continuum permeability. We assumed no heterogeneity in the y-direction across the 15 m thickness because all boreholes in the same vertical fan share the same y-coordinate, and consequently the pressure data for air injection can give no information as to the heterogeneity structure in the y direction. The same 19 block heterogeneity structure was applied to the borehole cluster 185–186, even through this cluster has only eight borehole injection and observation zones. The 19 blocks covered the portion (about 40 x 40 m^2) of the simulation domain occupied by the boreholes. (They will be referred to as "local blocks.") In addition, the rock elements immediately surrounding a borehole zone were assigned permeability that was different from that of the local blocks. (They will be referred to as "borehole blocks.") The rationale for these borehole blocks is as follows: The 19 local blocks shown in Figure 6 are relatively large, having dimensions of 6 m to 10 m in length. The permeability assigned to these large blocks controls the borehole-to-borehole flow characteristics—that is, the crosshole pressure response of the monitoring zones. However, the pressure response in the borehole zone into which air is injected will be most sensitive to the permeability immediately surrounding the borehole. It is therefore appropriate in the simultaneous inversion scheme to be able to assign permeability (that may not have the same value as that of the corresponding local block) to a smaller borehole block in the immediate vicinity of each borehole zone.

Three large material blocks were assigned to rock outside of this (40 × 40 m^2) zone containing the boreholes. Although the three zones are much larger than the 19 local blocks around the boreholes, and although the scale of heterogeneity in these zones is expected to be no different from those around the borehole, each of these large zones was given a homogeneous permeability. We did not implement a small-scale heterogeneity structure for these blocks of background rock, because there were no pressure-response data from these zones. Also, these zones were far enough away from the test boreholes that their permeability variations were not expected to significantly impact the interference-pressure data in the boreholes. In other words, they play little role in constraining the calibration process of the inverse modeling.

3.3. Simultaneous Inversion by Numerical Modeling

Numerical simulations were performed with TOUGH2 for a given set of permeability values assigned to each of the local blocks and borehole blocks. There were 38 per-

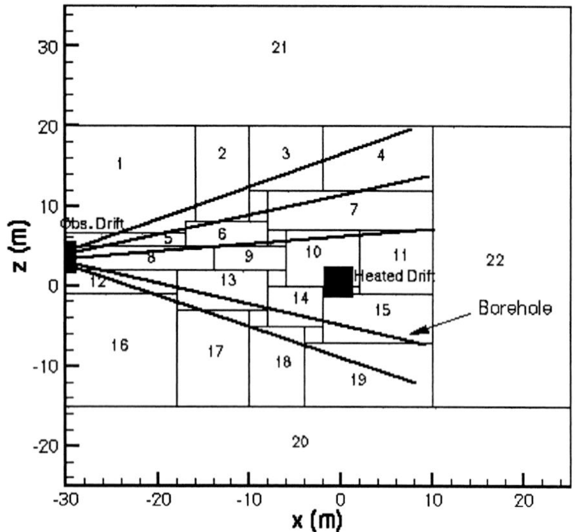

Figure 6. Schematic diagram of assigned heterogeneous material blocks for inverse modeling.

meability values in total (27 for borehole cluster 185–186) that could be adjusted for each simulation. Again, we emphasize that each simulation included all 19 injection tests in a borehole cluster (eight for borehole cluster 185–186). Our objective was to simultaneously fit the set of 19 × 19 (8 × 8 for boreholes cluster 185–186) simulated pressure responses from all borehole zones to the measured data. Our trial-and-error approach was to group the pressure responses for each injection test in descending order of magnitude, match the simulated results to the pressure response in the injection hole, and then successively fit those interference responses in order of decreasing magnitude. We also considered it important to reproduce with our simulations those data where pressure response was very small. Permeability values were adjusted by trial and error in a forward numerical model. This process proceeded until the simulated pressure changes in the injection borehole and most observation boreholes agreed well (by visual inspection) with the measured data.

The first-guess values of local permeability around each injection zone (for both the local blocks and the borehole blocks) were obtained from an analytical solution to a constant-flow-rate injection in a finite line source:

$$k = \frac{P_{SC} Q_{SC} \mu \ln\left(\frac{L}{r_w}\right) T_f}{\pi L (P_2^2 - P_1^2) T_{SC}}. \quad (1)$$

Equation (1) expresses the permeability around the injection zone in terms of the steady-state pressure response to injection. The notations are as follow:

Q_{SC}	steady-state injection flow rate at standard conditions, (m^3/s)
P_{SC}	pressure at standard conditions, 1.013×10^5 Pa
T_{SC}	temperature at standard conditions, 273.15 °K
T_f	temperature of the rock formation
P_1 and P_2	initial and final steady-state pressures
L	length of the air-injection zone
r_w	radius of the borehole
μ	dynamic viscosity of the air $(1.81 \times 10^{-5}$ Pa s)

An expression similar to Equation (1) has been used by *Rasmusson et al.* [1993] and *Guzman et al.* [1996] for the analysis of single-hole injection tests in fractured tuff at Apache Leap Research Site, Arizona. It was adapted from the steady-state analytical solution for ellipsoidal flow of incompressible fluid from a finite line source [*Hvorslev*, 1951] in an infinite medium, as shown by *LeCain* [1995]. The derivation of Equation (1) requires the assumption that air is the only mobile phase within the rock near the test interval and that it obeys the ideal gas law, so that the compressibility is inversely proportional to pressure.

4. RESULTS AND DISCUSSIONS

After intensive trial-and-error permeability adjustments for all material blocks ("local" and "borehole"), we arrived at a permeability set for which the calculated results closely approximate the observed data for all injection tests in boreholes in the same vertical zone. Because of space, not all the results can be presented here, so we shall focus on the simulated results for borehole cluster 57–61.

4.1. Borehole Cluster 57–61

Plate 1 shows the comparison of measured data (Plate 1a) with the simulated pressure changes in response to air injection (Plate 1b). The simulated (solid lines) and observed (dashed lines) pressure changes agree well for all the injection tests, for both transient time evolutions and steady-state magnitudes. Though TOUGH2 can successfully model the pressure response to the sharp rise and fall of the injection during the automatic flow adjustment at the initiation (first 10 to 20 s) of each injection test [*Huang et al.*, 1999], for simulations here we assumed the subsequent constant injection rate from the start. This assumption has no impact on the analysis presented here.

Whereas Plate 1 gives an overall view of the fitting results for all the tests, Figure 7 gives the simulated results for a few individual injection tests in more detail. Figure 7 shows the comparison of simulated pressure responses to

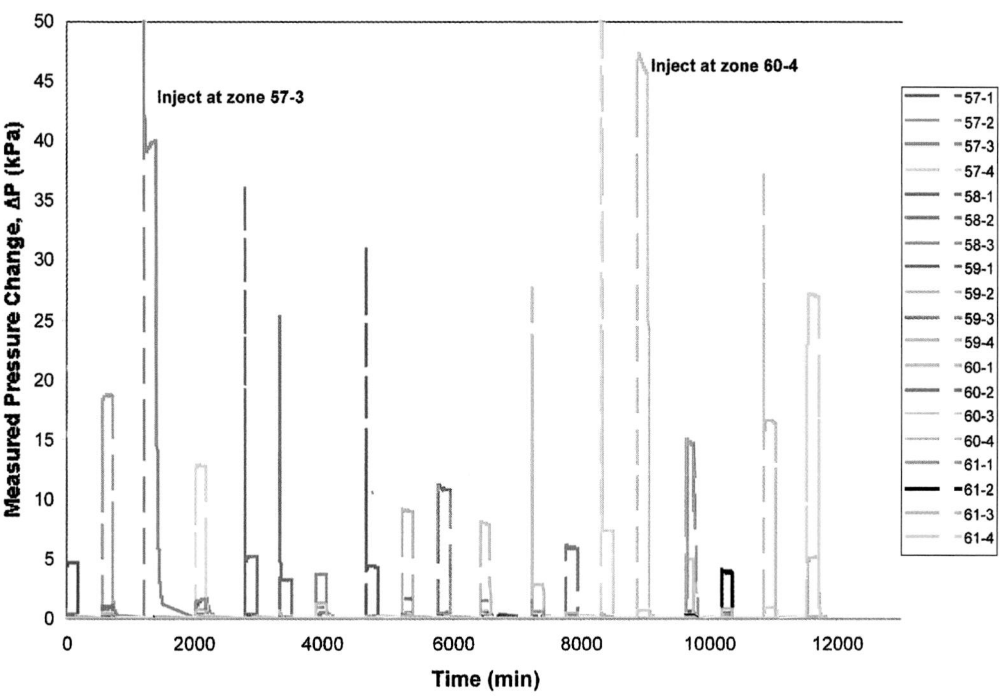

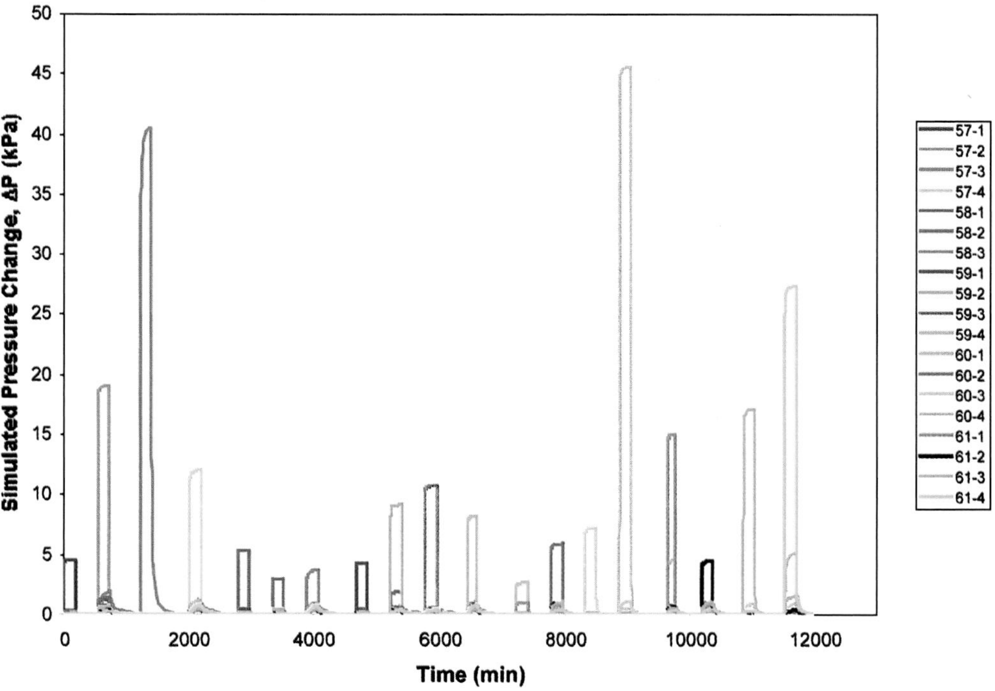

Plate 1. Measured data (a) and calculated pressure changes (b) for the air-injection tests at borehole cluster 57–61 (color).

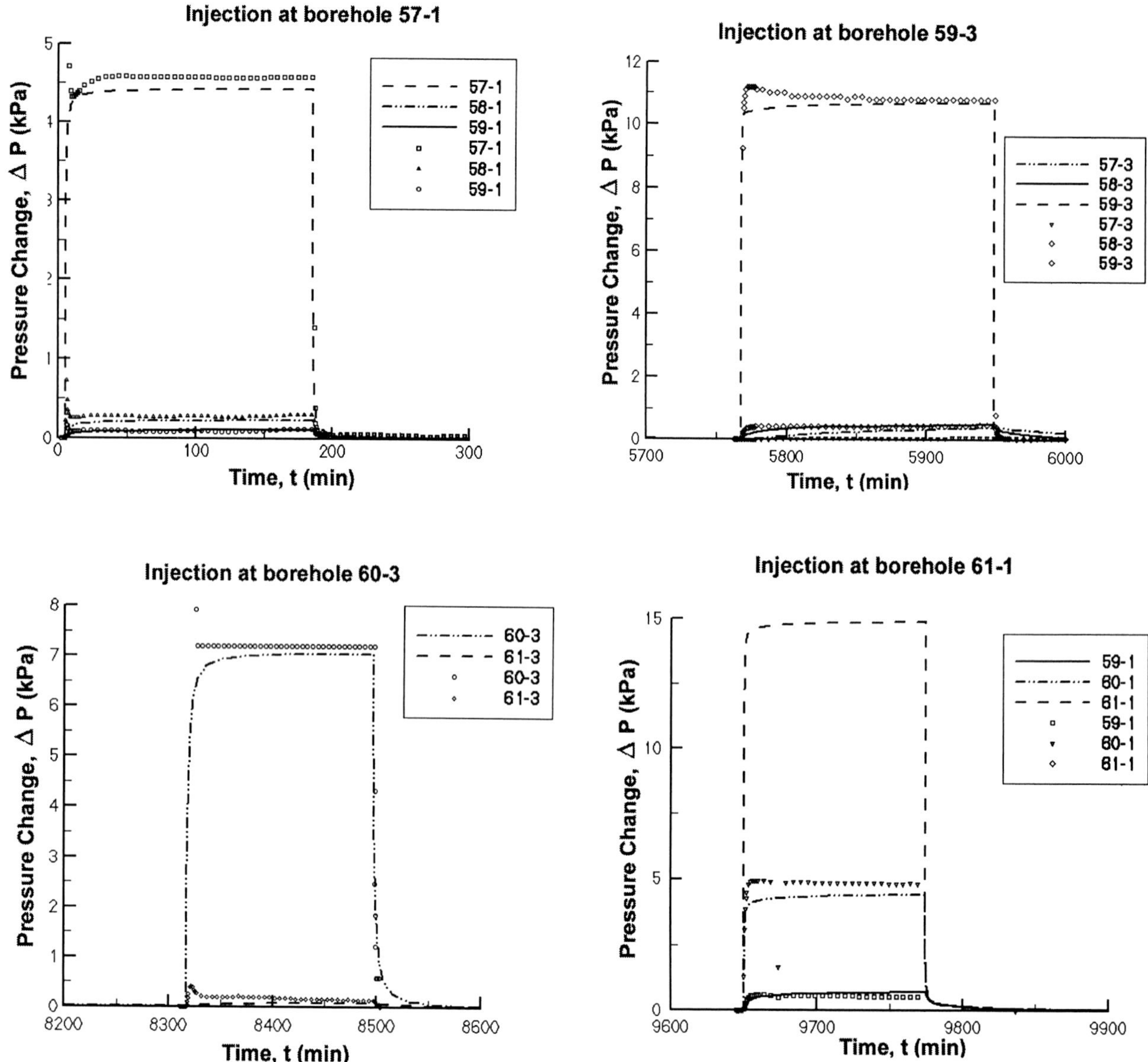

Figure 7. Comparison of simulated pressure changes (lines) with measured data (symbols) for selected injection tests at borehole cluster 57–61.

the observed data for injection experiments conducted at borehole zones 57-1, 59-3, 60-3 and 61-1. To avoid over-cluttering, only the pressure data for the injection borehole and a few of the 18 observation packed zones with appreciable pressure changes are shown. For injection at zone 57-1 (Figure 7a), most of the observation zones registered very small pressure increases. Even for those zones that are closest to it, 58-1 and 59-1 (at distances of 0.4 and 1.5 m from 57-1 respectively), the steady-state pressure changes are less than 0.25 kPa or less than 1/10 of the pressure rise in the injection zone itself. On the other hand, the steady-state pressure rise in the injection zone is not particularly large: 4.5 kPa at an injection rate of 20 L/min, indicating that permeability immediately surrounding the injection zone 57-1 is reasonable large. The small crosshole pressure responses in 58-1 and 59-1 therefore indicate that pneumatic connection between these zones and 57-1 is poor, i.e., few permeable fractures (running in the vertical direction) connect them (note that we are using a continuum approximation for flow through the fractures). This implies that the permeability for injection zone 57-1 must arise from horizontal rather than vertical fractures. A small pressure buildup in the injection zone (~ 3–5 kPa) was also observed from injection tests conducted at zones 58-1, 59-1, and 60-1 (see Figure 3). This is consistent with the hypothesis that the presence of horizontal (and not vertical) fractures around zones 57-1, 58-1, 59-1, and 60-1 accounts for their relatively large permeability.

Pressure response to injection in zone 61-1 (Figure 7d) is strikingly different from that in zones 57-1 (Figure 7a), 58-1, 59-1 and 60-1. First, the higher pressure rise of ~ 15 kPa in the injection zone originated from a higher injection rate of 100 L/min for this test. The injection in zone 61-1 also induced a large pressure change in zone 60-1. As noted earlier, the pressure response in the injection zones is mainly governed by the permeability assigned to the 19 "borehole blocks" immediately surrounding the borehole, while the 19 larger local blocks controlled the flow from injection zones to observation zones. In general, simulation is carried out by assigning initial permeability values for these blocks and iteratively adjusting the permeabilities until the simulated results for all the injection borehole zones match the measured data and, simultaneously, the agreement between simulations and data for all the observation zones is fairly good.

This approach worked for most of the tests. However, for the injection tests at zone 61-1, we were unable to obtain satisfactory pressure responses for both injection borehole zones and observation zones, even after many iterations and changes in permeabilities. The failure to accomplish a good fitting for both injection and observation zones indicates that the pre-assigned heterogeneity structure (Figure 6) near these borehole zones may not be correct. The large pressure response in zone 60-1 to injection in 61-1 indicates the presence of a strong pneumatic connection. This phenomenon may be accommodated by assigning a higher permeability value (e.g., 10^{-11} m^2) to some discretization elements between zones 60-1 and 61-1 so as to model a fast flow channel connecting these two zones. Such a fast flow channel was incorporated, which accounts for the good agreement between simulated and measured pressure response in 60-1 to injection in 61-1 (as shown in Figure 7d). This also highlights the fact that discrete-fracture effects still do surface in a small fraction of the crosshole pressure data.

Another example of heterogeneity being on a scale smaller than that of the pre-assigned heterogeneity structure of ~10 m (Figure 6) was evident in the pressure data pertaining to injection in 60-2. Although the pressure response in the injection zone indicates that the zone is rather permeable, the small pressure response from the adjacent zones (separated from the injection zone by inflatable packers) indicates minimal fracture connections between them. We could reproduce the pressure data in both the injection and monitoring neighboring zones in the same borehole by further refinement of the heterogeneity structure. This was done by dividing the borehole block into three different material zones and assigning lower permeability values to the discretization elements at the connecting end of the neighboring zones immediately adjacent to the impermeable packers.

The need to further refine the heterogeneity structure—by addition of fracture elements (e.g., injection in 61-1 in Figure 7d) and subdivision of the existing borehole block (e.g., injection in 60-2) to match the crosshole pressure response—underlines the inherent inadequacy of the continuum conceptualization for discrete fractures. The original assignment of heterogeneous permeability structure (Figure 6) is based on the configuration of the air-injection tests; that is, the length scale of the heterogeneity structure is of the same order as the length of the borehole zones isolated by impermeable packers. This structure is consistent with the continuum approximation that each packed-off borehole zone is identified by its own permeability, implying that flow is distributed uniformly over the length of the borehole zone. In fact, even in a densely spaced fracture network, flow would seek the least resistive paths and take place in only a small fraction of the continuum space. These discrete-fracture phenomena manifest themselves in the pressure response data as discussed above, and we were able to incorporate the effect of discrete fractures in our continuum conceptual model by further refinement (in an ad hoc manner) of the pre-assigned heterogeneity structure. For example, we showed that the addition of a high-permeability fracture channel connecting 60-1 and 61-1 brought the simulated pressure changes into agreement with data, irrespective of the exact spatial location of the channel. In other words, the data indicate only the presence of a fracture channel; they cannot provide constraint on its location. It is obvious that if air-injection tests were performed on a much smaller scale (that is, if the boreholes were packed off in submeter rather than 8–10 m intervals), then the interference- pressure data could surely provide constraint on the spatial location of fracture-flow connection. On the other hand, we would also expect the data on that much smaller scale to invalidate the continuum approximation; we would expect nonzero crosshole pressure response from only selected monitored zones [*Cook*, 1999], in contrast to the universal response of the Drift Scale Test data.

4.2. Borehole Clusters 74–78 and 185–186

Similar level of agreement between simulations and data as shown above were obtained for the vertical fans con-

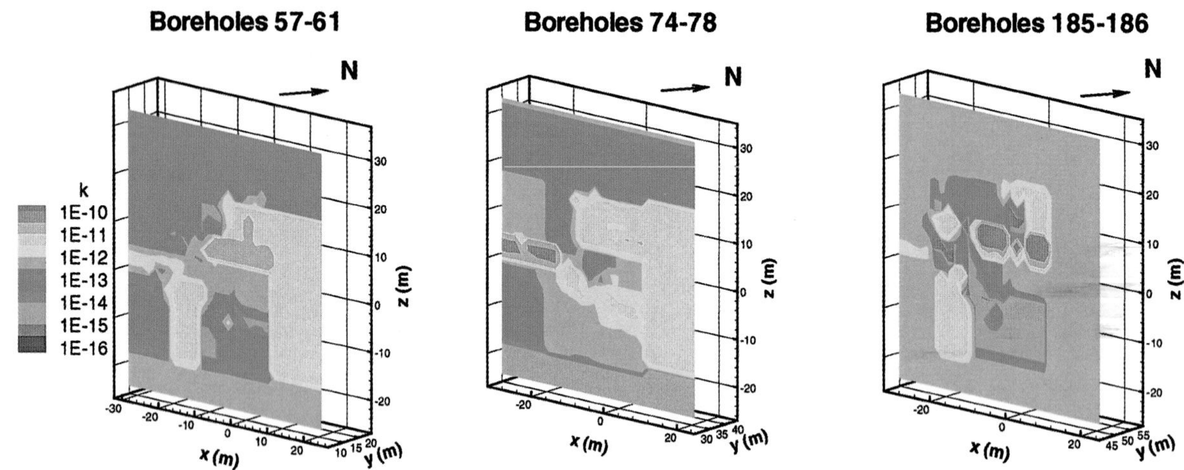

Plate 2. Three-dimensional view of the estimated permeability distribution for the test block (color).

taining boreholes 74–78 and 185–186. The results will not be shown here because of paper length constraints.

4.3. Heterogeneity Structure

The simulations presented above were done through a trial-and-error process, involving an adjustment of permeabilities of the local zone blocks and the borehole zone blocks until we obtained a "visual best fit." That is, we aimed to arrive at a set of permeability values for which the simulated pressure changes at all the injection borehole zones and most of the observation boreholes agreed with the measured data, simultaneously, for all the tests in a cluster. The permeability values finally used to obtain the best-fit results were assumed to be the estimated permeability distribution for the simulated rock block. The best-fitted permeabilities for the borehole blocks of all borehole zones are listed in Table 1, and those for the local material blocks are in Table 2. These two tables show that the local permeability varies from one borehole zone to another, spanning a range from 10^{-15} m^2 to 10^{-11} m^2. Permeability for most rock blocks is, however, on the order of 10^{-13} m^2, comparable to the values estimated for the nearby Single Heater Test block [*Huang at al.*, 1999]. The permeability values are also consistent with those used for the predictive and interpretative modeling of the Single Heater Test [*Birkholzer and Tsang,* 1996; *Tsang and Birkholzer,* 1999] and the Drift Scale Test [*Birkholzer and Tsang*, 1997; *Tsang et al.*, 1999b].

A detailed 3D view of the estimated permeability structure is shown in Plate 2. It shows the permeability in three vertical cross zones, each representing the simulated permeability distribution in respective planes of the three borehole clusters. The reader is referred to Figure 6 and reminded that the borehole clusters occupy the rock volume from x = –30 to 10 m and z = –15 to 20 m. Also, the permeability of local blocks 20, 21, and 22 would be insensitive to (and therefore not constrained by) the injection test data because of their distance from the cluster of boreholes. For this reason, the reader should not assign much significance to the apparent high permeability of local block 22 extending from x = 10 to 15 m for borehole clusters 57–61 and 74–78. However, the spatial heterogeneity in the rock volume containing the cluster of boreholes is a result of the simultaneous inversion of all the pressure response data. Plate 2 shows that permeability in the upper right part of the test block (toward the bottom of the boreholes and above the heater horizon) is generally high, ranging from 10^{-12} to 10^{-11} m^2. The rock around the Observation Drift (x = –30 to –24 m, z = –3 to 2 m) also exhibits a relatively high permeability, as illustrated from the plots for borehole clusters 57–61 and 74–78. These general trends of high permeability are consistent with the intensity and frequency of "open" fractures as observed from borehole videos.

Table 2. Estimated permeabilities ($\times 10^{-12}$ m^2) for all local blocks (see Figure 6) from simultaneous inversion.

Local block ID	Cluster 57–61	Cluster 74–78	Cluster 185–186
1	0.159	0.465	0.47
2	0.166	0.0612	0.226
3	0.00058	0.344	0.158
4	4.37	0.321	4.37
5	0.064	0.295	0.164
6	0.21	0.21	0.21
7	8.45	4.53	8.45
8	1.30	37.0	1.32
9	0.434	0.334	0.434
10	0.331	0.0676	0.331
11	0.969	0.894	9.69
12	0.285	0.385	0.385
13	3.98	0.898	0.00898
14	0.00135	0.635	0.00035
15	0.00985	2.94	0.00185
16	0.0533	0.0533	0.383
17	3.99	0.899	0.0271
18	0.00468	0.468	0.00734
19	0.0823	0.823	0.00868
20	0.2715	0.0715	0.8715
21	0.442	0.00242	0.842
22	6.42	5.42	0.842

4.4. Implication of Air-Injection Measurements for the Heater Test

The air-injection measurements described above were a crucial component of the preheat field characterization effort for the Drift Scale Test. The measurements provided information on the anticipated coupled thermal-hydrological processes for the heater test. With heat, water would vaporize, and moisture would be transported through the test block mainly in the gas phase as vapor. Over time, the rock mass in the vicinity of the heaters would become dry, while the liquid saturation in the rock mass in the cooler region would increase from condensation. Air-injection tests in the twelve boreholes identical to those described above are repeated quarterly, after the onset of heat, to track the moisture redistribution in the fractures. In those zones where fracture liquid saturation has increased as a result of condensation, resistance to airflow is anticipated to increase and the local air permeability to decrease. Indeed, we find the quarterly measurements of air permeability [based on Equation (1)] for the 46 packed-off zones agree well with the time evolution of condensation zones as predicted by the thermal-hydrological modeling of the Drift Scale Test [*Tsang et al.*, 1999b]. Where the model predicts an increase of liquid saturation because of condensation, the measured air permeability of those specific zones also declines, after heating, from its corresponding preheat value.

The configuration of the 12 boreholes was designed for studying thermal-hydrological coupling, specifically to monitor the drying and condensation processes after heating started. Zone lengths for conducting air-injection tests were commensurate with the scale of the drying and condensation zones, which is of the same order as the heat source (~10 m). On the other hand, to more properly assess the discrete fracture effects, we preferred to pack off the boreholes into consecutively smaller intervals, as discussed earlier. However, it was not feasible to isolate each borehole into more than four zones by packers, because each zone in the 12 holes was not only installed with pressure, packer inflation, and air-injection lines, but also with temperature and humidity sensors and their associated cables. Hence, four zones per borehole were the upper limit for the cable and tubing from all zones to fit into a 7.58 cm diameter borehole. The air-injection measurements on the current length scale of ~10 m are quite adequate to resolve the zones of drying and condensation. However, they are on too coarse a scale to resolve the exact location of discrete fractures important for flow, because the majority of fractures in the Topopah Spring welded tuff are less than a meter long. In a few of the 46 packed-off zones, water seepage occurred five months into heating. We attribute this phenomenon to preferential water drainage in fractures, which cannot be accounted for by the continuum approximation of fracture flow on the current scale of measurements.

5. SUMMARY AND CONCLUSIONS

It is a challenge to characterize the heterogeneous flow characteristics of a fractured rock formation. The Drift Scale Test, a large-scale thermal test in an underground facility at Yucca Mountain, provided a unique opportunity to address this problem. Because the dominant manifestation of thermal-hydrological coupled processes is the redistribution of moisture through vapor transport, the thermal test requires extensive preheat characterization to determine the flow characteristics and connectivity of the fractures. The preheat field characterization was carried out by air-injection tests in twelve 40 m long boreholes within the Drift Scale Heater test rock. These twelve boreholes were isolated by inflatable packers into 46 zones, each on the length scale of ~10 m. A constant flow rate of air was applied in each of the isolated zones, and pressure responses in each injection zone and all other 45 monitoring zones were monitored. We applied TOUGH2, a numerical code for multiphase flow and transport in forward modeling, to simultaneously invert the large set of pressure-change data from the air-injection tests. The densely fractured tuff was modeled as a continuum with a heterogeneous permeability field, with permeability treated as the adjustable parameter in the simultaneous inversion exercise. Through extensive trial-and-error efforts to adjust the local permeability values, we matched simulated air-pressure changes to the observed data for all 46 air-injection tests. Inversion is inherently nonunique, but the large number of data sets provided a tight constraint on the inversion. We could thus arrive at one set of local permeabilities that provides the best (visual and subjective) fit between the calculated and simulated pressure changes. We therefore believe that at the test scale, the permeability distribution obtained is quite reliable.

The estimated permeability distribution from the simultaneous inversion confirmed that the test block is highly heterogeneous, with permeability varying from 10^{-15} to 10^{-11} m^2. The estimated permeability distribution suggests that the lower part of the test block is in general less permeable than the upper part, with the upper northern portion of the test block the most permeable. A horizontal zone about 8 m long, extending from the observation drift to near the center of the block, appears to be a high-permeability feature (Plate 2). This zone extends discontinuously, with a gentle dip to the north side of the block, in the area where borehole cluster 74–78 is located. Around borehole cluster 57–61, a higher-permeability zone also exists in the lower part of the block, between the two drifts. This vertical zone, having a width of 6 m, extends from the middle ($z \sim 2$ m) to the bottom ($z \sim -20$ m) of the block.

In general, under the constant-rate injection, pressure change in the injection hole reaches steady state within minutes, with minimal well-storage effect. Interference pressure responses, though very slight, are almost instantaneous in the observation boreholes, suggesting good pneumatic connection among the boreholes. These data therefore support the assumption of the densely fractured rocks forming a continuum, on the current scale of measurement. In this simulation study, the test block was assigned a heterogeneous structure based on the test configuration. In other words, the scale of the heterogeneity structure is commensurate with the length of the zones of boreholes isolated by packers. Each block was assigned a different permeability. On a smaller scale, the rock elements in the immediate vicinity of each borehole zone were also assigned permeability values that were fitting parameters in the simultaneous inversion. In general, the trial-and-error process of adjusting the permeability values for all the blocks resulted in excellent agreement between the simulated and measured pressure changes. In instances where the interference-pressure response in certain monitoring borehole zones was exceptionally large, a good fit was nearly impossible to accomplish by simply varying the permeability values within the confines of the predesigned permeability structure. Instead, we modified the original permeability by incorporating explicit high-permeability pathways into the continuum model (for example, between the borehole zones 60-1 and 61-1). That is, although the continuum conceptual model did not account for discrete fracture effects, the large set of multizone data that we utilized to calibrate our model enabled us to incorporate the effects of discrete fractures on a scale smaller than the predesigned heterogeneity structure. This underscores the advantage of the simultaneous-inversion approach over the conventional two-well type-curve analyses of pressure data.

Acknowledgments. We thank Chin-Fu Tsang and Dan Hawkes for their careful review of this paper. We are grateful to Barry Freifeld for sharing his insight based on collection of these air-injection data in the field. We acknowledge the review comments and suggestions for improvement by AGU monograph referees: Gary LeCain, Walter Illman, and Paul Cook. This work was supported by the Director, Office of Civilian Radioactive Waste Management, U.S. Department of Energy, through Memorandum Purchase Order EA9013MC5X between TRW Environmental Safety Systems, and Berkeley Lab. The support was provided to Berkeley Lab through the U.S. Department of Energy Contract No. DE-AC03-76SF00098.

REFERENCES

Birkholzer, J. T., and Y. W. Tsang, *Forecast of Thermal-Hydrological Conditions and Air Injection Test Results of the Single Heater Test at Yucca Mountain*, Report LBNL-39789, Lawrence Berkeley National Laboratory, Berkeley, CA, December 1996.

Birkholzer, J. T., and Y. W. Tsang, *Pretest Analysis of the Thermal-Hydrological Conditions of the Drift Scale Test at Yucca Mountain*, Report LBNL-41044, Lawrence Berkeley National Laboratory, Berkeley, CA, 1997.

Cook, P., In situ pneumatic testing in Yucca Mountain, *Int. J. of Rock Mech. and Min. Sci.*, accepted.

Doughty, C., Investigation of conceptual and numerical approaches for evaluating moisture, gas, chemical, and heat transport in fractured unsaturated rock, *J. Contaminant Hydrology 38*(1–3), 69–106, 1999.

Dverstorp, B., Analyzing flow and transport in fractured rocks using the discrete network concept, doctoral thesis, R. Inst. Of Technol., Stockholm, 1991.

Finsterle, S., *ITOUGH2 User's Guide*, Report LBNL-40040, Lawrence Berkeley National Laboratory, Berkeley, Calif., 1999.

Finsterle, S., and P. Persoff, Determining permeability of tight rock samples using inverse modeling, *Water Resour. Res., 33*(8), 1803–1811, 1997.

Flint, L. E., *Matrix Properties of Hydrogeologic Unit at Yucca Mountain, Nevada, U. S. Geological Survey Open-File Report*, U. S. Geological Survey, Denver, Colo., 1998

Guzman, A. G, *Air Permeability Tests and Their Interpretation in Partially Saturated Fractured Ruffs*, Ph.D. thesis, Department of Hydrology and Water Resources, Univ. of Arizona, Tucson, Ariz., 1995.

Guzman, A. G., A. M. Geddis, M. J. Henrich, C. F. Lohrstorfer, and S. P. Neuman, *Summary of Air Permeability Data from Single-Hole Injection Tests in Unsaturated Fractured Tuffs at the Apache Leap Research Site, Results of Steady-State Test Interpretation*, Report Nureg/CR-6360, U.S. Nuclear Regulatory Commission, Washington, D.C., 1996.

Huang, K., Y. W. Tsang, and G. S. Bodvarsson, Simultaneous inversion of air injection tests in fractured unsaturated tuff at Yucca Mountain, *Water Resour. Res., 35*(8), 2375–2386, 1999.

Hvorslev, M. J., *Time Lag and Soil Permeability in Groundwater Observations, Bulletin 36*, U.S. Army Corps of Engineers, Water Ways Experimental Station, Vicksburg, Mich., 1951.

Illman, W. A., D. L Thompson, V. V. Vesselinov, G. Chen, and S.P. Neuman, *Single- and Cross-hole Pneumatic Tests in Unsaturated Fractured Tuffs at the Apache Leap Research Site: Phenomenology, Spatial Variability, Connectivity and Scale*, Report NUREG/CR-5559, U.S. Nuclear Regulatory Commission, Washington, D.C., 1998.

LeCain, G. D., *Pneumatic Testing in 45-Degree-Inclined Boreholes in Ash-Flow Tuff Near Superior, Arizona*, USGS Water Resources Investigation Report 95-4073, U.S. Geological Survey, Denver, Colo., 1995.

LeCain, G. D., *Results from Air-Injection And Tracer Testing in the Upper Tiva Canyon, Bow Ridge Fault, and Upper Paintbrush Contact Alcoves of the Exploratory Studies Facility, August 1994 through July 1996, Yucca Mountain, Nevada*, Water-Resources Investigations Report 98-4058, U.S. Geological Survey, Denver, Colo., 1998.

Long. J. C. S., J. S. Remer, C. R. Wilson, and P. A. Witherspoon, Porous media equivalents for networks of discontinuous fractures, *Water Resour. Res., 18*(3), 645–658, 1982.

Neuman, S. P., Calibration of distributed parameter groundwater flow models viewed as a multiple-objective decision process under uncertainty, *Water Resour. Res., 9*(4), 1006–1021, 1973.

Neuman, S. P., Stochastic continuum representation of fracture rock permeability as an alternative to the REV and fracture network concepts, *Rock Mechanics, Proc. of the 28th U.S. Symposium*, Balkema, Rotterdam, 533–561, 1987.

Peters, R. R., E. A. Klavetter, I. J. Hall, S. C. Blair, P. R. Heller, and G. W. Gee, *Fracture and Matrix Hydrologic Characteristics of Tuffaceous Materials from Yucca Mountain, Nye County, Nevada*, SAND84-1471, Sandia Nat, Laboratories, Albuquerque, N.M. 1984.

Pruess, K., *TOUGH2, a General-Purpose Numerical Simulator for Multiphase Fluid and Heat Flow*, Report LBL-29400, Lawrence Berkeley Laboratory, Berkeley, Calif., 1991.

Pruess, K., A. Simmons, Y. S. Wu, and G. Moridis, *TOUGH2 Software Qualification*, Report LBL-38383, Lawrence Berkeley Laboratory, Berkeley, Calif., 1996.

Rasmussen, T. C., D. D. Evans, P. J. Sheets, and J. H. Blanford, Permeability of Apache Leap tuff: Borehole and core measurements using water and air, *Water Resour. Res., 29*(7), 1997–2006, 1993.

Schwartz, F. W., and L. Smith, A continuum approach for modeling mass transport in fractured media, *Water Resour. Res., 24*(8), 1360–1372, 1988.

Sonnenthal, E. L., C. F. Ahlers, and G. S. Bodvarsson, Chapter 7: Fracture and fault properties for the UZ site-scale flow model, , in *The Site-Scale Unsaturated Zone Model of Yucca Mountain, Nevada, for the Viability Assessment*, edited by G. S. Bodvarsson, Report LBNL-39315, Lawrence Berkeley National Laboratory, Berkeley, Calif., 1997.

Sun, N., *Inverse Problem in Groundwater Modeling*, Kluwer Academic Publishers, 1994.

Tsang, Y. W., C. F. Tsang, F. V. Hale, and B. Dverstorp, Tracer transport in a stochastic continuum model of fractured media, *Water Resour. Res., 32*(10), 3077–3092, 1996.

Tsang, Y. W., and J. T. Birkholzer, Predictions and observations of the thermal-hydrological conditions in the single heater test, *J. Contaminant Hydrol.*, in press.

Tsang, Y. W., J. Apps, J. T. Birkholzer, B. Freifeld, M. Q. Hu, J. Peterson, E. Sonnenthal, and N. Spycher, *Yucca Mountain Single Heater Test Final Report*, Report LBNL-422537, Lawrence Berkeley National Laboratory, Berkeley, Calif., 1999a.

Tsang Y. W., J. Apps, J. T. Birkholzer, B. Freifeld, J. E. Peterson, E. Sonnenthal. N. Spycher, and K. H. Williams, *Yucca Mountain Drift Scale Test Progress Report*, Report LBNL-42538, Lawrence Berkeley National Laboratory, Berkeley, Calif., 1999b.

United States Department of Energy, *Site Characterization Plan: Yucca Mountain Site, Nevada Research and Development Area, Nevada*, Report DOE/RW-0199, Office of Civilian Radioactive Waste Management, U.S. Department of Energy, Washington, D.C., , 1988.

Yeh, W. W.-G., Review of parameter identification procedures in groundwater hydrology: The inverse problem, *Water Resour. Res., 22*(2), 95–108, 1986.

Y. W. Tsang, K. Huang, and G. S. Bodvarsson, Earth Sciences Division, Ernest Orlando Lawrence Berkeley National Laboratory, One Cyclotron Road, Berkeley, CA 94720

Estimation of Regional Recharge and Travel Time Through the Unsaturated Zone in Arid Climates

Alan L. Flint,[1] Lorraine E. Flint,[1] Joseph A. Hevesi,[1] Frank D'Agnese,[2] Claudia Faunt[3]

Studies are currently underway to determine the suitability of Yucca Mountain, Nevada, as the first high-level nuclear-waste repository in the United States. Values of net infiltration, estimates of travel time through the unsaturated zone, and recharge are useful for evaluating the expected performance of the potential repository as a waste-containment system. The current understanding of infiltration that has been developed at the Yucca Mountain site is presented, and is being extrapolated to the Death Valley region. A conceptual model of recharge as it relates to net infiltration is developed, where recharge is assumed to be near vertical, with a time delay that is a function of rate of net infiltration and the thickness of the unsaturated zone, and the effective travel-path porosity. From estimates of these three criteria, water travel time through the unsaturated zone can be calculated, providing additional understanding of spatial differences in recharge processes. Future climate change is expected to result in an increase in precipitation that may change the rate of net infiltration and therefore the travel time through the unsaturated zone. Although parts of the regional flow system may respond quickly to climate change, others may lag behind significantly. This variability may be significant in determining the rate and direction of groundwater flow. The approach presented in this paper is tested with data from the Yucca Mountain site, and compared to various other approaches for estimating recharge. The approach is extended to the regional scale for consideration as a useful tool for estimating recharge at various scales and under different climatic scenarios.

1. INTRODUCTION

The understanding of the role of fractured rock in unsaturated-zone hydrology has increased dramatically in the last decade. Interest in this field is partly because of the importance of the use of deep unsaturated-zones in arid and semiarid environments for the storage of nuclear waste products. These waste products decay over very long time periods, which require examination of the responses of the unsaturated zone to potential changes in future climate. Future climate change is expected to result in an increase in precipitation [*Spaulding*, 1985; *Botkin et al.,* 1991] that may increase the rate of net infiltration and therefore decrease the travel time of waste products through the unsaturated zone. In this paper, travel time is defined as the time it takes for water to travel from the soil-air boundary (land surface) to recharge the water table. It is also an important

[1]U.S. Geological Survey, Sacramento, California
[2]U.S. Geological Survey, Tucson, Arizona
[3]U.S. Geological Survey, San Diego, California

Dynamics of Fluids in Fractured Rock
Geophysical Monograph 122

Published 2000 by the American Geophysical Union

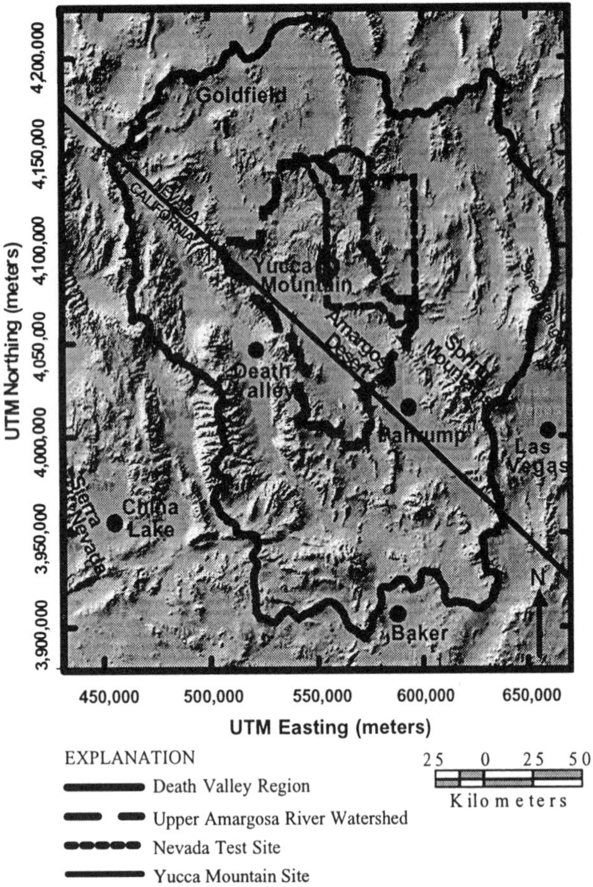

Figure 1. Elevation of the regional model domain.

calculation for water-resources management for timing of water availability.

The study of the Yucca Mountain site (Figure 1) as a potential high-level waste repository by the U.S. Geological Survey and the U.S. Department of Energy has provided a large set of field and laboratory data for the development of conceptual and numerical models of fluid flow in fractured rocks. The conceptual model presented here is a summary of the efforts to characterize the surface infiltration—that is, the process of water moving across the land surface (the air/soil or air/bedrock boundary) that results in infiltration into fractured bedrock. The model is based on 12 years (1984–95) of field measurements of soil/rock water content and water potential in the near-surface zone and in the zone above the soil/bedrock interface. Net infiltration is defined as a component of infiltration that moves below the near-surface zone affected by evapotranspiration and which can therefore become recharge to the underlying aquifer.

The evaluation of the conversion of net infiltration into recharge on a small scale commonly is done using unsaturated-zone flow models with the spatially distributed net-infiltration rates as the upper boundary condition. However, this method may not be applicable at large spatial scales owing to the intensity of the data requirements. This paper describes a method that estimates groundwater recharge at large spatial scales by utilizing estimates of net infiltration, effective flow-path porosity, the unsaturated-zone thickness, and an assumption of piston flow through a specified flow path. Perched water is not considered separately but simply as part of the flow path.

1.1. Study Area

The Death Valley Region (Figure 1) is primarily in the northern Mojave Desert, but extends into the southern Great Basin Desert, a subprovince of the Basin and Range physiographic province [*Grayson*, 1993], which is characterized by linear mountains and valleys with a distinct north to northwest trend. This physiography is predominantly the result of normal faulting in response to east-west extensional tectonics. The Death Valley Region is in the rain shadow of the Sierra Nevada, and elevations range from 86 m below sea level at the Death Valley basin itself to 3,600 m above sea level at Mount Charleston in the Spring Mountains. The basins in the region are filled with alluvium and some interbedded volcanic deposits [*Peterson*, 1981].

Weather patterns in the region vary seasonally. Summer precipitation primarily comes from the south and southeast. Winter winds carrying precipitation are from the west, resulting in a regional rain shadow east of the Sierra Nevada. Precipitation on the valley floors of the Amargosa Desert, Death Valley, and basins at lower elevations in the southern part of the region averages less than 150 mm/yr. Average precipitation in the mountain ranges is commonly 200 to 250 mm/yr and greater, and is as high as 500 to 750 mm in the Sheep Range and Spring Mountains (Figure 1), the highest ranges in the region.

The Death Valley Region consists of Precambrian and Cambrian clastic and crystalline rocks; Paleozoic clastic and carbonate rocks; Mesozoic Era clastic and intrusive rocks; varied Pliocene Epoch fluvial, paleic, and playa sedimentary deposits; Tertiary Period volcanic rocks, alluvium, and colluvium; and eolian deposits and alluvium of the Quaternary Period [*Waddell*, 1982]. Much of the region has undergone deformation, and some parts have been almost continuously tectonically active.

Soils in the Death Valley Region typically follow a pattern of lithosols (upland soils) on the mountains, medium- to coarse-textured soils on alluvial fans and terraces, and fine-grained, alluvial soils in the valley bottoms and stream channels. In general, the soils of the mountains and hills are shallow and are characterized by coarse texture and little moisture-storage capacity. The soils of the alluvial fans on the upper bajadas are also coarse textured but are much deeper, and thus infiltration rates can be relatively high. Infiltration rates of the alluvial basin soils are low because the downward movement of water is typically slowed by indurated calcium carbonate layers, fine-grained playa deposits and, more infrequently, silicified hardpans that form within the soils over time [*Beatley*, 1976]. These layers help to retain the water that is shallower in the profile and can more easily be lost to evapotranspiration. Soil thickness is related to topographic and geomorphologic positions as well as degree of slope.

Nine vegetation communities in the Death Valley Region are described as homogeneous units, although their natural distributions are commonly heterogeneous with variable species densities [*Munz*, 1974]. These communities are coniferous forests, pinyon-juniper woodland, sage-dominant areas, mixed shrub-transition, fan piedmont-mixed shrub, fan piedmont-creosote, fan skirt-creosote, alluvial flat-saltbush, and phreatophytes or agriculture.

1.2. Previous Approaches

Net infiltration and recharge have been estimated by previous investigators for the areas within the Death Valley Region. These estimations have used methods appropriate for arid environments such as water-balance techniques (e.g., basin-wide estimates of discharge or numerical models accounting for all significant components of the water balance), soil-physics techniques, geochemistry, and transfer equations based on other variables (such as precipitation).

Using a water-balance technique, *Winograd and Thordarson* [1975] estimated that 3 percent of the precipitation that falls in the watershed that discharges from springs at Ash Meadows (south of Yucca Mountain near the Nevada-California border) becomes recharge. This equates to an annual average recharge of 5 mm/yr over the entire watershed. *Maxey and Eakin* [1950] developed a method of estimating recharge to groundwater basins in Nevada, providing a baseline for the spatial distribution of recharge. This method uses average annual precipitation to classify a basin into 5 recharge zones. Each zone is characterized by a different percentage for average annual precipitation becoming recharge: 0 percent recharge for less than 203 mm/yr average annual precipitation, 3 percent for 203 to 304 mm/yr, 7 percent for 305 to 380 mm/yr, 15 percent for 381 to 507 mm/yr, and 25 percent for 508 mm/yr or greater. This method was applied to 167 basins in the Great Basin to estimate recharge for locations with an average annual precipitation in excess of 203 mm/yr (8 in).

The results of several studies were combined by *Hevesi and Flint* [1998] to help develop a modified Maxey-Eakin model for estimating recharge on a regional scale. An exponential model was used to fit the data presented by the original step function developed of *Maxey and Eakin* [1950], with a lower limit of recharge set to an average annual precipitation of 100 mm instead of the 203 mm/yr used by Maxey and Eakin. *Hevesi and Flint* [1998] used an estimate of regional precipitation (average annual precipitation of 175 mm/yr) and the modified model to estimate recharge of the Death Valley Region to be 2.9 mm/yr. A similar model was used by *D'Agnese et al.* [1997] to estimate recharge in their 3-D regional flow model of a subregion of the entire Death Valley Region (Figure 1), which is designated as the Death Valley regional groundwater flow system.

Net-infiltration and recharge estimates for basins in Nevada also have been obtained using chloride-mass-balance (CMB) calculations. This method compares chloride in recharge water and runoff to chloride deposited in source areas by precipitation and dry fallout. *Lichty and McKinley* [1995] provided an analysis of recharge for 2 basins in central Nevada using a 6 yr measurement period and 2 independent modeling approaches: water balance and CMB. Their results yielded recharge rates of 10 to 30 mm/yr for a drainage basin with an average annual precipitation of 270 mm, and 300 to 320 mm/yr for a drainage basin with an average annual precipitation of 640 mm. They determined that the CMB method was more viable than water balance methods for their study. *Dettinger* [1989] applied this method to 16 basins in Nevada; the estimates were close to those obtained using the Maxey-Eakin method and water-balance calculations. Dettinger concluded that the CMB method is practical at a reconnaissance level for estimating average rates of recharge for many desert basins of the western United States, but because the method assumes piston flow in porous media, it may not be as applicable for estimating recharge in fractured rock under shallow soils.

Several researchers have applied soil-physics techniques for estimating net-infiltration rates at locations with thick soil cover (alluvial fans and basins). These methods require data on ambient conditions; soil properties, specifically the soil moisture characteristics from which unsaturated hydraulic conductivity can be calculated; and an assumption of steady-state conditions. Using these techniques, *Winograd* [1981] estimated net infiltration through the thick soil valley fill at Sedan Crater, about 48 km northeast of Yucca Mountain, to be about 2 mm/yr. *Nichols* [1987] used water-potential measurements and a numerical model based on the Richards equation to perform water-balance calculations for the unsaturated zone near Beatty, Nevada, about 30 km west of Yucca Mountain. *Nichols* [1987] estimated a rate of net infiltration of 0.04 mm/yr in the thick soil materials, but the site potentially has lower precipitation rates than Sedan Crater. Both estimates are similar to values obtained using the modified Maxey-Eakin model.

In summary, average recharge values for the entire Death Valley Region range from 0.04 mm/yr to over 5 mm/yr, utilizing the various approaches discussed above. These approaches include the Maxey-Eakin model for percentage of precipitation estimates, chloride-mass-balance calculations, and water-balance models. Using these methods, researchers evaluated locations with differing average annual precipitation and thus developed correlations between precipitation and groundwater recharge. Sections 2 and 3 of this paper will discuss newer methods for estimating net infiltration and groundwater recharge. Section 4 will compare the results from earlier approaches to those determined using newer methods. Approaches that warrant consideration for use in distributing recharge estimates over regions are those that can use distributed properties or conditions, such as precipitation.

2. CONCEPTUAL MODEL OF RECHARGE

The current conceptual model for recharge in fractured rock [*Flint et al.*, 2000] was developed for conditions at Yucca Mountain, which is centrally located within the Death Valley Region. It identifies precipitation as the most significant environmental factor controlling net infiltration at Yucca Mountain. Precipitation at Yucca Mountain averages 170 mm/yr but is temporally and spatially variable [*Hevesi et al.*, 1991]. The depth of infiltration into the soil/bedrock profile fluctuates on a seasonal basis but is greatest in the winter owing to lower evapotranspiration demands, higher amounts of precipitation, and slow melting of snow.

The second most significant environmental factor controlling net infiltration is soil depth. In years when there is sufficient precipitation to produce net infiltration, the spatial distribution is generally defined by the spatial variability of soil depth. Field measurements indicate that when the soil/bedrock interface reaches near-saturated conditions, fracture flow is initiated in the bedrock, thereby increasing the hydraulic conductivity by several orders of magnitude. Soils exceeding 3 to 6 m in thickness virtually eliminate infiltration of water to the soil/bedrock interface or deeper except beneath channels [*Flint and Flint*, 1995]. Storage capacity in thick soil profiles is large enough that most water from precipitation is held in the root zone and removed by evapotranspiration. Soils that are generally shallower than approximately 3 m do not have enough capacity to store the infiltrating volume that results from above-average winter precipitation and often allow near-ponding conditions to occur at the soil/bedrock interface, particularly when the soil depth is less than about 0.5 m.

The third factor controlling net infiltration is bedrock hydraulic conductivity. At Yucca Mountain, welded tuffs and nonfractured, nonwelded tuffs are the principal rock types present in surface exposures or directly under soils. The saturated hydraulic conductivity of the nonwelded tuff matrix is higher than that of the welded-tuff matrix [*Flint*, 1998]. However, the fractures in the welded tuff increase the saturated hydraulic conductivity of those rocks. The lower storage capacity of the fractured welded tuffs allows infiltrating water to penetrate more deeply than in the nonwelded tuffs. Hydraulic properties of fractures depend on fracture aperture and whether or not the fractures are open or filled. Fractures at the ground surface commonly are filled with calcium carbonate or siliceous materials, which hinder infiltration. Calculations of fracture porosity, saturated hydraulic conductivity, and aperture [*Kwicklis et al.*, 1999] indicate that significant flux in fractures occurs only under saturated or near-saturated conditions. Fracture densities and matrix permeabilities vary greatly between the geologic units at Yucca Mountain.

In a deep, unsaturated zone, under steady-state conditions, net infiltration becomes recharge, except when perched water is discharged in springs and lost to evapotranspiration. Travel time through the unsaturated zone is controlled by net infiltration, unsaturated zone thickness, and effective flow-pathway porosity. As climate changes, the travel time of infiltrating water through the unsaturated zone may change, as well as the spatial variability of recharge. Recharge occurring today is spatially variable due to alluvium thickness, subsurface features, unsaturated zone thickness, layering, and properties of geologic and sediment strata. Recharge is dependent on the local net infiltration that responds to the variables influencing recharge over a variety of time scales—from years to centuries. The travel time of infiltrated water through the unsaturated zone is an important component for estimating groundwater recharge under future climate conditions.

The largest uncertainty in estimating the travel time of infiltrated water through the unsaturated zone, and therefore water-table response, is caused by the variability of the effective porosity of the fractured rock. Flow pathways may be in rock matrix, which is generally well constrained, or in fractures, which are generally poorly constrained. If net infiltration rates those calculated using exceed the matrix hydraulic conductivity, the infiltrated water is expected to travel laterally until the flow is intercepted by vertical fractures or faults. Lateral flow may be significant before the infiltrated water moves downward to the water table. However, because the vertical distance through the unsaturated zone is short relative to the horizontal distance (hundreds of meters vs. kilometers) of the regional aquifer, net infiltration will become recharge in the same relative location on a regional basis. For high infiltration rates, the effective flow-path porosity is assumed to be the fracture porosity. If net-infiltration rates are less than the hydraulic conductivity of the bedrock matrix, then flow will be through the matrix and the effective flow-path porosity will be the volumetric water content of the matrix at a specific net infiltration rate. The overall effective flow-path porosity must be the depth-weighted, mean-flow-path porosity of all layers between the surface and the water table.

3. NUMERICAL MODEL OF RECHARGE

Process-based recharge models are only approximations of the natural system. If the processes are modeled appropriately and the physical features are correctly represented, then model application can provide valuable insight into the behavior of the system. In many cases, scientists are less concerned with the absolute accuracy of the model results and more concerned with the spatial distribution of the recharge and associated processes. In addition, regional process models can give us a clearer picture of how climate change may affect water availability, or they can indicate areas where more study is needed—such as improving estimates of bedrock permeability or obtaining water chemistry data in locations where recharge is determined to be high.

In general, the complexity of the conceptual model of the hydrologic system depends on the availability of data and the scale of application [*Hatton*, 1998]. The model being presented in this paper is a pseudo 3-D model because surface routing of runoff is in 2 dimensions and following runoff the model reverts to separate 1-D columns, where net infiltration is modeled using a "bucket" model approach. *Hatton et al.* [1995] compared TOPOG, a 3-dimensional Richards equation-based watershed model, with a distributed 1-D bucket model to assess the conditions under which lateral processes became important in recharge modeling. Overland flow was found to be the most significant process, which is done in this study using the 2-D runoff model. The less significant component was lateral subsurface flow, which only occurred when saturated conditions existed on sloping soils. For subsurface lateral flow to be significant, *Hatton et al.* [1995] found that the slope and hydraulic conductivity had to be sufficiently high and in areas that exceeded 600 mm/yr rainfall. The model presented here routes runoff in the surface layer, eliminating that concern. In addition, it was determined that lateral flow is a minor component over the large region and would only shift the recharge values laterally a very small amount.

The numerical model of infiltration, based on the conceptual model described above, is the basis for the spatial estimates of recharge for the Death Valley Region. It must be understood that in this model, recharge is defined simply as the movement of water from the unsaturated zone into the saturated zone. Recharge is assumed to be equal to net infiltration and is defined as water that penetrates below the depths described in the conceptual model of infiltration with no time lags due to subsurface processes or features. Calculations of the time lags through the unsaturated zone are discussed below in Section 4.2. In this model, water movement in aquifers is not considered.

The numerical model used to simulate infiltration at Yucca Mountain (INFIL) is developed using components of the mass-balance equation (Figure 2). The numerical model is quasi three dimensional and uses quasi-mass-balance processes, real or stochastically simulated precipitation, the physical setting, and hydrologic properties of the site to approximate the actual conditions at Yucca Mountain and vicinity. (At the watershed or site scale, grid blocks are 30 m x 30 m. At the regional scale, grid blocks are 278 m x 278 m.)

A mass-balance equation calculating net infiltration for the near-surface zone can be written as:

$$I = P - \Delta W_s - ET - R_{\text{off}} \ , \qquad (1)$$

where I is the net infiltration, P is precipitation, ΔW_s is the change of soil water storage, ET is evapotranspiration, and R_{off} is surface runoff.

Net infiltration in the INFIL model is not solved directly using Equation 1. The model uses the mass-balance approach to estimate the total water stored in the profile. P is determined daily; ET is simulated daily using a solar radiation subroutine and a modified Priestley-Taylor equation [*Flint and Childs*, 1991]; R_{off} is routed for all grid locations; and ΔW_s is determined using an infiltration subroutine and the physical and hydrologic properties of the soils, such as soil-moisture conditions and properties, soil depth, and bulk bedrock hydraulic conductivity (saturated hydraulic conductivity of the matrix and fractures combined) of the underlying geologic formation. Following an occurrence of precipitation, the water is stored in the first layer of the soil profile until field capacity is exceeded. Field capacity is defined as the soil water content of the near-surface zone (i.e., the root zone) at which drainage becomes negligible.

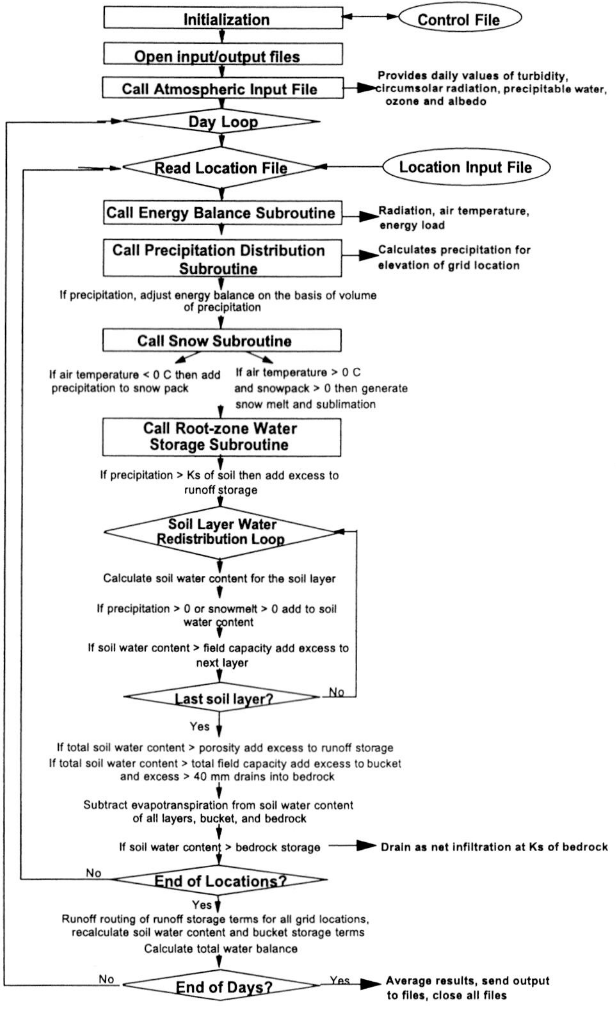

Figure 2. Flow diagram of numerical model.

Once the first soil layer has reached field capacity, excess water enters the next soil layer until this layer reaches field capacity, and so forth. If there is more water than the total soil can hold at field capacity, the water drains to the soil/bedrock interface, where it will accumulate in a storage component, or "bucket." If total soil water-storage capacity (porosity times depth) is exceeded, then runoff is generated. Runoff is also generated if the hydraulic conductivity of the soil is exceeded by the rate of precipitation for an event. Water that has accumulated in the bucket infiltrates into the underlying bedrock at a rate equal to the saturated hydraulic conductivity (assuming a unit hydraulic gradient) and is lost to evapotranspiration at a rate dependent on the available energy and the soil thickness. Therefore, net infiltration is calculated as the bulk hydraulic conductivity of the underlying bedrock during the time periods when saturated conditions exist at the soil/bedrock interface.

For locations with thin soils (<0.5 m), the root zone is extended into a fractured bedrock layer having an effective root-zone water storage capacity of 0.04 m^3/m^3 up to a maximum depth of 2 m below the ground surface water. Redistribution is averaged for every 24 hr period for all soil layers. For the thin soil that covers most of the mountains, this is probably acceptable during all types of precipitation events. For deeper soils, this method tends to move water deeper in the profile than would actually occur because it is averaging increases in water content that occurred to shallow depths over a deeper soil profile. When averaged over the profile, the volumes are small and the method is not likely to result in increases in net infiltration that would ordinarily occur due to low evapotranspiration and eventual drainage of the profile. In summary, this method does not directly solve the water-balance equation but determines if field capacity is exceeded by precipitation. If so, then the water drains to the soil/bedrock interface, where it becomes evapotranspiration or net infiltration. If the volume of infiltrating water exceeds the total soil water-storage capacity, it becomes runoff.

The model was calibrated by matching simulated daily mean discharge rates to those measured from episodic runoff events at calibration basins. In addition to measured stream discharge, the model was calibrated using volumetric water content data from neutron boreholes used to calculate flux. In this manner, the results of the model substantiate the original conceptual model of infiltration that was based on physical setting and hydrologic properties. The 4 infiltration zones are correlated strongly with the features that correspond most closely with the resultant estimates of net infiltration, namely, higher precipitation at higher elevations and shallower soil depths.

4. RESULTS

4.1. Net Infiltration

The net infiltration (recharge) for each of 10 calibration basins at Yucca Mountain is compared (Figure 3) to the other methods discussed in the introduction. Figure 3 shows recharge to be a volume calculated per basin area as a function of average annual precipitation volume. Estimates of recharge that were determined using the Maxey-Eakin method are from *Harrill and Prudic* [1998]. Those estimates determined using a modified Maxey-Eakin method are from *D'Agnese et al.* [1997] and *Hevesi and Flint* [1998]. The recharge estimates determined using the chloride-mass-balance method are from *Dettinger* [1989] and *Lichty and McKinley* [1995]. There is generally good agreement between methods (Figure 3), although direct

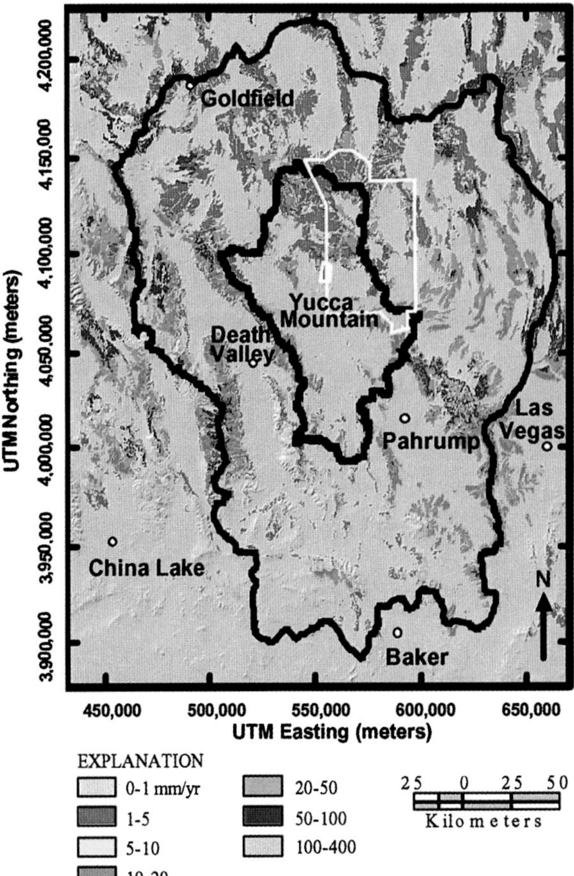

Plate 2. Simulated net infiltration for 1980–1995 for the Death Valley Region.

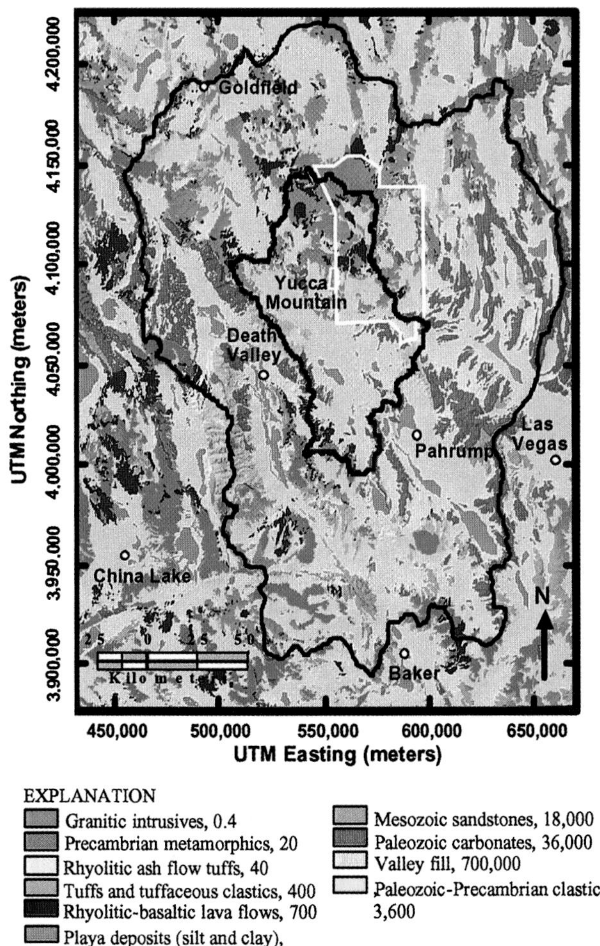

Plate 1. Bedrock geology and corresponding saturated hydraulic conductivity, in mm/year, of the Death Valley Region.

comparison cannot be made. Precipitation input used in the modified Maxey-Eakin method and INFIL model are estimated daily values for the 1980–1995 model calibration period, whereas the precipitation estimates used in the CMB and original Maxey-Eakin models are based on estimates of longer-term average precipitation rates. For a direct comparison, the modified Maxey-Eakin model [*Hevesi and Flint*, 1998] and the INFIL model were used to estimate net infiltration for 20 basins from *Harrill and Prudic* [1998] using the same precipitation estimate. The results illustrate that estimates from these two models generate similar relations between recharge and precipitation, with two distinct differences (Figure 4): the INFIL model predicts higher recharge in the basins that are dominated by high-permeability carbonate rock and lower recharge in basins dominated by valley fill and low-permeability volcanic rock. This analysis indicates that bedrock geology plays a significant role in controlling recharge, which is not accounted for in the modified or original Maxey-Eakin models.

In order to calculate net infiltration and recharge for the Death Valley Region, a series of Geographic Information System (GIS) coverages were assembled. The base maps were the regional digital elevation model (DEM) (Figure 1) and the bedrock geology map (Plate 1). These were used to distribute the energy-load calculations, runoff, and bedrock properties. The saturated hydraulic conductivity of the bedrock in the Death Valley region is also indicated on Plate 1. It can be seen that lower bedrock permeabilities correlate with the granitic rocks in the southwest part of the region and with the rhyolitic ash-flow tuffs and metamorphics, and that higher permeabilities correlate with areas of Mesozoic sandstones and Paleozoic carbonates.

Distributions of soil thickness and vegetation coverage were also used to calculate net infiltration and recharge. In order to produce a soil-thickness map, soil thickness was

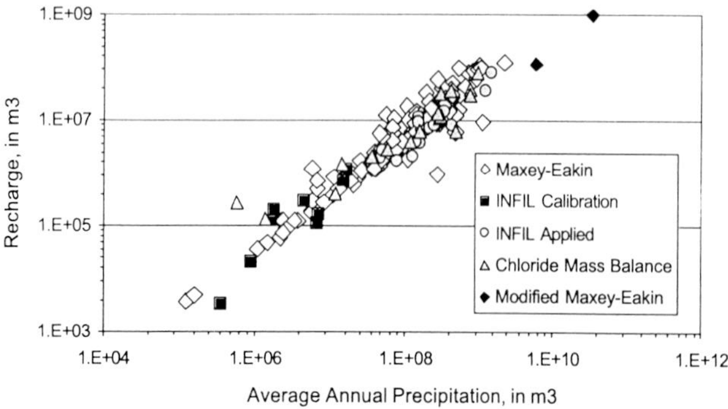

Figure 3. Comparison of various methods to estimate recharge in the Death Valley Region with model results from INFIL, as a function of average annual precipitation.

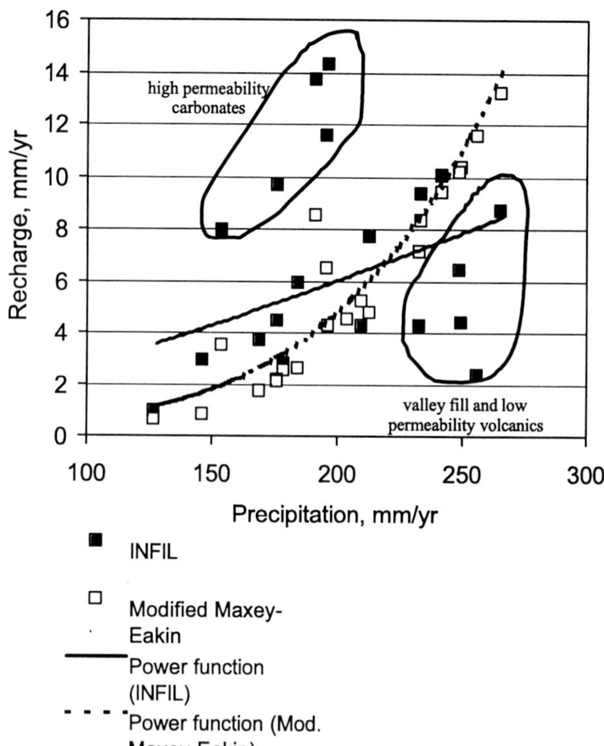

Figure 4. Comparison of recharge estimated using INFIL and a modified Maxey-Eakin model. Circled points indicate high-permeability carbonate bedrock or valley fill with low-permeability volcanics.

classified into five categories (0.1 m, 0.3 m, 0.5 m, 1.0 m, and 6.0 m), based on mapped soils and degree of slope. Soil properties are defined on the basis of field and laboratory analyses that were conducted on the soils around Yucca Mountain. The vegetation type has not been incorporated into the calculations for water use, but the percent of plant coverage is used in the estimates of the water balance to determine total water lost to evapotranspiration.

Regional estimates of average annual precipitation were made and range from <100 mm/yr in Death Valley to >400 mm/yr in the Spring Mountains. For these simulations, the daily precipitation was estimated for the central area in the model (Yucca Mountain) and extrapolated to the rest of the modeling domain by the correlation between precipitation and elevation. This method is generally acceptable for winter precipitation, which is very widely distributed. Errors in estimated precipitation increase with increasing distance from Yucca Mountain. In particular, the model overestimates precipitation input for the northern and western sections of the model domain. In addition, the method results in some errors in extrapolating summer precipitation, owing to monsoonal storms. On the basis of the Yucca

Mountain data for the period modeled (1980–1995), there were few monsoonal events that lead to net infiltration. Most recharge comes during the winter months of an El Niño weather pattern.

Using the coverages and the INFIL numerical model, net infiltration was simulated using rainfall records for the region that were developed into a daily precipitation record for 1980–1995. These simulations are for net infiltration, which is defined as water that has moved below the root zone at the location where it has infiltrated. Deeper unsaturated-zone processes may move the water vertically or laterally to the water table to become recharge or may move the water laterally until it reaches the surface and is removed by evapotranspiration processes, such as spring discharge, before it can become recharge. Simulated net infiltration is highest in the locations with high rainfall and shallow soils (Plate 2). In those locations, bedrock hydraulic conductivity dictates the net infiltration. Almost no infiltration (less than 0.01 mm/yr) is simulated in the locations with valley fill, which generally is thicker than 3 to 6 m.

4.2. Estimation of Regional Recharge

To evaluate the appropriateness of the approach to estimate recharge, calculations of unsaturated-zone travel time (UZTT) were compared with measured isotope data that yield estimates of residence time of groundwater in the unsaturated zone at Yucca Mountain. This exercise relies on a conceptual model (described above), available data for estimates of effective porosity (1 minus the water-filled porosity), unsaturated zone thickness, and estimates of net infiltration. Data for fracture densities are also fairly well understood. The UZTT is calculated as

$$\mathrm{UZTT} = \frac{\phi_{\mathrm{eff}} \cdot Z_{\mathrm{uz}}}{I_{\mathrm{net}}} \qquad (2)$$

where ϕ_{eff} is effective unsaturated zone porosity (m/m), Z_{uz} is the thickness of the unsaturated zone (m), and I_{net} is net infiltration (m/yr). At Yucca Mountain, the lithostratigraphy consists of an upper formation of densely welded, fractured tuff (Tiva Canyon Tuff), a lower formation of densely welded, fractured tuff (Topopah Spring Tuff), and a nonwelded, bedded sequence of rocks (nonwelded Paintbrush Group, PTn). These are sandwiched between the two welded-tuff formations. Estimates of average properties and thickness for those three layers and an estimate of average net infiltration are used to calculate UZTT as

$$\begin{aligned} \mathrm{UZTT} &= \mathrm{UZTT(Tiva)} + \mathrm{UZTT(PTn)} + \mathrm{UZTT(Topopah)} \qquad (3) \\ &= \frac{(0.002\,\mathrm{m/m})\cdot 75\,\mathrm{m}}{0.004\,\mathrm{m/yr}} + \frac{(0.23\,\mathrm{m/m})\cdot 75\,\mathrm{m}}{0.004\,\mathrm{m/yr}} + \frac{(0.002\,\mathrm{m/m})\cdot 75\,\mathrm{m}}{0.004\,\mathrm{m/yr}} \\ &= 38 + 4025 + 200 \\ &= 4263\,\mathrm{yr}. \end{aligned}$$

A fracture-porosity estimate of 0.002 m/m is based on ongoing field infiltration experiments in the fractured Tiva Canyon Tuff and is valid for both Tiva Canyon and Topopah Spring Tuffs. Because the rocks are less than saturated, the 0.23 m/m porosity value is based on current estimates of volumetric water content in the nonwelded Paintbrush Group sequence (PTn) [*Flint*, 1998].

The value of UZTT calculated at Yucca Mountain using this approach is supported, at least in order of magnitude, by groundwater ages based on carbon-14 age dating of three perched water zones at the base of the Topopah Spring Tuff. The age of water in these zones is estimated to be between 4,000 and 7,000 yr [*Yang and Peterman*, 1999], 2,000 to 7,000 yr [*Yang et al.*, 1996], and 1,340 to 5,800 yr [*Yang et al.*, 1998]. These measurements bracket our estimates of UZTT and provide support that this approach is viable. They also suggest that the estimates of effective porosity, which are the largest uncertainty, are fairly reasonable.

This approach can be extended to regional estimates of recharge using Equation 2. Estimates of unsaturated-zone thickness for the region are based on the difference between digital elevation maps and groundwater elevations (Plate 3). Thickness of the unsaturated zone for the region ranges from 0 to 2,000 m. Layering in the unsaturated zone of the region is currently being investigated. Once layers in the unsaturated zone are identified and their hydrologic properties estimated, the calculation can be made for each layer. Saturated matrix permeability and bulk permeability, which incorporate the permeability of the fractures, also are needed. At this time, a demonstration of the method to determine UZTT for the Death Valley Region can be shown using a one-layer model, simulated net infiltration, and estimates of effective flow-path porosity.

If net infiltration is greater than the saturated bulk hydraulic conductivity, then fracture flow is dominant and a small effective porosity is used. Two examples are presented to show a range of travel times based on the assumption of effective porosity: a matrix-dominated system (0.10 m/m, Plate 4), and a fracture-dominated system with high fracture porosity (0.03 m/m, Plate 5). If an even lower estimate of effective porosity (0.001) is used, an extreme example can be generated in which the majority of the mountain ranges have travel times of less than 100 years;

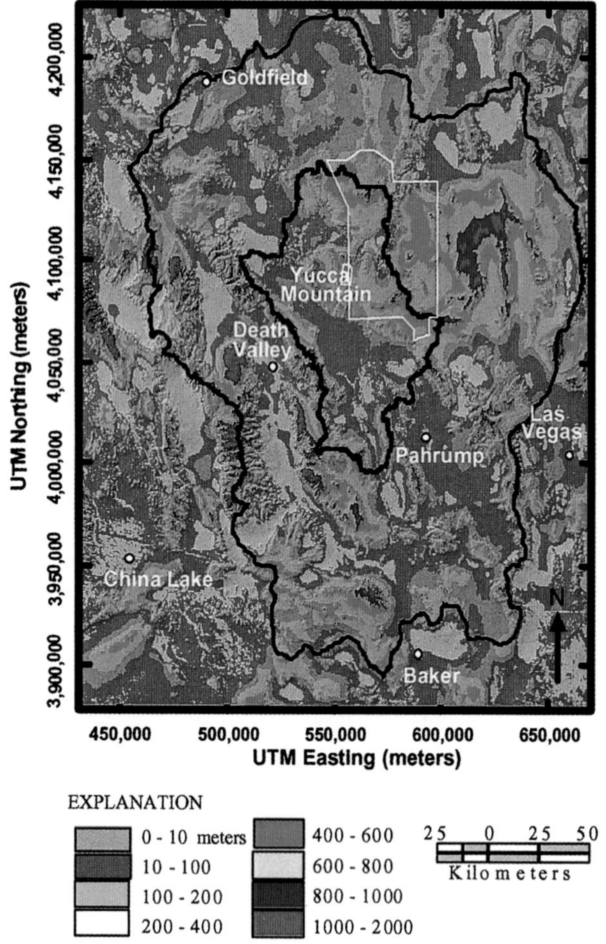

Plate 3. Thickness of the unsaturated zone for the Death Valley Region.

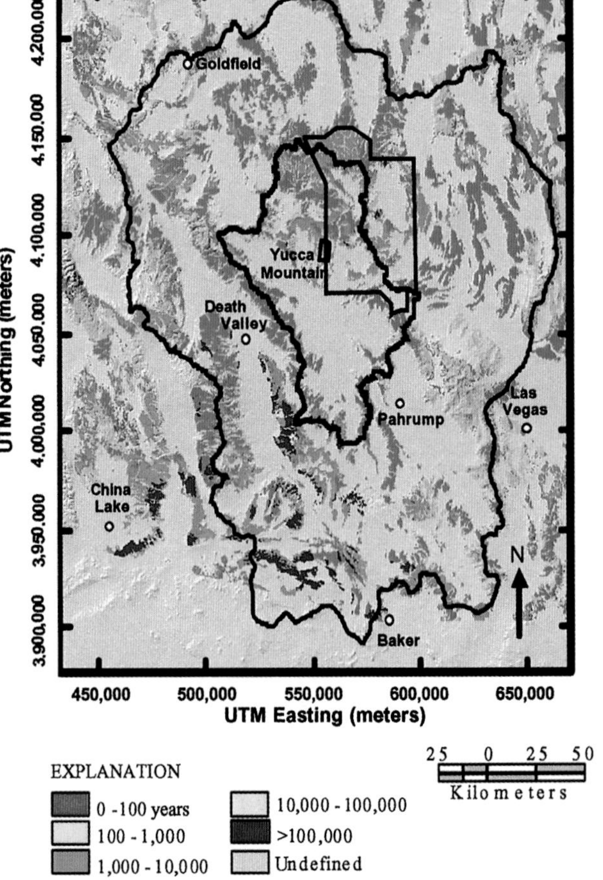

Plate 4. Calculated vertical unsaturated-zone travel time for an effective unsaturated-zone porosity of 0.1 cm^3/cm^3.

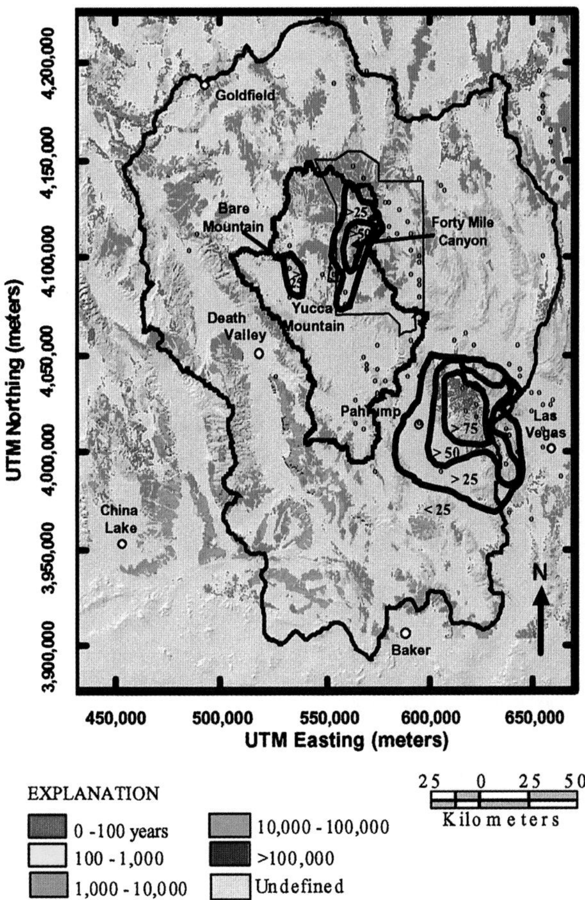

Plate 5. Calculated vertical unsaturated-zone travel time for an effective unsaturated-zone porosity of 0.03 cm^3/cm^3 and contours indicating percent modern carbon. Locations of C-14 measurements are shown as points.

this represents a fracture-dominated system with low fracture porosity. The most likely scenario is that all combinations exist in different parts of the region. The alluvial valleys are generally matrix-dominated systems, but with no recharge they become undefined in this analysis. It is unlikely that a thick unsaturated zone would be dominated by low-porosity fractures, and the fast travel times over most of the modeling domain represented by an effective porosity of 0.001 would not be expected. A fracture-dominated system with higher porosity (associated with high fracture density) would be more likely for areas such as the Paleozoic carbonates in the Spring Mountains between Pahrump and Las Vegas (Plate 5).

Testing or calibration of this model for the region can be approached by using available measurements of water-residence times at the base of the unsaturated zone or at the water table. These values could be employed inversely with the simulated net infiltration values to better refine estimates of effective flow-pathway porosity.

A comparison of the distribution of UZTT estimates using the combined matrix and fracture model, assuming 0.03 m/m effective porosity (Plate 5), was made with estimates of young water ages. These estimates are based on carbon-14 measurements of groundwater in southern Nevada [*Davisson et al.*, 1999] and indicate relatively high recharge zones. As groundwater mixing and water-rock interaction dominate most of the observed carbon-14 concentrations and bias the carbon-14 content of groundwater-dissolved inorganic carbon toward the younger end of the water age spectrum, only values of percent modern carbon (pmc) that exceed 25 are considered. Contour lines taken from *Davisson et al.* [1999] are included on Plate 5. Contour lines distinguish among carbon-14 values <25 pmc (the oldest water), 25–50 pmc, and >50 pmc (the youngest water). Areas within contour lines indicate zones of modern recharge. In the Spring Mountains the carbon-14 values systematically increase with elevation, showing the increase in modern recharge to be at high elevations in this area. These data correspond well with estimates of UZTT of between zero and 1,000 yr for the calculations using 0.03-m/m effective porosity. The other two high-recharge areas noted by the contour lines on Plate 5 are explained by a very shallow water table of less than 100 m in Fortymile Canyon just to the east of Yucca Mountain, corresponding with calculations of UZTT that are relatively fast, and high-permeability Paleozoic carbonate bedrock and high precipitation at Bare Mountain just to the west of Fortymile Canyon. Although there is a much more extensive estimate of recent recharge (0–1,000 yr) shown in Plate 5, there is not enough field data to test the model. If water wells are available in other locations than those analyzed by *Davisson et al.* [1999], our analysis certainly suggests areas where high or low values of carbon-14 might be expected. This analysis is in general agreement with their analysis.

5. SUMMARY AND CONCLUSIONS

A conceptual model of net infiltration, unsaturated-zone travel time (UZTT), and recharge for the Death Valley region has been presented. Net infiltration is a function of precipitation, soil thickness, and bedrock permeability. Present-day net infiltration is assumed to be equivalent to potential future recharge on a regional basis. Although the vertical component of the model is one-dimensional, it is assumed to be appropriate on the regional scale. The thickness of the unsaturated zone is no greater than 2,000 m and averages 500 m over the study area using the 278 m horizontal grid cells; therefore, it is unlikely that recharge would occur at a large distance laterally away from each cell.

Regional estimates of recharge were made using the net-infiltration model, and the timing of recharge was computed by using the thickness of the unsaturated zone and an assumption of effective flow-pathway porosity of the unsaturated zone. This porosity is used to estimate UZTT and is derived by knowing the bulk saturated hydraulic conductivity, the saturated hydraulic conductivity of the matrix, and net infiltration. If net infiltration exceeds the matrix permeability, then fracture flow predominates in the system and UZTT is much faster than if matrix flow were predominant. The combination of net infiltration and effective flow-pathway porosity control UZTT and, eventually, recharge. Knowledge of layering between the surface and the water table may be required to make better estimates of UZTT. This approach can be further refined as groundwater ages and hydrologic properties of the unsaturated zone become better known.

Sensitivity analyses of the numerical model used at Yucca Mountain indicate that as the climate becomes wetter, channel locations will experience a greater relative increase in net infiltration and the effect of bedrock hydraulic conductivity will become increasingly more significant. There is the likelihood of an increase in precipitation in the future, and the importance of attaining reliable estimates of the distribution of infiltration and recharge in the Death Valley region is further emphasized by the importance of the groundwater resource in the region.

The model presented here is supported by other measurements and estimates in the same and in similar basins. In terms of the estimated recharge volume relative to the input precipitation volume, when applied to basin parameters, this model agrees well with Maxey-Eakin, modified

Maxey-Eakin, and the chloride-mass-balance methods. Major deviations from the empirical Maxey-Eakin precipitation-recharge relation are caused by the bulk hydraulic conductivity of the dominant bedrock of a given basin. The major difference with this modeling approach is that the additional processes modeled give an estimate of the spatial distribution of recharge within a basin. Recharge estimates using this model and the calculations of unsaturated-zone travel time provide insight into the areas of the basin that are the most likely to contribute to recharge and respond quickly to changes in climate.

REFERENCES

Beatley, J. C., Vascular plants of the Nevada Test Site and central-southern Nevada, Ecological and geographical distributions, U.S. Department of Energy Research and Development Administration Report TID-26881, 308 p., U.S. Department of Energy, Washington, D.C., 1976.

Botkin, D. B., R. A. Nisbet, S. Bicknell, C. Woodhouse, B. Bentley, and W. Ferren, Global climate change and California's natural ecosystems, in *Global Climate Change and California, Potential Impacts and Responses*, edited by J. B. Knox and A. F. Scheuring, pp. 123–146, University of California Press, Berkeley, Calif., 1991.

D'Agnese, F. A., C. C. Faunt, A. K. Turner, and M. C. Hill, Hydrogeologic evaluation and numerical simulation of the Death Valley regional ground-water flow system, Nevada and California, U.S. Geological Survey Water-Resources Investigations Report 96-4300, 124 p., U.S. Geological Survey, Denver, Colo., 1997.

Davisson, M. L., D. K. Smith, J. Quenelle, and T. P. Rose, Isotope hydrology of southern Nevada groundwater: Stable isotopes and radiocarbon, *Water Resour. Res.*, *35*(1), 279–294, 1999.

Dettinger, M. D., Reconnaissance estimates of natural recharge to desert basins in Nevada, U.S.A., by using chloride-balance calculations, *J. Hydrol.*, *106*, 55–78, 1989.

Flint, A. L., and S. W. Childs, Modification of the Priestley-Taylor equation to estimate evapotranspiration for soil water limited conditions, *J. Agricul. & Forest Meteorol.*, *56*, 247–260, 1991.

Flint, L. E., Characterization of hydrogeologic units using matrix properties, Yucca Mountain, Nevada, U.S. Geological Survey Water-Resources Investigations Report 98-4243, 64 p., U.S. Geological Survey, Denver, Colo., 1998.

Flint, L. E., and A. L. Flint, Shallow infiltration processes at Yucca Mountain—Neutron logging data, 1984–93, U.S. Geological Survey Open-File Report 95-4035, 46 p., U.S. Geological Survey, Denver, Colo., 1995.

Flint, A. L., L. E. Flint, G. S. Bodvarsson, and E. M Kwicklis, Evolution of the conceptual model of unsaturated zone hydrology at Yucca Mountain, Nevada, in *Conceptual Models of Flow and Transport in the Fractured Vadose Zone*, National Academy Press, Washington, D.C., in press.

Grayson, D. K., The desert's past—A natural prehistory of the Great Basin, 356 p., Smithsonian Institution Press, Washington, D.C., 1993.

Harrill, J. R., and D. E. Prudic, Aquifer systems in the Great Basin region of Nevada, Utah and adjacent states—Summary Report, U.S. Geological Survey Professional Paper 1409, 66 p., U.S. Geological Survey, Denver, Colo., 1998.

Hatton, T. J., Catchment scale recharge modeling, Part 4, in *The Basics of Recharge and Discharge*, 19 p., CSIRO Publishing, Collingwood, Victoria, Australia, 1998.

Hatton, T. J., W. R. Dawes, and R. A. Vertessy, The importance of landscape position in scaling SVAT models to catchment scale hydroecological predictions, in *Space and Time Scale Variability and Interdependencies for Various Hydrological Processes*, edited by R. A. Feddes, University Press, Cambridge, Mass., 1995.

Hevesi, J. A., and A. L. Flint, Geostatistical estimates of future recharge for the Death Valley Region, *Proc. of the 9th International High-Level Radioactive Waste Management Conference, Las Vegas, Nev., May 11-15*, pp. 173–177, American Nuclear Society, LaGrange Park, Ill., 1998.

Hevesi, J. A., A. L. Flint, and J. D. Istok, Precipitation estimation in mountainous terrain using multivariate geostatistics—II. Isohyetal maps, *J. Appl. Metrol.*, *31*(7), 661–676, 1991.

Kwicklis, E. M., F. Thamir, R. W. Healy, and D. Hampson, Numerical simulation of water- and air-flow experiments in a block of variably saturated, fractured tuff, U.S. Geological Survey Water-Resources Investigations Report 97-4274, U.S. Geological Survey, Denver, Colo., 1999.

Lichty, R. W., and P. W. McKinley, Estimates of ground-water recharge rates for two small basins in central Nevada, U.S. Geological Survey Water-Resources Investigations Report 94-4104, 31 p, U.S. Geological Survey, Denver, Colo., 1995.

Maxey, G. B., and T. E. Eakin, Ground water in White River Valley, White Pine, Nye, and Lincoln Counties, Nevada: Nevada State Engineer, Water Resources Bulletin no. 8, 59 p., 1950.

Munz, P. A., A flora of southern California, 1,086 p., University of California Press, Berkeley, Calif., 1974.

Nichols, W. D., Geohydrology of the unsaturated zone at the burial site for low level radioactive waste near Beatty, Nye County, Nevada, U.S. Geological Survey Water-Supply Paper 2312, 52 p., U.S. Geological Survey, Denver, Colo., 1987.

Peterson, F. F., Landforms of the Basin and Range province defined for soil survey, Reno, Nevada, Agricultural Bulletin, no. 28, 53 p., University of Nevada, Reno, Nev., 1981.

Spaulding, W. G., Vegetation and climates of the last 45,000 years in the vicinity of the Nevada Test Site, south-central Nevada, U.S. Geological Survey Professional Paper 1329, 55 p., U.S. Geological Survey, Denver, Colo., 1985.

Waddell, R. K., Two-dimensional, steady-state model of ground-water flow, Nevada test Site and vicinity, Nevada, California, U.S. Geological Survey Water-Resources Investigations Report 82-4085, 72 p., U.S. Geological Survey, Denver, Colo., 1982.

Winograd, I. J., Radioactive waste disposal in thick unsaturated zones, *Science*, *212*(4502), 1457–1464. 1981.

Winograd, I. J., and W. Thordarson, Hydrogeologic hydrochemical framework, south-central Great Basin, Nevada-California, with special reference to the Nevada Test Site, U.S. Geological Survey Professional Paper 712-C, 126 p., U.S. Geological Survey, Denver, Colo., 1975.

Yang, I. C., and Z. E. Peterman, Possible source and residence time of perched water, in *Hydrogeology of the unsaturated zone, north ramp area of the Exploratory Studies Facility, Yucca Mountain, Nevada,* edited by J. P. Rousseau, E. M. Kwicklis, and D. C. Gillies, U.S. Geological Survey Water-Resources Investigations Report 98-4050, pp. 178–180, U.S. Geological Survey, Denver, Colo., 1999.

Yang, I. C., G. W. Rattray, and Pei Yu, Interpretation of chemical and isotopic data from boreholes in the unsaturated zone at Yucca Mountain, Nevada, U.S. Geological Survey Water-Resources investigations Report 96-4058, 58 p., U.S. Geological Survey, Denver, Colo., 1996.

Yang, I. C., Pei Yu, G. W. Rattray, J. S. Ferarese, and J. N. Ryan, Hydrochemical investigations in characterizing the unsaturated zone at Yucca Mountain, Nevada: U.S. Geological Survey Water-Resources Investigations Report 98-4132, 57 p., U.S. Geological Survey, Denver, Colo., 1998.

Alan L. Flint, Lorraine E. Flint, Joseph A. Havesi, U.S. Geological Survey, Placer Hall 6000 J Street, Sacramento, CA 95819-6129; Frank D'Agnese, U.S. Geological Survey, Tucson, AZ; Claudia Faunt, U.S. Geological Survey, San Diego, CA

Spatial and Temporal Instabilities in Water Flow through Variably Saturated Fractured Basalt on a One-Meter Field Scale

Robert K. Podgorney,[1] Thomas R. Wood,[1] Boris Faybishenko,[2] and Thomas M. Stoops[3]

Traditional approaches for predicting water flow and migration of contaminants in fractured rock vadose zones are often based on volume-averaged behavior of fluids in homogeneous porous media. However, this approach may overestimate the travel time through the vadose zone to underlying aquifers. This paper presents empirical data for nonideal infiltration and discusses factors and processes of three-dimensional spatial and temporal water-flow dynamics in variably saturated fractured basalt at the one-meter scale. Three small-scale ponded infiltration tests were conducted in fractured basalt during the summer of 1998 at the Hell's Half Acre field site in Idaho. A small reservoir (40 x 80 cm) was constructed on the surface exposure of a fracture at an overhanging basalt ledge. The tests were monitored with the purpose of determining factors that might generate nonideal flow behavior. Our major findings include: (1) infiltration and outflow were highly variable and apparently unrelated to the head in the reservoir; (2) the volume-averaged infiltration rate can be used to calculate an approximate range of the hydraulic conductivity, using Darcy's law for variably saturated flow; (3) measurements using tensiometers do not provide adequate three-dimensional resolution of the fracture and matrix capillary pressure for determining the exchange of water between the fracture and the basalt matrix; (4) the complex interactions of multiple, inherently nonlinear processes of water flow affected by gravity and capillary forces in both the fracture and matrix, as well as the fracture-matrix interaction and the effect of entrapped air, create temporal and spatial instabilities; and (5) under certain conditions, water dripping from an open fracture can exhibit deterministic chaotic behavior.

1. INTRODUCTION

Understanding the mechanisms of water flow and contaminant transport in a fractured rock vadose zone at the small field scale is critical for developing conceptual and mathematical models of migration of organic, inorganic, and radioactive contaminants and understanding the water dripping phenomenon in cavities, which could affect the integrity of waste disposal sites. A number of waste disposal sites are situated in geologic provinces with fractured rocks, including sites at the Idaho National

[1]Geosciences Department, Idaho National Engineering and Environmental Laboratory, Idaho Falls, Idaho

[2]Earth Sciences Division, Ernest Orlando Lawrence Berkeley National Laboratory, Berkeley, California

[3]Environmental Management Technical Integration Office, Idaho National Engineering and Environmental Laboratory, Idaho Falls, Idaho

Dynamics of Fluids in Fractured Rock
Geophysical Monograph 122

Published 2000 by the American Geophysical Union

Engineering and Environmental Laboratory (INEEL). Low levels of plutonium and other radioactive and organic contaminants have intermittently been detected in groundwater at several sites at INEEL. These detections were not expected or predicted using conventional volume-averaged models describing flow and contaminant transport through the vadose zone. For the past several years, INEEL and Ernest Orlando Lawrence Berkeley National Laboratory (Berkeley Lab) have investigated flow and transport through the fractured basalt vadose zone in southeast Idaho on scales ranging from tens—Box Canyon infiltration tests [*Faybishenko et al.*, 1998; 1999a,b]—to thousands of square meters—Large-Scale Infiltration Test [*Wood and Norrell,* 1996]. These tests showed some evidence of chaotic behavior for both water flow and contaminant transport. Experimental investigations, together with modeling studies [*Pruess et al.*, 1999], suggested that conventional volume-averaged approaches for calculating transport of contaminants through a fractured rock vadose zone may provide an unrealistic estimate of travel times.

Although numerous laboratory and field experiments have been conducted in porous materials, there are a limited number of data sets available for the validation of conceptual and numerical models for flow and transport in fractured rocks [*Evans and Nicholson,* 1987], implying a need for carefully controlled field investigations. Because of the complexity of flow and transport in fractured rocks, small-scale, detailed field investigations are of primary importance for developing the foundations for the basic theories governing flow and transport. The goals of this paper are to present empirical data of nonideal infiltration and discuss factors and processes affecting spatial and temporal water flow dynamics in fractured basalt using a series of infiltration tests at the one-meter scale.

To collect detailed data sets for analysis, we selected a filled site in the Hell's Half Acre (HHA) Lava Field of southeastern Idaho. This selection was the result of a careful site-evaluation process and the testing of many possible sites. The HHA test site consists of an overhanging ledge of basalt accessible on the top, front, and bottom surfaces. At the top surface there is an exposure of a single, small-aperture, vertical fracture that is the focus of this study.

2. REVIEW OF FIELD INVESTIGATIONS OF FLOW THROUGH VARIABLY SATURATED FRACTURED ROCKS

For a number of years, several experimental approaches have been used to investigate flow and transport phenomena in fractured rocks [*National Research Council,* 1996; *Wang and Narasimhan*, 1993]. Such investigations of fractured rock in the vadose zone have proven difficult, and have consisted of drilling and completion of boreholes, setting up instrumentation to provide access to fractures, and collecting water from fractures and matrix for chemical analysis. Despite the fact that field tests generate flow through both fractures and matrix, analysis of the flow rate through the matrix has often been neglected [*Kilbury et al.*, 1986].

Several field investigations of fractured rocks have obtained results that are difficult to explain using conventional approaches. A series of water flow and solute migration experiments were conducted in unsaturated fractured chalk in an arid area in the Negev Desert of Israel [*Nativ et al.*, 1995]. The researchers used anionic (chloride and bromide) and isotopic (tritium, oxygen-18, deuterium) tracer measurements to determine the amount of water flowing through the fractures and penetrating into the matrix across the fracture walls. *Dahan et al.* [1998] described an experimental field setup used to investigate flow and solute transport through a single natural fracture in the vadose zone. They designed a set of individually tagged percolation ponds, located above a chalk ledge and crossed by a fracture, and used a newly developed compartmental liquid sampler inserted in a horizontal borehole to collect infiltrating solution at several locations. This design allowed the authors to determine the spatial and temporal variations of the flow rate in the fractured rock. *Weisbrod et al.* [1998] studied flow processes in fractured chalk and determined that under variable moisture content conditions the fracture aperture, roughness, and flow channels varied with time. They also noted that under field conditions, the effective fracture aperture was more than one order of magnitude smaller than that expected from laboratory conditions.

In an attempt to characterize a fractured basalt vadose zone, several ponded infiltration tests were carried out in the Eastern Snake River Plain in Idaho. A large-scale infiltration test [*Wood and Norrell,* 1996] was conducted at INEEL (in a ponded area ~26,000 m^2), and several intermediate-scale infiltration tests were conducted at the Box Canyon site (in a ponded area 56 m^2) near Arco, Idaho [*Faybishenko et al.*, 1999a,b]. These tests adequately characterized the general volume-averaged characteristics of flow and transport in variably saturated basalt but could not be used to determine the specific mechanisms governing the flow and transport processes in individual fractures and fracture-matrix interactions. *Faybishenko et al.* [1999a] concluded that the infiltration rate through near-surface basalt fractures at the Box Canyon site in Idaho was limited by hydraulic properties of fracture infilling sediments. These results suggest that conventional approaches and concepts, such as a constant fracture aperture, may not be valid for characterizing unsaturated fractured rocks.

A number of laboratory-scale investigations were conducted using natural and artificial fractures to investigate factors and processes affecting flow and transport under controlled conditions at the centimeter-to-decimeter scale. For example, *Persoff and Pruess* [1995] injected a water-air mixture into a fracture replica under different flow rates and observed the temporal instabilities of water and air pressure at the entrance and exit of the fracture as well as the instabilities in the capillary pressure in the fracture. From their observations, the authors concluded that the redistribution of entrapped air created spatial and temporal flow instabilities. *Glass et al.* [1991], *Lenormand and Zarcone* [1989], *Nicholl et al.* [1993], and *Nicholl et al.* [1994] observed the phenomenon of preferential flow and channeling in fractures. *Su et al.* [1999] conducted a series of laboratory tests in which a dye-water mixture was supplied through a ceramic plate at the top of a transparent fracture replica with a variable aperture, allowing water to flow under unsaturated conditions. They observed that water moved down two main pathways, but the local flow geometry changed rapidly over time. *Geller et al.* [1998] studied the distribution of water in a fracture replica and determined that the water pressure and water-dripping frequency in fractures exhibited a deterministic chaotic behavior. *Or and Ghezzehei* [2000] studied water dripping into subterranean cavities in a fractured porous medium in order to improve estimates of dripping rates that could affect the integrity of waste disposal canisters placed in caverns.

Modeling methods have assumed volume averaging using effective-continuum [*Long et al.,* 1982; *Peters and Klavetter,* 1988; *Pruess and Narasimhan,* 1985], double-porosity [*Barenblatt et al.,* 1960; *Warren and Root,* 1963], dual-permeability [*Novakowski,* 1990], and multiple-interacting-continua [*Pruess and Narasimhan,* 1985] concepts, as well as fracture network models [*Cacus et al.,* 1990a,b; *Bear et al.,* 1993; *Dershowitz et al.,* 1992; *Kwicklis and Healey,* 1993; *Rasmussen,* 1987; *Smith and Schwartz,* 1993]. However, models developed for saturated conditions cannot describe the significant spatial and temporal variations of preferential flow caused by strongly heterogeneous, variably saturated flow conditions [*Birkholzer and Tsang,* 1997]. Numerical simulations conducted by *Pruess* [1999] showed that, in an unsaturated fracture with randomly distributed hydraulic parameters, a flow pattern could form a dendritic structure.

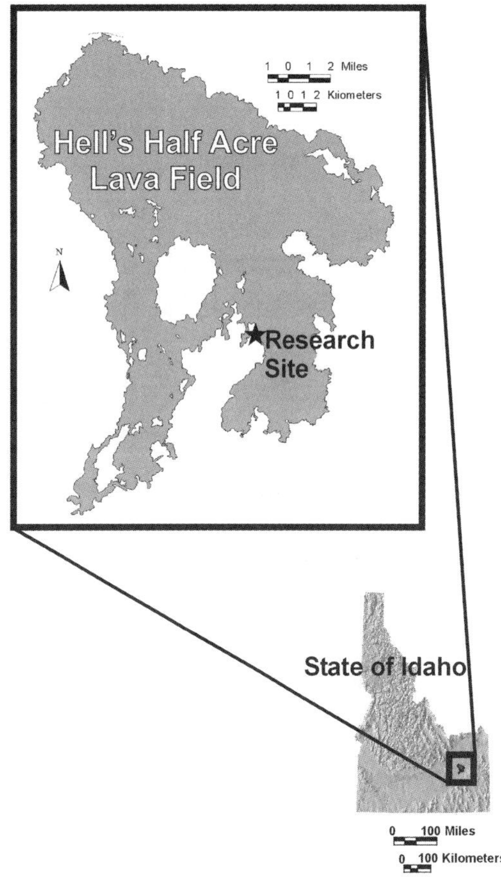

Figure 1. Location of the research site and the Hell's Half Acre lava field in Idaho.

Thus, our review of the literature indicates that there is a limited amount of data from small-scale field investigations directed at understanding the physics of flow in fractured rock vadose zones. However, we consider these tests to be important because they fill the gap between laboratory testing and large-scale characterization and predictive simulations. They have the advantage of being relatively inexpensive and easy to instrument in sufficient density for adequate characterization without the artificially constrained boundary conditions of laboratory experiments.

3. SITE LOCATION, TEST DESIGN, AND INSTRUMENTATION LAYOUT

The research site is located in the HHA Lava Field, approximately 30 km southwest of Idaho Falls, in Southeastern Idaho (Figure 1). The HHA Lava Field is the second largest holocene (~5,200 yr ago) lava field in the Eastern Snake River Plain, emplaced by a shield volcano approximately 20 km east of Atomic City, Idaho [*Kuntz et*

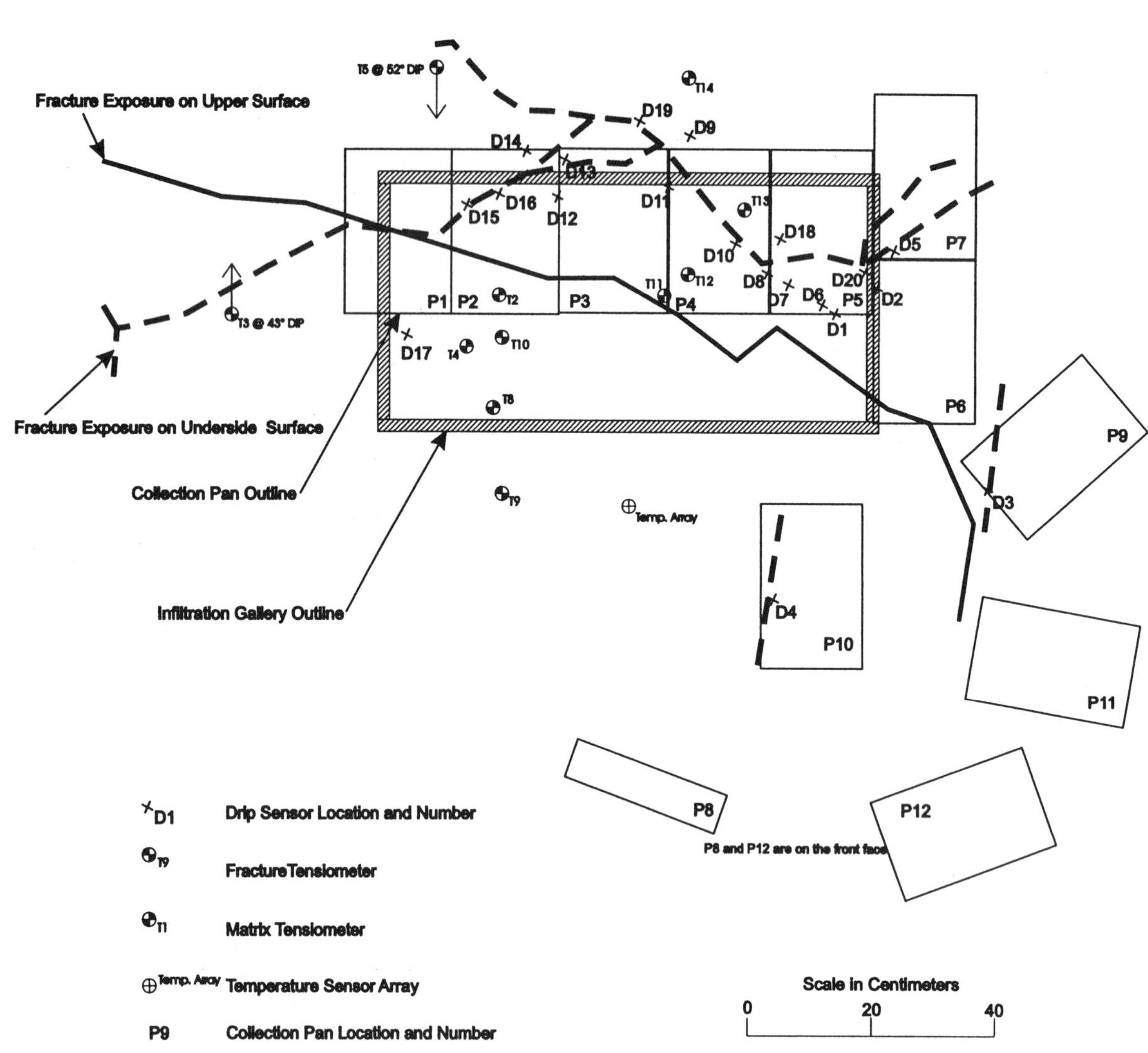

Figure 2. Plan view of the research area showing the fracture trace on the top and bottom surfaces of the basalt block, outline of the infiltration gallery, drip sensor locations, outflow collection pan locations, and tensiometer locations.

al., 1994]. The site consists of an overhanging basalt ledge or block, which is about a meter thick and is located on the edge of a collapsed lava tube. Access beneath the overhanging ledge is approximately 1 m high, 2 m wide, and 2 m deep. The basalt at the site ranges from moderately vesicular to dense, with a single vertical fracture exposed on the surface of the block. The fracture apparently bifurcates in the lower part of the block, resulting in two fracture traces on the underside of the overhanging ledge. A horizontal fracture is exposed on the face of the ledge, approximately 50 cm from the top of the block. Projection of the horizontal fracture into the block suggests that the vertical and horizontal fractures intersect. (This assumption was confirmed by the results of the infiltration tests described below.) The top, front, and undersides of the basalt block are exposed, allowing access for instrumentation and monitoring of discharge from the fracture.

A 40 × 80 cm infiltration reservoir was built on the surface of the basalt block over the exposed fracture. This water reservoir was used for maintaining water head conditions during the infiltration experiments. The walls of the infiltration reservoir were constructed of 1.9 cm plywood and painted with a sealant to prevent wicking.

The bottom of the reservoir was cut to the shape of the basalt surface and sealed with silicone sealant, with rock bolts inserted to apply downward pressure. The seal between the rock surface and the reservoir was assumed to be nearly unbroken because leaks were never visible at the perimeter of the reservoir. Figure 2 depicts the traces of the fracture on the surface and the underside of the basalt block, as well as the instrumentation layout. Figure 3 shows a cross-sectional sketch through the basalt ledge, including the location of the horizontal fracture, the location of the infiltration reservoir, and the water discharge collection system.

The site was instrumented to collect data of both temporal and spatial variations of the following parameters: water head and infiltration rates in the infiltration reservoir, water and rock temperature, capillary pressure (at 12 locations), barometric pressure, time intervals between drips (20 locations), and flow to a grid of 12 pans beneath the basalt ledge. Capillary pressures were monitored using the tensiometer design of *Hubbell and Sisson* [1998]. Tensiometers with 2.2 cm diameter ceramic cups were installed in 2.54 cm diameter holes drilled in basalt. Tensiometer ceramic cups were placed in wetted silica flour. Epoxy was used to fill the annulus between the tensiometer tube extended to the surface and basalt matrix. Twelve tensiometers were placed in and around the infiltration reservoir at depths ranging from 5 to 50 cm. Four intersected fractures (two in the vertical fracture, two in the horizontal fracture) and the remaining tensiometers were completed in the basalt matrix. Tensiometer locations are shown on Figure 2. As the tensiometers were in close proximity to one another at the research site and had similar responses throughout the tests, we will focus the discussion on only a few representative tensiometers. Tensiometers 2 and 4 are matrix tensiometers, installed at a depth of 25 cm, and located approximately 5 and 15 cm from the vertical fracture, respectively, within the footprint of the infiltration reservoir. Tensiometers 5 and 9 are called fracture tensiometers because they intersected both the fracture and basalt matrix. Tensiometer 5 was installed in a slanted borehole and intersected the vertical fracture at a depth of 30 cm in the vicinity of tensiometers 2 and 4. Tensiometer 9 intersected the horizontal fracture at a depth of 50 cm, outside the footprint of the infiltration reservoir.

The rock temperature was measured using an array of Campbell Scientific Instruments (CSI) CS-107B temperature sensors (with 15 cm vertical spacing) installed in a 2.5 cm hole drilled in the basalt block. After the sensor array was installed, the hole was filled with epoxy. The water dripping frequency was measured using specially designed sensors fabricated from 1.25 × 3.00 cm (0.50 × 1.18 in.) pieces of laminated piezoelectric film, which, when struck by a drop of water, emitted an oscillating voltage signal. Once a given piezoelectric drip sensor was placed under a drip location, its position remained fixed throughout the field season. Below the drip sensors was a grid of 20 × 30 cm pans. Each pan collected water from one or more drip locations from the area above it. The water was routed from the pans to bottles through tubes. The bottles hung from calibrated, weighing load cells. The load cells measured the weight of the bottles, thus monitoring the spatial and temporal variations in discharge from the fracture on a resolution of 20 × 30 cm.

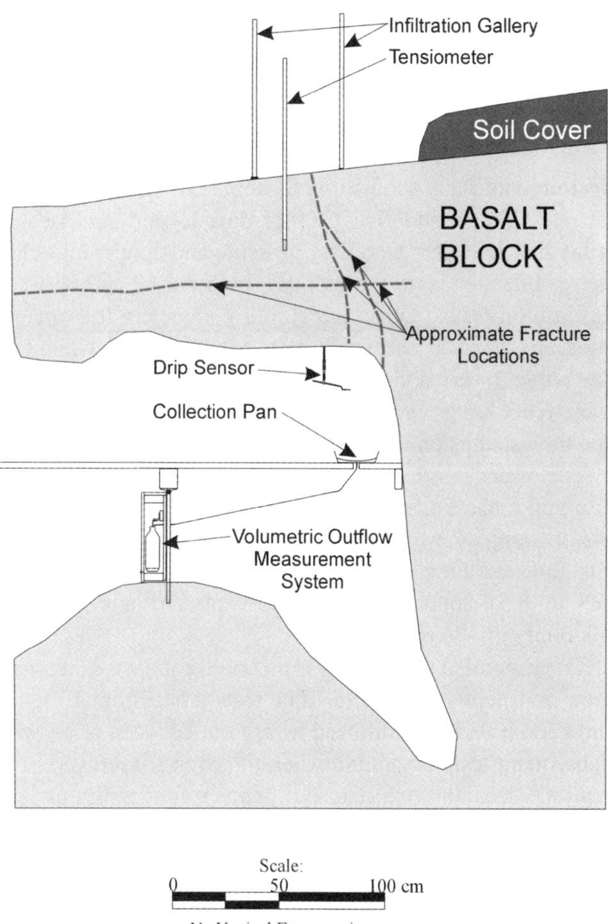

Figure 3. Cross-sectional schematic of the HHA research site.

Infiltration was calculated as the flow from the water supply tank into the reservoir. This was done by measuring the rate of water decline in the water supply tank using a pressure transducer and calculating the volume over time. During falling head tests, the infiltration was determined to be the decline of the water level times the cross-sectional area of the infiltration reservoir. A second pressure transducer was used to measure water levels in the infiltration reservoir. The accuracy of both water level transducers was +/– 0.1 cm.

All sensors were connected to a Cambell Scientific Instruments Data Acquisition System (CSI DAS) consisting of one CR21X and five CR10X data loggers and several relay multiplexers. Capillary pressure and temperature data were collected every ten minutes, both during and between infiltration tests. Other monitored parameters (barometric pressure, head in the infiltration reservoir, and weight of the bottles) were collected every minute. Drip-interval data were recorded as each drip fell, with a 0.125 s resolution of the time stamps on the drip events.

The water used for the infiltration tests was obtained from an Idaho Falls municipal water supply system. These water samples were representative of the waters of the Snake River Plain aquifer, with a pH of about 7.3 (range 6.5 to 8.5), total dissolved solids of 370 mg/l, and an alkalinity of 186 mg/l.

Three ponded infiltration tests were conducted between June and September, 1998. The tests lasted from 10 to 18 days. The rock was allowed to dry out between tests, with subsequent tests beginning when the overall average basalt matrix tension approached ambient conditions of approximately 80–100 mbars. We assumed that under this tension the fractures were empty and the matrix partially saturated. The first test, called the *constant head test*, was conducted under a constant water level of approximately 15 cm in the infiltration reservoir for nearly 17 days. The second test, called the *variable head test,* was conducted under variable water levels in the infiltration reservoir. The first part of the variable head test stepped from 15 cm to 7.5 cm and back to 15 cm. This was done to evaluate the relation of infiltration to head. The second part of the variable head test stepped repeatedly from 15 to 0 cm to evaluate the effects of entrapped air (air was allowed to enter the surface of the basalt when the reservoir was drained). The variable head test lasted nearly 10 days. The third test, called the *falling head test,* was performed from an initial constant head of 15 cm and then the water level in the reservoir was allowed to fall. This was repeated twice during the 16 day test.

4. RESULTS OF FIELD OBSERVATIONS

Before the test, we expected that after some early time fluctuations in infiltration and outflow, the system would approach steady state. However, steady state was never achieved for any appreciable length of time. We also assumed that individual tests would be repeatable. Yet despite our best attempts to maintain constant boundary conditions, the tests were not repeatable. Indeed, the variation from one identical test to the next could approach an order of magnitude difference in infiltration and/or outflow at any given time during each test.

4.1. Temporal Variations of Infiltration Rates

4.1.1. Constant head test. At the start of the constant head test, the infiltration reached a maximum of nearly 70 ml/min (Figure 4a), which corresponds to an infiltration flux (Darcy velocity) of 1.52×10^{-3} m/day. The infiltration flux was calculated as the ratio of the infiltration rate divided by the area of the reservoir. The infiltration then decreased to a minimum of less than 10 ml/min (0.22×10^{-3} m/day) on day 3. A gradual increase in the infiltration occurred over days 3 to 9 to a value of nearly 20 ml/min (0.43×10^{-3} m/day), at which time the rate abruptly increased to approximately 100 ml/min (2.17×10^{-3} m/day), the maximum observed during the test. It should be noted that these variations in infiltration occurred spontaneously under practically constant head conditions.

4.1.2. Variable head test. At the start of the variable head test, the infiltration was over 200 ml/min (3.4×10^{-3} m/day) (Figure 4b). Under the same boundary condition as the constant head test (head 15 cm), the infiltration was nearly three times higher. The flow rate then steadily declined for 4 days to about 30 ml/min (0.65×10^{-3} m/day). On day 4, the water level in the reservoir was reduced to 7 cm for 3 days. Figure 4b shows that, despite dropping the head by 50%, both the infiltration and outflow increased with time. This test concluded with three pulses in the head; by which the reservoir was filled to 15 cm for 12 hr and then left dry for 12 hr. This can be seen in Figure 4b as the three spikes in the water head, in the infiltration (inflow), and in the outflow, from approximately day 7 to the end of the test.

4.1.3. Constant-falling head test. At the start of this test, the head was set at a constant 15 cm, and the infiltration rate reached a maximum of approximately 80 ml/min (1.74×10^{-3} m/day) (Figure 4c). On days 2 through 4, the water supply into the reservoir was stopped and the water level allowed to fall. The basin was refilled to 15 cm from day 4 to day 9, after which it was again allowed to drain; the cycle was then repeated (Figure 4c). The falling head test was conducted to examine the conditions of falling head without a major perturbation to the system, as was done in the variable head test. Counterintuitively, the infiltration rate increased during the first falling head episode. After

to increase, and vice versa during periods of low flux. This may have occurred because the horizontal fracture, which was open to the atmosphere, formed a capillary barrier that was periodically overcome as water mounded above it. It can be seen in Plate 1c that, in contrast to pans 8 and 12, the flow and frequency fluctuation of the flux to other pans (6, 7, 9, 10, and 11) situated outside the footprint of the infiltration reservoir were small. We believe this characteristically lowered response reflects the fact that water contributions to flow came mainly from the matrix and fine fractures.

4.3. *Tensiometry*

Figure 7 illustrates the time variations in capillary pressure at four representative tensiometers. For comparison purposes, the head in the infiltration reservoir is also shown on Figure 7. These four tensiometers were located directly beneath the reservoir, with the exception of Tensiometer 9, which was located outside the reservoir area. Tensiometers 2 and 4 are "matrix" tensiometers, and 5 and 9 are "fracture" tensiometers (tensiometer locations are shown in Figure 2). Because assembly of the tensiometers was installed through the reservoir, there was concern that their annular seals might leak. However, the relative insensitivity of capillary pressure to changes in water level indicates that the seals were adequate (see Figure 7b). Figure 7 shows that low-amplitude fluctuations over periods of about 24 hours were diurnal effects caused by increases in air temperature and pressure during the day. These effects became less apparent during the third test (Figure 7c) as the contrast between day and night temperatures decreased at the end of the summer.

At the beginning of the infiltration tests, the capillary pressure in the fracture increased rapidly to about –25 to –15 cm because of the infiltration of water along the fracture. During the constant head test (Figure 7a), the initial increase in capillary pressure was followed by a decrease until day 2.5, which coincides with the decrease in the infiltration and outflow rates shown in Figure 4a. It is conventionally assumed that at the beginning of infiltration in a dry formation, capillary forces pull water into fractures and matrix at a high rate. Then, as the saturation of the fractures and matrix increases, both the capillary forces and the hydraulic gradient decrease, leading to a decline in the infiltration rate. We can hypothesize that flow in fractured basalt under ponded infiltration is affected by the same processes observed in laboratory and field infiltration tests in macroporous soils. Such processes include blocking of the porous space within the wetted zone by entrapped air, sealing of the soil surface by biofilms, and atmospheric air

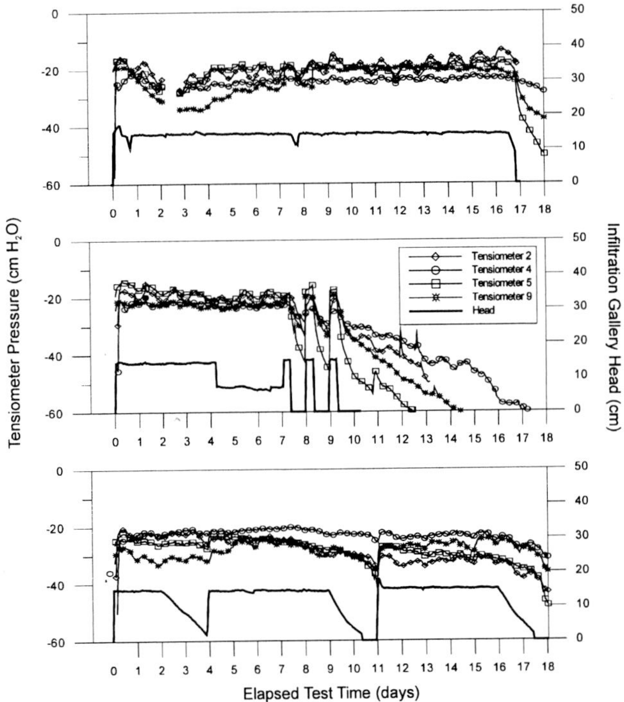

Figure 7. Tensiometry data from four typical locations for the three ponded infiltration tests: (a) constant head test, (b) variable head test, and (c) falling head test.

entering the soil profile when the water pressure falls below the air-entry pressure; consequently, the soils become unsaturated [*Faybishenko, 1995; 1999a*]. However, these effects may not be detected by tensiometers in fractured rocks because of water pressure averaging between the fracture and matrix over the surface area of the tensiometer porous tip.

From days 3 to 8 of the constant head test, the capillary pressure increased in the fracture tensiometers (Figure 7a). This corresponded to a slight increase in the flow rate (Figure 4a). For the rest of the test, capillary pressures remained practically constant, despite a spontaneous increase in the inflow and outflow rates on day 8 of the constant head test and variations in the flow rate that followed (Figure 4a). Thereafter, we assume the tensiometers did not respond to major fluctuations in flow because they were not located along the pathways leading to pans 8 and 12, which accounted for a large percentage of the total measured outflow.

Even though one would expect a positive capillary pressure in open fractures under conditions of ponded infiltration, the fracture tensiometers never showed positive pressure. This can be explained by the fact that the porous

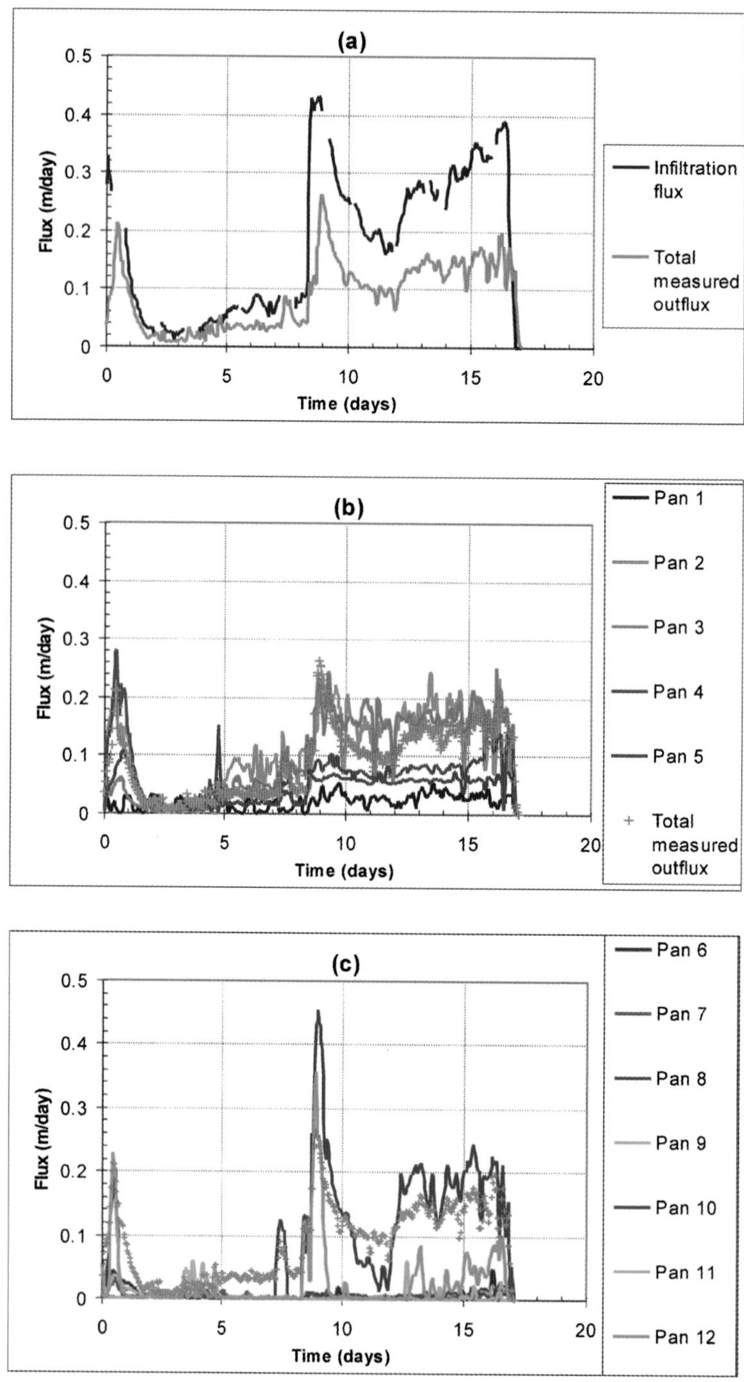

Plate 1. Time variations of (a) infiltration flux and total outflux, (b) outflux from pans located below the infiltration gallery, and (c) outflux from pans located outside of the footprint of the infiltration gallery.

tip intersected both the fracture and the matrix, and therefore the positive fracture pressure (developed over a small area of the fracture) and the negative matrix capillary pressure (developed over a larger area of the surrounding matrix) were averaged. This averaging generated a negative pressure value, which was then measured by the tensiometer [*Finsterle and Faybishenko*, 1999; *Faybishenko and Finsterle, 2000*].

After the constant head infiltration test was completed, the rock was allowed to dry out until the capillary pressure increased to approximately −100 cm. We assumed that under this pressure the fractures were practically empty and the matrix partially desaturated. However, more water was in storage in the basalt block at the start of the variable and falling head tests then at the start of the constant head test. Consequently, only a slight increase in capillary pressure was observed after the initial decrease during the second infiltration test.

This is in contrast to the constant head test, where the pressure declined during the low infiltration period between day 3 and day 8. We attribute this to the capillarity of the drier formation pulling water from the near field area into matrix storage. Capillary pressure seemed to be unaffected by the step in head from 15.0 to 7.5 cm during the variable head test (Figure 7b). However, when the basalt surface was exposed during the periodic flooding episodes, there was an immediate response in all of the tensiometers. We attribute this to atmospheric air entering the fracture and large, conductive basalt pores, which eliminated the effect of positive water pressure in the fractures on the tensiometers. As a result, the tensiometers only measured the matrix capillary pressure.

After the variable head test, the system was allowed to dry out again until the capillary pressure dropped to −100 cm, in preparation for the falling head infiltration test. Figure 7c shows that during the falling head test, the general trend of capillary pressure was correlated to the water-level variations in the pond. In general, capillary pressure remained insensitive to fluctuations of infiltration rate and changes in reservoir head, unless the bottom of the reservoir was exposed to the atmosphere.

4.4. Drip Intervals

Water flow in fractures includes the spreading and flowing of both water films and droplets on solid surfaces. This spreading and flowing is affected by a combination of surface tension, gravity, and inertial effects. Under such conditions, flow can exhibit nonperiodic, ordered patterns on a phase-space attractor that describes the evolution of the flow parameters (*Faybishenko,* 1999b). The only way to assess this behavior under field conditions is to evaluate a water dripping frequency using diagnostic parameters of chaos.

To determine whether the water dripping phenomenon exhibited chaotic behavior under a constant boundary condition, we analyzed the drip outflow rates obtained from the constant head test. Drip interval was considered to be the period of time between successive drips at a single location. The locations where drips left the ceiling of the overhanging basalt ledge remained constant over time and were usually associated with a low point at the edge of the fracture. The data analysis was conducted in phase space to determine the following six diagnostic parameters of a chaotic system (in otherwise random-looking time-series data): global embedding dimension (GED), local embedding dimension (LED), capacity (fractal) dimension, correlation dimension, Lyapunov dimension, and the largest Lyapunov exponent [*Tsonic,* 1992]. Surrogate analysis was also needed to determine whether the data set has generated a nonlinear system. This analysis can rule out stochastic processes in a time series.

4.4.1. Temporal variability. The drip intervals not only showed great variability within each test, but also between tests. Figure 8 contains some typical data for drip intervals during the constant head test (from Drip Point 10) over approximately 1.5 days (from day 11.6 to day 13.3). The horizontal axis is the drip number (n, the sequential number for the drip event), and the vertical axis is the time interval between the current and the previous drip events. Figure 8a shows that for Drip Point 10 the elapsed time between drip events is about 30 seconds, with a slight increase in the drip interval over time. At about n = 1200, there is a sudden decrease in the variation of the drip interval. The drip interval remains relatively constant through about n = 2600, at which time the drip interval again becomes erratic. The expanded view of the data for this change is shown in Figure 8b. The detailed plot of the drip interval shows repeated, low-amplitude, low-frequency fluctuations prior to the change at $n \sim 2560$. After $n \sim 2560$, the fluctuations become less regular and the amplitude and frequency increase.

An example of drip interval fluctuations with a different frequency and amplitude signature is presented in Figure 9a. It shows a series of drip intervals collected from Drip Point 9 during the constant head test (14.7 to 15.8 elapsed test days), with statistical parameters for the data presented in Table 1. We are interested in what caused these changes in the drip interval, but presently do not know what triggers the onset of this behavior. We have investigated environmental factors and have not found a correlation to rock temperature, air temperature, or barometric pressure.

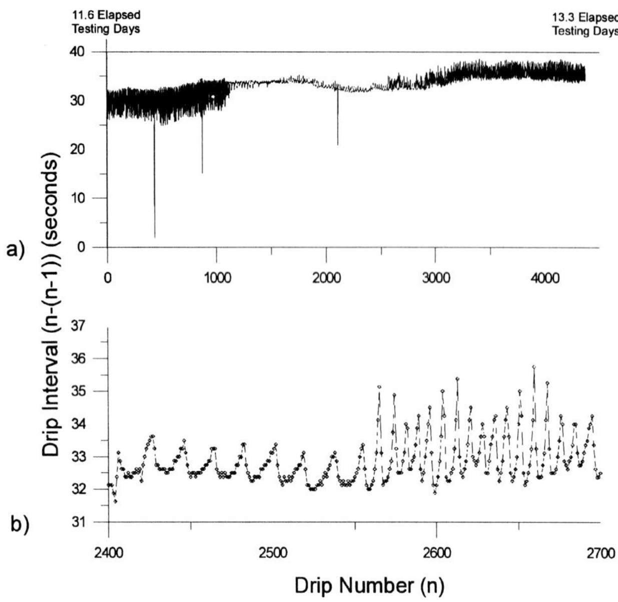

Figure 8. Time intervals between water dripping events at Drip Point 10, (a) from 11.6 to 13.3 elapsed testing days during the constant head test, and (b) an expended view for drip numbers from 2,400 to 2,700.

Ongoing and planned research will continue to explore this nonideal behavior.

A chaotic time-series analysis of data presented in Figure 9a was conducted using the Chaos Data Analyzer [*Sprott and Rowlands,* 1998] and the program CSPW [*Abarbanel,* 1996]. The results of this analysis are summarized in Table 2, and the three-dimensional attractor is shown in Figure 9b. The positive value of the largest Lyapunov exponent, the values of GED = 4 and LED = 4, and the presence of a well-defined structure of the 3-D attractor with a fractal dimension of 2.11 are evidence of low-dimensional (less than 5) deterministic chaos in the dynamics of water dripping from a fractured basalt. Some scattering of points on the attractor shown in Figure 9b indicates the presence of a stochastic component in the data set.

4.4.2. Spatial variability. The distribution of drips showed great variability, both within each individual test and between the tests, even under the same head conditions and similar infiltration rates. That is, the dripping would move from one area of the ledge ceiling to another during the tests, without an apparent correlation to head or infiltration rate. As a means of illustrating these variations, we calculated the percentage of drips recorded at each drip location by dividing the total number of drips recorded at each location by the sum of drips for all locations. The results are shown in Figure 10. The drip locations are plotted by their x-and y-coordinates under the basalt ledge, and the z-axis shows the percentage of drips. Discharge flow paths vary significantly between the tests. For example, drip location D15 (Figure 2), located near the leftmost edge of the fracture, accounted for over 6% of the total drips for the constant head test and less than 1% during the falling head test. From visual observations, there appeared to be a migration of the flow paths from the left to right side of the fracture as the testing progressed.

5. DISCUSSION OF PROCESSES AFFECTING WATER DYNAMICS IN FRACTURED BASALT

The infiltration tests were conducted on a single basalt ledge, having a few fractures and an areal extent of

(a)

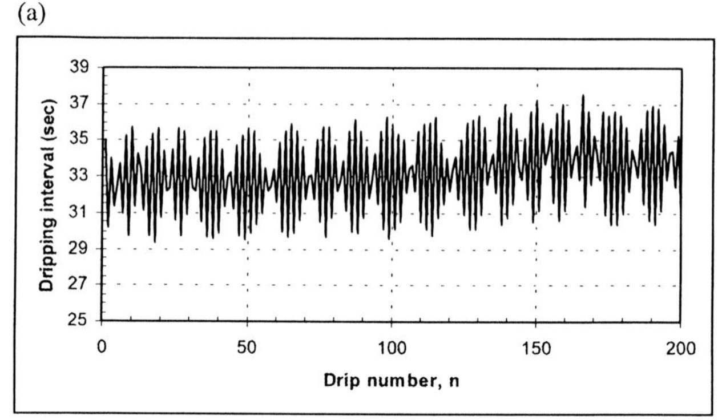

(b)

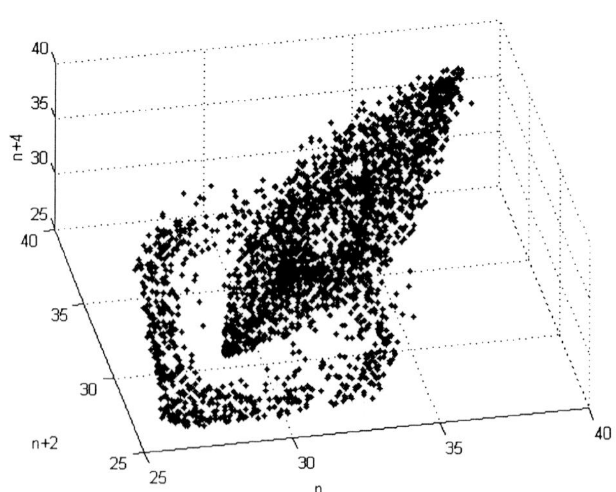

Figure 9. Time intervals between water dripping events at Drip Point 9, (a) from 14.7 to 18.3 elapsed testing days during the constant head test (first 200 points are shown), and (b) the three-dimensional chaotic attractor for this data plotted using the time delay $\tau = 3$ for the whole data set of 2,858 points.

approximately 0.5–1.0 m^2. The results clearly indicate a temporal instability of the infiltration and total outflow rates, as well as unstable, discrete outflow rates and chaotic behavior of dripping intervals. An important feature of flow in the basalt vadose zone is that the hydraulic system includes both variably saturated matrix and fractures [*Faybishenko,* 1999a]. However, it is evident from observations of dripping points that the saturated zones have a limited, local extent within flow channels in the fractures and cannot be distinctly identified using single probes such as tensiometers. Tensiometer porous tips are in contact with both the saturated fractures and unsaturated matrix.

Therefore, a tensiometer measurement represents a weighted average water potential of both the fracture and matrix water potentials, which may be quite different under transient water-flow conditions [*Finsterle and Faybishenko,* 1999; *Faybishenko and Finsterle,* 2000]. The initial rapid response of tensiometers (for example, tensiometer 9 located outside of the footprint of the infiltration reservoir), which was then followed by a slow decrease in pressure, supports the concept that under flooding conditions, the tensiometer response in fractured-porous media was determined by flow through fractures followed by matrix imbibition. Thus, tensiometers were insensitive to the main flow rate variations and could only measure approximate values of water pressure in fractured rock.

Table 1. Statistical parameters of the time-series of water dripping interval shown in Figure 9, Drip Point 9, constant head test.

Parameter	Value
Number of points	2858
Mean	32.102
Standard Error	0.060
Median	32.25
Mode	34
Standard Deviation	3.208
Sample Variance	10.290
Kurtosis	–0.7415
Skewedness	–0.1946
Range	13
Minimum	25.5
Maximum	38.5
Largest(1)	38.5
Smallest(1)	25.5
95.0% Confidence Level	0.1177

Table 2. Diagnostic parameters of a chaotic behavior of water dripping shown in Figure 9, Drip Point 9, constant head test.

Type of Parameter	Value of Parameter
Time delay, τ	2
Capacity (fractal dimension)	2.11
Global embedding dimension	4
Local embedding dimension	4
Lyapunov dimension	3.278
Largest Lyapunov exponent	0.22

At the beginning of the constant head infiltration test, the rock system was relatively dry. Immediately following the beginning of flooding, the infiltration and outflow rates increased because of the quick saturation of fractures and initially high rate of water imbibition into the dry basalt. We hypothesize that as water imbibed into the porous matrix under capillary forces, air was pushed out into the surrounding fractures. In fractures, air can become trapped in the small aperture zones, thus blocking some flow pathways. This leads to an overall decrease in the hydraulic conductivity and, consequently, a decrease in both the infiltration and outflow rates to minimum values after 2 to 4 days of ponding (Figures 4a and 7).

Despite the fact that we do not have direct measurements of the volume of entrapped air in the fractured basalt, the trend of the flow rate is identical to that observed in soils in the presence of entrapped air under field [*Faybishenko,* 1993; 1999a] and laboratory conditions [*Faybishenko,* 1995]. Using these results, we assume that the flow rate increases as entrapped air is removed in both free and dissolved phases. The simultaneous increase in the flow rate and tensiometric pressure (affected by fracture and matrix capillary pressures) after 4 to ~8 days of ponding (Figures 4a and 8a) supports the hypothesis of increasing rock saturation. The flow rates increased until the saturation reached a critical value, at which time additional flow paths opened and flow rates increased dramatically. The magnitude of the increase in flow rate at different locations along the fracture is quite varied and is accompanied by significant high-frequency flow rate fluctuations.

The other two tests showed an initial increase in flow rate because the rocks were initially wetter and water reached the underside of the ledge almost immediately. Of interest is that the outflux of pans 2, 3, 8, and 12 (which were collecting water from fractures) showed high-frequency fluctuations that were not evident in other pans or in the total volumetric outflow plot. Outflow into other

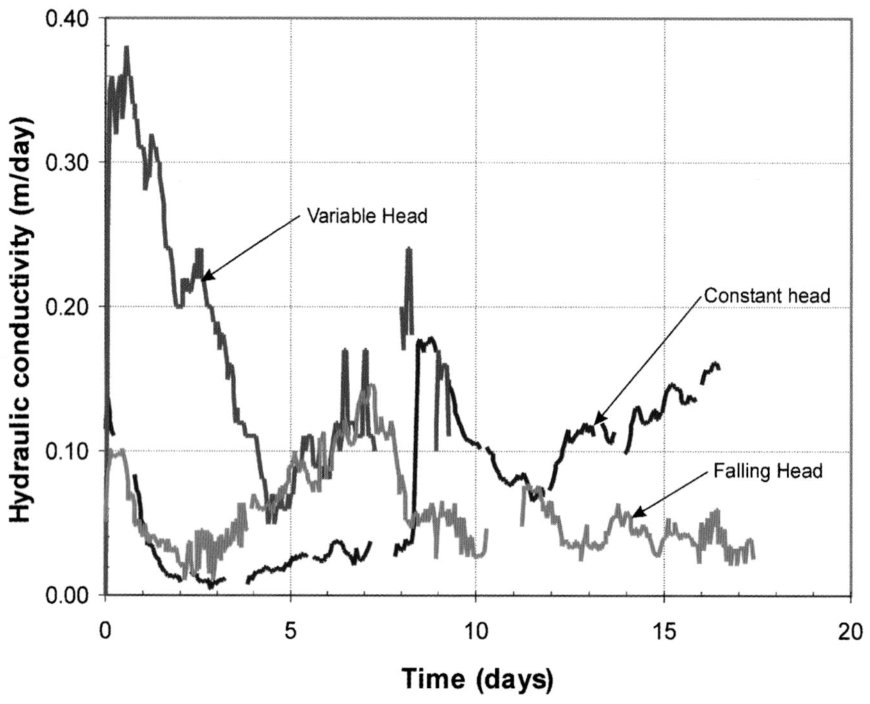

Plate 2. Time variations in the volume-averaged hydraulic conductivity of the near-surface 25 cm zone for three infiltration tests: (1) constant head test, (2) variable head test, and (3) falling head test.

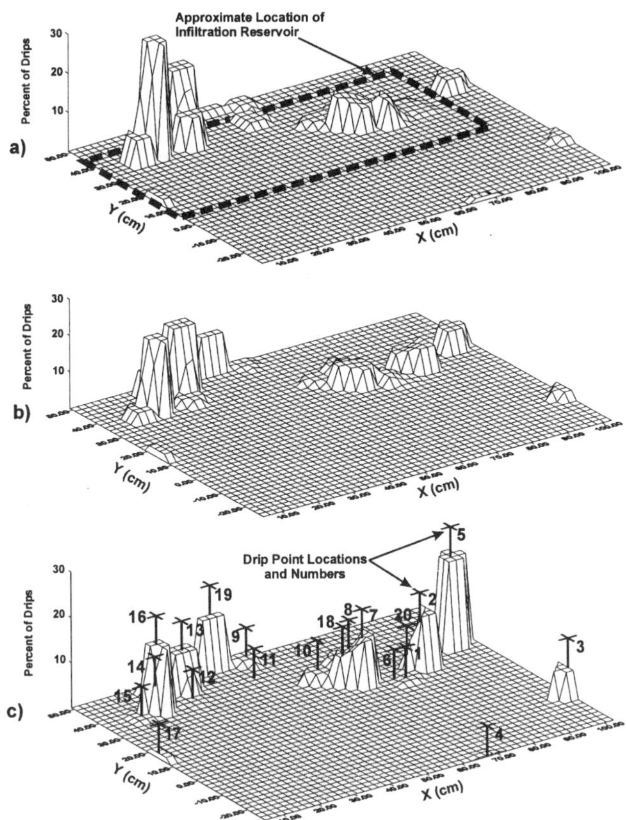

Figure 10. Spatial distribution of drip outflow for the (a) constant head test, (b) variable head test, and (c) falling head test shown as the percentage of cumulative drip events collected at each location for each test.

collection pans, which did not exhibit high-frequency fluctuations, was probably controlled by flow through the basalt matrix, which dampened the high-frequency fluctuations. The lower bound of the Darcy velocity in the matrix was approximately 0.02 m/day (which is the lowest value of the influx on Plate 1a). In order for water to reach the bottom of the ledge in less than a day, the effective matrix porosity should be a fraction of one percent. Thus, using collection pans, we were able to identify four main types of the outflow rate: (1) direct hydraulic connections between the top and the bottom of the fracture below the footprint of the reservoir (pans 2 and 3); (2) intermittent fracture flow as affected by episodic flow paths through a horizontal fracture (pans 8, 11, and 12); (3) fracture-matrix flow (pans 1, 4, and 5); and (4) mostly matrix flow (pans 6, 7, 9, and 10).

In general, it can be seen that small variations in the minimum values of the influx and outflux (between 2 and 4 days) led to a wide range of flux thereafter, which was quite unstable in time and space. The range of the flux was about three orders of magnitude—from <0.001 to 0.4 m/day. This range is evidence of chaotic behavior in flow through fractured rocks.

It is important to note that tensiometers are not able to reflect the rapid, high-frequency fluctuations that were observed for the flow rate. Furthermore, under field conditions tensiometers may not detect a positive water pressure, which is likely to develop locally within fractures under ponded conditions. Despite these limitations, the water pressure trends (Figure 7) combined with the flow rate data (Figures 4a and 7) indicate a rapid fracture saturation just after the beginning of flooding, water imbibition into the matrix, and fast water flow through fractures.

5.1. *Assessing the Range of the Volume-Averaged Hydraulic Conductivity*

Visual observations, flow rate, and water dripping measurements showed that flow in fractured basalt occurred through both the fractures and matrix. However, their contributions to the total flow rate cannot be separated in order to assess hydraulic properties of the fracture and matrix independently. To calculate the volume-averaged hydraulic conductivity over the area of the infiltration reservoir for the near-surface basalt, we assumed the validity of Darcy's law in the presence of entrapped air (*Faybishenko,* 1995) and used a one-dimensional equation:

$$Q = -K\,F \text{ grad } H \,, \tag{1}$$

where Q is the infiltration rate over the infiltration reservoir area F; K is the fractured rock hydraulic conductivity; and grad H is the hydraulic gradient determined as $grad\ H = (H_t - H_o)/z_t$. H_o is the hydraulic head at the surface ($z = 0$), determined as the water head in the infiltration reservoir, and H_t is the hydraulic head at a depth of z_t, determined as $H_t = P_t + z_t$. P_t is the water pressure (in cm) measured by a tensiometer at a depth of z_t. The hydraulic gradient was calculated using the actual water head in the infiltration reservoir and the pressure measured by Tensiometer 2 (at a depth of 25.4 cm) within the footprint of the reservoir. Because the negative water pressure developed below the infiltration reservoir (see Figure 7), the hydraulic gradient was between ~1.5 and ~2.5. Taking into account the area of the reservoir and the depth of Tensiometer 2, the effective basalt block for which the hydraulic conductivity was evaluated is approximately 81,300 cm^3. Plate 2 shows the results of calculations for the temporal variations of the hydraulic conductivity of basalt for all the tests. One can see from Plate 2 that the trends in the hydraulic conductivity are essentially the same on all tests. The initial values for the hydraulic conductivity were from 0.1 m/day (constant and falling head tests) to 0.38 m/day (variable head test), which correspond to a permeability range of 0.12 to 0.46 darcies, which probably characterizes the fracture permeability. The hydraulic conductivity then decreased to a range of 0.01 m/day (constant head test) to 0.06 m/day (variable head test), which corresponds to a permeability range of 0.012 to 0.072 darcies, typical for the basalt matrix [*Knutson et al.*, 1990; *Welhan and Reed*, 1997].

5.2. Chaotic Analysis of Water Dripping.

The motivation for using a dynamic chaotic approach to analyze the results of infiltration tests is twofold. First, contrary to continuous flow models based on Darcy's law and Richards' equation, the temporal regime of flow through an unsaturated fracture network can be characterized using the frequency of water dripping from a single observation point, assuming that flow is a combination of dripping faucets exhibiting chaotic behavior [*Shaw*, 1984]. Second, in the spatial regime, we recognize that the water funneling effect in the fracture system can combine with the lateral flow and capillary barrier effects at the fracture exits and in the overhang of the ledge. This combination of flow processes affected by capillarity, viscosity, and dripping effects can generate chaotic processes and result in chaotic attractors for dripping intervals.

For such a system, a dynamic chaotic approach can be used to describe flow in the fractured basalt vadose zone. In general, flow in a fractured basalt vadose zone can be considered a nonlinear, dynamic, deterministic-chaotic process, which is both temporally and spatially unstable. The chaotic nature of the flow is caused by the spatial and temporal variations in the fracture porosity, fracture-matrix interactions, hydraulic conductivity, fracture connectivity, and capillary barrier effects. When a new fracture flow path opens, water immediately flows from the reservoir. Therefore, small variations in flow parameters may lead to significant variations in predicted results for the fractured basalt.

6. CONCLUSIONS

Three ponded infiltration tests were conducted on a one-meter scale over a fracture in an overhanging basalt ledge at the Hell's Half Acre site in Idaho during the summer of 1998. Ponding was established using constant head, variable head, and falling head conditions. The ponded infiltration tests included measurement of reservoir water head, measurement of flow to the reservoir, temporal and spatial monitoring of water dripping from the undersurface of the ledge, measurement of the flux of water to a grid of pans beneath the overhanging ledge, and measurements of capillary pressure, barometric pressure, and temperature in the rock matrix, fractures, and atmosphere.

Conventional infiltration theories, such as by those propounded by Green and Ampt, Philip, and Kostyakov [*Jury et al.*, 1991] assume that the infiltration rate initially decreases and then approaches some steady state. However, steady state was never achieved for any appreciable length of time, and despite our best efforts to maintain constant boundary conditions, the tests were not repeatable. Indeed, the variation from one identical test to the next could approach an order of magnitude difference in infiltration and/or outflow at any given time during each test.

The infiltration rates observed during the tests were highly variable with time and were not sensitive to the head in the infiltration reservoir. The initial head conditions for the first four days of each test were very similar, yet for the variable head test, the infiltration during this initial period was about 5 times higher than that measured in the constant and falling head tests (Figures 4a, b, and c). As shown in Figures 4 and 7, the maximum values varied greatly, both within individual tests and between the tests, while the minimum value of infiltration for all tests was less than 10 ml/min (0.22×10^{-3} m/day). We hypothesize that a horizontal fracture (terminating in pans 8, 11, and 12, located outside of the footprint of the infiltration reservoir) served episodically as a capillary barrier, and when this flow path was open the infiltration rate increased because water discharged from the horizontal fracture.

The rapid appearance of infiltrating water at the bottom of the basalt ledge in both the fracture and matrix is evidence of fast, preferential flow. This phenomenon corroborates other studies by *Pruess* [1999], showing that volume averaging of flow parameters may lead to serious underpredictions of water travel time in the vadose zone.

Water dripping from a single location can be described using a chaotic model with some random component. At the same time, the volumetric outflow rates combined from several dripping locations exhibit spatial and temporal instability with primary low-frequency fluctuations, as well as secondary high-frequency fluctuations caused by local instabilities.

Acknowledgements. This research was supported in full under Grant No. 55359, Environmental Management Science Program,

Office of Science and Technology, Office of Environmental Management, United States Department of Energy (DOE). However, any opinions, findings, conclusions, or recommendations expressed herein are those of the authors and do not necessarily reflect the views of DOE. The participation in experimental studies of Douglas Whitmire, Matthew Peleschak, and Patrick Lebow; and reviews by Robert Trautz and Stefan Finsterle of Berkeley Lab, John Nimmo of the U.S. Geological Survey, and Mary Beckman of INEEL are greatly appreciated.

REFERENCES

Abarbanel, H. D. I., *Analysis of Observed Chaotic Data*, Springer-Verlag, Inc., New York, N.Y., 1996.

Barenblatt, G. E., I. P. Zheltov, and I. N. Kochina, Basic concepts in the theory of seepage of homogeneous liquids in fissured rocks, *J. Appl. Math, 24*(5), 1286–1303, 1960.

Bear, J., C.-F. Tsang, and G. de Marsily (eds.), *Flow and Contaminant Transport in Fractured Rock*, Academic Press, San Diego, Calif., 1993.

Birkholzer, J., and C.-F. Tsang. Solute channeling in unsaturated heterogeneous porous media, *Water Resour. Res., 33*(10), 2221–2238, 1997.

Cacas, M. C., E. Ledoux, G. DeMarsily, and B. Tillie, Modeling fracture flow with a stochastic discrete fracture network: Calibration and validation, 1. The flow model, *Water Resour. Res. 26*(3), 479–489, 1990a.

Cacas, M. C., E. Ledoux, G. DeMarsily, and B. Tillie, Modeling fracture flow with a stochastic discrete fracture network: Calibration and validation, 2. The transport model, *Water Resour. Res.* 26(3), 491–500, 1990b.

Dahan, O, R. Nativ, E. Adar, and B. Berkowitz, A measurement system to determine water flux and solute transport through fractures in the unsaturated zone, *Ground Water, 36*(3), 444–449, 1998.

Dershowitz, W. S., K. Redus, P. Wallmann, P. La Pointe, and C. - L. Axelsson, *The Application of Fractal Dimension in Hydrology and Rock Mechanics*, Swedish Nuclear Fuel and Waste Management Co. Tech. Report 92-17, Stockholm, Sweden, 1992.

Evans, D. D., and T. J. Nicholson (eds.), *Flow and Transport through Unsaturated Fractured Rock*, Monograph Series 42, American Geophysical Union, Washington, D.C., 1987.

Faybishenko, B. A., Two field experiments for ponded infiltration in foundation pits, *Proceedings of the 13th Annual Conference Hydrology Days*, pp. 139–148, Colorado State University, Fort Collins, Colo., 1993.

Faybishenko, B., Hydraulic behavior of quasi-saturated soils in the presence of entrapped air: laboratory experiments, *Water Resour. Res., 31*(10), 2421–2435, 1995.

Faybishenko, B., Comparison of laboratory and field methods for determination of unsaturated hydraulic conductivity of soils, *Proceedings of the International Conference on Characterization and Measurement of the Hydraulic Properties of Unsaturated Porous Media*, pp. 279–292, 1999a.

Faybishenko, B., Evidence of chaotic behavior in flow through fractured rocks, and how we might use chaos theory in fractured rock hydrogeology, *Proceedings of the International Symposium Dynamics of Fluids in Fractured Rocks: Concepts and Recent Advances*, pp. 207–212, Report LBNL 42718, Lawrence Berkeley National Laboratory, Berkeley, Calif., 1999b.

Faybishenko, B., C. Doughty, J. Geller, S. Borglin, B. Cox, J. Peterson Jr., M. Steiger, K. Williams, T. Wood, R. Podgorney, T. Stoops, S. Wheatcraft, M. Dragila, and J. Long, *A Chaotic-Dynamical Conceptual Model to Describe Fluid Flow and Contaminant Transport in a Fractured Vadose Zone*, Report LBNL 41223, Lawrence Berkeley National Laboratory, Berkeley, Calif., 1998.

Faybishenko, B. A., P. A. Witherspoon, C. Doughty, T. R. Wood, R. K. Podgorney, and J. T. Geller, *Multi-Scale Investigations of Liquid Flow in the Vadose Zone of Fractured Basalt*, Monograph Series, LBNL Report 42910, 1999a.

Faybishenko, B., C. Doughty, S. Steiger, J. Long, T. Wood, J. Jacobsen, J. Lore, and P. Zawislanski, *Conceptual Model of the Geometry and Physics of Water Flow in a Fractured Basalt Vadose Zone: Box Canyon Site, Idaho*, Report LBNL 42925, Lawrence Berkeley National Laboratory, Berkeley, Calif., 1999b.

Faybishenko, B. and S. Finsterle, On tensiometry in fractured rocks, in *Theory, Modeling, and Field Investigation in Hydrogeology: A Special Volume in Honor of Shlomo P. Neuman's 60th Birthday,* GSA, Boulder, Colo., in print, 2000.

Finsterle, S., and B. A. Faybishenko, What does a tensiometer measure in fractured rocks?, *Proceedings of the International Symposium Characterization and Measurement of the Hydraulics Properties of Unsaturated Media*, pp. 867–875, Riverside, Calif., 1999.

Geller, J. T., S. Borglin, B. A. Faybishenko, *Experimental Study and Evaluation Of Dripping Water in Fracture Models*, abstract (presented at the Chapman Conference of Fractal Scaling, Nonlinear Dynamics and Chaos in Hydrologic Systems, Clemson, S.C., May 12–15, 1998), American Geophysical Union, Washington, D.C., 1998.

Glass, R. J., T. S. Steenhuis, and J. Y. Parlange, Immiscible displacement in porous media: Stability analysis of three-dimensional, axisymmetric disturbances with application to gravity-driven wetting front instability, *Water Resour. Res., 27*(8), 1947–1956, 1991.

Hubbell, J. M. and J. B. Sisson, Advanced tensiometer for shallow or deep soil water potential measurements, *Soil Science, 163*(4), 271–276, 1998

Jury, W. A., W. R. Gardner, and W. H. Gardner, *Soil Physics, Fifth Edition,* Wiley & Associates, New York, N.Y., 1991.

Kilbury, R. K., T. C. Rasmussen, D. D. Evans, and A. W. Warrick, Water and air intake of surface-exposed rock fractures in situ, *Water Resour. Res., 22*(10), 1431–1443, 1986.

Knutson, C. F., K. A. McCormick, R. P. Smith, W. R. Hackett, J. P. O'Brien, and J. C. Crocker, *Vadose Zone Basalt Characterization, FY89 Report RWMC*, informal report prepared for the U.S. Department of Energy, Idaho Operations Office, DOE Contract No. DE-AC07-76ID01570, prepared by EG&G Idaho, Inc., July 1990.

Kuntz, M. A., B. Skipp, M. A. Lanphere, W. E. Scott, K. L.

Pierce, G. B. Dalrymple, D. E. Champion, G. F. Embree, W. R. Page, L. A. Morgan, R. P. Smith, W. R. Hackett, and D. W. Rodgers, *Geologic Map of the Idaho National Engineering Laboratory and Adjoining Areas, Eastern Idaho; U.S. Geological Survey Miscellaneous Investigation Map, I-2330, Scale 1:100,000,* U.S. Geological Survey, Boulder, Colo., 1994.

Kwicklis, E., and R. W. Healey, Numerical investigation of steady liquid water flow in a variably saturated fracture network, *Water Resour. Res., 29*(12), 4091–4102, 1993.

Lenormand, R., and C. Zarcone, Capillary fingering: percolation and fractal dimension, *Trans. Porous Media, 4*, 599–612, 1989.

Long, J. C. S., J. S. Remer, C. R. Wilson, and P. A. Witherspoon, Porous media equivalents for networks of discontinuous fractures, *Water Resour. Res. 18*(32), 645–658, 1982.

National Research Council, *Rock Fractures and Fluid Flow*, National Academy Press, Washington, D.C., 1996.

Nativ, R., E. Adar, O. Dahan, and M. Geyh, Water recharge and solute transport through the vadose zone of fractured chalk under desert conditions, *Water Resour. Res., 31*(2), 253–262, 1995.

Nicholl, M. J., R. J. Glass, and H. A. Nguyen, Wetting front instability in an initially wet unsaturated fracture, *Proceedings, Fourth High Level Radioactive Waste Management International Conference*, Las Vegas, Nev., 1993.

Nicholl, M. J., R. J. Glass, and S. W. Wheatcraft, Gravity-driven infiltration instability in initially dry nonhorizontal fractures, *Water Resour. Res., 30*(9), 2533–2546, 1994.

Novakowski, K. S., Analysis of aquifer tests conducted in fractured rock: A review of the physical background and the design of a computer program for generating type curves, *Groundwater, 28*(1), 99–107, 1990.

Or, D., and T. A. Ghezzehei, Dripping into subterranean cavities from unsaturated fractures under evaporative conditions, *Water Resour. Res., 36*(2), 381–393, 2000.

Persoff, P., and K. Pruess, Two-phase flow visualization and relative permeability measurement in natural rough-walled rock fractures, *Water Resour. Res., 31*(5), 1175–1186, 1995.

Peters, R. R., and E. A. Klavetter, A continuum model for water movement in an unsaturated fractured rock mass, *Water Resour. Res., 24*(3), 416–430, 1988.

Pruess, K., A mechanistic model for water seepage through thick unsaturated zones in fractured rocks of low matrix permeability, *Water Resour. Res., 35*(4), 1039–1052, 1999.

Pruess, K., B. Faybishenko, and G. S. Bodvarsson, Alternative concepts and approaches for modeling flow and transport in thick unsaturated zones of fractured rocks, *J. Contam. Hydrol., 38*, 281–322, 1999.

Pruess, K., and T. N. Narasimhan. A practical method for modeling fluid and heat flow in fractured porous media, *Soc. Pet. Eng. J., 25*(1), 14–26, 1985.

Rasmussen, T., Computer simulation model of steady fluid flow and solute transport through three-dimensional networks of variably saturated, discrete fractures, in *Flow and Transport through Unsaturated Fractured Rock*, Geophysical Monograph 42, edited by D. Evans and T. Nicholson, pp. 107–114, American Geophysical Union, Washington, D.C., 1987.

Shaw, R., *The Dripping Faucet as a Model Chaotic System*, Aerial Press, Santa Cruz, Calif., 1984.

Smith, L., and F. Schwartz, Solute transport through fracture networks, in *Flow and Contaminant Transport in Fractured Rock*, edited by J. Bear, C.-F. Tsang, and G. de Marsily, Academic Press, San Diego, Calif., 1993.

Sprott, J. C., and G. Rowlands, *Chaos Data Analyzer, The Professional Version 2. 1*, Physics Academic Software, Raleigh, N.C., 1998.

Su, G. W., J. T. Geller, K. Pruess, and F. Wen, Experimental studies of water seepage and intermittent flow in unsaturated, rough-walled fractures, *Water Resour. Res., 35*(4), 1019–1038, 1999.

Tsonic, A. A., *Chaos: From Theory to Applications*, Plenum Press, New York, N.Y., 1992.

Wang, J. S. Y., and T. N. Narasimhan, Unsaturated Flow in Fractured Porous Media, in *Flow and Contaminant Transport in Fractured Rock*, Academic Press, San Diego, Calif., 1993.

Warren, J. E., and P. J. Root, The behavior of naturally fractured reservoirs, *Soc. Pet. Eng. J., Transactions, AIME, 228*, 245–255, 1963.

Welhan, J. A., and M. F. Reed, Geostatistical analysis of regional hydraulic conductivity variations in the Snake River Plain aquifer, eastern Idaho, *GSA Bulletin, 109*(7), 855–868, 1997.

Weisbrod, N., R. Nativ, D. Ronen, and E. Adar, On the variability of fracture surfaces in unsaturated chalk, *Water Resour. Res., 34*(8), 1881–1887, 1998.

Witherspoon, P. A., J. S. W. Wang, K. Iwai, and J. E. Gale, Validity of cubic law for fluid flow in a deformable rock fracture, *Water Resour. Res., 16*(6), 1010–1024, 1980.

Wood, T. R., and G. T. Norrell, *Integrated Large-Scale Aquifer Pumping and Infiltration Tests, Groundwater Pathways OU 7-06, Summary Report, INEEL-96/0256*, Lockheed Martin Idaho Technologies Company, Idaho, 1996.

Robert K. Podgorney and Thomas R. Woods, Geosciences Department, Idaho National Engineering and Environmental Laboratory, Idaho Falls, ID 83415

Boris Faybishenko, Earth Sciences Division, Ernest Orlando Lawrence Berkeley National Laboratory, One Cyclotron Road, Berkeley, CA 94720

Thomas M. Stoops, Environmental Management Technical Integration Office, Idaho National Engineering and Environmental Laboratory, Idaho Falls, ID 83415

Overview of Preferential Flow in Unsaturated Fractures

Grace W. Su,[1,2,3] Jil T. Geller,[2] Karsten Pruess,[2] and James Hunt[1]

A number of laboratory, field, and theoretical studies have demonstrated that water flow through unsaturated rock fractures may proceed along localized preferential paths. We summarize results and observations from recent laboratory experiments conducted on analog fractures, transparent epoxy replicas of natural fractures, and actual rock fractures. Localized preferential flow channels were observed in all these experiments, with the exception of the experiments designed to examine film flow on rough fracture surfaces. In a number of the experiments, the flow channels underwent cycles of snapping and reforming, or intermittent flow, even in the presence of constant boundary conditions. Different modes of intermittent flow can occur during unsaturated flow in fractures, which have important implications for solute transport.

1. INTRODUCTION

Fractures in the unsaturated zone play an important role in water infiltration and contaminant transport. The transport of dissolved contaminants through fractures in the unsaturated zone and into the groundwater is a concern in many regions throughout the world, including Yucca Mountain, Nevada, and the Negev Desert in Israel. Nonaqueous phase liquids (NAPLs) have frequently been released into the unsaturated zone by leaking underground storage tanks and spills at industrial sites. Rock fractures may provide fast pathways for the migration of NAPLs through the unsaturated zone and to the groundwater table.

The conventional approach to describing seepage of liquids through unsaturated fractures involves using macroscale continuum concepts and volume averaging over a large scale [*Peters and Klavetter*, 1988]. Frequently, the fracture is modeled as parallel plates, and the aperture between the plates represents the average aperture of a natural rock fracture that contains variable apertures. This approach predicts the liquid to proceed with a spatially uniform infiltration front, subject to strong capillary imbibition effects from the rock matrix that draws the flowing liquid from the fracture [*Nitao and Busheck*, 1991; *Wang and Narasimhan*, 1993].

Field studies have provided strong evidence that water proceeds nonuniformly along fast flow paths through fractures in the unsaturated zone. At Rainier Mesa, Nevada, localized flow of water from fractures into drifts was observed [*Wang et al.*, 1993]. Tracers injected into a thick unsaturated zone of fractured chalk in the Negev Desert in Israel were observed to rapidly migrate across this zone [*Nativ et al.*, 1995]. At Yucca Mountain, elevated levels of bomb-pulse ^{36}Cl were measured at approximately 300 m depth, indicating that infiltrating water had reached those depths within only 50 years [*Fabryka-Martin et al.*, 1996]. These observations were explained by water migrating rapidly along localized preferential flow paths through the rock fractures. Preferential flow along fractures can accelerate contaminant transport to a rate well above what is predicted by the conventional models.

This paper discusses the mechanisms giving rise to preferential flow, presents a brief background on the theoretical

[1]University of California at Berkeley, Department of Civil and Environmental Engineering, Berkeley, California

[2]Ernest Orlando Lawrence Berkeley National Laboratory, Earth Sciences Division, Berkeley, California

[3]Now at U.S. Geological Survey, Water Resources Division, Menlo Park, California

Dynamics of Fluids in Fractured Rock
Geophysical Monograph 122

aspects of unsaturated flow in porous and fractured media, and gives an overview of the experimental studies of flow and transport in unsaturated fractures.

2. MECHANISMS FOR PREFERENTIAL FLOW

Preferential flow is the focusing of flow into narrow channels or fingers. Laboratory, field, and theoretical studies have demonstrated the occurrence of preferential flow in unsaturated fractures [e.g., *Scanlon*, 1992; *Nicholl et al.*, 1994; *Pruess*, 1998]. Preferential flow has also long been recognized to occur in saturated fractures [e.g., *Neretnieks et al.*, 1982; *Abelin et al.*, 1987; *Tsang and Tsang*, 1987; *Brown et al.*, 1998]. During two-phase flow, preferential flow or fingering can arise as a result of fluid instabilities created by density or viscosity differences between two immiscible fluids or because of heterogeneities in the geologic media [*Kueper and Frind*, 1988]. Unstable preferential flow has been studied extensively over a number of decades. During immiscible liquid invasion into homogeneous porous media or fractures, microscopic perturbations can form on the macroscopically planar front separating two fluid phases and develop into fingers when conditions for hydrodynamic instability are present. Viscous and gravitational forces can act as stabilizing or destabilizing forces for fingering. Viscous instability occurs when a less viscous fluid displaces a more viscous fluid and gravitational forces do not act to fully stabilize the interface. Gravitational instability occurs when a denser fluid displaces a lighter one from above, and viscous forces do not act to fully stabilize the front. Viscous instability has been studied primarily in the petroleum industry because of its importance for oil extraction, while gravitational instability has been examined extensively by soil scientists and hydrologists because of its significance in groundwater recharge and contamination.

A first-order linear stability analysis was performed by *Chuoke et al.* [1959] to determine the necessary and sufficient conditions for the onset of instability in homogeneous media. The fingering criteria are functions of the surface tension, interfacial velocity in the direction of flow, and the density and viscosity differences between the two fluids. The theoretical formulation agreed well with the experiments *Chuoke et al.* [1959] performed in parallel plates (Hele-Shaw cells) and two-dimensional cells filled with porous material. Linear stability analysis is still currently used to analyze results from laboratory fingering experiments. *Wang et al.* [1998a,b] generalized the analysis performed by *Chuoke et al.* [1959] to include the effect of wettability and successfully predicted the finger size and velocity for a range of conditions that gave rise to gravity fingering, including air entrapment, soil layering, water repellency, and surface desaturation. In nature, application of linear stability theory to predict the onset of fingering and the subsequent finger widths is difficult, since the conditions for this theory rarely, if ever, hold [*Glass and Nicholl*, 1996]. Even in nominally homogeneous media, fingers can meander and merge because of small-scale heterogeneity, and this is not described by linear stability theory [*Glass et al.*, 1989a,b; 1991].

In heterogeneous soils and fractures, fingering occurs because of the distribution of the heterogeneities rather than an unstable interface between the two fluids. When water moves through soils with discrete lenses of coarse and fine material, the lenses can act as barriers to flow and cause flow focusing [*Kung*, 1990a,b]. Flow can bypass the soil matrix and preferentially enter macropores, which are defined as large soil pores that have diameters ranging from 0.075 to 1.0 mm or more [*Luxmoore et al.*, 1990]. A similar phenomenon occurs in unsaturated fractured rocks, where the rock matrix is bypassed and flow preferentially occurs through the fractures. In macropores, water fills the entire cross section of the pore or flows as a film along the pore surface [*McCoy et al.*, 1994]. In unsaturated fractures, flow is more likely to proceed as localized fingers due to gravitational instabilities or variability in aperture sizes, rather than as a smooth sheet that fills the entire plane of the fracture. Film flow may also occur along the fracture surface driven by favorable wetting conditions and small-scale surface roughness [*Tokunaga and Wan*, 1997].

Flow of water through fractures while the rock matrix remains unsaturated is contrary to expectations from equilibrium-based capillary theory. As water enters an air-filled porous medium, capillary theory predicts that all pores smaller than a critical size will contain water, while larger pores remain air filled. According to this theory, the rock matrix should fill with water while fractures remain dry, since the matrix pores are small relative to the size of the fracture opening. Earlier conceptual models of flow in unsaturated, fractured rock assumed that most of the infiltrating water flowed through the rock matrix because the matrix has high suction pressures [*Wang and Narasimhan*, 1985; *Peters and Klavetter*, 1988]. Such models predict very low seepage rates through unsaturated fractured rock, rates that are inconsistent with the large seepage velocities measured in the field. Preferential flow through the rock fractures was hypothesized to be responsible for the large seepage velocities. Flow through the fractures in the presence of an unsaturated matrix can occur where the water flux becomes larger than the hydraulic conductivity of the matrix, resulting in ponded infiltration, or where local heterogeneity provides a direct supply of water [*Glass and*

Nicholl, 1996]. The preference for flow through the fracture instead of the matrix can also be explained using the simple model for infiltration derived by *Kao and Hunt* [1996], where the rate of infiltration is proportional to the pore size or aperture width. Since the fracture apertures are generally much larger than the pores in the matrix, the rate of infiltration into fractures should be greater than it is into the matrix.

3. APPROACHES FOR DESCRIBING UNSATURATED FLOW IN POROUS AND FRACTURED MEDIA

Petroleum engineers, soil scientists, and hydrologists have traditionally used a macroscopic approach to describe multiphase flow in porous media, involving scales much larger than the pore scale. In the unsaturated zone, liquids migrate under the combined action of gravitational, pressure, capillary, and viscous forces. At the macroscopic scale, the seepage of liquids through unsaturated porous media is usually described by Richards' equation [*Richards*, 1931], rewritten below in multiphase notation:

$$\frac{\partial(nS_l)}{\partial t} = \nabla \cdot \left[\frac{kk_{r,l}}{\mu_l} (\nabla P_l - \rho_l \mathbf{g}) \right] \qquad (1)$$

where k is the intrinsic permeability of the porous medium, $k_{r,l}$ is the relative permeability of the liquid, ρ_l is the liquid density, P_l is the liquid phase pressure, $\mathbf{g}$ is the gravitational acceleration vector, n is the porosity, and S_l is the saturation of the liquid. Relative permeability quantifies the interference of one phase with the other and varies between 0 and 1. Equation (1) assumes that the effect of displaced air during liquid infiltration can be neglected, flow is laminar, air pressure and temperature are constant, and liquid density and viscosity are also constant. Richards' equation has been applied to variable aperture fractures, based upon the conceptual model that views a fracture as a two-dimensional heterogeneous porous medium [*Pruess and Tsang*, 1990; *Pruess*, 1998]. Flow focusing, liquid ponding, and bypassing occurred in numerical simulations of water seepage in heterogeneous, unsaturated fractures [*Pruess*, 1998].

In the unsaturated zone, the capillary pressure is equal to the (constant) pressure in the air phase minus the pressure in the liquid phase ($P_c = P_{air} - P_l$). When Richards' equation is used to describe flow in porous media and fractures, constitutive relationships between the capillary pressure, saturation, and relative permeability are used. The constitutive relationships used in porous media have been applied to fractures, such as the curves derived by *Brooks and Corey* [1964] and *van Genuchten* [1980]. *Reitsma and Kueper* [1994] measured capillary pressure-saturation curves in horizontal rough-walled fractures and found that the measured curves were well represented by the Brooks and Corey curve. Relative permeability curves measured by *Fourar et al.* [1993] and *Persoff and Pruess* [1995] during two-phase flow in fractures were qualitatively similar to curves obtained in porous media. *Fourar et al.* [1993] obtained a family of relative permeability curves that depend on the liquid velocity rather than the single curve obtained in porous media.

Invasion percolation theory is also frequently used to describe preferential flow and has been successful in reproducing qualitative features of fingering observed in laboratory-scale experiments conducted in porous media and fractures [*Lenormand and Zarcone*, 1985; *Nicholl et al.*, 1994]. In percolation theory, a lattice is constructed made up of sites connected by bonds. The sites represent pore spaces while the bonds represent pore throats. The sites are given a probability of filling from a uniform probability distribution. All sites connected to a wetted site are available for filling, and the one with the highest assigned probability of filling is found and filled [*Glass*, 1993]. Simulations performed using invasion percolation are computationally faster than those using continuum, macroscale approaches [*Ewing and Berkowitz*, 1998], but it is not clear whether and how invasion percolation can be applied for flow and transport predictions in the field [*Pruess et al.*, 1999].

4. EXPERIMENTAL STUDIES ON UNSATURATED FLOW AND TRANSPORT IN FRACTURES

The ability of conceptual models to describe liquid seepage and solute transport in unsaturated fractures must be tested against experimental observations and measurements. Small-scale processes examined in laboratory experiments provide fundamental insight into mechanisms that may significantly impact larger-scale processes [*Glass et al.*, 1995]. This section summarizes laboratory work that has been conducted to study flow and transport in unsaturated fractures.

4.1. Observations of Preferential Flow Features

Schwille [1988] conducted a series of flow visualization experiments in unsaturated and saturated porous media and fractures. In the unsaturated fracture flow experiments, the fractures consisted of two flat glass plates, separated by a uniform aperture (Hele-Shaw cell) and were oriented vertically. Both smooth and rough (sandblasted) glass plates

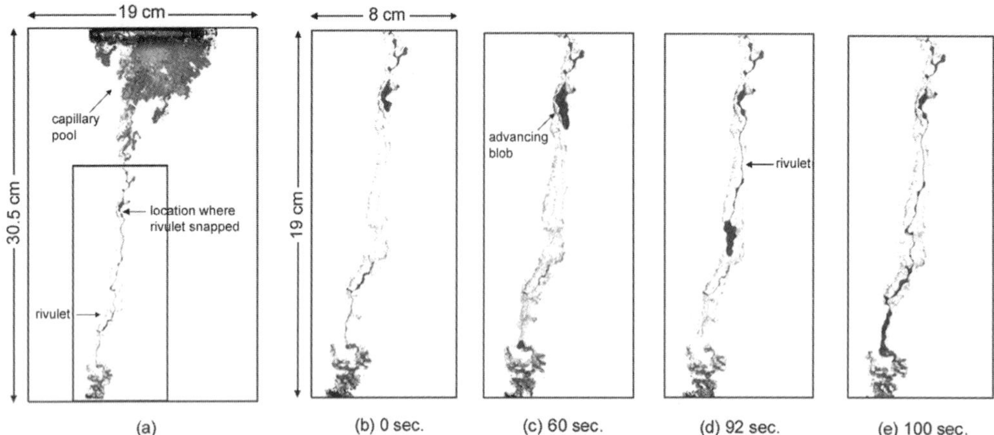

Figure 1. Liquid (dark) in an epoxy replica of a natural granite fracture. (a): full view of liquid distribution in fracture. (b)–(e): Enlargements of the boxed region in (a) where flow is intermittent. Water accumulates behind the advancing blob [(b)–(c)] and then forms a rivulet behind the blob [(d)–(e)]. Time is in seconds after the rivulet snapped [from *Su et al.,* 1999a].

were used. For the smooth plates, 0.1 mm and 0.2 mm apertures were used, and an aperture of 0.2 mm was used in the rough plates. Kerosene or trichlorethylene (TCE) was introduced as a point source into the initially dry plates at a constant flow rate. Both liquids formed one thin narrow liquid channel or rivulet in the 0.2 mm smooth plates. The distribution of the two liquids was different in the 0.1 mm smooth plates compared to the 0.2 mm plates because of the difference in the capillary force. The liquid spread laterally near the top of the 0.1 mm plates, forming a diffuse front. Below this diffuse front, multiple vertical rivulets formed, where the liquid front became gravitationally unstable. The liquid distribution was even more complex in the rough plates, where a network of fingers formed beneath the source. These fingers split and merged into each other and were much more tortuous than the rivulets that formed in the smooth plates.

Nicholl et al. [1994] used a less idealized fracture than *Schwille* [1984] in their studies of gravitational instability in inclined fractures. Their analog fracture consisted of two obscure glass plates, which had a statistically homogeneous aperture distribution. *Nicholl et al.* [1994] observed that several different boundary conditions produced gravitationally unstable fingers. Fingers immediately formed when a constant flux that was less than the saturated flux of the plates was applied to the inlet. Fingering also occurred when the plates were inverted so that the denser fluid rested on top of the lighter one. When a finite volume of water was injected uniformly across the width of the fracture, the water redistributed and fingers formed following the end of the stable infiltration. *Nicholl et al.* [1994] used the last condition to conduct a series of experiments where finger widths were measured and compared with predictions using linear stability theory. A range of angles of inclination and volumes of injected water were used. They found that linear stability analysis may provide a lower bound on their measured finger widths. The method of liquid application in their experiments caused finite perturbations along the air-water interface, however, and this may have controlled the width of the fingers that subsequently formed.

Linear stability theory is based on the assumption that an infinitesimal number of microscopic perturbations are present at the interface. This theory may not be applicable for the experimental results of *Nicholl et al.* [1994], since a finite number of perturbations formed. These results again illustrate the difficulty of maintaining conditions where linear stability theory is valid even when the system is nominally homogeneous. *Nicholl et al.* [1993] also examined the influence of initial moisture content on fingering instability in obscure glass plates and found that the fingers were faster, narrower, and longer compared to the initially dry case.

Flow visualization experiments were conducted by *Su et al.* [1999a] in a transparent epoxy replica of a natural granite fracture to examine how aperture variability affected the liquid distribution and flow behavior. A fracture replica with a realistic geometry was used to bring out the influences of heterogeneity on the flow behavior and liquid distribution that may not occur in the more idealized fractures used by *Schwille* [1988] and *Nicholl et al.* [1994]. A constant pressure or flow rate was applied to the fracture over a range of angles of inclination. *Su et al.* [1999a] observed highly localized and nonuniform flow channels consisting of broader, water-filled regions or capillary pools, connected by rivulets of liquid, as shown in Figure 1a. The capillary pools drained at higher angles of inclination because of the increase in the gravity force over the capillary force. Even in the presence of steady boundary conditions, the flow generally proceeded in an intermittent manner in all the experiments, where rivulets along the flow channel would snap and then reform. The sequence of an intermittent event along the flow channel is shown in Figures 1b–1e. The persistent occurrence of intermittent flow suggests that this is an important feature of flow through unsaturated fractures, but intermittent flow is not predicted by current conceptual models describing flow in unsaturated fractures. *Glass and Nicholl* [1996] also observed intermittent flow in their experiments conducted on an analog fracture.

Nonisothermal conditions can also significantly affect flow behavior in unsaturated fractures. *Kneafsey and Pruess* [1998] conducted a series of flow visualization experiments on natural and analog fractures to examine thermally driven liquid-flow behavior. Preferential flow paths created by vapor condensation were prevalent in these experiments. They exhibited a number of flow regimes including continuous rivulet flow, intermittent rivulet flow, and drop flow. Film flow and convective heat transfer by vapor-liquid counterflow (heat pipes) were also observed in their experiments.

Film flow along the fracture surface may also be an important factor contributing to fast flow in unsaturated fractures, where a film is defined as having one contact and one free surface. For cases where film flow in fractures is significant, conceptual models based on the aperture distribution may not correctly predict unsaturated flow in fractures. *Tokunaga and Wan* [1997] proposed and demonstrated that film flow could provide a mechanism for fast flow observed in unsaturated fractured rock. They measured average film thicknesses and velocities as a function of matrix potential on a rough fracture surface. The film velocities measured were on the order of 1,000 times faster than pore water under unit gradient saturated flow within the rock matrix.

The effect of matrix imbibition during seepage was not investigated in the experiments summarized in this section. In a homogeneous rock matrix, the total volume of water imbibed per cross-sectional area of the medium is linearly related to the square root of time. Imbibition was also found to follow a linear relationship with the square root of time even in a heterogeneous rock matrix [*Tidwell et al.*, 1995]. *Pruess* [1999] theoretically examined the significance of matrix imbibition for flow that occurred episodically and along localized preferential flow paths in fractures. *Pruess* [1999] found that the rate of matrix imbibition is small compared to rates that would occur if flow occurred as a spatially uniform sheet over long periods of time.

4.2. Intermittent Flow in Unsaturated Fractures

Flow intermittency is an important component of multiphase flow in porous and fractured media. In two-phase flow experiments using etched micromodels, designed to represent two-dimensional porous media, the non-wetting fluid snaps off during the imbibition of the wetting fluid [*Lenormand et al.*, 1983]. The geometric criterion for snap-off in a constricted cylindrical capillary was derived by *Roof* [1970]. Two-phase flow experiments conducted in horizontal fractures have demonstrated that unsteady flow may occur as a result of the interplay between capillary effects and pressure drop on account of viscous flow [*Persoff and Pruess*, 1995]. Flow oscillations have also been observed during gravity drainage of coarse sands [*Prazák et al.*, 1992]. *Kneafsey and Pruess* [1998] observed intermittent rivulet and drop flow in natural and analog fractures under conditions of partial saturation and thermal drive. Pulsating gas-water displacement in fractures was predicted in numerical simulations performed by *Thunvik and Braester* [1990]. Unsteady flow under two-phase conditions was observed in the field during the Stripa Validated Drift Experiments in Sweden [*Long et al.*, 1995].

Intermittent flow in unsaturated fractures was hypothesized by *Su et al.* [1999a] to evolve from liquid flowing through a sequence of small to large to small apertures, shown in Figure 2. Water advances under gravity through the top section until it encounters the capillary barrier at the interface of the top and middle section, which allows the formation of a rivulet through the large aperture section. When the advancing meniscus reaches the bottom section, the strong capillary force provided by the smaller aperture section pulls the rivulet through at a faster rate than the supply of water through the top section, thereby causing it to break. The flow capacity or permeability of the bottom section must be sufficient to pull the liquid away. Experiments were conducted on glass plates with a small to large to small aperture, and intermittent flow was successfully reproduced.

Three modes of flow were observed as a function of flow rate in these plates. At the lowest flow rates, the pulsating blob mode occurred, where the finger snapped before the advancing blob reached the bottom of the middle section. *Glass and Nicholl* [1996] also observed this mode of intermittent flow in their experiments conducted in analog obscure glass fractures. At intermediate flow rates, the snapping rivulet mode occurred, where the rivulet spanned the entire length of the middle section, remained connected for

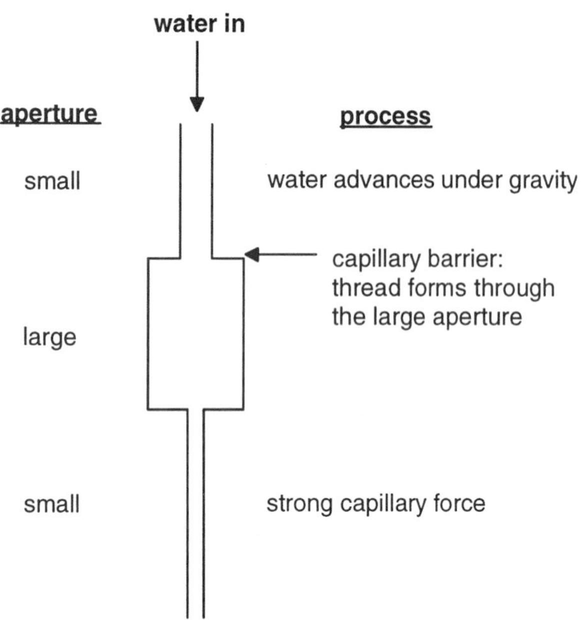

Figure 2. Cross section of the aperture sequence hypothesized to give rise to intermittent flow [from *Su et al.*, 1999a].

a period of time after the advancing blob drained into the bottom section, and then snapped. The snapping rivulet mode was the hypothesized flow mode occurring in the fracture-replica experiments. Finally, at higher flow rates, a steady thread formed in the middle section that did not snap. Images of the two modes of intermittent flow are shown in Figures 3a and 3b, with a five-second time interval between two subsequent images. A photograph of the steady liquid distribution is shown in Figure 3c.

Intermittent flow in unsaturated fractures is also strongly influenced by the wettability of the liquid on the fracture surface. The distribution of the preferential flow channels was similar for liquids with varying properties relative to water, but a strongly wetting liquid underwent much more frequent cycles of snapping and reforming than an intermediate wetting liquid [*Geller et al.*, 1998].

4.3. Effect of Preferential Flow on Solute Transport in Unsaturated Fractures

The effect of preferential flow on solute transport under unsaturated conditions has been primarily limited to studies in soils. The breakthrough curves (BTCs) measured in soils with macropores [*Wildenschild et al.*, 1994] and in heterogeneous sands [*Wildenschild and Jensen*, 1999] had a steep rise at early times and multiple peaks. *Birkholzer and Tsang* [1997] performed numerical simulations of flow and transport in unsaturated, heterogeneous porous media and obtained tracer breakthrough curves that also exhibited these features.

A number of field transport experiments have been performed in unsaturated fractured rocks [e.g., *Scanlon*, 1992; *Liu et al.*, 1995; *Nativ et al.*, 1995], but interpretation of field measurements is often difficult because detailed characterization of the subsurface is impossible, and flow and transport in unsaturated fractures is still not well-understood. Laboratory solute-transport experiments conducted to complement field and numerical studies can further the understanding of smaller-scale mechanisms that may affect measurements taken at a larger scale. *Geller et al.* [1996] conducted a laboratory-scale miscible dye tracer experiment in an epoxy replica of a granite fracture to examine the qualitative aspects of solute transport in an unsaturated fracture. The miscible dye tracer test was performed by introducing clear water into a preferential flow path already saturated with dyed water. Faster flowing regions were observed within the established preferential flow paths, even along the rivulets in the flow channel. Capillary pools along the flow path acted as reservoirs that diluted tracer concentration, as evidenced by the slow change in the dye concentration over time. Rivulets can potentially transport solute over great depths with minimal solute mixing.

Su et al. [1999b] conducted two laboratory experiments (A and B) on different epoxy fracture replicas to quantify solute transport along preferential flow paths in unsaturated fractures. Breakthrough curves of a nonsorbing chloride solute from these experiments were measured, and average travel times of the solute were obtained from the BTCs. At the laboratory scale, intermittent flow significantly affected the trend of the average velocities as a function of the relative strength of gravity. In Experiment A, the average travel times of the solute actually increased as the relative

Table 1. Summary of travel times from Experiments A & B.

Experiment	Q (ml/hr)	Angle of inclination (deg)	Average travel time (min)
		20	15.0
A	5	45	16.5
		80	19.5
A	3	20	22.8
		45	30.7
		20	16.4
B	5	45	10.7
		80	8.6

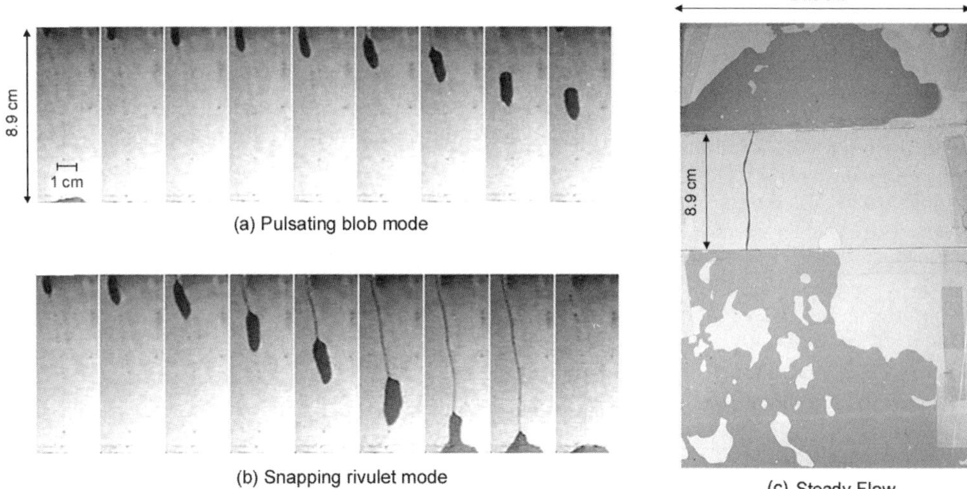

Figure 3. Flow models observed in glass plates with a small to large to small aperture sequence. Images of the finger forming to snapping in the middle section of the plates (large aperture) during the (a) pulsating blob mode and (b) snapping rivulet mode. Time between subsequent images in (a) and (b) is 5 sec. (c) Liquid distribution in the 3 sections when flow becomes steady.

strength of gravity increased, while in Experiment B the travel times decreased with increasing strength of gravity. The different trends in the measured travel times occurred because different modes of intermittent flow occurred in the two experiments.

In Experiment A, the snapping rivulet mode of intermittent flow occurred, while in Experiment B, the pulsating blob mode occurred. In Experiment A, the periods of steady flow were longer at the lower angle compared to the higher angles. The total time that the flow channel was disconnected increased at the higher angles, preventing the solute from reaching the outlet and thereby increasing the travel time. In Experiment B, however, the flow channel was always disconnected, so the travel time of the solute was controlled by how quickly the disconnected blob of water could reach the outlet. The rate at which the blob of water reached the outlet increased at the higher angles as a result of the increase in gravity force. A summary of the average travel times from these two experiments is presented in Table 1.

5. SUMMARY

Preferential flow occurs because of instabilities at the interface between two immiscible fluids or because of heterogeneities in the geologic media. Laboratory experiments conducted on analog and natural fractures with varying roughness and heterogeneity have demonstrated that infiltrating liquid proceeds along highly localized flow channels in unsaturated fractures. In addition, portions of these flow channels may undergo cycles of snapping and reforming, or intermittent flow. Intermittent flow occurs under isothermal and nonisothermal conditions and is not described by current conceptual models of unsaturated flow in fractures based on Richards' equation. Different modes of intermittent flow can occur, which have important implications for solute transport.

Future experiments should be conducted on three-dimensional fracture networks, since fracture intersections will significantly affect the seepage rates and liquid distribution of the flow channels. More detailed studies should also be performed on the interaction of the rock matrix with the flowing liquid in the fractures. Laboratory and field experiments are essential for providing direction in developing and testing conceptual models for flow in unsaturated fractures, and for obtaining a better understanding of the important flow processes that occur at different scales.

Acknowledgments. This work was supported by the Director, Office of Energy Research, Office of Health and Environmental Sciences, Biological and Environmental Research Program, of the U.S. Department of Energy under Contract No. DE-AC03-76SF000098. Thanks to Peter Persoff and two anonymous reviewers for comments and suggestions that improved this manuscript.

REFERENCES

Abelin, H., L. Birgersson, J. Gidlund, L. Moreno, I. Neretnieks, H. Widen, and J. Agren, *3-D Migration Experiment—Report 3: Performed Experiments, Results and Evaluation*, Stripa Project Technical Report 87-21, Swedish Nuclear Fuel and Waste Management Company (SKB), Stockholm, Nov., 1987.

Brown, S., A. Caprihan, and R. Hardy, Experimental observation of fluid flow channels in a single fracture, *J. of Geophy. Res., 103*, 5125–5132, 1998.

Birkholzer, J., and C.-F. Tsang, Solute channeling in unsaturated heterogeneous porous media, *Water Resour. Res., 33*(10), 2221–2238, 1997.

Brooks, R. H., and A. T. Corey, Properties of porous media affecting fluid flow, *J. Irrig. Drainage Div. Am. Soc. Civ. Eng., 92*(IR2), 61–88, 1966.

Chuoke, R. L., P. Van Meurs, and C. Van der Poel, The instability of slow, immiscible, viscous liquid-liquid displacements in permeable media, *Petrol. Trans. AIME, 216*, 188–194, 1959.

Ewing, R. P., and B. Berkowitz, A generalized growth model for simulating initial migration of dense non-aqueous phase liquid, *Water Resour. Res., 34*(4), 611–622, 1998.

Fabryka-Martin, J. T., A. V. Wolfsberg, A. V. Dixon, P. R. Dixon, S. Levy, J. Musgrave, and H. J. Turin, *Summary Report of Chlorine-36 Studies: Sampling, Analysis and Simulation of Chlorine-36 in the Exploratory Studies Facility*, Report LA-CST-TIP-96-002, Los Alamos National Laboratory, Los Alamos, N.M., 1996.

Fourar, M., S. Bories, R. Lenormand, and P. Persoff, Two-phase flow in smooth and rough fractures: Measurements and correlation by porous-medium and pipe flow models, *Water Resour. Res., 29*(11), 3699–3708, 1993.

Geller, J. T., G. Su, and K. Pruess, *Preliminary Studies of Water Seepage through Rough-Walled Fractures*, Report LBNL-38810, Lawrence Berkeley National Laboratory, Berkeley, Calif., 1996.

Geller, J. T., G. Su, H- Y. Holman, M. Conrad, K. Pruess, and J. C. Hunter-Cevera, Processes controlling the migration and biodegradation of non-aqueous phase liquids (NAPLs) within fractured rocks in the vadose zone, *FY97 Annual Report* (also, Report LBNL-41387), Lawrence Berkeley National Laboratory, Berkeley, Calif., 1998.

Glass, R. J., J.-Y. Parlange and T. S. Steenhuis, Wetting front instability, 1, Theoretical discussion and dimensional analysis, *Water Resour. Res., 27*(6), 1187–1194, 1989a.

Glass, R. J., J. Y. Parlange and T. S. Steenhuis, Wetting front instability, 2: Experimental determination of relationships between system parameters and two-dimensional unstable flow field behavior in initially dry porous media, *Water Resour. Res., 25*(6), 1195–1207, 1989b.

Glass, R. J., J. Y. Parlange, and T. S. Steenhuis, Immiscible displacement in porous media: Stability analysis of three-dimensional, axisymmetric disturbances with application to gravity-driven wetting front instability, *Water Resour. Res., 27*(8), 1947–1956, 1991.

Glass, R. J., Modeling gravity-driven fingering in rough-walled fractures using modified percolation theory, *Proceedings of the Fourth International High-Level Radioactive Waste Management Conference, Las Vegas, NV*, pp. 2042–2052, American Nuclear Society, La Grange Park, Ill., 1993.

Glass, R. J., M. J. Nicholl, and V. C. Tidwell, Challenging models for flow in unsaturated, fractured rock, through exploration of small scale processes, *Geophysical Research Letters, 22*(11), 1457–1460, 1995.

Glass, R. J., and M. J. Nicholl, Physics of gravity fingering of immiscible fluids within porous media: An overview of current understanding and selected complicating factors, *Geoderma, 70*, 133–163, 1996.

Kao, C. S., and J. R. Hunt, Prediction of wetting front movement during one-dimensional infiltration into soils, *Water Resour. Res., 32*(1), 55–64, 1996.

Kneafsey, T., and K. Pruess, Laboratory experiments on heat-driven two-phase flows in natural and artificial rock fractures, *Water Resour. Res., 34*(12), 3349–3367, 1998.

Kueper, B. H., and E. O. Frind, An overview of immiscible fingering in porous media, *J. of Contaminant Hydrology, 2*, 95–110, 1988.

Kung, K. J. S., Preferential flow in a sandy vadose zone, 1, Field observation, *Geoderma, 46*, 51–58, 1990a.

Kung, K. J. S., Preferential flow in a sandy vadose zone, 2, Mechanism and implications, *Geoderma, 46*, 59–71, 1990b.

Lenormand, R., C. Zarcone, and A. Sarr, Mechanisms of displacement of one fluid by another in a network of capillary ducts, *J. Fluid Mech., 135*, 337–353, 1983.

Liu, B., J. Fabryka-Martin, A. Wolfsberg, B. Robinson and P. Sharma, *Significance in Apparent Discrepancies in Water Ages Derived from Atmospheric Radionuclides at Yucca Mountain, Nevada*, Report LA-UR-95-572, Los Alamos National Laboratory, Los Alamos, N.M., 1995.

Long, J. C. S., O. Olsson, S. Martel and J. Black, Effects of excavation on water inflow to a drift, in *Proceedings of the Conference on Fractured and Jointed Rock Masses, Lake Tahoe, CA*, edited by L. R. Myer et al., pp. 543–549, A. A. Balkema, Brookfield, Vt., 1995.

Luxmoore, R. J., P. M. Jardine, G. V. Wilson, J. R. Jones, and L. W. Zelazny, Physical and chemical controls of preferred path flow through a forested hillslope, *Geoderma, 46*, 139–154, 1990.

McCoy, E. L., C. W. Boast, R. C. Stehouwer, and E. J. Kladivko, Macropore hydraulics: Taking a sledgehammer to classical theory, in *Soil Processes and Water Quality*, pp. 303–348, CRC Press, Boca Raton, Fla., 1994.

Nativ, R., E. Adar, O. Dahan, and M. Geyh, Water recharge and solute transport through the vadose zone of fractured chalk under desert conditions, *Water Resour. Res., 31*(2), 253–261, 1995.

Neretnieks, I., T. Eriksen, and P. Tahtinen, Tracer movement in a single fissure in granitic rock: Some experimental results and their interpretation, *Water Resour. Res., 18*(4), 849–825, 1982.

Nicholl, M. J., R. J. Glass, and S. W. Wheatcraft, Gravity-driven infiltration instability in initially dry nonhorizontal fractures, *Water Resour. Res., 30*(9), 2533–2546, 1994.

Nicholl, M. J., R. J. Glass, and H. A. Nguyen, Wetting front instability in initially wet fractures, *Proceedings of the Fourth*

International High-Level Radioactive Waste Management Conference, Las Vegas, NV, pp. 2023–2032, American Nuclear Society, La Grange Park, IL, 1993.

Nitao, J., and T. Buscheck, Infiltration of a liquid front in an unsaturated, fractured porous medium, *Water Resour. Res.*, *27*(8), 2099–2112, 1991.

Persoff, P., and K. Pruess, Two-phase flow visualization and relative permeability measurement in natural rough-walled rock fractures, *Water Resour. Res.*, *31*(5), 1175–1186, 1995.

Peters, R. R., and E. A. Klavetter, A continuum model for water movement in an unsaturated fractured rock mass, *Water Resour. Res.*, *24*(3), 416–430, 1988

Prazák, J., M. Sír, F. Kubík, J. Tywoniak and C. Zarcone, Oscillation phenomena in gravity-driven drainage in coarse porous media, *Water Resourc. Res.*, *28*(7), 1849–1855, 1992.

Pruess, K., and Y. Tsang, On two-phase relative permeability and capillary pressure of rough-walled rock fractures, *Water Resourc. Res.*, *26*(9), 1915–1926, 1990.

Pruess, K., On water seepage and fast preferential flow in heterogeneous, unsaturated rock fractures, *J. Cont. Hydrology*, *30*, 333–362, 1998.

Pruess, K., A mechanistic model for water seepage through thick unsaturated zones in fractured rocks of low matrix permeability, *Water Resourc. Res.*, *35*(4), 1039–1051, 1999.

Pruess, K., B. Faybishenko, G. S. Bodvarsson, Alternative concepts and approaches for modeling flow and transport in thick unsaturated zones of fractured rocks, *J. Cont. Hydrology*, *38*, 281–322, 1999.

Reitsma, S., and B. H. Kueper, Laboratory measurement of capillary pressure-saturation relationships in a rock fracture, *Water Resour. Res.*, *30*(4), 865–878, 1994.

Richards, L. A., Capillary conduction of liquids through porous medium, *Physics*, *1*, 318–333, 1931.

Roof, J. G., Snap-off of oil droplets in water-wet pores, *Soc. Pet. Eng. J.*, *10*, 85–90, 1970.

Scanlon, B. R., Moisture and solute flux along preferred pathways characterized by fissured sediments in desert soils, *J. of Cont. Hydrology*, *10*, 19–46, 1992.

Schwille, F., Spreading as a fluid phase in a fractured medium, in *Dense Chlorinated Solvents in Porous and Fractured Media*, translated by J. F. Pankow, pp. 61–72, Lewis Publishers, Chelsea, Mich., 1988.

Su, G., J. Geller, K. Pruess, and F. Wen, Experimental studies of water seepage and intermittent flow in unsaturated, rough-walled fractures, *Water Resour. Res.*, *35*(4), 1019–1037, 1999a.

Su, G., J. Geller, K. Pruess, and J. Hunt, Laboratory experiments on solute transport in unsaturated fractures, *Proceedings of Dynamics of Fluids in Fractured Rock: Concept and Recent Advances*, Report LBNL-42718, pp. 308–310, Lawrence Berkeley National Laboratory, Berkeley, Calif., 1999.

Thunvik, R., and C. Braester, Gas migration in discrete fracture networks, *Water Resourc. Res.*, *26*(10), 2425–2434, 1990.

Tidwell, V. C., R. J. Glass, and W. Peplinski, Laboratory investigation of matrix imbibition from a flowing fracture, *Geophysical Res. Letters*, *22*(11), 1405–1408, 1995.

Tokunaga, T., and J. Wan, Water film flow along fracture surfaces of porous rock, *Water Resour. Res.*, *33*(6), 1287–1295, 1997.

Tsang, Y. W., and Tsang, C.-F., Channel model of flow through fractured media, *Water Resour. Res.*, *23*(3), 467–479, 1987.

van Genuchten, M. T., A closed form equation for predicting the hydraulic conductivity of unsaturated soils, *Soil Sci. Am. J.*, *44*, 892–898, 1980.

Wang, J. S. Y, and T. N. Narasimhan, Hydrologic mechanisms governing fluid flow in a partially saturated, fractured, porous medium, *Water Resour. Res.*, *21*(12), 1861–1874, 1985.

Wang, J. S. Y, N. G. W. Cook, H. A. Wollenberg, C. L. Carnahan, I. Javandel, and C.-F. Tsang, Geohydrologic data and models of Rainer Mesa and their implications to Yucca Mountain, in *Proceedings, Fourth Annual International High Level Radioactive Waste Management Conference, Las Vegas, NV*, pp. 675–681, American Nuclear Society, La Grange Park, Ill., 1993.

Wang, Z., J. Feyen, and C. J. Rietsma, Susceptibility and predictability of conditions for preferential flow, *Water Resour. Res.*, *34*(9), 2169–2182, 1998a.

Wang, Z., J. Feyen, and D. E. Elrick, Prediction of fingering in porous media, *Water Resour. Res.*, *34*(9), 2183–2190, 1998b.

Wildenschild, D., K. H. Jensen, K. Villholth, and T. H. Illangasekare, A laboratory analysis of the effect of macropores on solute transport, *Ground Water*, *32*(3), 381–389, 1994.

Wildenschild, D. and K. H. Jensen, Laboratory investigations of effective flow behavior in unsaturated heterogeneous sands, *Water Resour. Res.*, *35*(1), 17–27, 1999.

Grace W. Su, U.S. Geological Survey, Water Resources Division, 345 Middlefield Road, MS 421, Menlo Park, CA 94025

Jil T. Geller and Karsten Pruess, Lawrence Berkeley National Laboratory, One Cyclotron Road, MS 90-1116, Berkeley, CA 94720

James Hunt, Department of Civil and Environmental Engineering, University of California, Berkeley, CA 94720

Fracture-Matrix Flow: Quantification and Visualization Using X-Ray Computerized Tomography

A.S. Grader,[1] M. Balzarini,[2] F. Radaelli,[2] G. Capasso,[3] and A. Pellegrino[2]

Natural and artificially induced fractures in a reservoir have a great impact on fluid flow patterns and on the ability to recover hydrocarbons. This paper focuses on multiphase flow in a rock sample that was artificially fractured and explores the application of x-ray computerized tomography (CT) imaging to determine the distribution of porosity and to track hydrocarbon loading of the sample. A method for establishing the relationship between porosity in the fractured zone and the fracture aperture is demonstrated. Converging and diverging flow patterns at the fracture tips are also shown. After saturating the sample with water, oil is injected and the loading process investigated. Displacement of the water out of the matrix layers is controlled by the fracture and by the individual properties of the layers. Water in the high-porosity and high-permeability layers is displaced first. The channeling of the flow by the fracture bypasses matrix fluid and increases the length of time needed to complete the hydrocarbon loading of the core with the displacing phase. Four-dimensional CT data presented in this paper form the basis for a fluid flow simulation that will be used to determine relative permeability and capillary-pressure distributions.

1. INTRODUCTION

The presence of fractures in a reservoir has a great impact on fluid flow patterns and on the reservoir's ability to transport fluids. Fractures affect both the recovery of dense nonaqueous phase fluids (DNAPLs) in groundwater contamination cases and the extraction of large volumes of oil and gas. Fractures can have either a positive or negative impact on oil and gas production. In tight formations, the naturally fractured system provides access to the hydrocarbon fluids stored in the matrix. In a geothermal formation, the fracture system provides access for heat mining. In low-productivity wells, hydraulically induced fractures provide the necessary connection between the wells and the reservoir to allow economical production. In some cases, however, fractures can have a negative effect on production or recovery operations. This occurs when they provide bypassing paths, especially in production-injection systems. For example, injected fluid may preferentially flow through the fractures, leaving behind inaccessible hydrocarbons. In geothermal injection-production schemes the fractures cause the water to short-circuit and arrive cold at the production wells. In secondary and tertiary oil recovery projects, injected fluid may channel through the fractured network and therefore not contact the fluid in the matrix. Fractures may also be nonconductive and form barriers to fluid flow. In other cases, they can be used to enhance the efficiency of displacement operations if the main flow directions are perpendicular to the orientation of the fractures. However, if the flow is parallel to the orientation of the fractures, displacement efficiency becomes poor.

[1]Pennsylvania State University, University Park, Pennsylvania
[2]ENI/AGIP, Italy
[3]Politecnico di Torino, Italy

Dynamics of Fluids in Fractured Rock
Geophysical Monograph 122

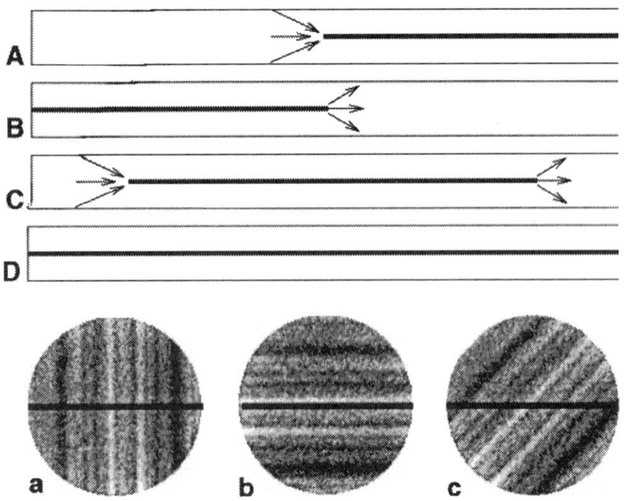

Figure 1. Fracture geometry and orientation.

Fractures have a strong effect on the pressure distribution in the flow domain. The heterogeneous nature of the flowing system and the presence of multiple phases create a difficult predictive environment [*Piggot et al.*, 1991; *Lichtner and Seth*, 1996; *Zimmerman and Bodvarsson*, 1996]. In this paper we present a preliminary study that explores the impact of an artificially induced fracture on fluid-flow patterns in a layered rock sample using x-ray computerized tomography (CT) imaging. The goals are to map the porosity distribution in the matrix and the fracture, quantify the flow in the sample, and describe the mass transport between the matrix and in the fracture in a two-phase flow environment. We focus on the CT methodology and possible use of 4D CT for understanding and simulating multiphase processes in fractured rocks.

2. EXPERIMENTAL PROCEDURE

2.1. Induced Fracture

The experiment described in this paper used a layered Berea sandstone sample with a length of 600 mm and a diameter of 53 mm. Berea is a calcite-cemented natural sandstone with about 5% clay content, typical porosities of 18%–22%, and an unconfined compressive stress ranging from 6 to 15 ksi [see *Clark*, 1966]. The specific sample used in this experiment had an average permeability of 200 md. The layers were parallel to the direction of flow. Figure 1 shows four possible fracture configurations (A, B, C, D). The work presented here focuses on case 1-C, where the fracture is fully internal and does not reach either end of the core. Figure 1 also shows three possible fracture orientations with respect to the layers in the sample: perpendicular, parallel, and diagonal (Figures 1a, 1b, and 1c, respectively). The perpendicular orientation was used in this experiment (1a); thus, the final configuration was C-a. During the experiment the sample was oriented in the imager with the fracture in a horizontal manner to minimize gravitational effects in the fracture during two-phase flow periods. The sample was fractured by a Brazilian-like test [see *Vukuturi et al.*, 1974]. In order to obtain a longitudinal planer fracture, the cylindrical sample was compressed between two parallel plates, thereby inducing a tensile state of stress in the center of the cylinder. We compressed the sample, putting it along the diagonal of the two square plates of the loading frame (Figure 2) so as to induce a fracture along most of the core. The two compression plates measured 420 mm × 420 mm with a diagonal of 590 mm.

This technique is a modification of the Brazilian test. In classical rock mechanics, this test is used with a thin disc loaded by uniform pressure applied radially over a short strip of the circumference at opposing sides (Figure 3). The underlying hypothesis of the Brazilian test is that the fracture starts from the center of the disc where, assuming that the material is homogeneous, isotropic, and linearly elastic, the value of the tangential stress σ_θ at the center of the disc is $\sigma_\theta = -F/(\pi r_o t)$, where F is the force applied, r_o is the radius of the disc, and t is the thickness of the disc. Values of the tangential stress σ_θ the radial stress σ_r, and the shear stress τ along the loaded diameter are given in Figure 3.

During compression, however, the conditions differ from the elastic analysis case. The concentration of stresses in

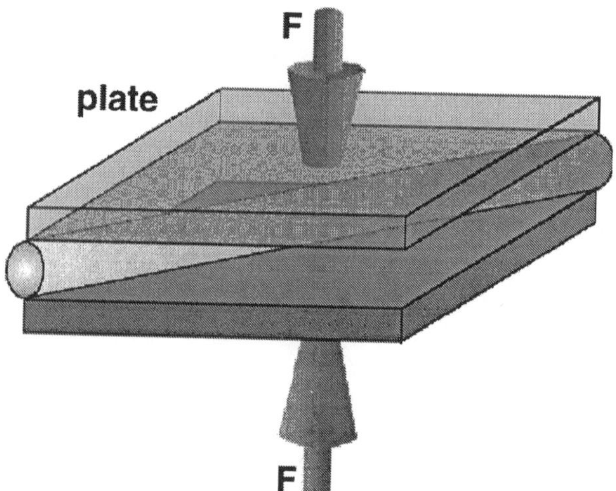

Figure 2. The modified Brazilian compression test.

the line of loading induces the development of high-shear stresses, which cause local crushing, thereby changing the loading conditions from a line to a strip. In the zone of contact, tangential stresses are developed, exerting a horizontal compressive restraint, which results in the formation of a small incipient shear wedge.

Using Brazilian-like compression techniques, we were able to create the desired fracture, almost perpendicular to the layers in the central part of the core. Two fractures were visible on the side of the core, confirming the formation of a shear wedge. A visual inspection of the fracture confirmed that the fracture fully opened across the core and was a good quality tensile fracture.

2.2. Apparatus

Figure 4 shows a schematic of the experimental apparatus, which has three main components: a fluid-flow system, a sample containment system, and a CT-imaging system. The fluid-flow system is designed to handle two-phase injection in a "once-through" mode, or in a circulating mode that minimizes errors in the measurements of fluid volume. The fluids used in the experiment were water and mineral oil. The sample was contained in a modified triaxial cell constructed out of aluminum that facilitated x-ray imaging, and was rated at a confining pressure of 20 MPa. The experiment was performed at a confining pressure of 3 MPa. The CT imager is a fourth-generation medical scanner. The images presented in this study were acquired at 130 kV at a 100-mA setting with a slice thickness of 2 mm and a scanning time of 4 s. The in-plane pixel resolution was 0.5 mm. A single CT pixel volume, termed voxel, has a dimension of 0.5 × 0.5 × 2.0 mm. The measured value of each voxel represents the average x-ray attenuation of the material in the volume. Hence, even features smaller than the physical resolution of the imager can be detected. In the case of a "flat" feature, such as a fracture, the large-scale topology can be determined. To enhance the attenuation contrast between oil and water, the water phase used in the experiment was tagged with 3% by weight potassium iodide (KI) to provide high x-ray attenuation. The oil was not tagged and had a low x-ray attenuation. The core assembly and the fluid flow system were mounted on traveling tables in the imaging unit, allowing the acquisition of tomographical images as well as digital radiographs.

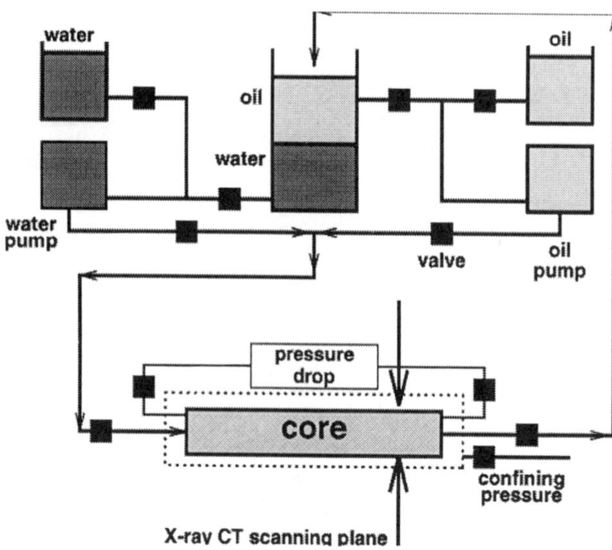

Figure 4. A schematic diagram of the experimental apparatus.

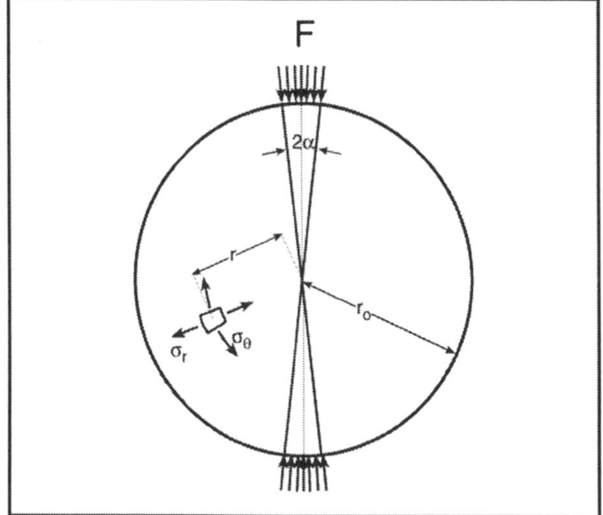

Figure 3. A schematic layout of the Brazilian test.

The main goals of the experiment were to obtain a three-dimensional porosity distribution and to study the oil-loading process. The experiment consisted of the following steps:

1. Assembling the sample in the core holder and mounting the sample and holder in the scanner
2. Applying vacuum to the sample
3. Calibrating the scanner
4. Scanning the core under dry conditions
5. Saturating the core with KI water
6. Scanning the core under water-saturated conditions.
7. Injecting oil into the core in several stages while acquiring images

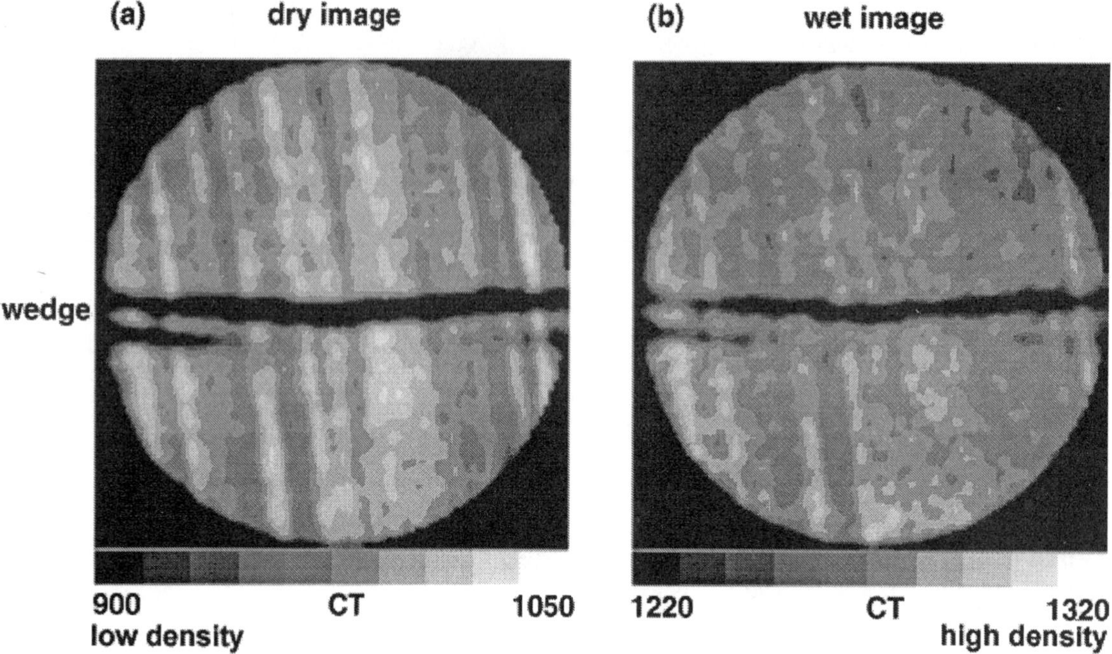

Figure 5. Cross-sectional images, 270 mm from the sample inlet.

3. POROSITY MAPPING

The porosity distribution in the core was determined using the CT-analysis method presented by *Vinegar and Wellington* [1987]. The total average porosity, Φ_{ave}, was determined during the initial saturation stage. A total of 58 "dry" and "wet" scans were acquired along the sample, at a spacing of 10 mm. Each CT image was in a matrix form, in which each element, a pixel, represented a volume of $0.5 \times 0.5 \times 2.0$ mm. In most medical applications, CT numbers are defined as: $CT = 1000(\mu_{object} - \mu_{water})/\mu_{water}$ (water is used as a calibration base since its density is close to the density of a human body). Hence, water has a CT value of 0 and air has a CT value of –1000. This definition of CT values is arbitrary. In the research presented in this paper, the base calibration was done with the dry Berea rock sample, yielding CT values close to 1000. The average CT value of all the dry images, $CT_{ave,dry}$, and the average CT value of all the wet images, $CT_{ave,wet}$, were computed. Since the rock sample is lithologically homogeneous, we can map the porosity distribution of the sample using the following formulation (the # symbol denotes a pixel-by-pixel value):

$$\Phi^{\#} = \frac{CT^{\#}_{wet} - CT^{\#}_{dry}}{CT_{ave,wet} - CT_{ave,dry}} \Phi_{ave} . \tag{1}$$

Figure 5 shows a dry image and a wet image. The fracture is visible as the dark region in the middle of the images. The layers in the matrix are also clearly visible. In the grayscale rendering of the images, the dark gray shades correspond to low-density material (low CT values) and the light gray shades correspond to high-density material (high CT values). The overall CT value of the wet image is higher than the average value of the dry image. They are displayed with different color windows so that the details of the layers are clearly maintained. The porosity distribution for the same location generated by Equation (1) is shown in Figure 6. Note that the dark layers in the two images in Figure 5 are now light-colored layers representing high porosity. The fracture appears as a white color. The layer porosities are between 15% and 22%. The five rectangular regions shown in Figure 6 are used to calculate the relationship between the porosity in the fracture and fracture width.

Figure 7 shows two porosity profiles: one vertical profile highlighting porosity in a single layer (b, marked as line B on the porosity image, Figure 7a), and one horizontal profile highlighting the variability of porosity between layers (c, marked as line C on the porosity image, Figure 7a). The porosity varies between the layers and also within each layer. It is expected that the high-porosity layers have

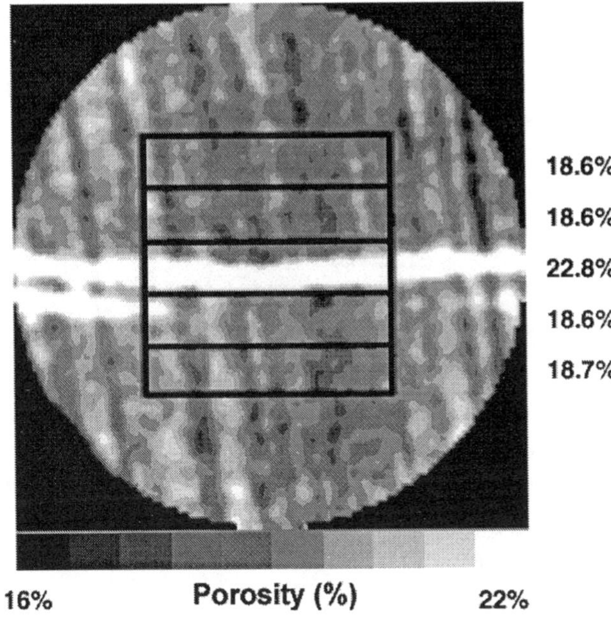

Figure 6. Porosity image, 270 mm from the sample inlet.

high permeabilities. A single-layer porosity distribution (d) shows that the average porosity is about 19%. The "tail" of the distribution at porosity values of 22% and above corresponds to the fracture. The impact of the fractured zone on the porosity profile shown in (b) is 5 pixels wide (2.5 mm). The fracture is not as wide, and is probably closer to 0.5 mm. The rise of the x-ray attenuation response is typical of wellbore logging measurements. The values recorded by the imager are averaged over widths greater than the width of the fracture and cannot be assigned directly to the fracture. The fracture process zone and the fracture itself have a width of about 1–2 mm, as can be seen upon physical inspection of the sample.

The fracture cannot be described as two parallel plates separated by an empty gap. The fracture meanders in space and has asperities on the fracture surfaces that keep the fracture from closing when confining pressure is applied. The correlation of porosity in the fracture with fracture width is generated using the rectangular sections shown in Figure 6. Each rectangle is 5 mm × 25 mm (10 × 50 pixels). The average porosities of the areas in the rectangles are shown to the right of the image. The four regions, not including the central one, have identical porosities. We assume that the central region had the same porosity as the other sections before the fracture was induced. The increase in porosity in the rectangle that contains the fracture is caused by the presence of the fracture. The same rectangular areas are used to calculate porosity images for the rest of the core. Two vertical (i.e., longitudinal) reconstructions are shown in Figure 8. The upper vertical reconstruction is a pixel-by-pixel projected average, generating an array of 58 pixels (length) × 50 pixels (height). Every element shown in the top reconstruction is an average of the stack of 50 pixels across the width of the sample, as shown in the shaded grid to the right. The reconstruction shows that the fracture is not perfectly planer.

Fracture porosity is usually defined as the ratio of the pore space in the fractures to the bulk volume of the rock [*Barenblatt and Zheltov*, 1960]. In this paper we focus on the porosity of the fractured zone, which is defined as the ratio between the pore volume of the fractured zone and the bulk volume of the fractured zone. In order to develop a relationship between the porosity in the fractured zone and the width of the fracture, we calculate the average porosity of each rectangle, leading to an array consisting of 58 × 5 elements, shown in the lower reconstruction in Figure 8. Since the core is largely homogeneous (with respect to average values over an area 5 mm × 25 mm) in areas that do not include the fracture, we calculate a correlation for each image along the core, relating porosity in the fracture and fracture width (Figure 9) for seven locations along the sample. The correlation is of the form where

$$h_{rec}(\Phi_{rec,\text{with fracture}} - \Phi_{ave}) = W_f(\Phi_f - \Phi_{ave}), \quad (2)$$

the fracture width W_f and the porosity in the fracture Φ_f are the unknowns, and h_{rec} is the height of the rectangle (5 mm). Consider the curve at 300 mm (uppermost curve). The curve demonstrates that if the fracture is fully open (100% porosity), its width is about 0.25 mm. As the porosity decreases (that is, more of the space between the fracture walls is filled with asperities), the fracture width increases. Effective fracture widths as a function of position along the core for various porosity values are shown in Figure 10. The fractured zone width is largest at the center of the core, as expected from the fracturing process. Also, the minimum computed width of the fracture is about 0.2 mm when its porosity is 100% (no fill material). Since the fracture width is on the order of the pixel resolution, it is not possible to directly measure the fracture aperture with this CT system. The distribution of width porosity will allow us to refine the flow model for this sample so that we may accurately simulate fluid-flow processes in the future. As shown below, the width and transmissivity of the fracture are heterogeneous.

We attempted to relate the porosity values measured by CT imaging to actual permeability values. At the end of the

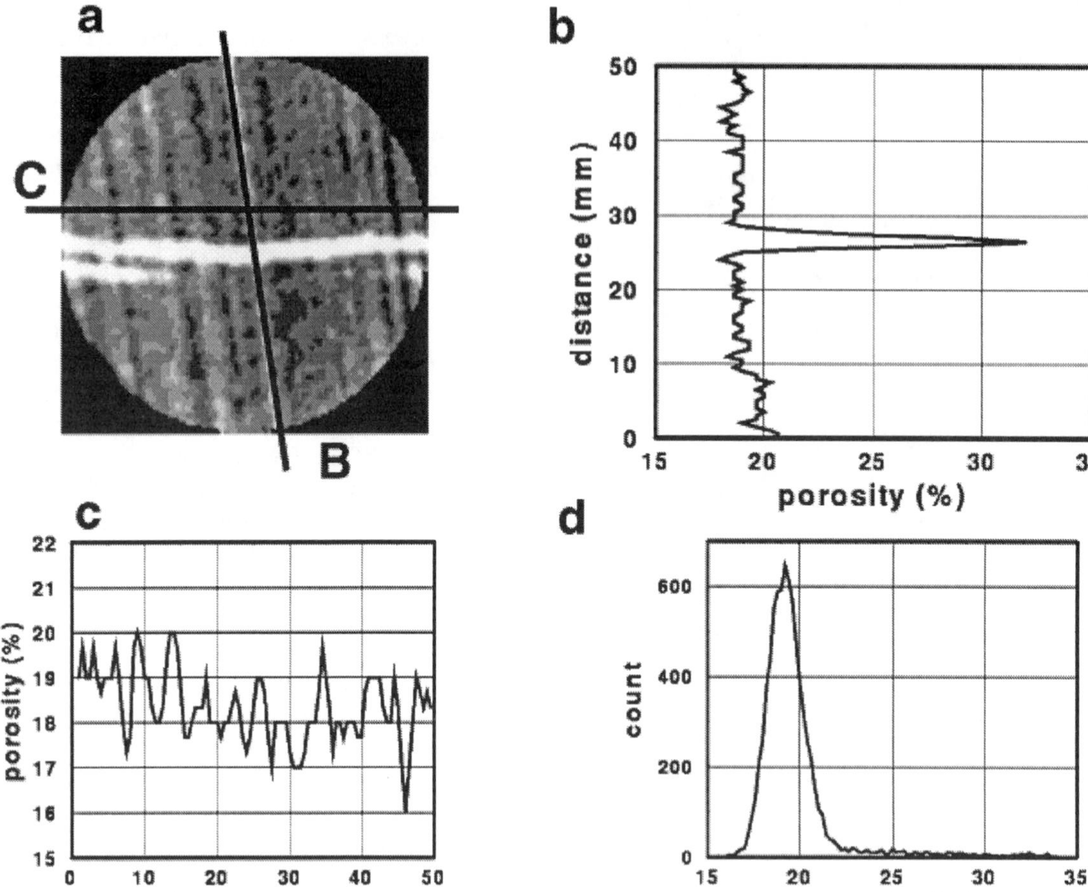

Figure 7. Porosity distribution at a specific location. (a) A porosity image identifying two profile locations. (b) A vertical profile including the fracture. (c) A horizontal profile not including the fracture. (d) A porosity distribution of the entire image shown in (a).

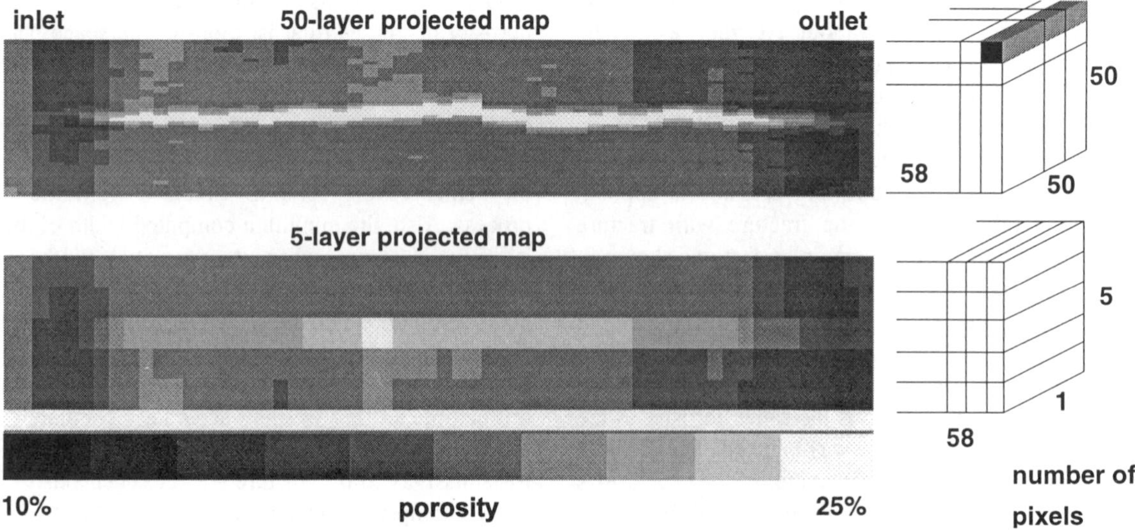

Figure 8. Projected vertical reconstructions including the fracture.

experiment, the sample was removed from the assembly and permeabilities were measured over two planes, as indicated in Figure 11. A mini-permeameter was used to cover a rectangular grid at each location. This technique is documented by *Eijpe and Weber* [1971], *Goggin et al.* [1988], *Deltaban et al.* [1989], and Jensen [1990]. The first surface was at the injection end of the sample, as shown in Figure 11. The mini-permeameter was programmed to collect permeability values on an 11 × 11 grid (35 mm × 35 mm). This grid is shown within the highlighted square in Figure 12a, which is a porosity map obtained from the first scan next to the inlet end of the sample. The highlighted square is isolated in Figure 12b, and is composed of a 70 × 70 matrix of porosity values. Since the permeability map is on an 11 × 11 grid, the porosity image was transformed by simple averaging so that each grid element was the same as for the permeability map (Figure 12c). Figure 12d shows the distribution of permeability values at the inlet of the core. We note that the formulation used for analyzing the mini-permeameter data is not correct in this instance because of sample heterogeneity. We are currently developing an inverse method suitable for our case. However, the mini-permeameter data are qualitatively correct and are used here to demonstrate the proposed methodology of linking permeability and porosity in a lithologically homogeneous sample. The maps in Figures 12c and 12d are similar. The layers are seen in both maps, with the high-permeability zones corresponding to the high-porosity zones. The reader is urged to view these two maps from a distance of 2 to 3 m and use the analogue ability of the human eye to match the patterns.

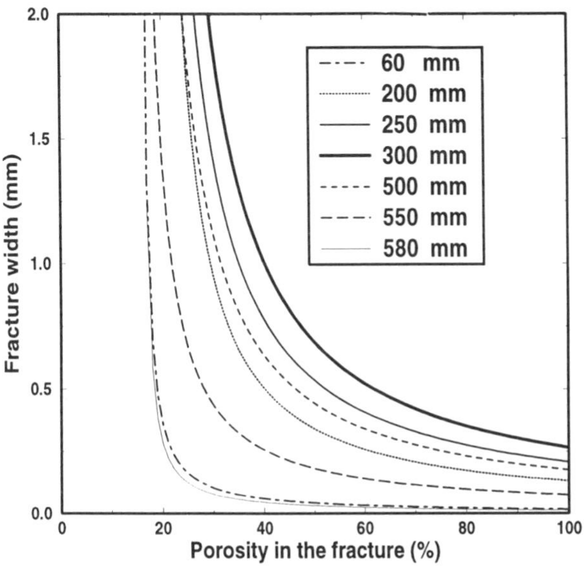

Figure 9. Apparent fracture width as a function of porosity in the fractured zone for various locations along the core.

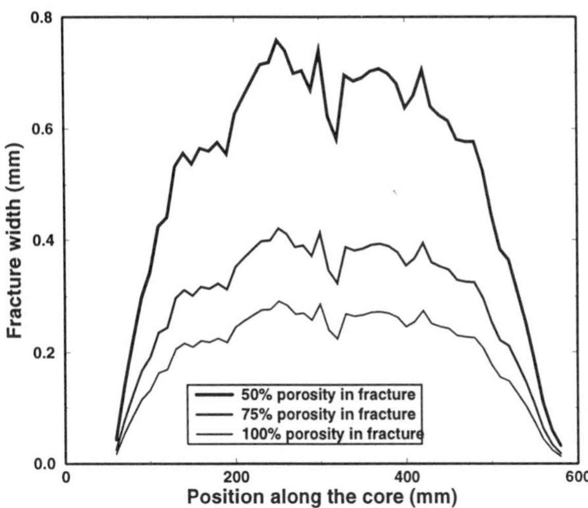

Figure 10. Apparent fracture width along the sample for different fixed values of porosity in the fractured zone.

The second set of mini-permeability measurements was made at the outlet end of the sample. The sample was cut parallel to and about 5 mm above the fracture, as shown in Figure 11. Measurements were done with the mini-permeameter over an area of 40 mm × 72 mm. Figure 13a shows a porosity reconstruction of the area. CT images were collected every 10 mm; there are only 8 scans through the designated permeability surface. The image in Figure 13a is a matrix of 8 × 80 porosity values. The mini-permeameter data, an array of 13 × 21 permeability values,

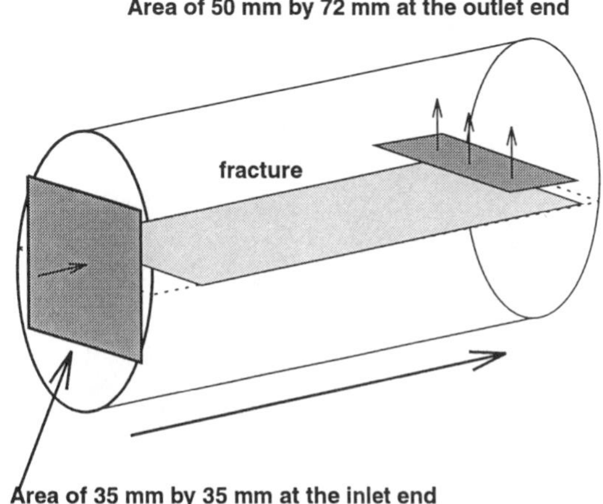

Figure 11. A schematic diagram identifying the minipermeability planes.

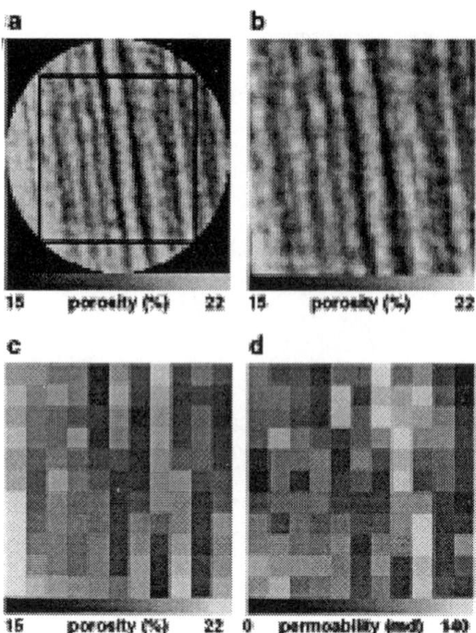

Figure 12. A comparison between porosity and permeability maps, obtained at the inlet end of the sample.

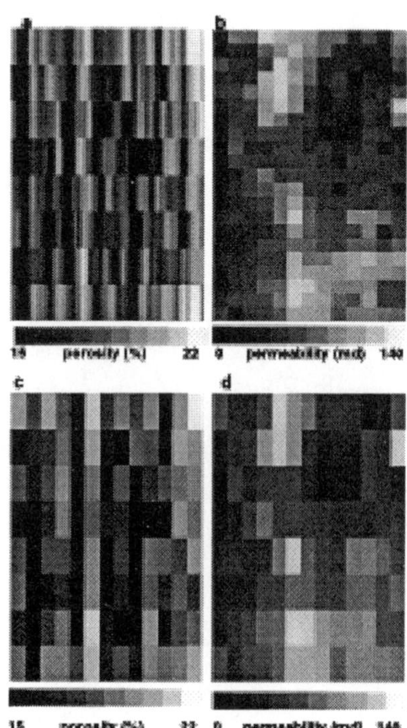

Figure 13. A comparison between porosity and permeability maps, obtained at the outlet end of the sample.

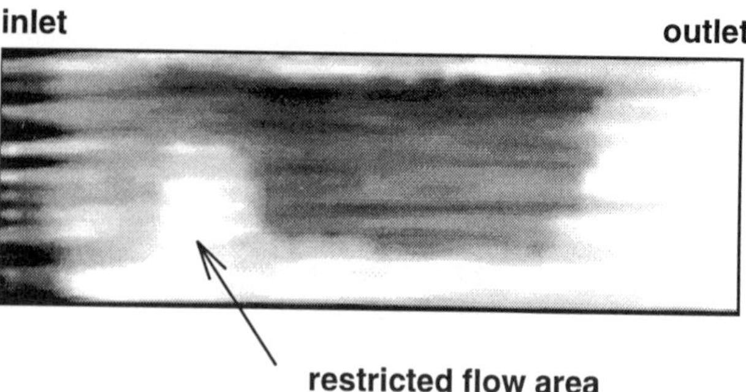

Figure 14. A horizontal projected reconstruction of oil in the fracture prior to breakthrough.

are shown in Figure 13b,. The upper images in Figure 13 are upscaled to arrays of 8 × 13 values and compared in the lower images in Figure 13. The relationship between porosity and permeability is not as strong in this case. However, the high-permeability and high-porosity layer in the middle appears in both images. Part of the reason for the poor match is the low axial sampling of the CT data. We could have collected CT values every 1 mm and improved the spatial resolution. Also, a mini-permeameter with better spatial resolution would be an advantage in developing a correlation between porosity and permeability.

The two examples presented here demonstrate a method for relating the quantitative CT work to mass transport in a layered and fractured sample. We expect that in the future, detailed mini-permeability work will be linked with quantitative statistics to allow us to collect a small number of miniperm data and, through the CT-derived porosity, map the permeability of the entire volume of the sample.

4. TWO-PHASE FLOW

After the core sample was saturated with water, a light mineral oil was injected and its distribution tracked using the x-ray imager. The total pore volume of the rock sample was 230 cc, and the overall porosity was 17.5%. After injecting 5 cc of oil (0.022 PV), we scanned the sample. Figure 14 shows the distribution of the oil phase in the fracture before oil breakthrough. This is a projected horizontal reconstruction that includes the fracture. We can compute a rough estimate of the fracture aperture using an assumed porosity in the fracture of 100%. About 2.5 cc are occupying one half of the area of the fractured zone. The "fracture" area (viewed from the top) is about 50 mm wide and 500 mm long, a total of 25,000 square mm. Hence, the average aperture of the fracture is 0.2 mm. This value

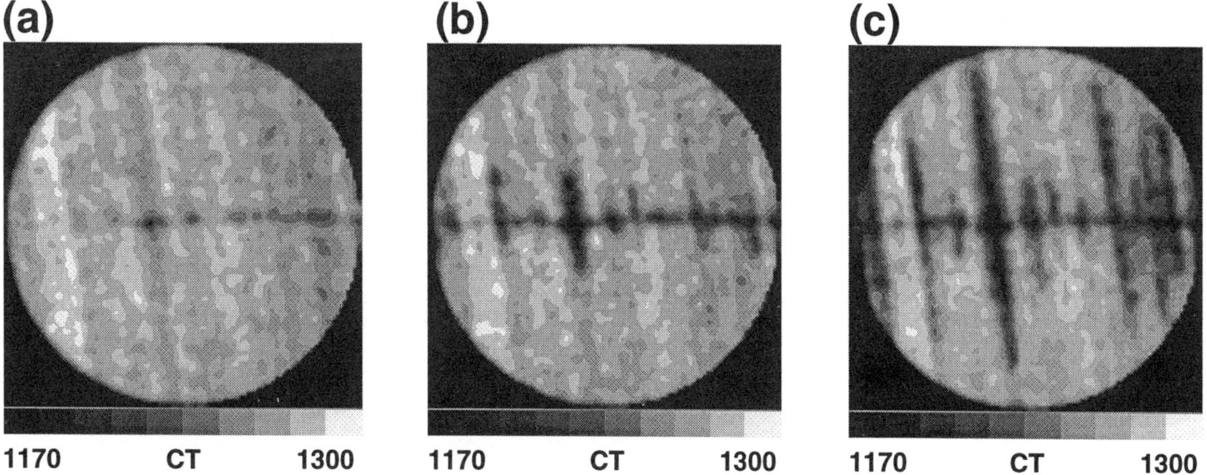

Figure 15. Oil displacing water at the outlet end of the sample.

matches well with the width shown in Figure 10 and is a first-order estimate.

Since the fracture did not extend all the way to the inlet end of the core, oil flow at the inlet is dominated by the layers shown as the dark and light alternating strips at the left of the figure. Once oil invasion reached the fractured region, the flow converged to the fracture and displaced the water out of the fracture. The oil advanced faster in the fracture sections that are adjacent to high-permeability layers than in those opposite the low-permeability layers. This can be seen at the leading edge of the oil in the fracture on the right side of Figure 14. This observation leads to a hypothesis that the effective fracture permeability, width, and porosity depend on the layers adjoining the fracture. If this hypothesis is true, it has an important impact on the interaction between the fracture and the matrix during displacement and recovery processes. Another observation supporting this hypothesis is presented immediately below.

Oil injection was resumed, and the scanner was positioned 10 mm from the outlet end of the core, where it collected images during the breakthrough period. As shown in Figure 10, the fracture is evident at the end of the core but is not as open as in the central regions of the core. Hence, as oil arrives at the outlet it is forced to displace water from the matrix. Figure 15 shows three images taken at the end of the core at three different times. The invasion of the oil phase into the layered matrix is denoted by the dark portions on the images (the invading oil phase has a low x-ray attenuation). In the early image (Figure 15a), the oil is only evident in the fracture as a dark color, and is concentrated opposite the high-permeability layers. In the second image (Figure 15b), acquired after injection of oil equal to 1% of the pore volume, the fracture is almost full of oil and the oil has started to displace water from the high-permeability layers. In the third image (Figure 15c), acquired after injection of oil equal to 2% of the pore volume, water in one of the high-permeability layers is almost fully displaced. The displacement of the water in the layers was also affected by gravity, since oil is lighter than water. The dark streaks extend farther in the upward direction than in the downward direction (Figure 15c).

The net amount of oil is calculated by image subtraction, and is shown in Figure 15. The oil presence in a binary form (locations that contain invading oil are shown in black) is shown in Figure 16. Both figures illustrate that the oil (the dark shade) was in the fracture at locations opposite high-porosity layers. As the oil arrived at the outlet end and was forced to invade the matrix, it displaced the water from the high-porosity and high-permeability layers first. The subtracted binary images shown in Figure 16 are essential as a proof that in this case there was significant oil invasion into the high-porosity layers and no invasion at all into the low-porosity layers early in the oil-injection process. However, as the injection continued, the low-porosity layers were also invaded.

Following breakthrough of oil, injection was continued, and the core was imaged five times at the following values of pore volumes injected (PVI): 0.035, 0.043, 0.056, 1.000, and 10.000. Figure 17 presents the average profiles at these PVI values. In this experiment, the CT profiles were not converted to saturation profiles since water injection did not follow the oil injection. The oil and water saturations can be computed based on overall external material

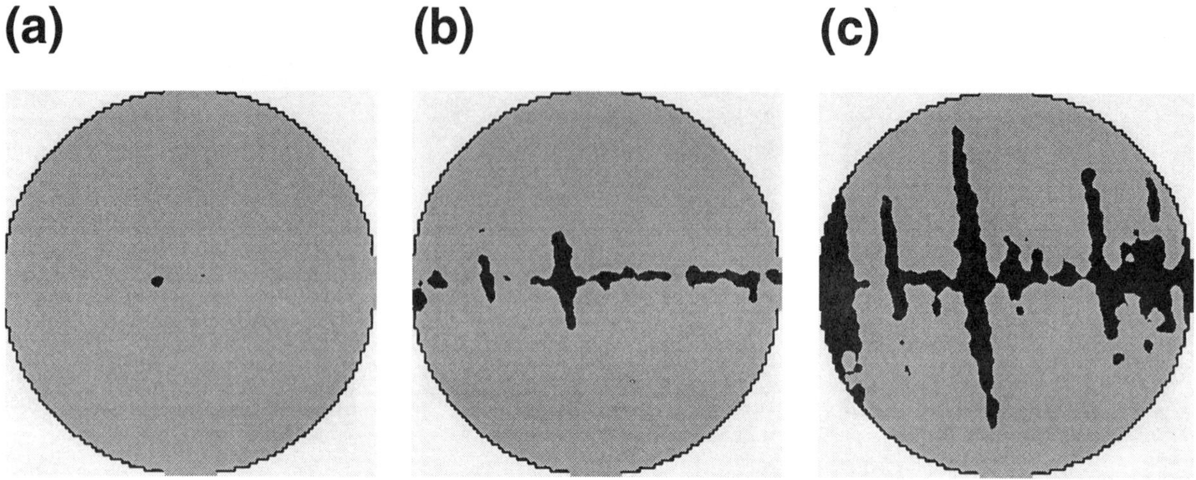

Figure 16. The net presence of oil at the outlet end of the sample during oil injection.

balance. However, these profiles serve the purpose of describing the displacement process. The upper dark solid line is the profile just before oil injection, where the core is fully saturated with water. The thin solid line denotes the profile after injecting 0.035 PV. This curve shows that oil had accumulated at the injection end, mainly in the high-permeability layers. Then the oil converged to the fracture and flowed mainly in the fracture, denoted by the small gap between the 100% water curve and the 0.035 PVI curve. The third profile was collected after injection of 0.043 PV and is denoted by the dotted line. The two curves (for 0.035 and 0.043 PVI) are practically identical in the injection side of the core. The oil had now fully filled the fracture and started to accumulate in the outlet end, as denoted by the difference between the dotted curve and the 100% solid curve. As the oil displacement extended, the water was

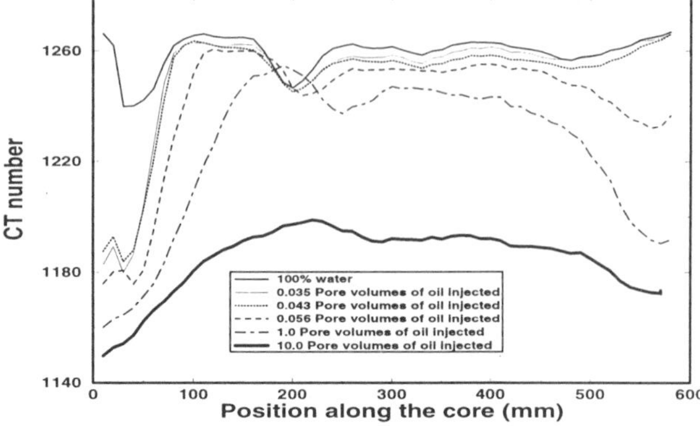

Figure 17. Average profiles along the sample for various values of pore volumes of oil injection.

displaced from the inlet end and the outlet end, and high-water saturation was left in the middle of the core.

The displacement of water from the middle of the core was slow, and was controlled by the presence of the high-permeability fracture. As shown in Figure 15c, the water was displaced from the high-permeability layers first, and then from the low-permeability layers. The presence of the fracture controlled the displacement order in the system. The CT values at the inlet end (corresponding to water saturation) were declining even after 10 pore volumes of oil injection, denoting that water was still being displaced, a typical *Buckley and Leverett* [1941] process. After injecting 10 pore volumes of oil, the average oil saturation in the core was only 46%. In a nonfractured Berea sample, the water saturation is expected to be between 20% and 30% after 1–2 pore volumes of oil injection. The goal of the oil-injection stage was to reduce the water saturation to a low (residual) value, simulating the geological conditions in the reservoir. After the oil-injection stage, the oil was produced from the sample by various recovery processes, such as water injection. The presence of the fracture increased the time required to bring the core to residual water saturation, and prepare it for primary and secondary recovery processes.

Figure 18 shows longitudinal reconstructions (images extracted from a 3D volume of CT data that is "cut" vertically in the longitudinal direction) through the highest porosity layer (Figure 7) at five injection stages. In reconstruction 18a, the fracture is not fully saturated with oil (dark color). At the injection end, the oil is flowing toward the fracture, creating a "cup" shape. In reconstruction 18c the oil cone at the end of the core is established. The oil leaves the fracture and flows into the high-porosity/permeability layer. The point of departure from the fracture moves in the direction of the injection end as time progresses, as seen in reconstruction 18d. In the last reconstruction, the trapped water in the middle of the core is represented by the light gray color. Figure 18 illustrates the channeling effects of the fracture, creating converging flow at the inlet and diverging flow at the outlet. The oil displacement within a single layer is not a piston-like process, but rather exhibits Buckley-Leverett frontal advance behavior. Figure 19 shows horizontal profiles obtained from the outlet end of the core for the five PVI values. The measurements show that the displacement of water by oil is highly variable in the rock matrix, and is closely associated with the layers.

The confining stress has an impact on the fracture aperture [*Gale*, 1982; *Barton et al.*, 1985; *Sundaram et al.*, 1987; *Gentier et al.*, 1997]. It was not possible to fully quantify this effect. However, CT images were taken at a fixed sample position at different confining stresses. Figure 20 shows two vertical profiles obtained for a strip 5-mm wide in the center of the sample, as shown in the inset. The light curve was acquired at a confining pressure of 0.5 MPa, and the thick curve at a confining pressure of 3 MPa. The increase in stress caused a reduction in fracture aperture, resulting in a slight increase in the CT values in the fracture region. The CT values in the rest of the sample did not change. In future work, we expect to be able to quantify the impact of changing stress on multiphase transport in the fracture and on the interaction between the fracture and the rock matrix that govern fluid flow in the sample [*Tsang and Witherspoon*, 1981, 1983].

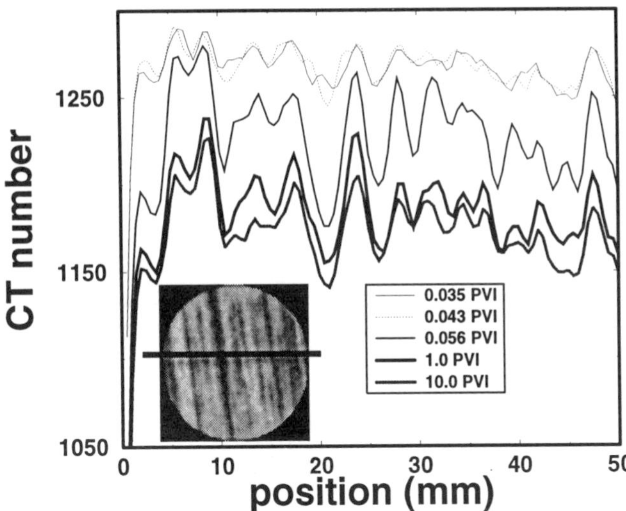

Figure 19. Horizontal profiles of CT values as a function of position, obtained at the end of the sample for different values of pore volumes of oil injection.

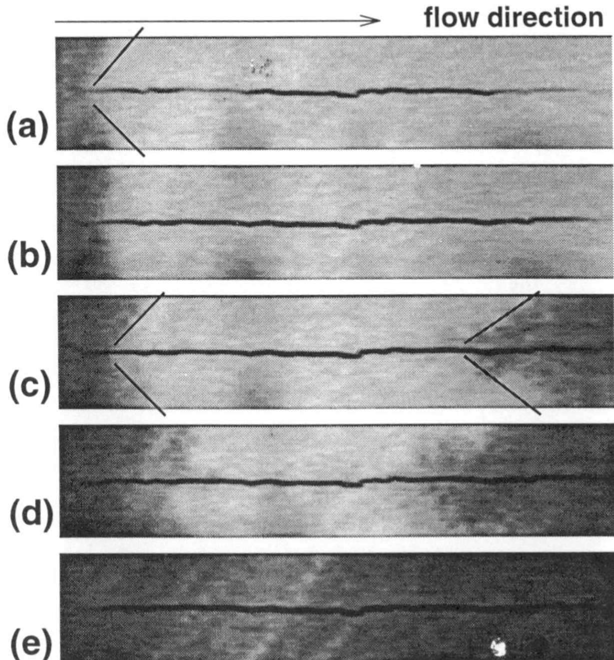

Figure 18. Longitudinal reconstructions at various values of pore volumes of oil injection: (a) 0.035, (b) 0.043, (c) 0.056, (d) 1.0, and (e) 10.0.

5. SUMMARY

For oil injected into a water-saturated sample, the presence of a nonterminating fracture channels the flow and

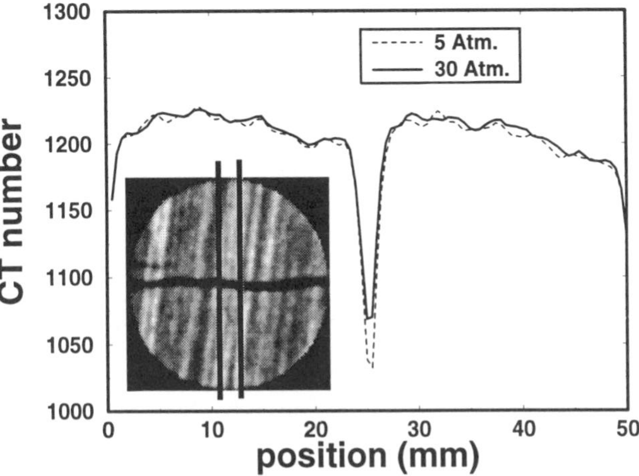

Figure 20. The effect of confining stress on CT values. The profiles are an average of the 5-mm strip shown in the inset image.

has a significant impact on the displacement process. The injected oil first invades the high-permeability and high-porosity layers. Then the oil converges toward the fracture tip and flows mainly in the fracture. The artificially induced fracture is not homogeneous. The water in the fracture does not get displaced with an even front. We hypothesize that the fracture permeability and porosity are high in areas where the fracture is adjacent to the high-porosity layers. Laboratory observations support this hypothesis, but more definitive proof is needed to confirm it. Higher resolution imaging or some type of injection-solidification process and destructive measurements may provide the necessary data. At the outlet end of the fracture the flow diverges preferentially into the high-porosity layers. The oil phase emerges progressively out of the fracture with continued injection.

The experiment demonstrates the usefulness of x-ray tomography for quantifying the porosity distribution in the sample and for generating a relationship between fracture aperture and porosity in the fracture. Volumetric nondestructive imaging permits us to collect 4D fluid distribution during multiphase flow processes. The quantitative values of the fluid saturations should allow the use of numerical modeling and simulation to determine the individual properties of each layer and the connectivity between the layers. Furthermore, simulation of the 4D CT data will shed light on the capillary pressure and relative permeabilities of the distinct layers in the core.

Acknowledgements. The authors thank ENI/AGIP for supporting this research project and for the permission to publish it. We acknowledge Adriano Figoni and Eduardo Fiorentino for the fracturing and miniperm work. We also thank Pennsylvania State University.

REFERENCES

Buckley, S. E., and M. C. Leverett, Mechanism of fluid displacement in sands, *Transactions AIME*, *146*, 107–116, 1942.

Barenblatt G. I., and Yu. P. Zheltov, Fundamental equation of homogeneous liquids in fissured rocks, *Dokl. Akad. Nauk SSR*, *132*(3), 545–48, 1960.

Barton, N., S. Bandis, and K. Bakhtar, Strength, deformation and conductivity coupling of rock joints. *Int. J. R. Mech.*, *22*, 121–140, 1985.

Clark, S. P., Jr., *Handbook of Physical Constants*, The Geological Society of America, Yale University, New Haven, Conn., 1966.

Deltaban, T. S., J. J. M. Lewis, and J. S Archer, Field mini-permeameter measurements—their collection and interpretation, in *Proc., 5th European Symposium on Improved Oil Recovery, Budapest*, pp. 671–682, 1989.

Eijpe, R., and K. J. Weber, K. J., Mini-permeameters for consolidated rock and unconsolidated sand, *AAPG Bulletin*, *55*(2), 307–309, 1971.

Gale, J. E., The effects of fracture type (induced versus natural) on the stress-fracture closure—fracture permeability relationships, in *Proc. 23rd U.S. Symposium on Rock Mechanics*, edited by R. E. Goodman and F. Heuze, pp. 290–298, 1982.

Gentier, S., and D. Hopkins, Mapping fracture aperture as a function of normal stress using a combination of casting, image analysis and modeling techniques, *Int. J. Rock Mech. & Min. Sci. 34*(3–4), Paper No. 132, (1997)

Goggin, D. J., R. L. Thrasher, and L. W. Lake, A theoretical and experimental analysis of mini-permeameter response including gas slippage and high velocity flow effects, *In Situ*, *12*(1–2), 79–116, 1998.

Jensen, J. L., A model for small-scale permeability measurement with applications to reservoir characterization, in *Proc., SPE/DOE 7th Symposium on Enhanced Oil Recovery, Tulsa, OK*, 1990.

Lichtner, P. C., and M. Seth, Multiphase-multicomponent non-isothermal reactive transport in partially saturated porous media, in *Proc. Int. Conf. on Deep Geological Disposal of Radioactive Waste, Winnipeg, Canada*, pp. 3-133–3-142, 1996.

Piggott, A. R., J. Xiang, and D. Elsworth, Inversion of hydraulic and electrical data for the determination of fracture aperture, in *Proc. U.S. Symposium on Rock Mechanics, Norman, Oklahoma*, pp. 1135–1144, 1991.

Sundaram, P. N., D. J. Watkins, and W. E. Ralph, Laboratory investigations of coupled stress-deformation-hydraulic flow in a natural rock fracture, in *Proc. 28th U.S. Symp. on R. Mechs.*, edited by I. Farmer, J. Daemen, C. Desai, C. Glass and S. Neuman, , pp. 585–592, Balkema, Rotterdam 1987.

Tsang, Y. W., and P. A. Witherspoon, Hydromechanical behavior of a deformable rock fracture due to normal stress, *J. Geophys. Res.*, *86*(B10), 9287–9298, 1981.

Tsang, Y. W., and P. A. Witherspoon, The dependence of fracture mechanical and fluid properties on fracture roughness and sample size, *J. Geophys. Res. 88*(B3), 2359– 2366, 1983.

Vinegar, H. J., and S. L. Wellington, Tomographic imaging of three-phase flow experiments, *Rev. Sci. Insts.*, January, 96–107, 1987.

Vukuturi, V. S., R. D. Lama, and S. S. Saluja, *Handbook on Mechanics Properties of Rocks, Volume I*, Trans Tech Publications, Switzerland, 1974.

Zimmerman, R. W., and G. S. Bodvarsson, Hydraulic conductivity of rock fractures, *Trans. Porous Media. 23*, 1–30, 1996.

A. S. Grader, Pennsylvania State University, 110 Hosler Building, University Park, PA 16802-5000

M. Balzarini, F. Radalini, and A. Pellegrino, ENI/AGIP, Italy

G. Capasso, Politecnico di Torino, Italy

Role of Fracture Geometry in the Evolution of Flow Paths under Stress

S. Gentier,[1] D. Hopkins,[2] and J. Riss[3]

This paper summarizes more than a decade's research at BRGM on the hydromechanical behavior of natural fractures in granite under normal and shear stress. The paper's emphasis is on the importance of understanding the role of fracture geometry in fluid flow and, in particular, the evolution of fracture flow paths with changes in stress. Experimental results were obtained by modifying classical hydromechanical tests to allow detailed analysis of fracture geometry under zero load and of the spatial organization of flow. Fracture-wall geometry is analyzed using profilometry; a casting methodology is used to determine the geometry of the fracture's void space. Tracer tests show that the decrease in transmissivity that occurs with increasing normal stress is associated with increasingly distinct channeling. This channeling is strongly linked to correlation lengths identified from geostatistical analysis of surface profiles and data obtained from the casts of fracture void space. Modeling results show that deformation of fracture surfaces with increasing normal stress causes substantial, nonuniform changes in void-space geometry that can change the flow regime. To better understand the mechanical behavior of fractures under shear stress, image analysis techniques are used to identify geometrical parameters that affect the micromechanical behavior and the evolution of damage zones during shearing. Laboratory experiments indicate that a fracture's mechanical response to shear stress can be broken down into at least five phases, which are shown to be associated with changes in flow. In general, application of shear stress induces an opening of the fracture, sometimes preceded by a closure phase, that causes a very large increase in global transmissivity that is associated with a reorientation of flow subperpendicular to the shear direction. Reorientation culminates just after peak shear stress is reached. During the subsequent softening and residual phases, flow tends to return to a more isotropic pattern.

[1]Bureau de Recherches Géologiques et Minières (BRGM), 3 Av. Claude Guillemin, B.P. 6009, 45060, Orléans cedex 2, France

[2]Ernest Orlando Lawrence Berkeley National Laboratory, MS 46A-1123, 1 Cyclotron Road, Berkeley, California

[3]Centre de Développement des Géosciences Appliquées (CDGA), Université Bordeauz 1, Av. des Facultés, 33405 Talance cedex, France

Dynamics of Fluids in Fractured Rock
Geophysical Monograph 122

1. INTRODUCTION

Any variation in stress on a fracture changes the geometry of its void space, which in turn changes its global permeability and preferential flow paths. The geometry of the fracture's void space depends on the geometry of its surfaces and the way in which these surfaces are matched under stress. The geometry of the fracture's void space also depends on whether or not there is "infill"—material present as a result of the fracture's chemical, tectonic, and hydraulic history.

In situ stresses on a fracture are subdivided into normal and shear components. Historically, research on fracture behavior first looked at hydromechanical behavior under normal stress [*Gale*, 1982; *Raven and Gale*, 1985; *Gentier*, 1987; *Pyrak-Nolte et al.*, 1990; *Iwano*, 1995] and only later progressed to studying behavior under shear stress [*Gale*, 1990; *Makurat et al.*, 1990; *Esaki et al.*, 1995; *Gentier et al.*, 1997; *Olsson*, 1998; *Yeo et al.*, 1998].

To understand a fracture's hydromechanical behavior, we first have to understand the relationship between flow in the general sense (spatial organization of flow paths) and the geometry of the fracture's void space (volume and spatial organization of the void space). Several numerical parametric studies have determined the effect of surface roughness and tortuosity on flow for specific flow models [e.g., *Brown*, 1987; *Tsang*, 1984; *Thompson and Brown*, 1991]. This work has led to the conclusion that an estimate of "representative hydraulic aperture" is necessary in order to use the cubic law to predict fluid flow in fractures. In addition, understanding the spatial organization of flow requires a precise knowledge of void-space geometry. Increasingly sophisticated hydromechanical experiments have given more and more precise data on fractures' hydraulic and mechanical parameters; at the same time morphological data have been gathered that describe surface roughness [*Gentier*, 1987], contact areas [*Pyrak-Nolte*, 1987], and aperture [*Hakami et al.*, 1995]. Collection and analysis of these data required development of new data-acquisition techniques (e.g., profilometry, casting, and dye injection) and new or modified data processing techniques (e.g., image processing algorithms, and geostatistical methods).

All of the research cited above followed from Paul Witherspoon's original work to determine the applicability of the cubic law to natural rock fractures under normal stress [*Witherspoon*, 1980]. Use of the cubic law is restricted to those cases where we can make strong assumptions about fracture geometry and flow regime; i.e., the cubic law is only valid when a fracture can be represented as two parallel planes and in cases where the fluid is incompressible and flow in the fracture is laminar and stationary. Other significant early work demonstrated the importance of surface roughness in determining the flow regime in reconstructed fractures [*Louis*, 1967, and *Lomize*, 1951 (cited in *Louis*, 1967)]. A consequence of this work is the knowledge that research on the hydromechanical behavior of fractures must be conducted in parallel with studies of the influence of fracture geometry on the flow regime.

Although research has clearly demonstrated that fracture geometry plays a significant role in flow and the evolution of flow paths with stress, much work remains in defining geometrical parameters and determining their roles in hydromechanical behavior. Inherent in this research is the problem that we have only partial information to work with. As a result, research is iterative. Laboratory techniques are developed to provide more detailed information about fracture geometry and hydromechanical behavior. This information, in turn, is used to refine and verify models. As we move toward a comprehensive conceptual model, it is critical to verify that the data we collect are consistent to ensure that we are not missing some important but as yet unknown phenomenon that is key to understanding fracture behavior.

In keeping with this progression, new techniques have been developed to provide increasingly detailed information. The results summarized here were obtained by modifying classical hydromechanical tests to allow detailed analysis of the spatial organization of flow, the geometry of fractures under zero load, and the evolution of contact areas under stress. This paper focuses specifically on approaches and results from current research designed to shed light on the evolution of flow paths in a fracture with changes in stress. The remainder of the paper consists of two major sections that describe the hydromechanical behavior of fractures under normal and shear stress. In each section, results of laboratory experiments and modeling studies are summarized and interpreted with respect to fracture geometry. The results presented are only for natural granite fractures with no infill or with a particular type of hard infill that is strongly bonded to fracture surfaces. All experiments were performed on samples that were approximately 12 cm in diameter.

2. HYDROMECHANICAL BEHAVIOR UNDER NORMAL STRESS

Laboratory study of hydromechanical behavior of fractures under normal stress (though this stress bears little resemblance to stresses in the field) is a first step toward understanding a fracture's response to those realistic and more complex stresses. From a modeling point of view, the phenomena that must be considered when studying fractures under normal stress are simpler than those for fractures in shear; the natural granite fractures that we have studied generally exhibit only elastic behavior under normal stress. In addition, the deformation that occurs for fractures under normal stress are relatively small compared to the large deformations and displacements that occur under shear stress.

2.1. Experimental Results

During the past decade, BRGM's research has led to the following key conclusions, illustrated in Figure 1:

- Intrinsic transmissivity decreases nonlinearly as normal stress increases. When normal stresses are small, intrinsic transmissivity decreases rapidly in response to stress. For intermediate normal stresses, intrinsic transmissivity decreases until it reaches a limit. The initial value of the intrinsic transmissivity is reduced by 1.5 to 10 times under increasing normal stress, depending on the type of fracture. The smallest reductions are observed for fractures with hard infill.
- The limit value of intrinsic transmissivity is reached at the stress above which further increases in stress do not produce a significant reduction in intrinsic transmissivity. The critical normal stress at which this limit occurs is between 8 and 15 MPa, depending on the type of fracture. The lower values correspond to fractures with hard infill.
- Hysteresis always exists in plots of intrinsic transmissivity versus normal stress; hysteresis is reproducible after several loading/unloading cycles, which are necessary to eliminate effects of sample preparation that affect the mating of the fracture surfaces [*Gentier*, 1987]. Note that the data plotted in Figure 1a were obtained after several cycles of loading and unloading where the hysteresis was reproducible.
- With increasing normal stress, the number of fluid exit points on the periphery of the sample decreases with a corresponding increase in the mean distance between exit points (Figure 1b). The number of exit points reaches a constant value at moderate levels of normal stress.

Intrinsic transmissivity is calculated, rather than transmissivity, because it is independent of the fluid used, allowing us to compare results from fluid-flow experiments that use different fluids. Intrinsic transmissivity is a global fracture property that can be calculated from flow rates and pressures measured in the laboratory. The results cited above for intrinsic transmissivity hold so long as fracture flow is laminar. In this case, we can use the classical formula for intrinsic transmissivity for radial divergent injection:

$$k_f.e = -(\mu/2\pi)\,Ln(r_i/r_e)\,Q/P\ , \qquad (1)$$

where $k_f.e$ is explicitly intrinsic permeability multiplied by aperture, μ is the dynamic viscosity, r_i is the radius of the injection well, r_e is the external radius of the sample, Q is the injection flow rate, and P is the stabilized injection pressure. This formula assumes that hydraulic pressure varies as a logarithmic function of the radius of the sample, that aperture is constant, and that flow in the fracture is laminar. If flow is laminar, the ratio between flow rate and

(a)

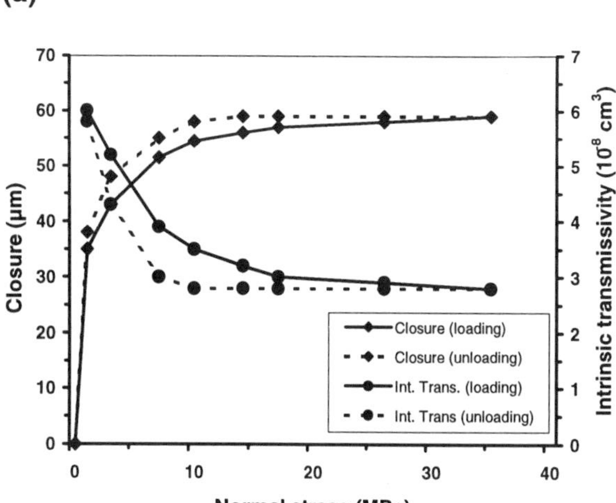

(b)

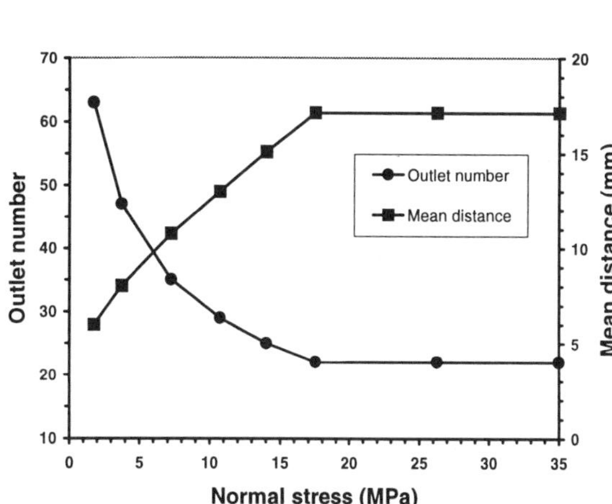

Figure 1. (a) Intrinsic transmissivity and global closure with increasing normal stress in a fracture without infill. (b) Number of fluid exit points (outlets) and mean distance between exit points with increasing normal stress measured on the fracture periphery during the same experiment in which fracture closure and intrinsic transmissivity were measured (a).

injection pressure (Q/P) for a constant geometry of the fracture remains constant. Figure 2 illustrates the evolution of this ratio Q/P during hydromechanical tests performed for a range of stresses between 1 and 10 MPa and a range of flow rates between 1 and 10 ml/min [*Gentier et al.*, 1996].

For low normal stresses (1 to 5 MPa), the ratio Q/P effectively remains constant for the range of flow rates used. In this range of stresses, the change in the void space that occurs with progressive closure does not lead to a modification of the fracture's global hydraulic regime. Only the

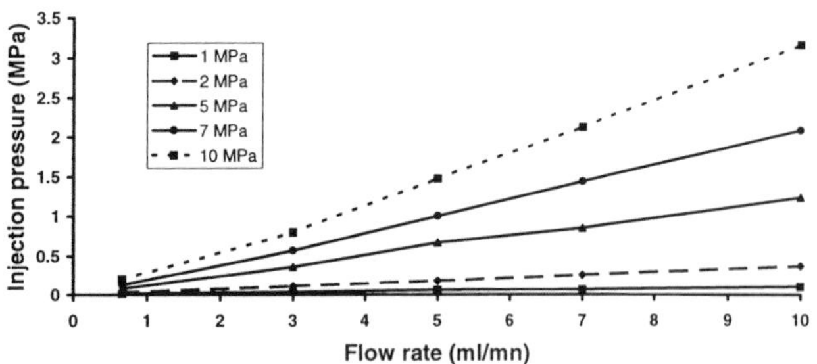

Figure 2. Injection pressure versus flow rate measured in a fracture without infill for 5 levels of applied normal stress.

ratio Q/P increases, which signifies a reduction in the fracture's global transmissivity. Over 5 MPa, the relationship between flow rate and pressure becomes nonlinear for the highest flow rates. This nonlinearity signifies a gradual evolution of the flow regime; the flow becomes less and less laminar and gradually changes to a turbulent regime. More complex evolution of the ratio Q/P has been observed in fractures with hard infill.

These results show that the flow regime evolves depending on stress and flow rate, which greatly complicates measurement and modeling. The difficulties lie in identifying flow regimes as a function of hydromechanical conditions and also, among other things, in determining the Reynolds number for fluid flow, which is inferred from flow velocity.

To fully understand the implications of these conclusions about hydraulic behavior, we must look at them in conjunction with the following conclusions about the mechanical behavior of fractures under normal stress:

- Measurements of fracture closure with linear variable differential transformers (LVDTs) arranged around the periphery of a sample show that closure does not increase linearly with increased applied normal stress; closure increases rapidly at low normal stresses and then approaches a constant value (maximal closure). Maximal closure is found to occur at the same value of normal stress at which the minimum value of intrinsic transmissivity occurs (Figure 1a). Maximum closure is typically 10–20% of the mean geometrical fracture aperture (measured from matched surface profiles or casts of the void space obtained before load is applied to the sample). Maximum-closure values equal to as much as 30–40% of the mean geometrical aperture have been observed.
- Measurements of normal displacement show that closure is not uniform across the fracture.
- Normal displacement is always accompanied by a small tangential displacement, resulting from the roughness of the fracture, which can be measured during the tests. Tangential displacement can represent from 5–10% of normal displacement for a fracture without infill and 20–30% for fractures with hard infill.
- Plots of fracture closure as a function of normal stress are, like plots of intrinsic transmissivity, reversible and reproducible after the first two or three loading cycles.

Although the correspondence between fracture closure and intrinsic transmissivity is clear, closure data give only partial information about the reduction in the void-space volume that occurs with increasing normal stress (i.e., closure data indicate reductions in aperture in the direction perpendicular to the average plane of the fracture). To obtain information not available from classical hydromechanical experiments, particularly on the evolution of the area of a fracture's surface accessible to flow, we injected fluid dye into fractures. The dye's exit points on the sample periphery were then identified. These exit points represent the intersection of the network of channels in the plane of the fracture and the sample's exterior surface. Fluid-flow experiments conducted at different normal stress levels show that:

- Under low normal stress, flow channels are numerous but identifiable.
- When normal stress increases, the number of channels decreases, rapidly in the beginning and then more slowly, becoming constant when the normal stress reaches the critical value defined above in conjunction with intrinsic transmissivity and maximum closure (Figure 1b).

These injection test results shed light on the role of channeling in fluid flow in the fracture plane. If the density of channels is sufficiently high, then under low normal stresses the flow can be considered quasi-homogeneous in the fracture plane. The density of channels diminishes with increasing normal stress, and flow becomes more heterogeneous. At some critical value of normal stress, the channel density approaches a constant value. Small reductions in intrinsic transmissivity observed at this point are probably a result of volume reduction in the channels rather than changes in the topology of the channel network.

Preferential flow paths in the fracture plane were identified by tracer tests used in conjunction with a fluid recovery system that allows measurement of the fluid exiting the fracture at the periphery of the sample [*Gentier et al.*, 1995]. The periphery was divided into eight sectors, and fluid leaving the exit points in each sector was collected and weighed. Breakthrough curves obtained for five of the eight sectors during a tracer test are shown in Figure 3. The curves show substantial differences in tracer concentration and arrival time for different sectors. In addition, most breakthrough curves exhibit multiple peaks, which probably correspond to different flow paths. The flow rate in this experiment was so slow (40 ml/h) that even if we used the tracer breakthrough time and the shortest distance between the injection well and exit point, we estimate both a very low velocity and a very low Reynolds number, which is characteristic of laminar flow. Even without the use of tracers, the partitioned flow-recovery system, which allows measurement of fluid in different sectors, provides very useful information on the heterogeneity of flow in fractures.

2.2. *Geometric Interpretation of Experimental Results*

Results presented above can be interpreted with respect to the fracture's initial geometry and its evolution under increasing normal stress. A complete description of the fracture's initial geometry requires characterization of the fracture surfaces and void space between the surfaces. The topography of the fracture surfaces is obtained by measuring surface profiles. The geometry of the void space is determined either from matched profiles of the two surfaces or directly, by making a cast of the void space [*Gentier et al.*, 1989]. Neither of these techniques provides a description of the fracture's geometry under load. To determine changes in void space geometry that occur with changes in stress, we must develop a model or technique for casting the void space under load. A casting methodology is under development, but it will only provide data for the stress level under which the cast was made.

2.2.1. Relationship between experimental results and the geometry of fracture surfaces. Geostatistical methods based on variograms can be used to determine surface components from profile data. The topography of the fracture walls (map of surface altitudes measured from a reference plane) can be described by the superposition of several components corresponding to different scales. Each component is characterized by a range (for stationary data, range corresponds to correlation length) and a sill (for stationary data, sill corresponds to variance). Figure 4a shows a variogram calculated for a surface profile, and the ranges of three components that describe the profile. The histogram of ranges obtained from all identifiable components on the profiles measured on a fracture surface is shown in Figure 4b. In this case, the sample was 12 cm in diameter, and the surface topography is best described by the superposition of three components whose ranges are 4–10 mm, 12–16 mm, and 18–22 mm, respectively. These three components are then superimposed onto structures with ranges greater than 30 mm, which can only be partially sampled because of the size of the specimen.

These geostatistical parameters were compared to the

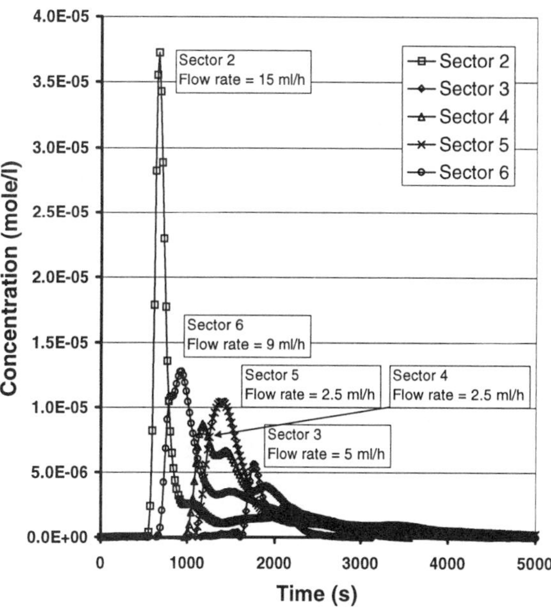

Figure 3. Breakthrough curves measured during a tracer test on a fracture under normal stress of 4 MPa. Fluid was collected at the fracture periphery, which was divided into 8 sectors; breakthrough curves are shown for 5 of the 8 sectors.

values of the average distance between exit points for dye injected at different levels of normal stress during hydromechanical tests. The average distance between the exit points for the fluid (Figure 1b) varied from 6 mm (corresponding to the range of the first component) to 18 mm (corresponding to the range of the third component). The correspondence between the distance between fluid exit points and the "correlation length" (range) for surfaces allows us to understand the progressive matching of the two fracture surfaces, which signifies a global closing of the fracture. For the specimen size studied, this closing is governed by the first and third components that constitute the surface geometry. This observation is valid for fractures with relatively symmetrical surfaces and no infill in the void space.

2.2.2. Relationship between experimental results and the geometry of the void space. To determine the void space geometry from either two- or three-dimensional maps of the fracture surfaces, we must be able to position the surfaces precisely. Once the surfaces of the fracture are aligned, geostatistical analysis of the fracture's void space shows that its geometry corresponds most closely to the surface components whose ranges are between 4 and 6 mm (i.e., the smallest range identified on surface profiles). Ranges corresponding to other components can also be identified, but less clearly. This result, obtained from a purely geometric analysis, is consistent with laboratory ex-

(a)

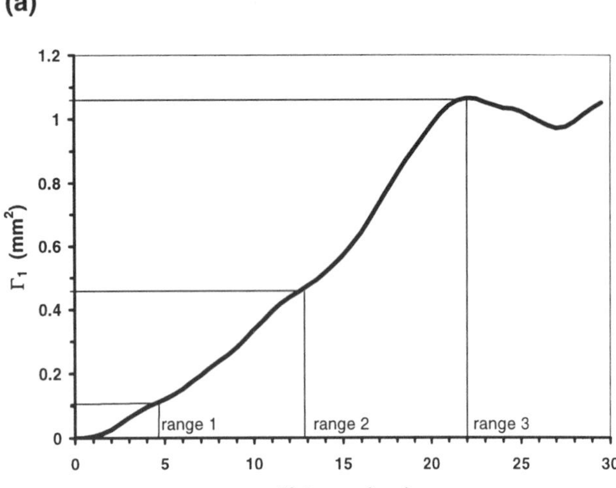

(b)

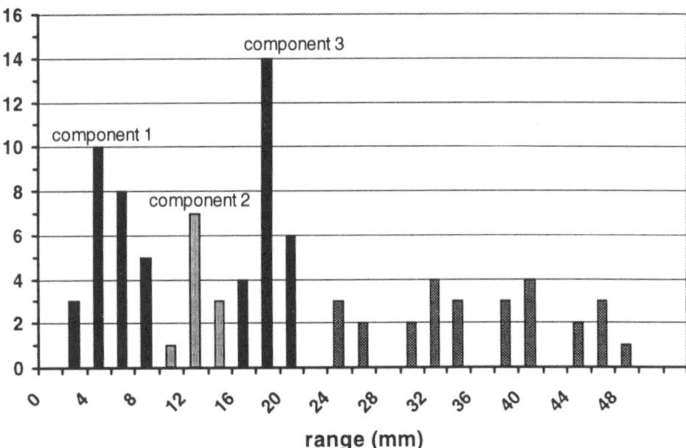

Figure 4. (a) Variogram calculated from a surface profile. In this case, the profile can be described by the superposition of 3 components with corresponding ranges indicated on the graph; for stationary data, range corresponds to correlation length. (b) Histogram of ranges corresponding to all identifiable components on profiles measured on a fracture surface 12 cm in diameter (ranges obtained from a single profile are shown in Figure 4a). In this case, the surface topography is best described by the superposition of 3 components, whose ranges are 4–10 mm, 12–16 mm, and 18–22 mm.

periments in which the average distance between the flow channels (exit points) on the fracture's periphery was found to correspond to the range of the first surface component (Figure 4b).

The above results were obtained from analyzing surface profiles. The same geostatistical analysis was applied to a map of the void space (Figure 5) obtained directly by casting a different granite fracture that contained hard infill [*Gentier and Poinclou*, 1996]. Ranges estimated from surface profiles and a void-space map in two perpendicular directions (Directions 1 and 2) are listed in Table 1. Analysis of the cast data led to the same conclusions reached from analysis of the surface profiles: a strong relationship exists between the components identified in the void-space data and those identified in the fracture-surfaces analysis. This is particularly clear for Direction 2, for which profile data are reported in Table 1. Along the Direction 2 profile, which is perpendicular to that of Direction 1, only three of the four components observed on the fracture surface are found in the void analysis. Comparing results from the two analyses shows that void-space data obtained from casting shows a much clearer relationship to surface data than void-space data obtained from matching two-dimensional profiles. We believe that the difference indicates the difficulty of accurately matching two-dimensional profiles obtained from the two fracture surfaces.

2.2.3. Relationship between experimental results and geometry of the contact area. Fracture closure with increasing normal stress is accompanied not only by a reduction in mean aperture but by an increase in contact area, which reduces the volume of void area through which flow can occur. Contact area is mapped for several stress levels by placing a very thin plastic film between the fracture surfaces. This film shows where contact occurs when normal stress is applied. The contact area indicated on the film, detected by image analysis, is larger than the actual contact area because of the film's thickness (20 μm) (i.e., areas with aperture smaller than 20 μm appear to be in contact). This technique shows that for a natural granite fracture:

- Contact area is not uniformly distributed on the fracture

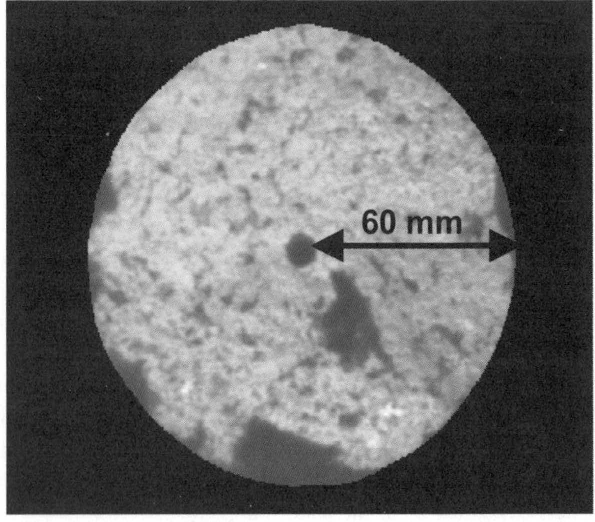

Figure 5. Map of fracture void space obtained by casting.

(a)

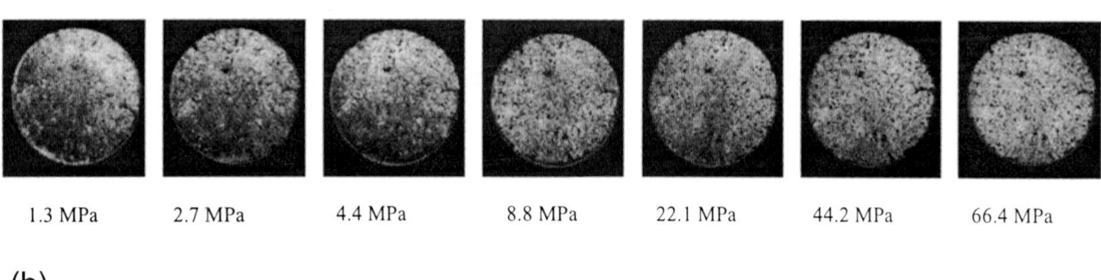

(b)

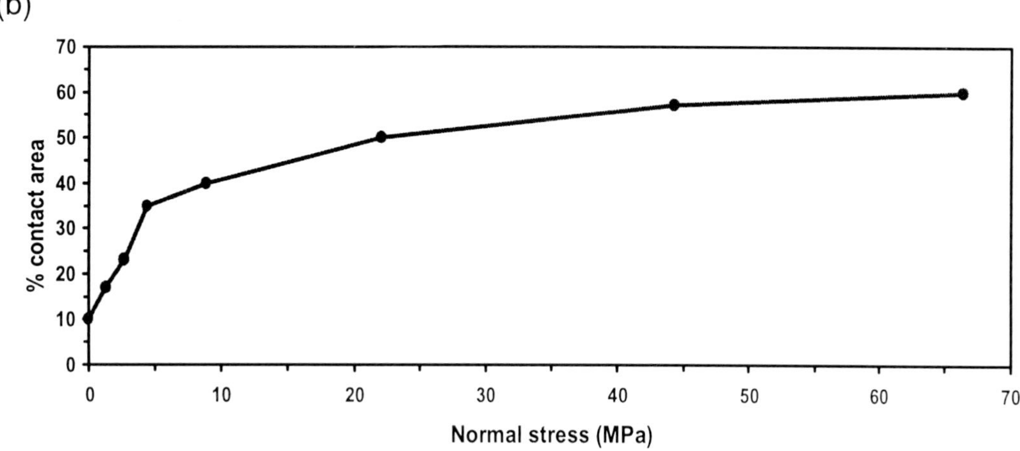

Figure 6. Evolution of contact area [white areas in (a)] with increasing normal stress estimated from image analysis of thin plastic sheets placed between surfaces of the same fracture. (b) The actual contact area for these areas is less than that indicated because the film's thickness (20 μm) means that apertures of less than 20 μm appear to be in contact.

surface (Figure 6a). This result is consistent with the heterogeneous closure described above.

- Contact area does not increase linearly with normal stress (Figure 6b). It increases rapidly at low normal stress, becoming constant at the critical normal stress observed at maximal closure and minimal intrinsic transmissivity.
- Contact area (the area with aperture smaller than 20 μm) does not exceed 40–60% of the nominal fracture area for normal stresses up to 65 MPa (Figure 6b).

2.3. *Hydromechanical Modeling*

Results from the hydraulic, mechanical, and geometrical analyses presented above are consistent with each other, permitting us to give a generalized description of a fracture's hydromechanical behavior. Modeling undertaken from the hydraulic and mechanical points of view is intended to reproduce the general characteristics of a fracture's behavior as observed in experiments.

2.3.1. Hydraulic modeling. Modeling of flow in a fracture has become its own research area as it has become clear that the cubic law, which originated in hydraulics for simple and well-defined geometries, cannot readily be applied to natural fractures in rock. This is because it is based on the assumption of laminar flow between two parallel planes. In an open, smooth-walled fracture in which asperi-

Table 1. Ranges corresponding to components identified on surface profiles and a cast of the void space between the surfaces in two perpendicular directions.

Ranges estimated from profiles measured on fracture surfaces (mm)		Ranges estimated from analysis of void-space map obtained from fracture cast (mm)	
Direction 1	Direction 2	Direction 1	Direction 2
37.0–46.0			
	27.0–30.0	20.0–30.0	25.0–35.0
17.0–18.0	13.5–19.0	15.0–17.0	12.0–18.0
7.0–10.0		8.0–11.0	8.0–10.0
	6.0–6.5	5.5–6.5	5.0–6.0
2.5–5.0	3.0–4.0	2.0–3.5	2.0–3.5

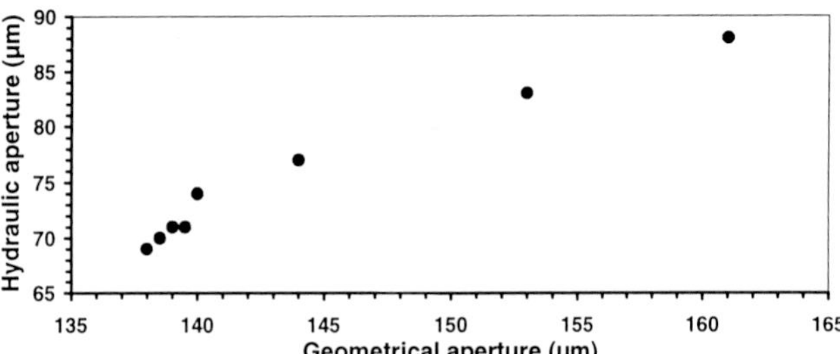

Figure 7. Relationship between geometric and hydraulic apertures for a fracture without hard infill under normal stresses between 1.5 and 35 MPa.

ties are small with respect to the mean aperture, the cubic law is still applicable. However, as soon as fracture geometry becomes more complicated, we cannot apply the cubic law because of the difficulty of determining whether to use hydraulic, mechanical, or geometric mean aperture values. Because it is difficult to determine which mean aperture is appropriate, *Louis* [1967] attempted to expand the range of fracture types to which the cubic law could be applied by introducing a coefficient of roughness and a reduction in flow surface:

$$k_f.e = (e^3/12)\,(\tau\,/C)\ , \qquad (2)$$

where:

($k_f.e$) is intrinsic transmissivity [intrinsic permeability (k_f) multiplied by aperture (e)],

τ is the separation ratio (the percentage of the fracture accessible to fluid),

C is a roughness coefficient [equal to 1 if relative roughness (k/D_h) is smaller than 0.033 and equal to $1 + 8.8\,(k/D_h)^{1.5}$ if relative roughness (k/D_h) is greater than 0.033],

k is the mean height of the asperities, and

D_h is the hydraulic radius of the fracture, usually taken as twice the mean aperture for the case of two parallel planes.

Louis's approach did not greatly extend the range of fractures to which the cubic law can be applied. This can be attributed to the impossibility of integrating the various roughness scales and the complexity of fracture flow paths into a single roughness coefficient. For example, using experimentally measured intrinsic transmissivities and assuming we can model the fracture as two parallel planes, we calculate the ratio between geometric and hydraulic aperture to be approximately 0.5, where the geometrical aperture is deduced from the geometrical matching of fracture-surface profiles. Figure 7 illustrates the relationship between these two apertures, calculated for different levels of normal stress. If we introduce geometric aperture values to the formula modified by *Louis* (Figure 8), we get ratios (τ/C) with values between 0.12 (35 MPa) and 0.17 (1.53 MPa). Taking into account contact area estimated by analyzing plastic film placed between fracture surfaces, and using *Louis's* equation, we calculate roughness coefficients (C) varying between 5.0 for a normal stress of 1.35 MPa and 3.0 for a normal stress of 35 MPa in one case and between 41 for a normal stress of 0.9 MPa and 175 for a normal stress of 13 MPa in another case. These last values of the roughness coefficient (C) lead to values of relative roughness (k/D_h) as high as 3 and 7, respectively. In other words, the estimated relative roughnesses are 3 and 7 times the aperture, respectively, which is not physically possible. Thus, to use this approach, we must rethink the definitions of roughness and aperture.

Current research focuses on better understanding the relationship between flow and the geometry of the fracture's void space. The difficulty of capturing fracture geometry in a single parameter requires finding new ways to incorporate realistic geometries into flow equations that adequately describe the complexity of the void space. The suggestions for describing flow in fractures proposed by *Louis* [1967] should be reevaluated in light of the more precise knowledge we now have about fracture morphology.

Modeling studies [e.g., *Neretnieks et al.*, 1982; *Bourke*, 1987; and *Tsang and Tsang*, 1989] have also led researchers to the conclusion that flow models must incorporate void-space geometries that are more realistic than parallel-plate models. *Billaux and Gentier* [1990] developed a flow model based on identifying a network of discrete channels. A map of void space obtained by casting was transformed into a network of possible flow channels using algorithms from mathematical morphology. Flow was thus calculated for a network of one-dimensional elements, assuming either a linear or cubic relationship between transmissivity and aperture. These calculations showed that the results of the hydromechanical experiments presented above (Section 2.1) are best described by a linear relationship between transmissivity and aperture. This approach explicitly takes into account flow channeling.

More recently, these same flow calculations were redone. A regular network of channels was superimposed on the pixels of the void-space map, which enabled us to calculate a map of flow rates and pressures in the fracture during radial-divergent-injection experiments (Plate 1). The map of flow rates presented here shows that flow is organized in channels (channels are defined as the areas where flow rates are higher than elsewhere in the fracture). The map of pressures shows some asymmetry around the injection well.

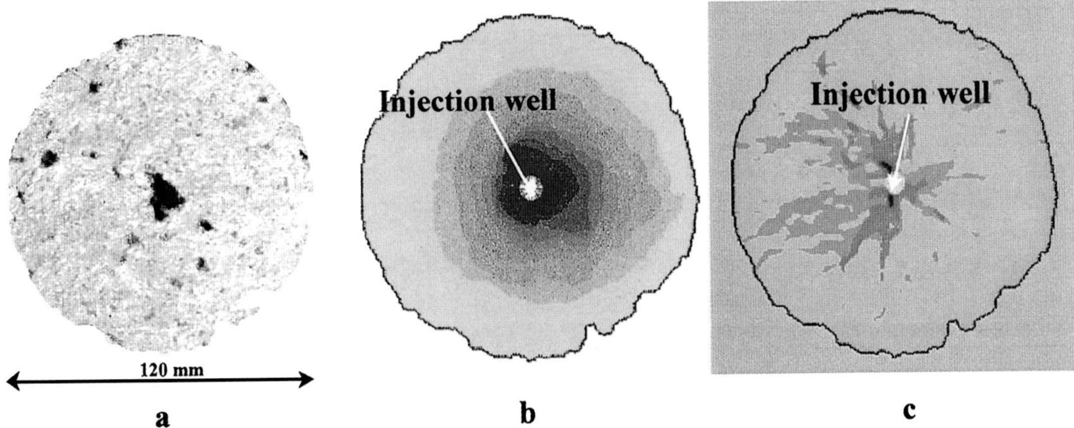

Plate 1. To calculate a map of pressures and flow rates in a fracture without hard infill during radial-divergent-injection experiments, a regular network of channels was superimposed on the pixels of the void-space map obtained by casting; dark areas correspond to high-aperture regions (a). The map of pressures shows some asymmetry around the injection well (blue corresponds to highest pressures) (b). The flow-rate map shows that flow is organized in channels (c), identified as areas of high flow rate (red/orange corresponds to highest flow rates).

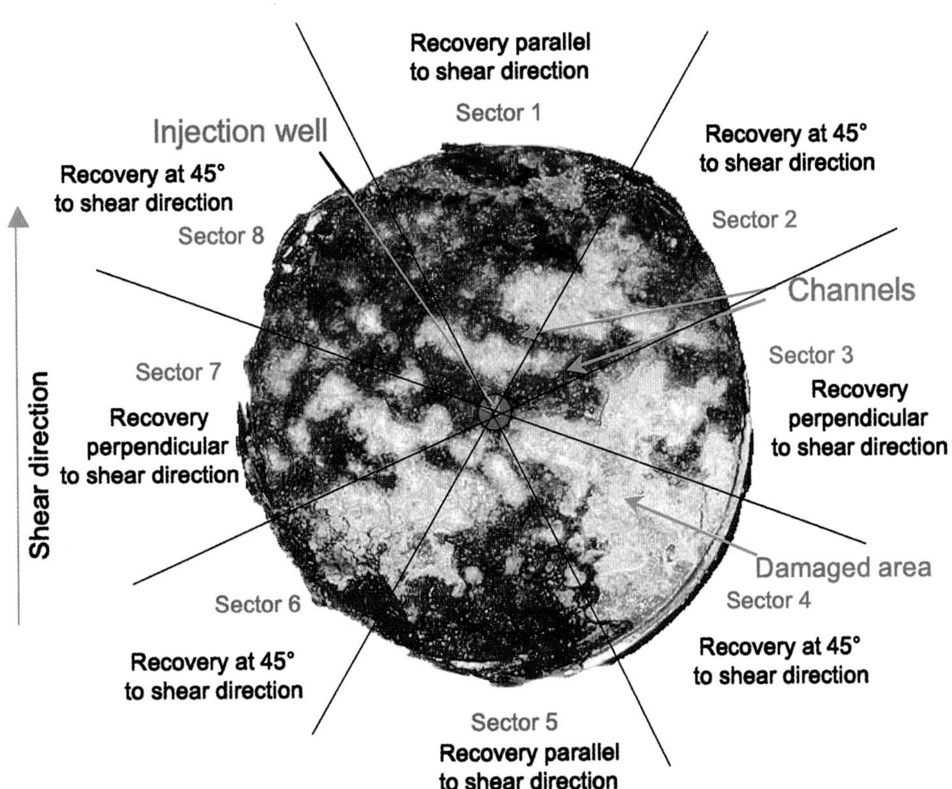

Plate 2. Photograph of a cast of the void space in place on the lower fracture surface after shearing. Channels (darkest areas) and damage areas (lightest areas) tend to form in directions subperpendicular to the shear direction.

Subsequent research has attempted to better model the physics of flow in channels [*Davias*, 1997].

2.3.2. Mechanical modeling. A fracture's hydraulic characteristics cannot be dissociated from the stresses applied to the fracture. All changes in stress cause a change in the geometry of the fracture's void space, a change that has direct effects on its global hydraulic properties. The most common approach so far has been to study the mechanical effects of stress on a fracture's hydraulic response. The mechanical approach allows us to understand the geometric changes produced by mechanical stress on a fracture and thus to attempt to understand the resulting perturbations in flow.

Our first approach to basic mechanical modeling of a fracture under normal stress is based on the hypothesis of uniform closure across the fracture [*Billaux and Gentier*, 1990]. Even though contradictory in principle to the experimental observations (nonuniform closure measured by LVDTs on the periphery of samples), this approach permits us to reproduce the variations in permeability measured in the laboratory for two values of average closure (once we calibrate the transmissivity for the first application of normal stress). However, assuming uniform fracture closure leads to overestimation of the contact area.

More realistic mechanical deformation of fracture surfaces with normal stress is accounted for in the model developed by *Hopkins* [1991]. In this model, the fracture is represented as asperities between elastic half spaces. A composite surface is constructed by summing the heights of the fracture walls measured from arbitrary reference planes. Asperities are modeled as cylinders that deform elastically. Deformation of the fracture surfaces is calculated based on solutions from the theory of elasticity. The model accounts for mechanical interaction among all contacting asperities so that asperities can be contiguous. Parametric studies using this model have clearly shown the role of both the contact area and the spatial distribution of the contact area in the fracture's global mechanical behavior [*Hopkins*, 2000]. This model was used to analyze deformation under normal stress for data obtained from a natural granite fracture. A geometric model of the fracture's composite surface roughness was calculated based on a void-space map obtained from a fracture cast [*Gentier and Hopkins*, 1996]. It was thus possible to map aperture and contact area across the fracture for any specified applied normal stress and to determine the normal stresses acting at each contact point.

Taking into account mechanical deformation of the fracture surfaces (using Hopkins's model) leads to estimates of contact area much smaller than those obtained assuming uniform closure across the fracture (Figure 9). Deformation of fracture surfaces also leads to changes in void-space geometry with changes in stress. These results have particularly important implications for calculating the flow and the volume accessible to fluid in a fracture [*Hopkins*, 2000]. Ongoing research, including similar work being performed by *Capasso et al.* [1999], will help us understand the micromechanical behavior of fractures under stress.

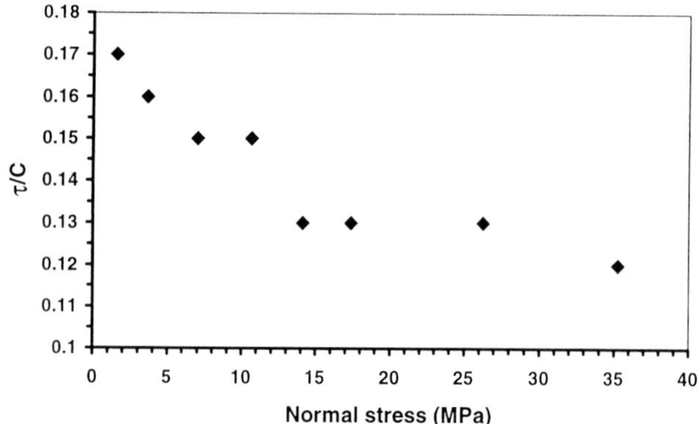

Figure 8. Relationship between the ratio of the percentage of the fracture accessible to fluid (τ) and roughness coefficient (C) with increasing normal stress (see Equation 2).

3. HYDROMECHANICAL BEHAVIOR IN SHEAR

Most fractures in situ are subjected to stresses that can be resolved into normal and shear components. The consequences of these stresses on a fracture's hydraulic properties are studied in the laboratory under constant normal stress. Much work has been done during the past 30 years to study the mechanical behavior of fractures in shear, with particular emphasis on problems of stability. However, very little theoretical or experimental work has been published to date on the hydromechanical behavior of fractures in shear.

3.1 Experimental Results

We performed tests of hydromechanical behavior in shear by using a classical shear machine adapted to permit divergent radial injection in the fracture. This allows us to maintain the same fracture geometry as in the tests of hydromechanical behavior under normal stress. Injected fluid is collected around the periphery of the sample for eight sectors. Because shear tests are destructive, they are performed on mortar replicas, which permits us to work with samples that have identical fracture geometry.

For a given normal stress (7 MPa) and a specific shear direction, shear produces (Figure 10):

- A global increase in intrinsic transmissivity with increasing shear displacement of typically two to three or-

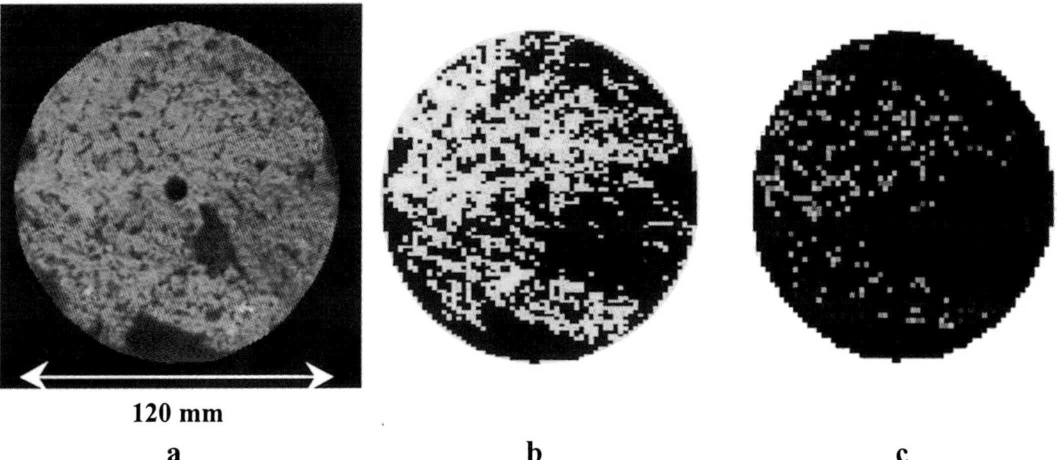

Figure 9. Map of void space obtained from a cast of a fracture with hard infill; dark areas correspond to high-aperture regions (a). Contact area calculated assuming uniform closure of the fracture (b) is much greater than that calculated using a model that takes account of deformation of the fracture surfaces (c).

ders of magnitude for a tangential displacement of about 1 mm (Figure 10a). This increase is directly related to the dilatancy of the fracture resulting from the roughness of the fracture surfaces. Values of dilatancy for the first millimeter of tangential displacement are about 15°. The increase in intrinsic transmissivity is sometimes preceded by a small reduction (about one order of magnitude) at the beginning of shearing, associated with an initial closure of the fracture, by as much as 10 to 15 percent of the initial aperture. Measurements indicate that dilatancy induces normal displacements of 2 to 3 times the initial aperture, when tangential displacements are between 2 and 2.5 mm.

- A change in flow pattern, illustrated by the three curves in Figure 10b, is associated with different phases of the fracture's mechanical behavior [*Gentier et al.*, 1997], with a particularly marked change at peak shear stress when the preferential flow paths reorient subperpendicular to the shear direction. This anisotropy tends to disappear during the post-peak softening stage with the appearance of residual shear behavior (no dilatancy).

The same basic behavior is found for all shear directions tested (0°, 90°, and 180°) for the same applied normal stress (7 MPa). In addition, the increased intrinsic transmissivity associated with dilatancy (after the initial decrease) is always of the same order of magnitude. However, shear direction affects the magnitude of the initial decrease in transmissivity associated with the initial closing of the fracture as well as the value of transmissivity during the residual phase of shear [*Gentier et al.*, 1997].

3.2. Geometrical Interpretation of Experimental Results

Analysis and interpretation of the results of hydromechanical experiments require understanding of the mechanical behavior—and the consequences of this behavior—on the geometry of the fracture walls and void space. In particular, the creation of damage zones during the course of shearing plays an important role in the evolution of flow paths. We use image-analysis and casting techniques (in conjunction with laboratory experiments) to better understand how fracture geometry and damage created during shear relate to changes in flow.

3.2.1. Relationship between damage zones and fracture geometry. The key phenomena in the hydromechanical behavior of fractures described above can be explained in part by the evolution of damage zones during shear. Damage zones vary in size and spatial distribution during the course of shearing. Damaged surfaces identified on images obtained at the end of each stage of shear show that damage zones increase in size and then coalesce with increasing tangential displacement [*Riss et al.*, 1997]. The spatial organization of damage zones is not random. They appear along lines subperpendicular to the shear direction. This observation is fundamental to understanding the reorientation of flow paths during shearing (Figure 11). Most of the damage occurs after the peak shear stress, during the softening and residual phases. Only a small percentage of the nominal area is damaged for each 1 mm of tangential displacement. Although the creation of damage zones may explain the reorientation of flow for the largest tangential displacements, it does not explain the reorientation of flow and increase in transmissivity that occur before the first millimeter of tangential displacement. In these first stages of shear, the maximum dilatancy (10 to 16° depending on shear direction) corresponds approximately to the mean value of the local gradient (13°) determined from surface profiles and from analysis of the entire surface reconstructed from the profiles (described above) using geostatistical methods. Moreover, the tangential displacement at which residual behavior is observed corresponds to half of the smallest range determined from surface profiles (see Section 2.2.1, above).

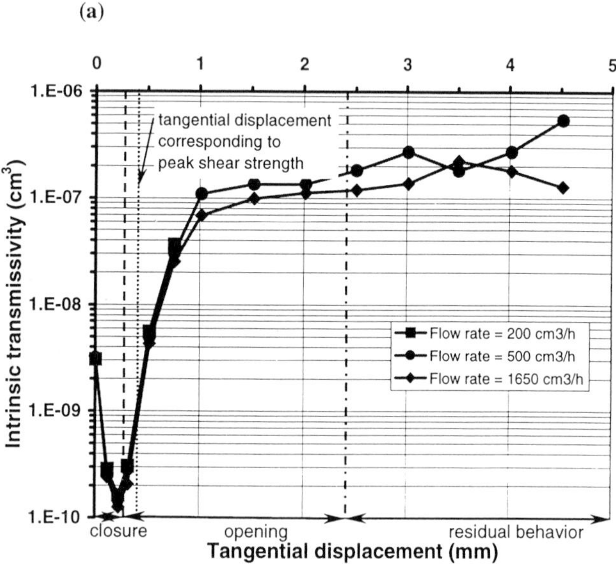

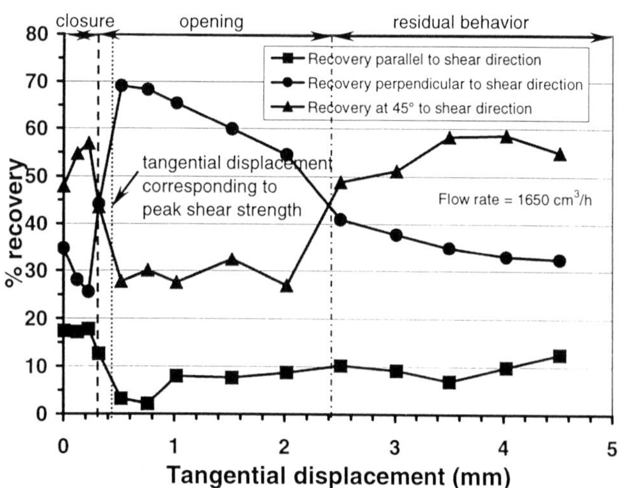

Figure 10. (a) Change in intrinsic transmissivity with increasing shear displacement for 3 flow rates. Marked changes in transmissivity correspond to closure, opening, and residual phases of mechanical behavior. (b) Change in percentage of fluid collected, by sector, with increasing shear displacement for sectors oriented parallel, perpendicular, and at a 45° angle to the shear direction.

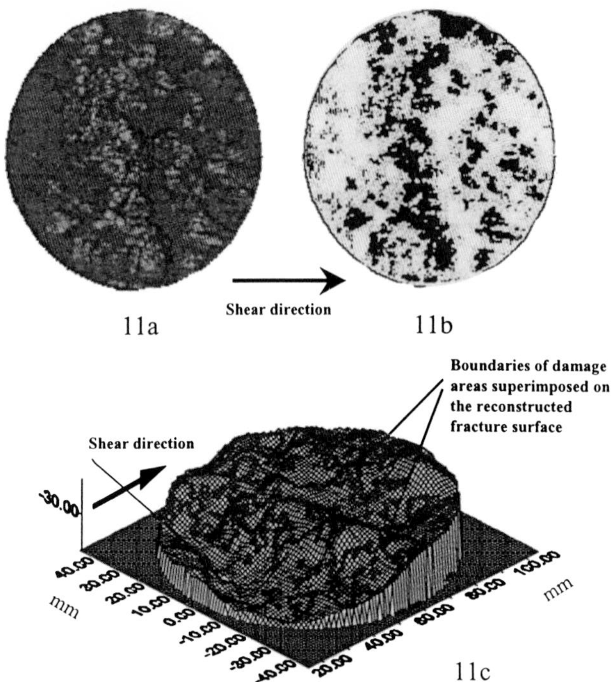

Figure 11. Grey-level image showing damage zones on a sheared fracture surface (a). The same data are displayed in a binary image (b) to isolate damage areas, which are then superimposed on fracture surfaces reconstructed from profile data using geostatistical methods (c). Damage zones generally appear along lines that are subperpendicular to the shear direction.

3.2.2. Relationship between hydromechanical behavior in shear and void geometry. The increase of transmissivity by several orders of magnitude observed during shear tests suggests that modifications in flow are not only a result of changes in the topology of the channel network. We must also pay attention to the global increase in the volume of void space accessible to fluid. An indirect measure of the volume accessible to fluid in a fracture is given by measuring the change in dilatancy that corresponds to the displacement of one surface relative to the other, in a direction perpendicular to the shear plane. Observed dilatancy values vary according to the shear direction but are of the same order of magnitude, which is consistent with transmissivities measured during the residual-shear phase.

Dilatancy gives merely an indication of the global change in fracture aperture with shear. To fully understand changes in flow during shearing, we must have complete information about the spatial organization of the void space and how it changes with shear displacement. A new methodology has been developed that allows us to cast a fracture's void space at the end of shearing. Techniques to analyze the cast are being developed to account for inclusion of particles in the resin cast; the particles are the result of damage during shearing. The casts are analyzed using image-analysis techniques. New techniques are required to analyze casts obtained post-shearing because particles in the cast show up on images with gray-levels that can be easily misinterpreted as areas of large aperture. A fracture cast obtained after shearing is shown in Plate 2 (this photograph is of the fracture cast in place on the lower fracture surface). Organization of channels in a direction subperpendicular to the shear direction (darkest areas) and damage areas (lightest areas) are clearly evident in the photograph. Qualitative analysis of images obtained from fracture casts after shearing shows that the apertures of the largest channels increase substantially during shearing. The next step in this research is to quantify the changes in aperture.

3.3. Hydromechanical Modeling

That we do not yet fully understand the mechanical behavior of fractures in shear, and the effect of shear on the geometry of the void space, impedes hydromechanical modeling. Recent work [e.g., *Archambault et al.*, 1996] indicates that mechanical behavior under shear can be broken down into at least five main phases, which can be summarized as follows:

1. The elastic phase, with reduction in aperture on the joint plane. Contact area and shear stiffness increase, and the void-space volume decreases.
2. The nonlinear-dilatancy phase (up to peak-shear stress) dominated by dilatancy and local redistribution of stress resulting from deformation and frictional slip at individual contacting asperities (isolated asperities may break).
3. The peak-shear-stress phase, with failure of asperities and a maximum dilatancy rate.
4. The postpeak phase, a progressive softening (unstable yielding) phase, during which the fracture surfaces progressively degrade (from microfracturing of asperities, crushing, and breaking of asperities, all depending on normal stress). The contact area between the two surfaces of the fracture increases, with a corresponding decrease in normal stress acting on the contact areas. Dilatancy still increases but at a lower rate than during Phase 3.
5. The residual-strength phase (stable sliding) in which shear and normal stresses are relatively stable, but degradation on joint surfaces still occurs.

This global view of shearing as a process that can be described by five mechanical phases is corroborated by the results of flow experiments in which fluid is collected and measured at the fracture periphery over the course of shearing. The flow patterns measured change more or less systematically during each phase (as described in Section 3.1). At first glance, these results suggest the possibility of a relatively straightforward mechanical process amenable to modeling. However, underlying the characteristic mechanical behavior exhibited by fractures in shear are complex micromechanical mechanisms that control the shear process. At present, the micromechanics of shearing are not well understood, making it difficult to develop a model that integrates micro- and macromechanical behavior. To take the next step and develop a model that couples hydraulic and mechanical behavior, we must determine whether flow is sensitive to micromechanical changes and, if so, how the micromechanics affect the evolution of the void-space geometry during each phase.

Laboratory experiments [*Archambault et al.*, 1996] indicate that little if any damage occurs during the first two phases of shearing prior to peak shear stress. This result simplifies mechanical modeling of the prepeak phases because it means that crushing and breaking of asperities do not have to be considered in these phases. Mechanical modeling of the prepeak dilatancy phase has been undertaken to study the stress redistribution that results from local slip at contacting asperities, which eventually leads to global slip after the peak shear stress is reached [*Hopkins*, 2000]. Because fracture surfaces are rough, sparse contact occurs between them, and the stress distribution across the contacting asperities is extremely heterogeneous. This nonuniform stress distribution leads to a progressive form of global slip (i.e., local slip occurs first at contacting asperities under the lowest normal stresses). Each local slip event results in a redistribution of stress to surrounding contact points. The stress redistribution leads to a stable state if neighboring asperities can accommodate the additional stress or to an unstable state if the stress redistribution triggers slip at other contact points. The next step in this modeling work is to calculate changes in the aperture distribution resulting from the stress redistribution that occurs during the prepeak dilatancy phase.

4. CONCLUSIONS

The hydromechanical behavior of fractures is far from perfectly understood. However, pieces of the puzzle are beginning to come together. The multiplicity of information from hydraulic, mechanical, and geometric studies is very helpful. Key results of research to date on natural granite fractures can be summarized as follows:

- The application of normal stress induces a normal closure that typically represents 10–20% of the initial geometric aperture. Because of surface roughness, application of normal stress also causes tangential displacement, resulting in closure that is typically 5–10%; this is based on experimental measurements. Associated with this closure is an increase in contact area. At maximum closure, contact area is at most 30–50% of the nominal area of the fracture surface.
- The consequence of the mechanical behavior on the hydraulic response of natural granite fractures is a decrease of one-half to one-tenth in the intrinsic transmissivity at maximum closure reached between 8 and 15 MPa of applied normal stress. This decrease in transmissivity is associated with increasingly distinct channeling as the normal stress increases. Channeling can be studied using tracer tests. These tests indicate that channeling is strongly linked to the various correlation lengths associated with the fracture walls and void space, and shows heterogeneity in the plane of the fracture.
- Applying a shear stress induces an opening of the fracture (due to dilation) marked by normal displacement that exceeds the initial aperture by 200–300%. The initial dilation appears to be very closely related to the mean value of the local slope in the shear direction. This prepeak phase is not associated with significant surface damage. Most damage occurs after the peak, when dilation is maximal. After peak-shear stress is reached, damaged areas increase in size. The maximum aperture is reached for a tangential displacement of 2.0 to 2.5 mm.
- The consequence of mechanical behavior on the overall hydraulic response of granite fractures is a very large increase (around two to three orders of magnitude) in the global intrinsic transmissivity. This increase results from maximal opening of the fracture, depending on shear direction. Closure at the beginning of shear, associated with small tangential displacements, leads to a decrease in average aperture of the same order as measured under normal stress. This initial closure does not affect the increase in transmissivity associated with dilatancy. Whatever the shear direction, this increase in global transmissivity is associated with an evolution in the topology of the flow channel network; in general, we observe a reorientation of the flow in a direction subperpendicular to the shear direction. This reorientation culminates just after the peak shear stress is reached. During the residual phase following the post-peak softening phase, flow tends to return to a more isotropic pattern.
- Some experiments indicate that flow does not remain laminar under increasing normal stress and flow rate. For fractures under shear stress, the global opening of the fracture induced by shearing likely leads to a more and more laminar flow, except during the initial closure period when very small tangential displacement is observed, which provokes an initial decrease in transmissivity.

It appears clear that the greatest increase in transmissivity observed during laboratory experiments on natural fractures occurs when the fractures are subjected to a shear stress. Even small tangential displacements can be sufficient to provoke large increases in transmissivity (after initial closure of the fracture). This result has particularly important implications for predicting the response of fractured rock in situ to changes in stress (for example, in designing production strategies for petroleum and geothermal reservoirs and ensuring stability of underground excavations). A better understanding of the relationship between transmissivity and changes in shear stress would also contribute greatly to improved production of fluids through hydraulic stimulation of fractured reservoirs.

Even though laboratory experiments are essential for understanding the phenomena that control the hydromechanical aspects of fracture behavior, prediction of behavior requires a good physical model. Modeling fracture behavior under normal stress permits us to take a first step toward understanding the evolution of flows paths and the fact that it seems impossible to completely close a fracture. The next step is to associate a closure model under normal stress with a flow model to verify consistency between experimental and modeling results. Shear is a very complex process and must be approached step by step, starting with small displacements that do not result in damage and progressing to understanding how slip/failure occurs at individual contact points and how failure progresses across the fracture plane. This work should lead to a better understanding of anisotropy, particularly the organization of flow in the fracture during different shear phases.

Much work remains. A better quantitative evaluation of the various phenomena related to hydromechanical behavior is necessary to further verify the consistency of the main results presented here, which are mostly qualitative. Moreover, all this work has been performed on surfaces corresponding to natural fractures in granite without infill or with hard infill. To extend the application of this work, we must study other kinds of fractures in various kinds of rock. Finally, we must determine how these results scale so that they can be applied to in situ field studies. This last point requires, in particular, identification of the most relevant parameters in hydromechanical behavior and techniques to estimate the values of these parameters at the field scale.

Acknowledgments. In addition to Paul Witherspoon, the authors are grateful to Bernard Feuga, Jane Long, Daniel Billaux, Guy Archambault, and Neville Cook, who encouraged and supported this work. The research described in this paper was made possible by the creativity of Lucien Bertrand, the laboratory genius of Christophe Poinclou and the geostatistical expertise of Jean-Paul Chilès. The authors thank Nan Wishner for her contributions to the paper, which include editing and translations of original work. The authors also gratefully acknowledge the support of BRGM, ANDRA, and LBNL.

REFERENCES

Archambault, G., S. Gentier, J. Riss, R. Flamand, and C. Sirieix, Joint shear behaviour revised on the basis of morphology 3-D modelling and shear displacement, *Proceedings, 2nd North American Rock Mechanics Symposium, Tools and Techniques*, Vol. 2, edited by M. Aubertin, F. Hassani, and H. Mitri, pp. 1223–1230, Balkema, Rotterdam, The Netherlands, 1996.

Billaux, D., and S. Gentier, Numerical and laboratory studies of flow in a fracture, *Proceedings, Int. Symp. Rock Joints*, edited by N. Barton and O. Stephanson, pp. 369–373, Balkema, Rotterdam, The Netherlands, 1990.

Bourke, P. J., Channelling of flow through fractures in rock, *GEOVAL'87*, 167–177, 1987.

Brown, S. R., Fluid flow through rock joints: the effects of surface roughness, *J. Geophys. Research*, *90*(82), 1337–1347, 1987.

Capasso, G., S. Gentier, C. Scavia, and A. Pellegrino, The influence of normal load on the hydraulic behaviour of rock fractures, *Proceedings of the 9th Int. Congress for Rock Mechanics, Paris, France, August 25–28, 1999,* vol. 2, edited by G. Vouille and P. Berest, pp. 863–868, 1999.

Davias, F., *Modélisation numérique d'écoulements en massif rocheux fracturé—Contribution à la modélisation hydromécanique de milieux fracturés*, Thèse de l'université Bordeaux 1, Mathématiques appliquées, 1997.

Esaki, T., K. Nakahara, Y. Jiang, and Y. Mitani, Effects of preceding shear history on shear-flow coupling properties of rock joints, in *Mechanics of Jointed and Faulted Rock*, edited by H. P. Rossmanith, pp. 501–506, Balkema, Rotterdam, The Netherlands, 1995.

Gale, J. E., The effects of fracture type (induced versus natural) on the stress-fracture closure-fracture permeability relationships, *Proceedings, 23rd U.S. Symposium on Rock Mechanics*, edited by R. Goodman and F. Heuze, pp. 290–298, Society of Mining Engineers of AIME, New York, N.Y., 1982.

Gale, J. E., Hydraulic behavior of rock joints, *Proceedings, Int. Symp. Rock Joints*, edited by N. Barton and O. Stephansson, Balkema, pp. 351–362, Rotterdam, The Netherlands, 1990.

Gentier, S., Lamontagne, E., Archambault, G. and Riss, J., Anisotropy of flow in a fracture undergoing shear and its relationship to the direction of shearing and injection pressure. *Int. J. Rock Mech. Min. Sci. (paper 094) 34*(3–4), 1997.

Gentier, S., P. Baranger, L. Bertrand, L. Rouvreau, J. Riss, Expérience d'écoulement dans une fracture pour la validation des modèles couplés chimie-hydro-thermo-mécanique en milieu fracturé, *Collection "Sciences et techniques nucléaires" de la Commission des Communautés Européennes*, EUR 17123 FR, 197 p., 1996.

Gentier, S., and C. Poinclou, *Comportement Hydromécanique sous Contrainte Normale d'une Fracture In Situ (FRABEX-Suède)—Morphologie à l'échelle Métrique et Décimétrique*, Rapport ANDRA B RP 0ANT 95-187/A, 86 p. and annexes, 1996.

Gentier, S., and D. Hopkins, *Comportement Hydromécanique sous Contrainte Normale d'une Fracture In Situ (FRABEX-Suède)—Modélisation du Comportement Mécanique des Essais en Laboratoire*, Rapport ANDRA B RP 0ANT 96-119/A, 54 p. and annexes, 1996.

Gentier, S., P. Baranger, and C. Poinclou, *Comportement Hydromécanique sous Contrainte Normale d'une Fracture In Situ (FRABEX-Suède)—Essais en Laboratoire de Comportement Hydromécanique et Hydrodynamique sous Contrainte Normale*, Rapport ANDRA B RP 0ANT 95-186/A, 79 p. and annexes, 1995.

Gentier, S., D. Billaux, and L. Van Vliet, Laboratory testing of the voids of a fracture, *Rock Mechanics and Rock Engineering 22*, 149–157, 1989.

Gentier, S., Comportement hydromécanique d'une fracture naturelle sous contrainte normale, *Proceedings, 6th ISRM Int. Congress for Rock Mechanics, Montreal*, pp. 105–108, Balkema, Rotterdam, The Netherlands, 1987.

Gentier, S., *Morphologie et Comportement Hydromécanique d'une Fracture Naturelle dans un Granite sous Contrainte Normale*, Thèse de l'université d'Orléans, Document du BRGM 154, 637 p., 1987.

Hakami, E., H. H. Enstein, S. Gentier, and M. Iwano, Characterisation of fracture apertures—Methods and parameters, *Proceedings, 8th Int. Congress on Rock Mechanics, Tokyo*, edited by T. Fujii, Balkema, pp. 751–754, Rotterdam, The Netherlands, 1995.

Hopkins, D. L., The implications of joint deformation in analyzing the properties and behavior of fractured rock masses, underground excavations, and faults, *Int. J. Rock Mech. Min. Sci.*, *37*(1–2), 175–202.

Hopkins, D. L., *The Effect of Surface Roughness on Joint Stiffness, Aperture, and Acoustic Wave Propagation*, Ph.D. dissertation, University of California, Berkeley, Calif., 1991.

Iwano, M., *Hydromechanical Characteristics of a Single Rock Joint*, Ph.D. dissertation, 321 p., Massachusetts Institute of Technology, Cambridge, Mass., 1995.

Louis, C., *Etude des Ecoulements d'Eau dans les Roches Fissurées et de Leurs Influences sur la Stabilité des Massifs Rocheux*, Thèse de l'université de Karlsruhe, 128 p., 1967.

Makurat, A., N. Barton, N. S. Rad, and S. Bandis, Joint conductivity variation due to normal and shear deformation, *Proceedings, Int. Symp. Rock Joints*, edited by N. Barton and O. Stephansson, Balkema, Rotterdam, pp. 535–540, 1990.

Neretnieks, I., T. Eriksen, and P. Tahtinen, Tracer movement in a single fissure in granitic rock: Some experimental results and their interpretation, *Water Resour. Res. 18*(4), 849–858, 1982.

Olsson R., *Mechanical and Hydromechanical Behaviour of Hard Rock Joints: A Laboratory Study*, Ph.D. dissertation, 195 p., Chalmers University of Technology, Gothenburg, Sweden, 1998.

Pyrak-Nolte, L. J., L. R. Myer, N. G. Cook, and P. A. Witherspoon, Hydraulic and mechanical properties of natural fractures in low permeability rock, *Proceedings, 6th ISRM Int. Congress for Rock Mechanics, Montreal*, pp. 225–232, Balkema, Rotterdam, The Netherlands, 1987.

Pyrak-Nolte, L. J., D. D. Nolte, L. R. Myer and N. G. W. Cook, Fluid flow through single fractures, *Proceedings, Int. Symp. Rock Joints*, edited by. N. Barton and O. Stephansson, pp. 405–412, Balkema, Rotterdam, The Netherlands, 1990.

Raven, K. G., and J. E. Gale, Water flow in a natural fractures as a function of stress and sample size, *Int. J. Rock Mech. Min. Sci. 22*, 251–261, 1985.

Riss, J., S. Gentier, G. Archambault, and R. Flamand, Sheared rock joints: Dependence of damage zones on morphological anisotropy (presented at the 36th U.S. Rock Mechanics Symposium, New York, N.Y.), *Int. J. Rock Mech. & Min. Sci., 34*(3–4), paper no. 258, 1997.

Thompson, M. E., and S. R. Brown, The effect of the anisotropy surface roughness on flow and transport in fractures, *J. Geophys. Research, 96*(B13), 21923–21932, 1991.

Tsang, Y., The effect of tortuosity on fluid flow through a single fracture, *Water Resour. Res. 20*, 1209–1215, 1984.

Tsang, Y. W., and C.-F. Tsang, Flow channeling in a single fracture as a two-dimensional strongly heterogeneous permeable medium, *Water Resour. Res. 25*(9), 2076–2080, 1989.

Witherspoon, P. A., J. S. Y. Wang, K. Iwai, and J. E. Gale, Validity of cubic law for fluid flow in a deformable rock fracture, *Water Resour. Res. 16*(6), 1016–1024, 1980.

Yeo, I. W., M. H. De Freitas, and R. W. Zimmerman, Effect of shear displacement on the aperture and permeability of a rock fracture, *Int. J. Rock Mech. Min. Sci.* 35, 1051–1070, 1998.

S. Gentier, Bureau de Recherches Géologiques et Minières (BRGM), 3 Av. Claude Guillemin, B.P. 6009, 45060, Orléans cedex 2, France

D. Hopkins, Ernest Orlando Lawrence Berkeley National Laboratory, MS 46A-1123, 1 Cyclotron Road, Berkeley, CA 94720

J. Riss, Centre de Développement des Géosciences Appliquées (CDGA), Université Bordeauz 1, Av. des Facultés, 33405 Talance cedex, France

Predicting Hydrology of Fractured Rock Masses from Geology

Paul R. La Pointe

Golder Associates, Inc., Redmond, WA 98052

Fracture network connectivity often dominates movement rate, flow volume, and mass transport through rock masses. These networks influence the effectiveness of petroleum reservoir development, safe disposal of nuclear waste, delineation of water supply or establishment of well-head protection plans, recovery from geothermal reservoirs, solution mining, construction of underground openings, and the remediation of contaminated rock. Well tests can provide a great deal of useful information on the hydraulic properties of fracture systems, but they are often expensive or logistically infeasible. These tests also may not provide an accurate description of the hydrologic properties of the rock volume under consideration. Methods to model fractured rock can be improved by quantifying the relation between geologic parameters and the hydrologically conductive fractures. This study illustrates the application of four statistical and pattern recognition methods—evaluation of correlation coefficients, contingency table analysis, multivariate regression, and neural net analysis. The data for the study consist of borehole and well-test information from eight boreholes used for characterizing a proposed low-level radioactive waste repository in Wake County, North Carolina. The analyses show that high localized flow rates are related to the presence of increased fracture intensity, and that this intensity is controlled by a complex interplay of structural geology and lithology. Some of the initial hypotheses concerning the relation of geology to hydrology were not substantiated by the data, leading to a refined conceptual model that differed in significant ways from the initial model. Although the techniques used are of general applicability, the precise nature of the correlation between geology and hydrology is site dependent.

1. INTRODUCTION

1.1 Overview

Fractures are geologic features that form networks capable of transporting fluids through rock over long distances. The rate of movement, the volume of flow, and the amount of mass transport through the system of interconnected fractures affect petroleum reservoir development, safe disposal of nuclear waste, delineation of water supply or establishment of well-head protection plans, recovery from geothermal reservoirs, the efficiency of solution mining, the construction of underground openings, and the remediation of contaminated rock.

It is common for engineers and hydrologists to employ hydrologic testing to directly characterize fracture network flow parameters. However, well tests are often expensive

Dynamics of Fluids in Fractured Rock
Geophysical Monograph 122

or logistically infeasible. For example, the number and spatial reach of wells drilled from an offshore oil platform are very much restricted. This may cause direct hydrologic test results to be very sparse for the volume of rock under consideration. It also presents a problem because hydrologic heterogeneity is often quite significant, so that and, therefore, a few tests may not provide an accurate description of the hydrologic properties of the large volume of rock under consideration. Such a situation makes it necessary to infer parameters from the well or boreholes to a much larger volume of rock.

There are two broad approaches to modeling flow and transport in fractured rock [*National Research Council,* 1996]: the Discrete Fracture Network (DFN) approach and the Stochastic Continuum (SC) approach. Both methods have advantages and disadvantages.

DFN models require the specification of the geometry, location, and hydraulic properties of the fractures that play a significant role in the rock permeability.

SC models need permeability and porosity values that reflect the local matrix and fracture systems' effective properties at the scale and shape of the numerical grid. SC models incorporate geology to delineate large-scale statistically homogeneous regions, often referred to as *domains* or *zones*. Parameter values within these domains can be assigned according to statistical distributions or conditioned to geologic parameters. Often, the relation between mappable geology and values of effective permeability is not well understood. Parameter assignment within each domain is carried out using a spatial statistical model or an inversion that matches known well test results, but is constrained elsewhere to statistical parameters only [*Zimmerman et al., 1998*]. Better geologic conditioning of the parameter values within zones and better definition of zones can greatly improve model accuracy [*La Pointe et al.,* 1996*; National Research Council,* 1996].

Over the past decade, DFN models [*Hudson and La Pointe,* 1980*; Long et al.,* 1982*; Dershowitz,* 1984*; Endo et al., 1984; Robinson,* 1984*; Smith and Schwartz, 1984*] have evolved to successfully address problems of regional fracture network connectivity [*Cacas et al.,* 1990a,b; *Dershowitz et al.,* 1992*; Swaby and Rawnsley,* 1996]. The models represent fractures as polygons with flow and transport properties. However, the specification of these models relies upon accurately describing the geologic context of the subset of fractures that contribute to large-scale flow. This subset of fractures constitutes the network of *conductive* fractures. As detailed surveys of wellbores prove, only a portion, often less than 10% [*National Research Council,* 1996] play a role in flow at the scale of contaminant dispersal, energy production, or containment of nuclear waste. It is the intensity, geometry, and fluid-flow properties of this conductive subset that control the important behavior of fracture-dominated flow systems.

Thus, better understanding of the geologic habitat of conductive fractures could lead to improved DFN modeling, as well as a better understanding of why some fractures play a significant role in regional flow, while others play a limited, or insignificant role.

The attractiveness of geologically conditioning DFN or SC models rests on the presumption that hydrology relates to geology. By understanding the geologic characteristics that are associated with hydrologic variability, it is possible to describe the conductive fracture network or assign values of permeability (or other properties) at unsampled locations based upon the geologic characteristics.

The geologic approach is very appealing, since it is much more flexible in handling hydrologic variability as a result of variability in underlying geology than any zonal statistical model. This is because the geologic approach need not conform to overly simplistic statistical models. Moreover, a demonstrated connection between mappable geology and flow modeling is important in many licensing applications for proposed waste repositories or other facilities. Another advantage is that the geology is often known with reasonably high accuracy and resolution throughout the site.

However, the attempt to relate flow in fractured rock masses to underlying geology has proven challenging, particularly in fracture-dominated flow systems. Fracture network connectivity often controls flow and transport. This means that local geologic conditions might be affected by the properties of individual fractures and may not correlate to the larger-scale fracture network connectivity.

The current study attempts to understand the geologic habitat of fractures that play an important role in regional flow and mass transport. The focus of the study is a preliminary site investigation of a proposed low-level nuclear waste repository in Wake County, North Carolina. Licensing of the site requires modeling of possible movement of radionuclides through the rock and soil.

The goal of this work was to investigate the interrelation of hydrologic behavior and geologic characteristics by using multivariate analysis or other techniques, as appropriate, to help prepare the initial site conceptual model and support the development of the preliminary groundwater flow simulations. The identification of key geologic parameters was also important. These parameters needed to be measured in a manner sufficient for subsequent hydrologic modeling and so as to identify how data collection protocols might be altered in order to obtain the necessary data.

The problem of relating geology to hydrology is one of pattern recognition. Powerful and sophisticated tools to recognize patterns and correctly classify new data into proper groups are in common use in many disciplines. These tools differ in mathematical assumptions and outcomes, but all share the ability to classify data into categories. Section 1.4 below describes the methodology employed in this study in greater detail.

1.2 Data

A thorough data set was obtained for this study from geochemical surveys, hydrophysical logging, packer tests, and geologic logging. The data came from a series of eight wells, aligned from west to east. The line formed by these wells transects a prominent fault north-striking a normal fault (Figure 1). Wells W206 through W208 lie to the east of the fault in the footwall of the normal fault. Wells W201 through W205 lie in the hanging wall. As shown in Figure 1, well W205 cuts through the fault plane. Each of these boreholes was carefully logged to record both the lithologic characteristics and the attributes of any fractures present. Subsequently, the core and the borehole imagery was interpreted to create a data set in which fractures were identified, and their measured depths, orientations, apertures, and types recorded. Lithologic information was also recorded. This included grain-size classification, ranging from conglomerate to claystone, as well as the orientation and measured depth of contacts separating strata of contrasting lithology.

Hydrophysical anomalies were identified as part of the testing and logging program carried out in the eight wells. Such anomalies were determined as follows. First, a borehole was filled with deionized water and a logging tool to measure electrical conductivity is placed in the borehole. Next, the nonconductive deionized water was slowly pumped out, which allowed the conductive groundwater to flow through fractures or other permeable pathways into the borehole and be sensed by the logging tool. Continuous recording of the conductivity made it possible to identify locations and rate of groundwater flow into the borehole. The term "strength," as used in this study, refers to the magnitude of the conductivity change between the nonconductive, deionized water and the inflowing groundwater. The magnitude of the anomaly is a function of flow rate, and is used as a surrogate for fracture transmissivity or local wellbore permeability. The tool used in this study has a 1 ft (0.3 m) resolution. Together with image logs, it is possible to relate hydrophysical anomalies to fractures or other geologic features within the 1 ft intervals. Figure 2 summarizes the interpreted lithology, and shows the location and geometric aperture of detected fractures, as well as the location and magnitude of flow for hydrophysical anomalies. Figure 3 and Table 1 show the directly measured and derived geologic parameters used in this study. Parameters related to the

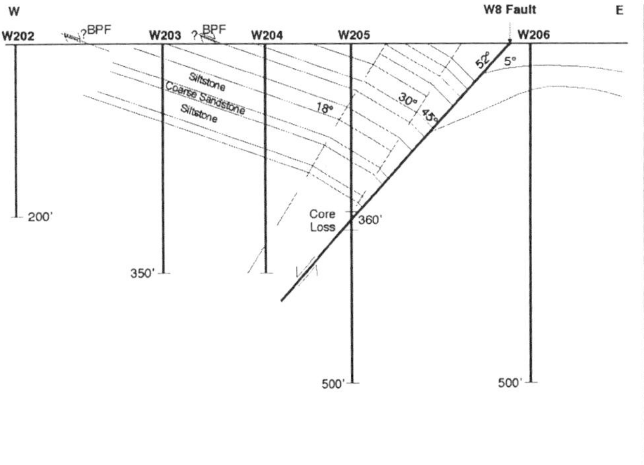

Figure 1. Geologic cross section along trench GM-1. Trench is oriented along an east-west axis. The normal fault, denoted as W8, strikes approximately north-south.

four nearest fractures were included to account for the 1 ft resolution of the hydrophysical logs, and for any small misregistration among the hydrophysical logs, the borehole imagery, and the core.

1.3 Initial Hypotheses

A preliminary focus of efforts at the Wake County site was to clarify the role of jointing and faulting in the site-scale hydrology. It was thought that fracture intensity at the site might be related to depth, weathering, structural deformation, and structural position. Thus, the study focused on evaluating some initial hypotheses concerning the relations between site geology and conductive flow features. These hypotheses are:

1. Hydrophysical anomalies are caused by fracture flow.
2. Fracture intensity is greater in the hanging-wall deformation zone.
3. Proximity to contacts between strongly contrasting lithological units leads to an increase in fracturing.
4. Fracture intensity relates to lithology.
5. Fracture intensity changes with depth/elevation because of weathering or lithostatic effects.
6. None of the above.

1.4 Methodology

A number of statistical and pattern-recognition techniques were applied to the data in order to investigate possible relations between geologic parameters and fracture flow. These studies were designed not only to test existing hypotheses, but also to uncover other unsuspected relations between geologic parameters and the

Table 1. Directly measured and derived geologic parameters.

Directly Measured Parameters	Symbol	Dimensions
Fracture Parameters		
Measured depth to fracture	MDF	Feet
Fracture dip and dip direction	DIP,DPDIR	Degrees
Fracture type: Shear or bedding	TYPE	
Fracture Aperture	AP	Inches
Lithologic Parameters		
Lithologic classification, based on grain size, gradated from 1 to 9 (1 being coarsest; 9 being finest): Conglomerate (Cnglom) = 1 Very coarse sandstone (VCrsSS) = 2 Coarse sandstone (CrsSS) = 3 Medium sandstone (MedSS)= 4 Fine sandstone (FineSS)= 5 Very fine sandstone (VFineSS)= 6 Siltstone (SiltS) = 7 Mudstone (MudStone) = 8 Claystone (Claystone) = 9	LITH	Grain size
Measured depth to contact between lithologic units	MDCON	Feet
Orientation of contact between lithologic units	CONDIP, CONDPDIR	Degrees
Layer (lithologic unit) thickness , distance measured vertically	UTHICK	Feet
Dip of contact above anomaly	DIP_A	Degrees
Dip of contact below anomaly	DIP_B	Degrees

hydrophysical anomalies. The following statistical techniques were used:

1. Evaluation of correlation coefficients among all parameters.
2. Contingency table analysis.
3. Multivariate regression.
4. Neural net analysis.

Correlation coefficients are very useful in examining the first-order relations between continuous variables such as the strength of the hydrophysical anomaly and its proximity to the boundary of a contrasting lithologic unit, or between the depth of a fracture and its aperture. These coefficients are also important in gaining an understanding of which variables may be redundant. Identifying redundant variables is important when data are sparse. When there are more variables—degrees of freedom—than hydrophysical anomalies, it is possible to predict the anomalies with a good degree of accuracy without actually determining any significant relations among the parameters.

Two-way contingency table analysis is a method for looking at the correlation among class or ordinal variables. It is also useful for examining relations between a parameter-like lithologic unit and a hydrophysical anomaly.

Contingency table analysis and correlation coefficients examine relations between pairs of variables. Multiple regression takes into account the combined relations among many variables. Multiple regression assumes that the dependent variable is a linear combination of independent (uncorrelated) variables.

Neural nets are the most complex of the methods used to investigate the relations among dependent variables and hydrophysical anomalies. This approach does not assume a simple or hypothesized model among the variables, nor does it require linear independence. Unlike multiple regression, it can also include class or ordinal variables such as lithologic type. Since the variable of interest, i.e., the hydrophysical anomaly flux rate, is a continuous variable, a Generalized Regression Neural Network

Table 1(continued).

Derived Parameters		
Distance from hydrophysical anomaly to four nearest fractures	D1, D2, D3, D4	Feet
Apertures of four fractures nearest to a hydrophysical anomaly	W1, W2, W3, W4	Inches
Lithology of four nearest fractures (ranging from conglomerate to mudstone, 1 to 8)	L1, L2, L3, L4	Grain size
Depth of four nearest fractures	MD1, MD2, MD3, MD4	Feet
Lithology contrast between layer containing hydrophysical anomaly lithology and layer above; grain-size class of layer above used for computation	LCA	Dimensionless
Lithology contrast between layer containing hydrophysical anomaly lithology and layer below; grain-size class of layer above below for computation	LCB	Dimensionless
Absolute contrast above; computed as ABS(LCA)	ALCA	Dimensionless
Absolute contrast below; computed as ABS(LCB)	ALCB	Dimensionless
Total contrast between lithological unit containing anomaly and units above and below; computed as LCA+LCB	TC	Dimensionless
Total absolute contrast between lithological unit containing anomaly and units above and below; computed as ALCA+ALCB	ATC	Dimensionless
Distance from anomaly to nearest upper contact; computed as distance perpendicular to contact interface	CDA	Feet
Distance from anomaly to nearest lower contact; computed as distance perpendicular to contact interface	CDB	Feet
Minimum distance to contact; computed as MIN(CDA, CDB	CMIN	Feet
Grain size contrast with nearest unit	C_NEAR	Dimensionless
Apparent distance to upper contact	APCDA	Feet
Apparent distance to upper contact	TDIST_MIN	Feet
Apparent distance to lower contact	APCDB	Feet
Angular difference between upper and lower contact; computed as DIP_B - DIP_A	UNCONF	Degrees
Absolute value of UNCONF	ABS(UNCF)	Degrees
Bed thickness	THICK	Feet

(GRNN), was employed. This type of network architecture has proven very useful for this type of application [*Ward*, 1996].

2. RESULTS

2.1 Correlation Coefficients

Table 2 shows the correlation coefficients for all of the independent variables with hydrophysical anomaly strength. This strength has been expressed in two ways: the arithmetic strength, which is the actual measured value, and the base-ten logarithm of the measured strength. The logarithmic transformation of a variable reduces the impact of extreme values, which may be outliers or spurious measurements.

Correlation coefficients vary between −1.0 (perfect anticorrelation) and +1.0 (perfect correlation). The statistically significant correlations at the 95% level in Table 2 are those in which the absolute value of the

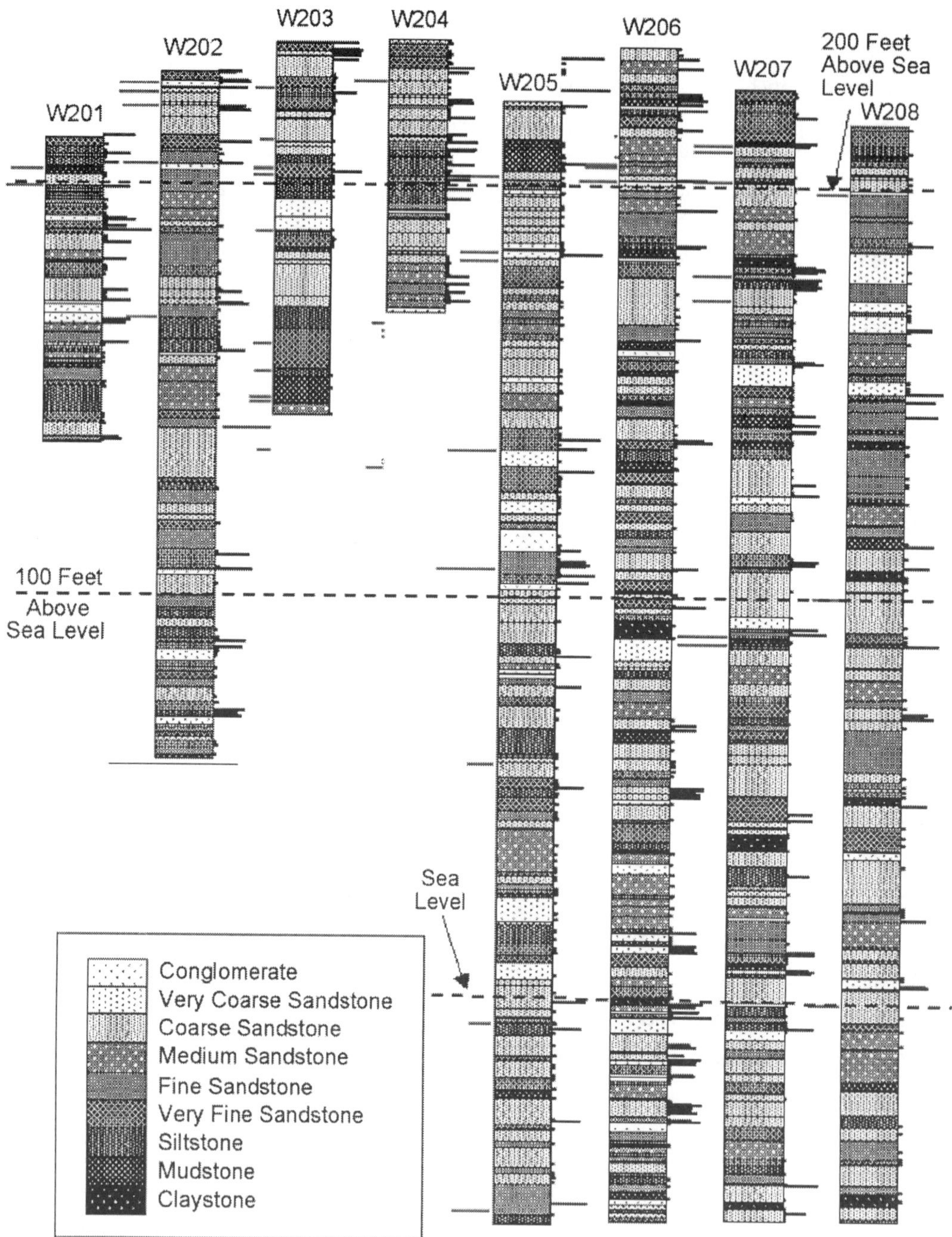

Figure 2. Summary of lithology, fractures, and hydrophysical anomalies. For each borehole, the location of hydrophysical anomalies and fractures are shown by horizontal lines. The length of the line to the left of the borehole is proportional to the logarithm of the hydrophysical anomaly magnitude, while the length of the line to the right of the borehole is proportional to the logarithm of the fracture aperture. In both cases the lines represent relative, not absolute lengths.

correlation coefficient is greater than or equal to 0.35.

This table shows that the anomaly strength and the log of the anomaly strength are positively correlated with the fracture width (aperture), the lithology of the unit in which the hydrophysical anomaly occurs, and the surrounding lithologic units, their contrast, and the dip of the contact between lithologic units. The positive correlation with lithology means that finer-grained lithological units have stronger anomalies. The positive correlation with the absolute total contrast (ATC) implies that stronger anomalies tend to be in lithologies, unlike the units immediately above and below. The correlation with contact dip may reflect structural disruption in the hanging wall of the fault, but other explanations may also be possible. The log of the anomaly strength is negatively correlated with the distance to the nearest fractures (in other words, the closer the fractures to the anomaly and the more there are of them, the stronger the anomaly).

Table 2. Correlation coefficients for flow anomaly and $\log_{10}$(anomaly) with geological parameters. Absolute values of the coefficient equal to 0.35 or greater are significant at the 95% level.

Variable	Anomaly	Log_{10} (anomaly)
W2	0.36	0.53
ALCA	0.31	0.48
ATC	0.20	0.47
W3	0.40	0.40
LITH	0.08	0.31
CONDIPB	0.47	0.27
CNEAR	–0.02	0.27
ALCB	0.00	0.23
W1	0.00	0.19
W4	–0.03	0.18
LCB	0.15	0.16
LCA	–0.25	0.15
LITHA	0.38	0.14
LITHB	–0.11	0.13
MDF	0.19	0.08
CONDIPA	0.23	0.01
APCDB	–0.01	0.00
CDB	–0.04	–0.02
CDA	0.17	–0.12
APCDA	0.19	–0.13
CMIN	0.08	–0.25
D1	–0.12	–0.34
D4	–0.15	–0.39
D3	–0.18	–0.40
D2	–0.18	–0.45

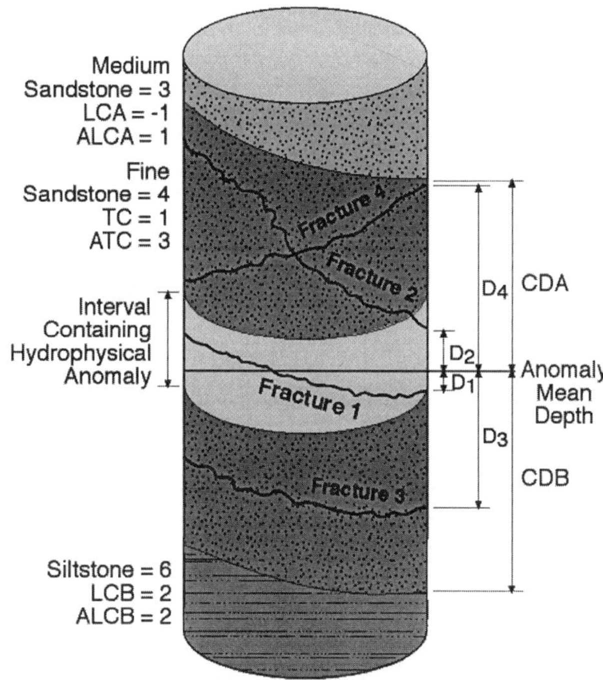

Figure 3. Derived parameters used in the statistical analyses. Table 1 provides additional explanation of these and other parameters.

2.2 Contingency Tables

Contingency tables were computed for various combinations of variables. This type of analysis is useful for determining whether a variable of interest shows up with unusually high or low frequency in some other variable class. For example, do siltstones have an unusually high frequency of fractures or hydrophysical anomalies that is out of proportion to their stratigraphic percentage?

Tables 3, 4, and 5 provide overall lithology and fracture intensity data for the eight wells. Table 3 shows how lithology varies among the eight wells. This Table illustrates that fine-grained lithological layers, particularly siltstone, increase eastward from W201 to W205, where such layers reach upwards of 45% of the total amount. They decrease abruptly in the three wells to the east of W205.

Tables 4 and 5 summarize the fracture intensity by lithology and well. These tables show that fractures in the western wells (W201 through W203) tend to reside in the coarse sandstone, while in the wells nearer to the fault, the fractures tend to occur in the siltstones. However, the

Table 3. Lithology statistics by well.

Total Thickness (inches) of Lithological Layers Within Wells										
	Lithology									
Well No.	1 Cnglom	2 VCrsSS	3 CrsSS	4 MedSS	5 FineSS	6 VFineSS	7 SiltSt	8 MudStone	9 ClayStone	Total
W201		50.25	146.74	218.29	108.29	74.42	177.11	79.44		854.54
W202			188.71	432.06	420.76	209.13	550.51	147.26		1948.43
W203		72.8	175.94	351.29	9.41	34.18	354.3	88.82		1086.74
W204		4.43	117.38	107.36	63.42	74.03	326.4	29.17		722.19
W205	30.71	220.76	479.08	1388.54	590.14	299.59	3403.86	648.01	70.87	7131.56
W206	66.35	186.64	304.61	912.62	484.35	611.72	1286.1	349.94	46.55	4248.88
W207	266.82	194.12	349.49	814.3	434.02	546.04	1814.42	402.63		4821.84
W208	110.12	67.26	189.86	474.12	1822.33	759.82	1584.26	429	43.26	5480.03
Total	474	796.26	1951.81	4698.58	3932.72	2608.93	9496.96	2174.27	160.68	26294.21
Percent of Thickness of Lithological Layers Within Wells										
	Cnglom	VCrsSS	CrsSS	MedSS	FineSS	VFineSS	SiltSt	MudStone	ClayStone	
W201		5.88	17.17	25.54	12.67	8.71	20.73	9.30		
W202			9.69	22.17	21.59	10.73	28.25	7.56		
W203		6.70	16.19	32.33	0.87	3.15	32.60	8.17		
W204		0.61	16.25	14.87	8.78	10.25	45.20	4.04		
W205	0.43	3.10	6.72	19.47	8.28	4.20	47.73	9.09	0.99	
W206	1.56	4.39	7.17	21.48	11.40	14.40	30.27	8.24	1.10	
W207	5.53	4.03	7.25	16.89	9.00	11.32	37.63	8.35		
W208	2.01	1.23	3.46	8.65	33.25	13.87	28.91	7.83	0.79	

greater number of fractures in siltstones near the faults may be due to a higher proportion of siltstone in these wells than in the western wells (Table 3), rather than a greater fracture intensity in siltstones near the faults.

Two series of questions have been formulated to specifically address fracture variability and hydrophysical anomalies among the wells. The first series of questions focuses on the fracturing itself. The correlation between hydrophysical anomaly strength and the number of fractures and their closeness to the anomaly indicates that

Table 4. Total number of fractures by lithology within wells, independent of aperture.

Total No. of Fractures by Lithology within Wells										
	Lithology									
Well No.	1 Cnglom	2 VCrsSS	3 CrsSS	4 MedSS	5 FineSS	6 VFineSS	7 SiltSt	8 MudStone	9 ClayStone	Total
W201		7	6	32	5	3	16	16		85
W202			22	30	9	2	19	10		92
W203		1	4	36	0	1	10	1		53
W204		0	16	19	11	16	37	2		101
W205	3	36	57	124	60	22	203	82	5	592
W206	2	10	24	54	19	17	82	54	7	269
W207	13	13	27	66	34	14	55	41		263
W208	5	3	7	37	46	40	69	43	13	263
Total	23	70	163	398	184	115	491	249	25	1718
Percent of Total Fractures by Lithology within Wells										
	Cnglom	VCrsSS	CrsSS	MedSS	FineSS	VFineSS	SiltSt	MudStone	ClayStone	
W201		8.24	7.06	37.65	5.88	3.53	18.82	18.82		
W202			23.91	32.61	9.78	2.17	20.65	10.87		
W203		1.89	7.55	67.92	0.00	1.89	18.87	1.89		
W204		0.00	15.84	18.81	10.89	15.84	36.63	1.98		
W205	0.51	6.08	9.63	20.95	10.14	3.72	34.29	13.85	0.84	
W206	0.74	3.72	8.92	20.07	7.06	6.32	30.48	20.07	2.60	
W207	4.94	4.94	10.27	25.10	12.93	5.32	20.91	15.59		
W208	1.90	1.14	2.66	14.07	17.49	15.21	26.24	16.35	4.94	

Table 5. Total number of fractures by lithology within wells for apertures > 0.03 in.

	Total No. of Fractures > 0.03 in. by Lithology within Wells									
	Lithology									
Well No.	1 Cnglom	2 VCrsSS	3 CrsSS	4 MedSS	5 FineSS	6 VFineSS	7 SiltSt	8 MudStone	9 ClayStone	Total
W201		2	2	4	2	0	2	6		18
W202			4	8	1	0	10	1		24
W203		0	0	5	0	0	4	0		9
W204		0	4	5	2	2	8	0		21
W205	0	3	3	9	6	1	27	9	1	59
W206	0	3	3	11	5	3	20	22	4	71
W207	4	3	7	20	9	2	7	18		70
W208	0	1	0	6	7	7	9	7	1	38
Total	4	12	23	68	32	15	87	63	6	310
	Percent of Total Fractures > 0.03 in. by Lithology within Wells									
	Cnglom	VCrsSS	CrsSS	MedSS	FineSS	VFineSS	SiltSt	MudStone	ClayStone	
W201		11.11	11.11	22.22	11.11	0.00	11.11	33.33		
W202			16.67	33.33	4.17	0.00	41.67	4.17		
W203		0.00	0.00	55.56	0.00	0.00	44.44	0.00		
W204		0.00	19.05	23.81	9.52	9.52	38.10	0.00		
W205	0.00	5.08	5.08	15.25	10.17	1.69	45.76	15.25	1.69	
W206	0.00	4.23	4.23	15.49	7.04	4.23	28.17	30.99	5.63	
W207	5.71	4.29	10.00	28.57	12.86	2.86	10.00	25.71		
W208	0.00	2.63	0.00	15.79	18.42	18.42	23.68	18.42	2.63	

the existence of multiple fractures could mean the presence of a strong hydrophysical anomaly.

If fractures are correlated with hydrophysical anomalies, then it is important to determine how fractures vary throughout the site, and what factors influence variations in fracture intensity. Tables 6 through 9 address the following four questions regarding fracture intensity and whether it varies among the wells or lithologic groups:

1. Are fractures distributed uniformly within all logged lithologic layers? If this is true, then it may imply that fracturing is relatively independent of lithological controls or structural position.
2. Are fractures distributed uniformly within each lithological layer in each well? If so, then fracture intensity may be controlled by lithology, and be relatively unaffected by structural position.
3. Are fractures with an aperture of >0.03 in. (0.76 cm) distributed uniformly within all lithologic layers? This question is similar to the first, except the emphasis here is on fractures with the largest geometric apertures. Such fractures may represent only a small subset of the most conductive fractures in a well, which may account for most of the flow.
4. Are fractures with an aperture of >0.03 in. distributed uniformly within each lithological layer within each well? This question is similar to the second, except it focuses on the fractures with the largest geometric apertures, which might in turn represent the most conductive fractures.

Table 6 shows that the p-value (the probability of observing the calculated Chi-Square statistic given that the null hypothesis is true) is essentially 0.0, implying that fracture intensity is not uniform among the wells. Wells W204, W205, and W208 are the most anomalous in terms of fracture intensity, as shown by their high Chi-Square values. Table 7 demonstrates that the frequency of fracture variation from well to well is influenced by lithology. The high Chi-Square value for siltstone and mudstone in well W205 shows that the fracture frequency is anomalous in these fine-grained lithologies in the well. The higher values for the coarser-grained layers in wells W202 and W203 confirm that fracturing is anomalous in these units. The p-values show that fracture intensity varies with lithology within individual wells. Tables 8 and 9 show similar results for fractures with apertures greater than 0.03 in. In summary, these tables indicate that fracture intensity varies among wells, and this variation is due not only to differences in the proportion of different lithologies. For example, siltstones in well W205 have a different frequency than siltstones in well W202.

With regards to the hydrophysical anomalies themselves, several questions arise:

1. Does the strength of the conductivity anomaly or number of anomalies vary by well?
2. Does the strength of an anomaly depend upon being located in a particular lithology?
3. Is the strength of an anomaly influenced by the thickness of the unit in which it occurs?
4. Is the strength of an anomaly influenced by the

Table 6. Evaluation of the degree to which fractures are distributed uniformly among the wells, using Chi Square criteria.

Well	Total Thickness of Layer (inches)	Total No. of Fractures	Expected No. of Fractures	Chi Square	p-value
W201	854.54	85	55.8	15.24	
W202	1948.43	92	127.3	9.79	
W203	1086.74	53	71.0	4.57	
W204	722.19	101	47.2	61.37	
W205	7131.56	592	466.0	34.09	
W206	4248.88	269	277.6	0.27	
W207	4821.84	263	315.0	8.60	
W208	5480.03	263	358.1	25.23	
Total	26294.21	1718	1718.0	159.16	4.82E-31

proximity to a contact with another lithologic unit?

5. Is the strength of an anomaly correlated to a sharp contrast in the lithologies of units immediately above and below the unit?

Figure 4 shows the percent of hydrophysical anomalies by lithology within each well. There is insufficient information to carry out meaningful contingency table analyses on these data, but the figure clearly illustrates that the anomalies occur in different lithologies in a manner that is not in proportion to the relative net thickness of the lithologic layers in the wells

Figure 5 illustrates the relation of the thickness of a unit containing a hydrophysical anomaly to the thickness of all units found in the eight wells. There is no obvious visual difference.

The hypothesis that lithologic contrasts are related to anomalies was tested by comparing the anomaly strength to:

1. The absolute strength of the contrast (parameter ATC).
2. The contrast with the nearest adjacent lithology (parameter C_NEAR).
3. The distance to the nearest adjacent lithology.
4. Whether anomalies are predisposed to occur more closely to lithologic unit contacts.

Figure 6 shows the relation between both total contrast (TC) and ATC with the strength of the anomaly. The

Table 7. Evaluation of the degree to which fractures are distributed uniformly with lithologic layer in individual wells.

Expected Number of Fractures											
Well No.	1 Cnglom	2 VCrsSS	3 CrsSS	4 MedSS	5 FineSS	6 VFineSS	7 SiltSt	8 MudStone	9 ClayStone	Total	
W201		5.00	14.60	21.71	10.77	7.40	17.62	7.90		85	
W202			8.91	20.40	19.87	9.87	25.99	6.95		92	
W203		3.55	8.58	17.13	0.46	1.67	17.28	4.33		53	
W204		0.62	16.42	15.01	8.87	10.35	45.65	4.08		101	
W205	2.55	18.33	39.77	115.26	48.99	24.87	282.56	53.79	5.88	592	
W206	4.20	11.82	19.29	57.78	30.66	38.73	81.42	22.15	2.95	269	
W207	14.55	10.59	19.06	44.41	23.67	29.78	98.96	21.96		263	
W208	5.28	3.23	9.11	22.75	87.46	36.47	76.03	20.59	2.08	263	
Total	26.59	53.13	135.73	314.47	230.75	159.14	645.52	141.76	10.91	1718	
Chi-Square Analysis											
	Cnglom	VCrsSS	CrsSS	MedSS	FineSS	VFineSS	SiltSt	MudStone	ClayStone	Total	p-Value
W201		0.80	5.06	4.87	3.09	2.62	0.15	8.30		24.90	0.00036
W202			19.23	4.52	5.94	6.28	1.88	1.34		39.19	2.18E-07
W203		1.83	2.45	20.78	0.46	0.27	3.07	2.56		31.41	0.00002
W204		0.62	0.01	1.06	0.51	3.08	1.64	1.06		7.98	0.23973
W205	0.08	17.05	7.47	0.66	2.48	0.33	22.40	14.79	0.13	65.39	4.05E-11
W206	1.15	0.28	1.15	0.25	4.44	12.19	0.00	45.77	5.57	70.81	3.39E-12
W207	0.17	0.55	3.31	10.49	4.51	8.36	19.53	16.51		63.42	3.13E-11
W208	0.02	0.02	0.49	8.92	19.65	0.34	0.65	24.40	57.48	111.96	1.51E-20

figure illustrates that there is a weak trend between increasing ATC and anomaly strength, but none between TC and anomaly strength. Figure 6 also compares anomaly strength to the contrast of the unit above (ALCA) and the unit below (ALCB). There is no obvious visual trend between these parameters and anomaly strength.

In addition, Figure 6 shows two parameters relating the anomaly strength to the contrast between the unit containing the anomaly (C_NEAR) and the distance to the nearest adjacent unit (CMIN). These data suggest no obvious trend.

The data shown in Figure 6 were tested to determine if hydrogeologic anomalies tend to be within 1 ft of the nearest contact. Does this imply that hydrophysical anomalies tend to occur close to lithology changes? This proposition was tested by selecting a random point in each unit penetrated by the eight wells and then computing the distance to the nearest contact. Figure 7 summarizes the result. Visually these two distributions appear very similar. A Chi-Square test does not reject this hypothesis, showing with an approximate 86% significance that these distributions are the same. This implies that anomalies do not appear to be preferentially located near major changes in lithology, although lithology contrasts have some importance.

The location and relative strength of each hydrophysical anomaly identified in the eight boreholes are indicated in Figure 2 by a line on the left-hand side of the borehole log. The length denotes the log of the flow rate. The figure indicates that the strongest hydrophysical anomalies are in W205, which intersects the fault.

2.3 Multiple Regressions

A series of step-wise multiple regressions was carried out using a data-mining application, SAS™. The results

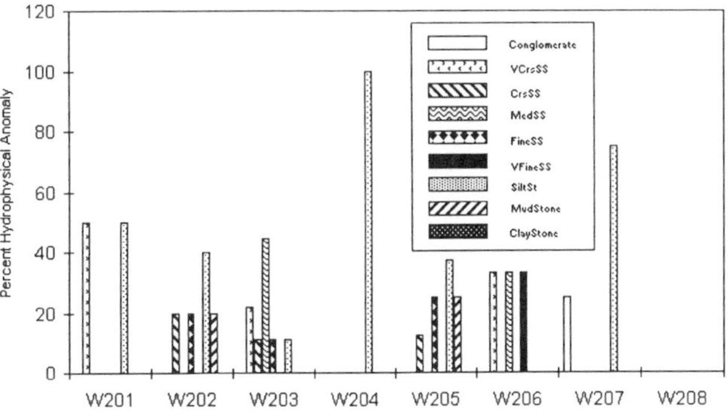

Figure 4. Percent of hydrophysical anomalies by lithology for individual wells. The height of the bars represents the percent of the logged interval of the well comprised by each lithology. Missing bars indicate the absence of a particular lithology in a well.

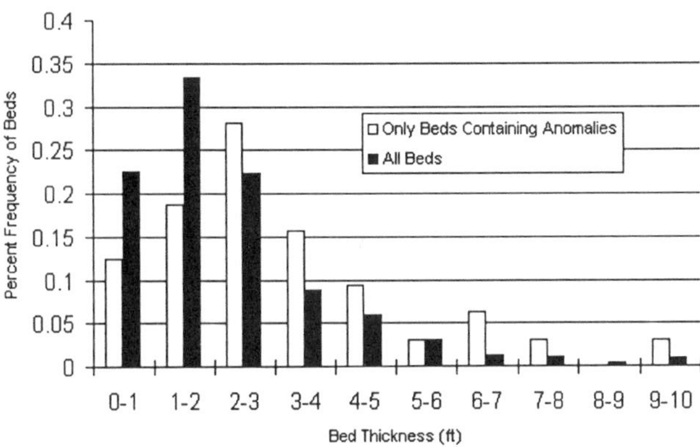

Figure 5. Bed thickness histograms. Open bars represent the histogram of bed thickness for all beds encountered by the wells. Black bars represent the histogram of bed thickness for only those beds containing hydrophysical anomalies. The vertical axis represents the percent of the beds belonging to a particular bed thickness class.

are summarized in Table 10. This table contains the results for the six multiple regressions with the highest R-square value for different numbers of independent variables.

Improvement in R-square decreased significantly after seven variables. Those variables that produced the best R-square values for a given number of parameters are shown in the table and marked by an "X." The table demonstrates that certain variables explain much of the variance in anomaly strength. These variables include the distance to the nearest fractures (D1, D2, D3), the aperture, or width of the fractures (W3), the absolute lithologic contrast (ATC), the dip of the unit below the anomaly (CONDIPB), and the contrast with the unit above (ALCA).

Thus, the multiple regression results suggest that the distances to the three nearest fractures are important, as well as the data on a strong lithology contrast and wider fracture apertures. These variables account for approximately 80% of the variability in anomaly strength.

2.4 Neural Net Analyses

All neural net analyses were carried out using NeuroShell II™ Version 3.0 [*Ward,* 1996]. A genetic adaptive GRNN architecture was used [*Specht,* 1991]. GRNNs are designed to predict a continuous variable. They are memory-based feed-forward networks based on the estimation of probability density functions. The classic GRNN contains only one adjustable parameter, termed the *smoothing factor*. The smoothing factor allows the GRNN to interpolate between patterns in the training data set. The genetic adaptive GRNN differs in that it has smoothing factors for each input parameter as well as a conventional overall smoothing factor. These smoothing factors are

Table 8. Fractures > 0.03 in. are distributed uniformly with lithologic layer.

Well	Total Thickness (inches)	Total Fractures	Expected Fractures	Chi Square	p-value
W201	854.54	18	10.1	6.23	
W202	1948.43	24	23.0	0.05	
W203	1086.74	9	12.8	1.13	
W204	722.19	21	8.5	18.31	
W205	7131.56	59	84.1	7.48	
W206	4248.88	71	50.1	8.73	
W207	4821.84	70	56.8	3.04	
W208	5480.03	38	64.6	10.96	
Total	26294.21	310	310.0	55.93	9.75E-10

optimized by iterative changes in the factors based on cross-validation to minimize the mean squared error of the outputs over the entire test set. The genetic adaptive GRNN is much more robust against noisy data and redundant parameters than traditional GRNNs, and is particularly well suited when the input parameters are of different types, and some may have markedly more impact on predicting the output variable than others.

When an observation, which consists of parameter values for all input parameters under consideration, is presented to the GRNN, it is compared to all of the patterns in the training set to determine how closely this observation corresponds to those patterns. The GRNN essentially computes a weighting factor for each training data-set observation, and then computes an overall output value based upon these weights and the input training set parameter values.

Application of a neural net to the hydrophysical anomaly dataset would be improved if the data set were larger. This is because a small training set and a small test set increase the variability in the results. Also, with a very small training set, it is possible that the training data would not span the *n*-dimensional parameter space and would thereby have difficulty interpolating to new observations with characteristics outside of the data training subspaces. Likewise, it is important to have fewer degrees of freedom

Table 9. Evaluation of the degree to which fractures > 0.03 in. are distributed uniformly with lithologic layer in individual wells.

All Fractures > 0.03 in. by Lithology within Wells											
Expected Number of Fractures											
	Cnglom	VCrsSS	CrsSS	MedSS	FineSS	VFineSS	SiltSt	MudStone	ClayStone	Total	
W201		1.06	3.09	4.60	2.28	1.57	3.73	1.67		14.89	
W202			2.32	5.32	5.18	2.58	6.78	1.81		10.40	
W203		0.60	1.46	2.91	0.08	0.28	2.93	0.74		5.05	
W204		0.13	3.41	3.12	1.84	2.15	9.49	0.85		2.47	
W205	0.25	1.83	3.96	11.49	4.88	2.48	28.16	5.36	0.59	5.73	
W206	1.11	3.12	5.09	15.25	8.09	10.22	21.49	5.85	0.78	67.51	
W207	3.87	2.82	5.07	11.82	6.30	7.93	26.34	5.85		51.47	
W208	0.76	0.47	1.32	3.29	12.64	5.27	10.99	2.97	0.30	15.45	
Total	6.00	10.02	25.73	57.80	41.30	32.48	109.91	25.10	1.66	172.96	
Chi-Square Analysis											
	Cnglom	VCrsSS	CrsSS	MedSS	FineSS	VFineSS	SiltSt	MudStone	ClayStone	Total	p-Value
W201		0.84	0.39	0.08	0.03	1.57	0.80	11.19		14.89	0.02111
W202			1.21	1.35	3.38	2.58	1.53	0.37		10.40	6.47E-02
W203		0.60	1.46	1.50	0.08	0.28	0.39	0.74		5.05	0.53790
W204		0.13	0.10	1.13	0.01	0.01	0.23	0.85		2.47	0.87224
W205	0.25	0.75	0.23	0.54	0.26	0.88	0.05	2.47	0.29	5.73	6.78E-01
W206	1.11	0.00	0.86	1.18	1.18	5.10	0.10	44.62	13.35	67.51	1.54E-11
W207	0.00	0.01	0.73	5.66	1.16	4.43	14.20	25.28		51.47	7.42E-09
W208	0.76	0.61	1.32	2.24	2.51	0.57	0.36	5.45	1.63	15.45	5.10E-02

(i.e., input variables) in the network than training cases; otherwise, the net "memorizes" the training set without uncovering useful relations.

The performance of a neural net is determined by how well it predicts the strength of the anomaly of a test set as quantified by R and R-square statistics. These statistics vary from 0.0 to 1.0; a perfect prediction has an R and R-square statistic of 1.0. The importance of an input parameter is quantified by its smoothing factor, which can vary from 0.0 (no importance) to 3.0 (very important).

Because of the small (approximately 30) number of hydrophysical anomalies available for use as training and test patterns, the results are sensitive to the random subset selected as the training pattern. For this reason, several random subsets were selected and processed.

Correlation shows up in the smoothing factors as a substitution of one correlated variable in random samples. For example, in one run, the distance to the third nearest fracture might be important. In a subsequent run, this distance might have little importance but the distance to the second nearest fracture would now be important. Highly correlated variables can substitute for one another since virtually all of the information in all of the variables can be contained in any one of the variables.

Neural net regression was much more successful for the logarithm of the anomaly than for the anomaly itself. A series of preliminary neural nets was constructed to look at different components of the data—in particular, fracture-related parameters and lithological parameters. In the preliminary processing that focused on fractures, several parameters played an important role. These included depth of the fractures, width of the fractures, distance from the anomaly to the nearest fractures, and the lithology of the

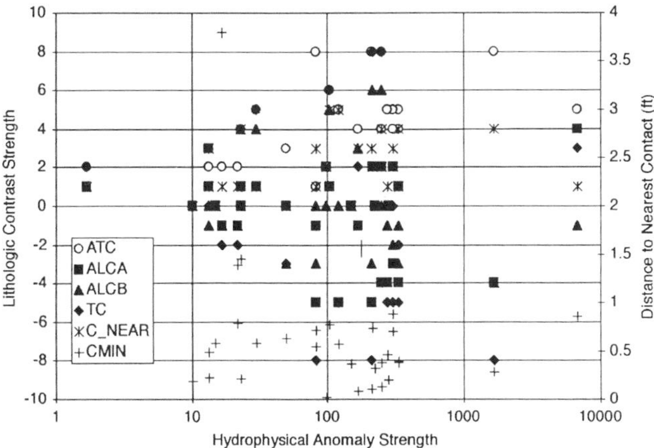

Figure 6. Anomaly strength vs. lithological contrast parameters. The vertical scale on the left-hand side of the graph pertains to the contrast parameters ATC, ALCA, ALCB, TC, and C_NEAR. The vertical scale on the right-hand side indicates the value of the distance parameter CMIN.

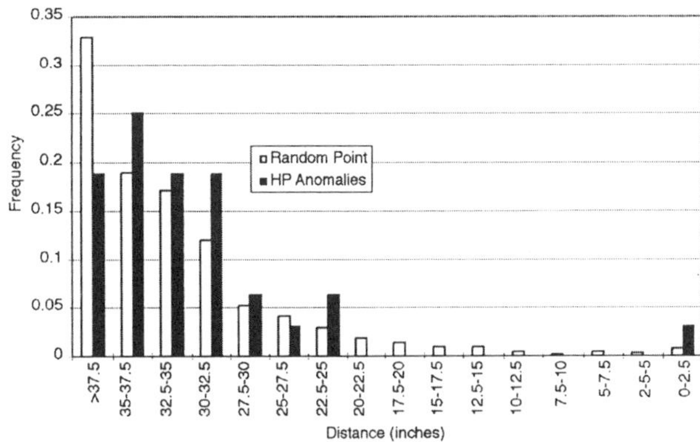

Figure 7. Percent frequency histogram of distance to nearest lithology change. Open bars represent distances between a location randomly assigned in the wells and the nearest contact. Black bars represent the distance from the locations of all hydrophysical anomalies to the nearest contact. The frequencies have been normalized to percentages in order to facilitate visual comparison.

nearest fractures. Depth was probably an artifact of the available dataset, however. Well W205, the deepest well, had the strongest anomalies. The other wells either were not drilled to the depth of W205, or else did not have all of the necessary parameters for the deeper portions of the borehole.

The lithologies above and below the hydrophysical unit did not appear to explain the hydrophysical anomaly. However, the lithologies of the four nearest fractures were quite informative. Although these lithologies correlated very strongly, surprisingly, at least three of them always seemed to have significant smoothing factors. This is reminiscent of the results from the step-wise regression, which showed that the distance to at least three of the four nearest fractures was important. If the information content in any one of these parameters were a good representation of the information content in any other, then it would be expected that only one of the lithology, depth, or aperture parameters would have a significant smoothing factor. The fact that more than one lithology parameter has a significant smoothing factor might be evidence that hydrophysical anomalies are in some way enhanced by the presence of multiple fractures near the anomaly, whether or not these fractures actually have an anomaly associated with them.

These preliminary runs guided the final runs in which redundant variables and depth measures had been eliminated. Tables 11 and 12 summarize those results.

Table 11 shows the mean of the smoothing factors for five randomly drawn training and test sets. The lithology of the fractures (L1 and L2), fracture width (W1 and W2), and proximity (D1 and D2) to fractures to the anomaly are

Table 10. Results of step-wise regression.

No. of Variables	R-square	ALCA	ATC	CDA	CONDIP A	CON DIPB	D1	D2	D3	D4	LCA	LITH	Log(W1)	Log(W2)	Log(W3)	W1	W2	W3
2	0.477	X												X				
2	0.438		X											X				
2	0.419	X															X	
2	0.402							X						X				
2	0.387							X									X	
2	0.383		X														X	
3	0.527	X						X						X				
3	0.513		X			X		X										
3	0.511	X											X	X				
3	0.508	X								X				X				
3	0.507		X					X						X				
3	0.506	X				X								X				
4	0.622						X	X	X									X
4	0.596		X					X	X						X			
4	0.591						X	X	X						X			
4	0.581	X						X	X					X				
4	0.580		X			X	X	X										
4	0.571	X				X							X	X				
5	0.717		X				X	X	X						X			
5	0.711		X				X	X	X									X
5	0.687						X	X	X							X		X
5	0.684						X	X	X					X				X
5	0.679	X					X	X	X									X
5	0.677		X			X	X	X	X									
6	0.802		X			X	X	X	X									X
6	0.783		X			X	X	X	X						X			
6	0.765		X				X	X	X					X				X
6	0.753	X					X	X	X					X				X
6	0.751		X				X	X	X					X	X			
6	0.750		X				X	X	X							X		X
7	0.828		X	X		X	X	X	X									X
7	0.823		X			X	X	X	X					X				X
7	0.820		X			X	X	X	X							X		X
7	0.817		X		X	X	X	X	X									X
7	0.816		X			X	X	X	X		X							X
7	0.815		X			X	X	X	X			X						X

always important. In particular, the distance to the second closest fracture is much more important than the distance to the closest fracture. Two other parameters, used only in the neural net analysis, also were important—UNCONF and ABS(UNCF). These variables were introduced to assess structural or stratigraphic disruption, an area also important in the neural net analyses. Absolute lithology contrast (ATC), lithology of the anomaly (LITH), and distance to the nearest contact (CMIN), have a moderate impact. Bed thickness (BED_TH) and the contrast between the anomaly unit and the nearest adjacent unit (C_NEAR) play very minor roles.

Table 12 shows the R-square and r-square statistics for both the training and test sets. R-square is the coefficient of multiple determination, whereas r-square is the more familiar coefficient of determination [*Ward*, 1996], which is the square of the correlation coefficient. The R-square value for the training sets averages nearly 92%, with a variation of 81% to 100%. The R-square value for the test sets averages nearly 77%, with a variation of 50% to 97%. Note that the lowest R-square for the test set comes from realization 2, which conversely has the highest value (1.0) for the training sets. The smoothing factors for this realization are also somewhat anomalous. The R-square values for both the training and test sets are higher than for the multivariate regression. This suggests that some of the input parameters have a nonlinear relation with anomaly strength. In general, the GRNN achieves about a 20% improvement over the multivariate regression in terms of R-square values.

Table 13 shows the predictions made by the neural net method for the ten test sets of observations. In general, the predictions are reasonably accurate, the average error being –0.0250 and the standardized mean squared error 0.34.

Thus, the neural net analysis confirms the results of the simpler statistical techniques and suggests that a GRNN offers greater accuracy than multivariate regression.

3. CONCLUSIONS

The statistical results for the Wake Co. site support the following conclusions:

1. There are statistically significant correlations between the geologic variables directly measured or derived from core data or borehole imagery and the strength of hydrophysical anomalies.
2. Hydrophysical anomalies do not appear to be located preferentially in particular lithologies for all eight wells. The frequency histograms and associated statistical test results show that hydrophysical anomalies are located in randomly selected subsets of all of the lithologic units.
3. Hydrophysical anomalies are not preferentially located near to boundaries between contrasting lithologic units. Frequency histograms of the proximity of anomalies to the nearest lithologic boundary are statistically indistinguishable from the distance from a random point located in every unit to its nearest boundary.
4. There is a weak trend between the magnitude of a hydrophysical anomaly and the strength of the contrast between the lithologic unit containing the anomaly and the units above and below it.
5. The trend noted in the conclusion above is independent of whether the anomaly occurs in a fine-grained unit surrounded by coarse-grained units or a coarse-grained unit surrounded by fine-grained units.
6. There is no correlation between the strength of the hydrophysical anomaly and the strength of the contrast with the lithologic unit above or below.
7. There is no correlation between the strength of the hydrophysical anomaly and the distance to either the upper or lower contact with the adjacent lithologic units.
8. The strongest hydrophysical anomalies occur in the fine-grained lithologies in Well W205, which intersected the fault and was drilled and logged deeper than most of the other wells. The anomalies in well W205 have three unique characteristics: they occur preferentially in siltstone or mudstone lithologies,

Table 11. Average smoothing factors for neural net analyses.

Average of 5 Realizations	**Smoothing Factor [0 to 3.0]**
Lithology to 1st Nearest Fracture	2.37
Distance to 2nd Nearest Fracture	2.24
Lithology to 2nd Nearest Fracture	2.09
Aperture of 1st Nearest Fracture	1.97
Absolute Value of Angular Difference between Upper and Lower Contacts	1.86
Difference between Upper and lower Contacts	1.59
Distance to Nearest Contact	1.28
Aperture of 2nd Nearest Fracture	1.21
Absolute Grain Size Contrast	1.15
Grain Size of Unit Containing Anomaly	1.12
Distance to 1st Nearest Fracture	1.08
Minimum Apparent Distance to Nearest Contact	0.76
Bedding Thickness	0.60
Dip of Contact with Unit Above	0.47
Grain Size Contrast with Nearest Unit	0.29

Table 12. Statistics for GRNN prediction of hydrophysical anomaly strength.

Training Set	#1	#2	#3	#4	#5	Average
R squared	0.91	1.00	0.99	0.81	0.88	0.92
r squared	0.94	1.00	0.99	0.90	0.88	0.94
Mean squared error	0.05	0.00	0.01	0.12	0.06	0.05
Mean absolute error	0.15	0.00	0.06	0.24	0.07	0.10
Min. Absolute error	0.00	0.00	0.00	0.00	0.00	0.00
Max. Absolute error	0.80	0.00	0.25	1.04	1.35	0.69
Correlation coefficient r	0.97	1.00	0.99	0.95	0.94	0.97
Percent within 5%	45.46%	100.00%	77.27%	27.27%	87.50%	67.50%
Percent within 5% to 10%	36.36%	0.00%	13.64%	22.73%	3.13%	15.17%
Percent within 10% to 20%	13.64%	0.00%	9.09%	31.82%	6.25%	12.16%
Percent within 20% to 30%	0.00%	0.00%	0.00%	9.09%	0.00%	1.82%
Percent over 30%	4.55%	0.00%	0.00%	9.09%	3.13%	3.35%
Test Set	**#1**	**#2**	**#3**	**#4**	**#5**	**Average**
R squared	0.86	0.50	0.97	0.85	0.67	0.77
r squared	0.96	0.53	0.98	0.87	0.73	0.81
Mean squared error	0.04	0.34	0.01	0.02	0.19	0.12
Mean absolute error	0.15	0.29	0.08	0.11	0.20	0.17
Min. Absolute error	0.00	0.00	0.00	0.03	0.00	0.01
Max. Absolute error	0.37	1.70	0.14	0.36	1.35	0.78
Correlation coefficient r	0.98	0.73	0.99	0.93	0.85	0.90
Percent within 5%	50.00%	60.00%	50.00%	80.00%	60.00%	60.00%
Percent within 5% to 10%	10.00%	0.00%	40.00%	0.00%	10.00%	12.00%
Percent within 10% to 20%	10.00%	30.00%	10.00%	10.00%	20.00%	16.00%
Percent within 20% to 30%	20.00%	0.00%	0.00%	0.00%	0.00%	4.00%
Percent over 30%	10.00%	10.00%	0.00%	10.00%	10.00%	8.00%

which is not true for anomalies in other wells; they occur at a greater depth than in the other wells; and the fractures nearest the anomalies have greater apertures than the fractures near anomalies in other wells. The depth correlation is probably an artifact.

9. Away from faults, hydrophysical anomalies are smaller on average and not as strongly associated with siltstone and mudstone. Near faults in the hanging wall, the strength of the anomalies is greater and they are correlated with fractures with larger aperture.
10. The results of our analyses confirm that a fault-related deformation produces new fractures or enhances the aperture of existing fractures in lithological units with lower resistance to shear—mudstones and siltstones—and thereby increases the strength of hydrophysical anomalies.

4. RECOMMENDATIONS

This section offers recommendations for using the results of this study to construct site-scale DFN and SC flow models. The results confirm that it is not correct to assign permeability values to grid cells or flow properties to fractures based on lithologic type alone. The boundaries between contrasting lithologic types do not focus strain. If they did, then the DFN model could have reproduced this effect by having large, bedding-parallel fractures at these interfaces with enhanced transmissivity values. SC models could have approximated this effect by having bedding-parallel layers of grid cells with enhanced subhorizontal permeability. While there is some correlation between lithological contrasts, the correlation is second order and should probably be ignored in preliminary flow models.

Table 13. Comparison of measured hydrophysical anomaly strength with strength predicted by GRNN for ten randomly selected anomalies not used in the GRNN calibration.[a]

#1		#2		#3		#4		#5	
Actual	GRNN	Actual	GRNN	Actual	GRNN	Actual	GRNN	Actual	GRNN
2.52	2.52	3.22	2.62	2.44	2.46	2.52	2.48	3.82	2.48
2.48	2.45	2.52	2.52	2.34	2.29	2.48	2.54	2.52	2.52
2.48	2.39	2.52	2.44	2.32	2.41	2.40	2.35	2.48	2.44
2.22	2.19	2.22	1.92	2.18	2.17	2.35	2.10	2.45	2.39
2.09	2.28	2.09	2.09	1.92	2.03	2.34	2.27	2.40	2.22
2.02	1.99	1.48	1.46	1.48	1.60	2.32	2.43	2.09	2.35
1.99	2.23	1.37	1.37	1.37	1.28	2.18	2.21	1.48	1.47
1.48	1.78	1.34	1.34	1.37	1.32	2.02	1.99	1.37	1.37
1.12	1.37	1.18	1.34	1.12	1.26	1.92	1.83	1.34	1.34
1.00	1.37	0.22	1.92	1.12	1.21	1.12	1.49	1.18	1.34

[a]The ten rows, corresponding to the randomly selected anomalies for each of the five test sets, were sorted by anomaly, from largest anomaly to smallest, to illustrate that the GRNN tends to overpredict the strength of small anomalies (calculations were carried out for five runs).

Fault proximity is a first-order control in and of itself, and lithology plays a significant role near faults. Deformation is greatest in the hanging-wall siltstones and mudstones within the deformation zone that extends from the fault eastward to the fold hinge that intersects the bottom of W204. This implies that fractures in these lithologies and structural positions should have larger aperatures, resulting in higher transmissivity. Alternatively, the grid cells within this region of a SC model would have higher layer-parallel permeability. Based on the data obtained, the footwall appears to be similar to the rock outside of the hanging-wall deformation zone.

The fact that anomalies do exist outside of the hanging-wall zone, but are smaller and not as strongly correlated with lithology, suggests simple "background" fracture and permeability models for the remainder of the rock mass. At least for the preliminary model, the background fracturing would be homogeneous throughout the model. This means that the discrete fractures in the DFN model would be distributed with the same size and orientation distributions, the same transmissivity and transport properties, and the same average intensity independent of depth, lithology, or location. In the SC model, the permeability, porosity, and transport parameter values would be assigned as random (Monte Carlo) draws from a single parent distribution for each parameter. Only in the vicinity of the hanging-wall deformation zone would properties be changed as previously described.

The next stage in refining these DFN or SC models would be to simulate field flow or transport experiments in order to assess whether the models are reproducing the first-order effects, and if so, whether certain second-order effects should be included to improve the modeling forecasts. There is considerable freedom in assigning transmissivity and transport properties either to the fractures themselves in the DFN formulation or as effective properties to the stochastic continuum models. Iterative refinement of these properties to match the field-flow or transport results is usually an important step in the evolution of models that will be used subsequently for forecast or design.

Acknowledgments. Portions of this work were performed under contract to Harding-Lawson Associates, whose support is gratefully acknowledged. The opinions expressed in this paper, however, remain those of the author and do not necessarily coincide with those of Harding-Lawson Associates or any other party involved in the North Carolina LLRW Facility Project.

REFERENCES

Cacas, M. C., E. Ledoux, G. De Marsily, B. Tillie, A. Barbreau, E. Durand, B. Feuaga, and P. Peaudecerf, Modeling fracture flow with a discrete fracture network: calibration and validation 1, the flow model, *Water Resour. Res., 26*(3), 479–489, 1990a.

Cacas, M. C., E. Ledoux, G. De Marsily, A. Barbreau, P. Calmesl, B. Gaillard, and R. Margritta, Modeling fracture flow with a discrete fracture network: calibration and validation 2, the transport model, *Water Resour. Res., 26*(3), 491–500, 1990b.

Dershowitz, W. S., *Rock Joint Systems*, Ph.D. thesis, Massachusetts Institute of Technology, Cambridge, Mass., 1984.

Dershowitz, W. S.N. Hurley, and K. Been, Stochastic discrete fracture modeling of heterogeneous and fractured reservoirs, in *Proceedings of the Third International Conference on the Mathematics of Oil Recovery,* Delft University Press, Delft, Holland, 1992.

Endo, H. K., J. C. S. Long, C. K. Wilson, and P. A. Witherspoon,

A model for investigating mechanical transport in fractured media, *Water Resour. Res.*, *20*(10), 1390–1400, 1984.

Hudson, J. A., and P. R. La Pointe, Printed circuits for studying rock mass permeability, *Int. J. Rock Mechanics and Mining Science and Geomechanics Abstracts*, *17*, 297–301, 1980.

La Pointe, P. R., P. C. Wallmann, and S. Follin, Continuum modeling of rock masses: Is it useful?, *Proceedings of the Int. Society of Rock Mechanics Symposium on Prediction and Performance in Rock Mechanics and Rock Engineering—EUROCK '96*, pp. 343–350, Balkema, Rotterdam, 1996.

Long, J. C. S., J. S. Remer, C. R. Wilson and P. A. Witherspoon, Porous media equivalents for networks of discontinuous fractures, *Water Resour. Res.*, *18*(3), 645–658, 1982.

National Research Council, Committee on Fracture Characterization and Fluid Flow, *Rock Fractures and Fluid Flow*, Chap. 6, pp. 307–404, National Academy Press, Washington, D.C., 1996.

Robinson, P., *Connectivity, flow and transport in network models of fractured media*, Ph.D. thesis, Oxford University, UK, 1984.

Smith, L., and F. W. Schwartz, An analysis of the influence of fracture geometry on mass transport in fractured media, *Water Resour. Res.*, *20*(9), 1241–1252, 1984.

Specht, D. F., A generalized regression neural network, *IEEE Trans. on Neural Networks, 2*, 568–576, 1991.

Swaby, P. A., and K. D. Rawnsley, An interactive 3D fracture modelling environment, *Proceedings of the Petroleum Computer Conference, SPE 36004*, Society of Petroleum Engineers, Dallas, Tex., 1996.

Ward, Inc., *NeuroShell 2 User's Manual*, Ward Systems Group, Inc., Frederick, Md., 1996.

Zimmerman, D. A., G. De Marsily, C. A. Gotway, M. G. Marietta, C. L. Axness, R. L. Beauheim, R. Bras, J. Carrera, G. Dagan, P. B. Davies, D. P. Gallegos, A. Galli, J. Gómez-Hernández, S.M. Gorelick, P. Grindrod, A. L. Gutjahr, P. K. Kitanidis, A. M. Lavenue, D. McLaughlin, S. P. Neuman, B. S. Rama Rao, C. Ravenne, and Y. Rubin, A comparison of seven geostatistically-based inverse approaches to estimate transmissivities for modeling advective transport by groundwater flow, *Water Resour. Res.*, *34*(6), 1373–1413, 1998.

Paul R. La Pointe, Golder Associates, Inc., 18300 NE Union Hill Rd. #200, Redmond, WA 98052

Fracture Spatial Density and the Anisotropic Connectivity of Fracture Networks

Carl E. Renshaw

Department of Earth Sciences, Dartmouth College, Hanover, New Hampshire

The permeability of fractured rock depends on both the average transmissivity of the fractures and on how the fractures connect together to form a network. The connectedness of a network is a function of the spatial density of the fractures, which can be measured directly at the surface or inferred from the subsurface using seismic travel times. Spatial density is related to network connectedness using a new connectivity model based upon the power-law fracture length distribution commonly observed in the field. Numerical simulations suggest that for effective spatial densities less than approximately one (which include the typical range of effective spatial densities observed on outcrops), the new model provides a reasonable estimate of the connectivity of power-law networks for typical values of the power-law exponent. A review of the typical range of fracture apertures determined from field and laboratory tests suggests that while the new model, in conjunction with field observations, constrains the connectivity of a network, in the absence of additional information on the average fracture transmissivity the new methodology does not significantly constrain the overall conductivity of fractured rock. Therefore, the practical application of indirect geologic and geophysical data to predict the conductivity of fractured rock requires new methods for estimating average fracture transmissivities from the structural, mechanical, and geophysical properties of fractured rock.

1. INTRODUCTION

The permeability of fractured rock depends on both the transmissivity of individual fractures, which is strongly controlled by the fracture aperture distribution [e.g., *Renshaw*, 1995], and how the individual fractures link together to form a connected network. It has been suggested that much of the error involved in current simulations of flow and transport through fractured rock is due to deficiencies in our understanding of these basic geometric properties of natural fractures and their impact on permeability [*USNCRM*, 1996]. This suggests that an understanding of the structural, mechanical, and geophysical properties of fractured rock that control the growth and geometry of rock fractures may help constrain appropriate models for the permeability of fractured rock.

The use of the structural or mechanical properties of fractured rock to constrain fractured rock permeability requires a methodology for relating these indirect data to fracture permeability. At present, techniques for using structural or mechanical data to constrain the transmissivities of fractures within a network remain limited. However, we demonstrate here that the spatial density of a fracture system, which can be directly mapped at the surface or inferred from seismic travel times in the

Dynamics of Fluids in Fractured Rock
Geophysical Monograph 122

subsurface, can be related to the interconnectedness of a fracture network using a new closed-form predictive model [*Renshaw*, in press]. Using this model, and the typical range of observed near-surface fracture densities, a sensitivity analysis reveals that once a network is connected, the primary source of uncertainty in predicting the permeability of fractured rock does not arise from uncertainty in the connectedness of the network, but rather from uncertainty in the transmissivities of the individual fractures, which arises from uncertainty in the fracture apertures, state of stress, and multiphase flow effects. Thus, while the interconnectedness of a network can be constrained, it does not significantly increase our ability to predict the permeability of fractured rock because it does not address the primary source of uncertainty. However, the conceptual understanding of fracture connectivity developed here may still be of use in understanding the transport and dispersion of contaminants through fractured rock.

2. FRACTURE DENSITY AND CONNECTIVITY

Because outcrop observations and seismic travel times do not directly depend on fracture network permeability, these data must be used to first constrain the statistical properties of fracture networks. The constraints on the statistical properties can then be used to constrain the bulk connectivity of the networks if the relationship between the statistical properties and network connectivity is known. In this work, connectivity is defined as the bulk permeability of the network assuming all fractures are equally transmissive. This definition is most directly applicable to describing the permeability of near surface flow systems where stress effects on fracture transmissivity are minimized [*Barton*, 1995].

In contrast to the large number of proposed models for the statistical properties of rock fractures [e.g., *Dershowitz and Einstein*, 1988], there exist relatively few closed-form predictive models relating the statistics of a network to its connectivity. Despite its uncertain applicability to natural fracture systems, the model of *Snow* [1969], in which a fracture is represented by two infinite parallel plates, remains a commonly used model. A limited number of alternative models have been developed to account for the finite lengths of fractures. However, even for simple networks, the connectivities predicted by these models are inconsistent [*Renshaw*, in press].

For example, consider the simplistic representation of the complex geometry of natural fracture networks as two-dimensional bond percolation networks (Figure 1a). The connectivity of these networks predicted by various models, over the range of fracture densities commonly encountered in geologic formations [*Renshaw*, 1997], is shown in Figure 1b. Note the inconsistent predictions of these models even for these simplistic networks, with some models predicting connected networks (for $\rho < 3.5$), and others not.

Fracture density has been variously defined as the number of fractures per unit length, area, or volume. These definitions, which have units of inverse length, inverse area, or inverse volume, are more accurately described as fracture frequencies. This paper follows the fracture mechanics literature and reserves the term fracture density to refer to the dimensionless excluded volume parameter introduced by *Bristow* [1960] and *Walsh* [1965]

$$\rho_{2D} = \frac{1}{A}\sum_{i=1}^{n} a_i^2 \qquad \rho_{3D} = \frac{1}{V}\sum_{i=1}^{n} \ell_i^3 \qquad \text{(1a, b)}$$

where the subscripts *2D* and *3D* indicate the dimension of the fractures, A and V are the area and volume of the sample containing n fractures, and a_i and ℓ_i are the half length and average radius of the *i*-th fracture, respectively. Significantly, unlike fracture frequencies, fracture density can be related to the effective elastic properties of a fractured medium [*Kachanov*, 1992]. Hence, in situ fracture densities can be determined from geophysical data sensitive to effective elastic properties, such as seismic travel times [for a complete discussion of the methodology and its limitations, see *Crampin and Zatsepin* (1997)]. Therefore, seismic travel times can constrain fracture connectivity if a consistent relationship exists between fracture spatial density and fracture connectivity.

The existence of a consistent relationship between fracture density and connectivity has been questioned based on field investigations indicating that fracture frequencies, which can be determined using borehole logging equipment, are poorly correlated with transmissivities determined from borehole packer tests [e.g., *Paillet*, 1998]. However, we stress that in contrast to fracture frequency, fracture density cannot be determined from borehole data because the determination of the density requires information on fracture lengths that is unavailable from borehole data. Further, because fracture density is a function of the fracture lengths, fracture density is only defined at scales significantly greater than the average fracture length. Since borehole packer intervals typically are only of the order of the average fracture length or less, individual packer test results cannot be meaningfully correlated with fracture densities as the fracture density is not defined at this scale. Consequently, the connectivities plotted in Figure 1b and discussed here are implicitly large-scale bulk values for the entire network. Rather than

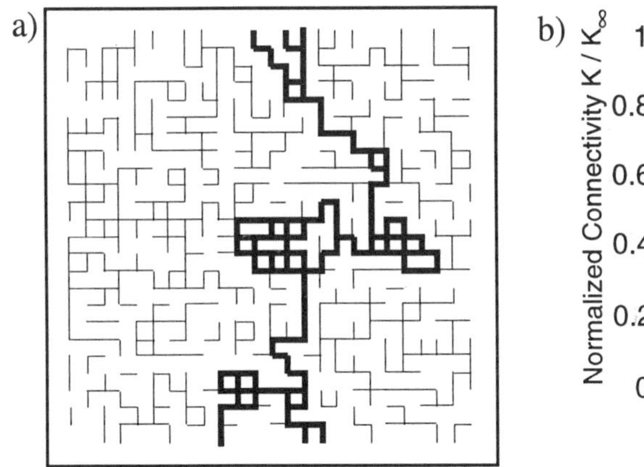

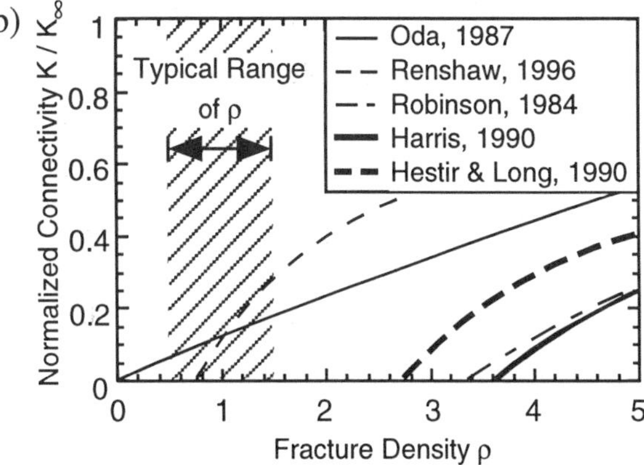

Figure 1. (a) Example of a bond percolation network created by defining a fixed probability for a connection between any two adjacent nodes. The connected backbone between the upper and lower boundaries is highlighted. (b) Curves show the connectivity of bond percolation networks predicted by various models as a function of the fracture density of the network. Shaded region indicates the range of fracture densities typically observed on surperficial exposures [*Renshaw*, 1997]. Connectivities are normalized by the connectivity predicted using the parallel plate model of *Snow* [1969].

representing the results from single packer tests, these connectivities can be thought of as representing the average connectivity determined from multiple packer tests at a given site or by calibrating regional flow models [e.g., *Guimerà et al.*, 1995; *Tiedeman et al.*, 1997].

Similarly, we note that the relationship between seismic travel times and fracture density also is only valid at scales significantly greater than the average fracture length. Thus, the fracture densities discussed here are bulk values. Locally the fracture density may be greater or less than the average bulk value inferred from seismic travel times.

3. POWER-LAW NETWORKS

The discrepancies between the various models for relating bulk fracture density to bulk connectivity (Figure 1a) arise, in part, from different assumptions regarding the underlying fracture length distribution. For example, whereas the fracture connectivity model of *Renshaw* [1996] is based on an exponential fracture length distribution, that of *Robinson* [1984] assumes a narrow range of fracture lengths. Since connectivity models are sensitive to fracture length distribution, the appropriate connectivity model for relating fracture density to fracture connectivity should be based on a fracture length distribution that is appropriate for fractured rock.

Fracture length distributions obtained from outcrop-scale maps often follow a log-normal distribution. However, outcrop-scale maps under-represent both smaller and larger scale fractures. When data from several different scales of observation are combined, the distribution of fracture lengths often follows a power-law, or fractal, distribution [e.g., *Gudmundsson*, 1987; *Marrett*, 1997; *Odling*, 1997; *Segall and Pollard*, 1983].

A power-law distribution of fracture lengths is defined by

$$n(l) = Cl^{-a} \tag{2}$$

where $n(l)$ is the number of fractures of length l, C is a constant that depends on the size of the network, and a is the power-law exponent. It can be shown that because observations indicate the fracture frequency decreases with increasing fracture length, and because the total fracture volume cannot exceed the sample volume, $0 < a < 3$ [*Renshaw*, in press]. Fracture networks with fracture-length distributions following the power-law distribution are referred to here as power-law networks.

Equation (2) is only valid over a given range of fracture lengths defined by the fracture length cutoffs. The power-law length distribution is not valid at scales less than those of microcracks (often of the order 10^{-5}–10^{-3} m) or at scales greater than that of the largest fracture, which may be of the order 10^{1}–10^{2} m or larger.

Within the range of fracture lengths for which Equation (2) is valid, the value of the power-law exponent may not be constant. Changes in the power-law exponent, known as scaling breaks, may occur across fundamental length scales that affect the growth of fractures, such as the grain size and the mechanical layer thickness in sedimentary rocks. However, here it is assumed that the power-law exponent is constant.

Values of the power-law exponent for geologic fracture networks can be estimated from fracture-length histograms derived from published fracture trace maps [*Renshaw*, in press]. For all the fracture trace maps considered in Figure 2, the estimated value of the power-law length exponent falls within the range $1 < a < 3$. Within this range the estimates are widely scattered and there is no consistent dependence on scale of the trace map. The mean exponent value for all maps, shown by the solid line in Figure 2, is 1.8, with a standard deviation, shown by the dashed lines, of 0.4.

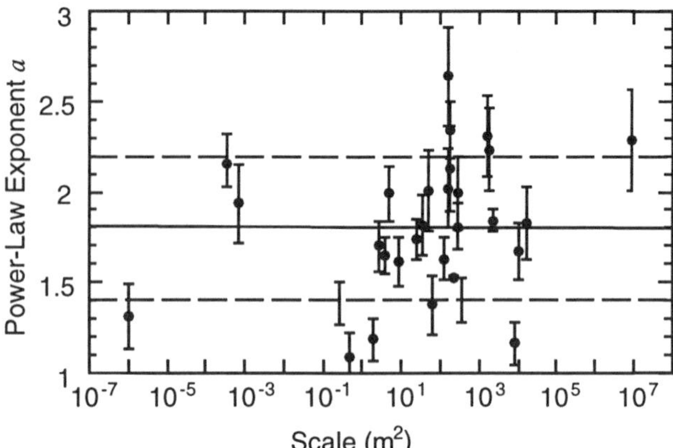

Figure 2. Values of the power-law exponent estimated from published fracture trace maps versus scale of the map. Error bars indicate +/– one standard deviation of the estimate. After *Renshaw* [1999].

4. CONNECTIVITY OF ISOTROPIC POWER-LAW NETWORKS

A number of investigators have developed fracture connectivity models based on percolation or equivalent medium theories. In these approaches, the connectivity of the network is typically given as a power-law function of the fracture frequency. Example results from these theories are shown in Figure 1. Most percolation-based approaches to network connectivity are problematic for fracture networks with power-law length distributions because they require a well-defined average fracture length. In the absence of a lower fracture-length cutoff, the average fracture length for a power-law network is only defined for $a < 2$ [*Renshaw*, in press]. Of course, all fracture networks have a lower fracture-length cutoff, but this length is often difficult to define and introduces an additional parameter, making the connectivity relation difficult to generalize.

An alternative model for the connectivity of power-law networks has been proposed by *Renshaw* [in press]. In this approach, the connectivities are normalized following the discussion of fractured rock permeability scaling in *Renshaw* [1998]. In this work it is noted that the observed scaling of fractured rock permeability near the percolation threshold is similar to that of a single fracture crossing the entire system. Accordingly, the normalizing connectivity K_* is given by the effective permeability of a single fracture spanning an $L \times L$ area

$$K_* = \frac{t_f}{L} \tag{3}$$

where t_f is the transmissivity of the fracture, defined as

$$t_f = b^3 \frac{\rho g}{12\mu} \tag{4}$$

where b is the hydraulic aperture, g the gravitational constant, and ρ and μ the fluid density and viscosity, respectively [*Renshaw*, 1995].

In addition, results from numerical investigations of the connectivity of power-law networks suggest that near the percolation threshold there is an inverse relationship between the connectivity of the network and the power-law exponent [*Renshaw*, in press]. This suggests an alternative measure of the fracture intensity, defined as the effective two-dimensional fracture spatial density

$$\rho_{eff} = \frac{1}{aL^2} \sum_{i=1}^{n} \left(\frac{l_i}{2} \right)^2 \tag{5}$$

where l_i is the length of the i-th fracture.

Figure 3 shows results from numerical experiments on the connectivity of power-law networks with power-law exponents within the typical range suggested by Figure 2 [*Renshaw*, in press]. When plotted as a function of the effective spatial density, the connectivities for the various values of the power-law exponent all plot along a single curve for effective spatial densities less than about 1.0. For reference, also shown on this figure is the typical range in effective fracture densities observed in surficial exposures of joint networks. This range is determined by dividing the spatial density data discussed in *Renshaw* [1997] by the power-law exponent determined for each network [*Renshaw*, in press]. Significantly, the typical range in effective spatial densities is included within the range over which the connectivities for the various values of the power-law exponent are approximately coincident. This suggests that a single curve can be developed that provides a reasonable estimate of the connectivities of these networks.

Following percolation theory, the appropriate measure of the effective fracture density, with respect to predicting connectivity, is not the absolute density, but rather the difference between the absolute effective density and the

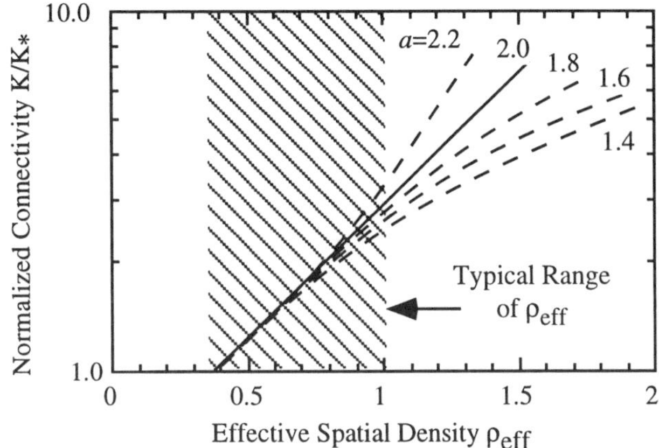

Figure 3. Normalized connectivity of isotropic fracture networks with power-law fracture length distributions as a function of the effective spatial density. Solid line indicates connectivity predicted by Equation (8). Dashed lines indicate connectivities determined from numerical simulation of power-law networks.

effective density that corresponds to a critical threshold. Accordingly, the effective spatial density offset Γ is defined as

$$\Gamma = \rho_{eff} - \rho_c \tag{6}$$

where ρ_c is the critical effective spatial density corresponding to $K/K_* = 1.0$. From Figure 3, $\rho_c \approx 0.4$ for two-dimensional power-law networks.

We then note that if a fracture is added to a very sparse (low spatial density) network it is likely that the fracture will, at most, intersect only one other fracture and thus not contribute to the overall network connectivity. In contrast, if the same fracture is added to a dense (high spatial density) network, then it is likely that the fracture will connect several fractures, possibly creating multiple new pathways for fluid flow and significantly increasing the overall network connectivity. Therefore, the change in the normalized connectivity of a network for a unit increase in effective spatial density is proportional to the existing normalized connectivity. Thus, the relationship between density and connectivity can be written

$$\frac{d\kappa}{d\Gamma} = \beta\kappa \tag{7}$$

where $\kappa = K/K_*$ and β is a constant that only depends on the dimension of the fractures. Integrating, we have

$$\kappa = C_1 e^{\beta\Gamma} \tag{8}$$

where C_1 is a constant. Since, by definition, $K/K_* = 1$ when $\Gamma = 0$, $C_1 = 1$. Thus, for two-dimensional isotropic networks, the relationship between normalized connectivity and effective density offset only depends on the parameter β.

The best-fit line ($\beta = 1.9$) of the form of Equation (8) is shown by the solid line in Figure 3. Within the typical range of effective fracture densities, this model provides a reasonable estimate of the connectivity for the range of power-law exponents typically observed. Within the typical range of effective fracture densities, equally good fits to the power-law model simulations are achieved by assuming that the change in normalized connectivity for a unit increase in effective spatial density is proportional to the existing effective spatial density. In that case, the relationship between spatial density and connectivity is a power law of the form $\kappa \propto \Gamma^t$, with a best-value $t \approx 1.2$. While connectivity relations for networks near the percolation threshold are more commonly described using power laws [*Stauffer and Aharony*, 1992], the exponential relation is used here because the requirement that $\kappa(\Gamma{=}0) = 1$ results in the exponential relation having a simpler form than the power law.

5. ANISOTROPIC POWER-LAW NETWORKS

The model developed above can be extended to anisotropic power-law networks by making the simple assumption that any increase in connectivity in one direction due to an anisotropy in the bulk spatial density is offset by a corresponding decrease in the connectivity in an orthogonal direction. However, for a given total bulk effective spatial density, the sum of the principal connectivities remains constant and is equal to the sum of the orthogonal connectivities in isotropic networks.

Implicit in the above assumption is that the network is composed of two orthogonal fracture sets or that the anisotropy of a network of arbitrarily oriented fracture sets can be shown to be equivalent to that of an orthogonal network. More specifically, implicit in this work is that the connectivity tensor is orthotropic. The anisotropic connectivity for arbitrary fracture orientations is then obtained by solving the eigenvalue problem for a second-rank tensor. Fortunately, *Kachanov* [1992] has shown that the crack density tensor, upon which the connectivity model is based, is fully orthotropic for any 2D fracture statistics and that the deviations from orthotropy in the 3D case are small and can usually be neglected. Therefore, the spatial density of any fracture set can be reduced to its principal components and these components treated as representing an equivalent orthogonal network.

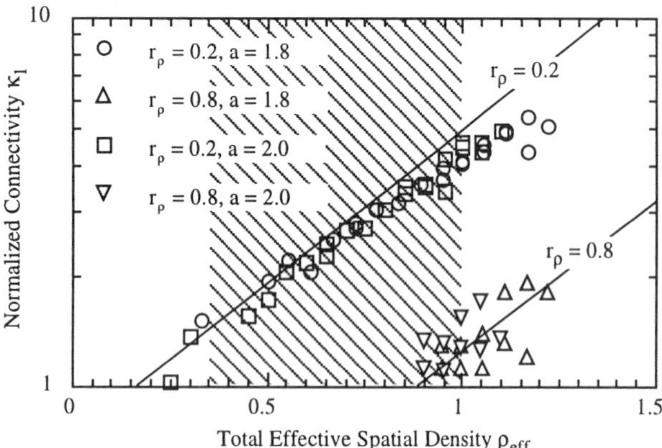

Figure 4. Results from numerical simulation of the normalized connectivity of anisotropic fracture networks with power-law fracture length distributions as a function of the total effective spatial density. Each point represents the average connectivity from at least 50 realizations. Solid lines indicate connectivities predicted by anisotropic connectivity model [Eq. (10)].

In particular, let κ_1 and κ_2 be the principal normalized connectivities in a fracture network having corresponding principal effective spatial densities ρ_1 and ρ_2. In this case, an effective spatial density anisotropy ratio $r_\rho = \rho_1/(\rho_1+\rho_2) > 0.5$ will result in a normalized connectivity anisotropy ratio $r_\kappa = \kappa_1/(\kappa_1+\kappa_2) < 0.5$ as increasing the spatial density in one direction, say ρ_1, increases the connectivity orthogonal to this direction κ_2. At a given total effective spatial density $\rho_{\text{eff}} = \rho_1 + \rho_2$, the sum of the principal connectivities is equal to the sum of the orthogonal connectivities in an isotropic network, given by Equation (8) as

$$\kappa_1 + \kappa_2 = 2e^{\beta(\rho_1 + \rho_2 - \rho_c)} \tag{9}$$

where, from Figure 3, $\beta \approx 1.9$. Either principal normalized connectivity can be determined from the additional requirement that $r_\kappa = 1 - r_\rho$, yielding

$$\kappa_1 = 2(1 - r_\rho)e^{\beta(\rho_1 + \rho_2 - \rho_c)}$$
$$\kappa_2 = 2r_\rho e^{\beta(\rho_1 + \rho_2 - \rho_c)} \tag{10a, b}$$

Figure 4 compares the anisotropic connectivities predicted using Equation (10) to those simulated numerically for various values of the power-law exponent and the effective spatial density ratio. Over the typical range of effective spatial densities there is a good match between the simulated connectivities and those predicted by the anisotropic connectivity model.

6. MODEL IMPLICATIONS

Figures 3 and 4 demonstrate that the average connectivity of a power-law fracture network can be estimated from the bulk spatial density of a fracture network. Given that the typical range of fracture spatial densities in the geologic formation is constrained by mechanical limitations [*Renshaw*, 1997], we now explore whether knowledge of the fracture spatial density can be used to predict the permeability of fractured rock.

As noted in the introduction, the permeability of fractured rock depends on both the connectivity of the network and the transmissivities of the fractures. If the average transmissivity of the fractures is known, then Equation (10) provides a lower bound estimate of the large-scale bulk conductivity of the network. The conductivity predicted by Equation (10) is a lower bound estimate because it was developed for two-dimensional networks, whereas three-dimensional flow in natural fracture networks will increase the connectivity of the network. Despite this limitation, Figure 5 demonstrates that where data on the average transmissivity of the fractures are available, either from well testing, model calibration, or direct measurement of fracture apertures, the connectivity model developed here [Eq. (8)] can provide correct order-of-magnitude estimates of the actual conductivity of the fractured rock. All the data in Figure 5 come from near-surface studies. Predictions for deeper flow systems are

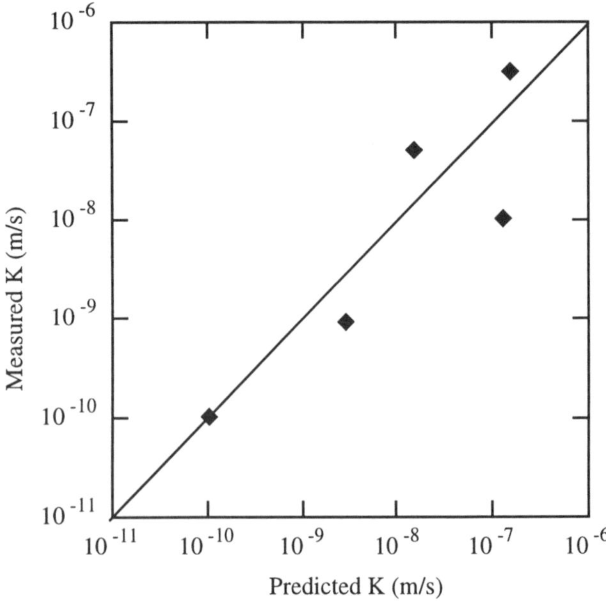

Figure 5. Measured versus predicted values for the bulk conductivity of fractured rock. Measured conductivities come from a variety of field sites around the world. After *Renshaw* [1999].

Table 1. Measured Apertures of Single Fractures.

Source	Rock Type	Mean Aperture µm	Source	Rock Type	Mean Aperture µm
[*Banwart et al.*, 1994]	granite	100–130	[*Novakowski and Lapcevic*, 1994]	shale	245
[*Engelder and Scholz*, 1981]	quartzite	3–35	[*Raven et al.*, 1988]	gniess	118
[*Folger et al., 1996*]	granite	110–380	[*Raven and Gale*, 1985]	granite	6–110
[*Guimera et al.*, 1995)	granite	5–50	[*Schrauf and Evans*, 1986]	grano-dioritie	210–590
[*Himmelsbach et al.*, 1998]	granite	100–270 430–580	[*Shapiro and Nicholas*, 1989]	dolomite	2,150–2,700
[*Hinsby et al.*, 1996]	glacial till	13–120	[*Suzuki et al., 1998*]	granite	0.2–1.2
[*Hsieh et al.*, 1993]	granite	11–240	[*Witherspoon et al.*, 1980]	granite	4–240
[*Klavetter and Peters* 1986]	volcanic tuff	3–355	[*Witherspoon et al.*, 1980]	marble	4–280
[*Kumar et al.*, 1995]	limestone	265	[*Witherspoon et al.*, 1980]	basalt	9–220
[*Long and Billaux*, 1987]	granite	24–321			

more problematic due to the strong impact of state of stress on the permeability of individual fractures [*Barton*, 1995].

Even for near-surface systems, data on the average transmissivities of fractures are usually unavailable and, at present, it is uncertain how to estimate these transmissivities from indirect structural or mechanical data. However, we can estimate the typical range in fracture transmissivities from the numerous field and laboratory investigations of flow through single fractures (Table 1). These investigations suggest that fracture apertures generally range from a few microns (10^{-6} m) to a few millimeters (10^{-3} m). Using the cubic law [Eq. (4)], this range of fracture aperture corresponds to fracture transmissivities ranging from 10^{-12} to 10^{-3} m^2/s. Since the bulk conductivity of a network is linearly related to the average fracture transmissivity, this range of transmissivities results in a similar nine-order-of-magnitude range in the bulk conductivity of the network. In contrast, Figure 1 indicates that fracture densities typically only range over one order of magnitude (from 10^1 to 10^0), resulting in a similar order of magnitude uncertainty in fracture conductivity once the network is connected. Consequently, while knowledge of the fracture spatial density can be used to constrain the connectivity of a network, once the network is connected, this constraint will not significantly decrease the uncertainty in estimates of the bulk conductivity of the network unless the transmissivities of the fractures can be determined.

These results suggest that the application of structural and mechanical data to the prediction of permeability requires new methods for inferring fracture transmissivities from these data. However, the connectivity model developed here may still have practical value. For example, once the bulk conductivity of a fracture network is measured, the average aperture of the fractures can be determined from the spatial density of the network. Knowledge of the average aperture and density, along with an understanding of the compliance of fractures, can then be used to predict changes in bulk conductivity induced by drawdowns or other changes in effective stress. We also note that contaminant transport and dispersion through fractured rock may be more sensitive to the detailed geometry of the network than bulk conductivity. Thus the conceptual understanding of fracture connectivity developed here may be of use in understanding the transport and dispersion of contaminants through fractured rock.

7. SUMMARY

Existing models for the connectivity of fracture networks yield inconsistent results as each assumes a different distribution of fracture lengths. We present a new connectivity model based upon the power-law fracture length distribution commonly observed in the field. Numerical simulations suggest that for effective spatial densities less than about one (which includes the typical range of effective spatial densities observed on outcrops), the new model provides a reasonable estimate of the connectivity of power-law networks for typical values of the power-law exponent. A review of the typical range of fracture apertures determined from field and laboratory tests suggests that while the new model provides a methodology for using geophysical data to constrain the connectivity of a network, in the absence of additional information on the average fracture transmissivity the new methodology does not significantly constrain the conductivity of fractured rock. Therefore, the practical application of geophysical data to predict the conductivity of fractured rock requires new methods for estimating average fracture transmissivities from the structural, mechanical, and geophysical properties of fractured rock.

REFERENCES

Banwart, S., E. Gustafsson, M. Laaksoharju, A.-C. Nilsson, E.-L. Tullborg, and B. Wallin, Large-scale intrusion of shallow water into a vertical fracture zone in crystalline bedrock: Initial hydrochemical perturbation during tunnel construction at the Äspö Hard Rock Laboratory, southeastern Sweden, *Water Resour. Res.*, *30*, 1747–1763, 1994.

Barton, C. A., M. D. Zoback, and D. Moos, Fluid flow along potentially active faults in crystalline rock, *Geology*, *23*, 683–686, 1995.

Bristow, J. R., Microcracks and the static and dynamic elastic constants of annealed and heavily cold-worked metals, *British J. Appl. Phys.*, *11*, 81–85, 1960.

Crampin, S., and S. V. Zatsepin, Modelling the compliance of crustal rock—II. Response to temporal changes before earthquakes, *Geophys. J. Int.*, *129*, 495–506, 1997.

Dershowitz, W. S., and H. H. Einstein, Characterizing rock joint geometry with joint system models, *Rock Mech. and Rock Engr.*, *21*, 21–51, 1988.

Engelder, T., and C. H. Scholz, Fluid flow along very smooth joints at effective pressures up to 200 Megapascals, in *Mechanical Behavior of Crustal Rocks*, edited by N. L. Carter, M. Friedman, J. M. Logan, and D. W. Stearns, pp. 147–152, American Geophysical Union, Washington, D.C., 1981.

Folger, P. F., E. Poeter, R. B. Wanty, D. Frishman, and W. Day, Controls on 222Rn variations in a fractured crystalline rock aquifer evaluated using aquifer tests and geophysical logging, *Ground Water*, *34*, 250–261, 1996.

Gudmundsson, A., Tectonics of the Thingvellir fissure swarm, SW Iceland, *J. Struct. Geol.*, *9*, 61–69, 1987.

Guimerà, J., L. Vives, and J. Carrera, A discussion of scale effects on hydraulic conductivity at a granite site (El Berrocal, Spain), *Geophys. Res. Letts.*, *22*, 1449–1452, 1995.

Himmelsbach, T., H. Hötzl, and P. Maloszewski, Solute transport processes in a highly permeable fault zone of Lindau fractured rock test site (Germany), *Groundwater*, *36*, 792–800, 1998.

Hinsby, K., L. D. McKay, P. Jørgensen, M. Lenczewski, and C. P. Gerba, Fracture aperture measurements and migration of solutes, viruses, and immiscible creosote in a column of clay-rich till, *Ground Water*, *34*, 1065–1075, 1996.

Hsieh, P. A., A. M. Shapiro, C. C. Barton, F. P. Haeni, C. D. Johnson, C. W. Martin, F. L. Paillet, T. C. Winter, and D. L. Wright, Methods of characterizing fluid movement and chemical transport in fractured rock, in *Field Trip Guidebook for Northeastern United States*, edited by J. T. Chaney, and J. C. Hepburn, Contribution #67, Department of Geology and Geography, University of Massachusetts, Amherst, MA, 1993.

Kachanov, M., Effective elastic properties of cracked solids: critical review of some basic concepts, *Appl. Mech. Rev.*, *45*, 304–335, 1992.

Klavetter, E. A., and R. R. Peters, *Estimation of Hydrologic Properties for an Unsaturated Fractured Rock Mass*, SAND84-2642, Sandia National Laboratories, Sandia, New Mex., 1986.

Kumar, A. T. A., P. D. Majors, and W. R. Rossen, Measurement of aperture and multiphase flow in fractures using NMR imaging, in *SPE Annual Technical Conference and Exhibition*, SPE Paper 30558, Soc. Petrl. Eng., Dallas, Tex., 1995.

Long, J. C. S., and D. M. Billaux, From field data to fracture network modeling: an example incorporating spatial structure, *Water Resour. Res.*, *23*, 1201–1216, 1987.

Marrett, R., Permeability, porosity, and shear-wave anisotropy from scaling of open fracture populations, in *Fractured Reservoir Characterization and Modeling*, edited by T. E. Hoak, A. L. Klawitter, and P. K. Blomquist, pp. 217–226, Rocky Mtn. Assoc. Geol., Denver, Colo., 1997.

Novakowski, K. S., and P. A. Lapcevic, Field measurement of radial solute transport in fractured rock, *Water Resour. Res.*, *30*, 37-44, 1994.

Odling, N., Scaling and connectivity of joint systems in sandstones from western Norway, *J. Struct. Geol.*, *19*, 1257–1271, 1997.

Paillet, F. L., Flow modeling and permeability estimation using borehole flow logs in heterogeneous fractured formations, *Water Resour. Res.*, *34*, 997–1010, 1998.

Raven, K. G., and J. E. Gale, Water flow in a natural rock fracture as a function of stress and sample size, *Int. J. Rock Mech. Min. Sci. Geomech. Abstr.*, *22*, 251–261, 1985.

Raven, K. G., K. S. Novakowski, and P. A. Lapcevic, Interpretation of field tracer tests of a single fracture using a transient solute storage model, *Water Resour. Res.*, *24*, 2019–2032, 1988.

Renshaw, C. E., On the relationship between mechanical and hydraulic apertures in rough-walled fractures, *J. Geophys. Res.*, *100*, 24,629–24,636, 1995.

Renshaw, C. E., Influence of sub-critical fracture growth on the connectivity of fracture networks, *Water Resour. Res.*, *32*, 1519–1530, 1996.

Renshaw, C. E., Mechanical controls on the spatial density of opening mode fracture networks, *Geology*, *25*, 923–926, 1997.

Renshaw, C. E., Sample bias and the scaling of hydraulic conductivity in fractured rock, *Geophys. Res. Letts.*, *25*, 121–124, 1998.

Renshaw, C. E., Connectivity of joint networks with power-law length distributions, *Water Resour. Res.*, in press.

Robinson, P. C., Connectivity, flow and transport in network models of fractured media, Atomic Energy Research Authority, Harwell, United Kingdom, TP 1072, 1984.

Schrauf, T. W., and D. D. Evans, Laboratory studies of gas flow through a single natural fracture, *Water Resour. Res.*, *22*, 1038–1050, 1986.

Segall, P., and D. D. Pollard, Joint formation in granitic rock of the Sierra Nevada, *Geol. Soc. Amer. Bull.*, *94*, 563–575, 1983.

Shapiro, A. M., and J. R. Nicholas, Assessing the validity of the channel model of fracture aperture under field conditions, *Water Resour. Res.*, *25*, 817–828, 1989.

Snow, D. T., Anisotropic permeability of fractured media, *Water Resour. Res.*, *5*, 1273–1289, 1969.

Stauffer, D., and A. Aharony, *Introduction to Percolation Theory*, 181 pp., Taylor and Francis, London, 1992.

Suzuki, K., M. Oda, M. Yamazaki, and T. Kuwahara, Permeability changes in granite with crack growth during immersion in hot water, *Int. J. Rock Mech. Min. Sci.*, *35*, 907–921, 1998.

Tiedeman, C. R., D. J. Goode, and P. A. Hsieh, Numerical simulation of ground-water flow through glacial deposits and crystalline bedrock in the Mirror Lake area, Grafton County, New Hampshire, *U.G. Geological Survey Professional Paper*, *1572*, 1997.

USNCRM, *Rock Fractures and Fluid Flow*, 551 pp., National Academy Press, Washington, D.C., 1996.

Walsh, J. B., The effect of cracks on the compressibility of rocks, *J. Geophys. Res.*, *70*, 381–389, 1965.

Witherspoon, P. A., J. S. Y. Wang, K. Iwai, and J. E. Gale, Validity of cubic law for fluid flow in a deformable rock fracture, *Water Resour. Res.*, *16*, 1016–1024, 1980.

Carl E. Renshaw, Department of Earth Sciences, Dartmouth College, Hanover, NH 03755

Fluid Flow in Rock Fractures: From the Navier-Stokes Equations to the Cubic Law

Robert W. Zimmerman

T. H. Huxley School of Environment, Earth Sciences, and Engineering, Imperial College of Science, Technology, and Medicine, London, United Kingdom
and
Earth Sciences Division, Lawrence Berkeley National Laboratory, Berkeley, California

In-Wook Yeo

Department of Geological Sciences, University of Colorado, Boulder, Colorado

The mathematical analysis of the flow of a single-phase Newtonian fluid through a rough-walled rock fracture is reviewed, starting with the Navier-Stokes equations. By a combination of order-of-magnitude analysis, appeal to available analytical solutions, and reanalysis of some data from the literature, it is shown that the Navier-Stokes equations can be linearized if the Reynolds number is less than about 10. Further analysis shows that the linear Stokes equations can be replaced by the simpler Reynolds lubrication equation if the wavelength of the dominant aperture variations is about three times greater than the mean aperture. However, this criterion does not seem to be strongly obeyed by all fractures. The Reynolds equation (i.e., the local cubic law) may therefore suffice in estimating fracture permeabilities to within a factor of about 2, but more accurate estimates will require solution of the Stokes equations. Similarly, estimates of mean aperture based on inverting transmissivity data may have errors of a factor of two if any version of the local cubic law is used to relate transmissivity to mean aperture.

1. INTRODUCTION

Determining single-phase flow through a rough-walled rock fracture is the most basic problem in fractured-rock hydrology, and is the starting point for studies of more complex issues such as two-phase flow, flow through fracture networks, solute and tracer transport, etc. Although this problem has been the subject of many theoretical, computational, and experimental investigations, it is as yet far from being thoroughly understood. Mathematically, it would be desirable to have a clearer understanding of the conditions under which the governing Navier-Stokes (N-S) equations can be replaced by simpler and more tractable governing equations such as the Stokes equations or the Reynolds lubrication equation. These mathematical issues are closely akin to the practical question of how to relate the fracture transmissivity to geometric properties such as mean aperture, fracture roughness, and contact area.

Dynamics of Fluids in Fractured Rock
Geophysical Monograph 122

There are several important practical implications of these mathematical considerations. For example, if the flow is governed by the Stokes equations, a linear relationship is implied between mean pressure drop and mean flowrate, i.e., Darcy's law, whereas this relationship will not in general be linear if the flow is governed by the full Navier-Stokes equations. Furthermore, whereas the lubrication equation can be solved numerically for realistic fracture profiles, the Navier-Stokes and Stokes equations have not yet been solved for realistic fracture geometries derived from actual aperture measurements.

In this paper we will review the equations that govern single-phase fluid flow through a rough-walled rock fracture and investigate the conditions under which the various levels of mathematical simplification are possible. We will then discuss several models used to relate fracture transmissivity to fracture void space geometry. Several closely related issues, such as two-phase water-air flow and tracer transport, will also be explored. Finally, the relationship between mean aperture, hydraulic aperture, and "tracer aperture" will be discussed.

2. FROM THE NAVIER-STOKES EQUATIONS TO THE STOKES EQUATIONS

Fundamentally, flow through a rock fracture is governed by the Navier-Stokes (N-S) equations, a set of three coupled nonlinear partial differential equations which, in steady-state, can be written as [*Batchelor*, 1967]:

$$\rho(\mathbf{u}\cdot\nabla)\mathbf{u} = -\nabla P + \mu\nabla^2\mathbf{u}\,, \tag{1}$$

where ρ is the density, $\mathbf{u}$ is the velocity vector, μ is the viscosity, and P is the reduced pressure defined by $P = p - \rho g\zeta$, where p is the pressure and ζ is the coordinate pointing in the direction of the gravitational acceleration. The first term in Equation (1) is the advective acceleration, the second term is the pressure gradient, and the third term represents the viscous forces.

To simplify the discussion, we will usually consider a "one-dimensional" fracture whose aperture varies in only one direction, the x-direction, which we take to be the direction of flow; the direction perpendicular to the fracture plane will be z (Figure 1). If we need to consider a fully two-dimensional fracture plane, the other coordinate within the fracture plane will be y. The N-S equations must always be supplemented by the equation for conservation of mass, which for an incompressible fluid takes the form:

$$\nabla\cdot\mathbf{u} = 0\,. \tag{2}$$

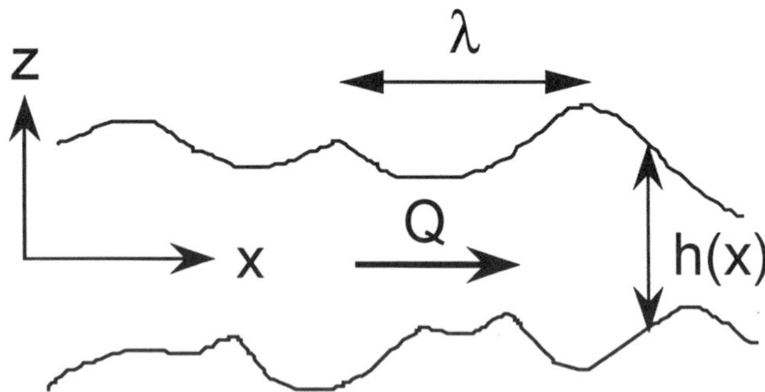

Figure 1. Generic rough-walled fracture, with aperture that varies only in the x-direction. The local aperture is $h(x)$, and λ is a characteristic wavelength of the aperture variation.

Equations (1) and (2) form four coupled equations for the four variables $\{\mathbf{u},P\}$.

The starting point for all discussions of the fracture flow problem is the special case of a fracture bounded by smooth, parallel walls that are separated by a distance h. This is, in fact, the only fracture geometry for which the N-S equations can be solved exactly. The simplification of the N-S equations in this case arises from the fact that the velocity vector points in the x-direction, but it varies only in the z-direction. Hence, the velocity and the velocity gradient are orthogonal to each other, and the nonlinear term on the left-hand side of Equation (1) vanishes identically. The velocity vector for flow between smooth, parallel plates is given by [*Batchelor*, 1967] as:

$$u_x = -\frac{1}{2\mu}\frac{dP}{dx}[z^2 - (h/2)^2],\, u_z = 0\,. \tag{3}$$

This velocity profile can be integrated across the fracture, to find the overall flowrate in the form [*Zimmerman and Bodvarsson*, 1996]

$$Q_x = \int_0^w \int_{-h/2}^{h/2} u_x(z)dz\,dy =$$

$$-w\int_{-h/2}^{h/2} \frac{1}{2\mu}\frac{dP}{dx}\left[z^2 - (h/2)^2\right]dz = \frac{-wh^3}{12\mu}\frac{dP}{dx}, \tag{4}$$

where w is the depth of the fracture, perpendicular to the flow direction. This result is usually written in terms of a transmissivity T, defined by $Q_x = -(T/\mu)(dP/dx)$, in which case Equation (4) leads to the "cubic law" [*Witherspoon et al.*, 1980]:

$$T = wh^3 / 12 . \quad (5)$$

For fractures that are not bounded by smooth, parallel walls, the transmissivity is frequently quantified in terms of the "hydraulic aperture," h_H, which is defined as the aperture that, if substituted into the cubic law (5), would yield the correct transmissivity.

Unfortunately, the full N-S equations are too difficult to solve, either analytically or numerically, for real, rough-walled fractures. This is also true for idealized fracture geometries such as a fracture bounded by sinusoidal or sawtooth-shaped walls. Therefore, various approximations are usually made that reduce the N-S equations to a more tractable form. The first level of simplification is to discard the acceleration terms in the Navier-Stokes equations, which yields the steady-state Stokes equations, a coupled set of three *linear* partial differential equations:

$$\nabla P = \mu \nabla^2 \mathrm{u} . \quad (6)$$

Equation (6) represents three equations, one for each component of the velocity vector. They are coupled through the pressure field, which appears in all three equations. As these equations are mathematically linear, they will *necessarily* lead to a linear relationship between the pressure gradient and the mean flowrate.

The reduction from the Navier-Stokes equations to the Stokes equations is possible only if the advective acceleration terms, $\rho(\mathbf{u} \cdot \nabla)\mathbf{u}$, are small compared to the viscous terms, $\mu \nabla^2 \mathbf{u}$. However, as it is not easy to accurately estimate the size of the various terms without actually finding the detailed solution to the problem, it is consequently difficult to arrive at sufficient *a priori* conditions for the N-S equations to be replaced by the Stokes equations.

The simplest order-of-magnitude analysis of the size of the various terms proceeds as follows. First, note that for the case of flow between two smooth, parallel walls, the velocity vector lies in the x-direction, whereas the velocity gradient lies in the z-direction, so the advective acceleration terms, $\rho(\mathbf{u} \cdot \nabla)\mathbf{u}$, vanish identically. In the general case of a spatially variable aperture, however, there will be a nonzero z-component of the velocity, and both velocity components will vary in the x-direction, as well as in the z-direction. In a component form, the Navier-Stokes equations (for two-dimensional, steady-state flow) can be written as:

$$\rho\left(u_x \frac{\partial u_x}{\partial x} + u_z \frac{\partial u_x}{\partial z}\right) = -\frac{\partial P}{\partial x} + \mu\left(\frac{\partial^2 u_x}{\partial x^2} + \frac{\partial^2 u_x}{\partial z^2}\right), \quad (7)$$

$$\rho\left(u_x \frac{\partial u_z}{\partial x} + u_z \frac{\partial u_z}{\partial z}\right) = -\frac{\partial P}{\partial z} + \mu\left(\frac{\partial^2 u_z}{\partial x^2} + \frac{\partial^2 u_z}{\partial z^2}\right). \quad (8)$$

It is reasonable to assume that the momentum-balance in the x-direction, i.e., along the direction of flow, is the more important of these two equations. If we let U_x and U_z be characteristic velocities in the x- and z-directions, λ be the characteristic length scale in the x-direction, and $\langle h \rangle$ be the mean aperture, then we can estimate the size of the terms as follows:

$$mag\left[\rho u_x \frac{\partial u_x}{\partial x}\right] \approx \frac{\rho U_x^2}{\lambda}, \qquad mag\left[\rho u_z \frac{\partial u_x}{\partial z}\right] \approx \frac{\rho U_x U_z}{\langle h \rangle}; \quad (9)$$

$$mag\left[\mu \frac{\partial^2 u_x}{\partial x^2}\right] \approx \frac{\mu U_x}{\lambda^2}, \qquad mag\left[\mu \frac{\partial^2 u_x}{\partial z^2}\right] \approx \frac{\mu U_x}{\langle h \rangle^2}. \quad (10)$$

Both inertia terms in Equation (9) are zero for the parallel-plate case, so it is difficult to judge which of the two terms will be largest; both must therefore be considered in the analysis. The relative magnitudes of the two viscous terms in Equation (10) depend on the ratio of aperture to wavelength, $\langle h \rangle / \lambda$. However, small wavelengths always correspond to small values of roughness [*Brown and Scholz*, 1985], and small-scale roughness at a high spatial frequency is known to be irrelevant for laminar flow [*Schlichting*, 1968, p. 580]. Then it seems that we need only consider $\lambda > h$, in which case the second term in Equation (10) is the larger one. Therefore, for both inertia terms to be negligible, we must have:

$$\frac{\rho U_x^2}{\lambda} << \frac{\mu U_x}{\langle h \rangle^2}, \text{and} \frac{\rho U_x U_z}{\langle h \rangle} << \frac{\mu U_x}{\langle h \rangle^2}, \quad (11)$$

which is equivalent to the two conditions

$$\frac{\rho U_x \langle h \rangle}{\mu} \cdot \frac{\langle h \rangle}{\lambda} << 1, \text{and} \frac{\rho U_z \langle h \rangle}{\mu} << 1 . \quad (12)$$

To interpret these conditions, we must choose values for the characteristic velocities, U_x and U_z. An obvious choice for U_x is the mean velocity in the x-direction. However, as the *mean* value of u_z is necessarily zero, we see that the mean value is *not* a meaningful choice for the

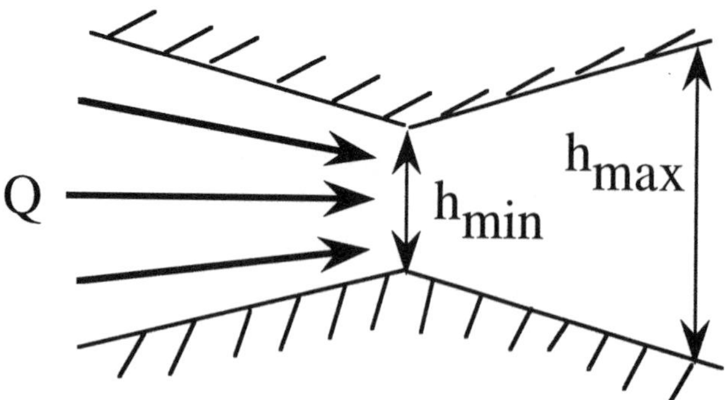

Figure 2. Sawtooth-type fracture profile. According to the Jeffery-Hamel solution for a converging or diverging channel, the streamlines in such a channel are radial lines that converge on the vertex; this fact allows us to estimate the ratio U_z/U_x.

"characteristic velocity" in the z - direction. In this instance, we need to estimate the mean of the *absolute* value of u_z. Consider the simple case of a sawtooth-type fracture profile (Figure 2). This geometry is similar to that of the flow between two smooth walls converging to (or diverging from) a point. This is the classical Jeffery-Hamel problem. The solution to this problem [*Schlichting*, 1968, p. 99] demonstrates that, at low Reynolds numbers, the fluid flows in straight paths toward (or away from) the vertex, which implies the following estimate:

$$\frac{U_z}{U_x} \approx \frac{h_{max} - h_{min}}{\lambda} \approx \frac{\langle h \rangle}{\lambda}. \tag{13}$$

Hence, both conditions in Equation (12) reduce to

$$\frac{\rho U_x \langle h \rangle}{\mu} \cdot \frac{\langle h \rangle}{\lambda} = Re \frac{\langle h \rangle}{\lambda} << 1, \tag{14}$$

where Re is the Reynolds number, and $Re\langle h \rangle / \lambda$ is a dimensionless parameter often referred to as the "reduced Reynolds number" [*Schlichting*, 1968, p. 109].

This seems to be as far as one can go with this type of order-of-magnitude analysis. However, because it is based on *a priori* estimates of the sizes of the terms in the governing equations, it is instructive to test the applicability of criterion (14) by comparing it to any analytical solutions that may be available. The only relevant solution to the N-S equations that we are aware of is the second-order perturbation solution found by *Hasegawa and Izuchi* [1983] for flow through a channel bounded by one flat wall and one sinusoidal wall (Figure 3). The aperture in their model of the fracture geometry is described by:

$$h(x) = \langle h \rangle [1 + \delta \sin(2\pi x / \lambda)]. \tag{15}$$

Hasegawa and Izuchi [1983] also found a few terms in the perturbation solution for the flow field, using Re and $\langle h \rangle / \lambda$ as their "small" parameters. Written in the present notations, the transmissivity was found to be given by:

$$T = \frac{w \langle h^{-3} \rangle^{-1}}{12}$$

$$\left[1 - \frac{3\pi^2 (1-\delta^2)\delta^2}{5(1+\delta^2/2)} \left(1 + \frac{13}{8085} \frac{(1-\delta^2)^5}{(1+\delta^2/2)^2} Re^2 \right) \left(\frac{\langle h \rangle}{\lambda} \right)^2 \right]. \tag{16}$$

The term in front of the square brackets represents the transmissivity that follows from the Reynolds lubrication equation [*Tsang and Witherspoon*, 1981; *Zimmerman et al.*, 1991]. The term inside the square brackets is therefore a correction term that represents the discrepancy between the predictions of the Reynolds equation and the Navier-Stokes equations. This term depends on the normalized magnitude of the roughness, δ, the normalized spatial frequency, $\varepsilon = \langle h \rangle / \lambda$, and the Reynolds number, Re. The term proportional to Re^2 therefore represents the discrepancy between the Navier-Stokes solution and the Stokes solution. The fact that the deviation depends on Re to the *second* power has been demonstrated by *Mei and Auriault*

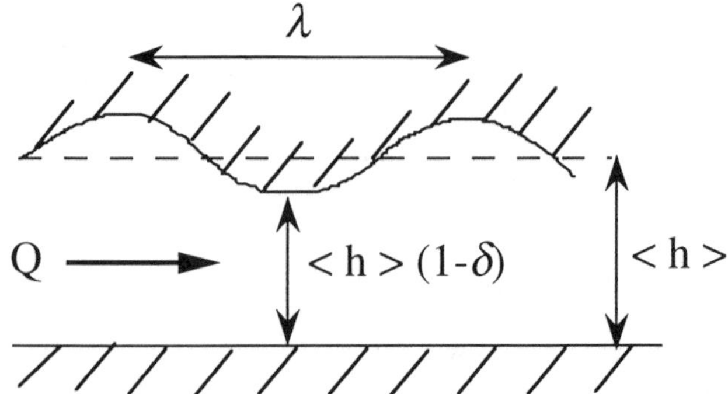

Figure 3. Fracture bounded by one smooth wall and one sinusoidally varying wall. The mean aperture is $\langle h \rangle$, the wavelength is λ, and the ratio of the amplitude of the aperture variation to the mean aperture is δ.

[1991], for porous media of arbitrary geometry, using homogenization theory.

We now consider the conditions under which the inclusion of the high Reynolds number effects will alter the transmissivity by less than, for example, 10%. For this to be the case, we need:

$$\frac{39\pi^2}{40{,}425}\frac{(1-\delta^2)^6\delta^2}{(1+\delta^2/2)^3}Re^2\frac{\langle h\rangle^2}{\lambda^2} < 0.1, \qquad (17)$$

which is equivalent to the condition:

$$\frac{(1-\delta^2)^3\delta}{(1+\delta^2/2)^{3/2}}Re\frac{\langle h\rangle}{\lambda} < 3.24\ . \qquad (18)$$

If $\delta = 0$, we see that a Reynolds number can have any value; this corresponds to the fact that the parabolic velocity profile holds for flow between smooth, parallel walls at any value of *Re*. (At values of *Re* greater than about 2000, the parabolic velocity profile is unstable, and turbulent flow will occur. This phenomenon is entirely unrelated to the effects being discussed in this paper, and is not of much relevance to flow through rock fractures, except in certain exceptional circumstances).

The prefactor in Equation (18) that depends on δ, but not on the Reynolds number or wavelength, is somewhat insensitive to δ over a broad range of values that are relevant to rock fractures. For $0.1 < \delta < 0.6$, this factor varies only between 0.096 and 0.216. Bearing in mind that very low values of δ are "trivial" in the present context, and values of $\delta > 0.6$ are rare, as they would correspond to aperture variations in which $h_{max} / h_{min} > 4$ [see *Yeo et al.*, 1998, Figure 8], for practical purposes Equation (18) effectively reduces to

$$Re^* = Re\frac{\langle h\rangle}{\lambda} < 15\,. \qquad (19)$$

Equation (19) illustrates that the reduced Reynolds number can be as high as 15 in order for the deviations because of high *Re* to be "negligible" for most engineering purposes; criterion (14) is therefore seen to have been much too conservative. This may be a result of the fact that the order-of-magnitude estimates are in a sense local, "worst-case" estimates, whereas the transmissivity is an integrated property, in which local deviations may partially cancel out. Furthermore, as pointed out by *Oron and Berkowitz* [1998], deviations from the Poiseuille-type parabolic velocity profile tend to occur near the upper and lower walls of the fracture, which are regions that carry little flow in any event. Most of the fluid is transmitted through the center of the channel, where the transverse z-component of the velocity is small.

If the wavelength is long enough, Equation (19) shows that the nonlinear effects will be negligible at *any* value of *Re*. On the other hand, since appreciable roughness will not be expected to occur at wavelengths less than the mean aperture, the worst case in Equation (19) will be when $\langle h\rangle \approx \lambda$, in which instance we see that the allowable upper limit on *Re* is about 15. Hence, we can say that, regardless of the details of the fracture geometry, Reynolds number effects should be negligible when $Re < 15$. This critical value of *Re* is comparable to the value of 10 suggested by *Oron and Berkowitz* [1998].

Skjetne et al. [1999] have solved the Navier-Stokes equations for a simulated one-dimensional fracture, at values of *Re* ranging from 0 to 52. They used a "self-affine" fracture with a Hurst roughness exponent of 0.8, as suggested by a number of experimental measurements on fractures in various materials. It should be noted, however, that the midplane of their fracture exhibits large tortuosity at wavelengths of about $\lambda \approx 10\langle h\rangle$, with amplitudes on the order of $\langle h\rangle$; such features are not usually observed in real rock fractures [*Hakami*, 1995; *Yeo et al.*, 1998]. They found that the fracture transmissivity was decreased by about 10% at about $Re \approx 7$. This critical value of *Re* is of the same order of magnitude as the value 15 predicted above; the slight discrepancy may result from the macroscopic tortuosity of their fracture.

Our criterion is also in rough agreement with the experimental finding of *Iwai* [1976] for flow through a tension fracture in granite. Iwai plotted his data in the form of friction factor, ψ, *vs.* Reynolds number, Re. Iwai used a converging radial flow configuration, and evaluated his Reynolds number at the inner radius. However, as the overall flow is indeed dominated by the properties near the inner wellbore radius, this seems appropriate, as long as we bear in mind that the Reynolds number varies with the radius in these experiments. As plotted by Iwai, deviations from linearity would manifest themselves as a departure from a straight line of slope −1; such a departure point is difficult to determine from this type of plot.

We have replotted Iwai's data in the form of the normalized transmissivity, $h_H^3/\langle h\rangle^3$, vs. the Reynolds number, *Re* (Figure 4a). To do so, we first note that his Reynolds numbers must be divided by 2 in order to be converted to the present definition, given by Equation (14).

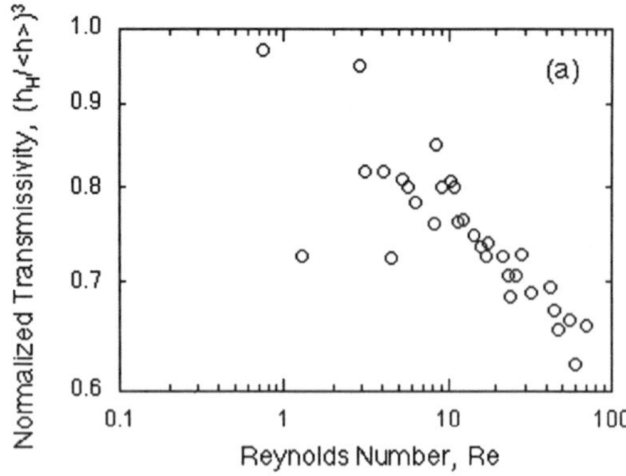

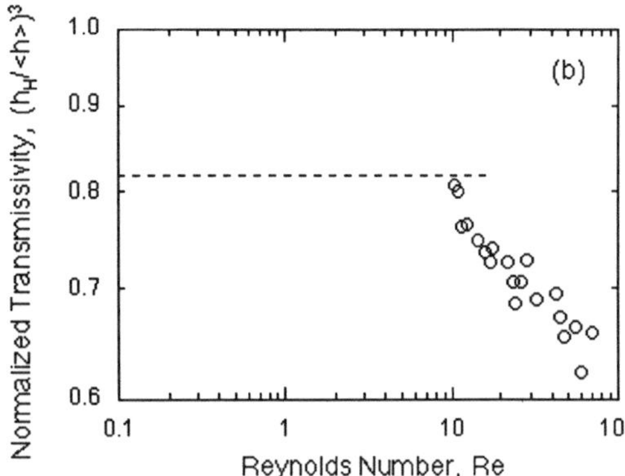

Figure 4. (a) Data from *Iwai* [1976], plotted as the ratio of the measured transmissivity to the cubic-law transmissivity calculated using the mean aperture. Data for lower Reynolds numbers have large error bars, due to difficulty in measuring low flow rates and small pressure drops. We expect the normalized transmissivities to converge to some constant value (< 1) as $Re \to 0$. (b) Data replotted, with all normalized transmissivities for $Re < 10$ replaced by their mean value, 0.817. Data now show that systematic deviations from linearity begin at about $Re \approx 10$.

Also, we note that $h_H^3/\langle h\rangle^3$ can be shown to be equal to $96/\psi Re$. Figure 4a demonstrates that the normalized transmissivity begins to decrease as a function of *Re* when *Re* is greater than approximately 10, as we expect. This trend is somewhat obscured by the great degree of scatter in the data at low values of *Re,* which is explained by a difficulty in accurately measuring the flowrates and the pressure drops in this regime.

Since we expect $h_H^3/\langle h\rangle^3$ to level off to some constant value as $Re \to 0$, we can replot the data in the following way (Figure 4b), which better illustrates the deviations from linearity. We first average out all 12 data points that have $Re < 10$, to find an asymptotic value of 0.817 for the normalized transmissivity at low Reynolds numbers, and then plot this asymptote along with all data for $Re > 10$. This method of displaying the data is, admittedly, somewhat circular in its logic. Nevertheless, the data do show a systematic deviation from the asymptotic value of 0.817, starting at a Reynolds number of about 10. This is in rough agreement with the estimate given by Equation (15), as well as with the theoretical estimate of *Oron and Berkowitz* [1998] and the numerical simulation results of *Skjetne et al.* [1999]. Hence, it seems that a Reynolds number of about 10 marks the onset of nonlinearity in the relationship between transmissivity and flowrate.

3. FROM THE STOKES EQUATIONS TO THE LUBRICATION EQUATION

Although they are linear, the Stokes equations are nevertheless very difficult to solve, and do not yet seem to have been solved for *any* realistic fracture geometry, either analytically or numerically. *Mourzenko et al.* [1995] and *Brown et al.* [1995] solved the Stokes equations numerically for fractures with synthetically generated aperture fields. *Cao and Kitanidis* [1998] used finite elements to solve the Stokes equations for a fracture bounded by two out-of-phase sinusoidal surfaces. However, solutions utilizing *measured* aperture fields do not yet seem to have been obtained. So, up until this time, it has been common to attempt to further simplify the Stokes equations before attempting to solve them.

The next level of simplification often used is to replace the Stokes equations with the Reynolds lubrication equation [*Brown*, 1987]. To accomplish this, we start with the two Stokes equations for our "one-dimensional" fracture, Equations (7) and (8), which can be explicitly written as

$$\frac{\partial P}{\partial x} = \mu\left(\frac{\partial^2 u_x}{\partial x^2} + \frac{\partial^2 u_x}{\partial z^2}\right), \tag{19}$$

$$\frac{\partial P}{\partial z} = \mu\left(\frac{\partial^2 u_z}{\partial x^2} + \frac{\partial^2 u_z}{\partial z^2}\right). \tag{20}$$

All terms in Equation (20), which represents the momentum balance in the direction perpendicular to the fracture plane, are very small, so we ignore this equation. Using the estimates of the magnitudes of the two viscous terms in Equation (19) that were given in Equation (10), we see that $\partial^2 u_x/\partial x^2$ will be smaller than $\partial^2 u_x/\partial z^2$ by a factor of about ten, if:

$$\langle h\rangle^2 / \lambda^2 < 0.1, \quad \text{or} \quad \langle h\rangle / \lambda < 0.3. \tag{21}$$

It is noteworthy that this condition does not depend on the variance of the fracture aperture; indeed, large aperture variations are allowable, as long as they occur gradually [*Langlois*, 1964].

If condition (21) is satisfied, the governing equations reduce, in the more general case of a fracture whose aperture varies in both the x and y directions, to

$$\frac{\partial P}{\partial x} = \mu \frac{\partial^2 u_x}{\partial z^2}, \qquad \frac{\partial P}{\partial y} = \mu \frac{\partial^2 u_y}{\partial z^2}. \tag{22}$$

These equations can be integrated twice with respect to z to find an in-plane velocity vector that is parabolic in z, but which is directed parallel to the *local* pressure gradient [*Schlichting*, 1968]. If the equation of the parabolic velocity profile is substituted into the conservation of mass equation, Equation (2), and integrated again in the z-direction, we arrive at the Reynolds lubrication equation [*Brown*, 1987; *Zimmerman and Bodvarsson*, 1996]:

$$\frac{\partial}{\partial x}\left(h^3 \frac{\partial P}{\partial x}\right) + \frac{\partial}{\partial y}\left(h^3 \frac{\partial P}{\partial y}\right) = 0, \tag{23}$$

which can be viewed as a local version of the cubic law. The Reynolds equation can be solved by finite-difference or finite-element techniques for actual fracture aperture distributions [*Brown*, 1987; *Yeo et al.*, 1998]. These solutions allow the estimation of transmissivity in terms of the statistics of the aperture distribution. Moreover, since Equation (23) is equivalent to the equation that governs conduction in a two-dimensional medium with spatially variable conductivity, all of the methods that have been used for that more general problem can be used to estimate, or to bound, the fracture transmissivity [i.e., *Dagan*, 1989].

It was argued above that the reduction from the Stokes equations to the Reynolds equation should be permissible if condition (21) holds. However, this conclusion was based on an *a priori* estimate of the size of the various terms in the equation, rather than on *solutions* to the equations. It is therefore instructive to again compare the criterion obtained by order-of-magnitude analysis with the predictions of the Hasegawa-Izuchi solution for the smooth/sinusoidal fracture. If we ignore the terms that depend on Re, the transmissivity of this fracture is given to the second-order in $\langle h\rangle^2 / \lambda^2$, as

$$T_{\text{Stokes}} = T_{\text{Reynolds}}\left[1 - \frac{3\pi^2(1-\delta^2)\delta^2}{5(1+\delta^2/2)}\left(\frac{\langle h\rangle}{\lambda}\right)^2\right]. \tag{24}$$

It should be noted that *Kitanidis and Dykaar* [1997] computed the perturbation solution to the *Stokes* equations for flow between two out-of-phase (i.e., unmated) sinusoidal walls, the aperture of which is still represented by Equation (15), and found precisely the same expression for the ratio $T_{\text{Stokes}}/T_{\text{Reynolds}}$. They also found the next term in the solution, which is proportional to $\langle h\rangle^4 / \lambda^4$. However, we are mainly interested in the conditions under which the Reynolds transmissivity begins to diverge from the Stokes transmissivity, and the criterion for this to occur can be found by examining only the first perturbation.

So, from Equation (24), the criterion by which the Reynolds and Stokes transmissivities differ by no more than 10% is:

$$\frac{3\pi^2(1-\delta^2)\delta^2}{5(1+\delta^2/2)}\left(\frac{\langle h\rangle}{\lambda}\right)^2 < 0.1. \tag{25}$$

The prefactor term that depends on the roughness parameter δ never exceeds 1.08, so the approximate condition for the Reynolds approximation to hold can be written as:

$$\frac{\langle h\rangle}{\lambda} < 0.3, \tag{26}$$

which agrees exactly with the results of the order-of-magnitude analysis, Equation (21).

However, it is not clear that replacement of the Stokes equations by the lubrication equation is justifiable for real fractures, since there may be substantial roughness at wavelengths smaller than that required by (26). Furthermore, several researchers have compared Stokes and Reynolds simulations on simulated fracture aperture distributions [*Mourzenko et al.*, 1995; *Brown et al.*, 1995] and concluded that the Reynolds approximation does *not* apply. *Yeo* [1998] measured apertures and transmissivities of a sandstone fracture and found that the Reynolds equation overestimated the transmissivity by a factor of 40-100%, despite the fact that the fracture was not particularly rough.

Examples of the fracture apertures observed by *Yeo* are shown in Figure 5. The top row shows a typical "smooth" region, the bottom row shows an extremely rough section of the fracture, and the middle row shows a typical

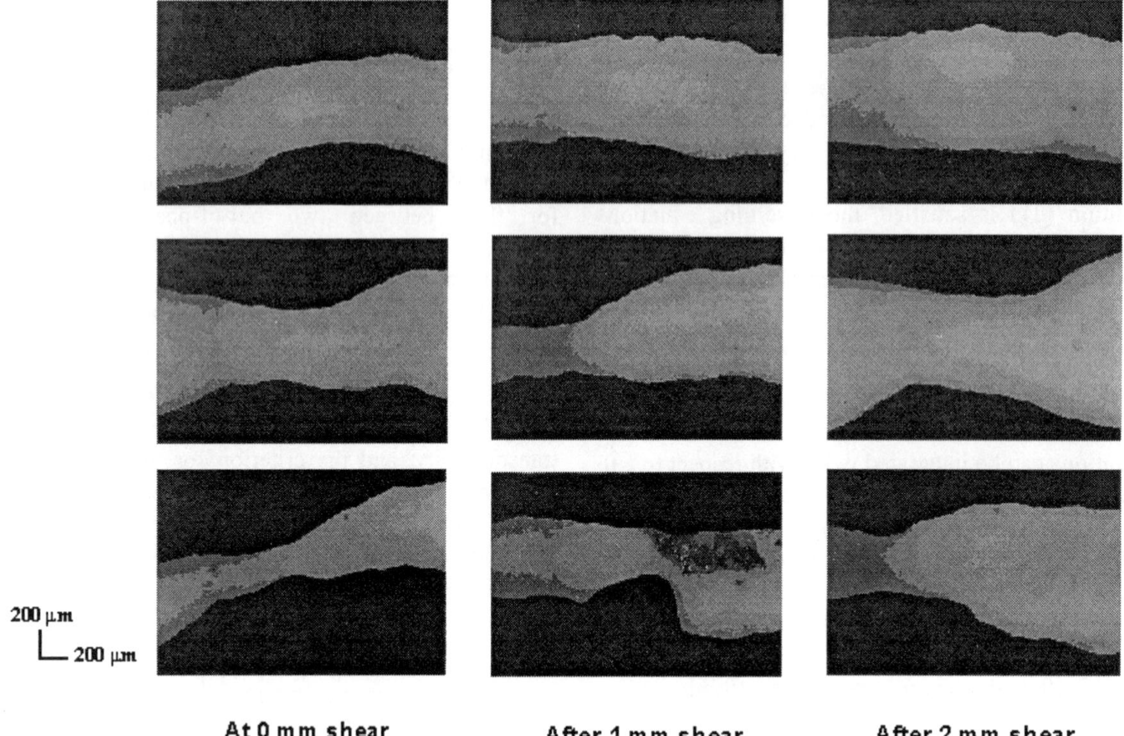

Figure 5. Examples of the fracture apertures observed by *Yeo* [1998]. Top row shows a typical "smooth" region, bottom row shows an extremely rough section, and the middle row shows a typical intermediate case. Different columns correspond to differing levels of shear displacement.

intermediate case. The different columns correspond to differing levels of shear displacement. It seems that the regions in which the aperture varied most rapidly correspond to about $\langle h\rangle/\lambda \approx 0.5$. Furthermore, although relative roughness increased slightly as the shear displacement increased, the values measured by *Yeo* tended to be in the range of $\sigma/\langle h\rangle \approx 0.3$. In order to use Equation (24), we must relate $\sigma/\langle h\rangle$ to the parameter δ. The sinusoidal aperture distribution has $\delta = \sqrt{2}\sigma/\langle h\rangle$, so it seems that the "effective" roughness parameter for the fracture investigated by *Yeo* would be $\delta \approx \sqrt{2}(0.3) \approx 0.42$. Equation (24) then predicts that the Stokes (low Reynolds number) transmissivity will be about 20% less than that predicted by the Reynolds equation. In fact, the transmissivities measured by Yeo were about 20–60% less than those predicted from numerical solution of the Reynolds equation, using the measured aperture fields. As the Hasegawa-Izuchi equation is based on the solution for a one-dimensional, sinusoidal fracture, precise agreement should not be expected. The salient point seems to be that the H-I equation does correctly predict that there will be substantial discrepancies between the Reynolds and Stokes predictions for a fracture having this degree of roughness.

4. IMPLICATIONS OF THE LUBRICATION THEORY APPROXIMATION

As mentioned above, lubrication theory Equation (23) for flow through a fracture is identical to the governing equation for any conductive-type process (groundwater flow through porous media, heat conduction, electrical conduction, etc.) through an inhomogeneous medium, with h^3 playing the role of the local conductivity. As such, the entire extensive array of methods that have been developed to estimate the effective macroscopic conductivity of an inhomogeneous medium may be employed. The results of any such analysis can be written in terms of an expression for the effective hydraulic aperture, as a function of the statistics of the aperture distribution.

If the aperture is lognormally distributed, then the local conductivity h^3 will also be lognormally distributed. In such cases, *Dagan* [1993] has shown that the effective conductivity is essentially equal to the geometric mean of the local conductivity, which is defined by $k_G = e^{\langle \ln k \rangle}$, where $\langle x \rangle$ denotes the mean value of the random variable x. This result has frequently been applied to rock fractures [*Piggott and Elsworth*, 1993], although the appropriateness of a lognormal distribution for fracture apertures has occasionally been questioned on theoretical grounds [*Oron and Berkowitz*, 1998]. Note that, for instance, the aperture distributions measured by *Yeo et al.* [1998] were fit much more closely with normal distributions than with lognormal distributions.

It therefore seems sensible to recast the effective conductivity into a form that reduces to the geometric mean in the case of a lognormal distribution, but which depends explicitly only on the mean and variance of the local conductivities. If k is distributed lognormally, it can be shown that [*Aitchison and Brown*, 1957]:

$$k_G = \langle k \rangle e^{-\sigma_Y^2/2}, \qquad \sigma_k^2 = \langle k \rangle^2 [e^{-\sigma_Y^2} - 1], \tag{27}$$

where $Y = \ln k$. So, Dagan's result for the effective conductivity in the case where the local conductivity is lognormally distributed can be recast in the form:

$$k_{eff} = \langle k \rangle \left[1 + \sigma_k^2 / \langle k \rangle^2 \right]^{-1/2}. \tag{28}$$

For small values of the log-variance, Equation (28) reduces to:

$$k_{eff} = \langle k \rangle \left[1 - \sigma_k^2 / 2\langle k \rangle^2 + \ldots \right], \tag{29}$$

which is a classical result that can been derived under the assumption of smoothness of the conductivity distribution, without requiring lognormality [e.g., *Landau and Lifshitz*, 1960].

Equation (28) could also be expressed directly in terms of the mean and standard deviation of the aperture, rather than in terms of the moments of $k = h^3$. However, the result would be an unwieldy expression that contained terms up to $\sigma_h^6 / \langle h \rangle^6$. *Zimmerman and Bodvarsson* [1996] simplified this result by dropping higher-order terms; however, these higher terms are *not* negligible for realistic values of $\sigma_h / \langle h \rangle$, such as the value 0.3 observed by *Yeo* [1998], or the values measured by *Hakami* [1989], which were higher yet. Hence, it seems advisable to use Equation (28) without introducing further approximations, with σ_k and $\langle k \rangle$ calculated directly from the aperture distribution using the identity $k = h^3$.

Direct comparison of these results with experimental data is problematic. *Zimmerman and Bodvarsson* [1996] used the first-order expression alluded to above,

$$h_{eff}^3 = \langle h \rangle^3 \left[1 - 1.5\sigma_h^2 / \langle h \rangle^2 \right], \tag{30}$$

and were able to achieve reasonably good matches to the transmissivities measured by *Hakami* [1989] on fractures in granite cores from Stripa, Sweden. However, two different effects are present in the comparison between the theory and data as presented by *Zimmerman and Bodvarsson* [1996]: the effect of roughness, which causes a discrepancy between the Reynolds and Stokes predictions; and the fact that, even if the Reynolds equation applies, expressions such as (28–30) for the effective transmissivity are merely approximations.

5. EFFECT OF CONTACT AREA

Another factor influencing the transmissivity of a rock fracture is the contact area between the two fracture walls in some locations, which blocks off flow. The fluid must flow around these regions, thereby introducing a tortuosity effect, which necessarily reduces the overall transmissivity. The reduction depends on both the shape and dimensions of the contact area (viewed as projections onto the nominal fracture plane).

A real fracture consists of contact areas in which the aperture is effectively zero, surrounded by areas of varying but nonzero aperture. One way to treat this situation is to use the sort of analysis that was described above for the "open" regions in order to first find an effective hydraulic aperture of the open regions. The open regions can then be replaced with a constant-aperture fracture having the appropriate aperture, and the effect that contact regions would have on the transmissivity can then be studied for this case. This latter problem becomes one of finding the overall conductivity of a two-dimensional medium of initial transmissivity T_O, with impermeable obstacles imbedded in it—a standard problem in effective medium theory that was in fact solved by *Maxwell* [1873] for the analogous three-dimensional problem of a medium containing impermeable spherical obstacles.

Walsh [1981] used Maxwell's approach to show that if the contact areas are circular in planform, the fractional decrease in fracture conductivity will be given by

$$\frac{T}{T_o} = \frac{1-c}{1+c}, \tag{31}$$

where c is the fractional contact area. It can be shown, by appealing to the Hashin-Shtrikman bounds of effective medium theory [*Hashin and Shtrikman*, 1962], that noncircular contact regions will always decrease the transmissivity by an amount that is greater than that given by Equation (31). In the specific case of randomly oriented elliptical contact regions, *Zimmerman et al.* [1992] found that:

$$\frac{T}{T_o} = \frac{1-\beta c}{1+\beta c}, \quad \text{where} \quad \beta = (1+\alpha)^2 / 4\alpha, \tag{32}$$

and $\alpha = 1$ is the aspect ratio of the elliptical contact areas. As an example, a large (but not unreasonable) value of 30% contact area [*Nolte et al.*, 1989; *Durham and Bonner*, 1994], arranged as ellipses of aspect ratio 0.5, would cause the transmissivity to decrease by about 50%. Other models have been developed to apply near the percolation threshold, where the contact area is sufficiently high as to nearly cut off the flow entirely [*Nolte et al.*, 1989; *Walsh et al.*, 1997], but there is a lack of experimental evidence showing that this situation is of importance in real rock fractures.

6. WATER FLOW UNDER UNSATURATED CONDITIONS

Another important issue is *unsaturated* flow through rock fractures (as in, for example, the characterization of the hydrologic system at Yucca Mountain, a possible site for an underground radioactive waste repository). In regions above the water table, the water phase cannot completely fill the void space of the fracture, and the remainder of the fracture is filled with air. The degree of liquid saturation is a function of the thermodynamic potential of the water. Consequently, the relative permeability of the fracture to water will also be a function of the water thermodynamic potential.

In the conceptual model for unsaturated fracture flow [*Pruess and Tsang*, 1990; *Murphy and Thomson*, 1993], the regions having aperture $h < h^*$ are filled with water, whereas the regions having aperture $h > h^*$ are filled with air. The critical aperture is given by the Young-Laplace equation as $h^* = 2\gamma / P_c$, where γ is the surface tension between the air and water, and $P_c = P_{air} - P_{water}$ is the capillary pressure. Liquid-phase flow is then modeled by assuming that the water obeys the local cubic law as it flows through the tortuous pathway that connects the water-filled regions of the fracture. The discussion given above, however, implies that use of the local cubic law may not be appropriate for the water phase, as it will probably overestimate the local transmissivity. In fact, this error may be *accentuated* in unsaturated flow, as it seems likely (see Figure 6) that the regions of smaller mean aperture, which are filled with water, will have larger values of relative roughness, $\sigma_h / \langle h \rangle$, and hence larger deviations from the cubic law.

A complementary conceptual model of unsaturated flow in rough-walled rock fractures has been developed by *Tokunaga and Wan* [1997]. They pointed out that, as long as the fracture plane is not perfectly horizontal, water can flow down the fracture in the form of a thin "film" that does not need to bridge the gap between the lower and upper rock surfaces. Hence, the criterion $h < h^*$ does not need to be satisfied in order for water to exist (and flow) at a certain location in the fracture plane. This film will "fill in" the regions of small-scale roughness on the lower fracture surface, in a manner that leads to a mean film thickness that decreases as the capillary pressure becomes larger [*Philip*, 1978].

Unfortunately, it is difficult to relate this model to the considerations of previous sections of this paper. By definition, the type of film flow hypothesized by *Tokunaga and Wan* [1997] depends on the roughness of the fracture wall, but *not* on the distance between the two walls. Although the aperture variation is related to the surface roughness, it also depends on the larger-scale "matedness" of the two fracture surfaces [*Kumar and Bodvarsson*, 1990], and so there is no simple relationship between wall roughness and aperture roughness. Another point is that, whereas small-scale roughness is not of great importance for saturated flow, such roughness is crucial to film flow under high capillary pressures. In particular, whereas fracture transmissivity under fully saturated conditions is primarily controlled by the mean aperture, with the small-scale roughness playing a perturbative role, the transmissivity under unsaturated conditions, particularly at very high suctions, will be independent of mean aperture and completely controlled by small-scale roughness. A quantitative theory of film flow along rough-walled fractures has yet to be developed, and it is not altogether clear how such a theory will relate to the models for saturated flow that have been discussed above.

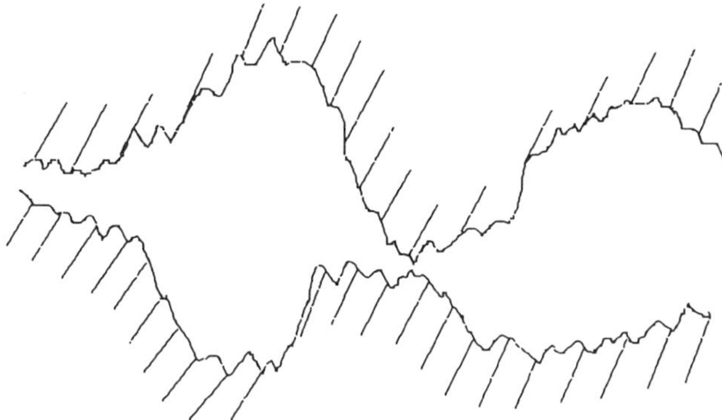

Figure 6. Drawing illustrating that if fracture-wall roughness is statistically uniform, regions of smaller mean aperture will have larger values of *relative* roughness, $\sigma_h / \langle h \rangle$.

7. TRACER TRANSPORT, MEAN APERTURE, AND FRACTURE POROSITY

Another important aspect of flow through fractured rock is the transport of tracers, which are often used to infer the total fracture porosity [*Silliman*, 1989]. Tracers are injected into the fractured rock mass at an injection well, and collected at a collection well. The tracer particles advect with the mean flow, but also diffuse through the fluid; this process is modeled by an advection/diffusion equation [*Bear et al.*, 1998]. The concentration of the tracer in the effluent, as a function of time, is then fit to a solution of the advection/diffusion equation, with the porosity of the fracture network as one of the fitting parameters. But this step rests on the assumption that the entire void space of the fracture takes part in the flow. In fact, circumstantial evidence exists to show that fractures contain numerous small dead-end zones that do not take part in the macroscopic flow [see *Raven et al.,* 1988]. The existence of such zones is indicated in the numerical simulations of *Cao and Kitanidis* [1998] for flow in channels containing sinusoidal bulges. Furthermore, the fact that measured fracture transmissivities are substantially less than those predicted by the Reynolds lubrication equation also gives indirect evidence of zones of recirculation. So, it seems that tracer measurements will actually yield an effective fracture that will be less than the actual physical porosity.

A similar error will occur if fracture porosity is estimated from, for example, gas-permeability measurements made between boreholes. In such a case, the measurements directly yield the permeability of the fracture continuum. If the fracture spacings can be estimated, it is then possible to estimate the mean transmissivity of the individual fractures. This value is then converted into a mean fracture aperture, using some relationship between hydraulic aperture and mean aperture [*Huang, et al.*, personal communication]. However, the discussion given above shows that existing relationships, all based on the local cubic law in one way or another, will overestimate the transmissivity for a given fracture geometry. Hence, if used in an inverse manner, the mean aperture will be underestimated (for a given transmissivity). If we assume an error in the local cubic law of approximately a factor of 2, as found above, and note that the aperture will still be proportional to transmissivity to the 1/3 power (with merely a different numerical factor in front), the error in mean aperture will only be about 25%. In many cases, such an error will be less than that involved in the estimation of fracture densities, etc.

8. CONCLUSIONS

The mathematical analysis of the flow of a single-phase Newtonian fluid through a rough-walled rock fracture has been reviewed, starting with the Navier-Stokes equations. By a combination of order-of-magnitude analysis, appealing to the analytical solution for a sinusoidal fracture developed by *Hasegawa and Izuchi* [1983], and replotting of the classic data of *Iwai* [1976], it was shown that the Navier-Stokes equations can be replaced by their linearized version, the Stokes equations, if the Reynolds number is less than approximately 10. At higher Reynolds numbers, the transmissivity is no longer independent of flowrate, but rather decreases with flowrate.

Further order-of-magnitude analysis of the Stokes equations, and the *Hasegawa-Izuchi* solution, showed that the Stokes equations could probably be replaced by the simpler Reynolds lubrication equation if the wavelength of the dominant aperture variations is about three times greater than the mean aperture. Analysis of the fracture aperture images collected by *Hakami* [1995] and *Yeo* [1998] showed that many fractures do *not* strongly satisfy this criterion. Indeed, the Reynolds equation simulations of *Yeo et al.* [1998], made using measured aperture profiles, tended to over-predict the measured transmissivity by about 25–100%. This discrepancy could be accounted for, but only qualitatively, by using the Hasegawa-Izuchi equation (16).

Hence, it seems that the Reynolds equation (i.e., the local cubic law) may suffice in estimating fracture permeabilities to within a factor of about 2, but more accurate estimates may require utilization of the Stokes equations. Conversely, if transmissivity measurements are used as a basis of estimating the mean aperture, errors on the order of about 2 may be expected if any version of the local cubic law is used such as, for example, using the geometric mean to estimate the hydraulic aperture.

REFERENCES

Aitchison, J., and J. A. C. Brown, *The Lognormal Distribution*, Cambridge University Press, New York, N.Y., 1957.

Batchelor, G. K., *An Introduction to Fluid Dynamics*, Cambridge University Press, New York, New York, N.Y., 1967.

Brown, S. R., Fluid flow through rock joints: the effect of surface roughness, *J. Geophys. Res.*, *92*, 1337–1347, 1987.

Brown, S. R., and C. H. Scholz, Broad bandwidth study of the topography of natural surfaces, *J. Geophys. Res.*, *90*, 12575–12582, 1985.

Brown, S. R., H. W. Stockman, and S. J. Reeves, Applicability of the Reynolds equation for modeling fluid flow between rough surfaces, *Geo. Res. Letts.*, *22*, 2537–2540, 1995.

Cao, J., and P. K. Kitanidis, Adaptive finite element simulation of Stokes flow in porous media, *Adv. Water Resour.*, *22*, 17–31, 1998.

Dagan, G., *Flow and Transport in Porous Formations*, Springer–Verlag, Berlin, 1989.

Dagan, G., Higher–order correction for effective permeability of heterogeneous isotropic formations of lognormal conductivity distribution, *Transp. Porous Media*, *12*, 279–290, 1993.

Durham, W. B., and B. P. Bonner, Self–propping and fluid–flow in slightly offset joints at high effective pressures, *J. Geophys. Res.*, *99*, 9391–9399, 1994.

Hakami, E., *Water Flow in Single Rock Joints*, Licentiate dissertation, Lulea Univ. Tech., Lulea, Sweden, 1989.

Hakami, E., *Aperture Distribution of Rock Fractures*, Ph.D. dissertation, Royal Inst. Tech., Stockholm, Sweden, 1995.

Hasegawa, E., and H. Izuchi, On the steady flow through a channel consisting of an uneven wall and a plane wall, *Bull. Jap. Soc. Mech. Eng.*, *26*, 514–520, 1983.

Hashin, Z., and S. Shtrikman, A variational approach to the theory of the effective magnetic permeability of multiphase materials, *J. Appl. Phys.*, *33*, 3125–3131, 1962.

Huang, K., Y. W. Tsang, and G. S. Bodvarsson, Simultaneous inversion of air-injection tests in fractured unsaturated tuff at Yucca Mountain, *Water Resour. Res. 35*, 2375–2386, 1999.

Iwai, K., *Fundamental Studies of Fluid Flow through a Single Fracture*, Ph.D. dissertation, University of California, Berkeley, Calif., 1976.

Kitanidis, P. K., and B. B. Dykaar, Stokes flow in a slowly varying two–dimensional periodic pore, *Transp. Porous Media*, *26*, 89–98, 1997.

Kumar, S., and G. S. Bodvarsson, Fractal study and simulation of fracture roughness, *Geophys. Res. Letts.*, *17*, 701–704, 1990.

Landau, L. D., and E. M. Lifshitz, *Electrodynamics of Continuous Media*, Pergamon, New York, N.Y., 1960.

Maxwell, J. C., *A Treatise on Electricity and Magnetism*, Clarendon Press, Oxford, England, 1873.

Mei, C. C., and J.-L. Auriault, The effect of weak inertia on flow through a porous medium, *J. Fluid Mech.*, *222*, 647–663, 1991.

Mourzenko, V. V., J. F. Thovert, and P. M. Adler, Permeability of a single fracture – validity of the Reynolds equation, *J. Phys. II*, *5*, 465–482, 1995.

Murphy, J. R., and N. R. Thomson, Two-phase flow in a variable aperture fracture, *Water Resour. Res.*, *29*, 3453–3476, 1993.

Nolte, D. D., N. G. W. Cook, and L. J. Pyrak-Nolte, The fractal geometry of flow paths in natural fractures and the approach to percolation, *Pure Appl. Geophys.*, *131*, 111–138, 1989.

Oron, A. P., and B. Berkowitz, Flow in rock fractures: the local cubic law assumption re-examined, *Water Resour. Res.*, *34*, 2811–2824, 1998.

Philip, J. R., Absorption and capillary condensation on rough surfaces, *J. Phys. Chem.*, *82*, 1379–1385, 1978.

Piggott, A. R., and D. Elsworth, Laboratory assessment of the equivalent apertures of a rock fracture, *Geophys. Res. Letts.*, *20*, 1387–1390, 1993.

Pruess, K., and Y. W. Tsang, On two-phase relative permeability and capillary pressure of rough-walled fractures, *Water Resour. Res.*, *26*, 1915–1926, 1990.

Schlichting, H., *Boundary-Layer Theory*, 6th ed., McGraw–Hill, New York, N.Y., 1968.

Silliman, S. E., Interpretation of the difference between aperture estimates derived from hydraulic and tracer tests in a single fracture, *Water Resour. Res.*, *25*, 2275–2283, 1989.

Skjetne, E., A., Hansen, and J. S. Gudmundsson, High–velocity flow in a rough fracture, *J. Fluid Mech.*, *383*, 1–28, 1999.

Tokunaga, T. K., and J. Wan, Water film flow along fracture surfaces of porous rock, *Water Resour. Res.*, *33*, 1287–1295, 1997.

Tsang, Y. W., and P. A. Witherspoon, Hydromechanical behavior of a deformable rock fracture subject to normal stress, *J. Geophys. Res.*, *86*, 9287–9298, 1981.

Walsh, J. B., S. R. Brown, and W. B. Durham, Effective medium theory with spatial correlation for flow in a fracture, *J. Geophys. Res.*, *102*, 22587–22594, 1997.

Walsh, J. B., The effect of pore pressure and confining pressure on fracture permeability, *Int. J. Rock Mech.*, *18*, 429–435, 1981.

Witherspoon, P. A., J. S. Y. Wang, K. Iwai, and J. E. Gale, Validity of cubic law for fluid flow in a deformable rock fracture, *Water Resour. Res.*, *16*, 1016–2024, 1980.

Yeo, I. W., *Anisotropic Hydraulic Properties of a Rock Fracture under Normal and Shear Loading*, Ph.D. dissertation, Imperial College, London, England, 1998.

Yeo, I. W., M. H. deFreitas, and R. W. Zimmerman, Effect of shear displacement on the aperture and permeability of a rock fracture, *Int. J. Rock. Mech.*, *35*, 1051–1070, 1998.

Zimmerman, R. W., and G. S. Bodvarsson, Hydraulic conductivity of rock fractures, *Transp. Porous Media*, *23*, 1–30, 1996.

Zimmerman, R. W., D. W. Chen, and N. G. W. Cook, The effect of contact area on the permeability of fractures, *J. Hydrol.*, *139*, 79–96, 1992.

Zimmerman, R. W., S. Kumar, and G. S. Bodvarsson, Lubrication theory analysis of the permeability of rough–walled fractures, *Int. J. Rock Mech.*, *28*, 325–331, 1991.

Robert W. Zimmerman, T.H. Huxley School of Environment, Earth Sciences and Engineering, Imperial College of Science, Technology and Medicine, London, UK, and Earth Sciences Division, Lawrence Berkeley National Laboratory, One Cyclotron Road, Berkeley, CA 94720

In-Wook Yeo, Department of Geological Sciences, University of Colorado, Boulder, CO 80302

Multiphase Flow in Fractured Rocks—Some Lessons Learned from Mathematical Models

Karsten Pruess

Earth Sciences Division, Lawrence Berkeley National Laboratory, Berkeley, California

1. INTRODUCTION

Fractured rock formations encompass an enormous variety of hydrogeologic properties [*Bear et al.*, 1993; *National Research Council*, 1996]. For the recovery of resources such as oil, gas, and geothermal energy from fractured reservoirs, we are primarily interested in systems with well-connected fracture networks of high permeability and with good matrix permeability and porosity. For the purposes of underground waste disposal, we generally prefer media with the opposite characteristics, i.e., sparse and poorly connected fractures and low matrix permeability. Multiphase flow processes of interest in fractured media include two-phase flows of water-gas, water-NAPL (nonaqueous phase liquid), and water-steam, and three-phase flows of oil, water, and gas. Water seepage through the vadose zone is a special kind of multiphase flow process that is an essential component of the hydrologic cycle. It may often be described in approximate fashion by considering the gas phase as a passive bystander. Multiphase flows may be complicated by strongly coupled heat transfer effects, as in geothermal production and injection operations, in the thermally enhanced recovery of oil and of volatile organic contaminants, and in the geologic disposal of heat-generating high-level nuclear wastes.

This paper presents a critical discussion of different approaches for modeling multiphase flows in fractured media with respect to oil, gas, and geothermal production, and vadose zone hydrology. We limit ourselves to methods that are based on the sound principles and well-established continuum field theories of classical theoretical physics [*Morse and Feshbach*, 1953; *Narasimhan*, 1982a,b] in which conservation of the active system components (water, air, chemical constituents, heat) is expressed by means of integral or partial differential equations (PDEs) for space-and-time varying fields of phase saturations, pressures, temperatures, solute concentrations, etc. Mass and heat fluxes are expressed through phenomenological relationships between intensive variables that drive flow such as multiphase extensions of Darcy's law for phase fluxes, Fick's law for mass diffusion, Scheidegger's hydrodynamic dispersion, and Fourier's law for heat conduction. Alternative approaches such as invasion percolation [*Glass*, 1993], lattice gas automata [*Stockmann et al.*, 1997], and chaos theory [*Pruess et al.*, 1999] have been applied for modeling the detailed spatial and temporal structure of multiphase flows in fractures, but are outside the scope of this article.

Dynamics of Fluids in Fractured Rock
Geophysical Monograph 122

Published 2000 by the American Geophysical Union

2. VOLUME-AVERAGED CONTINUUM APPROACHES

The study of fractured multiphase flow systems began in the context of oil and gas recovery (for a recent review, see *Kazemi and Gilman*, 1993). The groundbreaking concept on which most later work was based is the "double-porosity" method (DPM), formulated by *Barenblatt et al.* (1960) and introduced into the U.S. petroleum literature by *Warren and Root* [1963]. The basic idea is to associate each "point" in a fractured reservoir domain with not just one, but two sets of hydrogeologic parameters and thermodynamic state variables. Formally this is accomplished by attaching a sphere of "suitable" volume, which contains many fractures, to each point in the reservoir. The volume of the sphere is then partitioned into two subdomains, one for the fractures, the other for

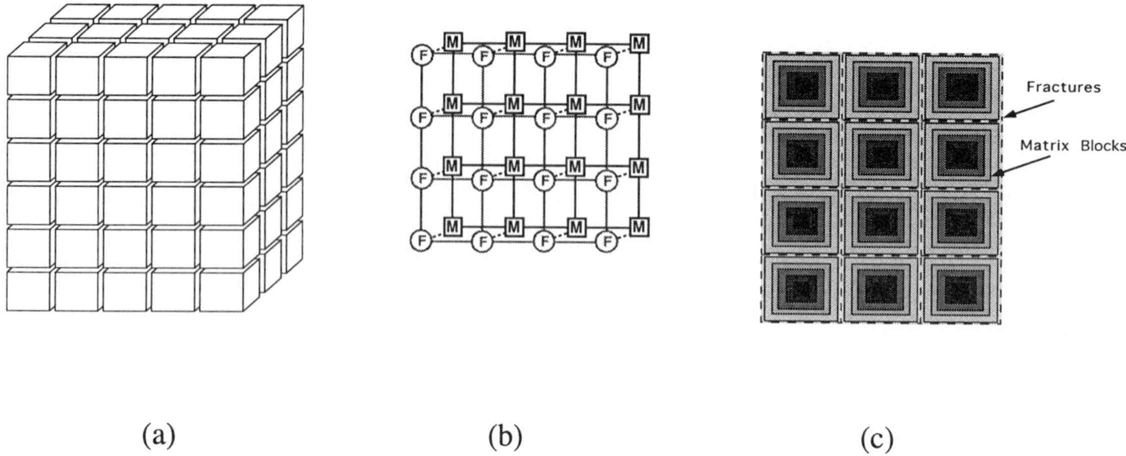

Figure 1. Illustration of concepts used for modeling of multiphase flow in fractured rocks: (a) double-porosity concept (DPM), after *Warren and Root* [1963], in which global flow occurs exclusively through a network of interconnected fractures, while fractures and matrix may exchange fluids and heat locally; (b) dual permeability model (DKM), with global flow in both fracture (F) and matrix continua (M); (c) MINC subgridding for resolution of gradients in the matrix blocks [after *Pruess and Narasimhan*, 1982, 1985].

unfractured "matrix rock." Volume averages for hydrogeologic data and thermodynamic parameters are considered separately for the two domains. The fractures are viewed as a porous continuum that carries the global flow in the reservoir and is characterized by customary porous medium-type parameters (absolute and relative permeability, porosity, capillary pressure, and compressibility). The matrix blocks provide storage and exchange fluid with the fractures locally. This "interporosity flow" is assumed to be "quasi-steady," occurring at rates that are proportional to the difference in fluid pressures. A schematic illustration of the double-porosity method is given in Figure 1a.

The early double-porosity work emphasized single-phase flow and closed-form analytical solutions, while later developments used numerical simulation to study processes such as water flooding of fractured petroleum reservoirs, where water injected into the fracture system is imbibed into matrix blocks by capillary force, expelling oil [e.g., *Kazemi et al.*, 1976, 1989, 1993; *Thomas et al.*, 1983]. It was recognized that under some conditions, for example in the gas cap of a fractured petroleum reservoir, there may be capillary continuity between matrix blocks, and global flow of the wetting phase may proceed via the matrix continuum [*Firoozabadi and Ishimoto*, 1994]. This led to an extension of the double-porosity approach, commonly referred to as "dual permeability model" (abbreviated "DKM;" Figure 1b), where global flow may occur in both fracture and matrix continua. In some cases the characteristic length of time for fracture-matrix exchange can be very long, so that the "quasi-steady" approximation for interporosity flow is no longer valid. This may occur when fracture spacing is large and/or when diffusivity in the matrix continuum is small. Examples for the latter include systems with tight matrix blocks, multiphase flows with large compressibility and/or large relative permeability changes, heat exchange between matrix and fractures, and diffusive migration of solutes.

When perturbations in the fracture system slowly invade the matrix blocks, it is necessary to resolve the temporal evolution of the gradients (of pressure, saturation, temperature, component concentrations) that drive flow at the matrix-fracture interface. This can be accomplished with the method of "multiple interacting continua" (MINC) [*Pruess and Narasimhan*, 1982, 1985], which partitions matrix blocks into several continua based on the distance of matrix material from the fractures (Figure 1c). Exchange between these matrix continua is usually treated by numerical simulation, although analytical and semi-analytical methods can also be used in certain cases. When implemented through a numerical approach, the MINC method can deal with nonlinear multiphase and non-isothermal processes, as well as with systems in which hydrogeologic properties of the matrix rock may not be homogeneous, but may change as a function of distance from the fractures [*Xu et al.*, 1999].

The double-porosity method and its extensions has primarily been used for studies of oil recovery mechanisms and geothermal production-injection operations in idealized systems, while site-specific modeling of actual fractured reservoirs has usually employed single porous medium approaches. Modeling studies of flow mechanisms have given interesting insights into the interplay between global fracture flow and local fracture-matrix exchange. For example, it was found that there is a general tendency for global flow to compensate for perturbations in local flow, and vice versa, making these types of flow systems "forgiving" in terms of required accuracy of hydrogeologic parameters and numerical discretization schemes. As an example, consider cold water injection into the fracture network of a geothermal reservoir. Let us suppose that the rate at which heat is transferred from the matrix blocks to the fluid near the injection point is underpredicted in a model, either because fracture spacing was chosen inappropriately large, resulting in too small of a heat transfer area, or because of space truncation errors from coarse numerical discretization that underestimates the initially large temperature gradients near the matrix block surfaces. The injected fluid will then have too low a temperature as it sweeps past downstream matrix blocks, which will tend to enhance heat transfer from these blocks, compensating for the upstream errors [*Pruess and Wu*, 1993]. Similar arguments apply for fracture-matrix exchange of fluids or chemical constituents, indicating a relative insensitivity to changes in the fracture-matrix interaction. This has both favorable and unfavorable aspects. It reduces the need for very detailed characterization data, making model predictions more robust, but it also limits the accuracy with which the modeler is able to determine in situ conditions.

3. ABSOLUTE AND RELATIVE PERMEABILITY

For a parallel-plate fracture with aperture b, absolute permeability is $k_f = \frac{b^2}{12}$, so that single-phase flow rate Q for a given pressure gradient is proportional to the cube of the aperture, $Q \propto b^3$ ("cubic law") [*Witherspoon et al.*, 1980]. Many studies have shown that the idealized parallel-plate model is inadequate for understanding flow and transport behavior of fractures on a field scale. The effective fracture aperture as determined from tracer tests can exceed the "hydraulic" aperture derived from pressure drop in viscous flow by as much as 2 to 3 orders of magnitude [*Neretnieks*, 1993]. The large deviation from the cubic law arises from the spatial variability of apertures in real rough-walled fractures. Generally speaking, it is the small apertures (bottlenecks) that control permeability, while it is the large apertures that contribute most to the void volume to be swept by solute tracer. Thus, fracture permeability and fracture aperture, in the sense of void volume per unit fracture wall area, are essentially independent parameters for real rough-walled fractures.

Modeling of multiphase flow behavior with DPM, DKM, or MINC approaches requires specification of absolute and relative permeabilities for a continuum formed by many intersecting fractures. From the mid-60s to the mid-80s the prevailing view in the petroleum literature was that, for fractures, relative permeabilities of wetting and nonwetting phases should sum to 1 regardless of saturation, $k_{rw} + k_{rn} \approx 1$. Often, the even more sweeping assumption was made that relative permeabilities should be equal to the respective phase saturations, $k_{rw} \approx S_w$, $k_{rn} \approx S_n$ (the so-called "X-curves"). These notions about fracture-relative permeabilities can be traced back to laboratory experiments by *Romm* (1966). Romm's experiments used artificial fracture assemblies of parallel plates lined with sheets of celluloid and polyethylene film, or waxed paper, which tended to minimize interference between the flowing phases. Experimental and theoretical work during the last ten years has questioned whether simplistic notions of "fracture-relative permeabilities" are applicable to realistic, rough-walled natural fractures, although the issue remains far from settled at the present time.

Considerable efforts have been made to determine permeability characteristics of individual fractures and of fracture networks in two and three dimensions, for both single-phase [*Long et al.*, 1982] and multiphase conditions [*Pruess and Tsang*, 1990; *Kwicklis and Healey*, 1993; *Karasaki et al.*, 1994]. Studies of individual fractures generally have employed a conceptualization of fractures as two-dimensional heterogeneous media (see below), while the fracture network studies have specified parameters such as spacing, length, orientation, and permeability of individual fractures by means of stochastic distributions. In either case, flow and transport behavior is predicted from postulated geometric characteristics, which conceptually is a very straightforward approach. It should be pointed out, however, that the aspects of fractured media that are most important for flow and transport behavior, namely, fracture connectivity and areal coverage of flow in the fracture plane, tend to be elusive in field observations. It appears that the geometry-based approach is more useful for gaining conceptual insight than for representing flow and transport behavior at specific sites.

The single-phase work has clarified the interplay between geometric characteristics of the fracture network (spacing, length, orientation) and permeability, and the approach to porous medium-like behavior for well-connected networks. The multiphase studies have considered the relative permeability of individual fractures, or fracture networks, to two phases flowing simultaneously. It was found that interference between phases is strong, causing the sum of wetting and nonwetting phase-relative permeabilities to be small at intermediate saturations [*Pruess and Tsang*, 1990]. This was confirmed in laboratory experiments [*Persoff and Pruess*, 1995], and is consistent with insights gained from percolation theory for the connectivity of two-dimensional lattices. However, the issue of "fracture-relative permeabilities" remains controversial. In a recent paper, *Horne et al.* [in press] presented steam-water flow experiments in fractures assembled from roughened glass plates, and stated that observations could only be matched by simulation when X-curve relative permeabilities were used.

Most theoretical and experimental studies have examined individual fractures on a relatively small scale, and their practical implications are not clear. For field-scale flow processes it is conceivable that the wetting phase may flow in the "small" fractures and the nonwetting phase in the "large" fractures, with minimal phase interference. In other words, the problem of two-phase flow in individual fractures may not be relevant to multiphase flow behavior in a field-scale fracture network. Simultaneous flow of two phases in a single fracture, if it does occur, may take place primarily at high rates in large fractures that feed wellbores; the quasi-static capillary-based phase occupation scheme postulated in the mathematical modeling of fracture-relative permeability may not apply for these conditions. For water seepage through fractured unsaturated zones, permeability itself may be an irrelevant parameter. In fact, it has been shown that in unsaturated fractured media the rate at which seeps advance downward may be larger in media with a smaller average permeability (see below).

4. HIGH-RESOLUTION FINITE DIFFERENCES

In thick unsaturated zones in fractured rocks of (semi-) arid regions, water seepage may proceed through highly localized preferential pathways. In such systems, much of the fracture volume does not participate in flow and large-scale volume averages may be completely meaningless. Continuum approaches may still be applicable to these systems, however, if applied on the actual scale where the flow processes occur. A key concept that has provided much useful insight into multiphase flow behavior is the view of fractures as "two-dimensional heterogeneous porous media." This conceptualization comes in two "flavors," a more microscopically oriented one in which a fracture is described in terms of a spatially variable aperture, and a more macroscopic model in which a fracture is represented by means of spatially variable permeability. The description in terms of apertures is appropriate for fundamental studies of flow on small spatial scales (on the order of 10^{-3} to 10^{-1} m). When using this approach investigators have either attempted to approximately solve the Navier-Stokes or Reynolds equations in the irregular pore space [*Brown*, 1987; *Glover et al.*, 1998], or have introduced the simplifying assumption that the fracture can be represented locally by a parallel plate model, so that flow can then be described by Darcy's law [*Pruess and Tsang*, 1990].

For flow processes on a somewhat larger scale, fractures are discretized into subregions of order 0.1 m or larger, and the customary continuum concepts of absolute and relative permeability and capillary pressure are applied [*Pruess*, 1998]. Justification for this is provided by laboratory experiments that have shown that, for "slow" flows in "small" fractures, continuum concepts are indeed applicable on a scale of order 0.1 m [*Persoff and Pruess*, 1995]. An areally extensive fracture is modeled as consisting of spatially correlated subregions with different permeability and capillary pressure characteristics (Figure 2). Aspects of heterogeneity in the fracture plane that are believed to be essential for replicating natural features include (a) regions of zero permeability, representing asperity contacts where the fracture walls touch; (b) a more or less gradual change towards larger apertures away from the asperities; (c) finite spatial correlation length for permeability; and (d) nonzero irreducible water saturation, representing water films held by capillary force in fracture wall roughness [*Tokunaga and Wan*, 1997]. In the continuum approach, effects of water held by capillarity and adsorption on fracture walls can be modeled by means of appropriate relative permeability and suction pressure relationships.

Simulation studies of water seepage in synthetic fractures with highly resolved heterogeneity have produced useful insights into hydrogeologic mechanisms in thick unsaturated zones in fractured rock. Fracture flow was found to proceed not in smooth sheets, but in dendritic patterns along localized preferential paths, giving rise to such features as ponding and bypassing (Figure 2). As long as fluxes are small compared to saturated hydraulic conductivity, unsaturated seepage may be dominated by flow funneling into localized pathways, due to subhorizontal barriers that may be formed by asperity contacts or fracture terminations. Flow funneling effects and localized seepage flux will increase with increasing length of subhorizontal barriers, while average vertical fracture permeability, as could be measured by monitoring

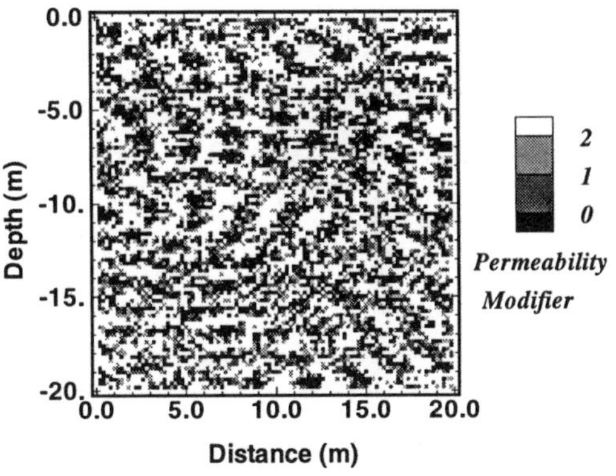

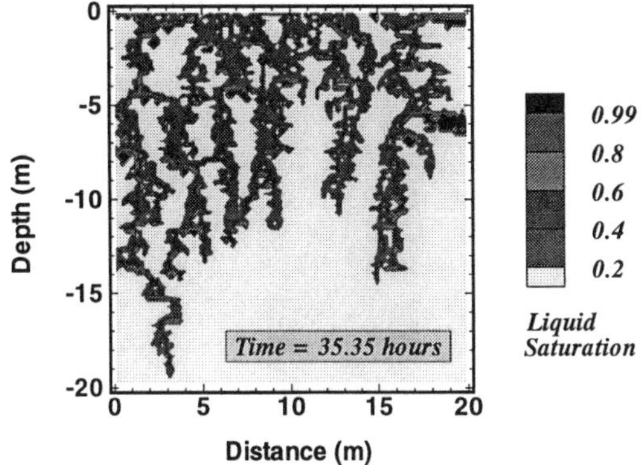

Figure 2. Stochastic permeability field (left) and seepage pattern (right) at the time of breakthrough at a depth of –19.5 m for water injection at a constant rate of 10^{-3} kg/s over the entire top of the fracture.

the propagation of gas pressure disturbances, would decrease [*Pruess*, 1999]. This is illustrated in Figure 3, which shows simulated saturation distributions for water injected uniformly over a 10-m-wide region at the top of a subvertical fracture (tilt angle of 80°). For a homogeneous fracture, water breakthrough at a 100-m depth occurs after 456.0 days. When a sloping impermeable obstacle is introduced into the fracture, flow is funneled into a narrow region, resulting in larger fluxes and accelerated breakthrough. At the same time, average permeability in the vertical direction becomes smaller when a longer obstacle is placed into the fracture. Thus, we have the remarkable situation in which unsaturated seepage can actually proceed faster in media with a lower average permeability (Figure 4). This seemingly paradoxical result emphasizes aspects that are unique to unsaturated flow in fractured media, and suggests that "average permeability" may not be a meaningful parameter for this process.

WATER INJECTION INTO VAPOR-DOMINATED GEOTHERMAL RESERVOIRS

Extensive steam production from the fractured vapor-dominated geothermal reservoirs at Larderello, Italy, and The Geysers, California, has caused a decline of reservoir pressures and well flow rates, and has led to an underutilization of installed electric generating capacity. These reservoirs are beginning to run out of fluid, while heat reserves in place are still enormous. Vapor-dominated geothermal reservoirs are naturally water-short systems. Fluid reserves tend to get depleted during exploitation much more quickly than heat reserves. Injection of water is the primary means by which dwindling fluid reserves can be replenished, and field life and energy recovery be enhanced.

Injection of cold water into homogeneous porous vapor zones entails rock-fluid heat transfer on a local (grain) scale, which is a rapid process, so the approximation of instantaneous local equilibrium between rocks and fluids is well justified. The process involves partial vaporization of the injected water and gives rise to two sharp fronts, a phase front at a temperature T_f, less than original reservoir temperature T_{res}, where conditions change from single-phase liquid to superheated steam and, closer to the injection point, a thermal front at which the temperature jumps from injection temperature T_{inj} to T_f [*Pruess et al.*, 1987]. When cold water is injected into hot fractures, heat transfer from the rocks to the fluids occurs slowly (conductively). Instead of sharp fronts, we then obtain very broad zones where fluid temperatures and saturations change gradually. In subvertical fractures, injection plumes evolve through a complex interplay of heat transfer, boiling and condensation phenomena, gravity effects, and two-phase flow. Vaporization dominates in the hotter portions of the plume, away from the injection point, while vapor tends to flow towards cooler lower-pressure regions near the injection point where it condenses. The counterflow of liquid away from the injection point and the flow of vapor towards it constitute a very efficient heat transfer system known as heat pipe, which tends to diminish temperature variations throughout the injection plume. Because vapor has a much lower density than liquid water, it has larger kinematic viscosity and acts as the more viscous fluid. Very considerable vapor pressure gradients may be generated during vaporization, which may be comparable in magnitude to a gravitational body force on the liquid, providing a mechanism for lateral flow of liquid, with associated potential for early breakthrough at neighboring production wells [*Pruess*, 1997].

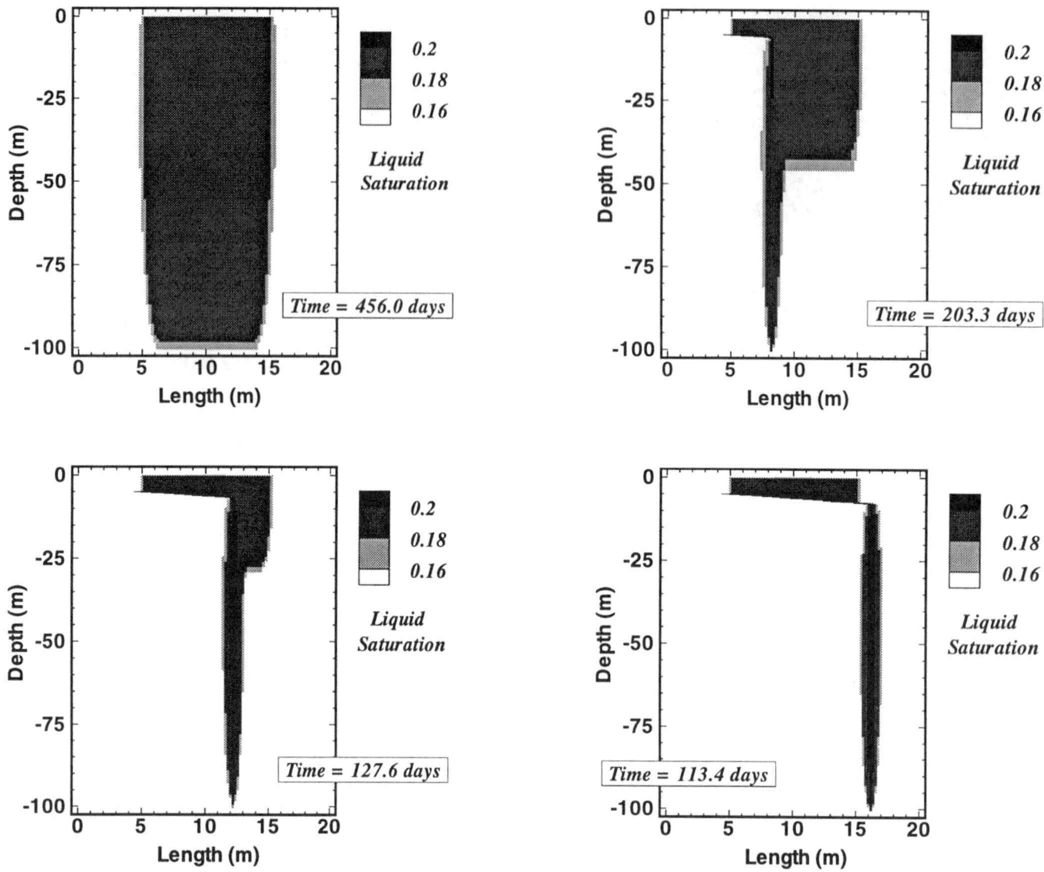

Figure 3. Simulated water saturations at time of breakthrough at a –100-m depth for seepage in subvertical (80°) homogeneous fractures with a single embedded subhorizontal obstacle of rectangular shape and variable length. The impermeable obstacle starts at the left boundary at a depth of –4 m and slopes downward to the right at an angle of 14.0°. Different cases were simulated in which the length interval blocked by the obstacle was l = 0 (no obstacle, top left), l = 8 m (top right), l = 12 m (bottom left), and l = 16 m (bottom right).

6. NUCLEAR WASTE DISPOSAL

Mathematical models have been extensively used in investigations of the thick (≈600 m) fractured vadose zone at Yucca Mountain as a potential site for a high-level nuclear waste repository. Numerical simulations of flow and transport at Yucca Mountain have generally emphasized large-scale spatial averages, and have employed fracture continuum approaches, such as DPM, DKM, MINC, and single effective continuum models (ECM) [*Wu et al.*, 1999]. High-resolution models with explicit discretization of fractures have also been used to study basic mechanisms of fluid and heat flow in this unusual hydrogeologic environment [*Birkholzer and Tsang*, 1998]. Large-scale volume-averaged models have been very successful at describing the propagation of barometric or artificial pressure pulses, and for describing temperature evolution during heater tests. Gratifying as this success is, it is not unexpected. This is because gas flow and heat conduction, being described by parabolic partial differential equations, are subject to strong internal averaging mechanisms. Water seepage in fracture networks at rates far below saturated hydraulic conductivity, however, is described by a hyperbolic PDE. In this case internal averaging mechanisms are essentially absent, and volume averages are not enforced through physical processes in the flow system, but are formal constructs of the analyst. Predictions of water seepage from volume-averaged continuum models must therefore be interpreted with a great deal of caution.

Recent observations of environmental tracers at Yucca Mountain have provided direct evidence that water can

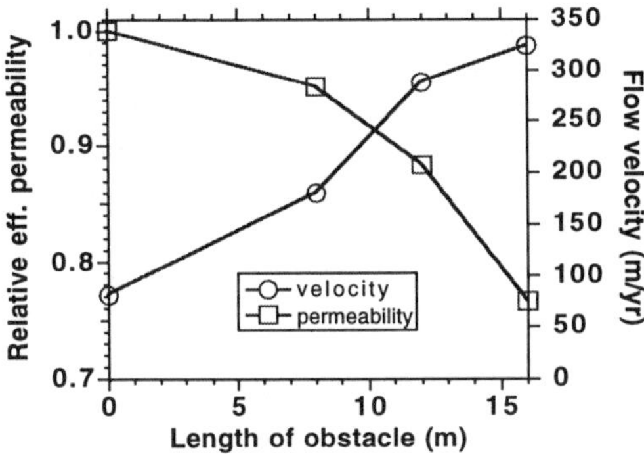

Figure 4. Average flow velocity for seepage in fractures with embedded obstacles of different length, as shown in Figure 3. Effective vertical permeability for single-phase flow is also plotted, and is seen to decrease with length of obstacle, while average water seepage velocity increases.

flow through unsaturated fractured rocks over vertical distances of several hundred meters at velocities of an order of 10 m/yr or more [*Yang et al.*, 1995; *Fabryka-Martin et al.*, 1996]. These observations came as a surprise to many, even though early work by U.S. Geological Survey scientists had suggested that the well-connected fracture network at Yucca Mountain would provide pathways that could rapidly drain away episodic infiltration [*Roseboom*, 1983]. Capillary theory would appear to suggest that the strong suction from the unsaturated rock matrix, of order $P_m \approx -3$ bar, at Yucca Mountain [*Rousseau et al.*, 1997] would quickly remove free water from the fractures, making it difficult to understand "how fractures could remain sufficiently saturated to act as fast paths in the face of high matrix suction" [*Cook*, 1991]. Matrix imbibition indeed would be a very strong process if water were flowing down fractures in the form of area-filling sheets [*Nitao and Buscheck*, 1991]. However, recent mathematical modeling has demonstrated and quantified several mechanisms that could drastically diminish water imbibition into the rock matrix, including (1) funneling of flow into localized preferential pathways, which reduces the wetted area where imbibition occurs; (2) the episodic nature of infiltration, which reduces the time available for removing water from the fractures; and (3) mineral coatings on fracture walls, which reduce imbibition fluxes. Based on numerical simulation experiments, it was suggested that the total wetted fracture-matrix interface area is comparable in magnitude to the land surface area beneath which it is present, and that spacing between major localized seeps may be on the order of 50 m or more [*Pruess*, 1999]. Flow of liquid films held on rough fracture walls may provide a mechanism for fast transport [*Tokunaga and Wan*, 1997], but total percolation flux carried in films is expected to be very small under the suction conditions that prevail at Yucca Mountain [*Pruess*, 1999].

Emplacement of heat-generating high-level nuclear wastes in thick unsaturated zones of fractured tuff at Yucca Mountain would give rise to complex multiphase fluid flow and heat transfer processes. For strongly heat-driven flows, water held in tight matrix pores will be vaporized as temperatures approach and exceed the boiling point at ambient pressures. The resulting pressurization will expel vapor from the matrix, which subsequently will flow away from the heat sources in the fracture network. Upon encountering cooler wall rock, the vapor will condense. The condensate will flow in the fractures under the combined action of gravity, pressure, and capillary pressure forces, and be partially imbibed into the rock matrix. Numerical simulations of this process using simplified geometric descriptions, large-scale volume averaging, and more or less homogeneous media have predicted that over time the rock in the vicinity of the heat sources will dry out [*Buscheck and Nitao*, 1993; *Pruess and Tsang*, 1993]. This observation has led some workers to propose a repository concept called "extended dry," in which high thermal loading would be used to effectively protect waste packages from being contacted by liquid water. However, critics have pointed out that liquid water can migrate considerable distances through fractured rock that is at above-boiling temperatures and be only partially vaporized [*Pruess and Tsang*, 1994; *Pruess*, 1997]. An added concern is that large repository heat loads would increase rates of vaporization and condensate formation, thereby promoting nonequilibrium matrix-fracture flow effects that could enhance localized and intermittent water flow near the waste packages.

7. SCALING RELATIONSHIPS

Laboratory experiments using transparent replicas of natural rock fractures, or artificial fracture assemblies made from materials such as roughened glass plates, have provided much useful qualitative and quantitative insight into multiphase flows under ambient conditions [*Su et al.*, 1999], as well as under thermal drive [*Kneafsey and Pruess*, 1998]. However, the significance of flow phenomena on a laboratory scale for the much larger spatial dimensions in field-scale problems is uncertain. Mathematical models can be very useful for evaluating relationships between flow processes on different space and time scales. Let us consider a plane heterogeneous fracture, with coordinates x in the horizontal and z in the (sub-) vertical direction. Applying the following simultaneous transformation of space and time coordinates

$$\begin{aligned} t \to t' &= \lambda_t \cdot t \\ x \to x' &= \lambda_x \cdot x \ , \\ z \to z' &= \lambda_z \cdot z \end{aligned} \tag{1}$$

it can be shown that the Richards' equation for unsaturated flow in the fracture remains approximately invariant when

$$\lambda_t = \lambda_x^2 = \lambda_z \,. \tag{2}$$

Rates of external sinks/sources scale by λ_x. Thus, (sub-) vertical-length scale and time need to be stretched by the square of the horizontal scale factor. The validity of the scaling invariance given in Equation (2), as well as its limits of applicability, were confirmed by numerical simulation [*Pruess*, 1998]. It is even possible to obtain an approximate scaling invariance for vaporizing water flow down hot rock fractures. In addition to the relations given in Equation (2), this requires scaling of the thermal diffusivity of the wall rock by a factor $\lambda_\theta = 1/\lambda_z$. It may appear as though this approximate scaling invariance has little practical value, involving as it does a scaling of thermal diffusivity, which for rocks is a material parameter with little if any systematic dependence on scale. However, vaporization behavior in a "large" rock fracture in the field could be replicated through smaller-scale experiments in the laboratory if different fracture wall materials with larger thermal diffusivities were used. For example, the thermal diffusivity of cast iron is approximately 10 times larger than that of typical hard rocks, so that vaporization and flow behavior in a laboratory fracture of $x = 1$ m, $z = 1$ m size in cast iron should be similar to that of a rock fracture of $x = \sqrt{10} = 3.16$ m, $z = 10$ m size with a 3.16 times larger water injection rate and a 10 times slower time scale.

8. CONCLUDING REMARKS

Fractured flow systems have received increasing attention during the last several decades. They exhibit a tremendous diversity of fracture and rock matrix properties and flow and transport processes. Early work emphasized applications to oil and gas reservoirs and large-scale volume averaged approaches. More recent studies have often been motivated by applications related to waste disposal, and to environmental protection and remediation, which typically involve higher spatial resolution of small-scale processes. Depending on the nature of the fractured flow system under study, and the engineering or geoscientific interest and purpose in dealing with the system, different approaches will be employed for characterization and modeling. It is well to remember that models represent idealizations and simplifications of real systems, and their formulation (governing equations, system parameters) typically is valid only for certain space and time scales, and for a limited range of physical, chemical, etc., conditions. These limitations are seldom made explicit in the formulation of models; in fact, they may often be poorly known. At best, mathematical models may be able to identify and quantify the key processes and parameters that determine the behavior of the flow system under study, for the conditions and space and time scales of interest to the analyst. For a mathematical modeling effort to be successful, perhaps the single most important prerequisite is to have very clear and specific objectives.

Most modeling approaches rely on volume averaging to some extent. This works well for systems in which physical averaging mechanisms are present, usually described by parabolic partial differential equations (PDEs). This includes diffusive processes such as heat conduction, gas flow in unsaturated zones, capillary-driven liquid flow, and molecular diffusion of solutes. Volume averaging can generate misleading results when internal averaging mechanisms are absent or weak (hyperbolic PDEs), as in episodic water seepage through highly permeable fracture networks in thick unsaturated zones. The absence of internal averaging mechanisms greatly complicates flow modeling.

Much useful insight into multiphase flow behavior and mechanisms in fractured formations has been gained through the study of idealized systems. Examples include oil recovery from fractured reservoirs through water- and steam-flooding, injection of cold water into fractured geothermal reservoirs, and water seepage in unsaturated rock fractures. Applications to site-specific predictive modeling have been more difficult to achieve, as they raise difficult issues of characterization and model calibration, and applicability of conceptualizations for processes operating on different space and time scales. A general problem with modeling of flow in fractured media arises from the geometric complexity of individual fractures and fracture networks. Fracture geometry on different scales is a very natural starting point for flow and transport modeling, but geometric features that are crucial for flow behavior, such as fracture connectivity, are very difficult to determine in the field.

The presence of fractures generally makes flow and transport behavior more complex than it is in homogeneous porous media, but fractures also allow for some unique simplifications. For example, flow in "small" fractures in hard rocks with low matrix permeability is essentially a two-dimensional process. Compared to three-dimensional porous media, flows in two-dimensional heterogeneous fractures are more easily modeled mathematically and are more amenable to direct observation and visualization on a laboratory scale. Thus, fractures can provide convenient systems for learning about flow in more general heterogeneous porous media.

Acknowledgments. The author appreciates comments and suggestions made by Yushu Wu, Boris Faybishenko, and two anonymous reviewers. This work was supported, in part, by the Assistant Secretary for Energy Efficiency and Renewable Energy, Geothermal Division, and by the Director, Office of Energy Research, Office of Health and Environmental Sciences, Biological and Environmental Research Program, of the U.S. Department of Energy under Contract No. DE-AC03-76SF00098.

REFERENCES

Barenblatt, G. E., I. P. Zheltov, and I. N. Kochina, Basic concepts in the theory of seepage of homogeneous liquids in fissured rocks, *J. Appl. Math, 24* (5), 1286–1303, 1960.

Bear, J., C. F. Tsang, G. de Marsily (eds.), *Flow and Contaminant Transport in Fractured Rock,* Academic Press, San Diego, Calif., 1993.

Birkholzer, J. and C. F. Tsang, Solute channeling in unsaturated heterogeneous porous media, *Water Resour. Res., 33*(10), 2221–2238, 1997.

Brown, S. R., Fluid flow through rock joints: The effects of surface roughness, *J. Geophys. Res., 92*(B2), 1337–1347, 1987.

Buscheck, T. A., and J. J. Nitao, The analysis of repository-heat-driven hydrothermal flow at Yucca Mountain, *Proceedings, Fourth High Level Radioactive Waste Management International Conference, Las Vegas, Nev., April 26–30, 1993,* 1993.

Cook, N. G. W., I. Javandel, J. S. Y. Wang, H. A. Wollenberg, C. L. Carnahan, K. H. Lee, *A Review of Rainer Mesa Tunnel and Borehole Data and Their Possible Implications to Yucca Mountain Study Plans*, Lawrence Berkeley Laboratory Report LBL-32068, Berkeley, Calif., December 1991.

Fabryka-Martin, J., A. V. Wolfsberg, P. R. Dixon, S. Levy, J. Musgrave, and H. J. Turin, *Summary Report of Chlorine-36 Studies: Sampling, Analysis and Simulation of Chlorine-36 in the Exploratory Studies Facility,* Los Alamos National Laboratory Report LA-CST-TIP-96-002, Los Alamos, New Mex., August 1996.

Firoozabadi, A., and K. Ishimoto, Theory of reinfiltration in fractured porous media: Part I—One-dimensional model, *Advanced Technology Series, 2*(2), 35–44, Society of Petroleum Engineers, Richardson, Tex., 1994.

Glass, R. J., Modeling gravity-driven fingering in rough-walled fractures using modified percolation theory, *Fourth Annual International High-Level Radioactive Waste Management Conference, Las Vegas, Nev.*, pp. 2042– 2052, American Nuclear Society, La Grange Park, Ill, 1993.

Glover, P. W. J., K. Matsuki, R. Hikima, and K. Hayashi,. fluid flow in synthetic rough fractures and application to the Hachimanti geothermal hot dry rock test site, *J. Geoph. Res.,. 103*(B5), 9621–9635, 1998.

Horne, R. N., C. Satik, G. Mahiya, K. Li, W. Ambusso, R. Tovar, C. Wang, and H. Nassori, Steam-Water Relative Permeability, manuscript submitted for presentation at World Geothermal Congress 2000, Kyushu-Tohoku, Japan, Stanford University, Stanford, Calif., in press.

Karasaki, K., S. Segan, K. Pruess, and S. Vomvoris, A study of two-phase flow in fracture networks, *Proceedings, Fifth Annual International High-Level Radioactive Waste Management Conference, Las Vegas, Nev.*, Vol. 4, pp. 2633–2638, American Nuclear Society, La Grange Park, Ill., 1994.

Kazemi, M., L. S. Merrill Jr., K. L. Porterfield, and P. R. Zeman, Numerical simulation of water-oil flow in naturally fractured reservoirs, *Soc. Pet. Eng. J.,* 317–326, 1976.

Kazemi, H. and J. R. Gilman. Multiphase flow in fractured petroleum reservoirs, Proceedings, Advanced Workshop on Heat and Mass Transport in Fractured Rocks, Laboratorio Nacional de Engenharia Civil (LNEC), Lisbon, Portugal, June 1989, 1989.

Kazemi, H. and J. R. Gilman. Multiphase flow in fractured petroleum reservoirs, in *Flow and Contaminant Transport in Fractured Rock*, edited by J. Bear, C. F. Tsang, and G. de Marsily , pp. 267–323, Academic Press, San Diego, Calif., 1993.

Kneafsey, T. J., and K. Pruess, Laboratory experiments on heat-driven two-phase flows in natural and artificial rock fractures, *Water Resour. Res., 34*(12), 3349–3367, 1998.

Kwicklis, E. M., and R. W. Healy, Numerical investigation of steady liquid water flow in a variably saturated fracture network, *Water Resour. Res., 29*(12), 4091–4102, 1993.

Long, J. C. S., J. S. Remer, C. R. Wilson, and P. A. Witherspoon, Porous media equivalents for networks of discontinuous fractures, *Water Resour. Res., 18*(3), 645–658, 1982.

Morse, P. M., and H. Feshbach. *Methods of Theoretical Physics*, McGraw-Hill, New York, 1953.

Narasimhan, T. N., Physics of saturated-unsaturated subsurface flow, in *Recent Trends in Hydrogeology*, edited by T. N. Narasimhan, Special Paper 189, The Geological Society of America, Boulder, Colo., 1982a.

Narasimhan, T. N., Multidimensional numerical simulation of fluid flow in fractured porous media, *Water Resour. Res., 18*(4), 1235–1247, 1982b.

National Research Council, *Rock Fractures and Fluid Flow,* National Academy Press, Washington, D.C., 1996.

Neretnieks, I., Solute transport in fractured rock—Applications to Radionuclide Waste Repositories, in *Flow and Contaminant Transport in Fractured Rock* , edited by J. Bear, C. F. Tsang, and G. de Marsily, pp. 39–127, Academic Press, San Diego, Calif., 1993.

Nitao, J. J., and T. A. Buscheck, Infiltration of a liquid front in an unsaturated, fractured porous medium, *Water Resour. Res., 27*(8), 2099–2112, 1991.

Persoff, P., and K. Pruess, Two-Phase Flow Visualization and relative permeability measurement in natural rough-walled rock fractures, *Water Resour. Res., 31*(5), 1175–1186, 1995.

Pruess, K., On vaporizing water flow in hot sub-vertical rock fractures, *Transport in Porous Media, 28*, 335–372, 1997.

Pruess, K., on water seepage and fast preferential flow in heterogeneous, unsaturated rock fractures, *J. Contam. Hydr., 30*(3–4), 333-362, 1998.

Pruess, K., A Mechanistic model for water seepage through thick unsaturated zones in fractured rocks of low matrix permeability, *Water Resour. Res., 35*(4), 1039–1051, 1999.

Pruess, K., B. Faybishenko, and G. S. Bodvarsson, Alternative concepts and approaches for modeling unsaturated flow and transport in fractured rocks, *J. Contam. Hydr., 38*(1-3), 281–322, 1999.

Pruess, K., and T. N. Narasimhan, On fluid reserves and the production of superheated steam from fractured, vapor-dominated geothermal reservoirs, *J. Geophys. Res., 87*(B11), 9329–9339, 1982.

Pruess, K., and Y. Tsang, Modeling of strongly heat-driven flow processes at a potential high-level nuclear waste repository at Yucca Mountain, Nevada, *Proceedings, Fourth International High Level Radioactive Waste Management Conference, Las Vegas, NV, April 26–30, 1993*, 1993.

Pruess, K., and T. N. Narasimhan, a practical method for modeling fluid and heat flow in fractured porous media, *Soc. Pet. Eng. J., 25*(1), 14–26, 1985.

Pruess, K., C. Calore, R. Celati, and Y. S. Wu, An analytical solution for heat transfer at a boiling front moving through a porous medium, *Int. J. of Heat and Mass Transfer, 30*(12), 2595–2602, 1987.

Pruess, K., and Y. W. Tsang, On two-phase relative permeability and capillary pressure of rough-walled rock fractures, *Water Resour. Res., 26*(9), 1915–1926, 1990.

Pruess, K., and Y. S. Wu, A new semianalytical method for numerical simulation of fluid and heat flow in fractured reservoirs, *SPE Advanced Technology Series, 1*(2), 63–72, 1993.

Pruess, K. and Y. Tsang, *Thermal Modeling for a Potential High-Level Nuclear Waste Repository at Yucca Mountain, Nevada*, LBL-35381, Lawrence Berkeley National Laboratory, Berkeley, Calif., 1994.

Romm, E. S., *Fluid Flow in Fractured Rocks*, (translated by W. R. Blake, Bartlesville, Okla., 1972), Nedra Publishing House, Moscow, Russia, 1966.

Roseboom, E. H., *Disposal of High-Level Nuclear Waste Above the Water Table in Arid Regions*, Circular 903, U. S. Geological Survey, Denver, Colo., 1983.

Rousseau, J. P., E. M. Kwicklis, and D. C. Gillies (eds.), *Hydrogeology of the Unsaturated Zone, North Ramp Area of the Exploratory Studies Facility, Yucca Mountain, Nevada*, Water Resources Investigations Report 98-4050, U.S. Geological Survey, Denver, Colo., 1997.

Stockman, H. W., C. H. Li, and J. L. Wilson, A lattice-gas and lattice Boltzmann study of mixing at continuous fracture junctions: importance of boundary conditions, *Geoph. Res. Lett., 24*(12), 1515–1518, 1997.

Su, G., J. T. Geller, K. Pruess, and F. Wen, Experimental studies of water seepage and intermittent flow in unsaturated, rough-walled fractures, *Water Resour. Res., 35*,(4), 1019–1037, 1999.

Thomas, L. K., T. N. Dixon, and R. G. Pierson, Fractured reservoir simulation, *Soc. Pet. Eng. J.*, 42–54, 1983.

Tokunaga, T. K., and J. Wan, Water film flow along fracture surfaces of porous rock, *Water Resour. Res., 33*(6), 1287–1295, 1997.

Warren, J. E., and P. J. Root, The behavior of naturally fractured reservoirs, *Soc. Pet. Eng. J., Transactions, AIME, 228*, 245–255, 1963.

Witherspoon, P. A., J. S. Y. Wang, K. Iwai, and J. E. Gale, Validity of cubic law for fluid flow in a deformable rock fracture, *Water Resour. Res., 16*(6), 1016–1024, 1980.

Wu, Y. S., C. Haukwa, and G. S. Bodvarsson, A site-scale model for fluid and heat flow in the unsaturated zone of Yucca Mountain, Nevada, *J. Contam. Hydr.*, in press.

Xu, T., S. P. White, K. Pruess, and G. Brimhall, Modeling of pyrite oxidation in saturated and unsaturated subsurface flow systems, *Transport in Porous Media*, in press.

Yang, I. C., G. W. Rattray, and P. Yu, *Chemical and Isotopic Data and Interpretations, Unsaturated Zone Boreholes, Yucca Mountain, Nevada,* Water Resources Investigation Report, U.S. Geological Survey, Denver, Colo., 1995.

Karsten Pruess, Earth Sciences Division, Lawrence Berkeley National Laboratory, One Cyclotron Road, Berkeley, CA 94720

Physical Considerations in the Upscaling of Immiscible Displacements in a Fractured Medium

Y. C. Yortsos

Department of Chemical Engineering and Petroleum Engineering Program, University of Southern California, Los Angeles, California

The upscaling of two-phase flow in a fractured medium remains a problem still largely unresolved, despite its practical importance in many applications. In this paper, we present a physical approach to upscaling immiscible displacement and discuss the various regimes that may emerge as a function of process and structure variables. The relevance of these regimes to the various existing models—such as the equivalent continuum model (ECM), the double porosity model (DPM), or the dual permeability model (DKM)—and fracture-matrix interaction are emphasized. Open problems and potential approaches for their solution are outlined.

1. INTRODUCTION

Two-phase flow in a fractured medium is controlled by the structure of the medium and by three main forces: capillary, viscous, and gravitational. There are two key structural characteristics of a fractured medium: the existence of (1) multiple length scales and (2) multiple connectivity within the space occupied by the fluids.

Length scales include the pore scale of the matrix, the correlation length of the matrix, the aperture of the fracture, the correlation length of the fracture aperture, the fracture length (or, equivalently, the matrix block size), and the correlation length of the fracture network. Geometric quantities at these scales are, in general, characterized with probability density functions. Spatial correlations over many scales and fractal aspects (e.g., of the fracture aperture or the fracture network) are also not unusual [*Acuna and Yortsos*, 1995; *NRC*, 1996]. The different length scales give rise to different characteristic times for advection or diffusion, which can have important implications for upscaling.

Multiple connectivity is important with respect to percolation and transport features of displacements [*Feder*, 1988]. For a fractured medium, this includes the pore-network connectivity of the matrix, the fracture aperture, and the fracture network. Adjacent matrix blocks may be connected through bridges at asperity contacts. They can be important in providing capillary continuity for imbibition.

In many cases, large-scale anisotropy of the fracture network is prevalent. Fractured media are an extreme case of highly heterogeneous porous media (streaks, layers, etc.), and approaches used for the upscaling of heterogeneous media could be useful in the upscaling of fractured systems.

Two different types of immiscible flows in porous media exist: (1) transient displacements, classified as drainage or imbibition (primary or secondary) and (2) steady-state flows [*Dullien*, 1992]. There are important differences between these processes. Displacements involve the development of patterns [*Lenormand*, 1990], the propagation of fronts, and the existence of saturation/desaturation gradients. Steady-state flows are characterized by a mean steady-state saturation, around

Dynamics of Fluids in Fractured Rock
Geophysical Monograph 122

which fluctuations occur [*Constantinides and Payatakes*, 1996]. Although conventionally treated by simple macroscopic equations, steady-state immiscible flows are still not fully understood [*Valavanides and Payatakes*, 2000]. In this paper, we will discuss transient processes. Although these can be modeled with relative success, many unresolved questions still exist even for simple, relatively homogeneous media. For the sake of simplicity, here we will consider only primary displacements.

The forces controlling transient immiscible displacements in porous media at low rates can be expressed in dimensionless form using three parameters: the capillary number (Ca), the gravity Bond number (B_{gx}), and the viscosity ratio (M), namely:

$$Ca = \frac{q\mu_{nw}}{\gamma\cos\theta},\ B_{gx} = \frac{kg_x\Delta\rho}{\gamma\cos\theta},\ M = \frac{\mu_w}{\mu_{nw}}, \qquad (1)$$

which express the relative magnitude of viscosity or gravity over capillary forces, and the ratio in viscosities, respectively. In the above, q denotes flow velocity, μ is viscosity, γ is the interfacial tension between the two fluids, θ is the contact angle measured from the side of the wetting phase, k is permeability, g_x is the component of gravity in the direction, x, of displacement, $\Delta\rho$ is the density difference between displaced and displacing fluids, and subscripts nw and w denote nonwetting and wetting phases, respectively. The expressions within Equation (1) containing the contact angle θ should not be used near the limit $\theta = \pi/2$, because as θ approaches $\pi/2$, capillarity becomes negligible, which is not the case for noncylindrical pore surfaces. The Bond number, B_{gx}, can be positive (e.g., when a lighter fluid displaces a heavier fluid downdip) or negative (e.g., when a lighter fluid displaces a heavier fluid updip). In the following analysis, we will assume that gravity acts predominantly in the direction of displacement. Thus, gravity segregation, which can further complicate the process, will not be included [see *Zhang et al.*, 2000]. Together with the geometrical description of the medium, the three parameters in (1) fully characterize the displacement in terms of the distributions of the various quantities (pore size or permeability) at the various scales and correlation lengths.

The most common method for describing upscaled single-phase flow in fractured media makes use of double porosity continuum models (DPM) [e.g., see *Warren and Root*, 1963]. Under certain conditions, well-defined volume-averaged quantities (e.g., pressure, saturation, etc.) can be assigned to the two continua—fracture and matrix. Even though substantial progress has been made in modeling steady and transient single-phase flows in fractured media [e.g., *Arbogast*, 1988; *Noetinger and Estebenet*, 1998], including fractal fracture networks, the state of the art in modeling two-phase flows is still evolving. A variety of models exist. They include the equivalent (or effective) continuum model (ECM) [*Peters and Klavetter*, 1988], extensions or modifications of the double (or dual) porosity model [*de Swann*, 1978; *Arbogast et al.*, 1988; *Bourgeat and Panfilov*, 1998], the multiple interactive continua (MINC) model [*Pruess and Narasimhan*, 1985] and the dual permeability model (DKM). The first is based on the postulate that matrix and fractures are in capillary equilibrium. The van Genuchten-type expressions [*van Genuchten*, 1980] for relative permeability and capillary pressure, borrowed from soil science literature, are often used to construct large-scale averages for the effective continuum. The other models postulate Darcy-type equations for the fracture continuum (and the matrix continuum in the case of DKM) and an exchange term to describe their interaction. Discrete fracture network (DFN) models are also used. The applicability and relevance of these models for representing two-phase immiscible displacements in fractured media are not well defined. Specifically, the validity of these models for different problems has not been determined. Important issues, such as the effect of large-scale connectivity, and the nonlinearities and instabilities involved in two-phase flow, are also not properly recognized [see also *Glass et al.*, 1995].

The goal of this paper to further the discussion of key physical issues that must be considered in developing upscaled transient displacements in fractured media, specifically the various flow regimes that may develop as a function of process and structural variables. A similar approach was advocated in *Glass et al.* [1995]. The analysis is intended to provide a better understanding of the effect of nonlinearities and instabilities, associated with immiscible displacement, on the distribution of flowing phases in a fractured medium and on the upscaled models. Because many of these problems need additional investigation, directions for future research are also suggested.

To begin with, we will briefly review the state-of-the-art for upscaling immiscible displacements in porous media. Such a review will help delineate the key issues in the upscaling of two-phase flows in fractured systems and provide a common background. Then, we will discuss the upscaling of fractured media by considering two different cases: (1) an impermeable matrix—essentially a problem of displacement in a fracture network, and (2) the upscaling of displacements in fractured media with full matrix participation. Key to this analysis is the classification of the displacement process according to the relative magnitude of viscous and gravitational forces, and whether these forces act to stabilize or destabilize the displacement. Drainage and imbibition are separately analyzed. In all cases, displacements by injection at constant rate are considered.

2. BRIEF REVIEW OF UPSCALING ISSUES IN IMMISCIBLE DISPLACEMENTS IN POROUS MEDIA

In nonfractured media (which, in the context of this paper, also include a single matrix block or an isolated fracture plane), upscaling typically involves the passage from: (1) pore scale to macroscopic scale, namely from Stokes to Darcy flows [e.g., *Whitaker*, 1986], to characterize pore-scale heterogeneity; and (2) from macroscopic scale to a larger scale (for example, the gridblock size of typical field simulators) to characterize permeability heterogeneity [e.g., *Bourgeat*, 1984]. In either case, the upscaling process requires extending Darcy's law for single-phase flow to multiphase flow by introducing relative permeability and capillary pressure functions or equivalent pseudofunctions [e.g., *Durlofsky*, 1998] for the respective processes. A key issue is determining the dependence of the upscaled functions on the variables of flow processes, specifically on the phase saturations but also on the flow velocity or other parameters, in a form that may possibly depend on initial and boundary conditions (particularly for the case of pseudofunctions). In principle, knowledge of the pattern occupied by a given phase enables us to compute the conductance (and the relative permeability) of that phase. The problem of determining the relative permeabilities thus becomes one of determining the displacement patterns. In the general case, however, these two problems are coupled, since relative permeabilities will affect the displacement pattern itself. Furthermore, upscaling usually necessitates the introduction of volume-average quantities, independent of the size of the averaging volume (the representative elementary volume or REV). Thus, well-posed and rigorous upscaling procedures are possible only if the underlying pattern is stationary over the averaging volume (i.e., if there are no significant gradients of saturation over the REV). The latter also requires that the REV scale is sufficiently larger than the correlation length of the heterogeneity. Then, length scales are separated, and the local pattern becomes uncoupled from the global pattern [*Amaziane and Bourgeat*, 1988]. Under these conditions, rigorous upscaling procedures can be developed.

In immiscible displacement, such rigorous upscaling is possible when the process is locally controlled by capillarity. When capillarity dominates, the governing process is invasion percolation, based on which the displacement pattern can be constructed. In early works, *Heiba et al.* [1982, 1983] used ordinary percolation theory in Bethe lattices to obtain expressions for relative permeabilities and capillary pressures of two-phase and three-phase flows in homogeneous media [see also *Wilkinson*, 1984, 1986]. More realistically, invasion percolation (IP) [*Wilkinson and Willemsen et al.*, 1983] and invasion percolation with trapping (IPT) [*Dias and Wilkinson*, 1986, *Yortsos and Sharma*, 1986] should be used to describe these displacements. IP-based approaches can also account for situations where the pore structure is anisotropic [*Xu and Yortsos*, 1994] or has large-scale correlations [e.g., *Du et al.*, 1996]. When the assumption of capillary control over the microscale breaks down and viscous or gravitational forces are not negligible, the occupancy patterns will be affected, possibly giving rise to saturation gradients, becoming nonstationary. The conventional upscaling approach then becomes questionable. However, the validity of this problem was understood only recently, as discussed below.

In upscaling from the pore scale to a square pore network of size $N \times N$, in the absence of gravity, *Lenormand* [1989, 1990] identified the various limiting regimes that may develop. In the case of drainage they are IP, piston-like displacement (PD), and viscous fingering (VF) of the diffusion-limited aggregation (DLA) type. *Lenormand* [1989] also delineated the regions of their validity in the *Ca-M*-plane phase diagram. The boundaries separating these regimes were evaluated as a function of the network size, the capillary number, and the viscosity ratio (see below). *Yortsos et al.* [1997, 1998] and *Xu et al.* [1998] further extended this approach to 3-D, fully developed drainage, where lattice size is no longer a relevant parameter. They characterized fully developed displacement in terms of two global regimes: a stabilized displacement (SD) regime (which includes both IP and PD patterns), and a capillary viscous fingering (CVF) regime (Figure 1). The first regime consists of a traveling wave saturation profile, involving the front of a finite extent (where IP concepts apply) followed upstream by a compact pattern of the PD type. The CVF regime is characterized by viscous fingering at the pore-network scale, in general not of the DLA type. The two regimes are separated by a curve in the capillary number-viscosity ratio plane, which at small capillary numbers is a power law of the viscosity ratio (see Figure 1).

Now, the conventional concept of relative permeabilities

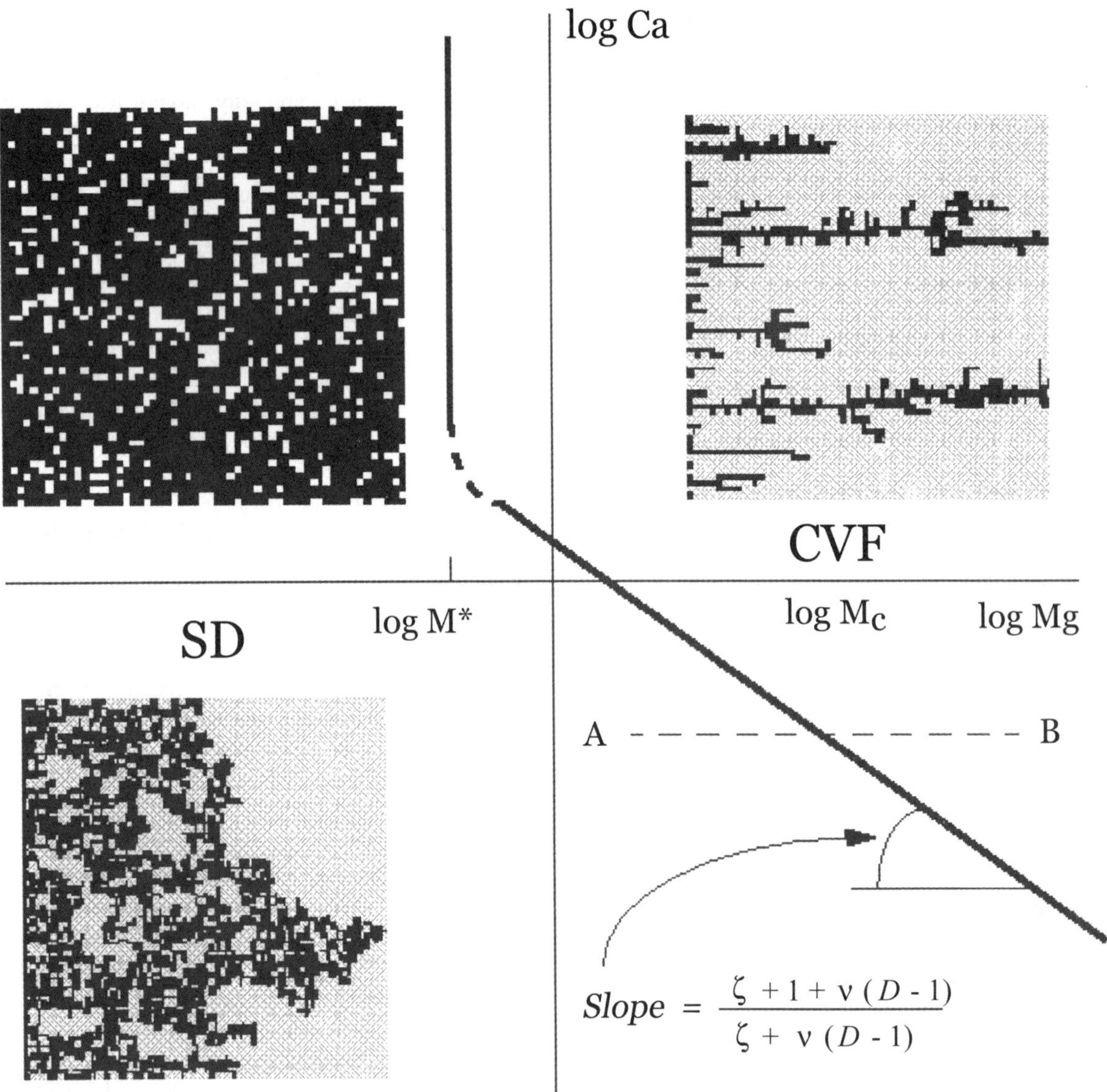

Figure 1. Phase diagram of fully developed drainage. From *Yortsos et al.* [1997].

dependent only on saturation is valid under conditions of an SD regime and at sufficiently small rates. Namely, the regime must be of the SD type, but also the capillary number must be sufficiently small such that the saturation profiles are almost flat at the pore-network scale. For uncorrelated networks, and away from percolation, this requires the condition (*Xu et al.*, 1998)

$$Ca \leq \varepsilon \left(\frac{\delta}{\Sigma} \right)^{2/3} \tag{2}$$

where ε and δ are small positive constants (of order 10^{-3}) and Σ is the standard deviation of the pore-size distribution normalized by the mean pore size. The condition of Equation (2) is satisfied at smaller *Ca* for larger disorder, Σ, and corresponds to the lower left region of the diagram in Figure 1. Under this condition, standard IP theories [see *Xu et al.*, 1998] can be applied for calculating rate-independent relative permeabilities, capillary pressure, and residual saturations, volume-averaged over well-defined REVs.

When the displacement is SD, but the rates are sufficiently large so that Equation (2) is violated, a variation of gradient percolation [*Gouyet et al.*, 1988], called invasion percolation in a stabilizing gradient (IPSG), can be used to model displacement patterns. These patterns have a frontal structure of a finite extent, σ (dimensionless with respect to the mean pore size), that decreases with increasing *Ca* following the power-law

$$\sigma \sim Ca^{-\alpha} \quad , \qquad (3)$$

where σ is a positive exponent that is a combination of percolation exponents showing a piston-like displacement at high capillary numbers. For such processes, upscaling in the conventional sense is still possible, although it has not yet been implemented in a formal way. One possible approach would be to define upscaled variables by averaging in a direction *transverse* to the direction of displacement, as done for example in miscible displacement models [see *Yang and Yortsos*, 1996] instead of using volume averages. Application of such an upscaling approach would yield relative permeabilities, capillary pressures, and residual saturations that are capillary-number dependent [see also *Blunt and King*, 1991].

Likewise, the problem is open when the regime of fully developed drainage is of the CVF type (large *Ca* and/or *M*, upper-right corner of Figure 1). This regime shares common properties with Invasion Percolation in a Destabilizing Gradient (IPDG) [*Birovljev et al.*, 1992]. Based on IPDG, we have conjectured that the finger thickness will scale as a negative power of the capillary number [see *Xu et al.*, 1998]. However, many of the properties of this regime are not known at present. We would expect some similarities between this regime's upscaling and viscous fingering models of miscible displacements [e.g., *Yang and Yortsos,* 1996], even though the role of capillarity is different from diffusion, and trapping of the displaced immisicible phase is also possible here.

Capillary control at the macroscopic scale is also the necessary condition for the rigorous upscaling of immiscible displacements in macroscopically heterogeneous porous media, for example, by using homogenization methods [see *Bourgeat*, 1984]. However, it must be pointed out that typical homogenization (and volume-averaging) methods are inherently unable to provide the correct distribution of phases under capillary-control conditions. Indeed, as in the passage from the pore scale, upscaling in macroscopically heterogeneous media under capillary control requires the use of IP. A large-scale version of IP, denoted as large-scale invasion percolation (LSIP), was already developed [*Yortsos et al.*, 1993]. It will be discussed in more detail below. Flow regimes analogous to Figure 1 and their extensions in the case of correlated heterogeneity [*Yortsos et al.*, 1998] are expected from the upscaling of macroscale heterogeneities. Such work remains to be done.

The previous analysis can also be readily extended to problems involving gravity (acting only in a direction parallel to the displacement) and correlated fields [*Yortsos et al.*, 1998]. When gravity is dominant, a condition analogous to Equation (3) holds, where $|B_{gx}|$ replaces *Ca*. When both gravitational and viscous forces compete, gravity acts in a generally additive way to viscous forces, and many of the previous results hold by replacing *M* by M-G_{gx}, where G_{gx} is the gravity number

$$G_{gx} = \frac{\Delta\rho g_x k}{q\mu_{nw}} \equiv \frac{B_{gx}}{Ca} \quad . \qquad (4)$$

The presence of large-scale spatial correlations gives rise to new regimes and in some cases makes relative permeabilities into stochastic functions. Finally, an approach similar to IP in a gradient (of the percolation probability) can also be applied to describe displacements when gravity acts in a direction perpendicular to the main displacement direction, as in the case of gravity segregation [see *Zhang et al.*, 2000].

3. UPSCALING IN FRACTURED MEDIA

The previous sections provide the necessary background for describing the upscaling of processes in fractured media. This problem is qualitatively different depending on whether the matrix is impermeable or impermeable. For this reason, the two different cases will be discussed separately below.

3.1. Impermeable Matrix (DFN)

If the matrix is impermeable (as a result, for example, of mineral deposits lining the fractures), the problem is one of displacement in a network of fractures, each of which can be viewed as a quasi-2-D pore-network (where the fracture aperture maps to an equivalent pore size). The overall problem becomes one of displacement in a DFN. Treating fracture planes as equivalent pore networks has been a useful approximation in the study of fractures [e.g., *Pruess and Tsang*, 1990]. In this section, we will focus on drainage. In principle, a similar approach is possible for imbibition, although some aspects of the latter are not well understood, even for a single fracture [see *Tokunaga and Wan,* 1997].

Consider, first, a process under capillary control over the correlation length of the fracture network, λ_f. This can be modeled as IP in a network of fracture planes. If the correlation length of the fracture aperture, λ_a, is much smaller than the linear fracture plane dimension, *L*, each of

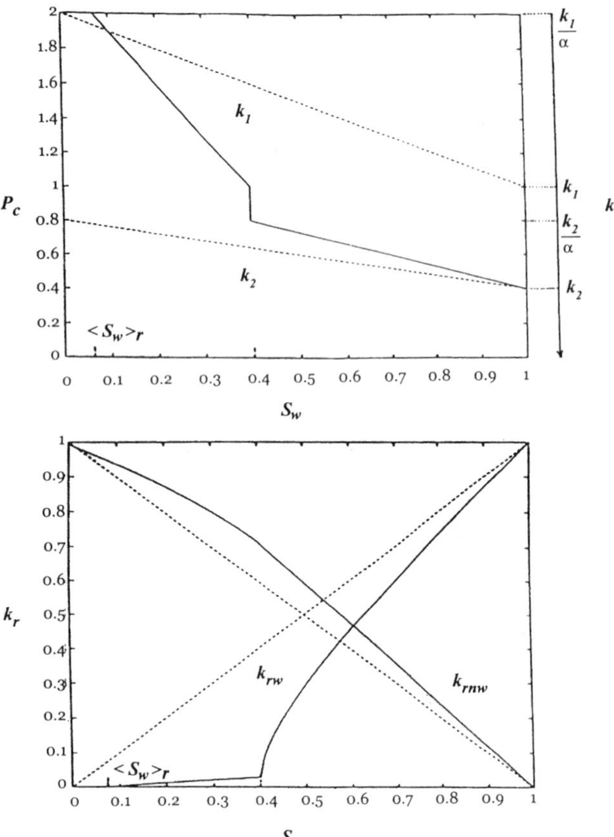

Figure 2. Large-scale capillary pressure curves (top) and relative permeabilities (bottom) for drainage in a 2-D uncorrelated heterogeneous porous medium. P_c is capillary pressure, S_w is saturation of the wetting phase, $k1$, k_2 refer to permeabilities of different media, and α is a factor of the capillary pressure curve. The medium consists of a bimodal permeability distribution, with $k_1 = 1$, $k_2 = 6.25$, and a relative fraction of 0.4. Occupancy is based on LSIP, capillary pressures on the Leverett J function, and relative permeabilities on Effective Medium Approximation. Dashed lines indicate the local. From *Yortsos et al.* [1993].

the fractures has well-defined percolation thresholds, relative permeabilities, capillary pressure and residual saturations. If not, these properties can be described as stochastic functions [*Du et al.*, 1996]. In either case, the problem is modeled by LSIP, where each element to be percolated is a fracture plane. Although to our knowledge, LSIP in a network of fractures has not yet been addressed, an analogous large-scale problem has already been solved. *Yortsos et al.* [1993] considered the upscaling of drainage in heterogeneous porous media at conditions of capillary control. To apply an IP approach, they represented the heterogeneous medium as a 3-D lattice of blocks, each with different permeability and hence different capillary pressure curves, expressed using the Leverett relationship

$$P_c = \frac{\gamma \cos\theta \sqrt{\phi}}{\sqrt{k}} J(S) \tag{5}$$

where φ is porosity, $J(S)$ is the Leverett J-function, and S is saturation. Equation (5) shows that the heterogeneity of k depends on heterogeneity of the capillary thresholds, the degree of saturation in any given block (or, in the present context, of a fracture), and the effective relative permeability of a block (or fracture). The LSIP problem becomes analogous to IP in a pore network: As in the latter, a site (e.g., block, fracture plane) is invaded when its capillary threshold is exceeded. Here, however, the saturation of the site is not unity as in the pore network, but is variable, dictated instead by its capillary pressure curve through Equation (5) and by its accessibility. As in the pore network, accessibility is determined from the IP aspects of the process and lattice connectivity. Because of limited accessibility, LSIP will lead to large-scale trapping and bypassing or incomplete filling of a finite fraction of the medium (or of the fracture network). Thus, part of the medium may become, after some time, inaccessible or "inactive." From the occupancy sequence of the porous medium, large-scale capillary pressure curves and relative permeabilities can be constructed. An example for a bimodal heterogeneity is shown in Figure 2. Under this condition of capillary control at the small scale, the large-scale averages represent rigorously upscaled functions. Note in Figure 2 the existence of large-scale trapped saturations, even though no trapping at the small scale was allowed. The application of an analogous approach based on LSIP for upscaling processes in a network of fractures should be straightforward. The description resulting from such an application would be an ECM of the fracture network, where the large-scale relative permeabilities and capillary pressure are calculated based on LSIP algorithms. Pronounced large-scale trapping is anticipated because of the generally limited connectivity of the network (small coordination number).

The applicability of the above approach rests on two assumptions: (1) the process is capillary-controlled in individual fracture planes, and (2) viscous and/or gravitational gradients are negligible over length scales exceeding the REV scale of the fracture network. A condition analogous to Equation (2) applies in the latter case. To delineate the conditions for the validity of the first assumption, we can extend the work of *Yortsos et al.* [1998] to obtain

$$Ca \left| cN^{\tau} - N^{\frac{\nu}{\nu+1}} (M - N_{gx}) \right| \leq \varepsilon \quad , \tag{6}$$

where c is a dimensionless constant, ε is a small parameter, N denotes the dimensionless length of the square lattice (the length of the fracture plane normalized with the mean fracture aperture), and τ and ν are exponents of percolation [*Stauffer and Aharony*, 1992]. The condition of Equation (6) generalizes the *Lenormand* [1989] condition of variable viscosity in the presence of gravity.

When viscous and/or gravitational gradients are important at the small scale, upscaling is a more difficult task. If the condition of Equation (6) applies, so that displacements in single-fracture planes are capillary controlled, then upscaling of the fracture network problem can be tackled using the approach for drainage described in the previous section. For this approach, we need to make the analogy used in LSIP between fracture planes and sites, and apply the theory of *Yortsos et al.* [1997], *Xu et al.* [1998] and *Yortsos et al.* [1998]. In essence, this is an extension of the previous pore-network scale theory to large-scale heterogeneous media. As in the upscaling of pore networks, we expect that two different large-scale regimes will emerge: a global SD regime and a global CVF regime. For specific results, additional work is needed—particular attention must also be placed on the possible emergence of *anisotropic* large-scale properties, in the presence of gravity, given that the pattern of occupancy will be different, depending on the orientation of the fracture planes with respect to the gravity vector.

Finally, if viscous and/or gravitational gradients within a single fracture plane are sufficiently large such that the condition of Equation (6) is violated, a number of possible flow regimes may arise (for example, local SD at the fracture scale, global SD at the large scale; global CVF at the large scale; and local CVF at the fracture scale). The properties of these regimes and their corresponding effective parameters are unknown. Certainly, DFN models would be most appropriate for their investigation.

3.2. Permeable Matrix

Consider next the more general and interesting problem, where the matrix is permeable and affects the displacement process. As before, we need to discuss two different cases, one in which capillarity is controlling, and another in which viscous or gravitational gradients are not negligible.

3.2.1. Capillary control. In the case of a fractured medium under capillary control, drainage will occur first by LSIP of the fracture network. Onset of matrix penetration will occur long after the large-scale percolation threshold of a fracture plane has been exceeded, at which time most fractures will be occupied by some amount of nonwetting fluid (except, possibly, for those that lack accessibility because of large-scale trapping). Assuming that the defending wetting phase can escape, penetration of the matrix will follow. From this point on, assuming capillary equilibrium between fracture and matrix, the overall process is described by an LSIP model, including both the fracture network and the matrix continuum with different sets of capillary properties. As in *Yortsos et al.* [1993], we anticipate that this process can be simulated in a relatively straightforward manner. Under this condition, the system is accurately simulated by ECM models.

In previous ECM models [*Peters and Klavetter*, 1988], large-scale capillary pressure and relative permeabilities were derived without taking into consideration the limited accessibility and large-scale trapping associated with an invasion process. In essence, this assumption is equivalent to the often-used representation of a pore network as a bundle of parallel capillaries [*Dullien*, 1992]. Thus, it suffers from the same limitations. The latter include unrealistically small percolation thresholds, lack of trapping and of residual saturations, and large relative permeabilities. We suggest that a more realistic approach for upscaling in the context of an ECM should be based on the LSIP method outlined above.

For the imbibition process, the pattern is such that matrix blocks are invaded first. Capillary continuity at asperity contacts will allow for the successive penetration of adjacent matrix blocks. Fracture penetration will occur at the end of this process unless the displaced phase is incompressible or cannot escape. Again, the problem best described by an LSIP model, with possible large-scale trapping. As in the process of drainage, an LSIP model can derive well-defined ECM properties.

We must note that implicit to the above is the assumption that the time for reaching equilibrium in a fracture-matrix composite over the REV scale is much faster than the characteristic time for changes in the capillary pressure (and the saturation) of the effective continuum. In the case of negligible gradients in capillary pressure, this condition is satisfied. If not, then models with a fracture-matrix interaction must be developed. These are discussed in the sections that follow.

3.2.2. Gradients. Consider the full problem, in which viscous or gravity gradients are not negligible and need to be considered. Because of the different roles played by fractures and matrix, we need, first, to distinguish between the different processes of drainage and imbibition. Second, we must differentiate the processes on the basis of whether the gradients are stabilizing or destabilizing (or, in the previous notation, whether the process is of the IPSG or the IPDG type).

Drainage—IPSG. With the drainage process, the relevant question is, first, whether there would be penetration into the matrix and, second, how to describe the ensuing fracture-matrix interaction. If the process in the

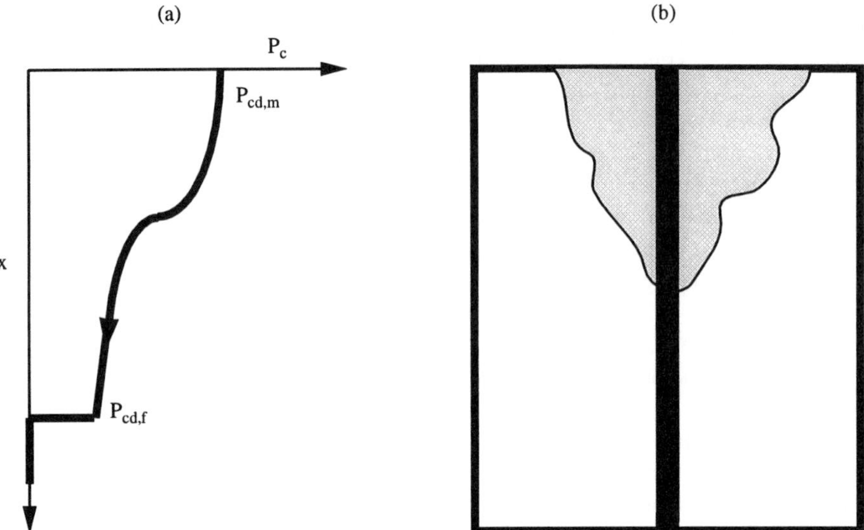

Figure 3. Schematics of drainage in a fractured medium under conditions of IPSG: (a) Profile of capillary pressure, (b) schematic of fracture-matrix occupancy of the displacing phase.

fracture is IPSG (conditions of stabilized displacement), the capillary pressure in the fracture plane, P_{cf}, decreases with distance, as shown in the schematic of Figure 3a. If P_{cf} is smaller than the matrix capillary threshold, $P_{cf} < P_{cmd}$, matrix penetration will not occur. For sufficiently large pressure gradients, however, P_{cf} will be sufficiently large, penetration of the matrix will occur somewhere upstream of the displacement, and the occupancy pattern will have the shape shown in Figure 3 (where for simplicity the fracture network is not shown).

A 2-D version of this problem was treated by *Haghighi et al.* [1994a]. These authors conducted drainage experiments in glass micromodels in the absence of gravity, in a geometry mimicking a single-fracture-matrix block system, as shown in Figure 4. The experiments were then simulated using a pore network simulator. As expected, matrix penetration is *Ca*-dependent. For the 2-D geometries, the critical value of *Ca* for which penetration did not take place was shown to be adequately approximated by the expression

$$Ca_d \approx 0.245 \frac{k_f}{L\sqrt{k_m F_m}} \quad , \qquad (7)$$

where L is the dimensional fracture length, k_i $(i = f,m)$ is the permeability of fracture or matrix, and F_m is the formation factor of the matrix. In Equation (7), we used the Katz and Thompson [1986] expression to relate capillary thresholds to permeability. An analogous expression involving the Bond number rather than the capillary number can be used when gravity effects dominate. Equation (7) shows that the critical capillary number depends on the geometrical-structural properties of the system, particularly on the matrix block length, with larger blocks being penetrated more readily. When the threshold for penetration is exceeded, experiments and simulations showed that the matrix was progressively penetrated, until a steady state was reached (Figure 4) in which the matrix saturation is a function of the capillary number, *Ca* (Figure 5). The partial matrix occupancy results from the fact that as more penetration occurs, the flow rate in the fractures decreases, the capillary pressure decreases, and an equilibrium is eventually reached at which further penetration ceases. In this geometry, and at steady state, the problem is effectively controlled by the fracture and matrix contributions being a rate-dependent flow resistance. *Haghighi et al.* [1994a] discussed the effect of various parameters, such as the mobility ratio, and defined large-scale composite relative permeabilities for the combined matrix-fracture system (which are *Ca*-dependent).

For the general problem of drainage in a 3-D fracture network, similar phenomena will take place. For flow rates exceeding a critical *Ca*, the matrix will be penetrated at a position sufficiently upstream of the displacement. However, the condition for matrix penetration, $P_{cf} > P_{cmd}$, will now also depend on a number of factors missing from Equation (7) (e.g., the relative permeability of the two fluids in the fractures, the flow partition in the fracture network and the spatial location). Thus it cannot be expressed by the simple condition shown in Equation (7). Following penetration, the transient exchange from the fracture continuum to the matrix blocks will occur at a rate that, in principle, can be calculated as a function of the *local Ca* number (and/or B_{gx} if gravity is important). As before, we anticipate the partial penetration of matrix blocks and the attainment of a steady state, given the constant rate processes assumed here. The penetration of the matrix would further stabilize the IPSG displacement assumed in the fracture plane. We should mention that, as in the single-block geometry in Figure 4, the fractures (and, here, specifically the fracture network) control the process. Indeed, since the matrix continuum is not interconnected, nonwetting-phase flow occurs through the fractures.

Regarding upscaling, we first note that under these conditions the dependence on the capillary number shown above makes the process nonlocal. It follows that in general the corresponding upscaled properties would also be nonlocal, depending on various other factors in addition to the local saturation. Appropriately constructed DPMs, with an exchange term describing the fracture-matrix interaction (which is also nonlocal), are usually necessary. The exchange term is a function of *Ca* and/or B_{gx}, and its computation requires considering two different cases. (1) If gradients are strong [e.g., as in *Haghighi et al.*, 1994a], the system behaves as an effective continuum, with the effective relative permeabilities of fracture plane-matrix block units time and capillary-number dependent. This situation may call for transversely averaged upscaling. Computing the effective permeabilities is in principle possible by solving a convection-diffusion type equation in the matrix. As mentioned, in general, the upscaled model involves *Ca-dependent* properties. (2) On the other hand, if gradients are weak in the fracture network, but not in the matrix, we have a double porosity case, where the fracture-matrix interaction can be approximated as a (nonlinear) diffusion term. Models of this type have been used to describe imbibition [e.g., *Zimmerman et al.*, 1993]. These models [e.g., *Arbogast et al.*, 1988; *Bourgeat and Panfilov*, 1998] take into account separation of scales. In the present context, this implies the existence of capillary equilibrium in the fracture network (although not in the matrix blocks) over the REV scale. In a sense, these models constitute the conventional ECM model, when gradients (in the matrix *but not* the fracture) become important. Additional comments on imbibition are also made below.

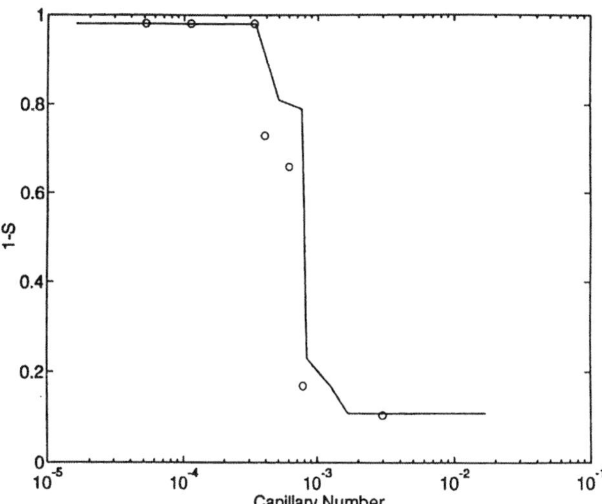

Figure 5. The wetting saturation in the matrix (fraction of pore volume of matrix occupied by the nonwetting phase) as a function of the capillary number for drainage in the 2-D single block-fracture geometry of Figure 4. Dots are experimental points, solid line is numerical. From. *Haghighi et al.* [1994a].

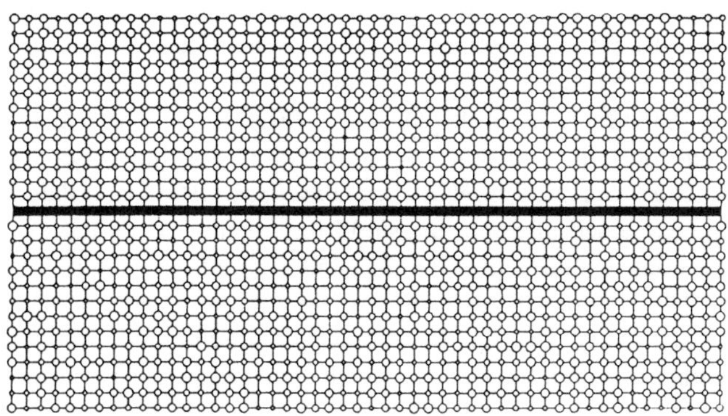

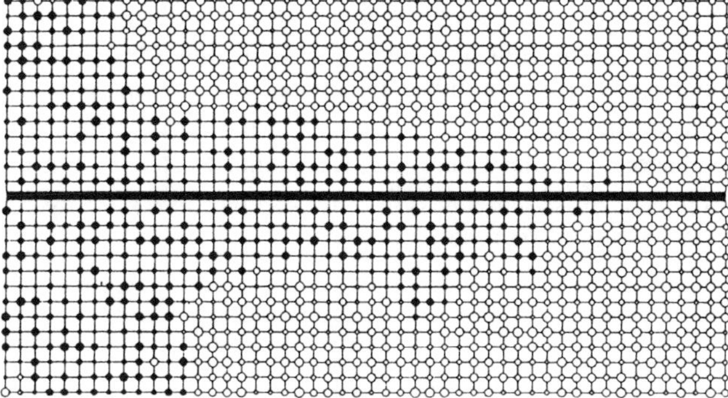

Figure 4. The occupancy pattern for drainage in a micromodel geometry of a 2-D single block-fracture geometry for $Ca = 3.25 \times 10^{-4} < Ca_d$ (top) and for $Ca = 3.9 \times 10^{-4} > Ca_d$ (bottom). From. *Haghighi et al.* [1994a].

Drainage—IPDG. When drainage in a single fracture plane is an IPDG process (for example, due to viscous or gravitational fingering), the capillary pressure in the fracture increases with distance in the direction of displacement, as shown schematically in Figure 6a.

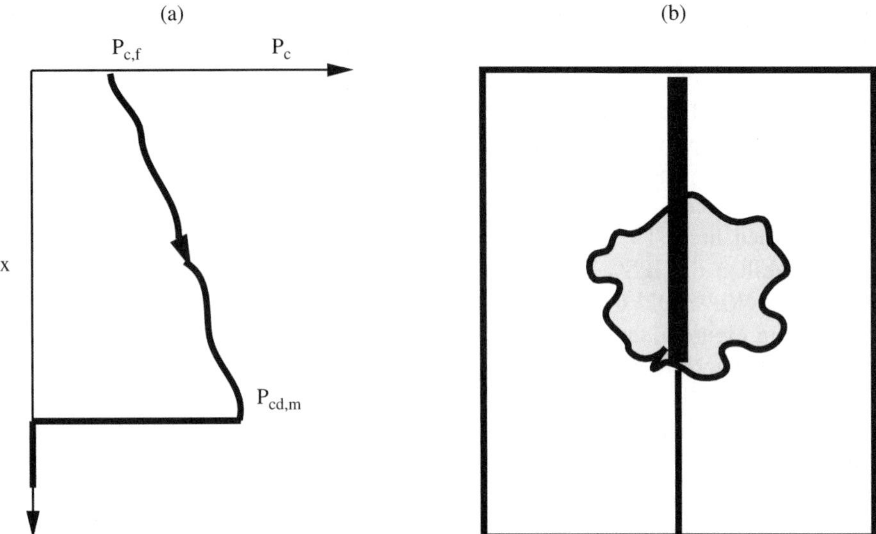

Figure 6. Schematics of drainage in a fractured medium under conditions of IPDG: (a) Profile of capillary pressure; (b) schematic of fracture-matrix occupancy of the displacing phase.

Sufficiently upstream of the displacement, the capillary pressure is sufficiently low and the displacement occurs only in the fracture. At some point downstream, however, the capillary pressure threshold of the matrix will be exceeded, resulting in penetration into the matrix (shown schematically in Figure 6b). In this case, therefore, conditions for the onset of penetration into matrix will exist at some point downstream of the displacement. The resulting exchange between fracture and matrix will somewhat mitigate the unstable displacement in the fracture and may result in further penetration into matrix. To our knowledge, this problem has not yet been analyzed. As in the previous case, penetration involves meeting a threshold condition that depends on a pressure distribution controlled by the fracture network. Computing the exchange term generally requires the solution of nonlinear convection and diffusion inside the matrix. For the most part, this term will be transient and depend on Ca (and/or B_{gx}). Remarks on the upscaling of this process similar to those in the above section apply here as well.

Imbibition—IPSG. Consider next the problem of imbibition. Two questions arise: is there flow in the fracture network? And, if so, what is the resulting fracture-matrix interaction?

Assume, first, that the imbibition process in the fracture plane is under conditions of stabilized displacement (IPSG). With that in mind, we would expect the same type of displacement in the matrix. Imbibition occurs first in the matrix. For the fractures to be invaded, the capillary pressure in the matrix must decrease to sufficiently low values, since the extent of imbibition increases with a decrease in the capillary pressure. For this to occur, capillary pressure gradients must develop in the matrix (resulting from viscous or gravitational forces), leading, under IPSG conditions, to an increase in capillary pressure in the direction of displacement (Figure 7). We expect, therefore, the existence of a threshold for the onset of fracture penetration. *Nitao and Buscheck* [1991] postulated the existence of a rate threshold for fracture penetration in the context of determining the limit of validity of the ECM. A 2-D version of this problem using micromodels, similar to D1 above and, in the absence of gravity, was studied both experimentally and through pore-network simulation by *Haghighi et al.* [1994b]. For an isolated matrix block, two critical values of the capillary number were found, such that for values above the critical capillary number, fracture penetration also takes place (at the same time as the first capillary number penetrates the matrix and the front in the matrix reaches the end of the matrix block for the second capillary number). Reasoning as in the case of drainage, we obtain the following estimate for the critical number in the first case:

$$Ca_i = \frac{q\mu_w}{\gamma\cos\theta} = \frac{\sqrt{k_m}}{ML}, \tag{8}$$

where L is the matrix block length. In the above equation, the capillary number was defined based on the viscosity of the wetting fluid. Note that Equation (8) for the critical

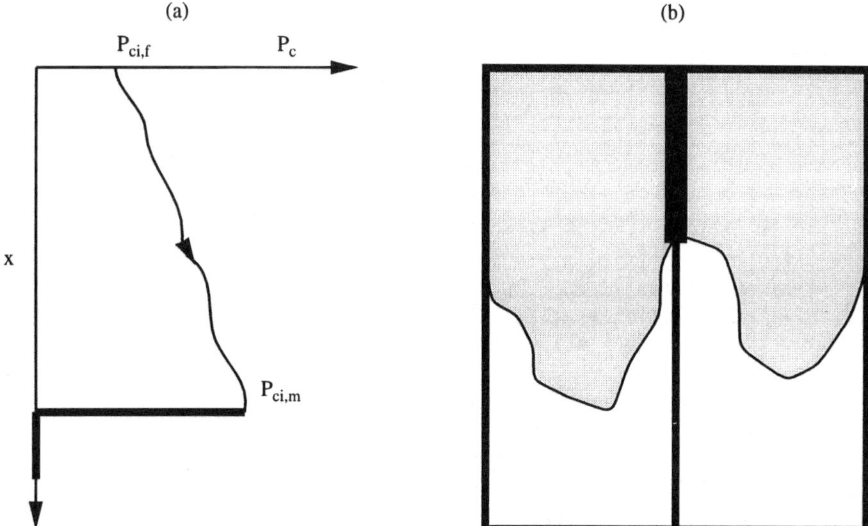

Figure 7. Schematics of imbibition in a fractured medium under conditions of IPSG: (a) Profile of capillary pressure; (b) schematic of fracture-matrix occupancy of the displacing phase.

imbibition rate is analogous to *Nitao and Buscheck's* [1991], where the corresponding threshold was found to be proportional to the capillary diffusivity for imbibition into the matrix [and which scales with the square root of the matrix permeability, consistent with Equation (8)].

Equation (8) is similar to Equation (7) for drainage under IPSG conditions. However, the critical value for fracture penetration is much smaller than in drainage because of the smaller value for matrix permeability. Hence, at first glance, imbibition into a fracture should be relatively easier than into matrix during drainage. On the other hand, the velocity in a fracture plane (for the case of flow in a fracture network only), would be much larger (roughly by the ratio of the respective porosities, ϕ_f / ϕ_m) than the velocity in the matrix (in the case of flow in the matrix only), for otherwise identical volumetric flow rates.

Following the onset of fracture penetration, the pattern development and the ensuing fracture-matrix interaction depends on how far the front penetrates the matrix at the onset of flow in the fracture. Because of the higher fracture permeability, the flow velocity in the fracture would be larger than in the matrix, thus leading to a faster frontal advance there. This mismatch would be mitigated somewhat by the generally smaller relative permeabilities in the fracture plane, as well as by the IPSG conditions assumed. Given sufficient fracture length, therefore, the front in the fracture will overtake that in the matrix, in which case the interaction will take the form discussed above. The possibility also exists that more flow in the fracture will lead to slower flow in the matrix and hence to an increase in the capillary pressure, which will further reduce flow penetration into the fracture.

In the micromodel study of *Haghighi et al.* [1994b], experiments and pore-network simulations were reported for a primary imbibition process, where water displaces air at a constant rate. In that study, the ratio in permeabilities and the matrix length was not sufficiently high for the front in the 1-D fracture to advance faster than in the matrix. Figure 8 shows two snapshots from the experiments corresponding to capillary numbers above and below the critical value (Ca_i), respectively. The features shown are consistent with the above interpretation. Next, consider 3-D fractured media. Here, the threshold for fracture penetration, $P_{cm} < P_{cfd}$, must be computed locally. Before the onset of fracture penetration, the fracture continuum is actually not connected, as far as the wetting phase is concerned. (However, this is not the case following the onset of flow and/or the nonwetting phase.) Thus, the pressure field is controlled by the properties of the matrix continuum and involves an exchange term for the fractures that needs to be computed as a function of Ca (and/or B_{gx}). The displacement in the matrix will be affected by the flow of the nonwetting phase in both the matrix and the fracture continua, and the permeabilities of both need to be accounted for. Thus, strictly speaking, Equation (8) does not apply. Nonetheless, strong rate effects are expected to determine the onset of fracture flow.

In upscaling this process, the difficulty is analogous to the drainage case D1. If gradients in the matrix are relatively weak over the fracture network, then penetration

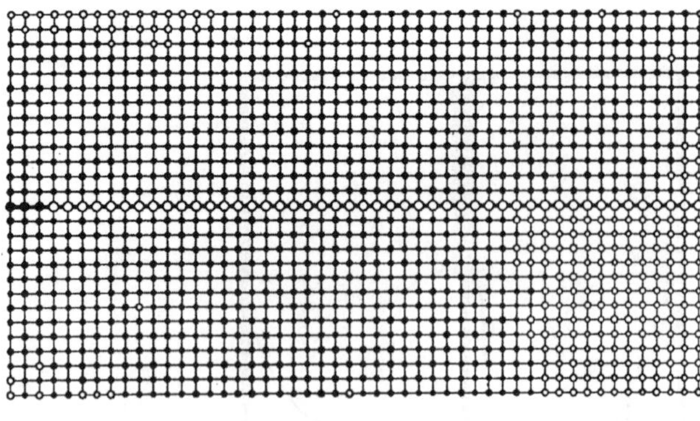

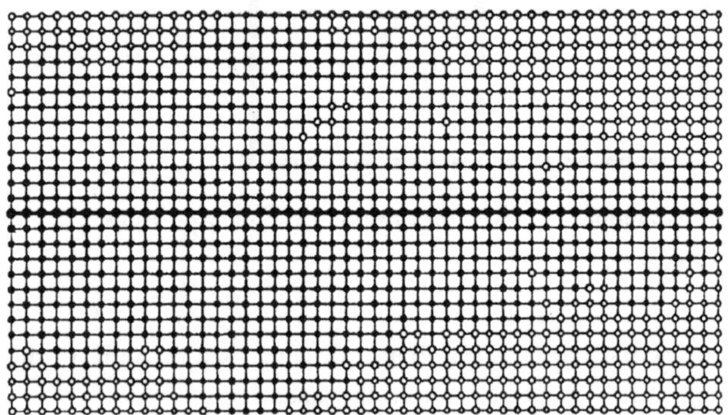

Figure 8. The occupancy pattern for water-air imbibition in a glass micromodel of a 2-D single block-fracture geometry, for $Ca = 2 \times 10^{-7} < Ca_I$ (top) and for $Ca = 2 \times 10^{-5} > Ca_I$ (bottom). From. *Haghighi et al.* [1994b].

of the fracture planes from the adjacent matrix blocks will occur almost uniformly. Because of the small volume contained in the fracture planes, equilibrium will be reached rapidly, and the relevant model is actually an ECM. Trapping the nonwetting phase displacement in the fracture network is possible and will additionally complicate the displacement. When the front in the fractures overtakes that in the matrix, the problem has the same features as the drainage problem previously discussed and also as imbibition under IPSG conditions (to be discussed in the following section). Here, further progress relies on whether there exists separation of scales in the matrix and the fracture, under which condition the double-porosity models of fracture-matrix interaction are applicable (see below). The situation depicted in Figures 7 and 8 represents an intermediate effective continuum with transient effective relative permeabilities, depending both on time and the local capillary number.

Imbibition—IPDG. The final case to be discussed involves primary imbibition under conditions of IPDG in the fracture (for example, involving viscous or gravitational fingering). Under such conditions, the capillary pressure decreases in the direction of displacement, as shown schematically in Figure 9. The same scenario is expected if the displacement is IPDG in the fracture and IPSG in the matrix. The existence of a negative gradient of capillary pressure in the fracture leads to an unstable process, promoting further displacement in the fracture. As pointed

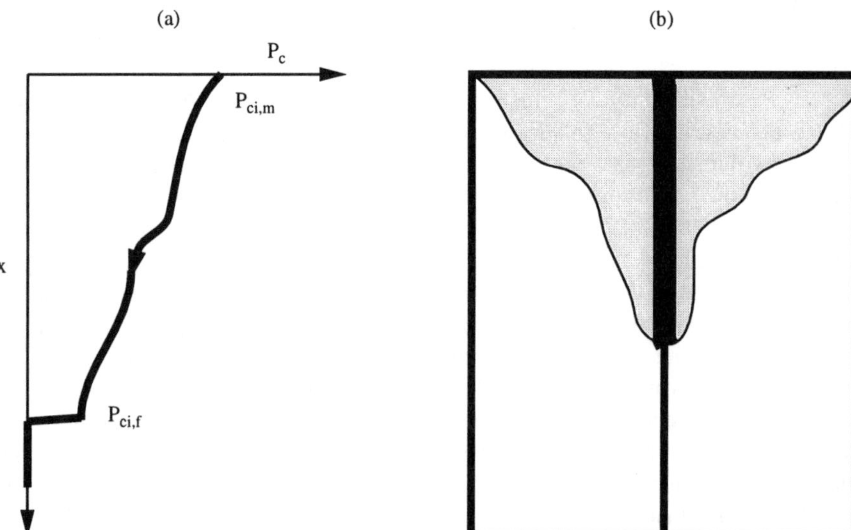

Figure 9. Schematics of drainage in a fractured medium under conditions of IPSG: (a) Profile of capillary pressure; (b) schematic of fracture-matrix occupancy of the displacing phase.

out above, in the case of imbibition with IPSG patterns, a front advancing faster in the fracture than in the matrix is also eventually expected,. During this process, the matrix block in contact with the wetting fluid in the fracture plane is progressively exposed to a lack of capillary equilibrium. As a result, invasion into the matrix will occur laterally from the fracture plane in a diffusion-like spontaneous imbibition. The ensuing diffusive exchange between fractures and matrix will slow down the unstable displacement in the fracture and, to a certain extent, mitigate the instability. Experimental support for such a scenario was provided by many authors in the past. In fact, most of the existing models for imbibition are based on such a conceptual model of the fracture-matrix interaction, regardless of the value of the capillary number or the nature of the process [e.g., *Arbogast et al.*, 1988; *Zimmerman et al.*, 1993], as pointed out above. We note again that as with drainage, the matrix continuum is disconnected with respect to the flow of the invading (now the wetting) phase.

In upscaling this process, we first note that, strictly speaking, separation of scales does not exist. However, if the rate of penetration in the fracture network is sufficiently fast, we may separately treat imbibition in the fracture and the matrix, and define an exchange term based on upscaled variables. Under this assumption, the problem is again represented by a DPM, with a transient term (accounting for imbibition in the matrix) to capture the fracture-matrix interaction. Quantifying the matrix-fracture interaction necessitates an adequate knowledge of the process of imbibition in a fracture plane under IPDG conditions—and also inclusion of the effects of film flow or flow in the roughness of the fracture surface [*Tokunaga and Wan*, 1997]. The wetting-phase distribution in the fracture planes will dictate the extent of the fracture-matrix interaction. This distribution needs to be clearly elucidated for use in upscaled models, where the contact area is a critical factor [*Ho*, 1997]. This extent is expected to be a function of Ca (and/or B_{gx}). If scale separation does not exist, however, then the rates of penetration in the matrix and the fracture are comparable, and we have an effective continuum with upscaled properties based on transverse averages, and which depend on Ca, space, and time.

4. CONCLUSION

In this paper we discussed various physical issues that arise when upscaling immiscible displacement in fractured media. We have emphasized the need to distinguish between stabilized and destabilized displacements, as these two different processes lead to both qualitative and quantitative changes in the flow process and the fracture-matrix interaction. We submit that upscaling approaches must recognize such flow regimes and the importance of structural parameters.

Because of the necessity to separate scales, the most rigorous model to date for upscaling is that based on capillary control, commonly known as the ECM in related contexts. We pointed out, however, that in deriving large-scale average functions (such as capillary pressure and relative permeabilities) an LSIP approach must be used to account for limited connectivity and/or accessibility. This is not the current practice. Furthermore, the applicability of ECM requires very small rates and the absence of gradients over relatively large scales, a limitation that could be difficult to overcome in realistic situations.

In the most general case, the effect of gradients during the invasion process is important and must be accounted for. This is typically attempted with a DKM. Upscaling will give rise to exchange terms between the fracture and matrix continua, which will depend on the type of displacement and the local competition between viscous, capillary and gravitational forces, as expressed through the appropriate dimensionless numbers, Ca and B_{gx}. This interaction is nonlinear, nonlocal in space and time, and directional. However, in most commonly used DKM models, the exchange terms are postulated to be only functions of the instantaneous saturations. In many of the situations analyzed, the process is controlled mostly by a single continuum (such as a DFN or a matrix-continuum), with exchange terms that describe the interaction with the other continuum.

Useful models of the double-porosity type, with transient terms to describe the fracture-matrix interaction, are possible when scale separation exists, such that gradients in one continuum (but not the other) are weak, over the REV scale. This could be the case in either drainage or imbibition. In essence, these models are transient versions of the ECM model, which they approach at steady state. Under conditions where gradients develop within a fracture plane, the concepts of volume-averaged saturation, relative permeabilities, etc., are called into question. It is suggested that in such cases, averages in a direction transverse to the main displacement direction are more appropriate. The existence of well-defined quantities transverse to the gradient will lead to directional dependence. Furthermore, such terms will be rate and/or gravity dependent. In general, when using such an upscaling approach, the local problem is coupled to the global problem.

Strictly speaking, the interaction between matrix and fracture will alter the relative permeabilities of the upscaled model, as can be seen from the analysis of the simpler ECM problem. When gradients are involved, the composite relative permeabilities and capillary pressures will certainly depend on *Ca* (and/or B_{gx}), in addition to saturations. Furthermore, in the case of strong gravity gradients, these properties are expected to be anisotropic or direction dependent.

For simplicity, our analysis was restricted to primary displacements, in which the medium is initially fully saturated with only one fluid phase. Secondary displacements (e.g., secondary imbibition), in which the medium initially contains the displacing phase at its residual (trapped) saturation, involve additional degrees of complexity. Many aspects of displacement are still not fully elucidated, even for macroscopically homogeneous media. For the same reasons, we did not discuss the important processes of steady-state two-phase flow, for which many unresolved questions still exist.

Acknowledgments. This work was partly supported by U.S. Department of Energy Contract DE-AC2699BC15211. The author would like to express his heartfelt appreciation to Paul Witherspoon for his generosity and guidance during several years of productive collaboration. Thanks are also due to Lang Zhan for technical assistance.

REFERENCES

Acuna, J., and Y. C. Yortsos, Application of fractal geometry to the study of networks of fractures and their pressure transient, *Water Resour. Res.*, *31*, 527, 1995.

Amaziane, B., and A. Bourgeat, Effective behavior of two-phase flow in heterogeneous reservoir, in *Numerical Simulation in Oil Recovery*, edited by M. F. Wheeler, IMA Volumes in Mathematics and Its Applications, *11*, 1, 1988.

Arbogast, T., The double porosity model for single-phase flow in naturally fractured reservoirs, in *Numerical Simulation in Oil Recovery*, edited by M. F. Wheeler, IMA Volumes in Mathematics and its Applications, *11*, 23, 1988.

Arbogast, T., J. Douglas, Jr., and J. E. Santos, Two-phase immiscible flow in naturally fractured reservoirs, in *Numerical Simulation in Oil Recovery*, edited by M. F. Wheeler, IMA Volumes in Mathematics and Its Applications, *11*, 47, 1988.

Birovljev, A., L. Furberg, J. Feder, T. Jossang, K. J. Maloy, and A. Aharony, Gravity invasion percolation in two dimensions: Experiments and simulation, *Phys. Rev. Lett. 67*, 584, 1991.

Blunt, M., and P. R. King, Relative permeabilities from two- and three-dimensional pore-scale network modeling, *Transport in Porous Media, 6*, 407, 1991.

Bourgeat, A., Homogenized behavior of two-phase flows in naturally fractured reservoir with uniform distribution, *Comp. Meth. Appl. Mec. Eng., 47*, 205, 1984.

Bourgeat, A., and M. Panfilov, Effective two-phase flow through highly heterogeneous porous media: Capillary non-equilibrium effects, *Comput. Geosc., 2*, 171, 1998.

Constantinides, G. N., and A. C. Payatakes, Network simulation of steady-state two-phase flow in consolidated porous media, *AICHEJ, 33*, 369,1996.

Dias, M., and D. Wilkinson, Invasion percolation with trapping, *J. Phys. A, 19*, 3131, 1986.

Du, C., C. Satik, and Y. C. Yortsos, Percolation in an fBm lattice, *AICHEJ, 42*, 2392,1996.

Dullien, F. A. L., *Porous Media: Fluid Transport and Pore Structure*, Academic Press, New York, N.Y., 1992.

Durlofsky, L., Coarse-scale models of two-phase flow in heterogeneous reservoirs: volume-averaged equations and their relationship to existing upscaling techniques, *Comput. Geosc., 2*, 73, 1998.

Feder, J., *Fractals*, Plenum, New York, N.Y., 1988.

van Genuchten, M. Th., Closed-form equation for predicting the hydraulic conductivity of unsaturated soils, *Soil Sci. Soc. Am. J., 44*, 892, 1980.

Glass, R. J., M. J. Nicholl, and V. C. Tidwell, Challenging models for flow in unsaturated fractured rock through exploration of small scale processes, *Geophys. Res. Lett., 22*, 1457, 1995.

Gouyet, J.-F., B. Sapoval, and M. Rosso, Fractal structure of diffusion and invasion fronts in three-dimensional lattices through the gradient percolation approach, *Phys. Rev. B, 37*, 1832, 1988.

Haghighi, M., B. Xu, and Y. C. Yortsos, Visualization and simulation of immiscible displacement in fractured systems using micromodels: I. Drainage, *J. Coll. Int. Sci., 166*, 168, 1994a.

Haghighi, M., B. Xu, and Y.C. Yortsos, Visualization and simulation of immiscible displacement in fractured systems using micromodels: II. Imbibition, in *DOE Annual Report*, Report DOE/BC/14899-17, U.S. Department of Energy, Washington, D.C., 1994b.

Heiba, A. A., M. Sahimi, L.E. Scriven, and H.T. Davis, Percolation theory of two-phase relative permeability, *SPE, 11015,* 1982.

Heiba, A. A., L. E. Scriven, and H. T. Davis, Percolation theory of three-phase relative permeability, *SPE, 12172,* 1983.

Ho, C. K., *Proc. ASME Fluids Eng. Div., Sixth Symposium on Multiphase Transport in Porous Media, FED-244, Dallas, Tex., Nov. 16–21, 1997*, p. 401, 1998.

Katz, A. J., and A. H. Thompson, Quantitative prediction of permeability in porous rock, *Phys. Rev. B, 34*, 8179, 1986.

Lenormand, R., Flow through porous media: limits of fractal patterns, *Proc. R. Soc. London A, 423*, 159, 1989.

Lenormand, R., Liquids in porous media, *J. Phys.: Condens. Matter, 2*, SA79, 1990.

Nitao, J. J., and T. Buscheck, Infiltration of a liquid front in an unsaturated, fractured porous medium, *Water Resour. Res., 27*, 2099, 1991.

Noetinger, B., and T. Estebenet, Upscaling of fractured media using continuous-time random walk methods, *Proceedings ECMOR VI*, Peebles, Scotland, September 8–11, 1998.

Peters, R.R., and E. A. Klavetter, A continuum approach for water movement in an unsaturated fractured rock mass, *Water Resour. Res., 24*, 416, 1988.

Pruess, K., and T. N. Narasimhan, A practical method for modeling fluid and heat flow in fractured porous media, *SPEJ, 25*, 14, 1985.

Pruess, K., and Y. W. Tsang, On two-phase relative permeability and capillary pressure of rough-walled fractures, *Water Resour. Res., 26*, 1915, 1990.

Stauffer, D., and A. Aharony, *Introduction to Percolation Theory*, Taylor & Francis, London, England, 1992.

Tokunaga, T. K., and J. Wan, Water film flow along fracture surfaces of porous rocks, *Water Resour. Res., 33,* 1287, 1997.

Valavanides, M. S., and A. C. Payatakes, True-to-mechanism model of steady-state two-phase flow in porous media, using decomposition into prototype flows, *Adv. Wat. Res.*, in print, 2000.

Warren, J. E., and P. J. Root, The behavior of naturally fractured reservoirs, *SPEJ, 3*, 245, 1963.

Whitaker, S., Flow in porous media II. The governing equations for immiscible two-phase flow, *Transport in Porous Media, 1*, 105, 1986.

Wilkinson, D. Percolation model of immiscible displacement in the presence of buoyancy forces, *Phys. Rev. A, 30*, 520, 1984.

Wilkinson, D., Percolation effects in immiscible displacement, *Phys. Rev. A, 34*, 1380, 1984.

Wilkinson, D., and J. F. Willemsen, Invasion percolation: a new form of percolation theory, *J. Phys. A, 16*, 3365, 1983.

Xu, B., and Y. C. Yortsos, Capillary pressure of anisotropic porous media, *DOE Annual Report*, DOE/BC/14899-17, U.S. Department of Energy, Washington, D.C., 1994.

Xu, B., Y. C. Yortsos, and D. Salin, Invasion percolation with viscous forces, *Phys. Rev. E, 57*, 739, 1998.

Yang, Z., and Y. C. Yortsos, Asymptotic regimes in miscible displacements in random media, paper *SPE, 35456*, 1996.

Yortsos, Y. C., and M. Sharma, Application of percolation theory of non-catalytic gas-solid reactions, *AIChEJ, 32*, 46, 1986.

Yortsos, Y. C., C. Satik, J.-C. Bacri, and D. Salin, Large-scale percolation theory of drainage, *Trans. Porous Media 10*, 171, 1993.

Yortsos, Y. C., B. Xu, and D. Salin, Phase diagram of fully developed drainage in porous media, *Phys. Rev. Lett., 79*, 4581, 1997.

Yortsos, Y. C., B. Xu, and D. Salin, Phase diagram of fully-developed drainage: A study of the validity of the Buckley-Leverett equation, *Soc. Pet. Eng. 49318*, 1998.

Zhang, Y., M. Shariati, and Y. C. Yortsos, The spreading of immiscible fluids in porous media under the influence of gravity, *Trans. Porous Media, 38*, 117, 2000.

Zimmerman, R. W., G. Chen, T. Hagu, and G. S. Bodvarsson, A numerical dual-porosity model with semianalytical treatment of fracture/matrix flow, *Water Resour. Res., 29*, *2127*, 1993.

Y.C. Yortsos, Department of Chemical Engineering and Petroleum Engineering Program, University of Southern California, Los Angeles, CA 90089-1211

Mixed Discrete-Continuum Models: A Summary of Experiences in Test Interpretation and Model Prediction

Jesus Carrera and Lurdes Martinez-Landa

Department of Geotechnics and Applied GeoScience, School of Civil Engineering, Technical University of Catalonia (UPC), Barcelona, Spain

A number of conceptual models have been proposed for simulating groundwater flow and solute transport in fractured systems. They span the range from continuum porous equivalents to discrete channel networks. The objective of this paper is to show the application of an intermediate approach (mixed discrete-continuum models) to three cases. The approach consists of identifying the dominant fractures (i.e., those carrying most of the flow) and modeling them explicitly as two-dimensional features embedded in a three-dimensional continuum representing the remaining fracture network. The method is based on the observation that most of the water flows through a few fractures, so that explicitly modeling them should help in properly accounting for a large portion of the total water flow. The applicability of the concept is tested in three cases. The first one refers to the Chalk River Block (Canada) in which a model calibrated against a long crosshole test successfully predicted the response to other tests performed in different fractures. The second case refers to hydraulic characterization of a large-scale (about 2 km) site at El Cabril (Spain). A model calibrated against long records (five years) of natural head fluctuations could be used to predict a one-month-long hydraulic test and heads variations after construction of a waste disposal site. The last case refers to hydraulic characterization performed at the Grimsel Test Site in the context of the Full-scale Engineered Barrier EXperiment (FEBEX). Extensive borehole and geologic mapping data were used to build a model that was calibrated against five cross-hole tests. The resulting large-scale model predicted steady-state heads and inflows into the test tunnel. The conclusion is that, in all cases, the difficulties associated with the mixed discrete-continuum approach could be overcome and that the resulting models displayed some predictive capabilities.

1. INTRODUCTION

1.1. Background and Motivation

Accurate simulation of flow and transport through porous media is, to say the least, complex. Numerous modeling approaches have been postulated. An excellent synthesis of most of them is provided in Chapter 6 of the *National*

Dynamics of Fluids in Fractured Rock
Geophysical Monograph 122

Research Council [1996] report on rock fractures and fluid flow. The abundance of methods may reflect the broad spectrum of situations encountered when dealing with fractured media. However, it may also reflect the lack of consensus in the hydrogeological community about how to conceptualize such media. In essence, most available approaches can be classified under one of two general headings: continuum and discrete fracture models.

Continuum models are based on assuming Darcy´s law to be valid in every point of the flow domain, such that mass conservation principles can be applied to derive conventional flow and transport equations. The most appealing feature of this approach is its apparent simplicity, although deriving continuum properties from fracture data may not be a simple task.

Continuum homogeneous models are normally used for the interpretation of hydraulic tests, at least for the preliminary interpretation. In fact, the methodology to obtain the equivalent hydraulic conductivity tensor from pump tests in 3-D is well established [*Hsieh and Neuman*, 1985; *Hsieh et al.*, 1985]. Recognizing that water in the best connected fractures can be more easily accessed than water in the matrix has led to multiple continuum concepts [*Gringarten*, 1984; *Barker*, 1988]. In fact, the latter provides solutions for noninteger dimensional domains, as a way of recognizing the variable role of fracture flow with the scale of the problem.

Within the heterogeneous models, stochastic continuum models [*Neuman*, 1988] are very appealing in that they apparently only require assuming that permeability be measurable at some scale. Since permeability is indeed measurable in boreholes at the few-meters scale, we might be tempted to conclude that conventional geostatistical tools could be applied to fractured media. In fact, they have been applied. It should be stressed, however, that neither the discontinuities observed in permeability logs nor the lateral-continuity of high conductivity fractured zones are consistent with conventional multigaussian geostatistics. Hence, unsurprisingly, experience with these types of models is very limited. In fact, best predictive capabilities have been obtained when the dominant fractures have been treated as distinct discrete features of the model, which falls within what we are terming *mixed discrete-continuum models* [*Gomez-Hernandez et al.*, 1999].

Discrete fracture network (DFN) models attempt to reproduce the geometry of fractured media as consisting of blocks separated by fractures. In general, the blocks are considered to be impervious, so that the hydraulic behavior of the system is entirely controlled by the properties of the fractures (transmissivity, orientation, extent, density). DFN models are essentially stochastic in that one cannot hope to characterize all fractures accurately. The only reasonable expectation is to characterize the statistical properties of the network. Then, realizations of fracture networks can be generated and flow simulated through each of them. Results then take the form of model-output statistics for all the realizations. Most DFN models are variations of this principle [*Long et al.*, 1982; *Dershowitz*, 1984; *Smith and Schwarz*, 1984; *Cacas et al.*, 1990; *Anderson and Thunvik*, 1986].

Unfortunately, the apparent fractal nature of fracture networks makes it very hard to include all fractures. In fact, it is difficult to simply define a threshold below which fractures can be neglected. Moreover, only a small fraction of all fractures effectively contribute to groundwater flow, and this fraction is difficult to identify because it correlates poorly with easy-to-measure parameters [*Jones et al.*, 1985]. In fact, one of the relevant findings of recent years is that only a small fraction of the fracture plane contributes to flow. This has led to the channeling concept [*Neretnieks*, 1983], wherein the fracture domain is represented by a network of channels [*Moreno and Neretnieks*, 1993]. This concept is needed in order to get a quantitatively consistent picture of flow and transport through fractures. In view of these difficulties, it is not surprising that the number of actual applications remain relatively small [*Dverstorp and Anderson*, 1989; *Cacas et al.*, 1990; *Dershowitz et al.*, 1991].

An intermediate approach that appears to share some of the advantages of DFNs and continuum models is the mixed discrete-continuum approach mentioned above. In essence, this approach consists of identifying the dominant fractures and including them in the model as 2-D features. They are embedded in a 3-D porous medium that represents the remaining fractures. This concept shares the simplicity of continuum models while presumably capturing the role of the most important fractures. The concept has been applied on a number of occasions [*Shapiro and Hsieh*, 1991; *Carrera and Heredia*, 1988; *Kimmeier et al.*, 1985].

1.2. Objective and Scope

To judge whether an approach is useful for modeling, it is probably best to define what a model is. A model is an entity that represents the behavior of a natural system. As such, it should help in integrating available data in a coherent manner and in making predictions that are not trivial consequences of the available data. From this perspective, the mixed approach may be more useful, albeit less elegant, than either the discrete fracture network or the continuum approaches. The objective of this paper is to argue for such a point.

This paper summarizes three cases in which the mixed discrete-continuum approach led to successful models. By successful models, we mean that they helped in integrating different types of data and they were capable of simulating the response of the fractured system to different conditions from those of calibration (that is, the models displayed predictive capabilities). We argue that by this standard, the mixed discrete-continuum approach is a successful (useful) modeling approach.

The methodology itself is described in Section 2. It must be stated at the start, however, that the proposed methodology is not particularly original and that it follows the steps of what can be considered good modeling practice. The three test cases (Chalk River Block, El Cabril, and FEBEX) are discussed sequentially. The descriptions are necessarily brief; the reader is directed to the appropriate references for details. The paper concludes with a summarizing discussion section.

2. METHODOLOGY

As stated in the introduction, the mixed approach does not require an essentially specific modeling methodology, except that significant emphasis must be placed on the identification and characterization of dominant fractures. The modeling procedure we advocate [*Carrera et al.*, 1993a] consists of using general scientific knowledge and site-specific data to derive a conceptual model. This model is discretized and calibrated against quantitative data. Often, the resulting numerical model fails to adequately fit existing data, making it necessary to revise the conceptual model in light of the discrepancies between measured and computed heads and concentrations. On the other hand, more than one model may fit the data. In such cases, additional measurements may be required to discriminate among alternative conceptual models or simply to reduce the uncertainty of estimated parameters. The methodology is outlined in Figure 1.

2.1. Conceptualization

Clearly, the most critical step in the above process is the definition of the conceptual model. Since this requires using all tools of hydrogeology, it would be pretentious to try an exhaustive description. The reader is directed to specific accounts [e.g., *Bear et al.*, 1993; *National Research Council*, 1996]. However, defining the dominant fractures is a difficulty specific to the mixed approach. Therefore, it is worth discussing the methods we have found most useful in determining the dominant fracture zones.

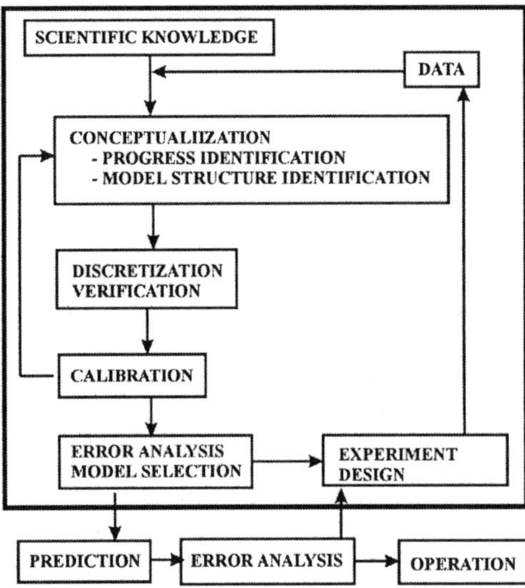

Figure 1. Schematic description of a generic modeling procedure.

A preliminary guess of the most important fractures can be obtained from both surface and downhole fracture mapping. Specifically, long fractures (or fracture zones) along directions of possible shear or tensile failure are likely candidates. Flowing fractures can be determined from fracture infilling

Regardless of the usefulness of indirect identification techniques, fracture characterization ends up requiring hydraulic testing. As discussed by *Meier et al.* [1998], conventional pumping tests yield effective transmissivity values even in highly heterogeneous media. Therefore, long-term pumping tests should always be performed. Clearly, getting a large number of observation intervals helps in defining the extent of the most important fracture zones.

In fact, early response at an observation interval to pumping elsewhere is probably the best way to identify connectivity [*Sanchez-Vila et al.*, 1999]. This issue will be discussed in some detail in Section 3 of this paper, which deals with the FEBEX case.

Despite all of this, the question remains whether we can unambiguously identify all important fractures. We do not think that this question can be answered satisfactorily in most cases. In fact, there may be cases in which the scale of fractures and data are such that no dominant fracture can be identified. In either case, uncertainties in the definition of these fractures fall in the realm of conceptual model uncertainties. As such, they are difficult to quantify (this is why the approach is not "elegant"). An idea about the level of conceptual model uncertainty can be gained by working simultaneously with all acceptable conceptual models. In fact, more than one model has been considered in all the cases discussed in this paper.

2.2. Mathematical Modeling

Equations governing fluid flow and solute transport in the type of medium described above are very similar to conventional flow and transport equations. Differences stem from the geometrical complexity associated with the fracture network. However, flow and transport mechanisms are no different from those of geometrically simpler porous media. The equations are listed here for the sake of completeness.

Let Ω be the flow domain and Γ its boundary. In what follows, we will assume Ω to be 3-D. Let Ω_{fi} be a 2-D domain representing the i-th fracture and contained in Ω. It is convenient to define a coordinate system, $\boldsymbol{u}_i$, for Ω_{fi}. Obviously, every fracture can be viewed as a discontinuity in Ω. As such, we may be interested in treating them as boundaries of Ω. In such cases, we will denote Ω_{fi}^{+} and Ω_{fi}^{-} the two faces of fracture i, and we will refer to them jointly as $\Omega_{fi}^{\pm}$. The boundary of the fracture is a 1-D domain Γ_{fi}. Fracture intersections are internal boundaries, denoted Γ_{ij}. Groundwater flow in Ω is governed by:

$$S_s \frac{\partial h}{\partial t} = \nabla \cdot (\boldsymbol{K} \cdot \nabla h) + w \qquad \boldsymbol{x} \in \Omega \qquad (1a)$$

subject to

$$h(\boldsymbol{x},0) = h_0(\boldsymbol{x}) \qquad \boldsymbol{x} \in \Omega, t = 0 \qquad (1b)$$

$$\boldsymbol{K} \cdot \nabla h \cdot \boldsymbol{n}_\Gamma = q_\Gamma(\boldsymbol{x},t) + \alpha(\boldsymbol{x},t) \cdot [H(x,t) - h(x,t)] \qquad \boldsymbol{x} \in \Gamma \qquad (1c)$$

$$\boldsymbol{K} \cdot \nabla h \, \boldsymbol{n}_{fi}^{+} = q_{fi}^{+}(\boldsymbol{x},t) \qquad \boldsymbol{x} \in \Omega_{fi}^{+} \qquad (1d)$$

$$\boldsymbol{K} \cdot \nabla h \, \boldsymbol{n}_{fi}^{-} = q_{fi}^{-}(\boldsymbol{x},t) \qquad \boldsymbol{x} \in \Omega_{fi}^{-} \,, \qquad (1e)$$

where h is head, S_s is specific storage, $\boldsymbol{K}$ is hydraulic conductivity tensor, w is a sink/source per unit volume, $h_0(\boldsymbol{x})$ is the initial head field, $\boldsymbol{n}_\Gamma$ is the unit vector normal to Γ and pointing outwards, q_Γ is a prescribed flux term, H is an externally prescribed head, and α is leakage factor. Variables $\boldsymbol{n}_{fi}^{+}$, $\boldsymbol{n}_{fi}^{-}$ are the equivalent to $\boldsymbol{n}$ for $\Omega_{fi}^{\pm}$, and q_{fi}^{+} and q_{fi}^{-} are the $\Omega_{fi}^{\pm}$ counterpart of q_Γ. However, the sign convention for (1d) and (1e) may be interpreted as opposite to that of (1c). Unit vectors $\boldsymbol{n}_{fi}^{+}$ and $\boldsymbol{n}_{fi}^{-}$ point outwards from the fracture and, consequently, towards Ω. As a result, the terms $\boldsymbol{K} \cdot \nabla h \, \boldsymbol{n}_{fi}^{+}$ are positive when water exits Ω. That is, q_{fi}^{+} and q_{fi}^{-} are defined with the sign convention of the i-th fracture: they are positive when water exits the block and enters the fracture, and *vice versa*. These terms, q_{fi}^{+} and q_{fi}^{-}, are unknown. Evaluating them requires solving flow in the fractures, which is governed by:

$$S_i \frac{\partial h_{fi}}{\partial t} = \nabla \cdot \left(\boldsymbol{T}_i \cdot \nabla h_{fi} \right) + r_i + q_{fi}^{+} + q_{fi}^{-} \qquad \boldsymbol{u}_i \in \Omega_{fi} \qquad (2a)$$

$$h_{fi}(\boldsymbol{u}_i,0) = h_{fi0}(\boldsymbol{u}_i) \qquad \boldsymbol{u}_i \in \Omega_{fi} \qquad (2b)$$

$$\boldsymbol{T}_i \cdot \nabla h \cdot \boldsymbol{n}_{fi} = q_{fi} + \alpha_{fi} \left(H_{fi} - h_{fi} \right) \qquad \boldsymbol{u}_i \in \Gamma_{fi} \qquad (2c)$$

$$q_{ij}^{+} + q_{ij}^{-} = q_{ji}^{+} + q_{ji}^{-} \qquad \boldsymbol{u}_i \in \Gamma_{ij} \,, \qquad (2d)$$

where h_{fi}, S_i, $\boldsymbol{T}_i$, and r_i are fractured head, storage coefficient, transmissivity tensor, and internal sink/source term (per unit surface area of fracture), respectively. The terms $\boldsymbol{n}_{fi}$, q_{fi}, α_{fi}, and H_{fi} are the i-th fracture equivalents of $\boldsymbol{n}_\Gamma$, q_Γ, α, and H in (1). Normally, q_{fi} and α_{fi} are zero. Internal boundary flux q_{ij}^{+} equals $\boldsymbol{T}_{fi} \cdot \nabla h_i \cdot \boldsymbol{n}_{ij}^{+}$ in parallel to (1d). The same can be said about q_{ij}^{-}. However, specific boundary conditions may be imposed when the fracture (Ω_{fi}) intersects the boundary (Γ). Solution of (1) and (2) requires imposing continuity:

$$h_{fi}(\boldsymbol{u}_i, t) = h(\boldsymbol{x}, t) \qquad Loc(\boldsymbol{x}) = Loc(\boldsymbol{u}_i) \,, \qquad (3)$$

where $Loc(\boldsymbol{x})$ and $Loc(\boldsymbol{u}_i)$ express the location of the point. That is, Equation (3) expresses that heads are identical as belonging to the fracture or as belonging to the block.

In summary, what Equations (1) through (3) express is that water may flow through either the fractures Ω_{fi} or the blocks in between. Continuity can only be established for h

[Equation (3)]. Fluxes are discontinuous at the fractures (as viewed from the blocks) and at fracture intersections (as viewed from the fractures). Mass balance is assured by forcing the condition that water exiting (entering) the blocks enters (leaves) the fractures [Equations (1d), (1e), and (2)].

Similarly, a mass balance must be imposed at fracture intersections (Equation 2d). Sometimes, a preferential flow path is conjectured to exist along fracture intersections. In such cases, a flow equation would have to be written for such intersections, and continuity conditions similar to those between block and fractures would have to be written instead of Equation 2d.

Two final remarks must be made. First, if fractures are viewed as 2-D entities, they are characterized by transmissivity ($\boldsymbol{T}_i$). Sometimes, hydrogeologists characterize fractures by their hydraulic conductivity ($\boldsymbol{K}_i$). In such cases, we must define a thickness such that, when multiplied by $\boldsymbol{K}_i$, it yields $\boldsymbol{T}_i$. The difference is unimportant in most cases. Keep in mind, however, that field measurements yield hydraulic conductivities of blocks or transmissivities of fractures (i.e., the Dupuit assumption usually holds for fractures).

Computer implementation of the above equations is much simpler than implied by the equations. In essence, it is sufficient to use 3-D elements to represent blocks and 2-D elements to represent fractures. Continuity in heads is ensured by using the same nodes for representing fractures and blocks. Mass balance is ensured by assembling together the 2-D and 3-D elements. The only numerical difficulty is related to discretization, which is discussed in Section 2.3.

2.3. Discretization and Calibration

Numerical solution of the flow problem requires a model that discretizes both the fractures and the porous blocks, unless the latter are homogeneous. In such a case, the use of a boundary method for solving the flow equation may be advantageous [*Shapiro and Anderson*, 1985]; otherwise, a complex grid would be needed. Note that flow in the matrix will tend to be perpendicular to fractures and large gradients are likely to occur during transient responses. Therefore, the 3-D grid must be refined close to the fractures. This may lead to very thin elements, which in turn may cause spurious effects. These are reduced by ensuring that discretization adjacent to the fracture is perpendicular to it.

Calibration of this type of model has few specific traits. We advocate the use of automatic calibration techniques [*Carrera and Neuman*, 1986; *Medina and Carrera*, 1996] to focus on the identification of the conceptual model (hand calibration may be time consuming). In fact, the mixed approach may require calibrating different combinations of fracture geometries. This can only be done realistically if each model can be calibrated in a short time. Model structure identification criteria [*Carrera and Neuman*, 1986] may help in selecting one model among a set of alternatives.

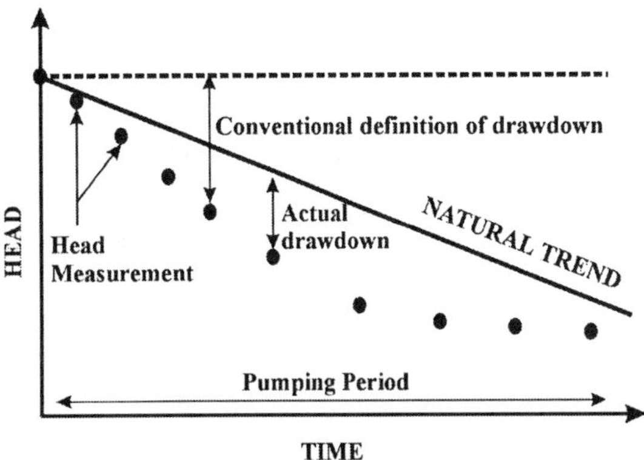

Figure 2. Filtering the natural trend out of head responses to pumping. Notice that ignoring natural trends (the conventional approach) may result in significant underestimation of drawdown.

An issue that may require special attention when modeling long-term pumping tests in low-permeability environments is the definition of drawdowns. They are usually taken as the difference between measured (during pumping) and initial head. However, this definition may not be appropriate when the response to pumping is weak, possibly smaller than natural head fluctuations. In such cases, heads must be observed for a long time both prior to the test and after recovery. This helps in separating natural trends from the response to pumping (Figure 2).

Another potentially important issue in low-permeability fracture media is the short-circuiting that may be caused by the observation wells. In some cases they have to be explicitly included in the model. This can be done with the help of 1-D elements embedded in the 3-D grid.

Further discussion on the calibration of this type of models is provided by *Carrera et al.* [1996], who also discuss some issues related to solute transport that are not treated here.

3. TEST CASES

3.1. Case 1—Chalk River Block

A week-long pump test was performed within the Chalk River Nuclear Laboratories (CRNL) in Canada. The flow

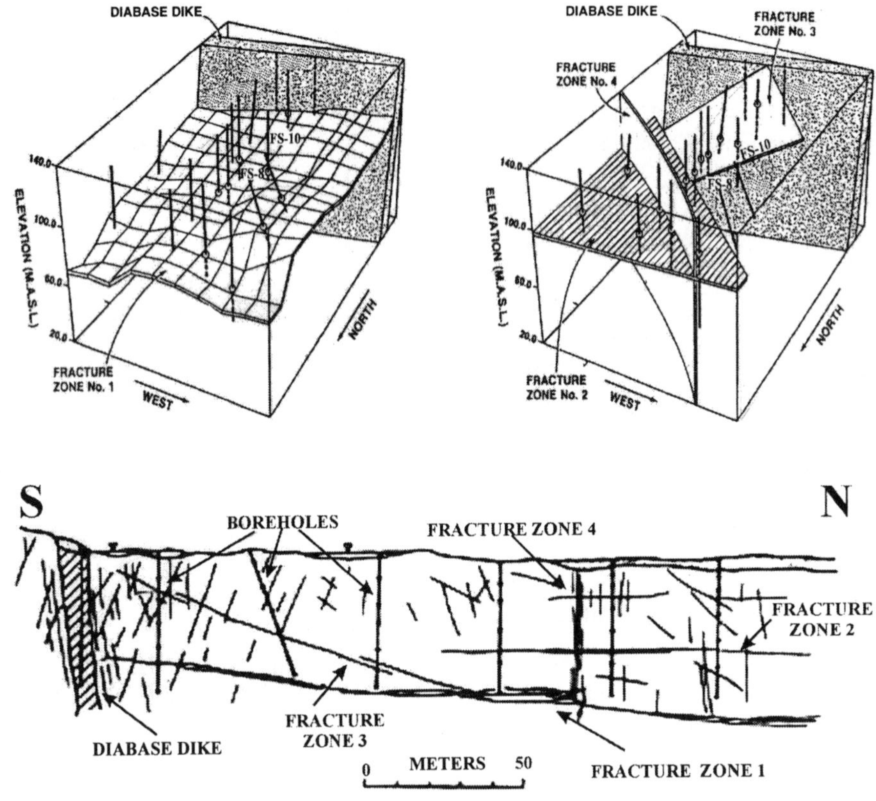

Figure 3. Schematic description of the Chalk River Block conceptual model, including the dominant fractures and the location of boreholes: (a) Block diagrams, (b) Vertical cross section.

region is well characterized by means of surface and downhole geophysics, surface mapping, and hydraulic testing [*Raven*, 1985]. The main rock unit is a folded sheet of quartz monzonite, overlain and underlain by paragneiss with inclusions of metagabbro, diabase, and pegmatite. Although the site is complex, both structurally and lithologically, *Raven* [1985] managed to construct a rather clear conceptual model for it, summarized in Figure 3. In essence, flow takes place through four fracture zones (three subhorizontal and one subvertical). The flow domain is bounded by a diabase dyke (assumed impervious) to the south and by a lake (assumed prescribed head) to the north. The eastern and western boundaries were chosen to coincide with natural flow divides. In all, the domain extends approximately 200 × 150 × 50 m in depth.

The following is a summary of the pumping test interpretation performed by *Carrera and Heredia* [1988] using the mixed continuum/discrete fractures approach. Details of this model are also provided by *Carrera et al.* [1990]. The method of *Carrera and Neuman* [1986] was used for automatic calibration, which was performed by matching the response at all observation intervals to pumping in well FS-10. The initial conceptual model (Model 1) was taken directly from *Raven* [1985] and *Resele et al.* [1986]. It assumed all four fractures to be homogeneous and treated the remaining (fractures and blocks) as a porous, homogeneous medium. This model is synthesized in Figure 4. Calibration of this model was satisfactory, yet analysis of residuals (differences between measured and computed heads) suggested that results could be improved by reducing the size of Fracture Zone 1. This led to Model 2. The actual choice of boundaries for the new fracture was done by carefully analyzing the residuals and trying to attribute them to boundary effects.

This type of residual analysis, together with a review of available lithological and structural descriptions, was useful in devising additional model improvements, which led sequentially to Models 3 and 4. Each of these models was an improvement over the previous one, both in terms of objective function and of model structure identification (MSI) criteria. Some statistics on model fit are shown in Table 1. A few of the head fits are shown in Figure 5, pointing out that indeed a very good model fit was obtained throughout the model region.

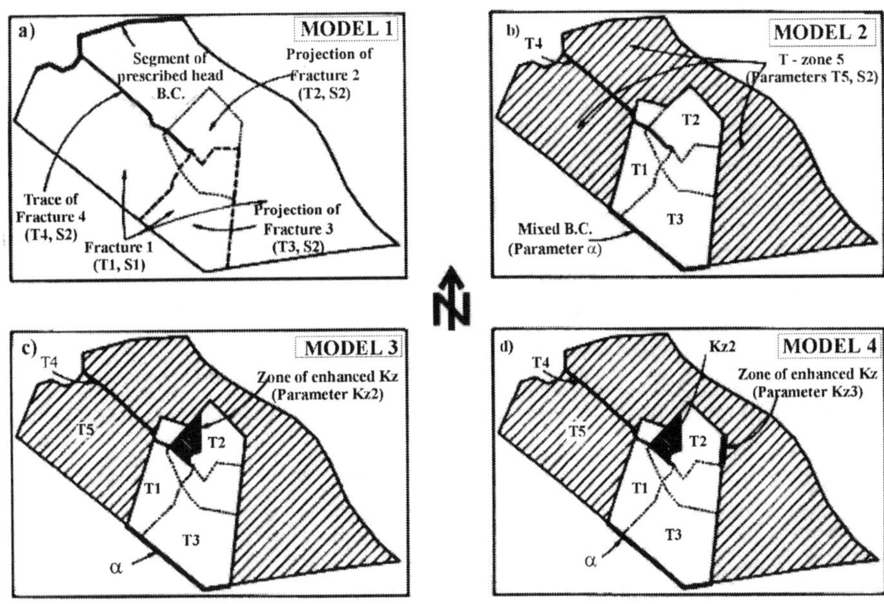

Figure 4. Plan view of the four conceptual models that were used for the Chalk River Block.

The most striking feature of the results is that the transmissivity of Fracture Zone 1 is 20 times larger than the geometric average of the nine single-hole T values available. This result is consistent with steady-state head values, which had not been used during calibration [*Carrera and Heredia*, 1988]. We attribute this discrepancy to channeling in the fracture plane, which may lead to effective T values much larger than the geometric average of point T values [*Sanchez-Vila et al.*, 1996; *Meier et al.*, 1998].

The four models were used to predict four pumping tests performed in different fractures (one in Fracture 1, another one in Fracture 2, and two in Fracture 3). In general, actual measurements were within confidence bounds, although uncertainty was slightly underestimated (Figure 6). Moreover, Model 4, which had been ranked as best by MSI criteria during calibration, was the one leading to fewest errors during prediction. On the other hand, the differences in predictive capabilities of all the models were not as marked as their calibration performances (Table 1). That is, Model 4 led to significantly better calibration results than Model 1, yet, its predictive capabilities were only slightly better than those of Model 1. We take this as indicative of the robustness of the approach. A modeler that had stopped at Model 1 would not have fared much worse than we did, provided that the dominant fractures had been properly included. It is unclear whether this conclusion can be extended to cases in which the system was less stressed, so that data are not as informative as they were here.

3.2. Case 2— El Cabril

El Cabril is the site for storage of low and intermediate level radioactive waste of Spain. Extensive characterization has been performed to control head fluctuations underneath the storage site, define an observation network to monitor possible transport pathways, and assess the retention properties of the geological medium. The site consists of layers of gneisses and meta-arkoses bounded to the west by quartzites. Available data in this case consisted of a very careful surface geological map, including detailed examination of fracture surfaces to seek indications of

Table 1. Summary of model fit and model prediction criteria for the Chalk River Block model.

Criterion	Model 1	Model 2	Model 3	Model 4
N_p	9	11	12	13
RSME (m)	0.38	0.45	0.23	0.21
S_l	1625	1322	1225	1224
d_k	1650	911	53	–14
Prediction				
RSSR	1.16	0.979	0.960	0.945
SRNK	12	12	9	7

N_p, number of parameters; RMSE, Root Mean Square Error; S_l, Support Function; d_k, Kashyap criterion [*Kashyap*, 1982] for Model Structure Identification; RSSR, Relative Sum of Squared Errors; SRNK, Sum of the prediction rankings for each test. See *Carrera et al.* [1990] for details.

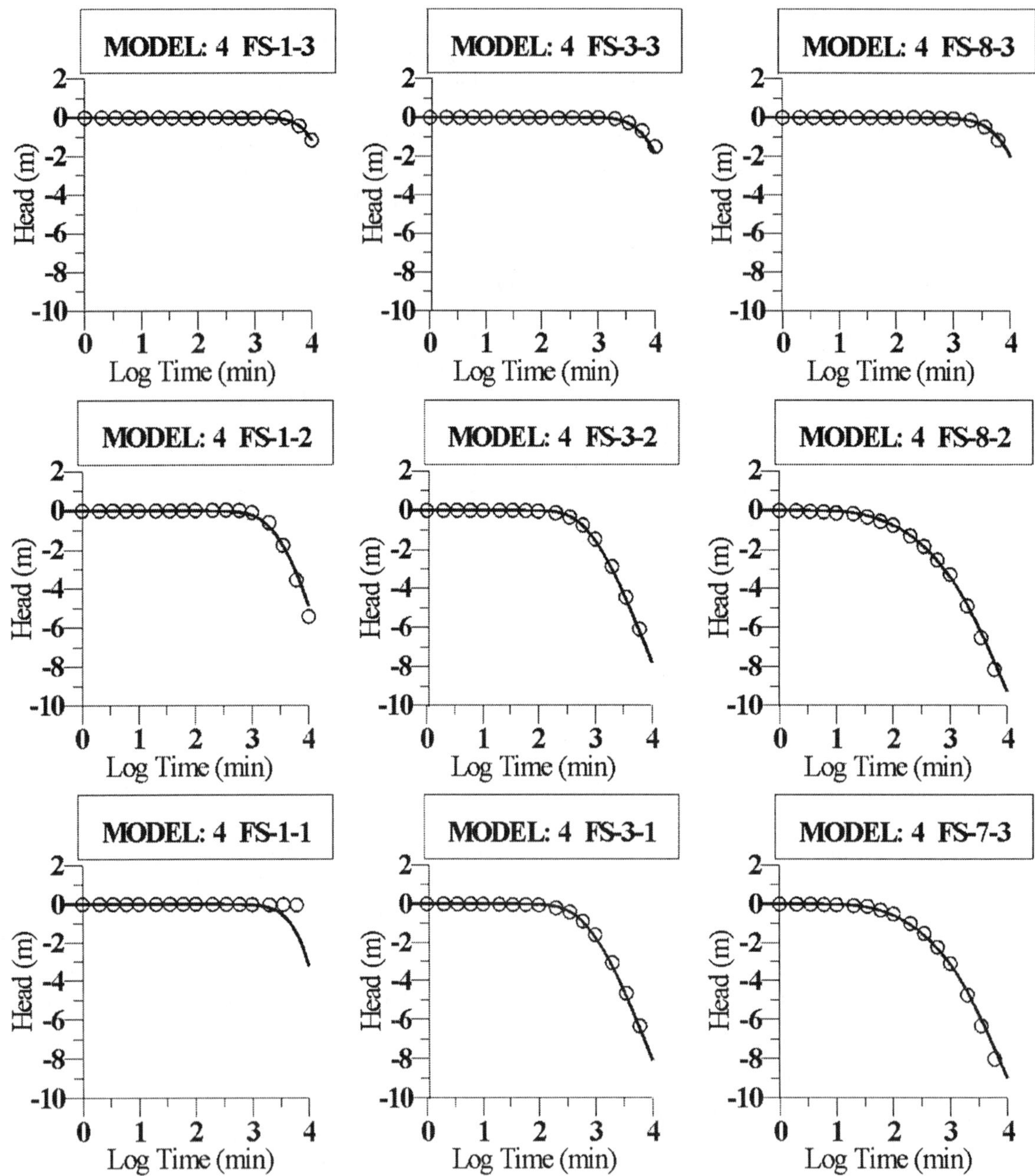

Figure 5. Calibration of FS-10 pump test using Model 4. Shown are the computed (continuous lines) and measured (dots) drawdowns versus time at selected observation points.

direction and type of displacement. This information, complemented with other information (fluid inclusions, fission tracks, break-outs, etc), helps in deriving the stress tensor. Approximately 100 boreholes were available. Short pumping or slug tests had been performed in most of them. Head data had been recorded in most of them for a few years, at a monthly or biweekly frequency.

The large scale (several km) of this model, together with a type of forcing (rainfall recharge), led to formulating the modeling procedure as that of conventional aquifers. That is, a significant effort was devoted to defining boundary conditions, natural recharge, etc. Discussing these issues falls outside the scope of the present work; details are given by *Carrera et al.* [1993b] and by *Sanchez-Vila et al.*

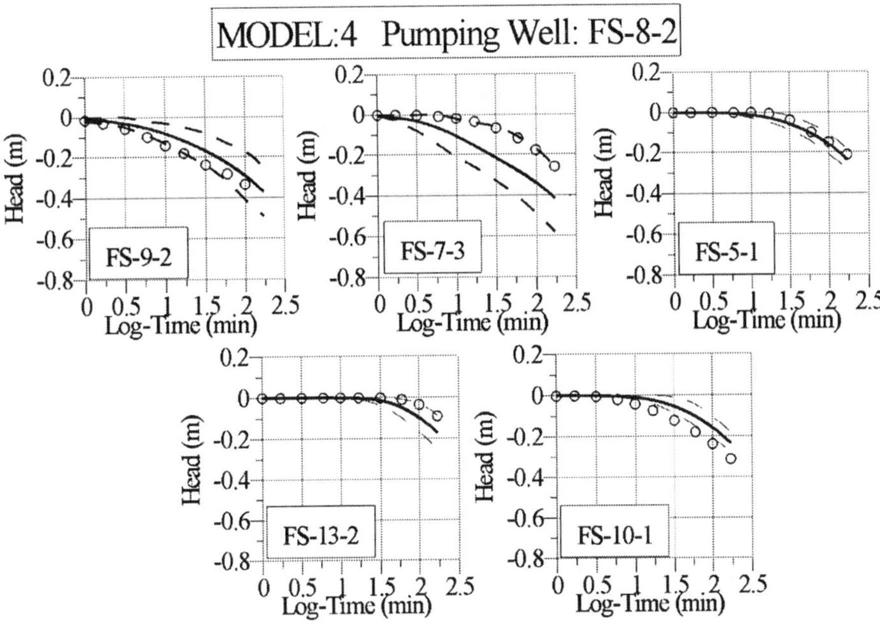

Figure 6. Predicted responses to two pumping tests performed at fractures different from the one pumped for calibration at the Chalk River Block.

[1993]. Note, however, that these issues are important when dealing with large-scale models.

For the purpose of this paper, it is relevant to discuss the way the medium was treated. In all, four models were considered: two were porous and two were mixed. Dominant fractures in the latter were parallel to the direction of schistosity (N-S), which was consistent with tectonic data, the results of a few pump tests, and the patterns of the hydrographs of some wells. All four models fitted historical drawdowns quite well. Figure 7 displays how one of the porous models fitted measured heads, but the fit was qualitatively similar in all models. The main

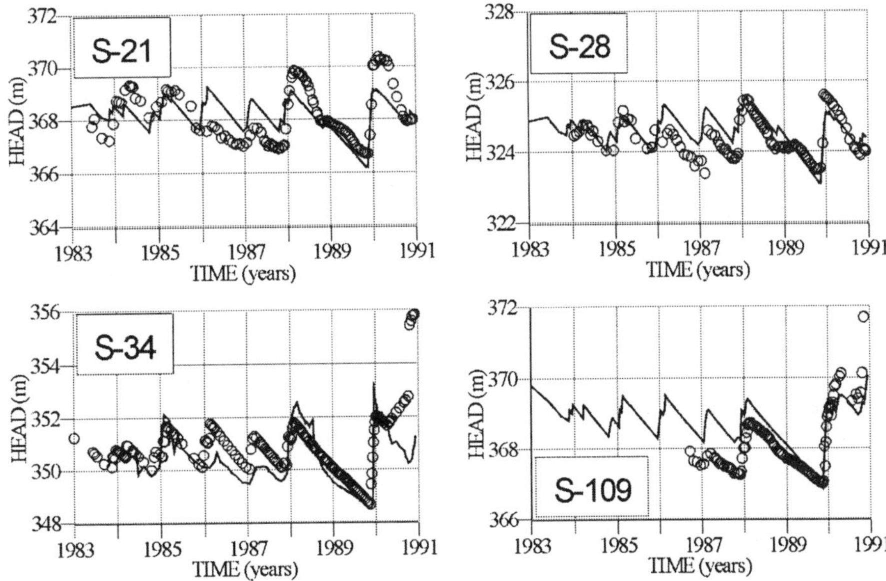

Figure 7. Calibration of El Cabril model. Computed (continuous lines) and measured (dots) evolutions of heads at selected piezometers.

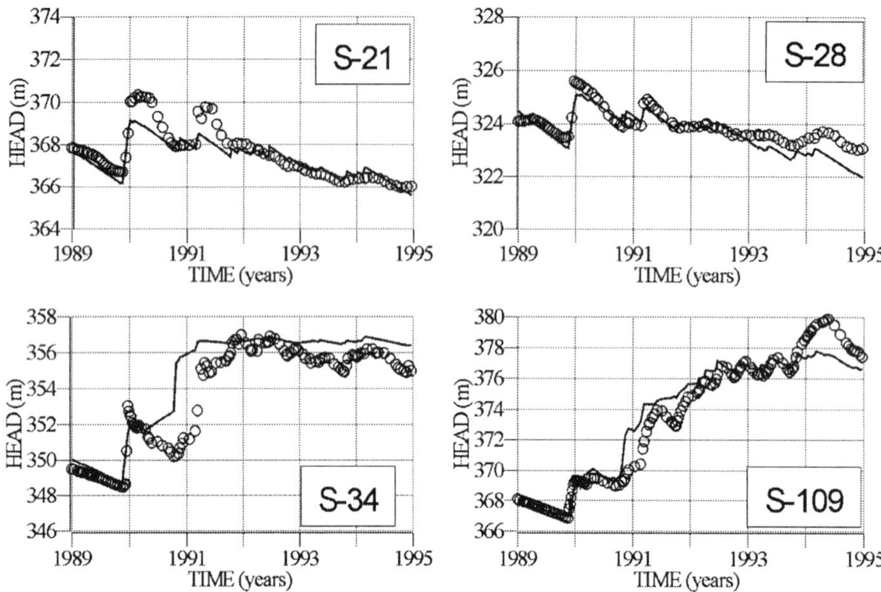

Figure 8. Response to increase in recharge rate at El Cabril. Evolution of measured and computed heads at selected piezometers.

difference in results lies in the fact that estimated conductivities in the porous models, lacking the highly transmissive N-S fractures, were significantly more anisotropic in that direction than the mixed models.

Construction of the storage facility led to eliminating the soil cover, which in turn resulted in a significant increase in recharge and in rising groundwater levels. All four models were able to reproduce the observed groundwater level rises with reasonable values of recharge. Figure 8 displays simulations obtained with the same model as Figure 7, but again predictions were similarly good for all four models. That is, response to changes in recharge was not appropriate for discriminating between the mixed and continuum models in this case, probably because the modified flow conditions were not very different from those during calibration.

On the other hand, response to a long-term pumping test led to significantly different responses in the porous and mixed models. Note (Figure 9) that the two porous models led to very similar results and that both clearly underestimated the response to pumping. The two mixed discrete-continuum models also led to similar predictions, but both led to fair simulations of the observed drawdowns. In summary, the mixed models displayed (in this case) better predictive capabilities than the continuum models. Note also that, as usual, the effect of conceptual model uncertainties is much more important than that of parameter uncertainties [*Carrera et al.*, 1993b].

3.3. Case 3—FEBEX Model

This model was built as part of the hydraulic characterization of the granite block around the FEBEX (Full-scale Engineered Barrier EXperiment), conducted at the Grimsel Test Site in Switzerland. This project aims at testing the viability of a bentonite barrier to encapsulate high-activity radioactive waste. In this context, the objective of the hydraulic characterization was to develop a numerical model of groundwater flow around the experiment tunnel so as to evaluate the distribution of water inflows and provide boundary conditions for hydrothermal modeling.

Available data included the following:

Geological mapping: Lithological and structural maps were made of the tunnels as well as detailed descriptions and photographs of well cores. Dominant features at the large-scale (100 m) are the lamprophyre dyke and the shear zones (Figure 10).

Geophysics: Seismic tomography and georadar measurements were carried between wells BOUS-1, 2, and 3 (See Figure 10). (Results indicated very weak correlation with actual hydrogeological observations.)

Water inflow: A method was developed to measure the spatial distribution of water inflows into tunnels excavated in low-permeability rocks. This method consists of evaluating the increase in weight per unit time for layers of absorbing material (cellulose) applied on the tunnel wall.

The method is described by *UPC* [1998]. In addition to the spatial distribution of water inflow, total inflow rate was also measured by stopping ventilation and collecting all the water seeping along the tunnel walls.

Hydraulic measurements at boreholes: Twenty-three boreholes with depths ranging from 7 to 150 m were available. Intact core was extracted from most of them. They were equipped with packers to isolate the most permeable fractures, which resulted in 42 observation intervals. The testing sequence for each interval included one or several pulse tests, measurement of the natural discharge rate when the interval was connected to the tunnel, and measurement of the pressure recovery after shutting the interval.

Pumping tests: Five crosshole pumping tests were performed in the area surrounding the last 20 m of the tunnel (what is denoted *experiment scale* in Figure 10).

Crosshole test data ended up being the most important set of data from the standpoint of hydraulic characterization. A preliminary interpretation was done by using the Theis model to match drawdowns at each observation well separately (one by one). Clearly, field conditions were far from meeting the Theis assumptions. One may argue that field conditions do not meet the Theis assumptions. Actually, the motivation for this interpretation is twofold. First, it provides an order of magnitude estimate of transmissivity. Second, it helps in identifying fracture connections, as discussed below.

Results of this interpretation are shown in Figure 11 for one of the tests (notice that the horizontal axis is t/r^2 in log scale, which should filter distance effects). That all drawdown curves show similar slopes but intercept the horizontal axis at different times explains why transmissivity estimates range over such a narrow interval (between 0.77 and 1.4 10^{-9} m^2/s), while estimated storage coefficients span an interval from 1.1 to 1500 × 10^{-8}. *Meier et al.* [1998] attribute this type of behavior to spatial variability of hydraulic conductivity. In fact, estimated transmissivities can be taken as large-scale representative values, virtually independent of local conditions around the pumping and observation wells.

On the other hand, the estimated storativity reflects how fast the observation well responds to pumping. This storativity may be related to the degree of hydraulic connection between pumping and the observation well. A low storativity (fast response) indicates a good connection. In fact, storage coefficients thus derived were used for defining the extent of the most conductive fractures. [Note that testing procedures were extremely careful (*Meier et al.*, 1995).] Therefore, trapped air or compliance effects can be discarded as an explanation for the variability of estimated storativities. Moreover, such effects would normally yield very high storativities, while those estimated are small.

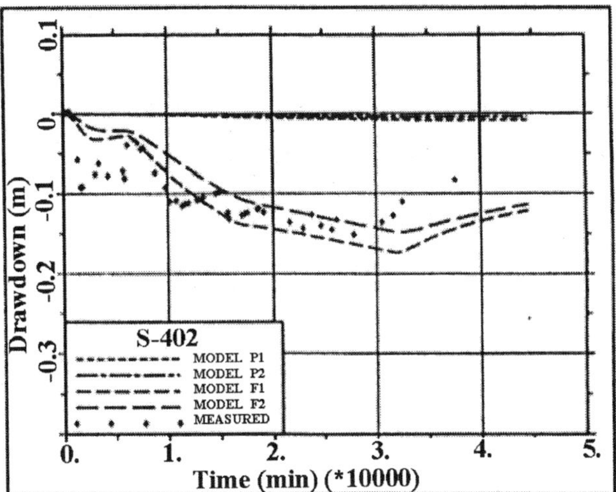

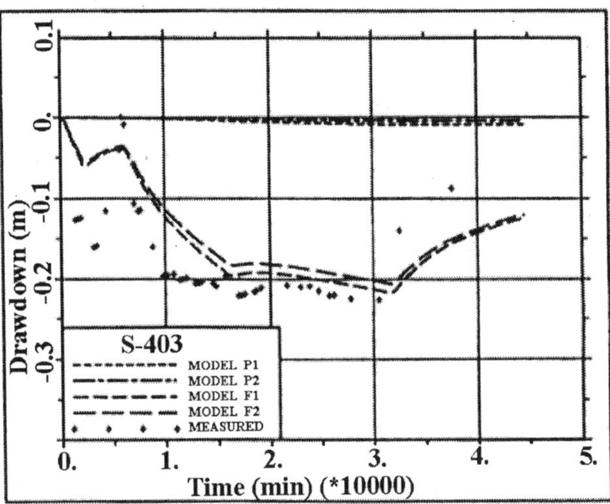

Figure 9. Response to pumping in well S-36 measured and computed with the four models. Notice that the two porous models (P1 and P2), which had properly modeled natural head fluctuations and the response to recharge increase, underestimate drawdowns. On the other hand, the mixed models (F1 and F2) produce reasonable predictions.

This approach for deriving connectivity patterns from the one-by-one separate interpretation of drawdown curves is simple and fast. The connectivity patterns were used to build 3-D mixed discrete-continuum models for interpreting each crosshole test (Figure 12). Here, we should stress that all drawdown curves are interpreted jointly, so that the parameters thus derived ought to be rather robust, even if the individual fits are not as good as those obtained by the separate interpretations. The values of transmissivity obtained for the dominant fracture of each

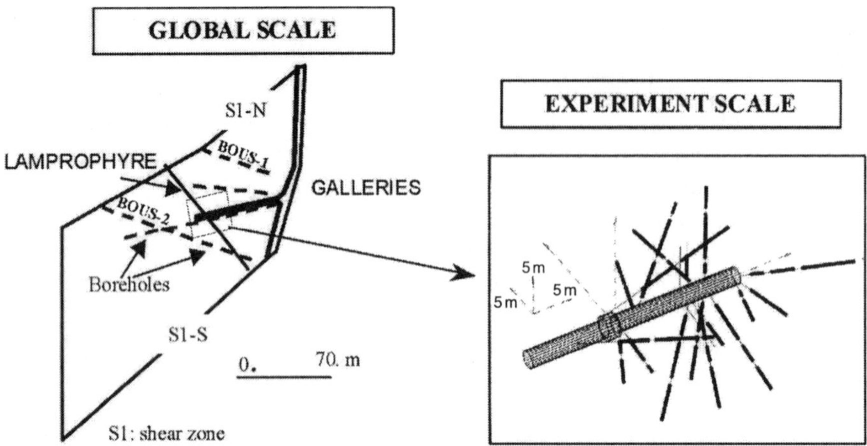

Figure 10. Schematic description of the FEBEX site showing boreholes and tunnels at the experiment and 100 m scales.

test were comparable to, but consistently larger than, the ones obtained by the Theis method (see Figure 13). The explanation for this bias is not entirely clear to us. It may reflect the fact that the fractures are not perpendicular to the borehole or that flow is not 2-D, as required by the Theis model.

A large-scale model was then built by geometric integration of all mapped fractures. Basically, tunnel maps, borehole logs, and (to some extent) geophysics were used to define shear zones, dykes, and dominant fractures. Patterns of tunnel inflows were also very useful. Values of transmissivity for fractures and fractured zones were taken from the crosshole models, as were the values of hydraulic conductivity for the blocks in between fractures.

The issue of density of fractures versus density of information is relevant to constructing the large-scale model. Four fracture zones were considered important around the experiment area when interpreting the crosshole test. Actually, there is nothing special about this area: a comparable amount of important fractures likely would appear in the remaining part of the domain. If they are not explicitly included in the model, then they are treated as porous. Obviously, the permeability of the resulting porous medium must be higher than the one in which the fractures were explicitly modeled. To address this issue in the FEBEX case, two regions were defined for the hydraulic conductivity of the porous blocks, one around the experiment area and one for the rest, where density of information was small. Obviously, hydraulic conductivity of the latter was much larger (actually, it was derived by

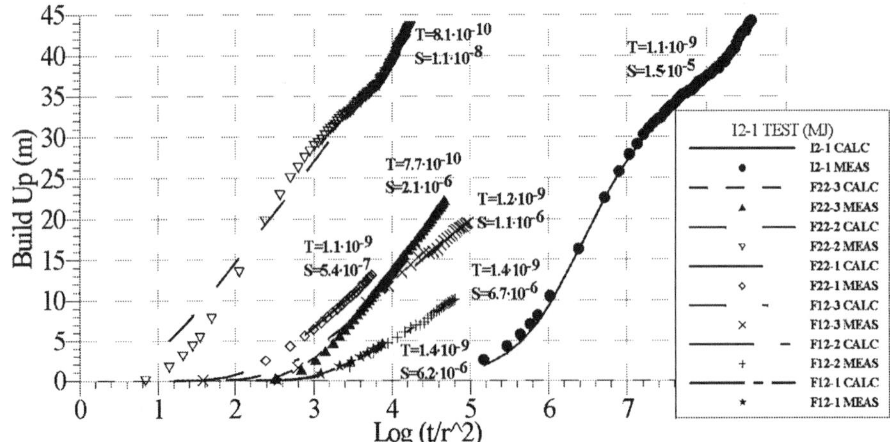

Figure 11. Separate (one by one) Theis model interpretation of drawdowns observed in response to pumping interval I2-1. Notice that t/r^2 is used as horizontal axis. Wellbore storage was taken into account in interpreting drawdowns at I2-1. Notice that T values range between 0.77 and 1.4 10^{-9} m^2/s, while S values range between 1.1 and 1500 10^{-8}.

simulating flow through the models used for interpreting the crosshole tests).

Lateral boundaries were chosen to coincide with shear zones, which are known to be rather transmissive. Outer boundary heads were fixed to values derived from a regional model [*Voborny et al.*, 1991]. Four models were built to accommodate uncertainties in the parameters and the geometry of boundary conditions.

The resulting models were capable of reproducing steady-state heads, both before and after excavation of the experiment tunnel. (Heads were available at four wells drilled from the access tunnel prior to excavating the experiment tunnel.) In itself, this result does not ensure model quality because outer boundary heads were modified to improve model fit. Still, some indication of good quality is derived from the fact that the two steady-state conditions (prior to and after the excavation) were properly reproduced. Furthermore, the model was capable of accurately simulating the overall water inflow into the experiment tunnel, a piece of information that had not been taken into account during calibration. Again, this result does not ensure model quality because the tunnel neighborhood had been intensely characterized. In fact, all things considered, we feel that confidence in model quality is best gained by analyzing the consistency that the model provides to all available measurements, as discussed below.

Prior to the model, it was felt that scale effects on hydraulic conductivity were being observed at this site. That is, K values derived from pulse tests suggested a value of effective permeability in the 10^{-11} to 10^{-10} m/s range. A slightly larger value (around 10^{-10} m/s) seems more

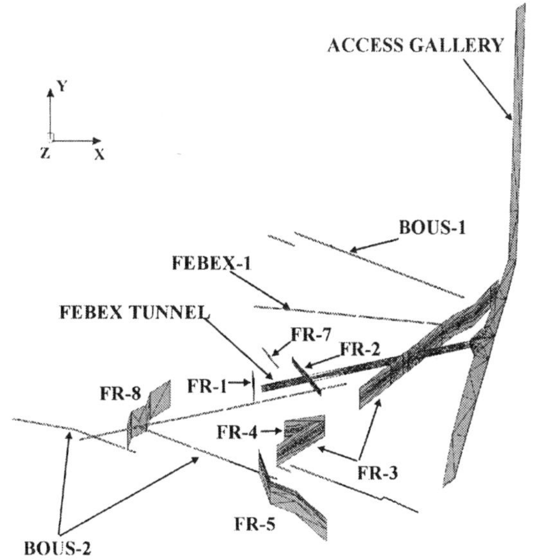

Figure 12. Fracture discretization and position at the large-scale model.

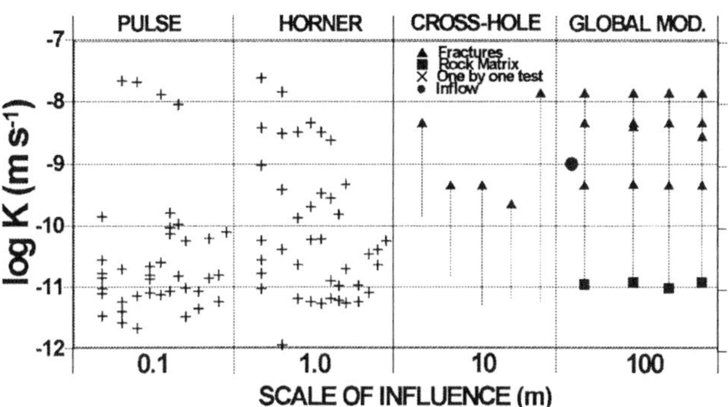

Figure 13. Schematic description of T values obtained from tests performed at different scales at the FEBEX site. Notice that the vast majority of small scale (0.1–1 m) measurements fall below 10^{-10} m/s, while the one-by-one interpretation of all crosshole tests yields values above 10^{-10} m/s. Tunnel inflow require conductivities of around 10^{-9} m/s. Modeling makes it apparent that large-scale behavior is controlled by fractures that occupy a small portion of the flow domain.

appropriate if one examines the Horner data, which average a slightly larger rock volume. Yet, K values derived from inflow measurements appear to be around 10^{-9} m/s. (All the values are summarized in Figure 13.) This suggests that the value of effective permeability is increasing with the model scale. As it turns out, the modeling exercise makes it apparent that what Horner and, especially, pulse tests tend to reflect is the large proportion of low-permeability areas. Large-scale behavior is controlled by the dominant fractures. These occupy a relatively small portion of the flow domain and thus affect only a few pulse tests, but their large lateral extent makes them important. Also, conventional pumping tests lead to consistent values of transmissivity that can be incorporated in large-scale models. Further details on this case are given in the UPC report [*UPC*, 1998].

4. SUMMARY AND CONCLUSIONS

The objective of this paper is to summarize the experiences of model predictions performed using the mixed continuum-discrete approach for fracture flow modeling. Three case studies are discussed. The proposed approach has been useful for integrating different types of data and has displayed some predictive capabilities in all three cases.

In the first case (Chalk River Block), the model was capable of simulating four pump tests performed in different fractures. We emphasize that the model had been calibrated against a much longer-lasting test performed in yet another fracture. Also, this test was calibrated using four sequential models, with each model an improvement over the previous one. Yet, while predictive performance improved with model calibration, it was not overly sensitive to model details. Certainly, differences in prediction capabilities were much smaller than in calibration. This suggests that the approach is rather robust (or, rather, that the data were sufficiently informative to lead to a robust model).

The second case study (El Cabril) consists of a much larger scale model (several km). In this case, the models were calibrated against natural head fluctuations, but extensive downhole and surface mapping data were also taken into account. Two continuum (porous) and two mixed (discrete fractures-continuum) models were calibrated. All four predicted quite well the head rises caused by increases in recharge rate. However, the two mixed models were clearly superior in predicting the response to a long-term pump test.

Finally, the third case study refers to the investigations of the FEBEX at the Grimsel Test Site. The models were calibrated against five crosshole tests. Much complementary information was also available, including geophysics, core descriptions, and tunnel mapping. The resulting models predicted quite accurately the heads around and inflows to the FEBEX tunnel. More importantly, the model was useful for explaining the shift in apparent conductivity from the single borehole scale to the tunnel scale.

To summarize, in all three cases the model could predict, within reasonable error bounds, the system response to flow conditions different from those used for calibration. The three cases corresponded to heavily tested media. Yet predictions were made under conditions and scales quite different from calibration. Moreover, the three cases span a broad range of model sizes and rock types. All this lends support to the proposed mixed continuum-discrete approach to simulate flow through fractured media.

Note that the only preconception of the mixed approach is that a large portion of the flow is concentrated in a few, discrete, fractures. As for the rest, the proposed approach consists of simply following what can be considered good modeling practices.

Acknowledgments. Case 1 was performed in the context of a contract between UPC and NAGRA. Cases 2 and 3 were performed in the context of contracts between UPC and ENRESA. Review and comments by W. A. Illman, V. V. Vesselinov, and B. Faybishenko are gratefully acknowledged.

REFERENCES

Anderson, J., and R. Thunvik, Predicting mass transport in discrete fracture networks with the aid of geometrical field data, *Water Resour. Res.*, *22*(13), 1942–1950, 1986.

Barker, J., A generalized radial-flow model for pumping tests in fractured rock, *Water Resour. Res*, *24*(10) 1796–1804, 1988.

Bear, J., C.-F. Tsang, and G. de Marsily, Flow and Contaminant Transport in Fractures Rocks. Academic Press, New York, N.Y., 1993.

Cacas, M. C., E. Ledoux, G. de Marsily, and B. Tillie, Modeling fracture flow with a stochastic discrete fracture network: calibration and validation. 1: The flow model, *Water Resour. Res.*, *26*, 497–489, 1990.

Carrera, J., and S. P. Neuman, Estimation of aquifer parameters under steady-state and transient conditions: I. Background and Statistical framework, *Water Resour. Res.*, *22*(2), 199–210, 1986.

Carrera, J. L., Vives, J. Heredia, B. Camacho, P. E. Martinez Alfaro, and S. Castano, Modelo de Gestion conjunta de recursos Hidraulicos Superficiales y Subterraneos. Aplicacion a la Cuenca del Rio Guadalhorce. MOPU, Aprovechamiento conjunto de recursos hidraulicos superficiales y subterraneos, *48*, 9–35, 1988.

Carrera J., and J. Heredia, *Inverse Modeling of the Chalk River Block*, HYDROCOIN LEVEL 2/Case5A. NAGRA TB 88-14, 117 pp., 1988.

Carrera, J., J. Heredia, S. Vomvoris, and P. Hufschmied, Modeling of flow with a small fractured monzonitic gneiss block, in *Hydrogeology of Low Permeability Environments, International Association of Hydrogeologist, Hydrogeology: Selected Papers, 2*, edited by S. P. Neuman and I. Neretnieks, pp. 115–167, Heise, Hanover, Germany, 1990.

Carrera, J., S. F. Mousavi, E. Usunoff, X. Sanchez-Vila, and G. Galarza, A discussion on validation of hydrogeo-logical models, *Reliability Engineering and System Safety*, *42*, 201–216, 1993a.

Carrera, J., X. Sanchez-Vila, J. Samper, F. J. Elorza, J. Heredia, J. A. Carbonell, and C. Bajos, Radioactive waste disposal on a highly heterogeneous fracture medium: 1. Conceptual models of groundwater flow, in *Hydrogeology of Hard Rocks*, pp. 203–214, International Association of Hydrogeologists, XXIVth Congress, Oslo, Norway, 1993b.

Carrera, J., L. Vives, P. Tume, M. Saaltink, G. Galarza, J. Guimera, and A. Medina. Interpretation of field tests in low permeability fractured media recent experiences, in *Parameter Identification and Inverse Problems in Hydrology, Geology and Ecology*, pp. 53–70, Kluwer, 1996.

Dershowitz, W. S., *Rock Joint Systems*, Ph.D. dissertation, Massachusetts Institute of Technology, Cambridge, Mass., 1984.

Dershowitz, W., P. Wallmann, J. E. Geier, and G. Lee, *Discrete Fractured Network Modeling of Tracer Migration Experiments at the SCV Site*, SKB Report 91-23, Swedish Nuclear Power and Waste Management Co, Stockholm, Sweden, 1991.

Dverstrop, B., and J. Anderson, Applicant of the discrete fracture network concept with field data: possibilities of model calibration and validation, *Water Resour. Res.*, *25*(3), 540–550, 1989.

Gomez-Hernandez, J. J., H. J. W. M. Hendricks, A. Sahuquillo, and J. E. Capilla, Calibration of 3-D transient groundwater flow models for fractured rock, in *MODELCARE99, Proceedings of the Int. Conf. on Calibration and Reliability in*

Groundwater Modeling, ETH, Zurich, Switzerland, 397–405, 1999.

Gringarten, A.C., Interpretation of tests in fissured and multi-layered reservoirs with double porosity behavior-theory and practice, *J. Petrol Technol.*, *36*(4), 549–564, 1984.

Hsieh, P. A., and S. P. Neuman, Field determination of the three-dimensional hydraulic conductivity tensor of anisotropic media. 1. Theory, *Water Resour. Res.*, *21*(11), 1655–1666, 1985.

Hsieh, P. A., and S. P. Neuman, G. K. Stiles, and E. S. Simpson. Field determination of the three-dimensional hydraulic conductivity tensor of anisotropic media. 2. Methodology and application to fractured rocks, *Water Resour. Res.*, *21*(11), 1667–1676, 1985.

Jones, J. W., E. S. Simpson, S. P. Neuman, and W. S. Keys, *Field and Theoretical Investigations of Fractured Crystalline Rock near Oracle, Arizona*, CR-3736. U.S. Nuclear Regulatory Commission, Washington, D.C., 1985.

Kashyap, R. L., Optimal choice of AR and Ma parts in autoregressive moving average models. *IEEE Trans. Pat. Anal. Intell.*, *PAMI-4*(2), 99–104, 1982

Kimmeier, F., P. Perrochet, R. Andrews, and L. Kiraly, *Simulation par Modele Mathematique des Ecoulements Souterrains entre les Alpes et la Foret Noire.* NAGRA Technischer Bericht, NTB 84-50, 1985.

Long, J. C. S., S. Remer, C. R. Wilson, and P. A. Witherspoon, Porous media equivalents for networks of discontinuous fractures, *Water Resour. Res.*, *18*(3), 645–658, 1982.

Long, J. C. S., P. Gilmour, and P. A. Witherspoon, A model for steady state flow in random, three dimensional networks of disk-shaped fractures, *Water Resour. Res.*, *21*(8), 1150–1115, 1985.

Medina, A., and J. Carrera, Coupled estimation of flow and solute transport parameters, *Water Resour. Res.*, *32*(10), 3063–3076, 1996.

Meier, P., P. Fernandez, J. Carrera, and J. Guimerà, *Results of Hydraulic Testing in Boreholes FBX-95.001, FBX-95.002, BOUS-85.001 and BOUS-85.002*, FEBEX Project, Phase 1. E.T.S.E.C.C.P., Barcelona, Spain, 1995.

Meier, P., J. Carrera, and X. Sanchez-Vila, An evaluation of Jacob´s method work for the interpretation of pumping tests in heterogeneous formations, *Water Resour. Res.*, *34* (5), 1011–1025, 1998.

Moreno, L., and I. Neretnieks, Fluid flow and solute transport in a network of channels, *J. Contam. Hydrol.*, *14*(3-4), 163–192, 1993.

National Research Council, *Rock Fractures and Fluid Flows Contemporary Understanding and Applications*, National Academy Press., Washington, D.C., 1996.

Neretnieks, I., A note on fracture flow mechanisms in the ground, *Water Resour. Res.*, *19*, 364–370,1983.

Neretnieks, I., Channeling effects in flow and transport in fractured rock: Some recent observations and models, *GEOVAL-87*, 315–335, 1987.

Neuman, S. P., E. S. Simpson, P. A. Hsieh, J. W. Jones, and C. L. Winter, Statistical analysis of hydraulic test data from crystalline rock near Oracle, Arizona, *International Association of Hydrogeologists, Memories, Vol. XVII*, 1985.

Neuman, S. P., Stochastic continuum representation of fractured rock permeability as an alternative to the REV and fracture network concepts, *Proceedings, NATO A.R.W on Advances in Analytical and Numerical Groundwater Flow and Quality Modeling*, Lisbon, Portugal, 331–362, 1988.

Raven, K. G., *Hydraulic Characterization of a Small Groundwater Flow System in Fractured Monzonitic Gneiss, A Report on Hydrogeologic Research Activities for Atomic Energy of Canada Ltd.,* Applied Geoscience Branch, Pinawa, Manitoba, Canada, 1985.

Resele, G., C. Wacker, and D. Job, *Chalk River Block: Application of Goodness-of-Fit Measures to the Calibration of Chalk River Block Flow System*, Unpublished Report, Motor Columbus Consulting Engineers Inc., Baden, Switzerland, 1986.

Sanchez-Vila, X., J. Carrera, J. Samper, F. J. Elorza, J. Heredia, J. A. Carbonell, and C. Bajos, Radioactive waste disposal on a highly heterogeneous fractured medium: 2. Numerical models of flow and transport, in *Hydrogeology of Hard Rocks,* pp. 203–214, International Association of Hydrogeologists, XXIVth Congress, Oslo, Norway, 1993.

Sanchez-Vila, X., J, Carrera and J. Girardi, Scale effects in transmissivity, *J. Hydrol.*, *183*, 1–22, 1996.

Sanchez-Vila, X., P. Meier, and J. Carrera, Pumping tests in heterogeneous aquifers: An analytical study of what can be obtained from their interpretation using Jacob's method. *Water Resour. Res.*, *35*(4), 943–952, 1999.

Shapiro, A. M., and J. Anderson, Simulation of steady state flow in three-dimensional fracture networks using the boundary element method, *Advances in Water Resources*, *8*(3), 1985.

Shapiro, A. M., and P. A. Hsieh, Research in fractured-rock hydrogeology: characterizing fluid movement and chemical transport in fractured rock at the Mirror Lake drainage basin, *Proceedings of the Technical Meeting of U.S. Geological Survey Toxic Substances Hydrology Program,* Monterey, Calif., March 11–15, edited by G. E. Mallard and D. A. Aronson (also, Water Resources Investigation Report 91-4034, U. S. Geological Survey), Reston, Va., 1991.

Smith, L. and F. W. Schwartz, An analysis of the influence of fracture geometry on mass transport in fractured media, *Water Resour. Res.*, *20*(9), 1241–1252, 1984.

UPC, *Hydrogeological Characterization and Modeling*, FEBEX. Report Prepared for ENRESA, School of Civil Engineering, Barcelona, Spain, 1998.

Voborny, O., P. Adank, W. Hürlimann, S. Vomvoris, and S. Mishra, *Grimsel Test Site: Modeling of Groundwater Flow in the Rock Body Surrounding the Underground Laboratory*, NAGRA Tech. Rep. 91-03, Switzerland, 1991.

Jesus Carrera and Lurdes Martinez-Landa, Department of Geotechnics and Applied GeoScience, School of Civil Engineering, Technical University of Catalonia, Barcelona, 08 034 Spain

A Discrete-Fracture Boundary Integral Approach to Simulating Coupled Energy and Moisture Transport in a Fractured Porous Medium

Stuart Stothoff

Stothoff Environmental Modeling, Houston, Texas

Dani Or

Utah State University, Logan, Utah

The high-level waste repository proposed for Yucca Mountain, Nevada, would be located in unsaturated, highly fractured, densely welded tuff. The spacing between fractures is sufficiently large relative to drift dimensions to render somewhat suspect predictions using continuum methods for simulating flow and energy transport. On the other hand, the spacing is sufficiently small to render standard discrete-fracture methods extremely computationally demanding. Using boundary integral approaches, discrete-fracture methodology is developed to overcome some of the limitations of standard computational methods. Fractures are discretized with standard finite volume methods, while each block between fractures is assumed to have piecewise-constant (although possibly time-varying) properties. With the assumption of piecewise-constant properties, the governing equation in the matrix blocks is transformed into a surface integral, obviating the need for a computational mesh within matrix blocks. Two formally equivalent alternative formulations may be used: discrete jump or multiple zone. Equations describing material-property changes, moving boiling fronts, and coupled fluid and energy transport in discrete fractures are presented, as well as a simple example demonstrating some effects of discrete fractures on unsaturated flow.

Two-phase flow in a fracture system adds considerable complexity relative to single-phase flow, and constitutive relationships are still relatively undeveloped. The liquid phase has the least developed theory, although it is of primary interest at Yucca Mountain. Ongoing efforts to extend constitutive theory for flow in a rough-walled fracture are discussed here, describing a fracture surface using partially connected pits and plateaus. Computational issues involved with routing film flow through wide-aperture fracture intersections are also discussed.

1. BACKGROUND

The high-level waste repository proposed for Yucca Mountain (YM), Nevada, would be located in the Topopah Springs welded (TSw) unit, a heavily fractured, densely welded tuff. The matrix has sufficiently low permeability

Dynamics of Fluids in Fractured Rock
Geophysical Monograph 122

that the fracture system is likely to be a strong participant in conducting fluxes through the unit, particularly under wetter climatic conditions. In addition to discrete-fracture pneumatic simulations using FracMan and the empirical WEEPS model, the fracture system has been incorporated into the Viability Assessment [*U.S. Department of Energy*, 1998] using continuum modeling approaches. While the assumption of a fracture continuum may be appropriate at certain scales, the applicability of fracture-continuum models is questionable when the scale of the problem is small compared to the size of the fracture spacing. In particular, a representative elementary volume may be larger than the grid blocks used in drift-scale simulations, motivating investigation of alternative simulation approaches.

Discrete-fracture methods are an alternative to fracture-continuum approaches. Standard finite volume and finite element methods can be used for simulating discrete fractures. When using discrete-fracture approaches, often the matrix is neglected and the fractures handled as conductive disks or converted into equivalent pipe networks [*Dershowitz and Fidelibus*, 1999]. In the YM system, however, the matrix is permeable enough relative to fluxes that matrix participation is likely to be significant. Disparities between time and space scales and between fractures and the matrix, however, place stringent gridding and time-stepping requirements that cause computational burdens for standard domain simulators.

Boundary integral methods have been used successfully to model saturated flow, unsaturated flow, and two-phase flow in porous media, as well as heat and electrical conduction problems. Boundary integral methods are particularly well suited to problems governed by Laplace's equation and can be used everywhere, except along discrete discontinuities, where the discontinuities may move or be governed by nonlinear equations. For suitable problems, two particular advantages of boundary integral methods are that the solution is analytic within the domain (subject to boundary discretization error) and the entire mesh is located only on discontinuities (e.g., fractures and boundaries) with no need to discretize the domain. Thus, the number of unknowns to be determined is greatly reduced and flexibility in describing geometries (e.g., fractures and drift walls) is greatly facilitated relative to domain methods. A significant disadvantage of boundary integral methods is the restriction on governing equations to potential-based problems with spatially invariant coefficients. Although more general situations may be considered, the additional work required to handle the complications (e.g., meshing the domain) renders the methodology much less attractive. The resulting linear equations produce an unsymmetric coefficient matrix that is fully populated, limiting the number of unknowns that may be solved for.

Considerations in using boundary integral techniques for solving discrete-fracture problems at the drift scale are presented here. The techniques might be applied across the range of conditions experienced in the near-field YM environment throughout the repository lifetime, such as ambient, thermal-pulse, and altered conditions. The discussion includes (i) boundary-integral formulations for coupled heat and moisture transport within the matrix, (ii) finite-volume fracture-flow formulations, and (iii) methods for reducing computational requirements. Particular emphasis is placed on movement of liquids in fractures under unsaturated conditions where classical constitutive theory is lacking. The emphasis is on presenting methods to extend present approaches. The effectiveness of discrete-fracture, boundary-integral methods for steady-state unsaturated problems is illustrated in a simple example.

2. BOUNDARY INTEGRAL METHODS

Boundary integral techniques are well known [*Jaswon and Symm*, 1977; *Banerjee and Butterfield*, 1981; *Liggett and Liu*, 1983]. Such methods were originally developed for and are most applicable to problems with a potential field; the following development honors this restriction. Boundary integral methods have also been successfully applied to unsaturated media with exponential conductivity [*Pullan and Collins*, 1987] and fracture networks [*Elsworth*, 1986; *Rasmussen*, 1987]. The boundary integral formalism has also been used to define so-called Green elements [*Taigbenu*, 1995], which are a type of mixed finite element. Green elements have been applied to a variety of linear and nonlinear problems, including transient unsaturated flow [*Taigbenu and Okyejekwe*, 1995].

There are several formulations for boundary integral problems. Formulations that work directly with physical quantities are preferred when investigating processes in discontinuities (e.g., fractures, changes in material properties). Two formally equivalent direct formulations may be used to characterize discontinuities: discrete jump or multiple zone. In both cases, compatibility constraints on potential and flux are imposed at the discontinuities. Steady state is assumed in the following development for simplicity, although transient problems could be approached by time stepping or, in the case of linear problems, through Laplace transform techniques.

In the discrete-jump approach, the singularity in the potential field represented by a discontinuity is removed from the domain by a line cut and replaced by a discretized set of jumps. A jump in potential might represent a thin barrier, whereas a jump in the gradient of potential might represent a source or a change in porous medium properties. Potential at a point is found using a generic discrete-jump formulation with

$$\sum_i \alpha_i u_i = \int_\sigma \left(G\frac{\partial u}{\partial n} - u\frac{\partial G}{\partial n} \right) d\sigma + \int_{\sigma^*} \left(-Gq_s - \Delta u \frac{\partial G}{\partial n} \right) d\sigma^* \quad (1)$$

where ϕ is a potential, $u = K\phi$ is also a potential (assuming K is piecewise constant), $\partial u / \partial n$ is the gradient of u perpendicular to the discontinuity, q_S represents a source, G is the Green's function for the operator, σ is the boundary, σ^* is the set of internal discontinuities, Δ represents a jump, and α is a weight that depends on whether the point at which the equation is evaluated is inside, outside, or on the boundary of the domain ($\alpha = 1/2$ on a smooth portion of the boundary). All boundary and internal discontinuities are discretized into elements, with u and $\partial u/\partial n$ interpolated within the elements. The N unknown values are simultaneously found by using Equation (1) at N distinct locations (usually at definition points for unknowns, such as at element centroids). The discretization process results in a fully populated unsymmetric matrix, which limits the number of unknowns that can be effectively determined relative to domain methods. Note that computational effort increases as N^3, and a set of linear equations with a fully populated unsymmetric 1000×1000 matrix requires on the order of one minute to solve on a PC. On the other hand, fewer unknowns are necessary, relative to domain methods, to achieve a given accuracy level.

The multiple-zone approach breaks a domain into smaller subdomains, usually splitting the domain along the same physical discontinuities considered by the discrete-jump method. Additional breaks can be applied to connect physical discontinuities, adding unknown values to be determined. Along the boundaries of the subdomains, compatibility of potential and flux between adjacent subdomains is enforced, which may require that additional unknown values be determined relative to the discrete-jump approach. Thus, each discrete jump is replaced with two unknown values, with compatibility requirements adding an additional equation. Each subdomain may be considered a type of finite element with high-order boundary discretization. By breaking the overall domain into subdomains, there is limited support for unknowns, and the resulting matrix has large empty blocks; however, appropriate solvers can significantly reduce computational effort. As the number of the subdomains increases, the multiple-zone approach reduces to a type of mixed finite-element method.

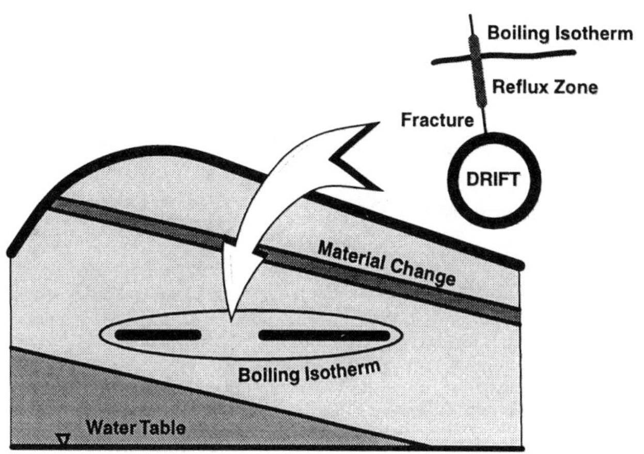

Figure 1. Example of a problem suited to the discrete-jump approach. A single discrete fracture intersecting a discrete shift is in the near field, with a far field that includes repository-scale sources and mountain-scale features. The quasi-steady behavior of refluxing liquids in the fracture may be examined with this model, assuming the boiling front moves with large inertia.

Both approaches are useful for YM applications. The discrete-jump approach is beneficial when there are relatively few discontinuities to consider and the problem is not conveniently broken into multiple zones. The multiple-zone approach is useful when there are numerous physical discontinuities, particularly if they form a regular pattern. When considering a drift-scale problem, one might use a multiple-zone approach for the heavily fractured near-field zone and a discrete-jump approach for the far field (considered as a continuum). Examples where a particular method might be applied are shown in Figures 1 and 2. In Figure 3, a simple example problem compares the quite different computational strategies the two approaches yield using the same discretization.

3. COUPLED MATRIX FLOW AND ENERGY

To effectively use a boundary integral method, one should avoid all domain integrals, typically by replacing them with surface integrals. As the processes in fractures are of primary interest at YM, it may be justifiable to fully emphasize the fracture processes while approximating the matrix processes. The most straightforward approximation is to assume that all coefficients are spatially invariant be-

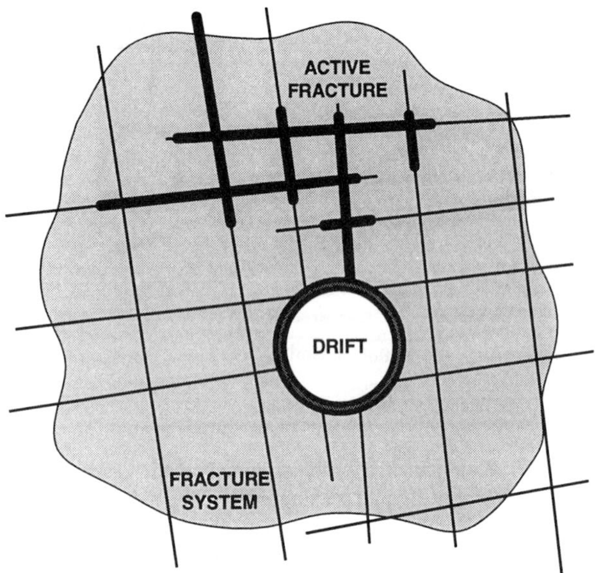

Figure 2. Example of a problem suited to the multiple-zone approach. Many discrete intersecting fractures form a topologically regular grid that is coarsely block banded. Either ambient conditions or thermally perturbed conditions may be examined.

tween discontinuities and dependent variables are not. In essence, the piecewise-constant approximation is equivalent to the assumption that a subdomain is a single finite volume, but with improved resolution of gradients and interface properties. The approximation is quite reasonable for small matrix blocks and becomes weaker as the size of the subdomain increases. Depending on the size of domain blocks, the piecewise-constant approximation is reasonable for the quasi-steady-state conditions assumed for the formulation, but may be unreasonable for highly transient problems. In cases where the matrix properties change in a boundary region near a discontinuity, the compatibility equations between subdomains can be augmented by including one or more layers of elements describing the near-fracture properties.

Following the assumption of piecewise-constant coefficients in a matrix subdomain, the governing equations for liquid, gas, and energy transport may be shown to reduce to a set of coupled Laplacians, with each equation in the form:

$$\langle c_1\rangle\nabla^2 P_l + \langle c_2\rangle\nabla^2\rho_g^2 + \langle c_3\rangle\nabla^2 T + \langle c_4\rangle\nabla^2 z = 0 \qquad (2)$$

where the variables are gas density (ρ_g), liquid pressure (P_l), temperature (T), and elevation (z). A piecewise-constant value of some variable c is denoted $\langle c\rangle$ (note that c may be a function of the dependent variables). A system of coupled total-mass, total-water, and total-energy equations is developed in the appendix.

A significant advantage of the piecewise-constant assumption is that the reduction of the governing equations to coupled Laplacians is particularly convenient. Each Laplacian may be transformed into a surface integral around the subdomains, so the governing equations may be solved with only surface integrals. Thus, there is no need for a mesh internal to any subdomain.

4. COMPATIBILITY EQUATIONS

The general boundary integral approach discussed in Section 2 must be specialized to handle particular situations. Specifically, along internal discontinuities there are twice as many unknown values as there are boundary integral equations available to evaluate them. For example, there is an unknown flux on either side of a discontinuity, but as these fluxes are evaluated at the same spatial location, a boundary integral equation may only be used once for the pair of fluxes. Additional compatibility equations, imposing continuity requirements on the pairs of unknowns, are used to close the equation set. Compatibility equations are also presented in this section that provide specializations appropriate for coupled flow and energy transport.

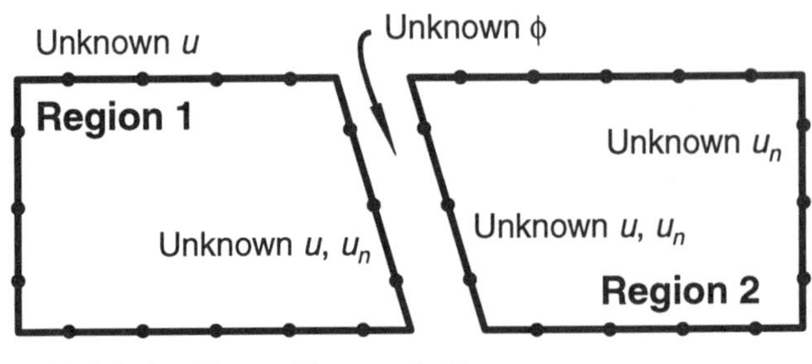

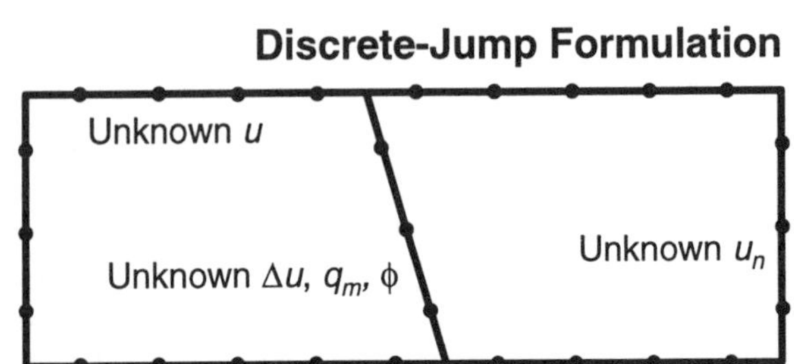

Figure 3. An example problem with formally equivalent multiple-zone and discrete-jump formulations. The notation u_n is shorthand for $\partial u/\partial n$.

In the multiple-zone approach, compatibility equations link potentials and gradients on opposite sides of a discontinuity, making the problem well posed. In the discrete-jump approach, the same compatibility equations are used to place restrictions on the left-hand side of Equation (1). Examples of these compatibility equations are presented in this section. Compatibility equations for fractures have also been developed (not shown) to account for modified resistances due to matrix-matrix contact via asperities and surface coatings on one or both faces. When the discontinuity represents a preferential pathway for flow (e.g., a fracture), it is necessary to add a supplemental set of equations describing flow along the discontinuity with an additional unknown (potential in the fracture). A finite-volume approach is convenient for solving the supplemental equations, consistent with piecewise-constant potentials and gradients in the boundary integrals. If interactions are being simulated that are local to the fracture region, the finite-volume region can easily be extended to include several elements within the matrix. The finite-volume equations are the same for both the multiple-zone and discrete-jump approaches.

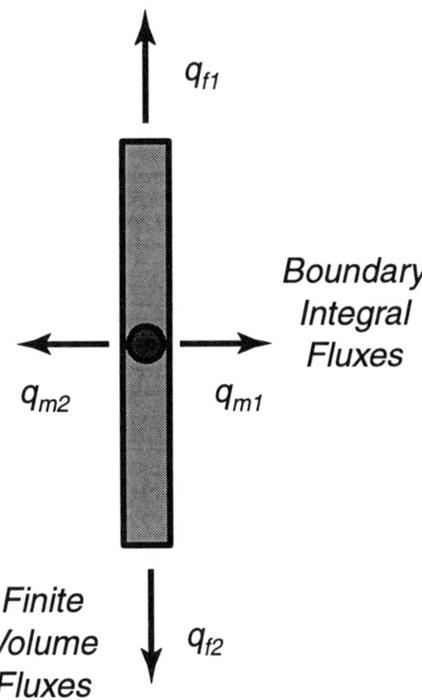

Figure 4. Interaction of one finite volume with matrix and fracture.

4.1 Single-Fluid Material Change (Multiple-Zone Approach)

Potential and flux must be continuous at material property changes (the two sides of the discontinuity are labeled 1 and 2 in Figure 3).

$$\begin{aligned} \phi_1 &= \phi_2 \\ K_1 \frac{\partial \phi_1}{\partial n} &= K_2 \frac{\partial \phi_2}{\partial n} \end{aligned} \tag{3}$$

where ϕ is potential and K is conductivity. Under saturated conditions, if there is a fracture between the two materials and the faces do not touch, an additional fracture equation is added linking lateral flow and interchange between the matrix and fractures. The equation set becomes

$$\begin{aligned} \phi_1 &= \phi_2 \\ K_1 \frac{\partial \phi_1}{\partial n} &= K_2 \frac{\partial \phi_2}{\partial n} + q_{mi} \\ \sum_j C_{ij}\left(\phi_i - \phi_j\right) &+ A_{mf} q_{mi} = 0 \end{aligned} \tag{4}$$

where C_{ij} is the conductance between fracture node i and a connected node j, A_{mf} is the area of exchange between the matrix and the fracture, and q_{mi} is the flux between the matrix and the fracture at node i [in Equation (1), q_s is the difference between q_m on the two sides of the fracture]. Note that local equilibrium between matrix and fracture is assumed. A schematic diagram of fluxes associated with a finite volume is shown in Figure 4, where fluxes with f subscripts are the fluxes exchanging between adjacent finite volumes.

4.2 Single-Fluid Material Change (Discrete-Jump Approach)

Potential and flux must be continuous at material property changes. Imposing compatibility pointwise allows the general form of Equation (1) to be retained, except for the left-hand side, which incorporates the compatibility requirements. The material-property-change compatibility requirements in Section 4.1 are restated by Δu and q_s as

$$\begin{aligned} \frac{K_1 + K_2}{2(K_2 - K_1)} \Delta u &= \int_\sigma \left(G \frac{\partial u}{\partial n} - u \frac{\partial G}{\partial n} \right) d\sigma \\ &+ \int_{\sigma^*} \left(-G q_s - \Delta u \frac{\partial G}{\partial n} \right) d\sigma^* \\ q_s &= 0 \end{aligned} \tag{5}$$

with inclusion of a fracture yielding

$$\frac{K_1+K_2}{2(K_2-K_1)}\Delta u = \int_\sigma \left(G\frac{\partial u}{\partial n} - u\frac{\partial G}{\partial n}\right) d\sigma$$

$$+\int_{\sigma^*}\left(-Gq_s - \Delta u\frac{\partial G}{\partial n}\right) d\sigma^* - \frac{q_{mi}}{2} \tag{6}$$

$$= \frac{\partial}{\partial n}\left[\int_\sigma \left(G\frac{\partial u}{\partial n} - u\frac{\partial G}{\partial n}\right) d\sigma + \int_{\sigma^*}\left(-Gq_s - \Delta u\frac{\partial G}{\partial n}\right) d\sigma^*\right]$$

$$\sum_j C_{ij}(\phi_i - \phi_j) + A_{mf}q_{mi} = 0$$

4.3 Coupled Fluid and Energy in Discrete Fractures

Use of the standard finite-volume approach to characterize flow within fractures provides a great deal of flexibility in modeling. The finite-volume equation for liquid, vapor, or energy as used in Sections 4.1 and 4.2 is extended to coupled equations using the form

$$\sum_j C_{ij}(\phi_i - \phi_j) + A_{mf}\, q_{mi} + Q_i = 0 \tag{7}$$

where Q_i generically represents sources, sinks, coupling terms, and nondiffusive fluxes within the fracture system. For example, Q_i might account for liberation of energy due to condensation and advective transport of energy due to liquid and vapor fluxes. In the solution process, changes to ϕ and q_m are solved for at each iteration, with the solutions to the other terms lagging by one iteration.

The finite-volume formalism straightforwardly allows for portions of the matrix near the fracture, as well as the fracture itself, to be discretized, leaving the bulk of the matrix block to be handled using the boundary integral approach. This procedure has the advantage that thin elements can be used without concomitantly requiring heavy subdomain discretization to maintain mesh integrity. For example, precipitation/dissolution near the fracture or fracture-dominated transport with strong matrix sorption can be simulated with a refined finite-volume mesh near the fracture.

4.4 Moving Phase Changes

When the temperature field adjusts rapidly relative to interface movement, a phase change (treated as a sharp interface) has a continuous temperature across the interface, but perhaps has discontinuous fluxes. Liquid saturation changes sharply across the interface. If a boiling isotherm marks the location of the phase change, requiring that the characteristic velocity of the interface be the same for liquid, vapor, and energy, it provides a compatibility constraint. The characteristic liquid velocity of the interface is

$$v_l = \frac{(q_{l2} - q_{l1})}{(\theta_2 - \theta_1)} \tag{8}$$

where q_l is liquid flux, θ is moisture content, and v_l is the characteristic velocity [*Stothoff and Pinder,* 1992]. Similar relationships hold for vapor and energy transport. With the requirement that all characteristic velocities are the same, rearrangement and simplification yield a compatibility constraint in the form

$$q_l X_l = -q_v X_v + v_e X_e \tag{9}$$

where q_v is vapor flux, v_e is the corresponding characteristic velocity of conductive heat flux, and the X factors account for porosity, heat capacities, latent heat, and saturation differences across the interface. If necessary, the sharp interface can be smeared into multiple moving interfaces. Such an approach can be quite accurate in one-dimension. However, experience [*Stothoff and Pinder,* 1992] suggests that, except in special cases, smearing a phase change using multiple moving interfaces is nightmarish to program in other than 1-D.

5. LIQUID FLOW IN FRACTURES

The computational approaches outlined previously rely on constitutive relationships to characterize fluxes as a function of state variables. The current theory is most poorly developed for unsaturated conditions, especially when the fracture aperture is sufficiently wide that film flow may occur. *Wang and Narasimhan* [1993] show that gravity effects dominate capillary effects for capillaries with radii greater than 3.8 mm, suggesting that capillary-based models are not valid for fractures with apertures greater than this width.

Discrete-fracture modeling places the emphasis on processes that occur within the fractures—precisely where theory is weakest. Efforts to augment present constitutive theory for wide-aperture fractures are discussed in this section. Two areas actively being investigated are the transition from unsaturated flow to film flow and flow routing under unsaturated conditions. These investigations are intended to help provide constitutive relationships to describe liquid flow in fractures when the fracture and surrounding matrix are not fully saturated, perhaps replacing the standard approaches that assume a fracture is a type of coarse porous medium.

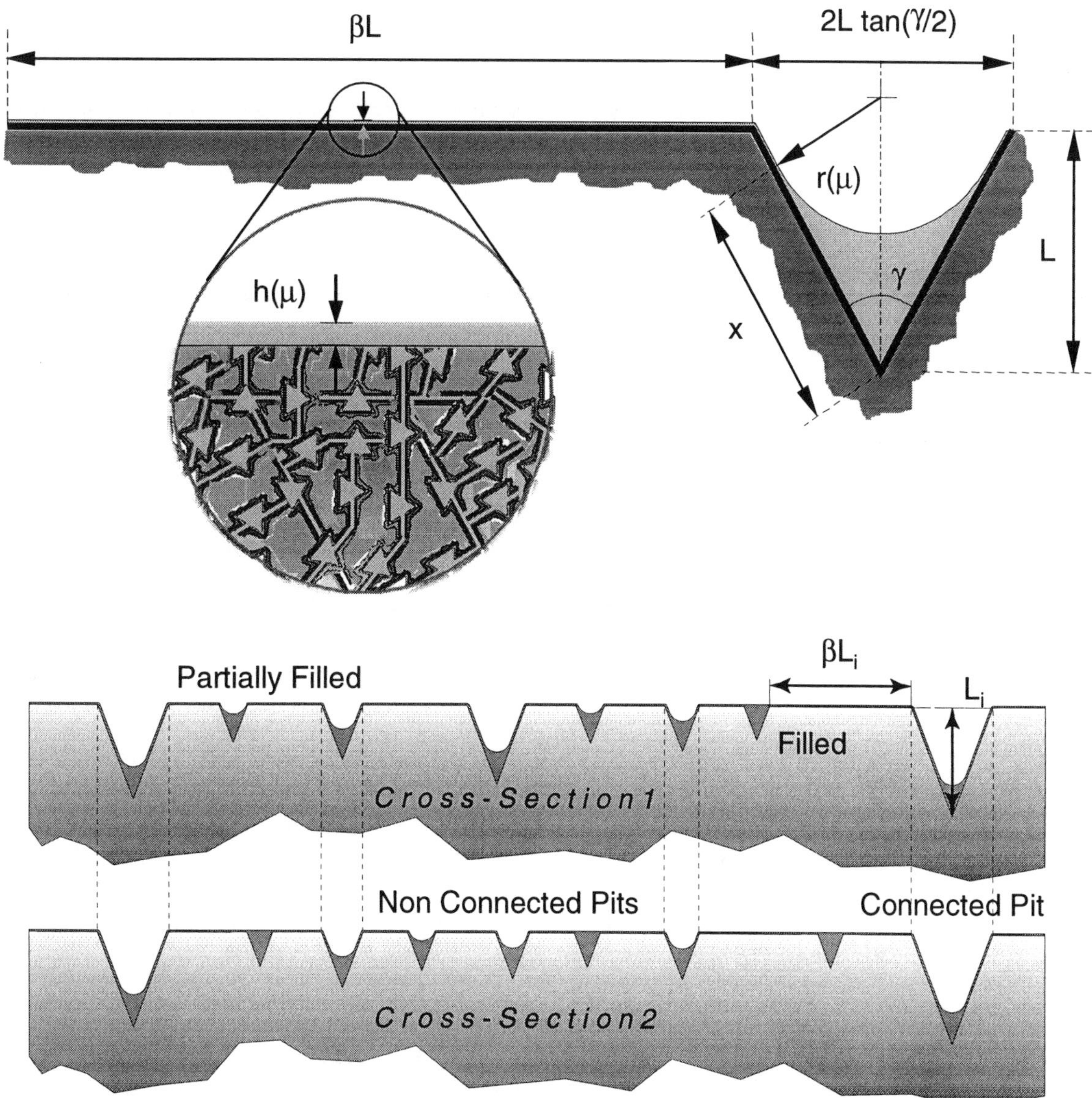

Figure 5. Conceptual model of a porous medium and fracture surface based on a combination of angular pores and slits (porous medium) or pits and flat surfaces (fracture surface). Capillary forces, adsorptive forces, and liquid/gas interfacial areas are accounted for. Pit connectivity affects hydraulic conductivity but not film thickness.

5.1 Constitutive Equations

Classical film-flow theory breaks down when the underlying matrix is unsaturated, so that pressures in the film are less than atmospheric. As shown by *Tokunaga and Wan* [1997], however, flows occur even under these conditions, especially on rough faces. One way to characterize the fluxes is to treat the fracture faces as a collection of half-tubes, using the classical capillary theory. Or and coworkers [*Or and Tuller*, 1999; *Or and Tuller*, 2000 in press; *Or and Ghezzehei*, 2000] recently developed a model for unsaturated flow in fractured porous media based on the idea of collections of both angular pores and slits, thereby explicitly accounting for the liquid-gas interface area and adsorptive forces. The classical porous-medium capillary theory is a limiting case of the expanded theory. The same approach is applied to unsaturated free-surface fracture flow—instead of pores and slits, there are pits and faces (half-pores and half-slits). A conceptual sketch for both porous medium and fracture faces is shown in Figure 5

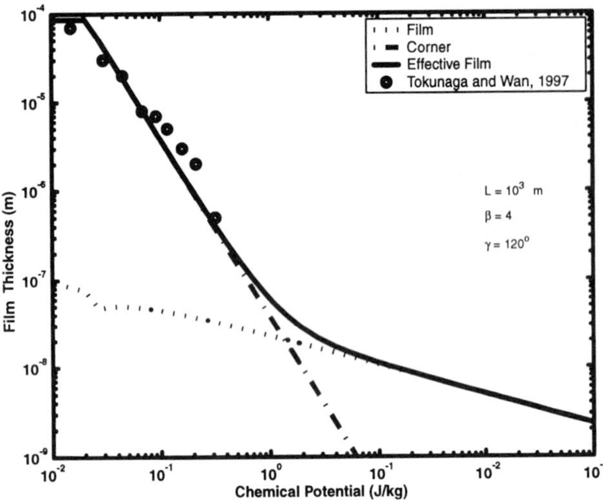

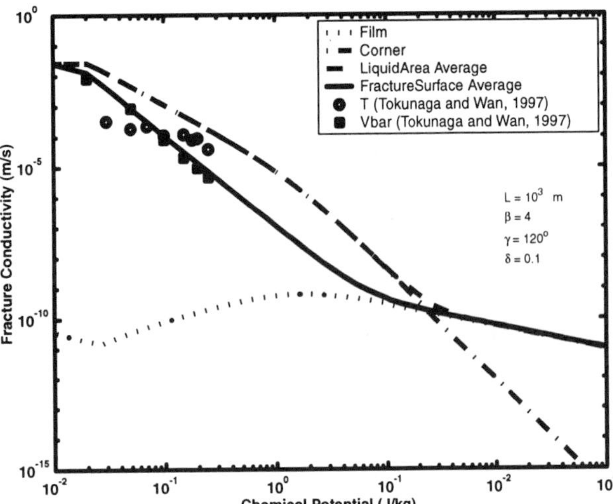

Figure 6. Comparison of observations with fracture-surface theory using a typical pit/face combination. The top figure demonstrates the contribution of corners (capillary forces) and faces (adsorptive forces) to average film thickness. Conductivity is demonstrated in the bottom figure using two methods of weighting the relative contributions of faces and pits. (Fracture-surface-average conductivity is the measurable quantity.)

[adapted from *Or and Tuller,* 2000 in press]. Note that the porous medium and fracture face are in local equilibrium. The unsaturated hydraulic conductivity of a fracture surface may be expressed by either the cross-sectional liquid area average (K_A) or the fracture-surface-length average (K_L):

$$K_A(\mu) = \frac{K_F(\mu)A_F + K_C(\mu)A_C\delta}{A_F + A_C}$$
$$K_L(\mu) = \frac{K_F(\mu)L_F + K_C(\mu)L_C\delta}{L_F + L_C} \qquad (10)$$

where μ is chemical potential; K is conductivity; subscripts F and C denote liquid film and pit corners; subscripts A and L represent cross-sectional area and projected fracture length and depend on μ and geometrical factors; and δ is the fraction of pits that are connected. Predictions of the model are compared with observations by *Tokunaga and Wan* [1997] in Figure 6 [also adapted from *Or and Tuller,* 2000 in press] for both film thickness and film conductivity using a single representative pit/face configuration (the theory has been extended to account for distributions as well). Results are relatively insensitive to model parameters. The theory becomes unstable for wetter conditions, implying that film movement is inherently unsteady and suggesting that rivulets may form.

5.2 Behavior at a Fracture Intersection

The flow behavior at a fracture network intersection is considerably more complex than a comparable saturated case and has seen relatively little study. Such complexity is caused by capillary forces. As shown in Figure 7, any completely unsaturated fracture intersection can have different behaviors on different faces, depending on the fracture orientations, offsets, and flux rates. For pure film flow, most configurations may be classified as either a split (independent flux paths) or a funnel (merging flux paths), with flow routing based on upstream fluxes. Lateral diversion before dripping on hanging-wall faces is flux dependent, with greater diversion at low fluxes and steep angles. Note that dripping is inherently unsteady; only time-averaged fluxes are compatible with the computational approach. The configuration usually considered in modeling studies, with vertical and horizontal fracture sets, is indeterminate for routing schemes that use geometry to determine upstream directions. If the typical configuration is rotated, like it is at YM, there is a potential for strong lateral diversion at lower flux rates. The situation becomes more complex when one or more fractures are liquid filled. This is because the assumptions required for routing fail. Each corner of the intersection may have a different potential for fully unsaturated conditions. When a fracture segment is full, both corners have the same potential. Capillary forces may preclude exit from a fracture segment into adjoining fractures.

Additional study is required to ascertain the conditions under which detailed analysis of flow at fracture intersections is important. Certainly the importance increases as the fracture apertures become larger and as film flow becomes more prevalent. In situations where flow routing is important, additional information is required on the distribution of offsets at fracture intersections. This information can be gathered during fracture-mapping exercises.

6. EXAMPLE PROBLEM

The utility of the boundary integral procedure is demonstrated by an example that uses a prototype simulator. This simulator was tested against a variety of analytic solutions for diffusion in a saturated porous medium in order to build confidence in predictions. Different flow rates yield cases that have various ratios of fracture conductivity to matrix conductivity. Under high pressures (low capillary pressures), unsaturated fractures are conduits; under low pressures (high capillary pressures), unsaturated fractures are barriers. Low and high source fluxes are selected to yield pressures such that fractures are primarily barriers or primarily conduits. Nondimensional units are used for all quantities.

The problem of interest is a square horizontal block. Fractures bound the domain on all four sides, and three evenly spaced parallel fractures run through the matrix in both directions, yielding 16 square matrix sub-blocks. One side of the domain is held at a constant pressure, while the remaining sides are no-flow boundaries. A constant flux is

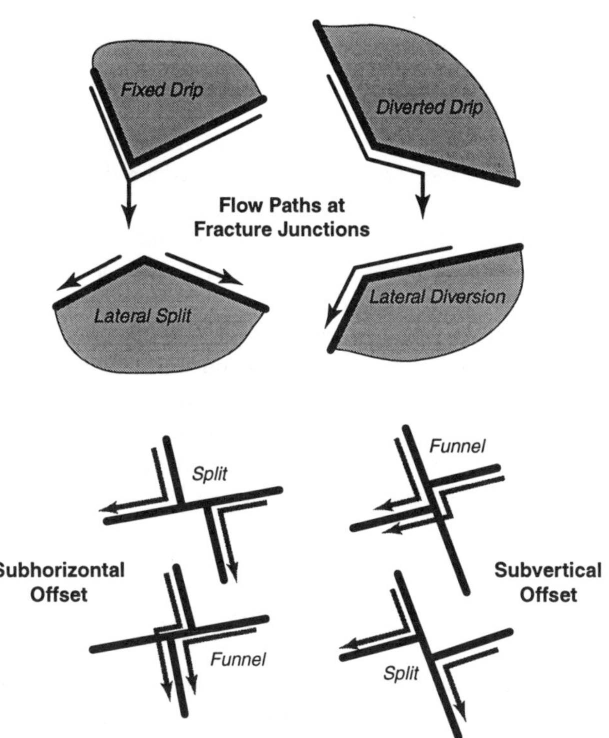

Figure 7. Classification of film flow and dripping behavior at fracture junctions. Lateral diversion on hanging walls is flux dependent. Routing into fractures is dependent on fracture capacity. Anisotropy from diversion and funneling may occur due to fracture offsets. Lateral diversion on hanging walls may convert funnel configurations into split configurations.

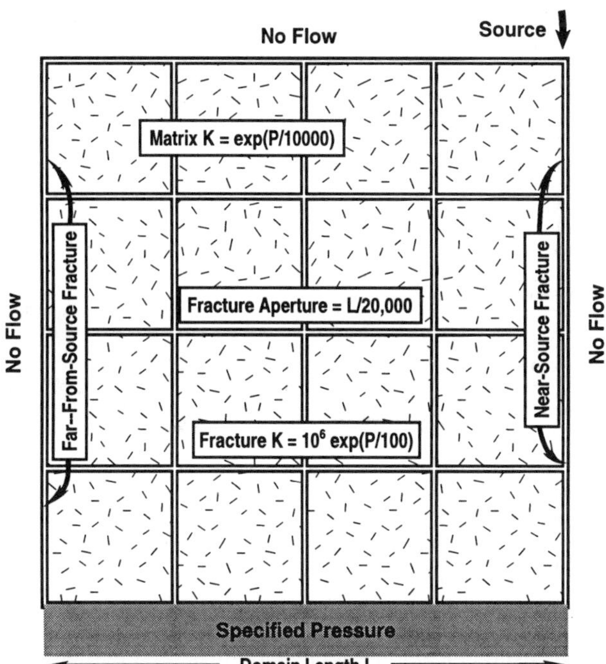

Figure 8. Layout of example problem with multiple matrix blocks and a source in one corner.

introduced into one element of the fracture at the corner opposite the pressure boundary. The low flux rate would produce a pressure drop of 2 across the domain if uniformly applied across the boundary opposite the pressure boundary, there were no fractures, and the matrix was saturated. The high rate would produce a pressure drop of 2000 under the same conditions. The problem description and matrix properties are shown in Figure 8.

Each fracture has a width 5×10^{-5} times the domain width and a saturated hydraulic conductivity 10^6 times greater than the matrix. A simple exponential relative permeability function for both matrix and fracture captures the essential characteristics of unsaturated flow. The fracture relative permeability function is much more sensitive to pressure than the matrix function, and fractures convert from conduits to barriers as pressures drop below about −1400. At the lower-boundary-specified pressure, the matrix hydraulic conductivity is about 3/4 of the saturated value, and the fracture conductivity is 7 orders of magnitude smaller than the matrix value. The across-fracture conductivity is assumed to be the same as the along-fracture conductivity.

The solution was produced using uniformly discretized fractures. Three cases were run for each flow, using 2, 4, and 8 elements per subdomain side (a total of 80, 160, and 320 elements, respectively). Material properties were up-

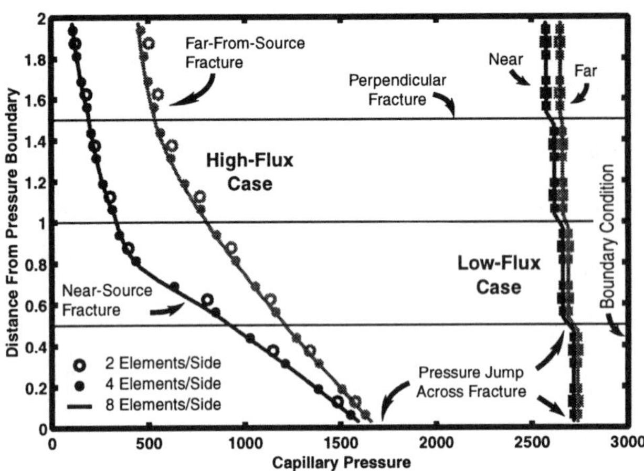

Figure 9. Pressure distribution in fractures at domain boundaries for both a high-flux and low-flux source in one domain corner. Note the large jump in pressure across the fracture next to the capillary-pressure boundary condition.

dated using Picard iteration, modified by limiting maximum change during an iteration. Tighter limits were imposed as iterations progressed. Picard iteration alone often failed to converge due to the nonlinearities in material properties; the limiting procedure was reasonably effective but not particularly efficient.

Selected results of the simulations are shown in Figure 9. Predicted pressures along the boundaries between the source and the specified pressure boundary are plotted. The solutions match quite well, even with as few as 2 elements per sub-block side. In the low-flow case, pressure within each matrix block is almost uniform, although there are jumps in pressure between each block. There is a total pressure difference of about 400 across the domain, although the total drop in the matrix is less than 3. In the high-flow case, there are no jumps between blocks except at the pressure boundary. The two rows of matrix blocks furthest from the pressure condition are dominated by fracture flow. The two rows nearest the pressure boundary are dominated by matrix flow, even though the fractures are more conductive than the matrix. This is because the cross-sectional area for flow in the fractures is so small.

These two cases demonstrate the strong control that unsaturated fractures provide on steady-state pressure distributions within the blocks outside a relatively narrow range of capillary pressures (roughly 500 to 2000 for this example problem). Under wetter conditions, the fractures carry most of the flux and little change in pressure is necessary for a large change in flux in the fracture network. For example, a capillary-pressure increase from 100 to 500 (as seen in the high-flux example) reduces the fracture conductivity by more than a factor of 50. Under drier conditions, the fractures are strong barriers restricting flow and, again, little change in pressure is necessary for a large change in flux across the fracture. The difference in matrix conductivity across the matrix-dominated range is only about 16 percent, and in the examples there is no matrix block that would have a difference in conductivity greater than about 6 percent. The result is acceptably close to the assumption of uniform hydraulic conductivity within the matrix block.

One strategy for discrete-fracture modeling using standard finite element methods is to insert reduced-dimension features between elements (e.g., line elements along triangle faces), thereby avoiding the issue of artificial anisotropy. Usually it is assumed that resistance to flow across fractures is negligible, so the matrix blocks and the fracture all have the same pressure. Across-fracture resistance is usually assumed to be negligible in continuum models as well. This assumption is questionable when the fractures are dry, as is evidenced in the low-flux case: the pressure drop from source to pressure boundary is roughly 200 times larger than is required for flux to pass through the matrix. Note that the neglect of fracture resistance is presumably less severe when fluxes can bridge dry fractures through matrix-block contact areas (which was not considered in this example).

Another strategy for discrete-fracture modeling is to artificially widen the aperture in order to reduce the range in element sizes, using equivalent properties for these elements. Flux-balance considerations dictate that along-fracture conductivity scales linearly with artificial aperture, while across-fracture conductivity scales inversely with artificial aperture. Thus, artificially broadening a fracture requires that the across-fracture conductivity is increased while the along-fracture conductivity is decreased, yielding numerically derived anisotropy even if the underlying fracture properties are isotropic. The discrete-fracture approach outlined here does not require or benefit from artificial widening.

7. SUMMARY

Boundary integral methods are well suited to the examination of steady-state or quasi-steady-state discrete-fracture processes in a fractured porous medium. Processes in the fractures are solved using standard finite-volume approaches, with the dimensionality of the fractures being one less than that of the problem domain (e.g., 1-D in a 2-D domain). The finite-volume approach allows general coupled equation formulations to be considered without conceptual difficulty. Given the approximation that coefficients are spatially constant within a subdomain (e.g., a matrix block), the governing equations may be transformed into a sum of Laplacians, enabling dependent variables to be analytically determined within the matrix (subject to discretization along discontinuities). Two formally equivalent formulations may be used to solve the boundary integral equations: a discrete-jump or a multiple-zone approach. The discrete-jump approach generates a set of linear equations with a fully populated coefficient matrix, with possible stability and accuracy advantages due to global support. The multiple-zone approach generates a larger coefficient matrix, but with blocks of zeros that may be taken advantage of to speed solution of the equations. The same compatibility equations at discontinuities apply in either case, but are implemented differently.

A prototype implementation of the multiple-zone boundary integral approach demonstrates the control that fractures can place on steady-state capillary pressure distributions within a matrix block. Outside a relatively narrow range of pressures, over which matrix conductivity varies little, fractures dominate the flow system, either as conduits or as barriers. In either case, pressure within a matrix block is essentially uniform. Under dry conditions, low-conductivity fractures may determine pressure gradients within a system; typically the barrier effect of dry fractures is neglected in continuum models. However, one can presume that the barrier effect of dry fractures may be abated when matrix asperities bridge the fracture, thus providing an alternate pathway for water to cross the fracture.

Constitutive behavior for gas-phase, energy, and saturated liquid-phase transport is relatively well understood in fractures. However, constitutive behavior for unsaturated liquid-phase transport (of particular interest at YM) is not as well understood. Therefore it is considered in greater detail in this paper. A new constitutive approach is presented here that treats both matrix and fracture surfaces using angular pores and connecting surfaces to consistently consider both capillary and adsorptive forces. Because of the flow in angular pits, the constitutive theory can be used to predict fracture flow when the matrix is unsaturated, even when film flow is detectable. For wetter conditions, the theory becomes unstable, suggesting that only dynamically steady-state conditions can exist. The physics of flow through fracture intersections can be much more complex under unsaturated conditions than under saturated conditions. Some of the factors affecting flow include fracture geometry, lateral diversion, and adsorption and capillary forces. Flow processes under these conditions may exhibit strong flow-dependent anisotropy, particularly with dipping planes such as exist at YM.

APPENDIX

The following development of equations for coupled liquid, gas, and energy transport within a subdomain assumes that steady state exists. The three unknowns are liquid pressure, the square of gas density, and temperature. Three governing equations are required to solve for the three unknowns: two mass balance equations and one energy balance equation. The mass balance equations are formulated in terms of total mass and total water mass, eliminating explicit consideration of transfer between phases. The equations are

$$\begin{aligned}
&\nabla\cdot(\rho_l q_l + \rho_g q_g) = 0\\
&\nabla\cdot(\rho_l q_l + \rho_v q_v) = 0\\
&\nabla\cdot[(\rho_l q_l C_{Pl} + \rho_g q_g C_{Pg})\Phi]\\
&\quad + \nabla\cdot(\rho_v q_v H_{lv}) - \nabla\cdot K_e \nabla\Phi = 0
\end{aligned} \tag{A1}$$

where ρ is density, q is flux, C_P is specific heat at constant pressure, H_{lv} is the coefficient of latent heat, K_e is thermal conductivity of the liquid/gas/solid mixture, Φ is reduced temperature ($T - T_0$), T and T_0 are absolute and reference temperatures, and subscripts l, g, and v denote liquid, gas, and vapor.

The mass flux terms are defined by

$$\begin{aligned}
\rho_l q_l &= -\rho_l k\lambda_l(\nabla P_l + \rho_l g\nabla z)\\
\rho_g q_g &= \rho_g k\lambda_g(\nabla P_g + \rho_g g\nabla z)\\
\rho_v q_v &= \rho_v q_g - \theta_g \tau D_v \nabla\rho_v
\end{aligned} \tag{A2}$$

where k is intrinsic permeability, λ is mobility (relative permeability divided by viscosity), P is pressure, z is elevation, g is acceleration due to gravity, θ is volume fraction, τ is tortuosity, and D_v is the diffusion coefficient for vapor in air.

Assuming the ideal gas law applies ($P = \rho RT$), where R is the gas constant, the following relationship holds for a gas

$$\rho\nabla P = R\rho(T\nabla\rho + \rho\nabla T) = \frac{RT}{2}\nabla\rho^2 + R\rho^2\nabla\Phi \tag{A3}$$

By the chain rule, the following relationship holds for the gradient of vapor density (which is an equilibrium function of capillary pressure and temperature in a porous medium)

$$\nabla\rho_v = \frac{d\rho_v}{dP_l}\nabla P_l + \frac{d\rho_v}{d\rho_g^2}\nabla\rho_g^2 + \frac{d\rho_v}{dT}\nabla T \tag{A4}$$

For Equations (A3) and (A4), each of the governing equations can be written in the form

$$\nabla \cdot (c_1 \nabla P_l + c_2 \nabla \rho_g^2 + c_3 \nabla T + c_4 \nabla z) = 0 \qquad \text{(A5)}$$

In the total mass equation, the coefficients are

$$\begin{aligned} c_1 &= \rho_l k \lambda_l \\ c_2 &= k \lambda_g RT/2 \\ c_3 &= \rho_g^2 k \lambda_g R \\ c_4 &= gk(\lambda_l \rho_l^2 + \lambda_g \rho_g^2) \end{aligned} \qquad \text{(A6)}$$

In the total water equation, the coefficients are

$$\begin{aligned} c_1 &= \rho_l k \lambda_l + \theta_g \tau D_v \frac{d\rho_v}{dP_l} \\ c_2 &= \tfrac{1}{2} \omega_g^v k \lambda_g RT + \theta_g \tau D_v \frac{d\rho_v}{d\rho_g^2} \\ c_3 &= \omega_g^v \rho_g^2 k \lambda_g R + \theta_g \tau D_v \frac{d\rho_v}{dT} \\ c_4 &= gk(\lambda_l \rho_l^2 + \omega_g^w \lambda_g \rho_g^2) \end{aligned} \qquad \text{(A7)}$$

In the energy equation, the coefficients are

$$\begin{aligned} c_1 &= \rho_l k \lambda_l C_{Pl} \Phi + \theta_g \tau D_v H_{lv} \frac{d\rho_v}{dP_l} \\ c_2 &= \tfrac{1}{2} \omega_g^v k \lambda_g RTC_{Pg} \Phi + \theta_g \tau D_v H_{lv} \frac{d\rho_v}{d\rho_g^2} \\ c_3 &= \rho_g^2 k \lambda_g RC_{Pg} \Phi \\ &\quad + \left(\omega_g^v \rho_g^2 k \lambda_g R + \theta_g \tau D_v H_{lv} \frac{d\rho_v}{dT} \right) H_{lv} + K_e \\ c_4 &= gk(\lambda_l \rho_l^2 C_{Pl} + \omega_g^w \lambda_g \rho_g^2 C_{Pg}) \end{aligned} \qquad \text{(A8)}$$

where $\omega_g^w = \rho_v / \rho_g$ is the mass fraction of vapor in the gas.

Each of the coefficients is nonlinear, with the strongest nonlinearity occurring due to the mobility terms. In the computational scheme based on the boundary integral approach, the coefficients are assumed to be piecewise constant over a subdomain. Representative coefficients could be obtained by using state variables evaluated at the subdomain centroid. Alternatively, representative coefficients could be obtained by considering the subdomain to be a single finite volume with averaged state variables, adding a set of balance equations describing the volume as a whole.

Acknowledgements. This paper documents work performed by the Center for Nuclear Waste Regulatory Analyses (CNWRA) for the Nuclear Regulatory Commission (NRC) under Contract No. NRC-02-97-009. The activities reported here were performed on behalf of the NRC Office of Nuclear Material Safety and Safeguards, Division of Waste Management. The prototype boundary integral simulator is not under CNWRA configuration control. This report is an independent product of the CNWRA and does not necessarily reflect the views or regulatory position of the NRC. The authors would like to acknowledge the suggestions and comments made by R. Fedors, B. Sagar, G. Ofoegbu, R. Green, and three anonymous reviewers, which tremendously improved the quality of the paper.

REFERENCES

Banerjee, P. K., and R. Butterfield, *Boundary Element Methods in Engineering Science*, McGraw-Hill, London, England, 1981.

Dershowitz, W. S., and C. Fidelibus, Derivation of equivalent pipe network analogues for three-dimensional discrete fracture networks by the boundary element method, *Water Resour. Res. 35*(9), 2685–2691, 1999.

Elsworth, D, A model to evaluate the transient hydraulic response of three-dimensional sparsely fractured rock masses, *Water Resour. Res. 22*(13), 1809–1819, 1986.

Jaswon, M. A. and G. T. Symm, *Integral Equation Methods in Potential Theory and Elastostatics,* Academic Press, London, England, 1977.

Liggett, J. A. and P. L.-F. Liu, *The Boundary Integral Equation Method for Porous Media Flow,* George Allen & Unwin, London, England, 1983.

Or, D., and T. A. Ghezzehei, Dripping into subterranean cavities from unsaturated fractures under evaporative conditions, *Water Resour. Res, 36*(2), 281–393, 2000.

Or, D., and M. Tuller, Flow in unsaturated fractured porous media—Hydraulic conductivity of rough fracture surfaces, *Water Resour. Res,* in press, 2000.

Or, D., and M. Tuller, Liquid retention and interfacial area in variably saturated porous media: Upscaling from single-pore to sample scale model, *Water Resour. Res., 35*(12), 3591–3605.

Pullan, A. J., and I. F. Collins, Two and three-dimensional steady quasi-linear infiltration from buried and surface cavities using boundary element techniques, *Water Resour. Res. 23*(8), 1633–1644, 1987.

Rasmussen, T. C, Computer simulation model of steady fluid flow and solute transport through three-dimensional networks of variably saturated, discrete fractures, in *Flow and Transport Through Unsaturated Fractured Rock, Geophysical Monograph 42*, edited by D. D. Evans and T. J. Nicholson, pp. 107–114, American Geophysical Union, Washington, D.C., 1987.

Stothoff, S. A., and G. F. Pinder, A boundary integral technique for multiple-front simulation of incompressible, immiscible flow in porous media, *Water Resour. Res. 28*(8), 2067–2076, 1992.

Taigbenu, A. E, The Green Element Method, *Int'l J. Numerical Methods Eng., 38*, 2241–2263, 1995.

Taigbenu, A. E., and O. O. Onyejekwe, Green element simulations of the nonlinear unsaturated flow equation, *Appl. Math. Modelling 19*, 675–684, 1995.

Tokunaga, T. K., and J. Wan, Water film flow along fracture surfaces of porous rock, *Water Resour. Res. 33*(6), 1287–1295, 1997.

Tuller, M., D. Or, and L. M. Dudley, Adsorption and capillary condensation in porous media—Pore scale liquid retention and interfacial configurations, *Water Resour. Res. 35*(7), 1941–1964, 1999.

U.S. Department of Energy, *Viability Assessment of a Repository at Yucca Mountain: Total System Performance Assessment*, DOE/RW–0508, U.S. Department of Energy, Office of Civilian Radioactive Waste Management, Las Vegas, Nev., 1998.

Wang, J. S. Y. and T. N. Narasimhan, Unsaturated flow in fractured porous media, in *Flow and Contaminant Transport in Fractured Rock*; edited by J. Bear, C.-F. Tsang, and G. de Marsily, pp. 325–394, Academic Press, San Diego, Calif. 1993.

Stuart Stothoff, Stothoff Environmental Modeling, Houston, TX 77019

Dani Or, Utah State University, Logan, UT 84322

Critique of Dual Continuum Formulations of Multicomponent Reactive Transport in Fractured Porous Media

Peter C. Lichtner

Los Alamos National Laboratory, Los Alamos, New Mexico

Subsurface flow processes may take place at many different scales. The different scales refer to rock pore structure, microfractures, distinct fracture networks (ranging from small to large fracture spacing), and even faults. Presently, there is no satisfactory methodology for quantitatively describing flow and reactive transport in multiscale media. Approaches commonly applied to model fractured systems include single continuum models (SCM), equivalent continuum models (ECM), discrete fracture models (DFM), and various forms of dual continuum models (DCM). The SCM describes flow in the fracture network only and is valid in the absence of fracture-matrix interaction. The ECM, on the other hand, requires pervasive interaction between fracture and matrix and is based on averaging their properties. The ECM is characterized by equal fracture and matrix solute concentrations, but generally different mineral concentrations. The DFM is perhaps the most rigorous, but would require inordinate computational resources for a highly fractured rock mass. The DCM represents a fractured porous medium as two interacting continuums with one continuum corresponding to the fracture network and the other the matrix. A coupling term provides mass transfer between the two continuums. Values for mineral and solute concentrations and other properties such as liquid saturation state may be assigned individually to fracture and matrix. Two forms of the DCM are considered, characterized by connected and disconnected matrix blocks. The former is referred to as the dual continuum connected matrix (DCCM) model and the latter as the dual continuum disconnected matrix (DCDM) model. In contrast to the DCCM model, in which concentration gradients in the matrix are allowed only parallel to the fracture, the DFM provides for matrix concentration gradients perpendicular to the fracture. The DFM and DCCM models can agree with each other only in the case where both reduce to the ECM. The DCCM model exhibits incorrect behavior as the matrix block size increases, resulting in reduced coupling between fracture and matrix continuum. The DCDM model allows for matrix gradients within individual matrix blocks in which the symmetry of the surrounding fracture geometry is preserved. However, the DCDM model breaks down for simultaneous heat and mass transport, and cannot account for significant changes in porosity and permeability caused by chemical reactions.

Dynamics of Fluids in Fractured Rock
Geophysical Monograph 122

1. INTRODUCTION

Fractured porous media, and more generally hierarchical media involving multiple length scales, play an important role in subsurface flow and transport processes. Fracture-dominated flow systems are involved in numerous subsurface geochemical processes including contaminant migration, ore deposition, and weathering. Practical applications involving fractured porous media include contaminant migration, oil recovery from fractured reservoirs, geothermal energy, degradation of cement, and potentially subsurface sequestration of CO_2, to mention but a few.

Considerable progress has been made in developing and applying reactive transport models to complex geochemical systems involving single-continuum porous media. [See *Lichtner et al.* (1996) for a general overview and references therein; and *Lichtner* (1998)]. However, subsurface flow processes may take place at many different scales: rock pore structure, microfractures, distinct fracture networks ranging from small to large fracture spacing, and faults. At present, no completely satisfactory methodology exists for describing quantitatively reactive flow and transport in multiscale media.

Because fractured porous media are characterized by bimodal distributions in physical and chemical properties (with generally distinct values associated with the fracture network and rock matrix), a description based on a single porous medium is generally unable to capture the unique features of a fractured system. Furthermore, present approaches used for describing fracture-matrix interaction are of limited use. This is especially true for transport of chemically reacting constituents and simultaneous flow of mass and heat. A prime example where present approaches may fail (and where more general methods are needed) is the proposed Yucca Mountain high-level nuclear waste repository, which is to be located in variably saturated fractured tuff rock. This paper reviews existing approaches for representing fractured media in continuum-based models applied to reactive flow and transport. Extension of these methods to hierarchical porous media is briefly considered. This paper is restricted to continuum-based formulations, leaving out discussions of other approaches such as algorithmic methods, including cellular automata and diffusion-limited aggregation (DLA), and network models. This is because the level of chemistry that can be incorporated into continuum models is on a par with the most sophisticated geochemical models. These models incorporate presently available thermodynamic and kinetic data for complex multicomponent systems.

2. CONTINUUM MODELS FOR REACTIVE FLOWS IN FRACTURED MEDIA

A number of different conceptual frameworks have been used to represent fractured porous media. They include the discrete fracture model (DFM), equivalent continuum model (ECM), variations of dual and multiple continuum models (DCM), and the representation of fractures as regions of high permeability/low porosity in heterogeneous media. Incorporating chemical reactions in models of fractured porous media requires newly considering the suitability and extension of basic techniques used to represent fluid flow (especially the appropriate length scale to account for the presence of reaction fronts). Furthermore, because equations for multicomponent systems require a much greater computational effort to solve, new numerical techniques are required. Finally, chemical reactions can dramatically alter the physical and hence hydraulic properties of a porous medium. Fractures may widen or become sealed as a result of chemical reactions. Alteration of the matrix surrounding fractures may affect the interaction between fracture and matrix.

A fractured porous medium can be thought of as composed of two distinct continuums, referred to as fracture and matrix, and represented by sub- and superscripts f and m. A representative elementary volume (REV) of bulk rock with volume V_b consists of the sum of fracture V_f and matrix V_m volumes:

$$V_b = V_f + V_m \,, \tag{2.1}$$

as illustrated in Figure 1. The fraction of volume occupied by fractures, denoted by ϵ_f, is defined by

$$\epsilon_f = \frac{V_f}{V_b} \,, \tag{2.2}$$

with $\epsilon_m = 1 - \epsilon_f$ representing the fraction occupied by the rock matrix. The fracture and matrix volumes may be further broken down into pore and solid fractions

$$V_\alpha = V_p^\alpha + V_{\mathrm{solid}}^\alpha \,, \quad (\alpha = f, m) \,. \tag{2.3}$$

Note that ϵ_f corresponds to the fracture porosity of the bulk rock volume for the case in which the fractures are not filled with solid $V_p^f = V_f$. In general, however, because of the presence of fracture-filling in the form of solids, the intrinsic porosity of the fracture is less than unity. Bulk and intrinsic fracture and matrix properties of some quantity Z are related by $\in_\alpha$:

$$Z_\alpha^b = \in_\alpha Z_\alpha, \tag{2.4}$$

where Z_α^b and Z_α denote the bulk and intrinsic properties, respectively.

Because of their small aperture and volume, fractures can be easily altered by chemical reactions. Consider, for example, the redistribution of silica between matrix and fracture as heat drives fluid from the matrix into the fracture network (as would the heat generated by the decay of high-level nuclear waste at the proposed Yucca Mountain repository). Imagine that the pore fluid in the rock matrix is brought to equilibrium with respect to a particular silica polymorph (such as amorphous silica at boiling conditions). Further, consider that as the fluid in the matrix boils, it escapes into the surrounding fracture network. As the matrix pore fluid is vaporized, its silica content is deposited in surrounding fractures, partially filling the fractures by precipitating silica polymorphs. At issue is the extent to which the fractures can be filled by the silica contained in the matrix pore water. To determine the volume fraction of solid precipitated in the fracture $\phi_{SiO_2}^f$, the expression

$$\phi_{SiO_2}^f = \frac{1-\in_f}{\in_f} \phi_m C_{SiO_2}^m \overline{V}_{SiO_2} \tag{2.5}$$

is evaluated, where $C_{SiO_2}^m$ denotes the concentration of silica in the matrix pore fluid at 100°C that is assumed to be in equilibrium with a particular silica polymorph of

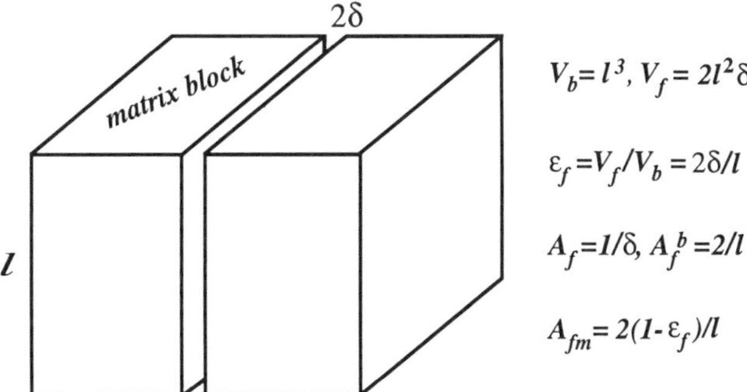

Figure 1. Illustration of geometric relations in a fractured porous medium with fracture aperture 2δ. See text for an explanation of symbols used in the figure.

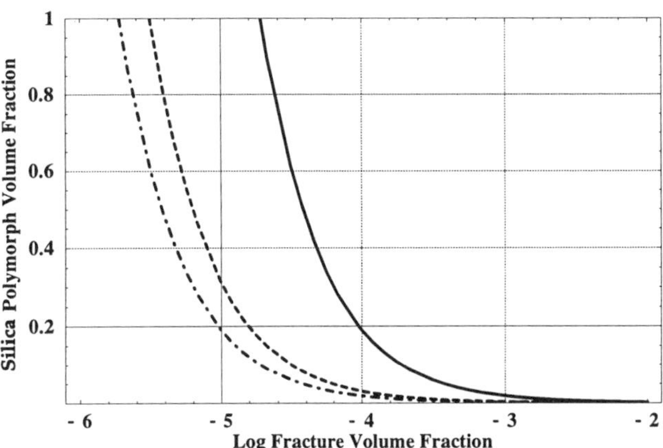

Figure 2. Volume fraction of quartz (solid curve), chalcedony (dashed curve), and amorphous silica (dash-dotted curve) plotted as a function of fracture porosity. A volume fraction of one represents complete filling of the fracture, assuming that the fracture was initially devoid of solid filling.

molar volume $\overline{V}_{SiO_2}$ and matrix porosity ϕ_m. This relation, derived from mass-balance considerations, is dependent on all matrix pore water flashing to steam in the fracture. If this is not the case—for example, if a drying front propagates inward into the matrix, depositing silica within the matrix—then Equation (2.5) provides an upper bound on the extent of fracture-filling. Other processes may also be possible, such as silica becoming remobilized from fracture coatings, which are not accounted for in this simple analysis. Results for a matrix porosity of $\phi_m = 0.1$ are shown in Figure 2 for quartz, chalcedony, and amorphous silica. From the figure, it is clear that for a given matrix porosity, the fracture's degree of sealing depends on the fracture volume fraction $\in_f$ and the particular silica polymorph that precipitates. Amorphous silica, with the highest solubility, gives the largest fracture filling of the silica polymorphs, followed by chalcedony and quartz. Complete sealing of the fracture $\left(\phi_{SiO_2}^f = 1\right)$ requires a very small fracture volume fraction. Moderate filling could lead to fracture coatings that armor the fracture and prevent or reduce imbibition into the matrix. Thus, very different consequences could result, depending on the extent of fracture-filling.

This example is but a highly simplified situation that could take place at the proposed Yucca Mountain nuclear waste facility. Heat from the waste is expected to lead to the formation of heat pipes, with resulting boiling and degassing of CO_2 (and an increase in pH and possible precipitation of salts and changes in silica solubility) as evaporation takes place [*Lichtner and Seth*, 1996].

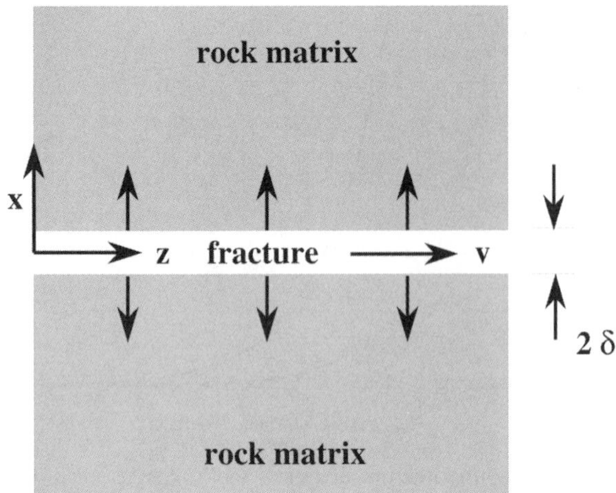

Figure 3. Discrete fracture model.

2.1 Discrete Fracture Model (DFM)

One approach is to treat fractures by explicitly taking into account coupling with the rock matrix through a mass-transfer term (Figure 3). This approach, referred to as the discrete fracture model (DFM), can be used to describe flow through a single fracture or an infinite number of equally spaced fractures. The DFM, however, rapidly becomes unwieldy for more than a few fractures if there is no simple geometric relation between them.

Several forms of the DFM are possible, depending on treatment of transport processes in the fracture and matrix. Here, a simplified form for the solute transport equations is considered, neglecting diffusion in the fracture and advection and diffusion parallel to the fracture in the matrix. This is a good approximation for sufficiently fast flow rates in the fracture. A single reacting species is considered obeying the reaction

$$A \leftrightarrow A_{(s)}, \tag{2.6}$$

with solid $A_{(s)}$ and aqueous species A. Transport equations for the DFM can be expressed as

$$\begin{aligned}\frac{\partial}{\partial t}(\phi_f C_f) + v_f \frac{\partial C_f}{\partial z} &= -k_f (C_f - C_{eq}) \\ &+ \frac{\tau_m \phi_m D}{\delta} \frac{\partial C_m}{\partial x}\bigg|_{x=\delta}\end{aligned}, \tag{2.7}$$

for the fracture and

$$\begin{aligned}\frac{\partial}{\partial t}(\phi_m C_m) + v_m \frac{\partial C_m}{\partial x} - \tau_m \phi_m D \frac{\partial^2 C_m}{\partial x^2} \\ = -k_m (C_m - C_{eq})\end{aligned}, \tag{2.8}$$

for the matrix, where z is the coordinate along the fracture and x is the coordinate in the matrix perpendicular to the fracture. Linear reaction kinetics are assumed with rate constants k_f and k_m for fracture and matrix, respectively, and equilibrium concentration C_{eq}. The solute concentration is denoted by $C_\alpha (\alpha = f, m)$, corresponding to fracture and matrix. Diffusivity is denoted by D, and fracture and matrix porosity and tortuosity by τ_α and ϕ_α, respectively. The fluid flow velocity in the fracture and matrix is represented by v_f and v_m, respectively. To complete the set of equations, initial and boundary conditions must be prescribed. At the fracture-matrix interface, the solute concentrations are presumed to be the same:

$$C_m(x = \delta, z) = C_f(z). \tag{2.9}$$

The fracture transport equation is coupled to the matrix equation by the last term on the right-hand side of Equation (2.7) representing the flux across the fracture-matrix interface.

Steefel and Lichtner [1998a,b] demonstrated the existence of a scaling relation between mineral alteration along a fracture and that within the rock matrix perpendicular to the fracture. Rate of alteration can be investigated by examining the stationary-state solution to the DFM transport equations. The stationary-state solution is useful for describing the time evolution of a reacting system. This system may be represented as a sequence of stationary states, with each stationary state corresponding to a different configuration of minerals along the flow path [*Lichtner*, 1988]. The stationary-state solution to the DFM transport equations can be expressed as [*Steefel and Lichtner*, 1998a]

$$C_f(z) = (C_f^0 - C_{eq}) e^{-z/\lambda_f} + C_{eq}, \tag{2.10}$$

for the fracture and

$$C_m(x; z) = (C_f(z) - C_{eq}) e^{-z/\lambda_m} + C_{eq}, \tag{2.11}$$

for the matrix. The term $\lambda_{f,m}$ represents equilibration lengths [*Lichtner*, 1988, 1998] in the fracture and matrix, respectively, defined by

$$\lambda_m = \sqrt{\frac{(\tau\phi D)_m}{k_m}}, \tag{2.12}$$

$$\lambda_f = \frac{\mathrm{Pe}\lambda_f^0 \lambda_m}{\lambda_f^0 + \mathrm{Pe}\lambda_m}, \tag{2.13}$$

where

$$\lambda_f^0 = v_f / k_f, \tag{2.14}$$

denotes the fracture equilibration length for pure advective transport [*Lichtner*, 1988]. The dimensionless Peclet-like number Pe is defined by

$$\mathrm{Pe} = \frac{v_f \delta}{(\tau\phi D)_m}. \tag{2.15}$$

According to these results, a wedge-shaped alteration front geometry is produced in the rock matrix with slope equal to the ratio of matrix and fracture equilibration lengths

$$\begin{aligned} \frac{dx}{dz} &= -\frac{\lambda_m}{\lambda_f} = -\left(\frac{1}{\mathrm{Pe}} + \frac{\lambda_m}{\lambda_f^0}\right) \\ &= -\frac{1}{\mathrm{Pe}}\left(1 + \frac{k_f \delta}{\sqrt{k_m (\tau\phi D)_m}}\right) \end{aligned}. \tag{2.16}$$

With increasing Peclet number, the slope approaches zero, and alteration becomes parallel to the fracture. As the Peclet approaches zero, alteration becomes perpendicular to the fracture. The second term in brackets is generally smaller than one, implying that the slope is independent of kinetics [*Steefel and Lichtner*, 1998]. A simple scaling relation exists between the concentration profile into the rock matrix C_m and along the fracture C_f of the form

$$C_m(x, z) = C_f\left(\frac{\lambda_f}{\lambda_m} x + z\right). \tag{2.17}$$

Similar scaling relations hold for other quantities such as reaction rates and mineral concentrations. Thus, by observing alteration in the matrix perpendicular to the fracture, it should be possible to predict mineralization along the fracture itself. Numerical analysis involving multicomponent systems with nonlinear reaction kinetics yield similar results [*Steefel and Lichtner*, 1998]. How well this relationship between fracture and matrix alteration is borne out in natural systems depends on strong communication between the fracture and matrix. Such communication could be significantly impeded by (for example) the presence of impermeable fracture coatings.

These results may be generalized to include an infinite set of equally spaced fractures with spacing d [*Lichtner*, 1998]. In this case, the stationary-state matrix solute concentration is given by

$$C_m(x; z) = \left(C_f(z) - C_{\mathrm{eq}}\right) \frac{\cosh\left[\dfrac{x - d/2}{\lambda_m}\right]}{\cosh\left[\dfrac{\delta - d/2}{\lambda_m}\right]} C_{\mathrm{eq}}. \tag{2.18}$$

This solution reduces to the previous-case infinite fracture spacing for $d \gg \lambda_m \gg \delta$. For finite fracture spacing that is small compared to the matrix equilibration length, the scaling relation between fracture and matrix concentration profiles no longer holds. If the fracture spacing is much smaller compared to the matrix equilibration length ($\delta \ll d \ll \lambda_m$), matrix concentration gradients disappear, and the solute concentrations in the fracture and matrix become equal. This is just the definition of the ECM, which is a limiting case of the DFM.

2.2 Dual Continuum Models: DCCM & DCDM Approaches

The dual continuum model (DCM) represents a fractured porous medium as two interacting continuums, with one continuum corresponding to the fracture network and the other to the matrix. A coupling term provides mass transfer between the two continuums. The fracture continuum is characterized by high permeability and low porosity compared to the matrix continuum. The DCM enables separate values of the field variables to be assigned to fracture and matrix continuums. Additional parameters are needed to represent the average matrix block size and fracture aperture or (equivalently) fracture volume, associated with a representative elemental volume (REV) of bulk medium. From these geometric quantities, the interfacial surface area between fracture and matrix can be computed.

In the field of reservoir engineering, DCMs have been in use for some time since their first introduction by *Barenblatt and Zheltov* [1960] and *Barenblatt et al.* [1960]. The approach put forth by *Barenblatt and Zheltov* [1960] represented a fractured reservoir as two distinct overlapping continuums. Flow equations were developed for each continuum, with a coupling term providing mass transfer between them. Shortly thereafter, *Warren and Root* [1963] published an alternative conceptual model in which the matrix was represented as a periodic array of identical blocks completely surrounded by fractures. *Pruess and Narisimhan* [1985] generalized the approach *of Warren and Root* [1963] to include multiple nodes within a matrix block, allowing for local gradients to be present within the rock matrix. These authors also provided for a fully transient description. In this approach, the matrix is discretized into concentrically nested blocks, spheres, or other geometric shapes. The outermost block is connected to the fracture continuum. The authors referred to their generalization as the MINC (Multiple Interacting Continuum) approach. The term MINC, however, is something of a misnomer. In their original paper, *Pruess and Narisimhan* [1985] associated different continuums with the properties of the grid, and not as a material property of the medium independent of the grid. If the matrix is considered as a single continuum in which provision is made for gradients in various field variables (such as pressure, temperature, saturation, and concentration), there are still only two continuums—fracture and matrix—rather than a "multiple" continuum.

The two distinct approaches to formulating DCMs may be conveniently distinguished by the connectivity of the rock matrix. (The fracture continuum is always considered to be connected in the following.) In the case of *Barenblatt and Zheltov* [1960], the matrix continuum is completely interconnected, with each matrix block connected to its neighboring blocks. In contrast, for the conceptual model used by *Warren and Root* [1963] and *Pruess and Narisimhan* [1985], the matrix continuum is partially disconnected, with each matrix block connected to surrounding fractures but not to other matrix blocks. In what follows, these two approaches are referred to as the dual continuum connected matrix (DCCM) and dual continuum disconnected matrix (DCDM) formulations of the DCM. As originally formulated, the DCCM model associates a single matrix node with each fracture node. This turns out to be a distinct disadvantage of this approach, since it does not allow for gradients within the matrix perpendicular to the fracture. An extension of the DCCM formulation to include more than one matrix node for each fracture node has been used, but only to a limited extent. This extension of the DCCM model to multiple matrix nodes is referred to as the multiple node dual continuum connected matrix (MDCCM) model. The structure of the MDCCM model is similar in many respects to the DFM, with the transport equation for the discrete fracture network replaced by a continuum formulation. As a consequence, it has computational requirements similar to the DFM. One must be careful, however, to ensure that the relation between matrix and fracture nodes is geometrically meaningful. The MDCCM model is not considered further in this critique.

In the case of sufficiently fast chemical reactions where the equilibrium length is on the order of pore scale or microscale, DCMs may not provide sufficient flexibility. In such instances, a hierarchical approach may be needed. In the past, "fast" heterogeneous reactions have often been represented by the local equilibrium. However, such reactions may in fact be much more complicated than surface-controlled kinetic reactions: they may result in local concentration gradients and hence become sensitive to pore and fracture geometry. An important unanswered question is how to scale such processes to the macroscale, where the continuum formulation is valid.

Terminology is not applied consistently in the literature when referring to these two different conceptual approaches. *Barenblatt et al.* [1960] used the term *double porosity*. However, other authors since then have attempted to distinguish between the case of a connected and disconnected matrix continuum, which they have called a *dual permeability* model [*Barenblatt and Zheltov*, 1960], versus *dual* (or *double*) *porosity* [*Warren and Root*, 1963] models. *Hill and Thomas* [1985] generalized the dual porosity model to include arbitrary connectivity referred to as a *dual permeability/dual porosity model*. Triple- and multiple-porosity models have also been considered [*Closmann*, 1975; *Abdassah and Ershaghi*, 1986; *Chen*, 1989; *Bai et al.*, 1993]. To confuse the issue, in the soil literature the connected matrix continuum approach is referred to as a double porosity model [*Gerke and Van Genuchten*, 1993; *Chittaranjan et al.*, 1997], rather than dual permeability, as is common in the oil reservoir literature. A summary of the acronyms and their definitions used here are listed in Table 1.

2.2.1. DCCM model. The DCCM formulation represents the fracture network and matrix as distinct from coexisting continuums. A coupling term provides exchange of mass and heat between the two continuums. In what follows, a multicomponent chemically reacting system consisting of N aqueous species and M minerals is considered. Homogeneous reactions within the aqueous phase and mineral precipitation/dissolution reactions take place represented by the reactions

$$\sum_j \nu_{ji} A_j \leftrightarrow A_i, \quad \sum_j \nu_{js} A_j \leftrightarrow M_s, \tag{2.19}$$

written in terms of a set of (nonunique) aqueous primary or basis species A_j, aqueous secondary species A_i, and minerals M_s. The quantities ν_{ji} and ν_{js} represent the stoichiometric reaction coefficients. These reactions take place simultaneously in the fracture and matrix continuums. Homogeneous reactions are presumed to be sufficiently fast, allowing for a local equilibrium description. Ion exchange and surface complexation reactions are not considered in the present treatment, although they are straightforward to include. Mass conservation equations for a multicomponent system for fracture and matrix continuums can be written in the form

$$\begin{aligned}\frac{\partial}{\partial t}\left(\epsilon_f\, \phi_f \Psi_j^f\right)+\nabla\cdot \epsilon_f\, \Omega_j^f \\ = -\epsilon_f \sum_s \nu_{js} I_s^f - \Gamma_j^{fm}\end{aligned}, \tag{2.20}$$

for the fracture continuum, for the matrix continuum as

$$\begin{aligned}\frac{\partial}{\partial t}\left(\epsilon_m\, \phi_m \Psi_j^m\right)+\nabla\cdot \epsilon_m\, \Omega_j^m \\ = -\epsilon_m \sum_s \nu_{js} I_s^m + \Gamma_j^{fm}\end{aligned}. \tag{2.21}$$

for aqueous primary species labeled j, and for minerals as

$$\frac{\partial \phi_s^\alpha}{\partial t} = \overline{V}_s I_s^\alpha, \quad (\alpha = f, m). \tag{2.22}$$

These equations are referenced to the bulk rock REV. For simplicity, a fully saturated system is considered. The quantities ϕ_α, ϕ_s^α, Ψ_j^α, Ω_j^α, and I_s^α, ($\alpha = f, m$) refer to intrinsic fracture and matrix properties corresponding to porosity, mineral volume fraction, total solute concentration and flux, and mineral reaction rate, respectively. The quantity $\overline{V}$ denotes the mineral molar volume. The total concentration Ψ_j^α is defined relative to an arbitrarily chosen set of primary species with concentrations C_j^α as

$$\begin{aligned}\Psi_j^\alpha &= C_j^\alpha + \sum_i \nu_{ji} C_i^\alpha, \\ C_i^\alpha &= \left(\gamma_i^\alpha\right)^{-1} K_i \prod_j \left(\gamma_j^\alpha C_j^\alpha\right)^{\nu_{ji}}\end{aligned}, \tag{2.23}$$

with C_i^α denoting the concentration of the ith secondary species derived from the primary species concentrations, through mass-action equations with equilibrium constant K_i and activity coefficients $\gamma_{i,j}^\alpha$ [see *Lichtner et al.* (1996) for more details]. The solute flux consisting of contributions from advection, dispersion, and molecular diffusion is defined by

$$\Omega_j^\alpha = -\tau_\alpha \phi_\alpha D_\alpha \nabla \Psi_j^\alpha + v_\alpha \Psi_j^\alpha, \quad (\alpha = f, m), \tag{2.24}$$

with tortuosity τ_α and diffusion coefficient D_α assumed to be the same for all species within each continuum. The mineral kinetic reaction rate I_s^α can be expressed as a sum over various parallel reaction mechanisms. It has the general form (based on transition state theory)

$$I_s^\alpha = A_s^\alpha \sum_l k_{sl} P_{sl}^\alpha \left[1-\left(K_{sl} Q_{sl}^\alpha\right)^{1/\sigma_{sl}}\right], \tag{2.25}$$

with kinetic rate constant k_{sl}, mineral surface area A_s^α, equilibrium constant $K_{sl}(T,p)$, Tempkin constant σ_{sl}, and prefactor P_{sl}^α consisting of products of primary and

Table 1. Summary of acronyms, and their definitions, used for continuum models describing flow and transport in fractured porous media.

Model	Definition	Description
SCM	Single Continuum Model	Fracture can be represented as a single continuum with no interaction with the matrix
ECM	Equivalent Continuum Model	Pervasive interaction between fracture and matrix
DFM	Discrete Fracture Model	Applicable to sparse, widely spaced fractures
DCM	Dual Continuum Model	Representation of fracture and matrix as separate continuums
DCCM	Dual Continuum Connected Matrix	DCM in which matrix continuum is connected and discretized by a single node
MDCCM	Multiple Node Dual Continuum Connected Matrix	DCM in which matrix continuum is connected and discretized by multiple nodes
DCDM	Dual Continuum Disconnected Matrix	DCM in which matrix continuum is disconnected

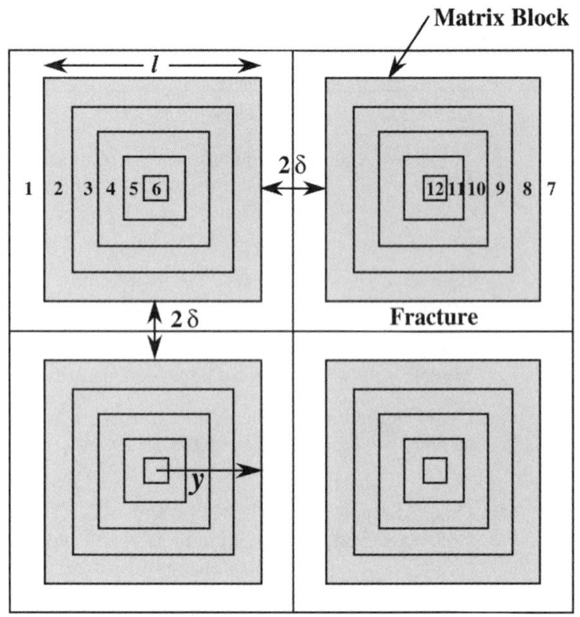

Figure 4. Geometry for the DCDM model for matrix block size l and fracture aperture 2δ indicating the 1-D coordinate y for each matrix block and possible node numbering scheme.

secondary species concentrations raised to respective powers n_{js} and n_{is} [*Lichtner*, 1998]

$$P_{sl}^{\alpha} = \prod_j \alpha_j^{n_{js}^l} \prod_i \alpha_i^{n_{is}^l} . \tag{2.26}$$

The ion activity product Q_{sl}^{α} is defined as

$$Q_{sl}^{\alpha} = \prod_j \alpha_j^{v_{js}^i} . \tag{2.27}$$

The surface area A_s^{α} is in general different for each continuum, and as a consequence so is the reaction rate I_s^{α} even in the limiting case of equal fracture and matrix aqueous concentrations.

The coupling term Γ_j^{fm} is equal to the product of the interfacial specific surface area A_{fm} between fracture and matrix continuums multiplied by the flux Ω_j^{fm} between fracture and matrix defined by

$$\Gamma_j^{fm} = A_{fm}\Omega_j^{fm} . \tag{2.28}$$

For the geometry shown in Figure 1, the interfacial specific fracture-matrix area A_{fm} is given by the expression

$$A_{fm} = \frac{2}{l}\left(1-\epsilon_f\right). \tag{2.29}$$

More complex geometries can also be represented, leading to more complicated expressions for the fracture-matrix surface area.

The aqueous and mineral mass conservation equations are coupled to one another through the reaction rate term and through changes in porosity, tortuosity, and permeability caused by chemical reactions. The latter effects are more difficult to incorporate into the conservation equations, requiring various phenomenological relations that relate changes in physical continuum properties to changes in mineral concentrations. One often-used approach is to relate porosity and mineral volume fractions for each continuum with the assumption that they add to unity

$$\phi_{\alpha} + \sum_s \phi_s^{\alpha} = 1 . \tag{2.30}$$

This relation, however, presupposes that the connected porosity and total porosity are equivalent. Other continuum properties such as permeability and tortuosity are then related to porosity through Archie's laws and the Carmen-Kozeny equation, for example. This approach needs to be tested in the field and in laboratory experiments.

2.2.2. DCDM model. An alternative formulation to the DCCM approach that circumvents the limitation of small matrix blocks is the DCDM approach. In this approach, it is assumed that each matrix block is completely surrounded by fractures (Figure 4). Different matrix blocks can only communicate with one another through the fracture network. In the DCDM formulation, the matrix is resolved into a set of nested rectangular or spherical regions forming an onion-skin-like nodal structure. Gradients across a single matrix block, caused for example by gravity (or thermal or concentration gradients), are thus impossible to describe in this formulation.

Mass transport equations for the DCDM model for fracture continuum have the following form

$$\begin{aligned} &\frac{\partial}{\partial t}\left(\epsilon_f \phi_f \Psi_j^f\right) \nabla \cdot \left(\epsilon_f \Omega_j^f\right) \\ &= -\epsilon_f \sum_s v_{js} I_s^f - \Gamma_j^{fm} \end{aligned} . \tag{2.31}$$

This equation may be of 1, 2, or 3 spatial dimensions. The matrix continuum transport equations, however, have the one-dimensional form

$$\frac{\partial}{\partial t}\left(\epsilon_m \phi_m \Psi_j^m\right)+\frac{\partial}{\partial y}\left(\epsilon_m \Omega_j^m\right) = -\epsilon_m \sum_s v_{js} I_s^m + \Gamma_j^{fm}, \tag{2.32}$$

written with generalized coordinate y representing the distance from the fracture to a point within the matrix. The coordinate y, for example, is a radial coordinate in the case of nested spheres or linear distance for a nested set of cubes. The fracture-matrix coupling term Γ_j^{fm} has the same form as given in Equation (2.28) for the DCCM formulation.

Fundamental difficulties occur when applying the DCDM approach to simultaneous heat and mass transport. This may be seen by considering a stack of matrix blocks, with each block surrounded by a fracture. The top and bottom of the stack are held at different fixed temperatures. In the absence of heat sources or sinks within the matrix blocks, temperature gradients clearly cannot exist within the matrix blocks at steady-state conditions. In fact, for steady-state conditions, the temperature of each matrix block must be the same as its surrounding fracture. As a consequence, heat conduction takes place through the fracture network only, not through the matrix blocks. This leads to an effective thermal conductivity determined by the fracture network $k_{\text{eff}} = k_f$. By contrast, in a layered medium with layer thicknesses l_i of alternating fracture and matrix properties, the effective thermal conductivity is given by the harmonic mean

$$k_{\text{eff}} = \frac{\sum l_i}{\sum \frac{l_i}{k_i}} = \frac{l+2\delta}{\frac{2\delta}{k_f}+\frac{l}{k_m}} \simeq k_m, \tag{2.33}$$

for $l \gg \delta$ and $k_m \ll k_f$, with half-fracture aperture δ and matrix block size l. Field observations suggest that heat conduction in a fractured porous medium is determined primarily by matrix conductivity and not by conductivity of the fracture network (which above the water table may be filled primarily with air). Presumably, this is because matrix blocks are not completely isolated from each other by fractures, but in fact are in direct contact over some fraction of the fracture interfacial area caused by asperities and an in situ stress field. Thus, the DCDM approach gives an incorrect value for the effective thermal conductivity of a composite medium such as fractured porous rock. However, for transient conditions with strong convective flow, as (for example) might occur for injection into a geothermal reservoir, the DCDM model can capture sharp temperature gradients within the matrix.

Other limitations exist to the DCDM approach as formulated here. It is restricted to a homogeneous matrix and cannot handle significant changes in porosity and permeability, which would alter the flow characteristics of the rock from a fracture-dominated system to one of porous flow, a common occurrence in chemical weathering [*Odling and Roden*, 1997]. A heterogeneous matrix block would break the symmetry of the nested matrix node structure. Significant changes in porosity and permeability can alter the physical properties of the porous medium altogether. Thus, during chemical weathering of a granitic rock, in the extreme case of a bauxite deposit formation, the weathering profile changes continuously with depth—from a lateritic layer near the surface (containing aluminum oxide ore), to a highly weathered saprolite zone (containing clay minerals), to the unweathered granite basement rock. The lateritic and saprolite layers are highly porous and have lost the fracture characteristics of the granite rock mass. The DCDM formulation, for instance, could not describe the continuous changes in material properties taking place with depth. As the medium becomes more porous and the fracture properties of the granite rock body are obliterated, the DCDM model would continue to impose an incorrect, relic symmetry on the medium corresponding to the initial fracture geometry. Moreover, the boundary between the two distinct media, as weathering proceeds, icontinuously changes, albeit slowly, with time. The issue of continuously joining a nonfractured porous medium to a fractured medium needs more study.

2.3 Equivalent Continuum Model (ECM)

The ECM representation of a fractured porous medium is based on a composite medium obtained by suitably averaging fracture and matrix properties. Concentrations of dissolved constituents are identical in the fracture and matrix. However, mineral concentrations and reaction rates may be, and generally are, distinct in each continuum. As is generally demonstrated below through scaling relations, the ECM represents the asymptotic limit of the DCCM model.

2.3.1. The ECM as a limiting case of the DCCM model. For conditions of sufficiently strong fracture-matrix coupling, the DCCM model reduces to the ECM. This may be seen by adding Equations (2.20) and (2.21) for fracture and matrix aqueous primary species. The fracture-matrix coupling term Γ_j^{fm} drops out, yielding the single equation

$$\frac{\partial}{\partial t}\left(\in_f \phi_f \Psi_j^f + \in_m \phi_m \Psi_j^m\right) + \nabla \cdot \left[\in_f \Omega_j^f + \in_m \Omega_j^m\right] = -\sum_s \nu_{js}\left(\in_f I_s^f + \in_m I_s^m\right) . \quad (2.34)$$

This equation reduces to the ECM, provided that the primary species concentrations in the fracture and matrix are identical:

$$C_j^f = C_j^m = C_j^{\mathrm{ecm}} . \quad (2.35)$$

In that case, $\Psi_j^{\mathrm{ecm}} = \Psi_j^f = \Psi_j^m$, and the accumulation term becomes

$$\in_f \phi_f \Psi_j^f + \in_m \phi_m \Psi_j^m = \phi_{\mathrm{ecm}} \Psi_j^{\mathrm{ecm}} , \quad (2.36)$$

where the ECM porosity ϕ_{ecm} is defined as an average over fracture and matrix porosities

$$\phi_{\mathrm{ecm}} = \in_f \phi_f + \in_m \phi_m . \quad (2.37)$$

Likewise, the solute flux reduces to an expression involving the ECM solute concentrations

$$\Omega_j^{\mathrm{ecm}} = \in_f \Omega_j^f + \in_m \Omega_j^m = -\tau_{\mathrm{ecm}} \phi_{\mathrm{ecm}} D \nabla \Psi_j^{\mathrm{ecm}} + q_{\mathrm{ecm}} \Psi_j^{\mathrm{ecm}} . \quad (2.38)$$

In this equation, the ECM tortuosity is related to intrinsic fracture and matrix properties by the expression

$$\tau_{\mathrm{ecm}} = \frac{\in_f \tau_f \phi_f + \in_m \tau_m \phi_m}{\phi_{\mathrm{ecm}}} , \quad (2.39)$$

and the Darcy flux q_{ecm} is given by the weighted sum of intrinsic fracture and matrix velocities:

$$q_{\mathrm{ecm}} = \in_f v_f + \in_m v_m . \quad (2.40)$$

For equal fracture and matrix concentrations, the mineral reaction rate reduces to

$$I_s^{\mathrm{ecm}} = \in_f I_s^f + \in_m I_s^m = -A_s^{\mathrm{ecm}} \sum k_{sl} P_{sl}^{\mathrm{ecm}} \left[1 - \left(K_{sl} Q_{sl}^{\mathrm{ecm}}\right)^{1/\sigma_{sl}}\right] , \quad (2.41)$$

where the ECM mineral surface area A_s^{ecm} is given by the weighted sum of intrinsic fracture and matrix surface areas:

$$A_s^{\mathrm{ecm}} = \in_f A_s^f + \in_m A_s^m . \quad (2.42)$$

In contrast to the ECM transport equations for solute species, the ECM equations for minerals generally cannot be reduced to a single bulk-averaged equation. Formally, mineral mass-transfer equations for the ECM can be derived that have the same form as the individual fracture and matrix continuums given by Equation (2.22):

$$\frac{\partial \phi_s^{\mathrm{ecm}}}{\partial t} = \overline{V}_s I_s^{\mathrm{ecm}} , \quad (2.43)$$

obtained by a weighted sum of Equation (2.22) written for fracture and matrix with weight factors $\in_f$ and $\in_m$. The mineral volume fraction ϕ_s^{ecm} in the ECM formulation is related to the intrinsic fracture and matrix volume fractions by the expression

$$\phi_s^{\mathrm{ecm}} = \in_f \phi_s^f + \in_m \phi_s^m . \quad (2.44)$$

The ECM porosity and mineral volume fractions satisfy the relation

$$\phi_{\mathrm{ecm}} + \sum_s \phi_s^{\mathrm{ecm}} = 1. \quad (2.45)$$

However, unlike solute concentrations, mineral concentrations in the ECM need not be, and generally are not, equal for fracture and matrix. This is because different mineral surface areas apply to each continuum, which, furthermore, may change with time as the reaction progresses. For example, fracture and matrix mineral surface areas may vary in reaction, with distinctly different dependencies on mineral volume fraction according to a relation such as

$$A_s^{\alpha} = A_s^{\alpha 0} \left(\frac{\phi_s^{\alpha}}{\phi_s^{\alpha 0}} \right)^{n_\alpha} , \quad (2.46)$$

where n_α is a constant that may be different for each continuum. One possible choice for the initial specific-mineral surface area in the matrix is to assume the surface area is inversely proportional to mineral grain size and directly proportional to the initial matrix mineral concentration:

$$A_s^{m0} = \frac{\phi_s^{m0}}{b_s^m}. \tag{2.47}$$

The initial fracture mineral specific surface generally has a different value proportional to the reciprocal of the half-fracture aperture δ for minerals located at the fracture wall, plus the ratio of initial fracture mineral concentration to grain size for fracture-filling minerals:

$$A_s^{f0} = \frac{1}{\delta} + \frac{\phi_s^{f0}}{b_s^f}. \tag{2.48}$$

The different mineral surface areas associated with each continuum lead to different reaction rates in fracture and matrix continuums. These different reaction rates lead to different mineral concentrations, even though the solute concentrations are the same for each continuum. For $n_\alpha \neq 0$, mineral volume fractions must be obtained directly from the individual mass-transfer equations for fracture and matrix continuums through Equation (2.22) and not the ECM equation (2.44). This is because it is not possible to express the ECM surface area, as defined by Equation (2.42), as a function of the ECM mineral volume fraction. Note the tacit assumption that ϵ_f remains constant for all time, which need not actually be the case.

Deciding what values to use for the reacting mineral surface areas is a formidable challenge. What makes specification of these parameters most difficult is that it is the hydrologically accessible surface area, the area in contact with the fluid, which is of interest. Accurately determining the surface requires in situ experiments and direct field measurements.

A consequence of averaging fracture and matrix properties in the ECM is that travel times of nonreacting tracer species are generally longer in the ECM compared to the other models describing transport in fractured media. Indeed, it follows that the ECM travel time for a tracer is given by

$$t_{\text{ecm}} = \frac{\phi_{\text{ecm}} L}{q_{\text{ecm}}}, \tag{2.49}$$

where L denotes the system length. Substituting for ϕ_{ecm} and q_{ecm} in terms of their intrinsic fracture and matrix properties gives (for the case of flow in the fracture network only, and $v_m = 0$)

$$t_{\text{ecm}} = \left(1 + \frac{\epsilon_m \phi_m}{\epsilon_f \phi_f}\right) t_f, \tag{2.50}$$

where the fracture travel time t_f is defined as

$$t_f = \frac{\phi_f L}{v_f}. \tag{2.51}$$

The travel time t_f applies to single and dual continuum formulations for the case when flow is absent in the matrix. As a consequence of Equation (2.50), the ECM is not conservative in predicting contaminant arrival times in the case of fracture-dominated flow.

2.3.2. Asymptotic limit of the DCCM model: scaling relations. As demonstrated by *Lichtner* [1993], through scaling relations, the reactive mass-transport equations based on a kinetic description of mineral reaction rates asymptotically approach the local chemical equilibrium limit. This asymptotic relation between a kinetic description and local equilibrium description provides an immediate understanding of local-equilibrium validity conditions. In addition, because the solution to the local equilibrium form of the reactive transport equations can be reduced to solving a set of algebraic equations, this relation also provides a way of checking the accuracy of the more complicated solution to the partial differential equations representing the kinetic formulation. The same considerations apply to the relationship between the DCCM formulation and the ECM. Applying the scaling transformation

$$\boldsymbol{r}_\sigma = \sigma^{-1}\boldsymbol{r}, \tag{2.52}$$

$$t_\sigma = \sigma^{-1} t, \tag{2.53}$$

with constant scale factor σ to the DCCM equations [Equations (2.20) and (2.21)] leads to the transformed equations

$$\begin{aligned}&\frac{\partial}{\partial t_\sigma}\left(\epsilon_\alpha \phi_\alpha \Psi_j^\alpha\right) + \nabla_\sigma \cdot \epsilon_\alpha \Omega_{j\sigma}^\alpha \\ &= -\sigma \epsilon_\alpha \sum_s \nu_{js} I_s^\alpha + \sigma\left(1 - 2\delta_{f\alpha}\right)\Gamma_j^{fm}, \; (\alpha = f,\, m),\end{aligned} \tag{2.54}$$

and

$$\frac{\partial \phi_s^\alpha}{\partial t_\sigma} = \sigma \overline{V}_s I_s^\alpha, \quad (\alpha = f, m), \tag{2.55}$$

where the transformed flux $\Omega_{j\sigma}^\alpha$ is given by

$$\Omega_{j\sigma}^{\alpha} = -\sigma^{-1}(\tau\sigma D)_{\alpha} \nabla_{\sigma}\Psi_{j}^{\alpha} + \boldsymbol{q}_{\alpha}\Psi_{j}^{\alpha} \quad . \qquad (2.56)$$

In the transformed flux, the diffusion/dispersion term is scaled, and ∇_{σ} represents the gradient operator with respect to the scaled spatial coordinates. Consequently, assuming that the boundary conditions imposed on the system are scale invariant, it follows that the solution to the solute and mineral conservation equations represented by the function $\mathcal{F}(\boldsymbol{r}, t|\{k\}, D, \boldsymbol{q}, A_{fm})$ scales according to the relation

$$\begin{aligned} &\mathcal{F}\left(\sigma\boldsymbol{r},\ \sigma t\middle|\{k\},\ D,\ q,\ A_{fm}\right) \\ &= \mathcal{F}\left(\boldsymbol{r},\ t\middle|\{k\},\ \sigma^{-1}D,\ \boldsymbol{q},\ \sigma A_{fm}\right) \ . \end{aligned} \qquad (2.57)$$

Taking the limit of the relation as $\sigma = \infty$ leads to the pure advective, local-equilibrium form of the ECM as the asymptotic limit of the DCCM equations.

3. DFM-DCCM MODEL COMPARISON

In this section, a comparison is made between stationary-state solutions to the DFM and DCCM model. The stationary-state DCCM transport equations for a single component system, expressed in terms of intrinsic properties for the solute species in fracture and matrix, have the form

$$\begin{aligned} &-\tau_{\alpha}\phi_{\alpha} D\frac{d^2 C_{\alpha}}{dx^2} + v_f \frac{dC_{\alpha}}{dx} \\ &= -k_{\alpha}\left(C_{\alpha} - C_{eq}\right) + \left(1 - 2\delta_{f\alpha}\right)\frac{\gamma}{\epsilon_{\alpha}}\left(C_f - C_m\right) \ , \end{aligned} \qquad (3.1)$$

where the Kronecker delta function $\delta_{f\alpha} = 1$ if $\alpha = f$, and zero otherwise. The fracture-matrix coupling term γ is defined by

$$\gamma = \frac{(\tau\phi D)_f (\tau\phi D)_m}{d_f (\tau\phi D)_m + d_m (\tau\phi D)_f} A_{fm} \, , \qquad (3.2)$$

where the notation d_f= and $d_m = l/2$ is introduced to refer to the perpendicular distances from the fracture and matrix node centers to their common interface. The kinetic rate constants k_{α} are effective rate constants, equal to the product of the intrinsic rate constant times the specific surface area for the fracture and matrix continuums, respectively. Thus they may differ significantly from each other. The coupling term is presumed to be linear for the difference in fracture and matrix concentrations at each node. The coupling strength γ has the same units as the kinetic rate constants (s^{-1}).

At large distances from the inlet, the solute concentration approaches the equilibrium concentration C_{eq} of the solid. The transport equations are subject to the following boundary conditions at the inlet and outlet to the fractured porous medium:

$$C_{\alpha}(0) = C_{\alpha}^{0}, \quad C_{\alpha}(\infty) = C_{eq} \, . \qquad (3.3)$$

To solve the stationary state transport equations, first note that the fracture transport equation may be solved for the matrix concentration C_m' to give

$$\begin{aligned} C_m' &= \left(1 + \frac{\epsilon_f k_f}{\gamma}\right) C_f' \\ &+ \frac{\epsilon_f v_f}{\gamma}\frac{dC_f'}{dx} - \frac{\epsilon_f (\tau\phi D)_f}{\gamma}\frac{d^2 C_f'}{dx^2} \ , \end{aligned} \qquad (3.4)$$

where

$$C_{\alpha}' = C_{\alpha} - C_{eq} \, . \qquad (3.5)$$

Substituting this expression into the matrix transport equation results in the following fourth-order ordinary differential equation with constant coefficients:

$$\begin{aligned} &a(\gamma)\frac{d^4 C_f'}{dx^4} - b(\gamma)\frac{d^3 C_f'}{dx^3} + c(\gamma)\frac{d^2 C_f'}{dx^2} \\ &\quad + d(\gamma)\frac{dC_f'}{dx} + e(\gamma) C_f' = 0. \end{aligned} \qquad (3.6)$$

The coefficients $a(\gamma)$, $b(\gamma)$, $c(\gamma)$, $d(\gamma)$, and $e(\gamma)$ are defined by

$$a(\gamma) = \frac{\epsilon_f \epsilon_m (\tau\phi D)_f (\tau\phi D)_m}{\gamma}, \qquad (3.7a)$$

$$b(\gamma) = \frac{\epsilon_m \epsilon_f}{\gamma}\left(v_m (\tau\phi D)_f + v_f (\tau\phi D)_m\right), \qquad (3.7b)$$

$$\begin{aligned} c(\gamma) &= \frac{\epsilon_f v_f \epsilon_m v_m}{\gamma} - \epsilon_m (\tau\phi D)_m \left(1 + \frac{\epsilon_f k_f}{\gamma}\right) \\ &- \epsilon_f (\tau\phi D)_f \left(1 + \frac{\epsilon_m k_m}{\gamma}\right), \end{aligned} \qquad (3.7c)$$

$$d(\gamma) = \epsilon_m v_m \left(1 + \frac{\epsilon_f k_f}{\gamma}\right) + \epsilon_f v_f \left(1 + \frac{\epsilon_m k_m}{\gamma}\right), \qquad (3.7d)$$

$$e(\gamma) = \gamma\left[\left(1 + \frac{\epsilon_f k_f}{\gamma}\right)\left(1 + \frac{\epsilon_m k_m}{\gamma}\right) - 1\right] \ . \qquad (3.7e)$$

The most general solution to Equation (3.6) for an infinite system subject to the boundary conditions at the inlet and outlet given by Equation (3.3) has the form

$$C_f(x;\gamma) = Ae^{-q_1 x} + Be^{-q_2 x} + C_{eq}, \tag{3.8}$$

where $q_1(\gamma)$ and $q_2(\gamma)$are the two non-negative roots of the characteristic fourth-order polynomial

$$p(q;\gamma) = a(\gamma)q^4 + b(\gamma)q^3 + c(\gamma)q^2 - d(\gamma)q + e(\gamma) = 0. \tag{3.9}$$

Because the coefficient $e(\gamma)$ is positive, there must always exist an even number of positive roots. Because $d(\gamma) \geq 0$, there can be only two non-negative roots. From Equation (3.4), the matrix concentration has the form

$$C_m(x) = w_1 A_1 e^{-q_1 x} + w_2 A_2 e^{-q_2 x} + C_{eq}, \tag{3.10}$$

where

$$w_i = 1 + \frac{\epsilon_f k_f}{\gamma} - \frac{\epsilon_f v_f}{\gamma} q_i - \frac{\epsilon_f \tau_f \phi_f D}{\gamma} q_i^2, \quad (i=1,\ 2). \tag{3.11}$$

The coefficients A_i are related to the boundary conditions imposed on the solution with the values

$$A_i = (-1)^{i+1} \frac{\left(C_m^0 - C_{eq}\right) - w_{3-i}\left(C_f^0 - C_{eq}\right)}{w_1 - w_2}, \quad (i=1,\ 2). \tag{3.12}$$

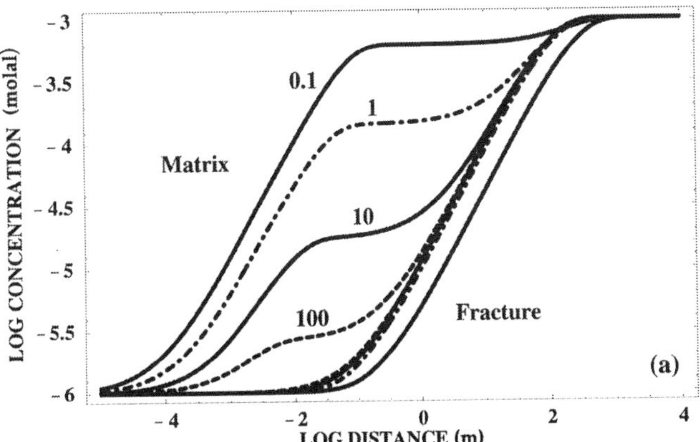

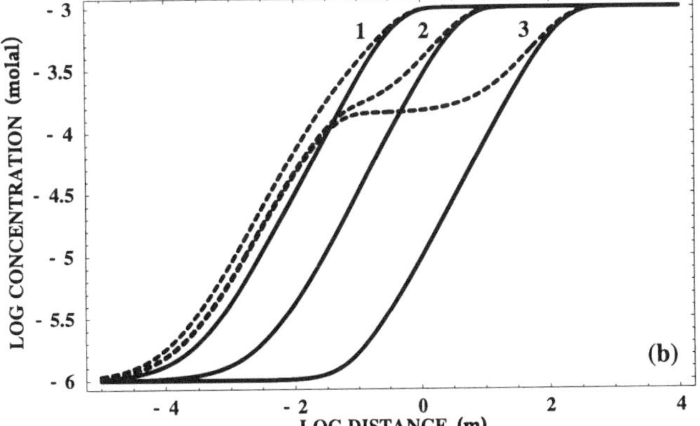

Figure 5. Stationary-state concentration profiles based on the analytical solution to the stationary-state transport equations: (a) Fracture-matrix surface area multiplied by factors of 100 (dashed), 10 (solid), 1 (dot-dashed), and 0.1 (solid); (b) Fracture (solid) and matrix (dashed) concentration profiles for kinetic rate constant equal to (1) 10^{-10}, (2) 10^{-11}, and (3) 10^{-14} moles cm^{-2} s^{-1}.

In the limit $\gamma \to 0$, the coupling term vanishes and the matrix and fracture continuums evolve independently of one another. The ECM is retrieved in the limit $\gamma \to \infty$.

Stationary-state profiles for fracture and matrix concentrations are illustrated in Figure 5, based on the analytical solution for a single component system. A fracture aperture of 1 mm, Darcy flow velocity of 1000 m yr^{-1}, matrix block size of 0.1 m, and matrix porosity of 0.05 are used in the calculations. As can be seen from Figure 5a, the matrix concentration plateau decreases as the fracture-matrix surface area increases. The ECM limit is recovered with increasing time. As the fracture kinetic-rate constant increases, the ECM limit is obtained at earlier times (as shown in Figure 5b).

Comparing the stationary-state solution for the DCCM formulation, Equations (3.8) and (3.10), with the DFM solution given by Equations (2.10) and (2.11), very different behavior for the solute concentration is observed. In particular, no scaling relation between fracture and matrix concentrations exists in the DCCM formulation—if for no other reason than only one matrix node exists for each fracture node. In fact, the DFM, DCCM, and ECM can only agree with each other when the ECM is valid. In the DCCM approach, concentration gradients are parallel to the fracture, whereas in the DFM (and DCDM), matrix gradients are perpendicular to the fracture. Although gradients parallel to the fracture could have been included in the DFM, they would not have made a significant difference in the qualitative behavior of the solution for sufficiently rapid fracture flow.

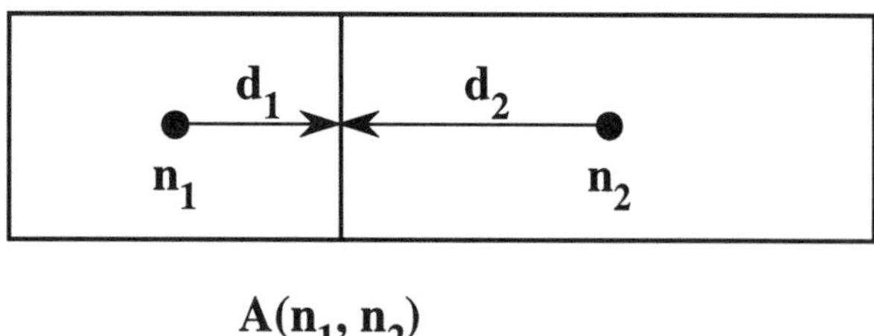

Figure 6. Integrated finite volume geometry.

4. NUMERICAL IMPLICATION

4.1 Integrated Finite Volume

To develop numerical techniques for solving the partial differential equations arising from the various formulations of the DCM, it is convenient to use an unstructured grid approach. In this approach, nodal connectivity, volumes, distances between connecting nodes, and surface areas can be specified arbitrarily (as illustrated in Figure 6) for a simple structured grid geometry with unequal spacing. The integrated finite volume equations for the primary species are expressed simply as

$$\frac{\left(\tau_n\phi_n\Psi_{jn}\right)_{t+\Delta t}-\left(\tau_n\phi_n\Psi_{jn}\right)_t}{\Delta t}+ \sum_{n'} A_{\langle n'n\rangle}\Omega_{j\langle n'n\rangle}^{t+\Delta t} = -\sum_s \nu_{js} I_{sn}^{t+\Delta t}, \tag{4.1}$$

for a fully implicit time discretization with time step Δt. The flux $\Omega_{j<n'n>}$ is defined as

$$\Omega_{j\langle n'n\rangle} = -(\tau\phi D)_{\langle n'n\rangle}\frac{\Psi_{jn}-\Psi_{jn'}}{d_n+d_{n'}}+v_{\langle n'n\rangle}\Psi_{j\langle n'n\rangle}. \tag{4.2}$$

The notation $<n'n>$ refers to the interface between nodes n and n' with interfacial area $A_{<n'n>}$ and distances to the interface denoted by d_n and $d_{n'}$. The sum in Equation (4.1) is over all nodes connected to the nth node. Note that there is no reference to fracture or matrix properties because that is handled automatically by the grid structure and its connectivity. Even explicit reference to the fracture-matrix coupling term has disappeared in the integrated finite volume form of the equations, with this term now included in the term containing the sum over fluxes. This approach offers greater flexibility in programming both the DCCM and DCDM methods, requiring only a change in preprocessor to invoke the appropriate geometrical relation between nodes. The internal part of the code can remain the same. In this formulation of the problem, the distinction between the different DCMs becomes lost.

Examples of integrated finite volume grids for the DCDM and DCCM formulations are illustrated in Figures 4 and 7, respectively. Note that the DCCM grid is in fact just a 2-D problem with two y-nodes. The difference between a true 2-D problem and the DCCM grid lies in the different assignment of areas at the fracture-matrix interface volumes and distances.. The node connections for the DCDM model corresponding to Figure 4 are listed in Table 2.

The flexibility of the unstructured grid framework of the various dual continuum formulations allows for practically arbitrary assignment of block connections and surface areas. However, the resulting finite volume equations must actually represent partial differential equations. The processes to be modeled must actually take place physically and cannot be merely an artifact of some artificially imposed grid structure.

4.2 DCCM: Harmonic Versus Arithmetic Averaging

An important consideration in the numerical implementation of the DCCM model is the computation of interface properties between fracture and matrix. This is especially true because of the often great difference in fracture aperture and matrix block size. In finite difference form, the coupling term [Equation (2.28)] is given by

$$\Gamma_j^{fm} = A_{fm}(\tau\phi D)_{fm}\frac{\Psi_j^f-\Psi_j^m}{d_f+d_m}, \tag{4.3}$$

where [as previously defined following Equation (3.2)], $d_f = \delta$ and $d_m = l/2$. Harmonic or arithmetic averages can be used to evaluate the product $\tau\phi D$ at the fracture-matrix interface. The harmonic average is more rigorously based by considerations of the steady-state flux across the interface [*Patankar*, 1980]. For $d_f \ll d_m$, the harmonic mean yields

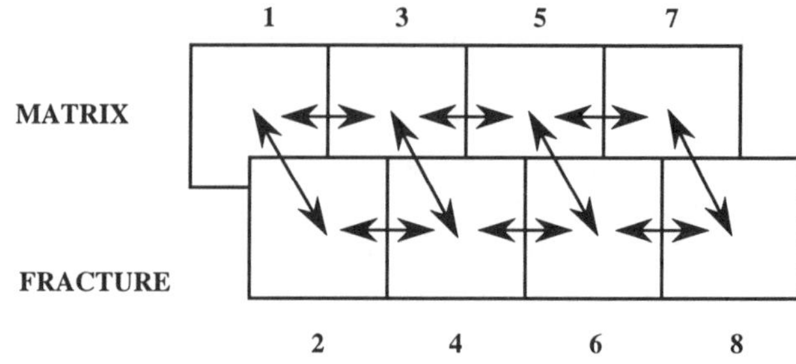

Figure 7. Nodes and their connections for the DCCM model.

$$(\sigma\tau D)_{fm}^{\text{harm}} = \frac{(d_f + d_m)(\tau\phi D)_f (\tau\phi D)_m}{d_m (\tau\phi D)_f + d_f (\tau\phi D)_m} \simeq (\tau\phi D)_m . \quad (4.4)$$

Thus, the harmonic mean yields a coupling term proportional to the effective matrix diffusivity. The arithmetic mean, however, gives for the interface property

$$(\phi\tau D)_{fm}^{\text{arith}} = \frac{d_m (\tau\phi D)_f + d_f (\tau\phi D)_m}{d_m + d_f} \simeq (\tau\phi D)_f , \quad (4.5)$$

yielding a coupling term proportional to the effective fracture diffusivity. Because the intrinsic fracture porosity ($\phi_f \sim 1$) is generally much larger than the matrix porosity ($\phi_m \ll 1$), harmonic averaging leads to a smaller coupling term compared to arithmetic averaging. Which approach is correct? Intuitively, one would expect that the flux across the fracture-matrix interface would be governed by diffusion in the matrix, and not the fracture, because of the very small fracture aperture. Hence, harmonic averaging is preferred.

The DCCM model appears to simulate the incorrect behavior as the matrix block size is increased. Evaluating the coupling-term equation (2.28), using harmonic averaging according to Equation (4.4) and inserting Equation (2.29) for the interfacial area, it is apparent from the finite-difference form of the coupling-term equation (4.3) that as the matrix block size increases, the coupling term decreases as d_m^{-2}

$$\Gamma_j^{fm} \sim (1 - \epsilon_f) \frac{(\tau\phi D)_m}{d_m^2} (\Psi_j^f - \Psi_j^m) . \quad (4.6)$$

Table 2. Node connections for the DCDM model with grid numbering as shown in Figure 4.

Node	Connecting Nodes			
1	1	2	7	
2	1	2	3	
3	2	3	4	
4	3	4	5	
5	4	5	6	
6	5	6		
7	1	7	8	13
8	7	8	9	
9	8	9	10	
	...		...	

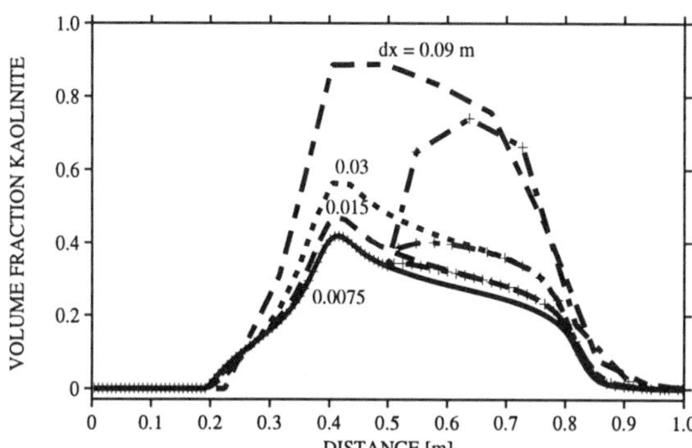

Figure 8. Profiles showing the volume fraction of kaolinite, which precipitates as K-feldspar is weathered. Shown are profiles for different grid spacing based on the harmonic mean. The calculations correspond to an elapsed time of 10,000 yr, with a fracture flow rate of 1000 m yr^{-1} and fracture volume fraction $\epsilon_f = 10^{-3}$. The intrinsic fracture porosity is unity and matrix porosity 0.1.

As a consequence, coupling between fracture and matrix decreases as the matrix block size increases. This behavior runs counter to that predicted by the DFM and what intuitively is to be expected. That is, the fracture-matrix interaction should be independent of the matrix block size, at least for times that are short compared to the transport time across the matrix block. The DCDM model does not have this limitation. It is able to describe narrow alteration halos surrounding fractures that result from sharp concentration gradients within the rock matrix.

Numerical difficulties arise when applying the DCCM model to cases where the fracture volume fraction ϵ_f is very small. As shown in Figure 8, variable grid spacing can lead to completely erroneous results for mineral concentrations. In this figure, the DCCM model is applied to formation of kaolinite resulting from the alteration of K-feldspar. Results are compared for uniform and variable grid spacing using harmonic averaging. Each pair of curves compare uniform grid spacing with a change in spacing 0.5 m from the inlet. Grid spacing varies from 0.0075 m to 0.09 m, as indicated in the figure. The figure shows that an erroneous jump in the kaolinite volume fraction is obtained at a change in grid spacing. The magnitude of the jump is also very sensitive to the absolute grid size. The DCDM model, on the other hand, does not suffer from this difficulty, since a small grid spacing on the order of the fracture aperture can be used to discretize the matrix in the neighborhood of the fracture.

Table 3. Model ore deposit giving primary ore and gangue mineral abundances, porosities, and associated mineral surface areas used in the calculations for dual, equivalent, and single continuum model values for the SCM are bulk properties.

Property	Volume Fraction				Surface Area (cm^{-1})			
	Fracture	Matrix	ECM	SCM	Fracture	Matrix	ECM	SCM
Chrysocolla	0.2	0.02	0.0253	0.0059	42.0	4.0	5.118	1.2350
Quartz	0.0	0.73	0.7085	0.0000	1.0	14.6	1.000	0.0294
Kaolinite	0.0	0.20	0.1941	0.0000	1.0	40.0	38.82	0.0294
Porosity	0.8	0.05	0.0721	0.0235				

4.3 DCDM: Decoupling Fracture and Matrix Transport Equations

Computationally, the DCDM model is generally much more expensive compared to the DCCM model. For a spatial domain consisting of N_f fracture nodes and N_m nodes within each matrix block, the DCDM model requires solving $N_c \times N_f \times N_m$ simultaneous equations for an N_c component system. whereas The DCCM model requires solving only $2 \times N_c \times N_f$ equations. However, perhaps surprisingly, it is possible to rigorously decouple the fracture and matrix equations in the DCDM model, reducing the system of equations to $N_c \times (N_f + N_m)$ in number. As noted by *Gilman* [1986] this is possible because of the one-dimensional form of the matrix equations in the DCDM model. This result is surprising given the strong, nonlinear coupling that is possible between fracture and matrix continuums. The procedure outlined by *Gilman* [1986] involves first a backward solution of the matrix equations, beginning with the innermost matrix node. This provides a relation between the concentration at the outermost matrix node and the concentration at the adjacent fracture node. Thus, the matrix concentration appearing in the coupling term in the fracture equations can be eliminated. As a result, the fracture equations are only a function of the fracture concentration and may be solved independently of the matrix equations. Once the fracture equations are solved, the solution of the matrix equations can be completed through a forward sweep of the matrix nodes, beginning with the outermost node. This approach also lends itself to parallel computing techniques [*Seth and Hanno*, 1995; *Smith and Seth*, 1999], in which the matrix equations can be solved in parallel, greatly reducing computation times and dramatically extending the capability of the DCDM to much larger numbers of nodes and chemical components than could be solved without these techniques.

5. EXAMPLE OF IN SITU COPPER LEACHING

The following example, involving the in situ leaching of a hypothetical copper ore body, illustrates and contrasts the various approaches previously discussed for describing transport in porous fractured media. [*Lichtner*, 1998]. Calculations were carried out using the computer code FLOTRAN [*Lichtner*, 1999]. A one-dimensional column contains the copper-bearing phase chrysocolla and gangue minerals in the form of kaolinite and quartz. A sulfuric acid solution with pH 1 is allowed to infiltrate into the column through a fracture network. The initial fluid in the column is assumed in equilibrium with chrysocolla, kaolinite, and quartz at a pH of 8. The composition of the host rock for the model ore deposit is listed in Table 3. For the model parameters listed in the table, the ore body has a copper grade of 0.90% and bulk rock density of approximately 2.44 g cm^{-3}. A matrix block size of 0.1 m is used in the calculations. A fracture aperture of 1 mm corresponding to a fracture volume fraction of $\epsilon_f = 2.941 \times 10^{-2}$ is used. A bulk Darcy velocity of 10 m yr^{-1} (corresponding to a fracture velocity of 340 m yr^{-1}) and an effective matrix diffusivity of 10^{-6} cm^2 s^{-1} is used in the calculations. No flow is allowed in the matrix. The fracture continuum was discretized into 100 nodes with equal spacing of 10 cm, giving a total length of 10 m for the leach column. In the DFM and DCDM models, the matrix was discretized into 10 grid blocks of variable spacing, with the smallest spacing equal to the fracture aperture neighboring the fracture. Secondary minerals that form during leaching are amorphous silica, gypsum, jurbanite, and alunite, and secondary copper minerals brochantite and antlerite.

Results for the copper breakthrough curves for the different models are shown in Figure 9. If diffusion is turned off, then the assumption of no flow in the matrix would require that the SCM give results identical to the DCCM and DCDM models. Thus, differences between breakthrough curves for these models stem from differences in how the interaction term between fracture and matrix is treated. The SCM breakthrough curve exhibits a single peak resulting from dissolution of chrysocolla in the fractures. Likewise, the ECM also exhibits a single peak, but it is delayed in time compared to the SCM—as expected from Equation (2.50), which predicts a retardation of approximately

$$1+\frac{\epsilon_f\,\phi_f}{\epsilon_m\,\phi_m}\simeq 3.02\,, \tag{5.1}$$

in agreement with the figure.

The width of the ECM peak is longer compared to the SCM: there is more copper to dissolve because the ECM incorporates copper from both the rock matrix and fractures. The breakthrough curves for the DFM, DCCM, and DCDM models show a bimodal distribution as a result of contributions from individual fracture and matrix copper sources. The shapes of the curves are somewhat different, with the DCDM model agreeing more closely with the DFM. Differences between the DFM and DCDM model can be attributed to different formulations of the matrix (treated as three-dimensional cubical blocks in the DCDM). The two models must give identical results for equal interfacial surface areas and nodal volumes. The DCCM curve follows the SCM curve closely during the early part of breakthrough, dominated by dissolution of copper in fractures. It then drops off to an almost constant value as the matrix becomes the dominant contributor. Clearly, the DCCM model is unable to give the proper behavior at longer times and overshoots the copper concentration (as predicted by the DFM and DCDM model) at early times.

Note that the peak copper concentration is quite high in these simulations compared to what might be expected from an actual five-spot leach field. This is an artifact of the one-dimensional form of the calculations.

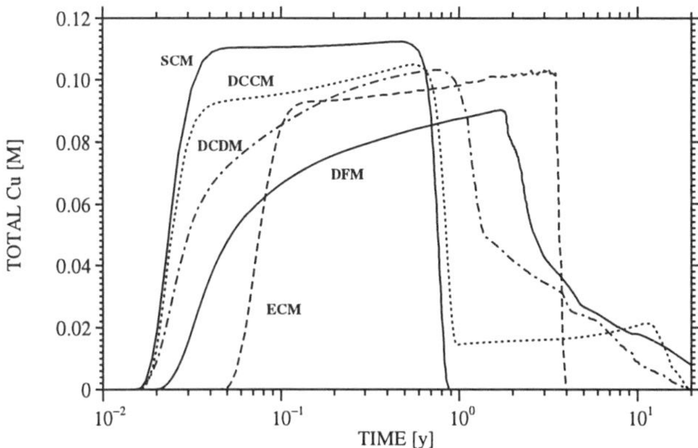

Figure 9. Copper breakthrough curves for the SCM, ECM, DFM, DCCM, and DCDM models.

6. CONCLUSION

Describing quantitatively reactive flow in fractured porous media presents a number of challenges that have yet to be resolved satisfactorily. DCMs attempt to account for the bimodal distribution in physical and chemical properties characteristic of fractured porous media. An equivalent porous medium description is generally unable to capture the unique features characteristic of fractured systems. DCMs are presumably applicable to highly fractured systems where the DFM becomes impractical, but not so highly fractured that the system can be described as an equivalent continuum. Whether a dual continuum (as opposed to a single continuum) representation of the fracture network is appropriate depends on the time scales of interest and the extent of interaction between the fracture network and rock matrix.

Two different DCM models were discussed in detail, characterized by the connectedness of the rock matrix. In the DCCM model, the matrix formed a connected continuum, with each fracture node associated with a single matrix node. The validity of the DCCM model is based on the absence of strong concentration gradients within the matrix perpendicular to the fracture. This is a consequence of representing the matrix by a single node for each fracture node. The DCCM model should be applicable to situations where the kinetic reaction rate varies smoothly over the matrix block or, equivalently, the characteristic chemical equilibration length scale is long compared to the matrix block size. Faster reactions imply shorter equilibration length scales, leading to steeper gradients and eventually the failure of the DCCM approach.

An alternative approach, the DCDM model, is applicable to situations where the fracture network segregates the matrix into disconnected blocks, linked with one another only through their common fracture interface. Within each matrix block, a fine grid may be used to capture arbitrarily sharp gradients, thereby removing one of the limitations of the DCCM model. However, in contrast to the DCCM approach, the DCDM model associates a single fracture node with each matrix block. The fracture node completely surrounds the matrix block. This symmetry imposes severe constraints on the DCDM model. It is not possible to account for gradients or reaction fronts across matrix blocks arising, for example, from gravity-driven flow. Furthermore, incorporation of heterogeneous matrix blocks would destroy this symmetry. Finally, it does not appear possible to describe simultaneous heat and mass flow within the DCDM framework in response to a geothermal gradient.

Many conceptual difficulties remain in providing a quantitative description of reactive flow and transport in fractured porous media. Although not discussed in any detail here, it is especially difficult to obtain the necessary data to apply the models to a particular field situation. Both the DCCM and DCDM approaches introduce additional parameters (such as matrix block size, fracture aperture, and fracture-matrix interaction parameters) that represent averages over distributions and are difficult to measure and characterize. In addition, these models require characterizing the reactive surface area of minerals separately for fracture and matrix continuums from experimental and field data, which may be difficult to obtain.

Acknowledgments. The author would like to thank Mohan Seth, Rajesh Pawar, and Daniel Tartakovsky for helpful discussions on this topic. Also, Tianfu Xu, Joel VanderKwaak, Daniel Tartakovsky, Bruce Robinson, and an anonymous reviewer deserve thanks for providing helpful comments, which greatly improved the manuscript.

REFERENCES

Abdassah, D., and I. Ershaghi, Triple porosity system for representing naturally fractured reservoirs, *SPE Form. Eval. 1*, 113–127, 1986.

Bai, M., D. Elsworth, and J.-C. Roegiers, Multiporosity/multipermeability approach to the simulation of naturally fractured reservoirs, *Water Resour. Res., 29*, 1621–1633, 1993.

Barenblatt, G. I., and Iu. P. Zheltov, Fundamental equations of filtration of homogeneous liquids in fissured rocks, *Soviet Physics-Doklady, 5*, 522–525, 1960.

Barenblatt, G. I., Iu. P. Zheltov, and I. N. Kochina, Basic conceptions the theory of seepage of homogeneous liquids in fissured rocks, *J. Appl. Math. Mech., 24*, 1286–1303, 1960.

Chittaranjan, R., T. R. Ellsworth, A. J. Valocchi, and C. W. Boast, An improved dual porosity model for chemical transport in macroporous soils, *J. Hydro., 193*, 270–292, 1997.

Closmann, P. J., An aquifer model for fissured reservoirs, *Soc. Pet. Eng. J., 15*, 385–398, 1975.

Gerke, H. H., and M. T. van Genuchten, A dual porosity model for simulating the preferential movement of water and solutes in structured porous media, *Water Resour. Res., 29*, 305–319, 1993.

Gilman, J. R., An efficient finite-difference method for simulating phase segregation in the matrix blocks in double-porosity reservoirs, *SPERE*, 403–413, 1986.

Hill, A. C., and G. W. Thomas, A new approach for simulating complex fractured reservoirs, *SPE, 13537*, 429–436, 1985.

Lichtner, P. C., Continuum model for simultaneous chemical reactions and mass transport in hydrothermal systems, *Geochimica et Cosmochimica Acta, 49*, 779–800, 1985.

Lichtner, P. C., The quasi-stationary state approximation to coupled mass transport and fluid-rock interaction in a porous media, *Geochimica et Cosmochimica Acta, 52*, 143–165, 1988.

Lichtner, P. C., Scaling properties of time-space kinetic mass transport equations and the local equilibrium limit, *Am. J. Sci., 293*, 257–296, 1993.

Lichtner, P. C., Modeling reactive flow and transport in natural systems, in *Environmental Geochemistry*, edited by G. Ottonello and L. Marini, Pacini Editore, Pisa, Italy, 5–72, 1998.

Lichtner, P. C., *FLOTRAN User's Manual*, Los Alamos National Laboratory, Los Alamos, N.M., 1999.

Lichtner, P. C., and M. S. Seth, Multiphase-multicomponent nonisothermal reactive transport in partially saturated porous media: Application to the Proposed Yucca Mountain HLW Repository, *Proceedings, International Conference on Deep Geologic Disposal of Radioactive Waste*, pp. 3-133–3-142, Canadian Nuclear Society, Winnipeg, Manitoba, Canada., 1996.

Lichtner, P. C., C. I. Steefel, and E. H. Oelkers (eds.) *Reactive Transport in Porous Media, Reviews in Mineralogy, 34*, 438 pp., 1996.

Odling, N. D., and J. E. Roden, Contaminant transport in fractured rocks with significant matrix permeability, using natural fracture geometries, *J. Contaminant Hydrology, 27*, 263–283, 1997.

Patankar, S. V., Numerical heat transfer and fluid flow, *Hemisphere Series on Computational Methods in Mechanics and Thermal Science*, 197 pp., 1980.

Pruess, K., and T. N. Narisimhan, A practical method for modeling fluid and heat flow in fractured porous media, *Soc. Pet. Eng., 25*, 14–26, 1985.

Seth, M. S., and M. Hanano, An efficient solution procedure for multiple interacting continuum flow, *Proceedings, World Geothermal Congress*, Florence, Italy, p. 1625, 1995.

Smith, E. H., and M. S. Seth, Efficient solution method for matrix-fracture flow with multiple interacting continuum, *Int. J. Numer. Anal. Meth. Geomech, 23*, 427–438, 1999.

Steefel, C. I., and P. C. Lichtner, Multicomponent reactive transport in discrete fractures: I. Controls on reaction front geometries, *J. Hydrology, 209*, 186–199, 1998a.

Steefel, C. I., and P. C. Lichtner, Multicomponent reactive transport in discrete fractures: II. Infiltration of hyperalkaline groundwater at Maqarin, Jordan, a natural analogue site, *J. Hydrology, 209*, 200–224, 1998b.

Thomas, L. K., T. N. Dixon, and R. G. Pierson, Fractured reservoir simulation, *SPEJ, 9305*, 42–54, 1983.

Warren, J. E., and P. J. Root, The behavior of naturally fractured reservoirs, *Soc. Petrol. Eng. J., 3*, 245–255, 1963.

Peter C. Lichtner, Los Alamos National Laboratory, Los Alamos, NM 87545.

On the Effective Continuum Method for Modeling Multiphase Flow, Multicomponent Transport, and Heat Transfer in Fractured Rock

Yu-Shu Wu

Earth Sciences Division, Lawrence Berkeley National Laboratory, Berkeley, California

Flow and transport through fractured porous media, which occurs in many subsurface systems, has received considerable attention in recent years because of its importance in the areas of underground natural resource recovery, waste storage, and environmental remediation. Among the methods of handling fracture-matrix flow and transport through geological media, the effective continuum method (ECM) has been widely used, and in some cases misused, because of its simplicity in terms of data requirements and computational efficiency.

This paper presents a rigorous, generalized effective continuum formulation, which has been incorporated as a special version of the TOUGH2 code [*Pruess*, 1991] for modeling multiphase, multicomponent, nonisothermal flow and transport in fractured rocks. Also included in this paper are discussions of the conditions under which the ECM approach applies and the procedures for evaluating the effective parameters for both flow and transport simulations. Three application examples—one multiphase flow, one heat flow, and one chemical transport problem—are presented to demonstrate the usefulness of the ECM method.

1. INTRODUCTION

The process of flow and transport through fractured porous media, which occurs in many subsurface systems, has received considerable attention in recent years because of its importance in the areas of underground natural resource recovery, waste storage, and environmental remediation. Since the 1960s, significant progress has been made in understanding and modeling fracture flow phenomena in porous media [*Barenblatt et al.*, 1960; *Warren and Root*, 1963; *Kazemi*, 1969; *Pruess and Narasimhan*, 1985]. Despite these advances, modeling the coupled processes of multiphase fluid flow, heat transfer, and chemical migration in a fractured porous medium remains a conceptual and mathematical challenge. The difficulty stems from the nature of inherent heterogeneity and uncertainties associated with fracture-matrix systems for any given field problem, as well as the computational requirements. Numerical modeling approaches currently used for simulating multiphase fluid flow, heat transfer, and chemical transport processes are generally based on methodologies developed for geothermal and petroleum reservoir simulations. They involve solving coupled multiphase fluid and heat flow, multichemical component migration formulations based on finite difference or finite element schemes with a volume averaging approach.

A key issue for simulating fluid and heat flow and

Dynamics of Fluids in Fractured Rock
Geophysical Monograph 122

Published 2000 by the American Geophysical Union

chemical transport in fractured rocks is how to handle fracture and matrix interactions under multiphase, nonisothermal conditions. The available methods for treatment of fracture and porous matrix interactions using a numerical model include: (1) an explicit, discrete-fracture and matrix model [*Sudicky and McLaren*, 1992]; (2) the dual-continua method including double- and multiporosity, dual-permeability, or the more general "multiple interacting continua" (MINC) method [*Pruess and Narasimhan*, 1985]; and (3) the effective continuum method (ECM) [*Wu et al.*, 1996a].

The discrete-fracture-modeling approach is, in general, computationally intensive and requires a detailed knowledge of fracture and matrix geometric properties and their spatial distributions, which are rarely known at a given site. For these reasons, it has found limited field application in modeling multiphase, nonisothermal flow and transport in fractured rock for large-scale problems. On the other hand, the dual-continua method is conceptually appealing and computationally much less demanding than the discrete-fracture-modeling approach and therefore has become the main approach used in handling fluid flow, heat transfer, and chemical transport through fracture-matrix systems. However, this approach also requires detailed knowledge of fracture and matrix geometric properties and their spatial distributions and is much less efficient computationally than the ECM method.

The classical double-porosity concept for modeling flow in fractured, porous media was developed by *Warren and Root* [1963]. In this method, a flow domain is composed of matrix blocks of low permeability, embedded in a network of interconnected fractures. Global flow and transport in the formation occurs only through the fracture system, described as an effective continuum. The matrix behaves as spatially distributed sinks or sources to the fracture system without accounting for global matrix-matrix flow. The double-porosity model accounts for fracture-matrix interflow, based on a quasi-steady-state assumption.

The more rigorous dual-continua method, MINC concept [*Pruess and Narasimhan*, 1985], describes gradients of pressures, temperatures, and concentrations between fractures and matrix by appropriate subgridding of the matrix blocks. This approach provides a better approximation for transient fracture-matrix interactions than using the quasi-steady state flow assumption of the Warren and Root model. Fluid flow and transport from the fractures into the matrix blocks can then be modeled by means of one- or multidimensional strings of nested gridblocks. In general, matrix-matrix connections can also be described by the MINC methodology. As a special case of the MINC concept, the dual-permeability model considers global flow occurring not only between fractures but also between matrix gridblocks. In this approach, fractures and matrix are each represented by one gridblock, and they are connected to each other. Because of the one-block representation of fractures or matrix, the interflow between fractures and matrix has to be handled using some quasi-steady- state flow assumption, as used with the Warren and Root model, and this may limit its application in estimating effects of gradients of pressures, temperatures, and concentrations within matrix. Under steady-state flow conditions, however, the gradients near the matrix surfaces become minimal, and the model is expected to produce accurate solutions.

The ECM approximation has long been used for modeling fracture-matrix flow problems. However, in both the literature and applications, a clear definition of the effective continuum or equivalent porous medium approach is lacking. Many different types of ECM models have been presented and used, based on different assumptions and approximations, such as for isothermal, unsaturated flow [*Peters and Klavetter*, 1988], coupled fluid and heat flow [*Pruess et al.*, 1988; 1990; *Nitao*, 1989], single phase flow and transport [*Kool and Wu*, 1991; *Berkowitz et al.*, 1988, and coupled multiphase fluid and heat flow and solute transport [*Wu et al.*, 1996a and 1996b].

This paper presents a generalized, rigorous ECM formulation for modeling multiphase, nonisothermal flow and solute transport in fractured porous media. The focus of this work is on the theoretical basis for the ECM methodology and the procedures to estimate effective parameters and constitutive relations in order to apply the method. Also included is a discussion of the conditions under which the ECM model can be used. In addition, three application examples (one multiphase flow, one heat flow, and one chemical transport problem) are given to demonstrate the usefulness of the ECM method.

2. FORMULATION

In concept, the ECM uses an "effective" porous medium to approximate a fractured-matrix system, and calculations for flow and transport are then simplified and performed by a single-porosity-continuum approach with a set of "effective" parameters. The ECM relies on the one critical assumption that approximate thermodynamic equilibrium (locally) exists between fractures and matrix at all times in the formation. This assumption implies that the local fracture-matrix interactions (fluid, heat, and concentration/ compnent exchanges) are simultaneously completed in a time scale that is short relative to the time needed for the global flow and transport occurring through surrounding fracture-fracture/matrix-matrix connections. Based on the local equilibrium condition, the ECM approach does not require detailed knowledge of distributions and interface areas of fracture and matrix geometric properties. It provides a simple, alternative conceptual model.

The favorable conditions for the porous-medium-like behavior occur when rock matrix blocks are relatively small and permeable, the fracture network is intensive and relatively uniformly distributed, and fracture-matrix interactions are rapid. The continuum method will be particularly suitable for a situation where a long-term, averaged, or steady-state solution is sought. However, the effective continuum approximation may break down under certain unfavorable conditions, such as for very large and low-permeability matrix blocks subject to rapid transient flow, because it may take a long time to reach a local equilibrium under such an environment, which violates the fundamental assumption of local equilibrium for the ECM.

Let us derive fluid-flow, heat-flow, and solute-transport governing equations starting from a dual-permeability conceptual model. At first, it is assumed that multiphase fluid-flow, multicomponent transport, and heat-transfer processes can be described using a continuum approach in both fractures and matrix, respectively, within a representative elementary volume (REV) of a formation. The condition of local thermodynamic equilibrium requires that temperatures, phase pressures, densities, viscosities, enthalpies, and component concentrations in fracture and matrix systems are the same locally at any REV of the formation. Therefore, governing equations for component mass and energy conservation can be greatly simplified by adding the fluxes of mass and heat through fractures and matrix, respectively. Darcy's law is still used to describe flow terms of a fluid phase. A mass component is transported by advection and diffusion/dispersion, and heat is transferred by convection and conduction mechanisms in fractures and matrix, respectively. This conceptualization results in a set of partial differential equations in the ECM formulation for flow and transport in fractured media, which may be written in the same form as such equations for a single-continuum porous medium [*Wu et al.*, 1996a; *Panday et al.*, 1995; and *Forsyth*, 1994]:

For transport of each species κ within a REV, the compositional equation can be written as:

$$\begin{aligned}&\frac{\partial}{\partial t}\left\{\phi\sum_{\beta}\left(\rho_\beta S_\beta X_\beta^\kappa\right)+(1-\phi)\,\rho_s\,\rho_w\,X_w^\kappa\,K_d^\kappa\right\}\\&+\lambda_\kappa\left\{\phi\sum_{\beta}\left(\rho_\beta S_\beta X_\beta^\kappa\right)+(1-\phi)\,\rho_s\,\rho_w\,X_w^\kappa\,K_d^\kappa\right\}\\&=-\sum_{\beta}\nabla\bullet\left(\rho_\beta\,X_\beta^\kappa\,\vec{v}_\beta\right)\\&+\sum_{\beta}\nabla\bullet\left(\rho_\beta\,\underline{D}^\kappa\bullet\nabla X_\beta^\kappa\right)+q^\kappa\end{aligned}\tag{1}$$

where β is an index for fluid phase [β = 1, ..., NP (total number of phases)]; κ is an index for components [κ = 1, 2, ..., NK (total number of components)]; and the rest of the symbols are defined below (also, see Nomenclature Table). The left-hand side of Equation (1) consists of (a) an accumulation term of component mass summed over all dissolved phases and adsorption on rock solids; and (b) a first-order decay term. The right-hand side of (1) is (a) an advection term contributed by all flowing phases; (b) diffusive and dispersive terms within all phases; and (c) a source/sink term.

The energy conservation equation is

$$\begin{aligned}&\frac{\partial}{\partial t}\left\{\sum_{\beta}\left(\phi\;\rho_\beta S_\beta U_\beta\right)+(1-\phi)\;\rho_s\;U_s\right\}=\\&-\sum_{\beta}\nabla\bullet\left(h_\beta\;\rho_\beta\;\vec{v}_\beta\right)\\&+\sum_{\beta}\sum_{\kappa}\nabla\bullet\left(\rho_\beta h_\beta^\kappa\underline{D}^\kappa\bullet\nabla X_\beta^\kappa\right)\\&+\nabla\bullet\left(K_{th}\nabla T\right)+q^E\end{aligned}\tag{2}$$

The left-hand of side of Equation (2) is an energy-accumulation term of summation over all phases and solids, and the right-hand side contains (a) a heat advection term contributed by all flowing phases; (b) diffusive and dispersive heat transfer terms within all phases; (c) a heat conduction term; and (d) a source/sink term.

Equations (1) and (2) contain many "effective" or "equivalent" parameters and constitutive correlations that need to be determined before the ECM approach can be used, as discussed in this and the following sections.

Terms and symbols in Equations (1) and (2) are defined as follows: ϕ is effective porosity, defined as

$$\phi = \phi_f + \phi_m \tag{3}$$

where ϕ_f and ϕ_m are effective porosities of fracture and matrix continua, respectively; ρ_β, ρ_s and ρ_w are densities of fluid β, rock solids, and water phase, respectively.
S_β is effective saturation of fluid β, defined as,

$$S_\beta = \frac{S_{\beta,f}\phi_f + S_{\beta,m}\phi_m}{\phi_f + \phi_m} \tag{4}$$

where $S_{\beta,f}$ and $S_{\beta,m}$ are saturation of fluid β, in fracture and matrix continua, respectively; X_β^κ is mass fraction of component κ in fluid β.

K_d^κ is the effective distribution coefficient of component κ between the water phase and rock solids of fractures and matrix, defined as,

$$K_d^\kappa = \frac{A_s K_d^{f,\kappa} + (1 - \phi_m - \phi_f) K_d^{m,\kappa}}{1 - \phi} \tag{5}$$

where A_s is the ratio of fracture surface areas to a bulk rock (fracture + matrix) mass, representing total fracture surface areas within unit mass of rock (L^2/M); $K_d^{f,\kappa}$ and $K_d^{m,\kappa}$ are distribution coefficients of component κ between the water phase and rock solids of fracture and matrix continua, respectively. It should be mentioned that here we use different definitions of solid-water partitioning coefficients, K_d, for fractures and matrix. has dimensions of ($M/L^2 \bullet L^3/M$) or (L), while $K_d^{m,\kappa}$ is a volume concept with a dimension of (L^3/M) [*Freeze and Cherry*, 1979].

$\vec{v}_\beta$ is the effective Darcy velocity of fluid β, defined as:

$$\vec{v}_\beta = -\frac{k\,k_\beta}{\mu_\beta}\left(\nabla P_\beta - \rho_\beta \vec{g}\right) \tag{6}$$

where P_β, μ_β, and $\vec{g}$ are pressure and viscosity of fluid β, and gravity vector, respectively; k is effective continuum permeability, defined as:

$$k = k_f + k_m \tag{7}$$

with k_f and k_m as absolute permeabilities of fracture and matrix continua, respectively; the effective relative permeability to fluid phase β, k_β, is defined as [*Pruess et al.*, 1990; *Peters and Klavetter*, 1988]:

$$k_\beta = \frac{k_f\, k_{\beta,f} + k_m\, k_{\beta,m}}{k_f + k_m} \tag{8}$$

$\underline{D}^\kappa$ is the combined fracture-matrix, hydrodynamic dispersion tensor accounting for both molecular diffusion and mechanical dispersion for component κ, weighted by total porosity,

$$\underline{D}^\kappa = \frac{\phi_f\, \underline{D}_f^\kappa + \phi_m\, \underline{D}_m^\kappa}{\phi_f + \phi_m} \tag{9}$$

where the fracture diffusion-dispersion tensor is:

$$\underline{D}_f^\kappa = \left(\alpha_{T,f}\left|\vec{v}_{\beta,f}\right|\right)\delta_{ij} + \frac{\left(\alpha_{L,f} - \alpha_{T,f}\right)\vec{v}_{\beta,f}\;\vec{v}_{\beta,f}}{\left|\vec{v}_{\beta,f}\right|} + \left(\phi_f\; S_{\beta,f}\tau_f\; d_f^\kappa\right)\delta_{ij} \tag{10}$$

and the matrix diffusion-dispersion tensor is:

$$\underline{D}_m^\kappa = \left(\alpha_{T,m}\left|\vec{v}_{\beta,m}\right|\right)\delta_{ij} + \frac{\left(\alpha_{L,m} - \alpha_{T,m}\right)\vec{v}_{\beta,m}\;\vec{v}_{\beta,m}}{\left|\vec{v}_{\beta,m}\right|} + \left(\phi_m\; S_{\beta,m}\tau_m\; d_m^\kappa\right)\delta_{ij} \tag{11}$$

where $\alpha_{T,f}$, $\alpha_{L,f}$, $\alpha_{T,m}$, and $\alpha_{L,m}$ are the transverse and longitudinal dispersivities, respectively, for fracture and matrix continua; τ_f and τ_m are tortuosities of fracture and matrix continua, respectively; d_f^κ and d_m^κ are the molecular diffusion coefficients in phase β of fracture and matrix continua, respectively; $\vec{v}_{\beta,f}$ and $\vec{v}_{\beta,m}$ are Darcy's velocities of fluid β of fracture and matrix continua, respectively; and δ_{ij} is the Kroneker delta function ($\delta_{ij} = 1$ for $i = j$, and $\delta_{ij} = 0$ for $i \neq j$).

λ_κ is radioactive decay constant of the chemical species;
q^κ and q^E are source/sink terms for component κ and energy, respectively;
h_β and h_β^κ are enthalpies of fluid phase β and of component κ in fluid phase β, respectively;
U_β and U_s are internal energy of fluid β and rock solids, respectively;
K_{th} is effective thermal conductivity, defined as,

$$K_{th} = \frac{\phi_f\, K_{th,f} + \phi_m\, K_{th,m}}{\phi_f + \phi_m} \tag{12}$$

where $K_{th,f}$ and $K_{th,m}$ are the thermal conductivities of fracture and matrix continua, respectively, and they may be functions of liquid saturations of each continuum. T is temperature.

3. EVALUATION OF EFFECTIVE PARAMETERS AND CONSTITUTIVE RELATIONS

In solving governing Equations (1) and (2) with the ECM approach, the primary variables selected in a numerical solution are normally fluid pressures, P_β, effective saturations, S_β, temperature, T, and mass fractions, X_β^κ, of chemical species. Once the primary variables are chosen, all of the secondary variables must be evaluated using the primary variables, effective parameters, and correlations to set up a set of solvable equations. Special attention needs to be paid to evaluating the secondary variables, effective constitutive relations, and parameters in this process. Many rock and fluid properties (such as porosity, absolute permeability, dispersivity, thermal conductivity, tortuosity, and diffusion and dispersion coefficients of fracture and matrix continua, fluid viscosity and density) should be determined from site-characterization studies. The key effective constitutive correlations, such as capillary pressures, relative permeability, dispersion tensor, and thermal conductivity relation for the ECM formulation, are discussed in this section.

The numerical implementation of the ECM scheme for evaluating the effective constitutive relations, as given in the previous section, is straightforward once we know fluid saturations in matrix and fracture, separately. This can be achieved by introducing a fracture-matrix combined (or composite) capillary pressure (P_c) curve (using tabulated values, based on the individual fracture and matrix P_c curves from the input data for a given rock type). Under local equilibrium conditions, the combined P_c curve is

$$P_c(S_\beta) = P_{c,m}(S_{\beta,m}) = P_{c,f}(S_{\beta,f}) \tag{13}$$

as a function of an effective or average liquid saturation, S_β, defined in (4).

During a Newton iteration in a numerical simulation, the liquid saturation obtained from the solution is exactly the effective saturation, as defined in Equation (4), as a primary variable. This effective saturation value can be used with the combined P_c curve to calculate the value of the effective capillary pressure function. The fracture and matrix saturations can then be determined by inverting the input capillary pressure functions of fractures and matrix, respectively.

Once fluid saturations, $S_{\beta,m}$ and $S_{\beta,f}$, in the fractures and matrix are separately determined, the effective relative permeabilities for the ECM calculations are evaluated using Equation (8), in which the relative permeabilities $k_{\beta,m}$ and $k_{\beta,f}$ also have to be evaluated at saturations $S_{\beta,m}$ and $S_{\beta,f}$, respectively.

Equation (5) can be directly used for estimating the effective K_d for adsorption terms of a component, contributed by both fractures and matrix. Similarly, Equation (12) should be used for calculating an effective thermal conductivity, in which the fracture and matrix thermal conductivities needed in Equation (12) can be handled as functions of fracture and matrix liquid saturations, respectively.

One of the primary difficulties in determining "effective" parameters for the ECM formulation is the evaluation of the effective dispersion tensor. In general, Equations (9), (10), and (11) can be used for this purpose. In addition to phase saturations in fracture and matrix continua, these equations require Darcy's velocities in fractures and matrix separately. These velocities can be approximated by:

$$\vec{v}_{\beta,f} = -\frac{k_f\, k_{\beta,f}}{\mu_\beta}\left(\nabla P_\beta - \rho_\beta \vec{g}\right) \tag{14}$$

for fractures, and

$$\vec{v}_{\beta,m} = -\frac{k_m\, k_{\beta,m}}{\mu_\beta}\left(\nabla P_\beta - \rho_\beta \vec{g}\right) \tag{15}$$

for matrix. The two velocities for a phase can be calculated per time step or per Newton iteration.

For 2D transport in a regular, two-orthogonal, parallel fractures system, *Kool and Wu* [1991] provide an alternative, approximate correlation for effective dispersivities as

$$\alpha_L = \frac{\phi_m\,(L_x - b)^2}{3\,\phi^2\,d_m}\left(\frac{R_m}{R_e}\right)^2 |v_x| \tag{16}$$

for effective longitudinal dispersivity perpendicular to the x-direction and

$$\alpha_T = \frac{\phi_m\,(L_y - b)^2}{3\phi^2\,d_m}\left(\frac{R_m}{R_e}\right)^2 |v_y| \tag{17}$$

for effective transverse dispersivity perpendicular to the x-direction. These two dispersivities were derived using a two-region type model of transport in an aggregated soil [*van Genuchten and Dalton*, 1986]. In Equations (16) and (17), L_x and L_y are the half spacings between parallel fractures in the x-direction and the y-direction, respectively; b is half the aperture of fractures; v_x and v_y are effective Darcy velocities in the x-direction and the y-direction, respectively. The retardation factors are then

$$R_m = 1 + \frac{(1-\phi_m)\rho_s K_d^m}{\phi} \tag{18}$$

for matrix and

$$R_e = \frac{\phi_f R_f + \phi_m R_m}{\phi} \tag{19}$$

for the effective continuum with

$$R_f = 1 + \frac{K_d^f}{b} \tag{20}$$

for fractures.

4. MODEL IMPLEMENTATION AND APPLICATION

The ECM formulation of Section 2 has been incorporated into a special version of the TOUGH2 code, a multiphase, multicomponent, nonisothermal reservoir simulator [*Pruess*, 1991]. It includes (a) the EOS3 and EOS4 modules for two phases (water and gas), two components (water and air), and heat; (b) the EOS9 module for two-phase, isothermal flow by solving Richards' equation; and (c) the EOS1G module for single gas flow in a two-phase condition with the aqueous phase serving as a passive phase [*Wu et al.*, 1996a]. In addition, a two-phase, three-component, nonisothermal version has been implemented as a module of a solute transport code, T2R3D [*Wu et al.*, 1996b; *Wu and Pruess*, 1998].

Because of the computational efficiency and simplicity of data requirements of the ECM method, the implemented modules of the TOUGH2 code and the T2R3D code have found a wide range of applications in field characterization studies at the Yucca Mountain, Nevada, site, a potential underground repository for high-level radioactive wastes. The ECM methodology has been used as a main modeling approach in 3-D, large-scale unsaturated zone model calibrations [*Wu et al.*, 1996c], in perched water studies [*Wu et al.*, 1997], and in ambient geothermal condition investigation [*Wu et al.*, 1998]. It also has been used for pneumatic data analyses [*Ahlers and Wu*, 1997], in modeling geochemical transport [*Sonnenthal and Bodvarsson*, 1997], and in thermal loading studies [*Haukwa and Wu*, 1996].

During these applications, several comparative and validation studies also were carried out to investigate the accuracy and applicability of the ECM approach to field problems [*Doughty and Bodvarsson*, 1997; *Wu et al.*, 1996a and 1996c]. These studies conclude that the ECM concept is adequate for modeling steady-state moisture movement and ambient heat flow, and transient gas flow in fractured unsaturated zones of Yucca Mountain, as long as there are strong fracture-matrix interactions in the system. However, the ECM approximation will introduce larger errors in cases where a strong nonequilibrium condition exists between fracture and matrix systems, such as within fast flow pathways along high-permeability flow channels.

Another important application of the ECM model in field studies is the use of its results as initial conditions for the dual-permeability simulations. This application has been proven to be extremely helpful in performing large-scale moisture flow simulations at Yucca Mountain [*Wu et al.*, 1998]. In this case, many 3-D steady-state dual-permeability simulations were needed to study the wide range in uncertainties for parameters ranging from surface infiltration rates to fracture-matrix properties. As a result of the more nonlinear nature of the dual-permeability formulation in handling fracture-matrix interactions and the doubling in size of the model grid relative to the ECM, a direct use of the dual-permeability modeling approach was impractical for completing all the simulations within the time constraint. Since the ECM and dual-permeability approaches give very similar results in terms of moisture distributions in fractures and matrix at steady state (as demonstrated in the following section), the initial condition for a dual-permeability run was estimated using the results of a corresponding ECM steady-state simulation. With a good initial guess, it was relatively easy for a dual-permeability simulation to converge to a steady-state solution. This approach provided an order of magnitude improvement in computer CPU times relative to those dual-permeability simulations without using ECM results as initial conditions.

5. APPLICATION EXAMPLES

Three application examples are given in this section for demonstrating the applicability of the ECM approach. They are:

(1) Comparison of the ECM with the more rigorous dual-permeability modeling approaches for two-phase flow through unsaturated, fractured rocks.
(2) Two-phase, 3-D nonisothermal fluid and heat flow at ambient geothermal conditions.
(3) Comparison of the result from a single-phase liquid, 2-

D flow and transport modeling approach with a result from a discrete-fracture model.

5.1. Example 1—Comparison with the Dual-Permeability Model Results

The first example is a one-dimensional vertical flow problem in the unsaturated zone of Yucca Mountain with the vertical column grid extracted directly from the 3-D site-scale model [*Wu et al.*, 1996c]. The grid consists of a set of 1-D vertical grid blocks, representing four hydrogeologic units (named TCw, PTn, TSw, and CHn), and an overlying boundary block and an underlying boundary block representing the atmosphere and the water table conditions, respectively. The fracture system in the unsaturated zone of the mountain is also subdivided into the same four units. The properties of fractures are taken from *Wilson et al.* [1994] for the four units, as listed in Table 1.

Since few fractures exist in the PTn unit, the effects of fractures on moisture flow in the PTn are ignored in this simulation, and the PTn formation is treated as a single-porous-medium rock. The same set of matrix and fracture properties, as listed in Table 2 [*Bandurraga et al.*, 1996], are used for both ECM and dual-permeability simulations.

In this table, α_m and α_f are van Genuchten's parameters [*van Genuchten*, 1980] of capillary pressure, and m_m and m_f are van Genuchten's parameters of relative permeability curves, respectively, for matrix and fracture systems.

The same type of boundary conditions are specified for both models (Dirichlet-type conditions), i.e., constant pressures, temperatures, and saturation. Also, the surface boundary is subject to a constant water infiltration of 3.6 mm/yr. However, the water infiltration on the ground surface is added as a source term only into the fracture elements on the top boundary for the dual-permeability model. It is distributed between fractures and matrix in the ECM model because, realistically, infiltration is expected almost entirely through the fractures on the land surface, but the continuum approach of the ECM uses equilibrium partitioning for the infiltration. Also, an isothermal condition was assumed to exist under two-phase flow conditions for this comparison.

Table 1. The Fracture-Continuum Porosity and Spacing Data Used for the Comparison Study [*Wilson et al.*, 1994].

Unit	Porosity	Spacing (m)
TCw	1.38e-3	0.618
PTn	4.12e-3	2.220
TSw	2.75e-3	0.740
CHn	9.98e-4	1.618

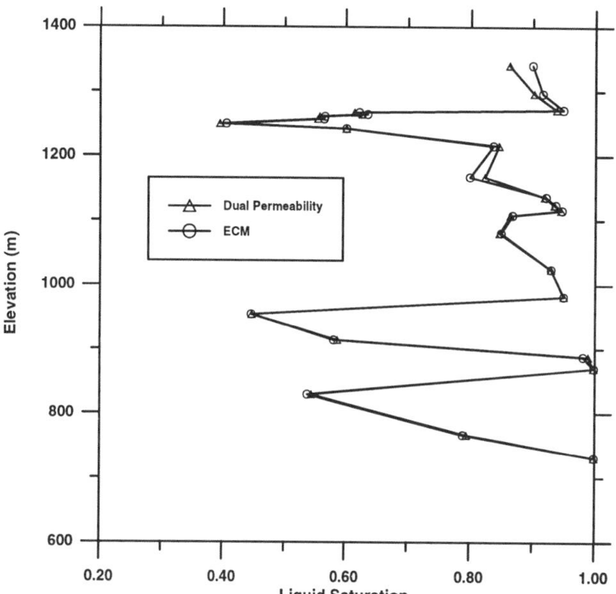

Figure 1. Comparison of the simulated matrix liquid saturation profiles using the ECM and the dual-permeability models.

Figure 1 presents the steady-state liquid saturation profiles for fractures and matrix, obtained using the steady-state solutions of both ECM and the dual-permeability models. Figure 1 shows that almost identical results for matrix saturations are obtained from the two modeling approaches. The only differences between the two solutions may be noticed near the top (TCw unit) or along the interfaces at an elevation of 1,200 m (between the PTn and the TSw). The reasons for these minor differences stem from the conceptual model, rather than from the modeling approaches. First, the water infiltration is imposed into fractures only on the top boundary in the dual-permeability simulation, while the ECM model puts the infiltration into both fracture and matrix systems. Second, the PTn unit is treated as a single-porous medium, which creates certain discontinuities in matrix-fracture vertical connections across the interfaces.

The comparison, as shown in Figure 1, indicates that as long as the local equilibrium condition is satisfied between the fracture and matrix systems, the ECM formulation will give accurate predictions of saturation distributions. We can create a situation under which the local equilibrium condition is not well satisfied, and the ECM approach will introduce considerable errors into the solution. Figure 2 shows such a comparison for simulated liquid saturation profiles also using the ECM and the dual-permeability methods.

Table 2. The Rock Properties of Matrix and Fractures Used in the ECM Comparison Study.

Unit/ Layer	k_m (m^2)	k_f (m^2)	α_m (Pa^{-1})	α_f (Pa^{-1})	m_m	m_f	ϕ_m	φ_f
tcw11	0.160E-18	0.910E-11	0.147E-04	0.518E-03	0.238	0.182	0.062	0.290E-03
tcw12	0.540E-15	0.575E-11	0.174E-05	0.977E-03	0.233	0.223	0.082	0.290E-03
tcw13	0.220E-16	0.575E-11	0.129E-05	0.123E-02	0.463	0.437	0.207	0.290E-03
ptn21	0.400E-12	n/a	0.244E-04	n/a	0.215	n/a	0.435	n/a
ptn22	0.240E-12	n/a	0.159E-04	n/a	0.310	n/a	0.222	n/a
ptn23	0.111E-12	n/a	0.445E-04	n/a	0.243	n/a	0.406	n/a
ptn24	0.880E-13	n/a	0.341E-04	n/a	0.295	n/a	0.499	n/a
ptn25	0.105E-11	n/a	0.138E-03	n/a	0.243	n/a	0.490	n/a
tsw31	0.261E-12	0.400E-11	0.237E-04	0.122E-02	0.206	0.203	0.048	0.243E-03
tsw32	0.194E-12	0.445E-11	0.172E-04	0.969E-03	0.248	0.249	0.156	0.243E-03
tsw33	0.796E-17	0.743E-11	0.573E-05	0.243E-03	0.247	0.250	0.154	0.243E-03
tsw34	0.100E-14	0.159E-11	0.754E-06	0.686E-03	0.321	0.325	0.110	0.243E-03
tsw35	0.423E-15	0.400E-11	0.234E-05	0.108E-02	0.231	0.226	0.130	0.243E-03
tsw36	0.776E-16	0.400E-11	0.522E-06	0.122E-02	0.416	0.416	0.112	0.243E-03
tsw37	0.316E-15	0.400E-11	0.804E-06	0.122E-02	0.368	0.368	0.036	0.243E-03
ch1vc	0.160E-11	0.723E-12	0.760E-04	0.122E-02	0.221	0.227	0.273	0.111E-03
ch2vc	0.550E-13	0.723E-12	0.980E-04	0.122E-02	0.223	0.227	0.344	0.111E-03
ch3zc	0.450E-17	0.100E-12	0.394E-05	0.388E-03	0.225	0.225	0.332	0.525E-04
ch4zc	0.210E-16	0.100E-12	0.150E-06	0.730E-03	0.475	0.470	0.266	0.525E-04
pp3vp	0.542E-14	0.300E-13	0.194E-04	0.122E-02	0.316	0.313	0.322	0.111E-03
pp2zp	0.269E-15	0.100E-12	0.126E-05	0.730E-03	0.311	0.312	0.286	0.525E-04

However, the interface areas between the fracture and matrix systems in this case were reduced by a factor of 10, 100, and 1,000, respectively, to reduce fracture-matrix interactions. The interface areas become 10%, 1%, and 0.1% of the geometric area of the fracture and matrix systems, as shown in Figure 2. As the fracture-matrix interface areas get smaller, the local equilibrium conditions are less well satisfied, and the figure shows more differences between the predictions using the ECM and the dual-permeability approaches in the certain units/layers (at elevations between 1,000 and 1,200 m). However, Figure 2 shows that even with several orders-of-magnitude reductions in the fracture-matrix interface areas, the comparisons are still reasonably close between the ECM and the dual-permeability results for the lower unit of the layers. The reason is that the matrix is relatively permeable in that unit, compared with the fractures, and the local equilibrium conditions are still reasonable. It should be mentioned that many 3-D comparisons between the ECM and dual-permeability model results have been made, with very similar results to this 1-D model [*Wu et al.*, 1996c].

5.2 Example 2—Comparisons with Measured Temperature Profiles

Several 3-D unsaturated-zone fluid and heat flow models have been developed using the ECM approach to estimate the geothermal conditions of the Yucca Mountain site [*Wu et al.*, 1998]. Measured temperature data [*Sass et al.*, 1988; *Rousseau et al.*, 1996] were compared with the 3-D simulation results for about 25 boreholes within and near the study area of the Yucca Mountain site. The 3-D model also uses the land surface and the water table as the top and bottom boundaries, with constant or spatially varying pressures, saturations, and temperatures described based on field data. In addition, a spatially varying, steady-state water infiltration map is specified at the top model boundary. The details on the model grid, parameters, and conditions are provided by *Wu et al.* [1998]. In general, it has been found that that the 3-D ECM models are able to match temperature data from all the boreholes. We present only one of the 25 comparisons conducted in the model calibration as a demonstration example. The selected borehole is H-5; the comparison between the modeled results and observations is shown in Figure 3.

5.3 Example 3—Single-Phase, 2-D Flow and Transport

This problem is used to test the ECM formulation in simulating a combined flow and transport case [*Kool and Wu*, 1991]. Effective dispersivities for transport are described by Equations (16) to (20) of Section 3 above. The vertical, 2-D domain, hypothetical waste site, depicted in Figure 4, is discretized into a rectangular element grid with a total of 1,768 elements. The ECM simulation was performed using a finite element code [*Huyakorn et al.*, 1991] and the ECM results were examined against the results from a discrete-fracture code [*Sudicky and McLaren*, 1992]. The detailed information on effective parameters and their evaluation, in addition to model conditions, is given in the report [*Kool and Wu*, 1991] for this problem. The flow and transport simulation was carried out by a decoupled approach, in which a steady-state flow run was conducted first, followed by the transient transport calculation. The model flow and transport boundary conditions are shown in Figures 5 and 6, respectively.

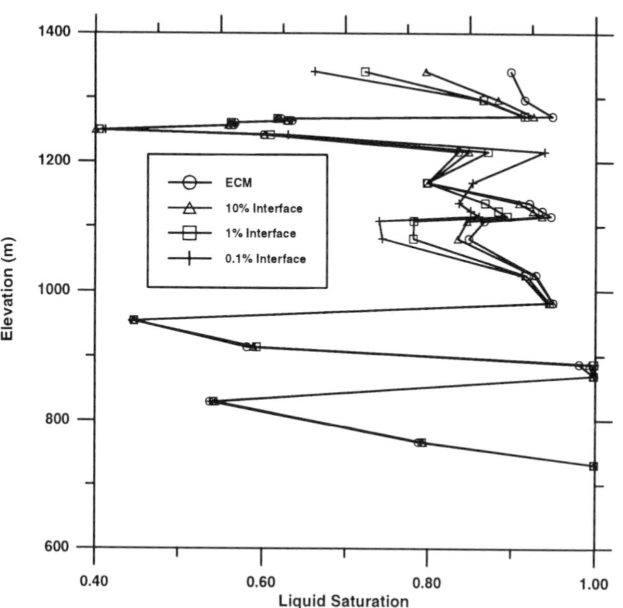

Figure 2. Comparison of the simulated matrix liquid saturation profiles using the ECM and the dual-permeability models with reduction of fracture-matrix interface areas.

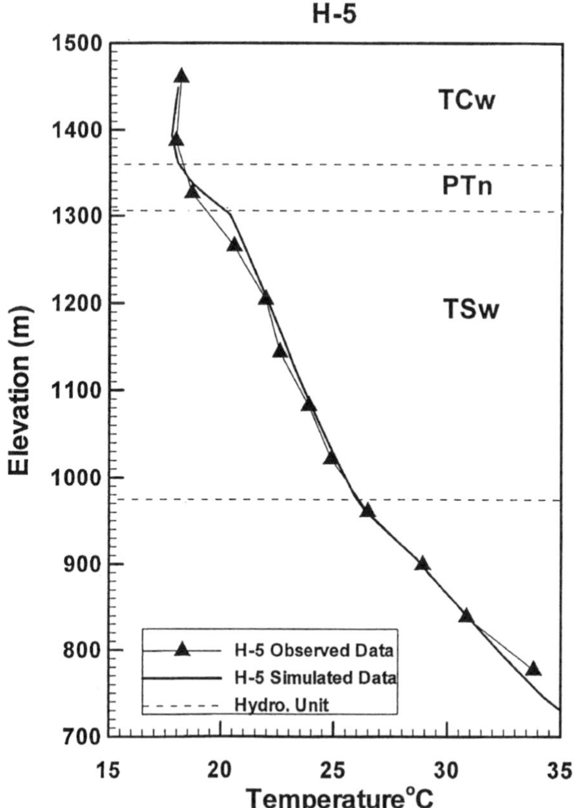

Figure 3. Comparison between measured and simulated temperatures for borehole H-5 in the unsaturated zone of Yucca Mountain.

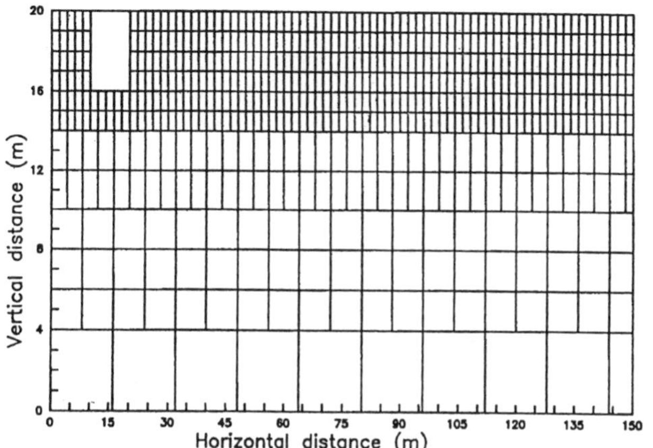

Figure 4. Fracture network and 2-D domain at hypothetical waste disposal site (note 5× vertical exaggeration).

Figure 5. Boundary conditions and coordinates used in the ECM flow simulation.

Figure 7 presents a comparison of simulated hydraulic head contours from the two models. The agreement is quite good, indicating the ECM approximation is reasonable in this case for describing the steady-state flow field. Results for transport simulation are shown in Figures 8 and 9, respectively, for a conservative and reactive species. Figure 8 shows that the comparison of the ECM and the discrete-fracture-model results is very favorable for this case. The extent of the contaminant plume, as indicated by the 0.005 relative concentration contour, is predicted well by the ECM model, as compared with the "true" solution by the discrete fracture model. However, the ECM model tends to overpredict the downward extent of the plume at this time.

The comparison of the ECM and discrete fracture model results for a sorbing species, displayed in Figure 9, also shows a similar pattern for the plume at the same time (2,500 yr). The large time values reflect the much slower contaminant migration rates with adsorption/reactive effects. The jaggedness in the discrete-fracture-model concentration contours results from fast-advection-transport effects through fractures and a nonequilibrium condition between fractures and matrix. The comparisons in Figures 8 and 9 are quite favorable with regard to the applicability of the ECM approach for modeling single-phase flow and transport in fractured rocks. However, this is a limited validation of the ECM method because the study system is very simple with known fracture characteristics. In general, the local equilibrium assumption for solute concentration between fractures and matrix may be a very serious limitation to the ECM application unless one is seeking a steady-state transport solution.

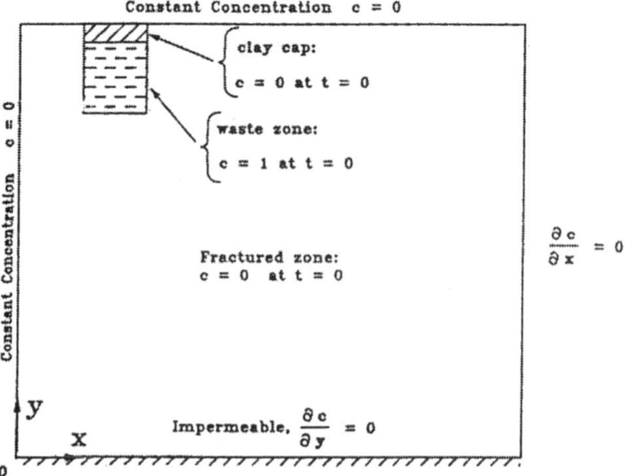

Figure 6. Boundary and initial conditions used in the simulation of conservative and reactive contaminant transport.

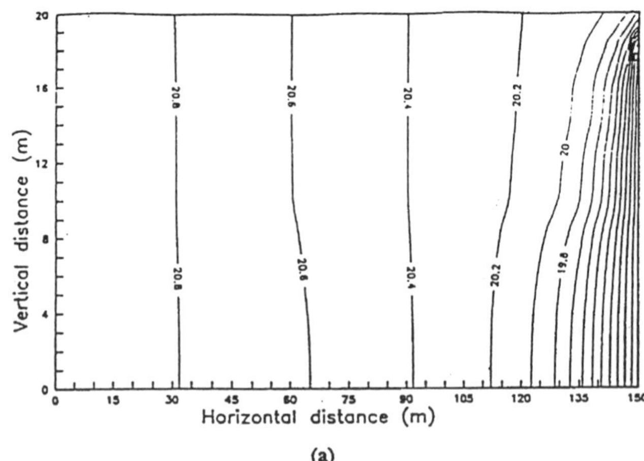

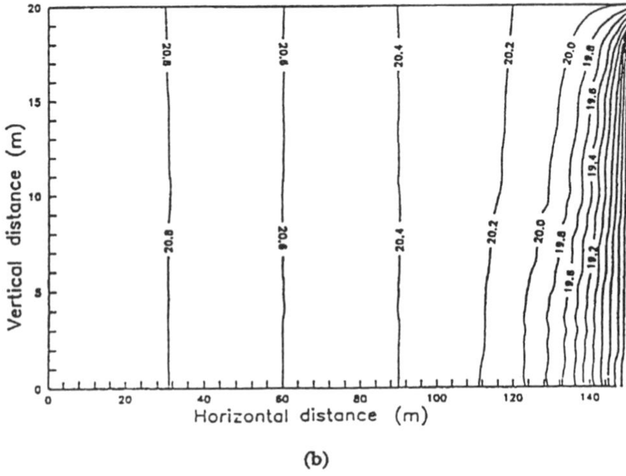

Figure 7. Steady-state hydraulic head distributions in the 2-D model domain simulated using the discrete fracture code [*Sudicky and McLaren*, 1992] and the ECM model [*Huyakorn et al.*, 1991].

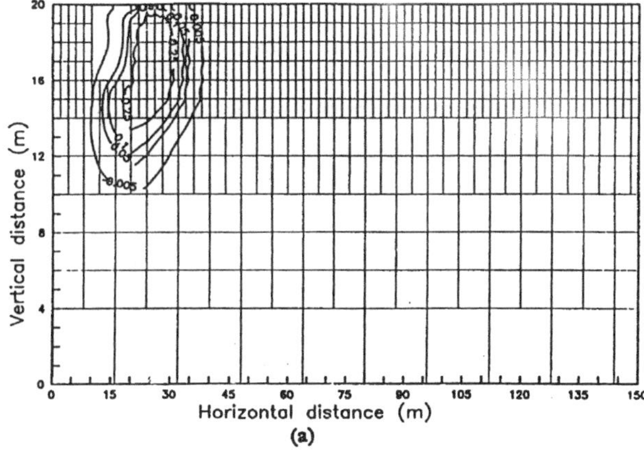

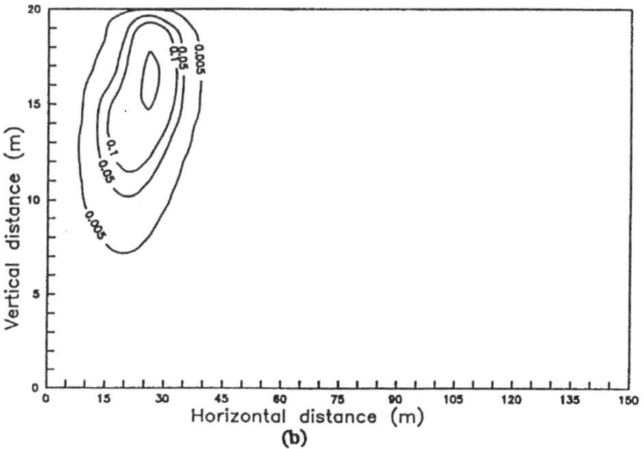

Figure 8. Concentration contours of conservative contaminant transport at t = 50 yr, simulated using the discrete fracture code [*Sudicky and McLaren*, 1992] and the ECM model [*Huyakorn et al.*, 1991].

6. CONCLUSIONS

This paper presents a rigorous derivation of the generalized ECM formulation and defines a complete set of the effective ECM parameters for modeling multiphase flow, multicomponent transport, and heat transfer in fractured rocks. Also included are discussions on implementation of the ECM formulation into a multidimensional, multiphase flow and transport reservoir simulator. In addition, three examples are provided for examining the ECM approach.

The ECM modeling approach is an efficient, alternative conceptual model for studies of flow and transport phenomena in fractured porous media. As discussed in this work, the ECM concept, relative to the dual-permeability approach, is built on only one additional but critical assumption—that there is approximate local thermodynamic equilibrium between fractures and matrix at all times in the formation. As long as this assumption is a reasonable approximation for a given application, the ECM approach will provide a very accurate solution, with substantial improvement in both computational intensity and requirement of detailed fracture geometric properties. The ECM approach is particularly suitable for the case in which a long-term, steady-state solution is sought for large time and spatial scales of concern. In a situation where no detailed fracture characteristic data are available (such as fracture network distributions and fracture-matrix interface areas) using a more rigorous modeling approach (the dual-permeability or other multicontinua method) may not gain more advantages than using a simple ECM model because of the uncertainties in fracture-matrix data. For these reasons, the ECM approach has been widely used as a practical tool in many field investigations of fracture flow problems.

However, the limitation of the ECM approximation must be recognized and examined before it is used. The single-continuum method may break down under certain unfavorable conditions, such as for very large and low-permeability matrix blocks subject to rapid transient flow and transport.

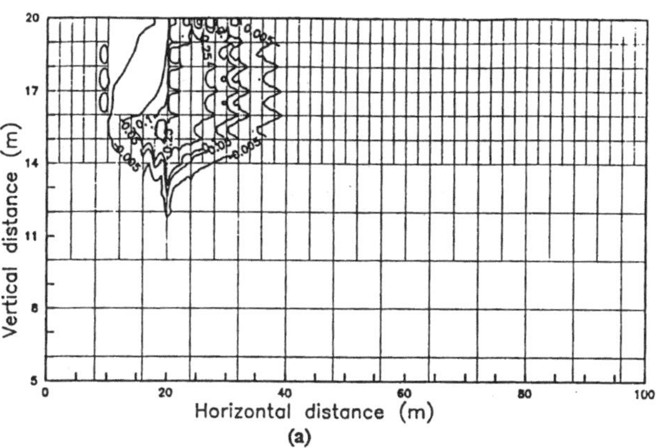

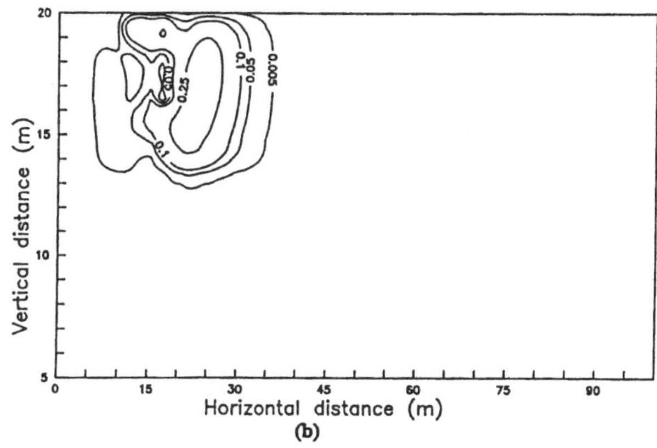

Figure 9. Concentration contours of reactive contaminant transport at t = 2,500 yr, simulated using the discrete fracture code [*Sudicky and McLaren*, 1992] and the ECM model [*Huyakorn et al.*, 1991].

Under such conditions, reaching local thermodynamic equilibrium between fractures and matrix systems may take too long. It is recommended that sensitivity and comparison studies be made using the ECM approach against multicontinua model results on a smaller-scale model before applying the ECM approach to a large-scale application.

NOMENCLATURE

A_s	Ratio of fracture surface areas to bulk rock mass, representing total fracture surface areas within unit mass of rock (m^2/kg)
b	Half of the aperture of fractures (m)
c	Concentration (kg/m^3)
d_m	Molecular diffusion coefficient (m^2/s) of a component in a fluid phase
d_f^κ, d_m^κ	Molecular diffusion coefficient (m^2/s) of component κ in a fluid phase in fractures and matrix, respectively
$\underline{D}^\kappa$	Effective hydrodynamic dispersion tensor accounting for both molecular diffusion and mechanical dispersion for component κ in phase β, defined in Equation (9) (m^2/s)
$\underline{D}_f^\kappa$	Hydrodynamic dispersion tensor accounting for both molecular diffusion and mechanical dispersion for component κ in phase β in fractures (m^2/s)
$\underline{D}_m^\kappa$	Hydrodynamic dispersion tensor accounting for both molecular diffusion and mechanical dispersion for component κ in phase β in matrix (m^2/s)
$\vec{g}$	Gravitational acceleration vector (m/s^2)
h	hydraulic head (m)
h_β	Specific enthalpy of phase β, (J/kg)
h_β^κ	Specific enthalpy of component κ in phase β, (J/kg)
k	Effective continuum permeability of fracture and matrix continua (m^2)
k_f	Absolute permeability of fracture continuum (m^2)
k_m	Absolute permeability of matrix continuum (m^2)
k_β	Effective relative permeability of fracture and matrix continua
$k_{\beta,f}, k_{\beta,m}$	Relative permeability to phase β of fractures and matrix, respectively
K_d^κ	Effective distribution coefficient of component κ between the fluid phase and rock solids of fractures and matrix, defined in Equation (5) (m^3/kg)
$K_d^{f,\kappa}$	Distribution coefficient of component κ between the fluid phase and rock solids of fractures (m)
$K_d^{m,\kappa}$	Distribution coefficient of component κ between the fluid phase and rock solids of matrix (m^3/kg)
K_{th}	Effective rock thermal conductivity, defined in Equation (12) (W/m °C)
$K_{th,f}$	Rock thermal conductivity of fracture continuum (W/m °C)
$K_{th,m}$	Rock thermal conductivity of matrix continuum (W/m °C)
L_x, L_y	Half spacing (m) between parallel fractures in the x-direction and the y-direction, respectively
P_c	Combined capillary pressure (Pa)
$P_{c,f}$	Capillary pressure in fracture continuum (Pa)
$P_{c,m}$	Capillary pressure in matrix continuum (Pa)
P_β	Pressure in phase β (Pa)
q^E	Source/sink terms for energy (W/m^3)
q^κ	Source/sink of mass for component κ (kg/s m^3)
R_e	Retardation factor for effective continua of fractures and matrix
R_f	Retardation factor for fracture continuum
R_m	Retardation factor for matrix continuum
S_β	Effective saturation of phase β, defined in Equation (4)
$S_{\beta,f}$	Saturation of phase β in fracture continuum
$S_{\beta,m}$	Saturation of phase β in matrix continuum
t	Time (s)
T	Temperature (°C)
U_β	Internal energy of phase β (J/kg)
U_s	Internal energy of rock solids (J/kg)
v_x, v_y	Effective Darcy's velocities in the x-direction and the y-direction, respectively (m/s)
$\vec{v}_\beta$	Effective Darcy's velocity of phase β (m/s)
$\vec{v}_{\beta,f}$	Darcy's velocity of phase β in fractures (m/s)
$\vec{v}_{\beta,m}$	Darcy's velocity of phase β in matrix (m/s)
X_β^κ	Mass fraction of component κ in phase β

Greek Symbols

α_L	Effective longitudinal dispersivity of fracture and matrix continua (m)
$\alpha_{L,f}$	Longitudinal dispersivity of fractures (m)
$\alpha_{L,m}$	Longitudinal dispersivity of matrix (m)
α_T	Effective transverse dispersivity of fractures and matrix continua (m)
$\alpha_{T,f}$	Transverse dispersivity of fractures (m)
$\alpha_{T,m}$	Transverse dispersivity of matrix (m)
δ_{ij}	Kronecker delta function. ($\delta_{ij} = 1$ for $i = j$, and $\delta_{ij} = 0$ for $i \neq j$)
ϕ	Effective porosity of fracture or matrix continua, defined in Equation (3)
ϕ_f	Effective porosity of fracture continuum
ϕ_m	Effective porosity of matrix continuum
λ_κ	Radioactive decay constant of the chemical species (s^{-1})
μ_β	Viscosity of fluid β (Pa•s)
ρ_β	Density of phase β at *in-situ* conditions (kg/m^3)
ρ_s	Density of rock grains (kg/m^3)
τ_f , τ_m	Tortuosity of matrix and fractures, respectively
ψ	Pressure head (m)

Subscripts

e	Effective
f	Fracture
L	Longitudinal
m	Matrix
T	Transverse
th	Thermal
β	Index for fluid phase

Superscripts

E	Energy
f	Fracture
m	Matrix
κ	Index for mass components

Acknowledgments. I would like to thank the three AGU Monograph reviewers, P. A. Witherspoon, C. Oldenburg, and R. W. Arnold, for their insightful and constructive comments and suggestions for improving this paper. The author is grateful to H. H. Liu, M. Cushey, and D. Hawkes for their careful review of this paper. Thanks are also due to J. Kool for his contributions to the 2-D flow and transport problem. This work was in part supported by the Assistant Secretary for Energy Efficiency and Renewable Energy, Office of Geothermal and Wind Technologies of the U. S. Department of Energy, under Contract No. DE-AC03-76SF00098.

REFERENCES

Ahlers, C. F., and Y. S. Wu, *Incorporation of Gas Flow Data into the UZ Model, Yucca Mountain Project*, Level 4 Milestone Report SP24UBM4, Lawrence Berkeley National Laboratory, Berkeley, Calif., 1997.

Bandurraga, T. M., S. Finsterle, and G. S. Bodvarsson, Chapter 3, Saturation and capillary pressure analysis, in *Development and Calibration of the Three-Dimensional Site-Scale Unsaturated-Zone Model of Yucca Mountain, Nevada,* edited by G. S. Bodvarsson and M. Bandurraga, Yucca Mountain Project Level 4 Milestone OBO2, Report LBNL-39315, Lawrence Berkeley National Laboratory, Berkeley, Calif., 1996.

Barenblatt, G. I., I. P. Zheltov, and I. N. Kochina, Basic concepts in the theory of seepage of homogeneous liquids in fissured rocks, PMM, *Sov. Appl. Math. Mech., 24*(5), 852–864, 1960.

Berkowitz, B., J. Bear, and C. Braester, Continuum models for contaminant transport in fractured porous formations, *Water Resour. Res., 24*(8), 1225–1236, 1988.

Doughty, C., and G. S. Bodvarsson, Chapter 5, Investigation of conceptual and numerical approaches for evaluating moisture flow and chemical transport, in *The Site-Scale Unsaturated Zone Model of Yucca Mountain, Nevada, for the Viability Assessment*, edited by G. S. Bodvarsson, M. Bandurraga and Y. S. Wu, Yucca Mountain Site Characterization Project Report, LBNL-40376, UC-814, Lawrence Berkeley National Laboratory, Berkeley, Calif., 1997.

Forsyth, P. A., Three-dimensional modeling of steam flush for DNAPL site remediation, *Int'l. J. Numerical Methods in Fluids, 19*, 1055–1081, 1994.

Freeze, R. A. and J. A. Cherry, *Groundwater*, Prentice-Hill, Englwood Cliffs, N.J., 1979.

Haukwa, C., and Y. S. Wu, Chapter 13, Thermal loading studies using the Unsaturated Zone Model, in *Development and Calibration of the Three-Dimensional Site-Scale Unsaturated-Zone Model of Yucca Mountain, Nevada*, edited by G. S. Bodvarsson and M. Bandurraga, Yucca Mountain Project Level 4 Milestone OBO2, Report LBNL-39315, Lawrence Berkeley National Laboratory, Berkeley, Calif., 1996.

Huyakorn, P. S., J. B. Kool, and Y. S. Wu, *VAM2D—Variably Saturated Analysis Model in Two Dimensions, Version 5.2 with Hysteresis and Chained Decay Transport, Documentation and User's Guide*, NUREG/CR-5352, Rev.1, prepared by HydroGeologic, Inc., for the U.S. Nuclear Regulatory Commission, 1991, Rockville, Md.

Kazemi, H., Pressure transient analysis of naturally fractured reservoirs with uniform fracture distribution. *SPEJ*, 451–62. *Trans., AIME, 246*, 1969.

Kool J. B., and Y. S. Wu, *Validation and Testing of the VAM2D Computer Code*, NUREG/CR-5795, prepared by HydroGeologic, Inc., for the U. S. Nuclear Regulatory Commission, 1991, Rockville, Md.

Nitao, J. J., *V-TOUGH—An Enhanced Version of the TOUGH Code for the Thermal and Hydrologic Simulation of Large-scale Problems in Nuclear Waste Isolation*, UCID-21954,

Lawrence Livermore National Laboratory, Livermore, Calif., 1989.

Panday, S., P. A. Forsyth, R. W. Falta, Y. S. Wu, and P. S. Huyakorn, Considerations for robust compositional simulations of subsurface nonaqueous phase liquid contamination and remediation, *Water Resour. Res., 31*(5), 1273–1289, 1995.

Peters, R. R., and E. A. Klavetter, A continuum model for water movement in an unsaturated fractured rock mass, *Water Resour. Res., 24*(3), 416–430, 1988

Pruess K., *TOUGH2—A General Purpose Numerical Simulator for Multiphase Fluid and Heat Flow*, Report LBL-29400, UC-251, Lawrence Berkeley National Laboratory, Berkeley, Calif., 1991.

Pruess, K., J. S. Y. Wang, and Y. W. Tsang, On the thermohydrologic conditions near high-level nuclear wastes emplaced in partially saturated fractured tuff, Part 2. Effective continuum approximation, *Water Resources Res., 26*(6), 1249–1261, 1990.

Pruess, K., J. S. Y. Wang, and Y. W. Tsang, *Effective Continuum Approximation for Modeling Fluid and Heat Flow in Fractured Porous Tuff*, Report SAND86-7000, Sandia National Laboratories, Albuquerque, N.M., 1988.

Pruess, K., and Narasimhan, T. N., A practical method for modeling fluid and heat flow in fractured porous media, *Soc. Pet. Eng. J., 25,* 14–26, 1985.

Rousseau, J. P., E. M. Kwicklis and D. C. Gillies (eds.), *Hydrogeology of the Unsaturated Zone, North Ramp Area of the Exploratory Studies Facility, Yucca Mountain, Nevada,* USGS-WRIR-98-4050 Yucca Mountain Project Milestone 3GUP667M, Denver, Colo., 1996.

Sass J. H., A. H. Lachenbruch, W. W. Dudley, Jr., S. S. Priest, and R. J. Munroe, Temperature, thermal conductivity, and heat flow near Yucca Mountain, Nevada: Some tectonic and hydrologic implications, USGS OFR-87-649, U.S. Geological Survey, 1988.

Sonnenthal, E.L., and G. S. Bodvarsson, Chapter 15, Modeling the chloride geochemistry in the unsaturated zone, *The Site-Scale Unsaturated Zone Model of Yucca Mountain, Nevada, for the Viability Assessment,* edited by G. S. Bodvarsson, T. M. Bandurraga, and Y. S. Wu, Yucca Mountain Project Level 4 Milestone SP24UFM4; Report LBNL-40376, UC-814, Lawrence Berkeley National Laboratory, Berkeley, Calif., 1997.

Sudicky, E. A., and R. G. McLaren, *User's Guide for Fractran: An Efficient Simulators for Two-dimensional, Saturated Groundwater Flow and Solute Transport in Porous or Discretely-fractured Porous Formations*, Groundwater Simulations Group, Institute for Groundwater Research, University of Waterloo, Waterloo, Ontario, Canada, 1992.

van Genuchten, M. Th., and F. N. Dalton, Models for simulating salt movement in aggregated field soils, *Geoderma, 38,* 165–183, 1986.

van Genuchten, M. Th., A closed-form equation for predicting the hydraulic conductivity of unsaturated soils, *Soil Sci. Soc. Amer. J, 44*(5), 892–898, 1980.

Warren, J. E., and P. J. Root, The behavior of naturally fractured reservoirs, *Soc. Pet. Eng. J.,* 245–255, 1963.

Wilson, M. L., J. H. Gauthier, R. W. Barnard, G. E. Barr, H. A. Dockery, E. Dunn, R. R. Eaton, D. C. Guerin, N. Lu, M. J. Martinez, R. Nilson, C. A. Rautman, T. H. Robey, B. Ross, E. E. Ryder, A. R. Schenker, S. A. Shannon, L. H. Skinner, W. G. Halsey, J. D. Gansemer, L. C. Lewis, A. D. Lamont, I. R. Triay, A. Meijer, and D. E. Morris, *Total System Performance Assessment for Yucca Mountain—SNL Second Iteration (TSPA-1993),* SAND-93-2675, 882 pp., Sandia National Laboratories, Albuquerque, N.M., 1994.

Wu, Y. S, and K. Pruess, A 3-d hydrodynamic dispersion model for modeling tracer transport in geothermal reservoirs, *Proc. Twenty-third Workshop, Geothermal Reservoir Engineering*, Stanford University, Calif., 139–146, 1998.

Wu, Y.S., A. C. Ritcey, C. F. Ahlers, J. J. Hinds, A. K Mishra, C. Haukwa, C., H. H. Liu, E. L. Sonnenthal, and G. S. Bodvarsson, *3-D UZ Site-Scale Model for Abstraction in TSPA-VA*, Yucca Mountain Project Level 4 Milestone Report SLX01LB3, Lawrence Berkeley National Laboratory, Berkeley, Calif., 1998.

Wu, Y.S., A. C. Ritcey, and G. S. Bodvarsson, Chapter 13,Perched water analysis using the UZ Site-Scale Model, in *The Site-Scale Unsaturated-Zone Model of Yucca Mountain, Nevada, for the Viability Assessment,* edited by G. S. Bodvarsson, M. Bandurraga and Y. S. Wu. Yucca Mountain Site Characterization Project Report, LBNL-40376, UC-814, Lawrence Berkeley National Laboratory, Berkeley, Calif., 1997.

Wu, Y. S., C. F. Ahlers, P. Fraser, A. Simmons, and K. Pruess, *Software Qualification of Selected TOUGH2 Modules*, Report LBNL-39490, Lawrence Berkeley National Laboratory, Berkeley, Calif., 1996a.

Wu, Y. S., S. Finsterle, and K. Pruess, , Chapter 4, Computer models and their development for the unsaturated zone model at Yucca Mountain, in *Development and Calibration of the Three-Dimensional Site-Scale Unsaturated-Zone Model of Yucca Mountain, Nevada*, edited by G. S. Bodvarsson and M. Bandurraga, Yucca Mountain Site Characterization Project Report, Lawrence Berkeley National Laboratory, Berkeley, Calif., 1996b.

Wu, Y. S., G. Chen, C. Haukwa, and G. S. Bodvarsson, Chapter 8, Three-dimensional model calibration and sensitivity studies, in *Development and Calibration of the Three-Dimensional Site-Scale Unsaturated-Zone Model of Yucca Mountain, Nevada*, edited by G. S. Bodvarsson and M. Bandurraga, Yucca Mountain Project Level 4 Milestone OBO2, Lawrence Berkeley National Laboratory, Berkeley, Calif., 1996c.

Yu-Shu Wu, Earth Sciences Division, Ernest Orlando Lawrence Berkeley National Laboratory, Berkeley, CA 94720, USA.

Effective-Porosity and Dual-Porosity Approaches to Solute Transport in the Saturated Zone at Yucca Mountain: Implications for Repository Performance Assessment

Bill W. Arnold

Sandia National Laboratories, Albuquerque, New Mexico

Hubao Zhang and Alva M. Parsons

Duke Engineering and Services, Albuquerque, New Mexico

The effective-porosity approach and the dual-porosity approach are examined as two alternative conceptual models of radionuclide migration in fractured media of the saturated zone at Yucca Mountain. Numerical simulations of one-dimensional radionuclide transport are performed for the domain relevant to repository performance assessment using the two alternative conceptual approaches. Dual-porosity solute transport modeling produces similar results to effective-porosity modeling for fracture spacing of less than approximately 1 m and greater than about 200 m, which corresponds to values of effective porosity equal to the matrix porosity and the fracture porosity, respectively. For intermediate values of fracture spacing, the dual-porosity approach results in concentration breakthrough curves that differ significantly from the effective-porosity approach and are characterized by earlier first arrival, greater apparent dispersion, and lower concentrations at later times. The effective-porosity approach, as implemented in recent performance assessment analyses of saturated zone transport at Yucca Mountain, is conservative compared to the dual-porosity approach in terms of both radionuclide concentrations and, generally, travel times.

1. INTRODUCTION

Yucca Mountain, Nevada, is being investigated as the potential site for the construction of a high-level radioactive waste repository. The repository system would consist of large, metal waste packages emplaced in drifts excavated within the unsaturated zone beneath the mountain. The potential repository elevation is approximately 300 m to 400 m below the land surface and about 300 m above the water table.

Radionuclides escaping a radioactive waste repository, following corrosion and breaching of the waste package, are conceptualized to travel downward, transported by groundwater in the unsaturated zone, to the water table where they will enter the saturated zone. Groundwater flow in the saturated zone below and directly downgradient of the potential repository at Yucca Mountain occurs in

Dynamics of Fluids in Fractured Rock
Geophysical Monograph 122

fractured volcanic rocks. The permeability of the rock matrix of this volcanic medium is generally several orders of magnitude lower than the bulk permeability of the fracture network, indicating that groundwater flow occurs predominantly in fractures. The contrasting values of fracture porosity and matrix porosity in the fractured tuffs (matrix porosity of two to four orders of magnitude higher than fracture porosity) suggest that the storage of groundwater and the potential storage of contaminants would occur primarily in the rock matrix.

Characterization of solute transport in the fractured media of the saturated zone has important implications for the performance assessment of a potential repository at the Yucca Mountain site. Rapid solute mass transfer from flowing groundwater in the fractures to the immobile groundwater in the matrix would result in a relatively long delay in the migration of contaminants to a point of likely groundwater withdrawal and release to the biosphere. Slower solute mass transfer from the fractures into the matrix would result in more rapid migration of contaminants and earlier arrival of radionuclides at the accessible environment, presumably from a pumping well. Radiological dose to human beings residing in the biosphere at the location of radionuclide release is the primary measure of repository performance considered in performance assessment calculations. Numerical modeling of these processes and other components of the repository system have recently been formally synthesized in the Total System Performance Assessment—Viability Assessment (TSPA-VA) [*CRWMS M&O*, 1998a].

1.1. Background

Various approaches have been investigated to simulate radionuclide transport in fractured media. One approach treats the fractured medium as an equivalent effective continuum at the scale of interest [*Berkowitz et al.*, 1988; *Schwartz and Smith,* 1988]. Work by *Robinson* [1994] indicates that the equivalent continuum approach, using an effective porosity equal to the matrix porosity, would be appropriate for the prediction of radionuclide migration in the saturated zone from Yucca Mountain. The equivalent continuum approach, assuming a range of uncertainty in the value of effective porosity, was utilized in saturated-zone flow and transport simulations in the TSPA-VA [*CRWMS M&O,* 1998a]. Alternatively, the physical process of solute diffusion into the matrix may be explicitly considered using the dual-porosity approach. A number of dual-porosity analytical solutions for contaminant transport in fractured media that assume idealized homogeneous geometries for the fracture network have been developed, including those by *Grisak and Pickens* [1981] and *Sudicky and Frind* [1982]. More recently, multiple rates of diffusion in heterogeneous, dual-porosity media have also been considered [*Haggerty and Gorelick*, 1995].

1.2. Objectives

The objectives of this study are numerical evaluation and comparison of the effective-porosity and dual-porosity approaches to radionuclide transport in that portion of the flowpath that resides in the fractured tuff in the saturated zone at Yucca Mountain. In addition, the implications of these alternative approaches for matrix diffusion in the saturated zone to repository performance assessment are analyzed. These objectives are addressed using numerical simulations of one-dimensional groundwater flow and transport, from the repository to a distance of 20 km.

2. YUCCA MOUNTAIN SATURATED-ZONE SYSTEM

The saturated-zone system at Yucca Mountain is important to repository performance because it is an important component of the natural system that constitutes a barrier between radionuclide releases from the unsaturated zone below the potential repository and the biosphere. Regulatory guidance [*NRC*, 1999] suggests that a significant mechanism for release of radionuclides to the biosphere would be groundwater discharge from hypothetical well(s) at a distance of approximately 20 km from the repository. Transport of radionuclides in the saturated zone would tend to dilute radionuclide concentrations (e.g., through dispersive processes) and delay the release (e.g., by matrix diffusion and/or sorption) of radionuclides to the biosphere.

2.1. Geology and Regional Flow System

The bedrock underlying Yucca Mountain consists of a stratified sequence of Miocene epoch ash-flow and ash-fall tuffs. Fractured carbonate rocks and clastic units of the Paleozoic era are present deeper in the saturated zone flow system in the area of Yucca Mountain. At distances greater than about 10 to 20 km downgradient from the repository, groundwater flows through interfingered alluvial and volcanic units.

The geological structures observed at Yucca Mountain are consistent with extensional tectonics of the Basin and Range Province. The geometry of the normal faults and strike-slip faults controls much of the topography at Yucca Mountain [*Luckey et al.*, 1996]. The normal faults generally are oriented north-south, dip steeply to the west, and typically have a component of oblique slip displacement. Displacements on the faults range from less than 1 m to more than 300 m.

Yucca Mountain is part of the Death Valley regional groundwater flow system. On the regional scale, groundwater flow occurs from areas of recharge in upland regions and mountain ranges to areas of discharge at wells, springs, and playas. Most of the regional-scale groundwater flow occurs in the Paleozoic carbonate rocks [*Winograd and Thordarson*, 1975].

2.2. Inferred Flow Pathways

Water-level measurements in wells indicate that the general direction of groundwater flow in the saturated zone is from north to south in the area downgradient of the potential repository. A relatively shallow hydraulic gradient extends from beneath the repository to the southeast for a distance of approximately 7 km, turning to the south and south-southwest for a distance up to 20 km (Figure 1). A significant upward hydraulic gradient from the deeper volcanic units and the Paleozoic carbonate aquifer [*Luckey et al.,* 1996] suggests that groundwater flow pathways from Yucca Mountain would remain relatively near the water table. Recharge to the saturated zone along Fortymile Wash [*Savard*, 1998] may result in deepening of the flow pathway toward the south.

The flow pathway in the saturated zone occurs in fractured tuffs for 10 to 20 km from the potential repository. Flow in the shallow saturated zone occurs in alluvium or valley-fill units further downgradient. Uncertainty in subsurface geology and in the horizontal location of the flowpath precludes exact determination of the point along the flowpath at which groundwater flow transits from fractured volcanic units to alluvium.

2.3. Previous Modeling Work

Previous modeling of radionuclide transport in the saturated zone for total system performance assessment analyses has employed dimensional and conceptual simplifications of saturated-zone transport processes. For TSPA-91 [*Barnard et al.*, 1992], one-dimensional transport of radionuclides in the saturated zone was based on two-dimensional flow modeling of the system [*Czarnecki and Waddell*, 1984]. For TSPA-93 [*Wilson et al.*, 1994], radionuclide transport simulations using a three-dimensional saturated-zone model were used to derive a distribution of travel times through the system. Saturated-zone transport simulations for TSPA-95 [*CRWMS M&O*, 1995] used the saturated-zone groundwater flow fields developed for TSPA-93. In all cases, the values of porosity used to simulate radionuclide migration were approximately equal to the matrix porosity of the fractured volcanic aquifer. This approach implicitly assumes equilibrium of solute concentrations in the fractures with the groundwater in the rock matrix.

2.4. Conceptual Model of Radionuclide Transport Processes

Groundwater transport of solutes in fractured volcanic media of the saturated zone at Yucca Mountain is governed by complex interactions among processes, including advection in fractures (and potentially in the matrix), dispersion in the fractures, sorption in fractures and matrix blocks, and diffusive transfer between groundwater in the fractures and in the matrix (Figure 2). Solute mass transfer between groundwater in fractures and rock matrix by molecular diffusion is the primary mechanism that influences the distribution of radionuclide travel times through the system. If radionuclide travel times are long relative to the half-life of a radionuclide, this diffusive process also reduces the concentration at the downstream end of the saturated zone system.

These radionuclide transport processes in the saturated zone influence repository performance by affecting the travel time of radionuclides to a location where they are likely to be pumped from a well and by changing their concentrations in groundwater. The radionuclide travel times in the saturated zone may be long relative to the 10,000 year time period of regulatory concern [*NRC*, 1999]. Therefore, we must evaluate which methods may provide more realistic and defensible representations of migration velocities.

3. ALTERNATIVE CONCEPTUAL APPROACHES

Two alternative conceptual approaches to the prediction of radionuclide transport in the fractured media in the saturated zone at Yucca Mountain are considered in this study. These approaches constitute conceptual simplifications of the groundwater flow and solute transport system that are dictated, in part, by lack of detailed knowledge of the physical system and by computational limitations in the numerical representation of the system. Primary uncertainties in the physical system include the geometry of the fracture network, the distribution of groundwater flow within the fracture network, and heterogeneities in the rock matrix. Explicit representation of these complexities is precluded at the scale of interest in performance assessment calculations (i.e., 20 km) and by our inability to characterize the system to this level of detail.

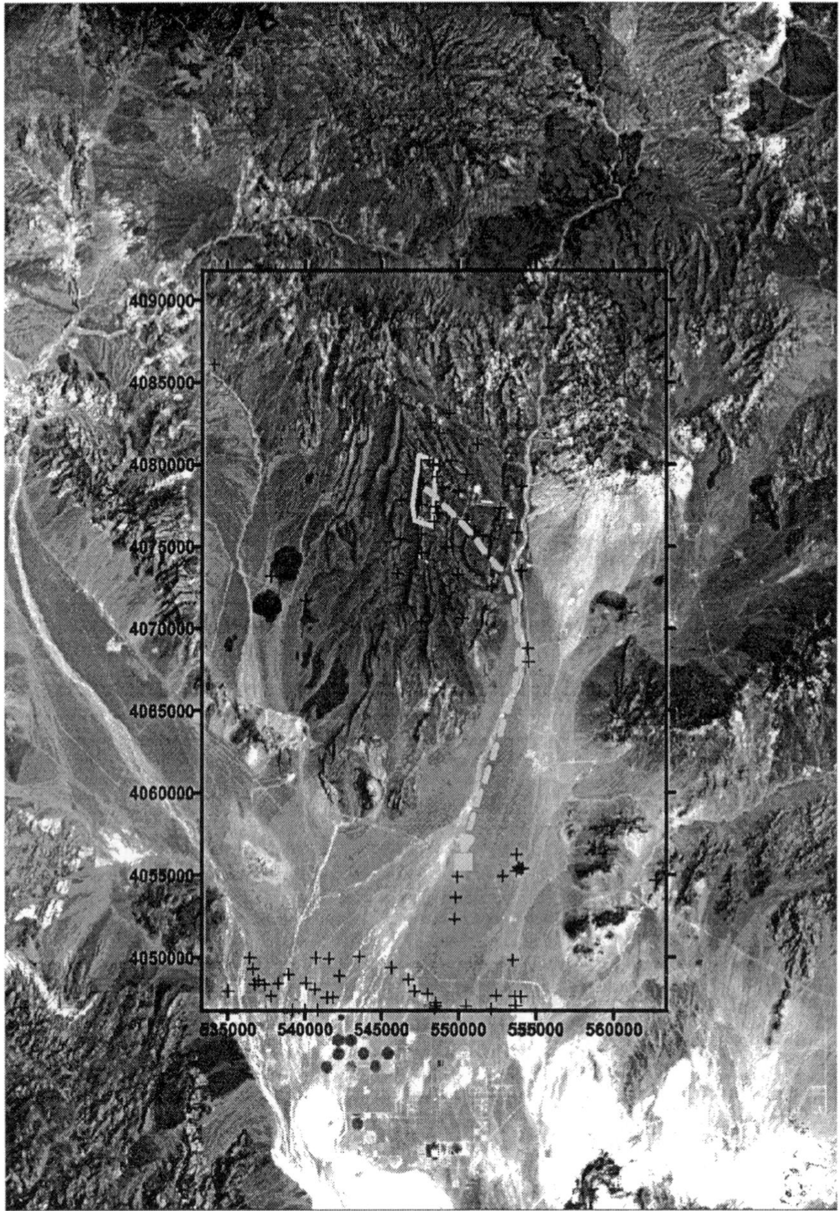

Figure 1. Region near the potential waste repository at Yucca Mountain, Nevada. The base map is an infrared satellite image. The outline of the potential repository is shown at the northern end of the dashed line. The inferred flow pathway in the saturated zone from beneath the repository to a hypothetical receptor at a distance of 20 km is shown by the dashed line. The locations of wells at which water level measurements have been made are shown by the + symbols. The outlined area of the saturated zone site-scale groundwater flow model (30 km × 45 km) is shown with the axes labeled in UTM coordinates (meters).

3.1. Effective-Porosity Approach

The effective-porosity approach assumes that some portion of the available matrix porosity in the fractured medium is immediately accessible to solutes in the fractures (Figure 3). If it is assumed that matrix diffusion will result in equilibrium in solute concentration between groundwater in the fractures and the matrix over the time of transport from the repository to the accessible environment, then a value of effective porosity equal to the matrix

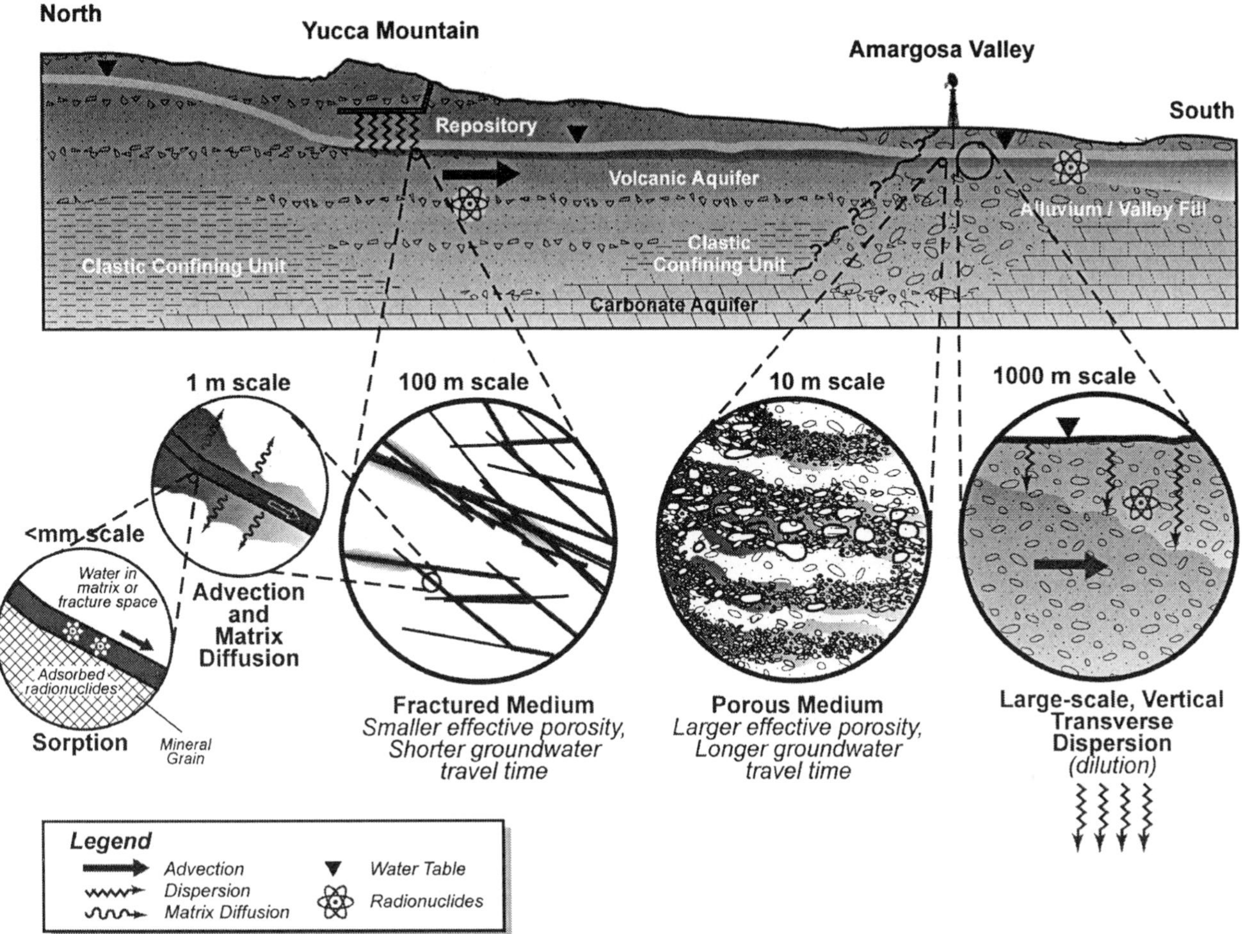

Figure 2. Conceptual model of radionuclide transport processes in the saturated zone. The cross section is a diagrammatic representation of the system along the inferred flow pathway in the saturated zone (as shown by the dashed line in Figure 1). Processes that are relevant to the transport of radionuclides at various scales are illustrated in the circular diagrams.

porosity would be used. If it is assumed that matrix diffusion will be extremely limited in the system of interest, then a value of effective porosity equal to the fracture porosity would be used. Intermediate values of effective porosity approximate the situation in which some fraction of the total solute storage capacity of the fractured medium has been filled by diffusion. The effective-porosity approach thus implicitly considers the effects of diffusion from fractures into the matrix; however, it must be applied in an ad hoc manner. In reality, the portion of the matrix porosity available for solute storage changes as a function of time. The effective porosity is thus a "lumped" parameter that incorporates uncertainty in underlying processes and results in an approximate solution for solute transport. The accuracy of the effective-porosity approximation, particularly for values of effective porosity intermediate between fracture porosity and matrix porosity, is dependent on several characteristics of the flow system including groundwater velocity, travel distance, and fracture spacing.

3.2. Dual-Porosity Approach

The dual-porosity approach explicitly considers the physical process of matrix diffusion of solutes from groundwater flowing in fractures. In this alternative simplification of the system, the fractured medium is conceptualized to consist of two continua. One continuum represents mobile groundwater in the fractures and the other continuum corresponds to immobile groundwater in the matrix. Advective transport is conceptualized to occur only in the fractures, and the matrix is available for the

Effective Porosity

Parameters:
- q (specific discharge)
- ϕ (effective porosity)
- R_f (retardation factor)

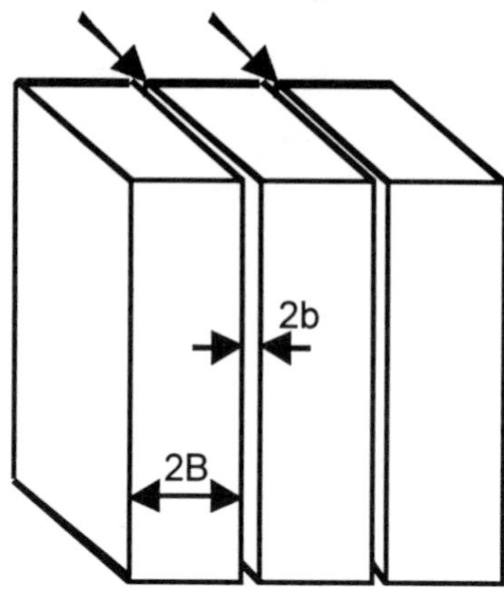

- 2b (fracture aperture)
- q (specific discharge)
- ϕ_m (matrix porosity)
- ϕ_f (fracture porosity)
- 2B (fracture spacing)
- D_m (matrix diffusion coefficient)
- R_f (retardation factor)

Figure 3. Schematic representation of the effective-porosity and dual-porosity approaches to solute transport in fractured media. The bold arrows at the top indicate volumetric groundwater flow rate. In the effective-porosity approach, the flow occurs in a single continuum in which the fraction of the total volume available to flow and solute transport is represented by the value of effective porosity. In the dual-porosity approach, flow occurs only within the volume represented by the fracture porosity, and the porosity within the matrix is available to solutes by diffusion.

storage of solute mass through the process of molecular diffusion.

Implementation of the dual-porosity conceptual model typically requires idealization of the fracture network geometry and groundwater flow within it, due to the continuum assumption. The continuum assumption is generally valid for a well-interconnected fracture network. The dual-porosity approach used in this study conceptualizes groundwater flow to occur through uniformly spaced, parallel fractures in the fractured medium (Figure 3). In this approach, homogeneous matrix blocks occupy the volume between the parallel fractures. The primary limitation of this approach is the extent to which it misrepresents the geometry of the system. Realistic fracture networks vary considerably in fracture spacing, orientation, and aperture; and divide the rock matrix into blocks of varying sizes. Consequently, there is significant uncertainty in the representative size of the matrix blocks (i.e., fracture spacing) in fractured media.

4. NUMERICAL APPROACHES

Numerical simulations of groundwater flow and solute transport are performed using a one-dimensional model of the saturated zone downgradient of Yucca Mountain. The one-dimensional model domain corresponds to the inferred groundwater flowpath at the site. Simulations using the effective-porosity approach and the dual-porosity approach are performed using the FEHM computer code [*Zyvoloski et al.*, 1995]. The FEHM code is a finite-element/finite-volume groundwater flow and solute transport simulator. Simulations of solute transport are obtained using the finite-element method. All simulations consist of first solving for a steady-state solution for the pressure distribution in the domain, then running transient solute transport simulations to obtain the concentration breakthrough curves at the downstream end of the system.

4.1. Model Domain

The model domain for the one-dimensional SZ flow and transport model consists of four hydrogeologic units corresponding to the approximate flowpath through the system. The respective, approximate lengths of these units within the one-dimensional model are taken from the results of particle tracking simulations with three-dimensional saturated zone flow modeling [*CRWMS M&O*, 1998a; TSPA-VA]. The three volcanic units are assumed to be fractured media, and the alluvium is assumed to be a porous medium. Because of the lack of subsurface data, there is significant uncertainty as to the location of the contact between the middle volcanic confining unit and the alluvium along the flowpath represented in the one-dimensional flow and transport model. This uncertainty in the subsurface geology of the saturated zone is incorporated into the effective-porosity simulations by varying the fraction of the flowpath through the alluvium. The length of the flowpath through alluvium is held constant at 2 km in the sensitivity studies of the dual-porosity approach to facilitate comparison between the conceptual approaches.

The one-dimensional model for flow and transport in the saturated zone represents a significant simplification of the three-dimensional system. The one-dimensional model does not account for transverse dispersion and consequently cannot simulate the dilution of radionuclide concentrations from this process. The one-dimensional model does provide an accurate simulation of radionuclide travel times through the saturated zone system.

Simulations for the effective-porosity approach are performed using a one-dimensional grid with a 5 m nodal spacing. Simulations for the dual-porosity approach are performed using a quasi-two-dimensional grid in which the second dimension represents length into the matrix continuum. The first column of nodes in the dual-porosity grid corresponds to the fracture. A high-resolution, exponentially spaced grid consisting of 50 nodes in the transverse direction is employed to accurately simulate diffusive movement of solute in the matrix. The fracture aperture (2b in Figure 3) is defined as the product of fracture porosity and fracture spacing.

4.2. Boundary Conditions

The groundwater boundary conditions applied to the effective-porosity and dual-porosity transport models consist of specified pressure at the downstream boundary and specified flux at the upstream boundary. In the dual-porosity model, the specified flux is applied at the node representing the fracture. A specific discharge value of 0.6 m/yr is used in all cases. As a consequence, the volumetric groundwater flow rate applied at the upstream boundary is a function of the fracture spacing (2B in Figure 3). The lateral boundaries of the quasi-two-dimensional dual-porosity model are no-flow boundaries. A constant, unit mass flux of a conservative radionuclide (^{99}Tc) solute is applied at the upstream boundary of the effective-porosity and dual-porosity transport models.

4.3. Input Parameters

The values of effective porosity applied to the four hydrogeologic units in the effective-porosity modeling approach are varied stochastically. Values ranging from those representative of fracture porosity to the matrix porosity are drawn independently for the units in multiple realizations of the flow and transport system [*CRWMS M&O*, 1998a]. The uncertainty distribution for effective porosity is log-triangular, with a mode value of 0.02. The length of the flowpath through the alluvium unit is randomly varied from 0 to 6 km in the effective-porosity realizations. The length of the middle volcanic confining unit is correspondingly varied to maintain the total length of 20 km. The values of longitudinal dispersivity, with a mean of 100 m, are also stochastically varied in the effective-porosity simulations. These values of longitudinal dispersivity are consistent with results of an expert elicitation regarding groundwater flow and radionuclide transport in the saturated zone at Yucca Mountain [*CRWMS M&O*, 1998b].

The parameter values used in the dual-porosity modeling are summarized in Table 1. The values of matrix porosity are held constant at the estimated value for each of the hydrogeologic units. Fracture porosity is held constant at a value of 1×10^{-4} for the fractured units. The value of fracture spacing (2B in Figure 3) is varied in sensitivity studies from 0.2 m to 200 m. The fracture aperture (2b in Figure 3) varies from 2.0×10^{-5} m to 2.0×10^{-2} m for the range of fracture spacing considered. Longitudinal dispersivity is held constant at 100 m in the dual-porosity simulations. The length of the flowpath in the alluvium unit is specified as a constant of 2 km in the dual-porosity simulations. A value of 3.2×10^{-11} m^2/s is assumed for the effective molecular diffusion coefficient of ^{99}Tc, based on laboratory measurements of diffusion in volcanic tuff [*CRWMS M&O*, 1998a]. In the dual-porosity simulations, groundwater flow is restricted to the fracture nodes by the large contrast in permeability (10 orders of magnitude) between the fracture nodes and the matrix nodes.

5. RESULTS

The results of the effective-porosity approach simulations are shown as concentration breakthrough curves in Figure 4. These are the results of 100 realizations for ^{99}Tc transport in the SZ system, as used in performance assessment analyses [*CRWMS M&O*, 1998a]. The median travel times for ^{99}Tc vary from less than 200 years to greater than 4,000 years for the ranges of uncertainty assessed in these realizations. Variation in the value of

Table 1. Parameter Values Used in Dual-Porosity Transport Model (CRWMS M&O, 1998a).

Hydrogeological Units	Matrix porosity	Fracture porosity	Fracture Spacing (m)	Fracture Aperture (m)
Upper volcanic aquifer	0.163	1.0×10^{-4}	0.2–200	2.0×10^{-5} – 2.0×10^{-2}
Middle volcanic aquifer	0.227	1.0×10^{-4}	0.2–200	2.0×10^{-5} – 2.0×10^{-2}
Middle volcanic confining unit	0.183	1.0×10^{-4}	0.2–200	2.0×10^{-5} – 2.0×10^{-2}
Alluvium/undifferentiated valley fill	0.25	N/A	N/A	N/A

longitudinal dispersivity results in significant variation in the apparent dispersion among the breakthrough curves.

The results of the sensitivity analysis for the dual-porosity approach are shown as the dashed concentration breakthrough curves in Figure 5. For comparison, the breakthrough curves for differing values of effective porosity are shown as the solid curves in the same figure. The longitudinal dispersivity and the length of the flowpath in alluvium are held constant at 100 m and 2 km, respectively, in the dual-porosity and effective-porosity simulations for the comparisons shown in Figure 5.

For a fracture spacing of 0.2 m with the dual-porosity approach, the concentration breakthrough curve corresponds to the breakthrough curve for the effective-porosity approach in which the values of matrix porosity are assigned to all hydrogeologic units. For a fracture spacing of 200 m with the dual-porosity approach, the midpoint of the concentration breakthrough curve corresponds approximately to the results of the effective-porosity approach in which the value of effective porosity in the fractured units is 0.005 or less. The breakthrough time for the midpoint of the curve (about 1,200 years) in this case indicates very rapid transport through the fractured units. Note that for the case of fracture spacing of 200 m with the dual-porosity approach there is some attenuation of the maximum concentration at later times. For intermediate values of fracture spacing (2 m to 20 m) with the dual-porosity approach, there are significant differences in the shapes of the concentration breakthrough curves between the dual-porosity simulations and the effective-porosity simulations. The dual-porosity simulations exhibit earlier ^{99}Tc breakthrough, greater apparent dispersion, and long tails for intermediate values of fracture spacing.

6. DISCUSSION

The results of the dual-porosity simulations of the saturated zone system at Yucca Mountain shown in Figure 5 indicate that for fracture spacing of about 1 m or less there would be nearly complete saturation of the matrix blocks within the fractured units along the flowpath of the contaminant plume. For this case of relatively closely spaced fractures carrying groundwater flow, diffusive solute mass transfer from the fractures to the matrix is dominant relative to the advective movement of solute in the fracture system. Consequently, the solute travel times to the 20 km compliance boundary for the repository are relatively long. In addition, the effective-porosity approach, using values of effective porosity equal to the matrix porosity of the fractured tuffs, would yield a relatively accurate solution for the concentration breakthrough curve and long travel time at these fracture spacings.

The modeling results for fracture spacing greater than 200 m indicate that there would be minimal interaction between contaminated groundwater flowing in the fractures and the groundwater contained in the matrix of the fractured units. For this case of widely spaced fractures, migration of contaminants in the system is dominated by advective movement in the fracture system relative to the solute mass transfer by diffusion from the fractures to the matrix. Transport of solutes to a distance of 20 km is much more rapid for large fracture spacing. The effective-porosity approach results in a relatively accurate solution for the concentration breakthrough curves, using values of effective porosity representative of fracture porosity at this large fracture spacing.

For intermediate values of spacing between fractures carrying groundwater, the dual-porosity modeling results indicate that the migration of contaminants is dominated by neither advection in the fractures nor diffusive mass transfer between fractures and matrix. The concentration breakthrough curves for these cases are characterized by the relatively early first arrival of solute at the 20 km boundary, but significant solute mass loss to the matrix and consequent lower concentrations at later times. The effective-porosity approach is inappropriate for these intermediate values of fracture spacing, in terms of providing an accurate solution for solute transport through the saturated zone system.

It should be noted that the conclusions with regard to the values of fracture spacing stated above are specific to the travel distance and specific discharge assumed in this study. Solute transport in the dual-porosity approach is a function of distance through the system and the groundwater velocity.

Comparison of Figures 4 and 5 indicates that the effective-porosity approach, as implemented in TSPA-VA analyses [(CRWMS M&O, 1998a], has potentially significant differences when compared with the dual-porosity approach. However, these inaccuracies produce generally conservative, and thus more easily defended, results from the perspective of regulatory assessments of repository performance. Conservatism is defined here as computational results that underestimate radionuclide travel times or overestimate radionuclide concentrations relative to other computational methods. The shortest median travel times among the TSPA-VA realizations using the effective-porosity approach shown in Figure 4 are shorter than the median travel times for the fastest breakthrough curves from the dual-porosity model shown in Figure 5. This is because the effective porosity in the alluvium was stochastically varied to include lower values in the TSPA-VA analyses shown in Figure 4 and was held constant at a higher value (0.25) in the dual-porosity simulations shown in Figure 5. The longest median travel times among the effective-porosity realizations are shorter than the longest

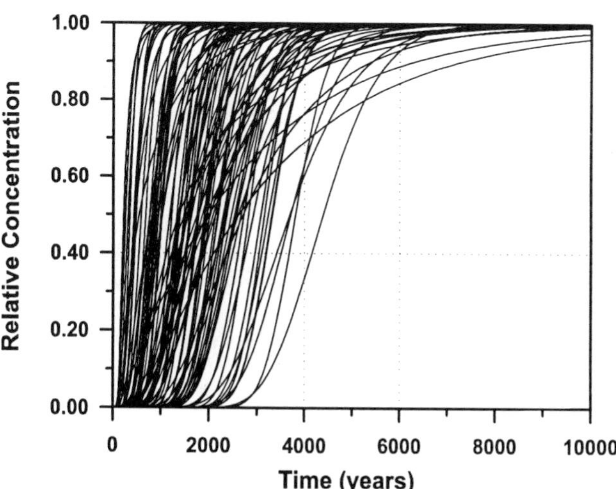

Figure 4. Concentration breakthrough curves at 20 km for ^{99}Tc from 100 realizations of the one-dimensional transport model using the effective-porosity approach.

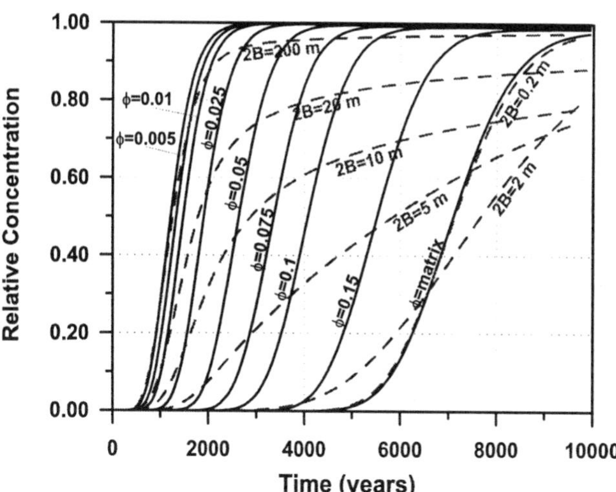

Figure 5. Concentration breakthrough curves for ^{99}Tc at 20 km from the one-dimensional transport model for several values of fracture spacing with the dual-porosity approach and for several values of effective porosity with the effective-porosity approach. Results for the dual-porosity approach are shown as dashed lines and results for the effective-porosity approach are shown as solid lines.

median travel times shown in Figure 5. This result occurred because none of the 100 TSPA-VA realizations shown in Figure 4 simultaneously sampled values of effective porosity near the matrix porosity values for all three fractured hydrogeologic units.

The simulated concentrations using the dual-porosity model shown in Figure 5 are significantly lower for intermediate values of fracture spacing at later times (e.g., 8,000 years) than the simulated concentrations from the effective-porosity approach. This result also indicates that the effective-porosity approach is conservative from the perspective of repository performance. Results from the dual-porosity modeling for intermediate values of fracture spacing indicate significantly greater spreading of the breakthrough curves (greater apparent dispersion) and earlier first arrival of solute than the effective-porosity approach as shown in Figure 5. The lower apparent dispersion from the effective-porosity approach is conservative in the context of TSPA calculations because longitudinal dispersion leads to attenuation of peak concentrations resulting from pulses of radionuclide mass at the source. The earlier first arrival of solute mass exhibited by the dual-porosity approach indicates that the effective-porosity approach is potentially nonconservative in terms of travel time. However, comparison of the first-arrival times shown for various fracture spacings in Figure 5 with the first-arrival times shown in Figure 4 indicates that the potential for early radionuclide arrival was represented in the TSPA-VA realizations. This result is a consequence of the uncertainty analysis in the effective-porosity modeling for TSPA-VA and is not inherent in the effective-porosity approach.

The effective-porosity approach as implemented in the TSPA-VA analyses of saturated zone transport is conservative compared to the dual-porosity approach from the perspective of both radionuclide concentrations and, generally, travel times. Future TSPA analyses of potential repository performance at Yucca Mountain may utilize more explicit modeling of matrix diffusion, which will have the effect of providing more realistic results than does the effective-porosity approach.

Acknowledgments. This work was supported by the Yucca Mountain Site Characterization Office as part of the Civilian Radioactive Waste Management Program, which is managed by the U.S. Department of Energy, Yucca Mountain Site Characterization Project. Sandia is a multiprogram laboratory operated by Sandia Corporation, a Lockheed Martin Company, for the United States Department of Energy under Contract DE-AC04-94AL85000.

REFERENCES

Barnard, R. W., M. L. Wilson, H. A. Dockery, J. H. Gauthier, P. G. Kaplan, R. R. Eaton, F. W. Bingham, and T. H. Robey, *TSPA 1991: An Initial Total-System Performance Assessment for Yucca Mountain,* SAND91-2795, Sandia National Laboratories, Albuquerque, N.M., 1992.

Berkowitz, B., J. Bear, and C. Braester, Continuum models for contaminant transport in fractured porous formations, *Water Resour. Res*, *24*(8) 1225–1236, 1988.

CRWMS M&O (Civilian Radioactive Waste Management System Management and Operating Contractor), *Total System Performance Assessment—1995: An Evaluation of the Potential Yucca Mountain Repository*, B00000000-01717-2200-00136, Rev. 01, TRW Environmental Safety Systems, Inc., Las Vegas, Nev., 1995.

CRWMS M&O, *Total System Performance Assessment—Viability Assessment (TSPA-VA): Technical Basis Document. Chapter 8 Saturated Zone Flow and Transport*, B00000000-01717-4301-00008, Rev. 01, TRW Environmental Safety Systems, Inc., Las Vegas, Nev., 1998a.

CRWMS M&O, *Saturated Zone Flow and Transport Expert Elicitation Project*, Yucca Mountain Project Deliverable SL5X4AM3, TRW Environmental Safety Systems, Inc., Las Vegas, Nev., 1998b.

Czarnecki, J. B., and R. K. Waddell, *Finite-Element Simulation of Ground-Water Flow in the Vicinity of Yucca Mountain, Nevada-California*, Water-Resources Investigations Report 84–4349, U.S. Geological Survey, Denver, Colo., 1984.

Grisak, G. E., and J. F. Pickens, An analytical solution for solute transport through fractured media with matrix diffusion, *J. Hydrol.*, *52*(1/2), 47–57, Elsevier Scientific Publishing Co., Amsterdam, 1981.

Haggerty, R., and S. M. Gorelick, Multiple-rate mass transfer for modeling diffusion and surface reactions in media with pore-scale heterogeneity, *Water Resour. Res.*, *31*(10), 2383–2400, 1995.

Luckey, R. R., P. Tucci, C. C. Faunt, E. M. Ervin, W. C. Steinkampf, F. A. D'Agnese, and G. L. Patterson, *Status of Understanding of the Saturated-Zone Ground-Water Flow System at Yucca Mountain, Nevada, as of 1995*, Water-Resources Investigations Report 96-4077, U.S. Geological Survey, Denver, Colo., 1996.

NRC (U.S. Nuclear Regulatory Commission), Disposal of high-level radioactive wastes in a proposed geologic repository at Yucca Mountain, Nevada; proposed rule, *Federal Register*, *64*(34), 8640–8679, 10 CFR Part 63, Office of the Federal Register, National Archives and Records Service, Washington, D.C., 1999.

Robinson, B. A., A strategy for validating a conceptual model for radionuclide migration in the saturated zone beneath Yucca Mountain, *Radioactive Waste Management and Environmental Restoration*, *19*(1–3, 73–96, Academic Yverdon, Switzerland, Harwood Publishers, International Publishers Distributor; Newark, N.J., 1994.

Savard, C. S., *Estimated Ground-Water Recharge from Streamflow in Fortymile Wash Near Yucca Mountain, Nevada*, Water-Resources Investigations Report 97-4273, U.S. Geological Survey, Denver, Colo., 1998.

Schwartz, F. W., and L. Smith, A continuum approach for modeling mass transport in fractured media, *Water Resour. Res.*, 24(8), 1360–1372, American Geophysical Union, Washington, D.C., 1988.

Sudicky, E. A., and E. O. Frind, Contaminant transport in fractured porous media: Analytical solution for a system of parallel fractures, *Water Resour. Res.*, *18*(6), 1634–1642, 1982.

Wilson, M. L., J. H. Gauthier, R. W. Barnard, G. E. Barr, H. A. Dockery, E. Dunn, R. R. Eaton, D. C. Guerin, N. Lu, M. J. Martinez, R. Nilson, C. A. Rautman, T. H. Robey, B. Ross, E. E. Ryder, A. R. Schenker, S. A. Shannon, L. H. Skinner, W. G. Halsey, J. D. Gansemer, L. C. Lewis, A. D. Lamont, I. R. Triay, A. Meijer, and D. E. Morris, *Total-System Performance Assessment for Yucca Mountain—SNL Second Iteration (TSPA-1993),* SAND93-2675, Sandia National Laboratories, Albuquerque, N.M., 1994.

Winograd, I. J., and W. Thordarson, *Hydrogeologic and Hydrochemical Framework, South-Central Great Basin, Nevada-California, With Special Reference to the Nevada Test Site*, Professional Paper 712-C, U.S. Geological Survey, Washington, D.C., 1975.

Zyvoloski, G. A., B. A. Robinson, Z. V. Dash, and L. L. Trease, *Models and Methods Summary for the FEHMN Application*, LA-UR-94-3787, Rev. 1, Los Alamos National Laboratory, Los Alamos, N.M., 1995.

B.W. Arnold and A.M. Parsons, Sandia National Laboratories, P.O. Box 5800, Albuquerque, NM 87185-0778

H. Zhang, Duke Engineering and Services, 1650 University Blvd., Suite 300, Albuquerque, NM 87102

Radionuclide Transport in the Unsaturated Zone at Yucca Mountain: Numerical Model and Preliminary Field Observations

Bruce A. Robinson and Gilles Y. Bussod

Earth and Environmental Sciences Division, Los Alamos National Laboratory, Los Alamos, New Mexico

The unsaturated zone at Yucca Mountain is one of the primary barriers to radionuclide migration from the potential repository to the accessible environment and, as such, has received great attention in site characterization activities. In this study, a dual-permeability flow and transport model has been developed that captures the disparate behavior of radionuclides, which depends on whether transport occurs in the fractures or the matrix. The model predicts stratigraphically controlled, abrupt changes in flow and transport— from fracture dominated to matrix dominated. A particle-tracking algorithm has been used to simulate radionuclide transport. This method handles a wide range of solute transport velocities due to fracture and matrix flow, as well as transport processes such as retardation due to sorption and matrix diffusion. Using sorption and diffusion data collected for the Yucca Mountain tuffs, the model predicts that the nonwelded vitric Calico Hills unit is the primary barrier to radionuclide migration in the unsaturated zone. This is because of the long percolation times in the matrix and the intimate contact of sorbing radionuclides with the host rock. Matrix diffusion is also shown to provide significant travel-time delay, especially for regions beneath the potential repository in which fracture transport occurs along the entire flow path to the water table. Because of the importance of the vitric Calico Hills to the performance of the unsaturated zone as a barrier to radionuclide migration, a field-scale Unsaturated Zone Transport Test (UZTT) is being conducted at Busted Butte, located 8 km southeast of the potential repository. Initial results indicate that the vitric Calico Hills unit does indeed exhibit matrix-dominated flow and transport under current and expected future hydrologic conditions at the site.

1. INTRODUCTION

Yucca Mountain, Nevada, is being studied by the U.S. Department of Energy as a potential site for the nation's first high-level nuclear waste repository. As part of the site characterization activities, the hydrologic and transport properties of the region near the potential repository are under investigation. The areas investigated include the unsaturated zone, where the potential repository is to be located, and the saturated zone along the likely pathways to the accessible environment. The present study focuses on the unsaturated zone, which consists of alternating layers of welded and nonwelded tuffs that are tilted, uplifted, fractured, and faulted. Infiltration rates into the unsaturated zone are thought to be low because of the low precipitation

Dynamics of Fluids in Fractured Rock
Geophysical Monograph 122

Published 2000 by the American Geophysical Union

and the large evapotranspiration rate [*Flint et al.*, 1996]. This percolating fluid is the primary carrier for most of the radionuclides of interest. For radionuclides to reach the accessible environment, the waste canisters must be breached and the radionuclides must become mobilized by dissolution into the percolating fluid. The radionuclides may then travel through the engineered barriers constructed within the emplacement drifts, in the unsaturated zone and, subsequently, toward the saturated zone within the groundwater flow system. In addition to the hydrologic processes that are important in controlling radionuclide migration, transport processes such as advection, dispersion, diffusion, and chemical reaction also affect the movement of radionuclides.

In this study, we present the development and results of a dual-permeability model for radionuclide transport. Factors such as the relative flow rates in the fractures and matrix, sorption, and matrix diffusion are examined to assess the important processes in unsaturated zone flow and transport at Yucca Mountain. The results presented are intended to simulate the range of representative flow and transport conditions characterizing the zone between the potential repository and the water table. In conjunction with the modeling, we present the preliminary results of the Unsaturated Zone Transport Test (UZTT) at Busted Butte, an experiment designed to demonstrate the applicability of the key assumptions in the transport model. Selected preliminary results from this field investigation relevant to the validity of the unsaturated zone flow and transport model are also reported.

2. DUAL-PERMEABILITY FLOW AND TRANSPORT MODEL

The complex nature of unsaturated zone flow and transport in fractured rock has been the subject of intensive study for the last two decades. It is clear from the field evidence and laboratory measurements of hydrologic and transport properties that the degree to which fractures play a role in hydrology depends on the hydrologic properties of the host rock. There is roughly a five order-of-magnitude variation in matrix hydraulic conductivity in the vertical direction from the repository to the water table, with variability closely correlated with geology and mineral alteration [e.g., *Altman et al.*, 1996]. Fractures in rocks of the low-permeability, welded tuffs of the Topopah Spring welded (TSw) unit are likely to transmit fluid under conditions of moderate or high infiltration. By contrast, the matrix rock of nonwelded units such as the vitric Calico Hills (CHnv) unit is permeable enough to transmit all of the percolating fluid, even under the highest flux values anticipated for the site.

This wide range of flow behavior, which is expected to occur in different types of rocks, places extreme demands on a numerical model. This model must capture the entire regime of flow—from fracture dominated to matrix dominated. We have developed dual-permeability flow and transport model formulations to handle this situation. Such models are a reasonable compromise between a discrete fracture representation, with its unwieldy structure and large data demands (fracture locations, lengths, apertures, etc.), and a single-continuum model, which cannot easily simulate mixed systems consisting of both fracture and matrix flow and transport processes. Details of the numerical formulation of the dual-permeability formulation can be found in *Zyvoloski et al.* [1997].

The "Residence-Time/Transfer Function" (RTTF) particle-tracking method is our cell-based particle-tracking algorithm for the simulation of solute transport in dual-permeability systems. Traditional particle-tracking models [e.g. *Kinzelbach*, 1988; *Tompson and Gelhar*, 1990] interpolate between points of known fluid potential or velocity to compute particle streamlines and travel times. While useful for structured grids and single-continuum models, this approach has two main disadvantages:

- Streamline particle-tracking is computationally intensive for unstructured grids
- The approximations associated with dual-permeability models (overlapping continua, fracture-matrix mass interchange) call into question the validity of and need for an interpolation of velocity streamlines within a computational grid cell

Instead, our cell-based particle-tracking method performs the computation of particle transport in only two steps:

- Cell-based particle-tracking holds the particle at a given cell for a prescribed time
- This model then determines where the particle is then moved, based on the relative flow rates exiting the cell

A detailed description of the first step (determining the particle residence time within a cell) is presented below. In the second step, the method makes no attempt to chart particle streamlines at a level of detail greater than the spacing of the finite volume grid. Instead, it opts for computational speed and ease of implementation in the dual-permeability formulation. In a dual permeability model, the connection to the node corresponding to the other continuum appears as an additional connecting node, which is treated identically to the other nodes (Figure 1). The method decides which cell the particle moves to probabilistically, with probabilities proportional to the flow rate to each adjacent cell. In this respect, the method is equivalent to the node-to-node particle routing method introduced by *Desbarats* [1990].

Returning to the first step, a transfer function methodology is used to probabilistically determine the holding time. First, for pure plug flow without retardation, the residence time in a cell is simply the fluid mass divided by the mass flux through the cell. From the solution of the flow field in a numerical model, the mass of fluid in the computational cell and the mass flow rate to or from each adjacent cell is computed. In the simplest case, the residence time of a particle in a cell, τ_{part}, is given by

$$\tau_{part} = \tau_f = \frac{M_f}{\sum \dot{m}_{out}} \quad , \tag{1}$$

where M_f is the fluid mass in the cell, and the summation term in the denominator refers to the outlet fluid mass flow rates from the cell to adjacent cells. In the absence of diffusion or other transport mechanisms, the transfer function describing the distribution of particle residence times is a Heaviside function (unit step function), which is unity at the fluid residence time, τ_f, because, for this simple case, all particles entering the cell will possess this residence time. Equilibrium, or linear sorption, is simulated by correcting the particle residence time by a retardation factor, R_f: $\tau_{part} = R_f \tau_f$, where R_f is given by

$$R_f = 1 + \frac{\rho_b K_d}{\phi S_f} \quad , \tag{2}$$

in which K_d is the equilibrium sorption coefficient (cm^3 fluid/g rock), ρ_b is the bulk rock density (g/cm^3), ϕ is the rock porosity, and S_f is the saturation of the phase in

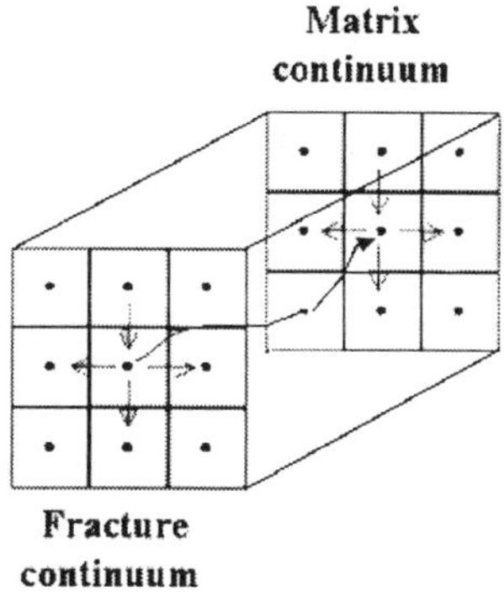

Figure 1. Schematic of the cell-based dual-permeability particle-tracking algorithm.

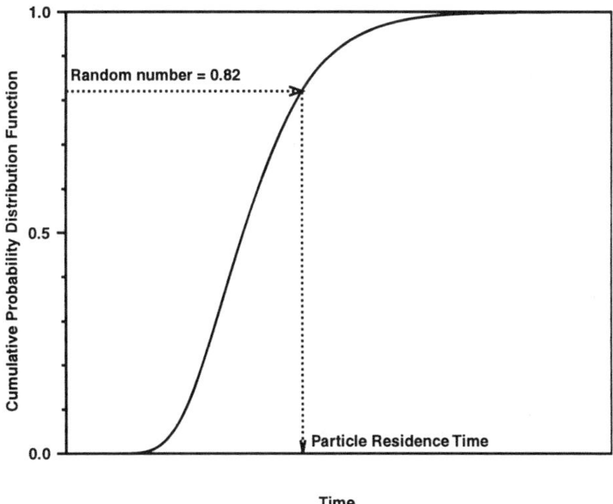

Figure 2. The residence time/transfer function (RTTF) particle tracking algorithm.

which the particle is traveling. Once again, in the absence of other transport processes, the transfer function is a Heaviside function.

For more complex processes such as matrix diffusion, the RTTF technique makes use of a transfer function (normalized cumulative residence time distribution curves) derived from an analytical solution to probabilistically determine the travel time of an individual particle within a cell, as shown conceptually in Figure 2. The matrix diffusion transfer function used in this study is derived from the following 1-D transport equation in a fracture and surrounding matrix:

$$R_{f,f} \frac{\partial C}{dt} = D_{eff} \frac{\partial^2 C}{dz^2} - v \frac{\partial C}{dz} - \frac{q}{b} \quad , \tag{3}$$

where t is time, $2b$ is the fracture aperture, $R_{f,f}$ is the retardation factor in the fracture, D_{eff} is the dispersion coefficient, and the flux at the fracture/matrix interface is given by

$$q = -\phi D \left. \frac{\partial C}{dx} \right|_{x=b} \quad . \tag{4}$$

Transport within the matrix is described by the 1-D diffusion equation

$$R_{f,m} \frac{\partial C}{\partial t} = D \frac{\partial^2 C}{\partial x^2} \quad , \tag{5}$$

where D is the effective diffusion coefficient and $R_{f,m}$ is the retardation factor in the matrix. In this model, D is treated as the fundamental transport parameter characterizing both the solute and the matrix. An analytical solution is given by

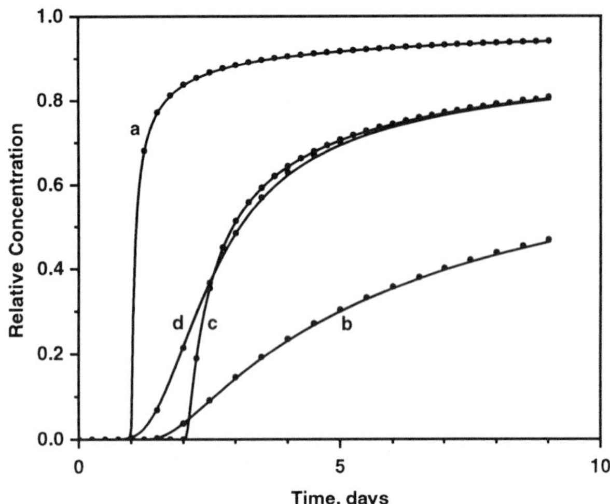

Figure 3. Breakthrough curves computed using the RTTF particle-tracking technique for the infinite spacing matrix diffusion model, compared to the analytical solution of *Tang et al.* [1981] (Solid Curves). (a) low matrix diffusion; (b) moderate matrix diffusion; (c) matrix diffusion and sorption on fracture and in matrix; (d) dispersion, matrix diffusion, and sorption on the fracture and matrix.

Tang et al. [1981] for the semi-infinite boundary condition $C = 0$ as $x \rightarrow B$. For the case of plug flow (no dispersion) in the fractures, the solution of *Starr et al.* [1985, Eq. 8b] can be used. Replacing *Starr's* term x / v_s with τ_f, as is called for in the RTTF technique, and ignoring radioactive decay, their solution for the normalized concentration C / C_0 at the outlet of the fracture is

$$\frac{C}{C_0} = erfc\left[\frac{\phi \tau_f \sqrt{R_{f,m} D}}{2b\sqrt{t - R_f \tau_f}}\right], \tag{6}$$

for $t > R_f \tau_f$, and $C / C_0 = 0$ for $t \leq R_f \tau_f$. The semi-infinite boundary condition between fractures limits the validity of either of these solutions to situations in which the characteristic diffusion distance for the transport problem is small compared to the fracture spacing B. As long as the solute has insufficient time to diffuse to the centerline between fractures, they solution provided by *Tang et al.* [1981] can be used as the transfer function for the particle-tracking technique. Alternatively, the analytical solution provided by *Sudicky and Frind* [1982] can be used to obtain the transfer function to handle finite fracture spacings; however, this model was not employed in the present study. To use Equation 6 as a transfer function, a numerical algorithm was developed to determine the inverse of the error function, that is, the value of x_d for a given value of y_d such that $y_d = erf(x_d)$. [Note that $erfc(x_d) = 1 - erf(x_d)$.] The numerical implementation of this method entails dividing the error function into piecewise continuous segments from which the value of x_d is determined by interpolation.

Several test cases have been developed to verify that the particle-tracking technique agrees with available analytical and numerical solutions. Most relevant to the modeling used in the present study are simulations of fracture transport with sorption and matrix diffusion. Figure 3 compares the results of a 1-D, particle-tracking solution for fracture flow, longitudinal dispersion, and matrix diffusion with the analytical solution of *Tang et al.* [1981]. The individual simulations consist of two different matrix diffusion coefficients—a case with sorption on the rock matrix, and a case with sorption both on the fracture surface and into the rock matrix. The close agreement of the particle-tracking results with the analytical solution illustrates that the RTTF technique provides an accurate transport solution for transport processes examined in the present study.

In the calculations that follow, a hybrid transport model is used in which a dual-permeability flow model captures flow in each continuum, with a fracture/matrix flow term to couple the continua. Particle transport occurs via advection within and between the fracture and matrix continua. In addition, for particles traveling in the fracture continuum, the matrix diffusion transfer function is used to approximate the diffusive interaction with the matrix. This set of assumptions is most appropriate for systems in which isolated fracture flow paths carry the water within the fractured, welded units such as the Topopah Spring unit. Then, in rock units with matrix-dominated flow, the fractures play no role, and the method reduces to a single-continuum transport model.

This particle-tracking method is used in the transport calculations that follow. Additional details about the numerical implementation of the particle-tracking method in the Finite Element Heat and Mass (FEHM) transfer computer code can be found in *Zyvoloski et al.* [1997].

3. FLUID FLOW AND SIMPLE TRANSPORT AT WELL SD-9

Although a full 3-D calculation is necessary to capture the complexity of the Yucca Mountain unsaturated zone, it is instructive to examine flow and transport mechanisms using a simplified model representative of the specific processes likely to be important. In this and the next section, we present 1-D dual-permeability flow and transport simulations that capture the range of behavior likely to be present in the unsaturated zone. All models extend from the ground surface to the water table, with several grid points within each hydrologic unit. Steady-state fluid-flow calculations are first performed at a fixed infiltration rate, followed by particle-tracking transport calculations to simulate radionuclide release at the potential repository horizon.

We begin with simulations for a 1-D model for an area around well SD-9, a test well at Yucca Mountain for which extensive characterization has occurred. This computation serves as a test bed for understanding the flow and transport mechanisms and the subsequent radionuclide transport simulations that follow. Hydrologic properties from *Bodvarsson et al.* [1997] are used for the various hydrostratigraphic units, and an infiltration rate of 3.6 mm/yr is applied as the surface condition [*Flint et al.*, 1996]. Agreement between the predicted and measured fluid saturation values (Figure 4, left) is excellent. The accompanying plot (Figure 4, right) of relative flow rates in the fractures and matrix illustrates the contrast in flow regime depending on the hydrostratigraphic unit. Fracture-dominated flow in the TSw transitions abruptly to matrix flow in the vitric Calico Hills unit (CHnv), a 10 m thick section at this location, followed by a more uniform distribution in the CHnz, a zeolite-bearing section of the Calico Hills unit.

The influence of this flow behavior on transport can be seen in Figure 5, which shows the predicted travel-time

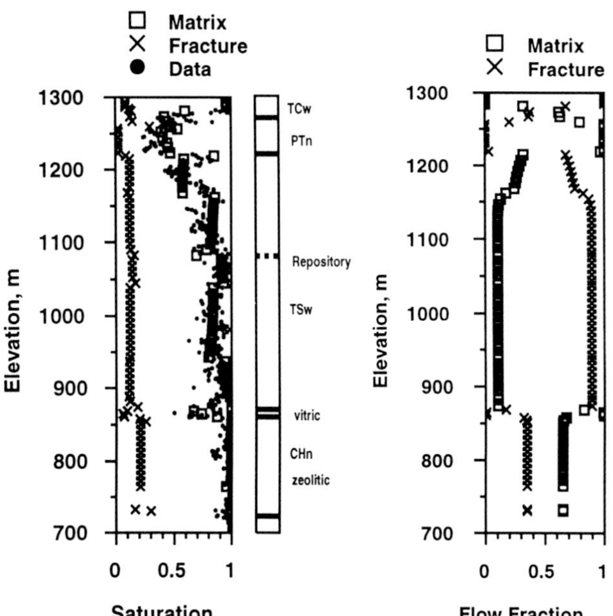

Figure 4. Flow and transport simulations at Well SD-9. Left: measured and predicted fluid saturation values; right: relative flow rates in the fractures and matrix; center: stratigraphy at SD-9.

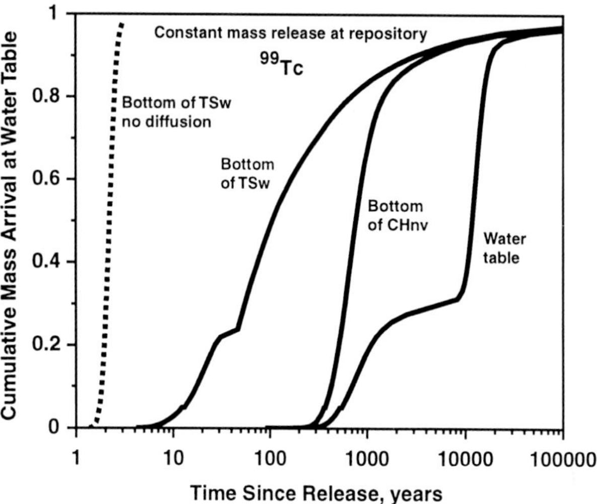

Figure 5. Predicted breakthrough curves of ^{99}Tc (a conservative species) from the potential repository to various stratigraphic units and the water table.

distributions of particles at well SD-9 from the potential repository to the bottoms of various hydrostratigraphic units and the water table. Transport to the bottom of the TSw occurs rapidly due to fracture flow. Travel times of less than 10 yr are predicted without matrix diffusion. With matrix diffusion, travel times are somewhat larger (of an order of 10 to 100 yr), but still short compared to the requirements of a repository to isolate waste for thousands of years. The CHnv, despite its thinness in this model calculation, contributes hundreds of years to the travel time because of the complete transition to matrix flow. Clearly, this unit is of critical importance to the unsaturated zone's ability to provide a barrier to radionuclide migration. For this reason, the field test presented later is sited in the CHnv unit. Finally, the CHnz unit has mixed fracture and matrix flow in this model, resulting in much longer travel times for a portion of the released radionuclides to the water table, while at the same time allowing a fraction of the mass to travel rapidly through fractures in this unit.

4. RADIONUCLIDE TRANSPORT CALCULATIONS

As the SD-9 test case shows, the hydrologic properties of the various tuff layers play a key role in controlling flow behavior and transport times. Therefore, examination of the range of possible unsaturated-zone transport conditions to be encountered by radionuclides released from the potential repository requires that variability in hydrologic and transport parameters across the region of interest be considered. In addition to flow properties, sorption is a key controlling process influencing travel time predictions within the un-

Table 1. Sorption coefficient K_d(cm^3) of radionuclides on the Yucca Mountain Tuffs [from *Triay et al.*, 1997].

Element	Devitrified	Vitric	Zeolitic	Notes
Non-Sorbing (Kd=0)				
Carbon	0	0	0	
Chloride	0	0	0	
Iodine	0	0	0	
Technetium	0	0	0	
Moderately Sorbing (1<Kd<10)				
Neptunium	1	1	4	DOE, 1998
Uranium	2	1	7	DOE, 1998
Selenium	3	3	2	
Strontium	10	0	500	Minimum Kd approach
Strongly Sorbing (Kd>20)				
Actinium	100	100	100	Expert judgement
Americium	100	100	100	Minimum *Kd* approach
Cesium	20	10	500	Minimum *Kd* approach
Lead	100	100	100	Minimum *Kd* approach
Nickel	100	50	100	
Niobium	100	100	100	Expert judgement
Plutonium	100	100	100	
Protactinium	50	50	50	DOE, 1998: uniform dist., 0 to 100
Radium	100	100	1000	Minimum *Kd* approach
Samarium	100	100	100	Expert judgement
Tin	20	20	100	Minimum *Kd* approach
Thorium	100	100	100	Expert judgement
Zirconium	100	100	100	Expert judgement

saturated zone. In this section, we examine the radionuclide transport behavior using a suite of 1-D, dual-permeability simulations at various locations within the potential repository. To set the stage for this modeling, we first discuss the data used to assign flow and transport parameters for the models.

4.1. Mineralogic Data

To perform radionuclide transport calculations that cover the range of possible behavior based on spatial variability in geologic and mineralogic properties, we selected two 1-D columns through the geologic and mineralogic models constructed for Yucca Mountain [*Chipera et al.*, 1997]. Hydrostratigraphic picks at locations near the northern and southern edges of the potential repository (called the "northern" and "southern" repository flow paths) were used to construct the 1-D columns. In the north, the Calico Hills unit is highly altered to zeolites, and little vitric tuff is present beneath the potential repository. More vitric tuff is encountered in the middle and southern portions of the repository.

As in the SD-9 calculations presented earlier, hydrologic properties from the calibrated flow model of *Bodvarsson et al.* [1997] were used. Site data suggest that there is a strong correlation between percent zeolitic alteration and permeability [e.g., *Altman et al.*, 1996]. *Robinson et al.* [1997] used a correlation that is consistent with the data, and based hydraulic conductivity directly on geostatistically rendered distributions of percent alteration. Unfortunately, this correlation has been determined through examination of unaltered rocks and rocks with zeolitic alteration of 60% and more, with no data at intermediate values. In the present study, we opt for a simpler technique of applying a cutoff alteration, above which the rocks are given the hydrologic properties of zeolitic tuffs. It is difficult to define a cutoff alteration fraction to capture what is undoubtedly a continuous process. Based on the more complex model of *Robinson et al.* [1997], a value of 30% alteration was chosen for these calculations. This aspect of the modeling should be considered tentative until a more complete data set becomes available. Using the 30% value, the following cumulative thicknesses of vitric and zeolitic sections within the Calico Hills tuffs were obtained for the 1-D models:

Northern:	Vitric: 0 m;	Zeolitic: 137 m
Southern:	Vitric: 74 m;	Zeolitic: 22 m

4.2. Transport Parameters

Transport parameters for Yucca Mountain tuffs have been studied extensively, and a large database exists for distribution coefficients K_d and diffusion coefficients for representative fluid chemical compositions. Table 1, from *Triay et al.* [1997], summarizes the K_d data for various radionuclides on the devitrified, vitric, and zeolitic tuffs from Yucca Mountain. Most values are based on laboratory batch sorption measurements; others are estimated based on analogies to other radionuclides of similar chemical characteristics in the Yucca Mountain fluids (designated as "expert judgement" in the table). To categorize the radionuclides with respect to sorption, we segregate the radionuclides into three groups: nonsorbing ($K_d = 0$ cm^3/g), moderately sorbing ($K_d = 1$ to 10 cm^3/g), and strongly sorbing ($K_d > 20$ cm^3/g). For many of the strongly sorbing radionuclides, sorption coefficients were determined on the least reactive minerals to provide an acceptable lower bound on K_d, which is labeled "minimum K_d approach" [after *Meijer*, 1992].

There is strong evidence, both at the laboratory scale [*Triay et al.*, 1997] and the field scale [e.g., *Reimus and Turin*, 1997], that matrix diffusion (the process of radionuclide transport from the fractures to the relatively stagnant fluid in the matrix) is likely to occur. Measured diffusion coefficients [from *Triay et al.*, 1997] in the various tuff types in rock beakers are the basis for the range of values used in these simulations, which account for matrix diffusion using the particle-tracking techniques described earlier.

4.3. Transport Model Results

In this section, we present 1-D, dual-permeability transport simulations from the two repository locations and the three radionuclide sorption categories. Results are plotted in the form of cumulative travel time breakthrough at the water table versus time. These breakthrough curves represent the response to a constant source of unit concentration released into fractures at the potential repository horizon. Through Figures 6a and 6b, we show the influence of sorption coefficient and release location on travel times, selecting a value of matrix diffusion coefficient of 10^{-10} m^2/s for the comparison. Simulations of nonsorbing processes were performed with $K_d = 0$ cm^3/g, moderately sorbing $K_d = 3$ cm^3/g, and strongly sorbing $K_d = 20$ cm^3/g. The southern repository location yields much longer travel times because of matrix-dominated flow and transport in relatively thick vitric and devitrified tuff sequences. The northern location exhibits fracture flow and transport from the repository to the water table. Travel times are correspondingly short. Sorption strongly retards radionuclide migration for both release locations.

In Figures 6a and 6b, we used a single value of matrix diffusion coefficient, chosen to be consistent with laboratory data [*Triay et al.*, 1997]. Since the system consists of alternating sequences of tuffs that exhibit fracture-dominated and matrix-dominated transport, it is impossible, from the data presented in Figures 6a and 6b alone, to separate the relative importance of matrix diffusion in fractured tuffs from matrix flow and transport in units such as the vitric Calico Hills unit. To clarify the interpretation, similar calculations were performed for the strongly sorbing radionuclides with the diffusion coefficient varied from 0 to 10^{-10} m^2/s (Figures 6c and 6d). The influence of larger diffusion coefficients is to increase travel times significantly for strongly sorbing radionuclides (Figure 6c). For example, even for the northern location, which exhibits fracture-dominated flow, matrix diffusion in the tuff dramatically increases radionuclide travel times. However, without matrix diffusion, extremely short travel times are predicted because of the persistent fracture transport from the repository to the water table.

5. UNSATURATED ZONE TRANSPORT TEST PRELIMINARY FIELD OBSERVATIONS

To summarize the modeling results just presented, there are two processes of unsaturated-zone flow and transport that contribute to most of the travel time delay for radionuclides that may escape the engineered barrier system:

1. Transition to matrix-dominated flow in nonwelded to partially welded tuffs such as the CHnv yields predicted travel times on the order of hundreds to thousands of years for nonsorbing radionuclides, and travel times in excess of 10,000 years for more strongly sorbing radionuclides. This behavior is likely in the southern and central portions of the potential repository, but may not occur in the north if fracture-dominated flow occurs from the repository to the water table, as this modeling suggests.
2. Matrix diffusion retards transport even if fracture flow is dominant. This was shown to be particularly relevant for strongly sorbing radionuclides.

It is critical to base models on field tests designed to demonstrate the validity of the conceptual models and laboratory-derived parameters used in site-scale models.

In this section, we report preliminary results of a field test designed to confirm important elements of the radionuclide transport model results presented. It is probably not possible to select a single site to test all processes important to transport. Therefore, the approach we used in this testing

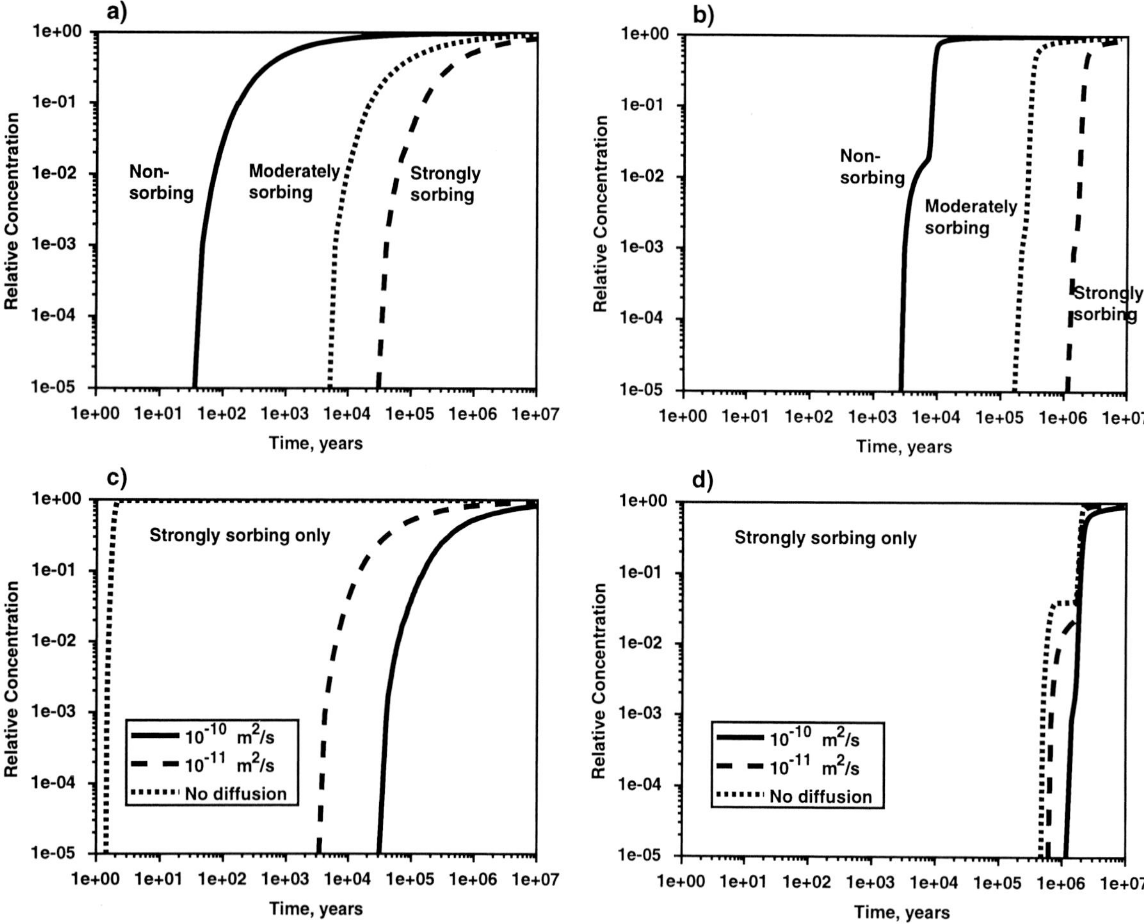

Figure 6. Predicted breakthrough curves at the water table for various radionuclide transport parameters. Infiltration rate = 5 mm/yr. (a and b) Influence of sorption coefficient. (c and d) Influence of matrix diffusion coefficient for a strongly sorbing radionuclide. (a) northern release location, (b) southern release location, (c) northern release location, (d) southern release location.

program was to select a site with access to the vitric Calico Hills tuff to test the conceptual model of matrix-dominated flow and transport in this unit. The importance of the transition to matrix-dominated flow is specifically being studied in the test. This test, called the Unsaturated Zone Transport Test (UZTT) at Busted Butte, is located in Area 25 of the Nevada Test Site (NTS), Nevada, 8 km southeast of the potential Yucca Mountain repository area. The site was chosen because of the presence of a readily accessible exposure of unsaturated rocks of the Topopah Spring/Calico Hills formations, stratigraphic units beneath the potential repository that are accessible at Busted Butte.

The principal objectives of the test are to evaluate the fundamental processes and uncertainties associated with flow and transport in the unsaturated zone site-scale models for Yucca Mountain, including the following:

1. The effect of heterogeneities on flow and transport under unsaturated conditions in the Calico Hills. In particular, the test aims to address issues relevant to the possible role of fractures in a tuff with a high matrix permeability, the nature of fracture/matrix interactions, and the role of permeability contrast boundaries.
2. The evaluation of the 3-D site-scale flow and transport process model (i.e., equivalent-continuum/dual-permeability/discrete-fracture-fault representations of flow and transport) used in performance assessment abstractions.
3. The validation, through field testing, of laboratory sorption experiments in unsaturated Calico Hills rocks.
4. The effect of scaling from laboratory scale to field and site scales.
5. The migration behavior of colloids in fractured and unfractured Calico Hills rocks.

Based on the needs and the time restrictions of the Yucca Mountain Project, an accelerated experimental design was

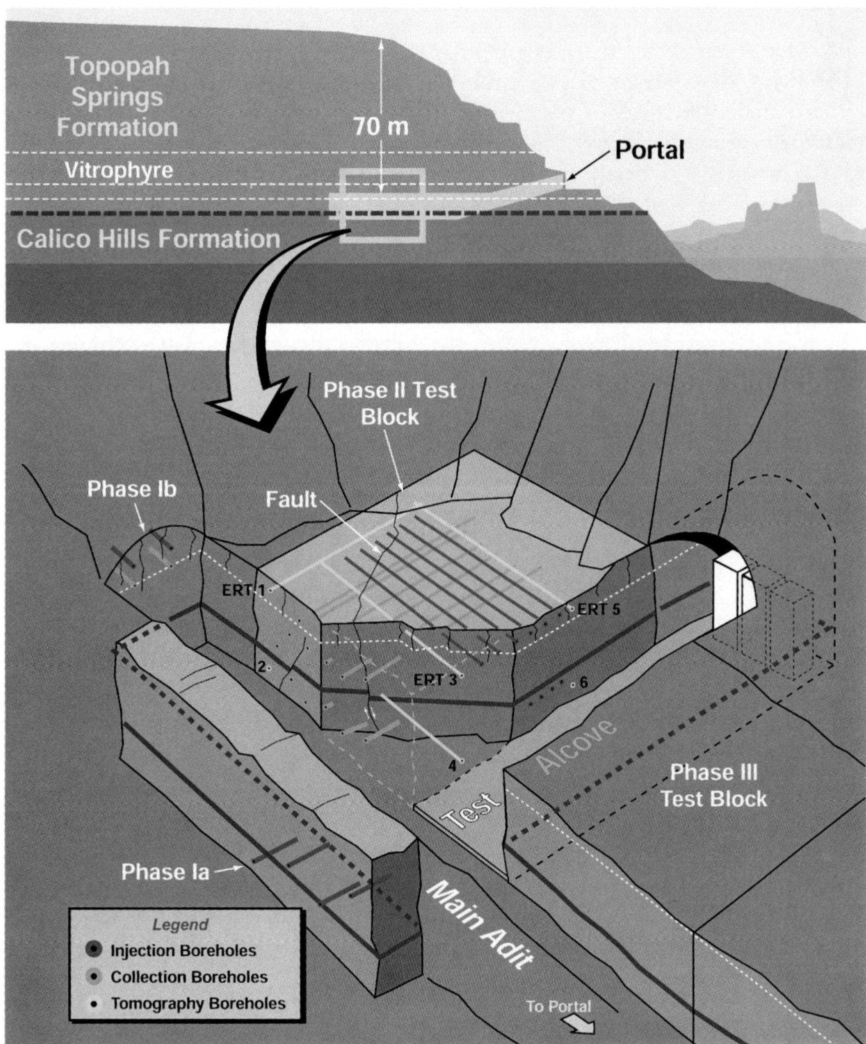

Figure 7. Schematic of the unsaturated zone transport test (UZTT) at Busted Butte, showing the locations of Phases 1A, 1B, and 2.

put in place that would maximize our ability to accomplish these objectives. In the present study, we focus on Objective 1 because initial test observations most directly address this issue. Examination of other test objectives awaits the collection and analysis of more data from the UZTT.

The underground experimental facility was sited, designed, and built from August 1997 to January 1998. A 3m x 5m portal located on the south, lower flank of Busted Butte allows access to a main adit (~70 m deep) and a test alcove perpendicular to it (~20 m deep). The underground test area consists of several areas in the exposed walls of the tunnel system and includes a large "block" 7 m high, 10 m deep, and 10 m wide, two sides of which are exposed underground (Figure 7). Prior to testing, numerous rock and pore water samples were collected and analyzed in preparation for the selection of hydrochemically suitable tracers and their injection into the tuffs. The large test "block" was itself located with respect to a normal fault offsetting at least 6 nearly horizontal (±6°) Miocene pyroclastic layers. The test block exposes the uppermost Calico Hills Formation, while the upper parts expose the basal two units (Tptpv1, Tptpv2) of the Topopah Spring Formation, which exhibits various degrees of fracturing associated with columnar jointing. Differences of approximately six orders of magnitude in the hydraulic conductivity of different layers (i.e., k = 10^{-6} to 10^{-12} m/s) are observed at the site. The UZTT comprises two main phases [*Bussod et al.*, 1999], briefly described below.

Phase 1 consisted of relatively simple, short-duration experiments intended to provide logistic and technical "scoping" for the design of the large block tests (see below). Six injection boreholes and two inverted-membrane collection boreholes, all 2 m long and 10 cm in diameter, were used to continuously inject a mixture of conservative tracers (bromide, fluorescein, pyridone, and fluorinated benzoic acids), a reactive tracer (lithium), and fluorescent polystyrene microspheres that tracked flow, reactive transport, and colloid migration, respectively. Phase 1A, initiated in the Calico Hills Formation, was a noninstrumented test consisting of four single-point injection boreholes that were excavated after injection in successive stages by mineback and auger sampling and photographed using strobe and UV lighting. Continuous injection at rates of 1 mL/hr and 10 mL/hr began on April 2, 1998, and was terminated on January 12, 1999. The Phase 1B experiments, in the Topopah Spring Formation (unit Tptpv2), consisted of two single-point injection boreholes and two collection boreholes, located directly below the injection boreholes. Each collection borehole contained an inverted membrane with a minimum of nine equally spaced collection pads. This test, which was located in relatively low permeability, fractured Tptpv2, was conceived to provide data on fracture-matrix interactions and to aid in the development of pretest predictions of Phase 2 transport. Continuous injection of 1 mL/hr and 10 mL/hr began on May 12, 1998, and was terminated on November 9, 1998.

Phase 2 testing involved a large 7 m high, 10 m wide, 10 m deep block with injection and collection in both Calico Hills (Tac) and Topopah Spring (Tptpv1,2) formations. Unlike the single-point injections in Phase 1, Phase 2 was designed to activate a large volume of the block to look at the effect of field-scale spatial heterogeneities (e.g., fractures, faults, and layering). A network of injection boreholes for this phase was distributed in two horizontal, subparallel planes that were drilled from the alcove (see Figure 7). Injection solutions contained the tracers from Phase 1, plus additional sorbing tracers (i.e., Ni^{2+}, Co^{2+}, Mn^{2+}, Sm^{3+}, Ce^{3+}, Rhodamine WT, and iodide), believed to represent a range of sorption behaviors expected of radionuclides present in the waste that would be disposed of at Yucca Mountain. Injection rates of 1, 10, and 50 mL/hr in various parts of the block were used to mimic fluid flux values ranging from present-day conditions to infiltration rates larger than would be expected even under wetter climates. Twelve collection boreholes, drilled from the main adit, are perpendicular to the injection boreholes. Each is 8.5–10 m long and contains a minimum of 15–20 collection pads evenly distributed on inverted membranes. In addition to direct sampling, several tomographic techniques (Electrical Resistance Tomography, Ground Penetrating Radar) and neutron logging were used to image the saturation of the block.

Both phases of the test have been carried out with a strategy of iterative numerical model predictions and subsequent comparison and updating of the model. In this study, we highlight initial experimental observations from the Phase 1 experiments, documenting important observations relevant to the flow and transport mechanisms in the Calico Hills unit.

After approximately nine months of continuous injection, direct observations of the Phase 1 test results were made using overcoring and successive mineback techniques. During the mineback, the extent of migration of fluorescein dye from the injection points was determined visually using digital photographs of the samples and mineback walls using UV illumination (Plate 1). Several key observations can be made from the Phase 1 tests:

- *Strong capillary forces exist that draw fluid upward as well as downward from the injection points within the vitric Calico Hills unit.* Plate 1, which represents a mineback face, 90 cm into the tunnel wall and illuminated by UV light, is an excellent illustration of the strong capillary-driven flow in this rock. Injection points for boreholes BH.2 and BH.3 are located at a 3 o'clock position. The borehole on the right (Plate 1, BH.3) represents a high-injection-rate experiment (10 mL/hr). The borehole on the left (Plate 1, BH.2) represents a lower injection-rate experiment (1 mL/hr). Both are representative of the migration of fluorescein dye over approximately eight months of continuous injection.
- *Lithologic boundaries, whether clay rich or silica rich, play a role in controlling either upward or downward flow, mostly by impeding it.* The buildup of dye at the Tptpv1/Tac boundary is clearly evident in Plate 1. In addition to impeding the downward migration of water and tracer from the high-injection borehole (BH.3), upward migration from the low-injection borehole was also arrested by this boundary (BH.2). In a transient flow scenario such as this injection test, the interface dampens the downward migration of the moisture. In a longer-term situation, such as water movement under natural conditions, we suggest that such interfaces would dampen pulses of infiltrating water and would probably spread channels of high flow over larger horizontal distances, thereby more uniformly distributing the flow. Transitions in flow regime might be expected to be strongly influenced by such boundaries, especially if fractures terminated there rather than extend through the interface.
- *Fractures within the vitric Calico Hills unit do not greatly enhance flow and are not behaving as classical "fast pathways."* Evidence for this result was obtained in both Phases 1A and 1B. In Plate 1 (Phase 1A), a high-angle fracture mapped during the mineback has been traced as a dashed line. Very slight distortions to

the otherwise uniform distribution of moisture and tracer can be seen at the edges of the plume, coincident with the fracture. However, there is very little evidence of this fracture serving as a preferential flow channel at this location. Rather, the distortions are probably the result of the differences in hydrologic properties in and immediately adjacent to the fracture.

In Phase 1B, more quantitative tracer breakthrough results were obtained from the collection boreholes. Figure 8a is a schematic of the experimental layout, in which water was injected at a vertical fracture intersecting the borehole and collection ports were situated in a second borehole drilled 28 cm below the injection borehole in the same vertical plane. Collection pads were placed at the fracture-borehole interface and at several locations away from the fracture within the collection borehole to sample the rock matrix. The tracer breakthrough results of Figure 8b indicate that, although the earliest breakthrough occurred at the location where the fracture intersected the borehole (130 cm from the collar), the first-arrival time was between 28 and 35 days, far longer than would be expected if flow had channeled only through the fracture. Furthermore, tracer subsequently broke through to the pads within the matrix, suggesting that within just 28 cm of the injection point, imbibition had drawn water from the fracture to the surrounding matrix. This interpretation was verified during overcoring operations of the Phase 1B experiments, where fracture-matrix interactions were clearly visible as a fluorescein plume on either side of the fracture walls.

This observation from the Tptpv2 unit, when combined with the Phase 1A observation after mineback, points to a conceptual model in which percolation of fluid through the rock matrix is the most likely scenario. Of course, in regions of extensive faulting, rock hydrologic properties may be significantly different from rock accessed at the UZTT site. Nevertheless, the observations lend credence to models of the unsaturated zone, such as those presented in this study, in which water (and released radionuclides) migrate relatively slowly through the rock, and indicate that fracture flow, the most severe type of preferential flow, does not appear to be important. Although data are not yet available to assess preferential flow in the rock matrix, initial Phase 1 observations obtained from the mineback operation suggest that the flow is relatively uniform. These flow conditions also would be favorable for sorption, a process that requires intimate contact of the radionuclide with the rock.

6. DISCUSSION AND CONCLUSIONS

A dual-permeability model of flow and transport in the Yucca Mountain unsaturated zone is presented in this study, along with illustrative 1-D model calculations to explore the transport mechanisms and estimate travel times of conservative and sorbing radionuclides. The rationale behind adopting the dual-permeability approach is to capture the disparate behavior of fluids and radionuclides, which is dependent on whether transport occurs in the fractures or matrix. Dual-permeability flow and transport models for the unsaturated zone allow us to simulate the stratigraphically controlled shifting of flow and transport from fracture dominated to matrix dominated. Along with the flow model results, we present the RTTF particle-tracking algorithm developed for transport in dual-permeability, unsaturated media. This method allows processes such as matrix diffusion to be predicted accurately using an analytical solution, while still retaining the relative simplicity of the dual-permeability formulation. This approach appears to circumvent the numerical problem of the inability to capture sharp concentration gradients into the matrix with a single grid block.

Using the model, we have demonstrated that the nonwelded, vitric Calico Hills unit should be an important barrier for many radionuclides because of the transition from fracture flow to matrix flow. This result is contingent upon the overall validity of 1-D simulations from the potential repository to the water table. Large-scale flow processes such as fault-controlled flow and transport could result in bypassing of the vitric Calico Hills unit. However, field evidence that this occurs on a large scale, such that radionuclides distributed over a several km^2 area would predominantly travel laterally to fault zones, is scarce. Strongly sorbing radionuclides are predicted to be retarded in this unit for longer than 10,000 years under current model assumptions. The field evidence is ambiguous with respect to whether certain regions beneath the potential repository will exhibit fracture flow from the repository to the water table. In the present study, modeling suggests that this could occur in the northern region, where extensive zeolitic alteration could give rise to low matrix permeability along the flow path, which, when coupled with high enough percolation rates, results in fracture flow through the Calico Hills tuffs. However, even if this occurs, matrix diffusion combined with sorption is predicted to result in significant radionuclide retardation. Therefore, we conclude that the most strongly sorbing radionuclides will be retarded for long times in the unsaturated zone, unless colloid-facilitated transport is shown to be an important mechanism for transmitting radionuclides quickly to the water table.

For weakly sorbing or conservative radionuclides such as ^{99}Tc, even the matrix-dominated transport scenario predicts arrival at the water table in less than 10,000 years (Figures 6a and 6b). Moderately sorbing radionuclides such as ^{237}Np

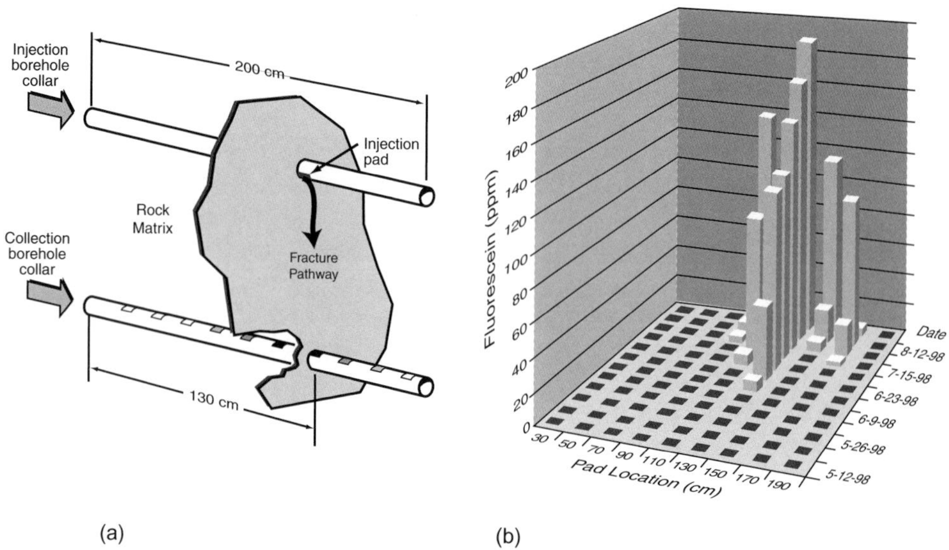

Figure 8. Phase 1B tracer experiment. (a) Schematic of borehole injection and collection system with fracture location. Boreholes are subparallel, 28 cm apart in the same vertical plane. Fracture surface is at 60° to the borehole axis and 130 cm from the borehole collars. (b) Fluorescein concentration versus time along the collection borehole. Sampling pad locations are marked as squares.

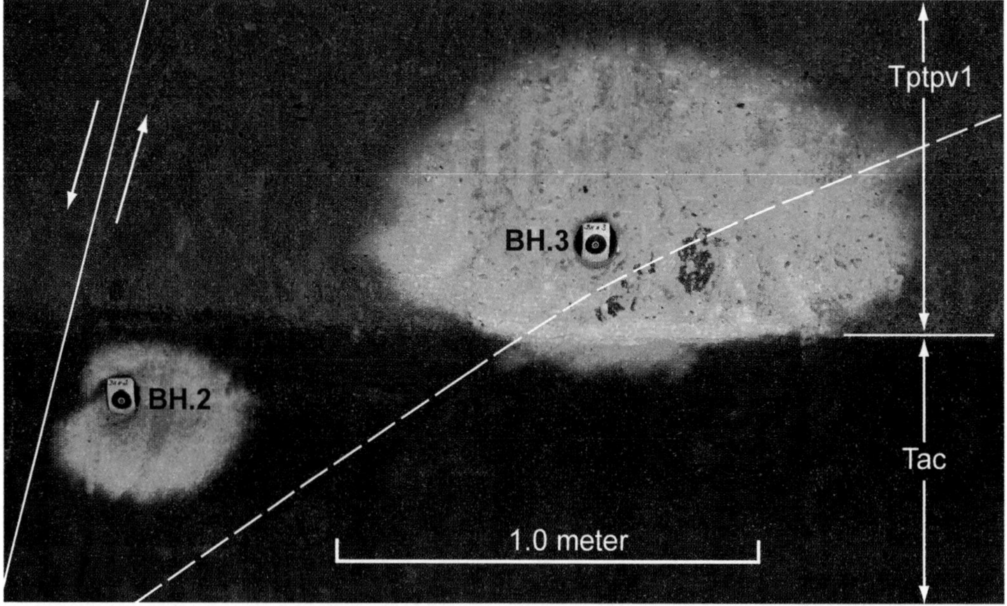

Plate 1. Digital photo of Phase 1A mineback face, 90 cm into the wall of the left rib of the main tunnel adit. The wall is illuminated by a UV light source. Visible are the traces representing the distance migrated by the fluorescein dye injected in boreholes BH.2 (left) and BH.3 (right). Mapped traces of a high-angle fracture (dashed line) and a normal fault with a 10 cm offset (solid line) are also represented. Both exhibit opal mineralization along their surfaces. The Tac-Tptpv1 contact is characterized by a 5–10 cm silicified zone of lower permeability than the rock above or below.

fall between, with good unsaturated zone barrier performance predicted in the southern region and possible travel times of between 1,000 and 10,000 years in the north. In summary, the modeling suggests that the unsaturated zone should be an important, although not complete, barrier to transport for a large portion of the inventory of radionuclide inventory.

The Unsaturated Zone Transport Test (UZTT) at Busted Butte is designed to test several important elements of the conceptual and numerical models used for performing transport predictions. Preliminary observations from the UZTT support the model findings for radionuclide transport in several important respects. Phase 1 observations demonstrate that capillary forces in the vitric Calico Hills tuff are strong and fracture-matrix interactions of fluid are considerable. Initial data from this test, although mostly qualitative, point to the likelihood of matrix-dominated flow and transport in the vitric Calico Hills tuff. Fractures within the region of injection, as in Phase 1A, or that actually intersect the injection borehole, as in Phase 1B, do not seem to behave as fast pathways. This behavior runs counter to observations of unsaturated fracture flow in some systems, including the welded, fractured tuff at Yucca Mountain. A possible explanation for the observation of matrix-dominated flow of the Calico Hills tuff is its large matrix permeability, which allows water to percolate readily through it. Furthermore, water residing in fractures tends to imbibe into the matrix under unsaturated conditions, thereby resulting in matrix flow. Further confirmation of this conclusion, along with quantitative comparisons of data and numerical models, is the subject of ongoing work.

Many of the modeling predictions point to the interplay of fluid flow and sorption as the determining factor in whether a radionuclide reaches the water table in less than 10,000 years. Whereas conservative radionuclides are predicted to reach the water table within a few thousand years, even with matrix flow in the Calico Hills, sorbing radionuclides are retarded for much longer times based on model calculations. Therefore, it is crucial to examine the model assumptions for sorbing solutes. Theoretical and laboratory studies have established a strong basis for sorption of many radionuclides, but an unresolved issue is the applicability of these results for large-scale transport under unsaturated conditions. The sorbing tracer tests in Phase 2 of the UZTT should provide important data for evaluating model predictions.

Acknowledgments. This work was supported by the Yucca Mountain Site Characterization Program Office as part of the Civilian Radioactive Waste Management Program. This project is managed by the U.S. Department of Energy, Yucca Mountain Site Characterization Project. The authors wish to thank Julie Canepa for her unwavering support during the course of design and construction of the UZTT. We would also like to thank the many colleagues who participated in the modeling effort and experimental program. Finally, special thanks go to Bill Carey for providing data and interpretations of the mineralogic model of Yucca Mountain, and several anonymous reviewers for their useful critiques of this manuscript.

REFERENCES

Altman, S. J., B. W. Arnold, C. K. Ho, S. A. McKenna, R. W. Barnard, G. E. Barr, R. E. Eaton, Flow calculations for yucca mountain groundwater travel time (GWTT-95), SNL Technical Report SAND96-0810, Sandia National Laboratories, Albuquerque, N.M., 1996.

Bodvarsson, G. S., T. M. Bandurraga, Y. S. Wu, *The Site-Scale Unsaturated Model of Yucca Mountain, Nevada, for the Viability Assessment*, Report LBNL-40378, Lawrence Berkeley National Laboratory, Berkeley, Calif., 1997.

Bussod, G. Y., H. J. Turin, and W. E. Lowry, *Busted Butte Unsaturated Zone Transport Test: Fiscal Year 1998 Status Report- Yucca Mountain Site Characterization Project*, Los Alamos Technical Report, LA-13670-SR, 300 pp., Los Alamos National Laboratory, Los Alamos, N.M., 1999.

Chipera, S. J., K. Carter-Krogh, D. L. Vaniman, D. L. Bish, J. W. Carey, Preliminary three-dimensional mineralogical model of yucca mountain, Nevada, Yucca Mountain Site Characterization Project Milestone SP321AM4, 1997.

Desbarats, A. J., Macrodispersion in Sand-Shale Sequences, *Water Resour. Res., 26*(1), 153–163, 1990.

DOE (U.S. Department of Energy), *Total System Performance Assessment. Volume 3 of Viability Assessment of a Repository at Yucca Mountain.* DOE/RW-0508, U.S. Government Printing Office, Washington, D. C., 1998.

Flint, A. L., J. A. Hevesi, L. E, Flint, *Conceptual and Numerical Model of Infiltration for the Yucca Mountain Area, Nevada*, U. S. Geological Survey Water Resources Investigation Report, MOL19970409. 0087, U.S. Geological Survey, Denver Colo., 1996.

Kinzelbach, W., The random walk method in pollutant transport simulation, *Groundwater Flow and Quality Modelling*, pp. 227–245. D. Reidel Publishing Company, Norwell, Mass., 1988.

Meijer, A., A strategy for the derivation and use of sorption coefficients in performance assessment calculations for the Yucca Mountain site, LANL Report LA-12325-C, Los Alamos National Laboratory, Los Alamos, N.M., 1992.

Reimus, P. W., H. J, Turin, Results, *Analyses, and Interpretation of Reactive Tracer Tests in the Lower Bullfrog Tuff at the C-wells, Yucca Mountain, Nevada*, Yucca Mountain Site Characterization Project Milestone SP2370M4, 1997.

Robinson, B. A., A. V. Wolfsberg, H. S. Viswanathan, G. Y. Bussod, C. W. Gable, A. Meijer, *The Site-Scale Unsaturated*

Zone Transport Model of Yucca Mountain, Yucca Mountain Site Characterization Project Milestone SP25BM3, 1997.

Starr, R. C., R. W. Gillham, and E. A. Sudicky, Experimental Investigation of Solute Transport in Stratified Porous Media, 2. The Reactive Case, *Water Resour. Res., 21*(7), 1043–1050, 1985.

Sudicky, E. A., and E. O Frind, Contaminant transport in fractured porous media: analytical solutions for a system of parallel fractures, *Water Resour. Res., 18*(6), 1634–1642, 1982.

Tang, D. H., E. O. Frind, and E. A. Sudicky, Contaminant transport in fractured porous media: analytical solution for a single fracture, *Water Resour. Res., 17*(3), 555–564, 1981.

Tompson, A. F. B., and L. W. Gelhar, Numerical simulation of solute transport in three-dimensional, randomly heterogeneous porous media. *Water Resour. Res., 26*(10), 2541–2562, 1990.

Triay, I. R (ed.), *Summary and Synthesis Report on Radionuclide Migration for the Yucca Mountain Site Characterization Project*, LANL Report LA-13262-MS, Los Alamos National Laboratory, Los Alamos, N.M., 1997.

Zyvoloski, G. A., B. A. Robinson, Z. V. Dash, and L. L. Trease, *Summary of the Models and Methods for the FEHM application—A Finite-Element Heat- and Mass-Transfer Code*, LANL Report LA-13307-MS, Los Alamos National Laboratory, Los Alamos, N.M., 1997.

Bruce A. Robinson and Gilles Y. Bussod, Earth and Environmental Sciences Division, Los Alamos National Laboratory, Los Alamos, New Mexico 87545.

Using Environmental Tracers to Constrain Flow Parameters in Fractured Rock Aquifers; Clare Valley, South Australia

Peter G. Cook

CSIRO Land and Water, Glen Osmond, South Australia

Craig T. Simmons

Flinders University of South Australia, Bedford Park, South Australia

In fractured rock aquifers, apparent groundwater ages obtained with environmental tracers (e.g., ^{14}C, CFC-12, and ^{3}H) usually do not represent the hydraulic age of the water. Diffusion of solute between the fractures and matrix results in apparent ages that are greater than hydraulic ages, and that may be different for different tracers. We use approximate analytical solutions and numerical simulations of tracer transport through fractured porous media to illustrate the dependence of ^{14}C and CFC-12 ages and ^{3}H concentrations on fracture and matrix properties. In the Clare Valley, South Australia, environmental tracer data are interpreted in conjunction with hydraulic data to constrain flow parameters in a fractured shale aquifer. Hydraulic conductivity, matrix porosity, fracture spacing, and groundwater age are measured, and a value for matrix diffusion coefficient is assumed. Equations describing tracer distribution and hydraulic properties of the system are solved simultaneously, to yield estimates of fracture aperture, vertical water velocity, and aquifer recharge rate. In particular, the recharge rate is estimated to be approximately 100 mm yr^{-1}. A sensitivity analysis showed that this value is most sensitive to the measured values of matrix porosity and groundwater age, and highly insensitive to the measured hydraulic conductivity and the assumed matrix diffusion coefficient. A major horizontal fracture at 37 m depth intercepts most of the vertical flow. The leakage rate to the deeper flow system is estimated to be less than 0.1 mm yr^{-1}.

1. INTRODUCTION

In simple porous-media aquifer systems, groundwater flow rates may be determined using Darcy's Law and estimates of head gradient and hydraulic conductivity. Alternatively, if the age of groundwater can be determined using environmental tracers [e.g., ^{14}C, ^{3}H (tritium), CFCs

Dynamics of Fluids in Fractured Rock
Geophysical Monograph 122

(chlorofluorocarbons)], then estimates of horizontal and vertical flow velocity can be directly determined [e.g., *Vogel*, 1967; *Robertson and Cherry*, 1989; *Cook et al.*, 1995]. In fractured porous media, neither method is so straightforward. Hydraulic conductivity is usually highly spatially variable, and so estimates of flow velocity derived from this are unreliable. Solute transport in these systems is characterized by rapid advection through fractures, with diffusive exchange between solute in the fractures and solute in the relatively immobile water in the matrix. The result is apparent groundwater ages obtained with environmental tracers that do not reflect the hydraulic age of the water [*Grisak and Pickens*, 1980].

Because of the great complexity of fractured porous media, the best approach may be to combine hydraulic and environmental tracer methods. However, while the hydraulics of groundwater flow has received considerable attention, environmental tracer methods have not been so extensively studied. Yet, because of the extreme spatial variability of hydraulic properties in fractured media, the inherent spatial averaging provided by environmental tracer methods should make them attractive, particularly when one is interested in determining average flow rates.

In the first section of this paper, a simplified model of groundwater flow and solute transport in fractured rocks is presented. Three of the commonly used environmental tracers (^{14}C, CFC-12, and ^{3}H) are briefly discussed, and their distribution within groundwater is related to fracture properties. In the second section of the paper, environmental tracer data are presented from the Clare Valley, South Australia. The distribution of the tracers within the aquifer is used in conjunction with available hydraulic data to estimate fracture parameters and average groundwater flow rates.

2. THEORY

2.1. Groundwater Flow through Fractured Rocks

Consider a system of planar, parallel, vertical fractures in a relatively impermeable rock matrix. The hydraulic conductivity of the medium can be expressed as

$$K = \frac{(2b)^3}{2B}\frac{\rho g}{12\mu} \tag{1}$$

where $2b$ is the fracture aperture, $2B$ is the fracture spacing, ρ is the density of water (1.00 g cm^{-3} at 20°C), g is acceleration due to gravity (9.8 m s^{-2}), and μ is the viscosity of water (1.00 mPa s at 20°C). The Darcy velocity, V_D, is then given by

$$V_D = K\frac{\delta i}{\delta z} \tag{2}$$

where $\delta i/\delta z$ is the hydraulic gradient. The Darcy velocity will be related to the water velocity within the fractures, V_w, by

$$V_D = V_w \theta_f = V_w b B^{-1} \tag{3}$$

where $\theta_f = b/B$ is the fracture porosity. In a vertical, one-dimensional flow field, the Darcy velocity is equal to the aquifer recharge rate, R.

2.2. Solute Transport, Matrix Diffusion and Equivalent Porous Media

Solute transport through this system is characterized by advection through the fractures, with diffusion into the immobile water in the matrix. Diffusion of solute from the fractures to the matrix (matrix diffusion) causes an apparent retardation of the solute tracer relative to the water. The result will be apparent solute velocities, V_a, that are less than water velocities in fractures ($V_a < V_w$) and apparent solute residence times, t_a, that are greater than water residence times, t_w ($t_a > t_w$).

If equilibration between fracture and matrix concentrations occurs, then the apparent age of the tracer, t_a, is related to the age of the water in the fractures, t_w, by the ratio of the total porosity, θ_t, to the fracture porosity, θ_f:

$$t_a = t_w \frac{\theta_t}{\theta_f} \tag{4}$$

The total porosity is equal to the fracture porosity plus the matrix porosity; $\theta_t = \theta_f + \theta_m$. (In practice, $\theta_m >> \theta_f$, and so θ_t and θ_m are often used interchangeably.) The tracer velocity will be equal to the Darcy velocity divided by the total porosity

$$V_a = V_w \frac{\theta_f}{\theta_t} = \frac{V_D}{\theta_t} \tag{5}$$

This condition is sometimes referred to as Equivalent Porous Media (EPM) for solute transport. *Van der Kamp* [1992] noted that EPM conditions would arise where $D\theta_m t/B^2$ was large, and *Cook et al.* [1996] suggested an approximate condition of $\exp[D\theta_m t/B^2] > 100$, where D is the diffusion coefficient for the aquifer matrix and t is the duration of the tracer "experiment." (In this paper, the diffusion coefficient in the matrix is defined by $D = D_0\tau$, where D_0 is the free solution diffusion coefficient for the particular solute species, and τ is the tortuosity.)

2.3. Environmental Tracers

Carbon-14. Carbon-14 (^{14}C), with a half-life of 5,730 years, is formed naturally in the upper atmosphere by cosmic ray bombardment of ^{14}N. In the atmosphere, most ^{14}C is in the form $^{14}CO_2$. It is incorporated into groundwater during recharge, and most dissolved inorganic carbon occurs as HCO_3^-. Where HCO_3^- behaves conservatively (i.e., does not react with the aquifer matrix or undergo other chemical changes that affect the ^{14}C activity), groundwater ages can be estimated from measured ^{14}C concentrations, usually by assuming an initial activity of 100 percent modern carbon (pmc).

$$t_a = -\lambda^{-1} \ln\left(\frac{c}{c_0}\right) \tag{6}$$

where c is the measured ^{14}C activity, c_0 is the initial activity (100 pmc), and $\lambda = 1.21 \times 10^{-4}$ yr^{-1} is the decay constant for ^{14}C.

Carbon-14 was also produced in large quantities by nuclear weapons testing during the 1950s and 1960s, with atmospheric concentrations peaking at approximately twice natural levels in 1963 (as shown in Figure 1). Concentrations have been declining since, but remain in excess of 100 pmc. Activities in excess of 100 pmc will produce apparent negative ages when dating assuming 100 pmc input. In some cases, chemical reactions also alter the ^{14}C activity of the groundwater, usually resulting in apparent increases in ^{14}C age, but detailed discussion of this is beyond the scope of this paper. Measurement of the stable isotope ^{13}C and other aquifer chemistry sometimes allows corrections to be made for variations in ^{14}C activities resulting from chemical processes [*Kalin*, 1999].

Analytical solutions for the transport of a constant-source, radioactive tracer through a system of planar, vertical fractures have been presented by *Tang et al.* [1981) for the case of a single fracture, and *Sudicky and Frind* [1982] for a series of parallel fractures, although the relevance to ^{14}C dating was not discussed by these authors. Under these conditions, and assuming constant input of ^{14}C and negligible dispersion, the apparent ^{14}C ages within fractures can be expressed

$$\frac{t_a}{z} = V_w^{-1}\left[1 + \frac{\theta_m D^{1/2}}{b\lambda^{1/2}} \tanh\left(BD^{-1/2}\lambda^{1/2}\right)\right] \tag{7}$$

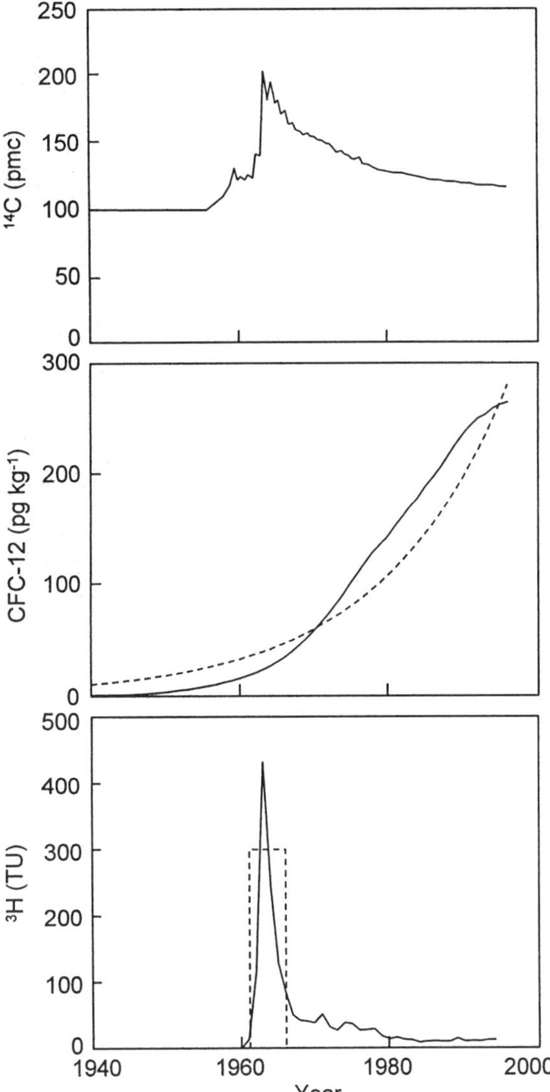

Figure 1. (a) Concentrations of ^{14}C in atmospheric CO_2 for northern hemisphere air (Kalin, 1999). (b) Concentrations of CFC-12 in water at a temperature of 16°C and pressure of 1 atmosphere (pg kg^{-1}). Atmospheric concentrations are as measured at Cape Grimm, Australia [*Cunnold et al.*, 1994]. The broken line represents an exponential growth rate of $k = 0.06$ yr^{-1}. (c) Concentrations of ^3H in rainfall at Vienna, Austria, corrected for radioactive decay to 1996. The broken line represents a square pulse approximation to the measured ^3H distribution.

[*Neretnieks*, 1981, Eq. (7); *Sudicky and Frind*, 1982, Eq. (43b); *Maloszewski and Zuber*, 1985, Eq. (31); *Sanford*, 1997, Eq. (12)]. Concentration profiles can be determined from (7) by substituting for t_a using (6). For large fracture spacings, defined by *Maloszewski and Zuber* [1985] as $BD^{-1/2}\lambda^{1/2} \geq 2$, $\tanh(BD^{-1/2}\lambda^{1/2})$ approaches 1. For small fracture spacings ($BD^{-1/2}\lambda^{1/2} \leq 0.25$), (7) reduces to (4). *Maloszewski and Zuber* [1991] extended this analysis to include the effects of carbonate sorption and exchange, although in this paper we will consider only cases where ^{14}C behaves as a chemically conservative species.

Input concentrations above 100 pmc within the past 50 years will mean that (7) will underestimate ^{14}C activities and overestimate ^{14}C ages for shallow groundwaters. It should reliably predict concentrations at greater depth, where the influence of the bomb pulse is reduced.

Chlorofluorocarbons. Chlorofluorocarbons (CFCs) are synthetic organic compounds that are produced for a range of industrial and domestic purposes. They have relatively long residence times in the atmosphere, and so their atmospheric concentrations are uniform over large areas and are steadily increasing [*Cunnold et al.*, 1994; Figure 1]. Concentrations of CFCs dissolved in groundwater can be used as indicators of groundwater age [*Busenberg and Plummer*, 1992]. In porous-media aquifers, dissolved CFC concentrations have been used to estimate vertical groundwater velocities and aquifer recharge rates [*Cook et al.*, 1995].

Cook et al. [1996] performed some numerical simulations of CFC-12 (CF_2Cl_2) transport in fractured porous media, although to date, quantitative interpretation of CFC ages in fractured aquifers have been limited to systems in which equivalent porous-media assumptions could be invoked. However, if the chlorofluorocarbon input functions can be approximated by an exponential growth curve (ae^{-kt}), then a simple analytical solution can be derived that expresses apparent CFC age as a function of the hydraulic age of the water, t_w, and fracture and matrix parameters. In fact, for a system of planar, parallel, vertical fractures, the solution is identical to (7), where the decay constant, λ, is replaced by the exponential growth rate, k. Figure 1 compares the measured atmospheric concentrations of CFC-12, converted to aqueous concentrations using the measured solubility at 16°C, with an exponential growth rate of $k = 0.06$ yr^{-1}. The exponential function provides a good fit to the actual input concentrations, although this will not be the case in the future as atmospheric CFC concentrations level off and then begin to decline [*Khalil and Rasmussen*, 1993].

Tritium. Tritium (^3H) is a radioactive isotope of hydrogen, with a half-life of 12.43 years. Nuclear weapons testing resulted in elevated concentrations of ^3H in rainfall (in the form $^3H^1HO$) during the 1960s (Figure 1). Mean annual ^3H concentrations reached several hundred times natural levels in the northern hemisphere, and between five and thirty times natural levels in the southern hemisphere. Several studies have measured the apparent velocity of the bomb pulse of ^3H in porous-media aquifers and used this information to estimate vertical flow velocities and groundwater recharge rates [e.g., *Robertson and Cherry*, 1989].

Lever and Bradbury [1985] present an analytical solution for the transport of a short 'top-hat' pulse of solute through a system of planar, parallel fractures. Where the fracture spacing is large (no interaction between adjacent fractures), the concentration within the fracture as a function of depth is given by

$$\frac{c(x,t)}{c_0} = \frac{T\tau^{1/2}}{\pi^{1/2}(t-t_w)^{3/2}} \exp\left[\frac{-\tau}{t-t_w}\right] \tag{8}$$

$$\tau = \frac{D\theta_m^2 t_w^2}{4b^2} \tag{9}$$

where c_0 is the concentration of tracer during the pulse input (zero at other times), T is the length of the pulse, and t is the time elapsed since the commencement of the pulse. Although it neglects radioactive decay, (8) can be used to approximate ^3H transport through fractured porous media, if fracture spacing is large ($B >> 12(D\theta_m)^{1/2}$). Analytical solutions for smaller fracture spacings and incorporating radioactive decay are available [*Sudicky and Frind*, 1982], but do not reduce to such simple algebraic expressions.

Because ^3H concentrations in rainfall did not reduce to zero after the 1960s, the analytical solution for the rectangular pulse will underpredict ^3H concentrations in shallow groundwaters. However, it should reliably predict concentrations below the bomb pulse.

3. SIMULATIONS

Figure 2 depicts simulated vertical profiles of ^{14}C, CFC-12, and ^3H within fractures, in an aquifer comprising planar, parallel, vertical fractures in an impermeable matrix. Concentration profiles have been simulated using the analytical solutions for idealized input functions represented by Equations (7) and (8) (broken lines), and using numerical solutions for actual input functions (solid lines). Numerical simulations were performed using FRAC3DVS [*Therrien and Sudicky*, 1996]. A rectangular grid was constructed that contained a single vertical fracture. Uniform recharge was applied to the surface of the model, with constant head on the bottom and no-flow conditions on all other sides. The width of the grid was chosen to be equal to the fracture half-spacing to simulate

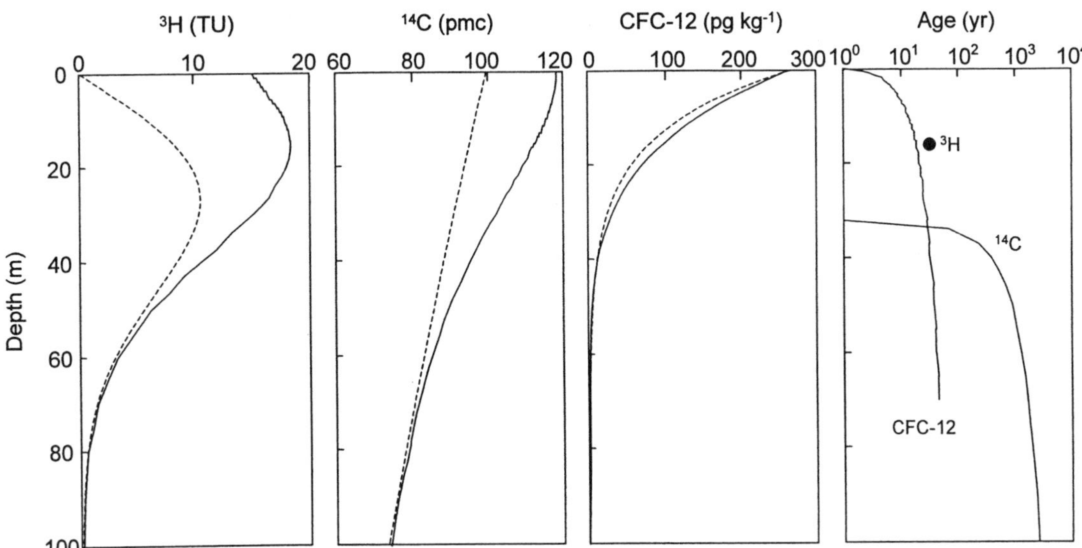

Figure 2. Simulations of ^{3}H, ^{14}C, and CFC-12 concentrations and apparent ^{3}H, ^{14}C, and CFC-12 ages in an aquifer comprising planar, parallel, vertical fractures in an impermeable matrix. Profiles depict concentrations within the fracture, as would have been measured in 1996. The simulations are for aquifer parameters $D = 10^{-3}$ m^2 yr^{-1}, $\theta = 0.02$, $b = 40$ μm, $B = 4$ m, $V_w = 50$ m yr^{-1} and $R = 0.5$ mm yr^{-1}. The solid lines are numerical simulations obtained using FRAC3DVS, with boundary conditions and matrix hydraulic conductivities chosen to simulate steady, vertical flow through the fractures and no advection within the matrix. Input concentrations are as shown in Figure 1. Apparent ^{14}C ages have been calculated assuming a constant input concentration of 100 pmc. The circle in the rightmost figure indicates the groundwater age of 32.5 years, indicated by the position of the maximum ^{3}H concentration in the profile. Broken lines represent approximate analytical solutions using (7) and (8), where ^{3}H input is represented as a square pulse defined by $T = 5$ yr, $t = 35$ yr, $c_0 = 300$ TU, and CFC-12 input is represented by an exponential growth rate of $k = 0.06$ yr^{-1}. Deviations between the approximate analytical solutions and the numerical solutions occur for ^{3}H and ^{14}C at shallow depths and are due to the idealized input functions used for the analytical solutions.

steady-state vertical flow through parallel fractures. For the numerical simulations, matrix conductivity was set to a very low value (10^{-9} m yr^{-1}) to simulate flow through fractures only, and dispersivity within the fracture was set to 1.0 m (the simulations are not sensitive to this value). Numerical simulations were also performed with the idealized input functions (broken lines in Figure 1), and the analytical solutions were accurately reproduced. Thus, differences between the numerical simulations using the actual input concentrations (solid lines in Figure 2) and the analytical solutions (broken lines in Figure 2) are attributable to differences in the input function for the two cases. As discussed above, the approximate analytical solutions for ^{3}H and ^{14}C reliably reproduce the distributions of these tracers at depth, but diverge at shallow depths because of differences between the idealized input function used by the analytical solution and the actual input of the tracers. For CFC-12, the approximate analytical solution provides a reasonable description of the tracer distribution at all depths.

The approximate analytical solutions predict that ^{3}H will be present to greater depths as the water velocity or fracture aperture increases, and as the matrix porosity and diffusion coefficient decreases. Similarly, younger CFC-12 and ^{14}C ages (higher concentrations) will result from greater water velocity and fracture aperture, and lower matrix porosity and diffusion coefficient. Increasing fracture spacing results in decreased CFC-12, ^{14}C, and ^{3}H concentrations at any given depth, until a level is reached such that there is no interaction between adjacent fractures. Beyond this point there is no change in tracer concentration with increasing fracture spacing, provided that V_w (and not R) is held constant. Conversely, as D becomes large, there will come a point where complete equilibration between fracture and matrix concentrations occurs, beyond which tracer concentration will be independent of D. For a given recharge rate, R, the tracer concentrations will also be independent of b, B, and V_w. (As D becomes large, the approximate analytical solution for ^{3}H given by (8) breaks down. This solution assumes no interaction between solute diffusing from adjacent fractures, which will not be the case at large diffusion coefficients.)

Although the recharge rate used in the simulations shown in Figure 2 is relatively low (R = 0.5 mm yr^{-1}), ^{3}H is still present in fractures to more than 80 m depth. CFC-12 is present to approximately 70 m, with CFC-12 ages increasing linearly with depth. Carbon-14activities are greater than 100 pmc above 32 m, giving negative apparent ^{14}C ages. Below 32 m, ^{14}C ages rapidly increase. Figure 2 clearly shows that the groundwater ages obtained with the various tracers will be different. In particular, between 40 and 70 m depth, CFC-12 and ^{3}H are present, with ^{14}C ages greater than 100 years. This apparent discrepancy results from enhanced retardation of ^{14}C by matrix diffusion relative to CFC-12 and ^{3}H. The maximum discrepancy between ^{14}C, ^{3}H, and CFC-12 ages that can be obtained under this model can be readily determined from the simplified analytical solutions. In the case of CFC-12 and ^{14}C ages, the greatest discrepancy is obtained with high matrix porosity, small fracture aperture, and large fracture spacing. It is easy to show that the maximum value the ^{14}C to CFC-12 age ratio can attain is $(k/\lambda)^{0.5}$ = 22.3. (The ^{14}C bomb pulse is not considered in this analytical calculation. It will have the effect of lowering the ^{14}C /CFC-12 age ratio at shallow depth, and so will not increase this maximum value.)

We can similarly determine the extent to which ^{3}H may be present in water having old ^{14}C ages. Figure 3 depicts the ^{14}C age measured in fractures near the leading edge of the ^{3}H pulse, determined from simultaneous solution of (7) and (8). In the northern hemisphere, we may define the presence of ^{3}H by c/c_0 = 0.001. (We can approximate the ^{3}H input function by a pulse of 5 years duration, with a concentration of 300 TU after radioactive decay. Thus, c/c_0 = 0.001 represents a measured concentration of 0.3 TU.) Thus, for a diffusion coefficient of 10^{-4} and large values of θ/b, ^{3}H may be present in water with ^{14}C ages up to 2500 years. At smaller values of θ/b, however, it may not be possible for ^{3}H to be present in water with such old ^{14}C ages. (Differences between measured ^{3}H concentrations in rainfall and the rectangular input pulse used for the analytical solution will not greatly affect these calculations. Differences between the real and idealized input distributions are most significant for the period since the 1970s, whereas these calculations are based on the position of the leading edge of the ^{3}H pulse in the subsurface.)

4. SITE DESCRIPTION AND METHODS

The Clare Valley, located approximately 100 km north of Adelaide, South Australia, consists of low-grade metamorphosed, folded, and faulted sedimentary rocks of Proterozoic age [*Morton et al.*, 1998]. The major lithologies are sandstone, shale, quartzite, and dolomite. In this paper, data are presented from three boreholes drilled within the Mintaro Shale. Two 200 mm diameter boreholes were drilled to 100 m depth, and a 250 mm boreholes to 26.4 m depth. (The three boreholes are spaced only 5–10 m apart.) A nest of six 50 mm PVC piezometers was installed in the 250 mm borehole and three 50 mm piezometers in one of the 200 mm boreholes. Screen lengths range from 0.5 to 5 m, being longer in the deeper piezometers. Gravel packs were used around the screens, which were separated using cement plugs. The nest of nine piezometers allows groundwater samples to be collected from depths between 8.5 and 95 m below the land surface. Mean annual rainfall is approximately 650 mm yr^{-1}, with approximately 450 mm falling in the six months between May and October. The position of the water table varies seasonally, from approximately 4 m depth in October–November to 7 m in July–August. The vertical head gradient is less than 10^{-3}.

During drilling, core samples were taken at selected intervals for laboratory determination of porosity and matrix hydraulic conductivity. The primary porosity was measured to be approximately 2% by helium porosimetry. The hydraulic conductivity of the same sample was 6 × 10^{-13} m s^{-1}. Fracture properties were determined by outcrop mapping, in an exposure approximately 70 m long and 6 m deep, located

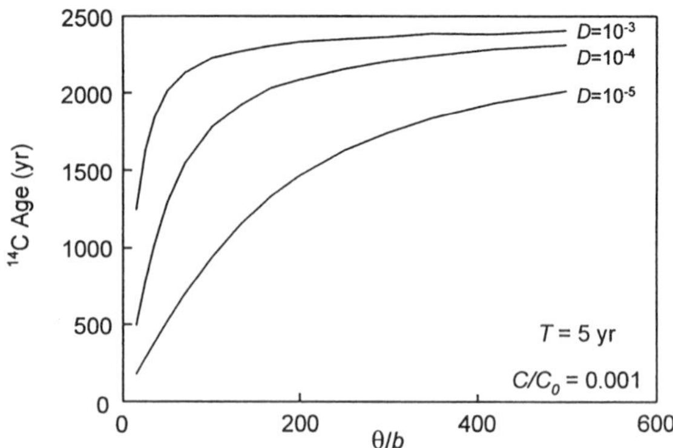

Figure 3. Concentrations of ^{14}C on the leading edge of the ^{3}H pulse, for transport of both tracers through a single, planar, vertical fracture, determined using the approximate analytical solutions represented by (6) and (7). The ^{3}H distribution in rainfall has been approximated by a square pulse of 5 years duration, commencing 35 years before present. The leading edge of the ^{3}H pulse in groundwater is defined by c/c_0 = 0.001. The diagram depicts maximum ^{14}C ages that could be expected on water samples also containing measurable ^{3}H, for diffusion coefficients between 10^{-3} and 10^{-5} mr yr^{-1}.

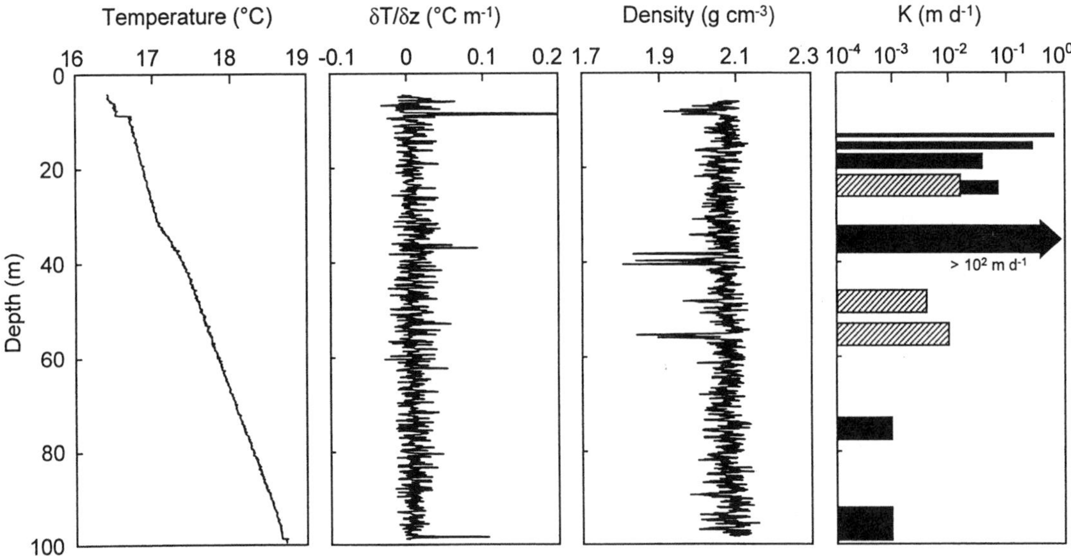

Figure 4. Temperature, density, and hydraulic conductivity measurements at Pearce Road, Clare Valley: (a) Ambient temperature profile; (b) Measured temperature gradient; (c) Density log; (d) Hydraulic conductivities measured on piezometers (solid bars) and using pump-packer equipment (hatched bars). Depths are below land surface.

approximately 10 km north of the borehole installations. Outcrop data indicate 5 sets of fractures are present, four steeply dipping sets and one gently dipping set. For this site, on the west limb of the Hill River Syncline, the aquifers are on end, with the majority of fractures and bedding planes oriented vertically. The bedding comprises one vertically dipping fracture set that strikes 151°. A second set of vertical fractures strikes 59°. A set of conjugate shear fractures has the same trend as the second set of vertical fractures, with dips of approximately 60°. The only shallow dipping fracture set has a strike of 345° and a dip of 27°. Fracture spacings are approximately log-normally distributed, with a mean fracture spacing of approximately 0.16 m.

Mean hydraulic conductivities of the aquifer were determined from single-well pumping tests carried out in each of the piezometers and also in the open hole using a pump-packer system. The hydraulic conductivity of the aquifer ranges between 10^{-3} and $>10^{2}$ m d^{-1} (Figure 4). Fracture mapping and pumping tests that monitored drawdown in piezometers above and below the pumping well indicate a vertical hydraulic conductivity generally much greater than the horizontal hydraulic conductivity.

Three well volumes were purged from each piezometer prior to sampling. We measured ^{14}C and ^{13}C after first precipitating the dissolved inorganic carbon as $BaCO_3$. The ^{14}C was analyzed using a liquid scintillation counter and the direct absorption method. Concentrations of ^3H were measured either by liquid scintillation counting after electrolytic enrichment or by helium ingrowth. CFC-11 and CFC-12 concentrations were measured by gas chromatography, although only CFC-12 concentrations are reported in this paper. Measured concentrations in groundwater have been converted to equivalent atmospheric concentrations using a recharge temperature of 16°C. All samples described in this paper were collected between August 1996 and May 1998.

5. RESULTS

Despite the large vertical variation in hydraulic conductivity, profiles of ^{14}C, CFC-12, and ^3H are remarkably smooth (Figure 5). Values for $\delta^{13}C$ range between –12‰ and –15‰, and show little variation with depth. The values are consistent with a $\delta^{13}C$ value for soil CO_2 of –22‰ and a +8‰ fractionation between gaseous CO_2 and dissolved bicarbonate. Thus, no corrections to the ^{14}C values for chemical reactions appear necessary at this site. Carbon-14 activities are between 90 and 100 pmc in the upper 20 m of the aquifer, decreasing to 65 pmc at 35 m, and 20–30 pmc between 70 and 100 m. Above-background concentrations of ^3H and CFC-12 are present to 40 m.

A major fracture at approximately 37 m depth is identified by anomalies in the temperature and density logs, and by a high hydraulic conductivity in the piezometer screened between 33.0 and 38.5 m depth (Figure 4). It appears to divide the aquifer into an upper flow system,

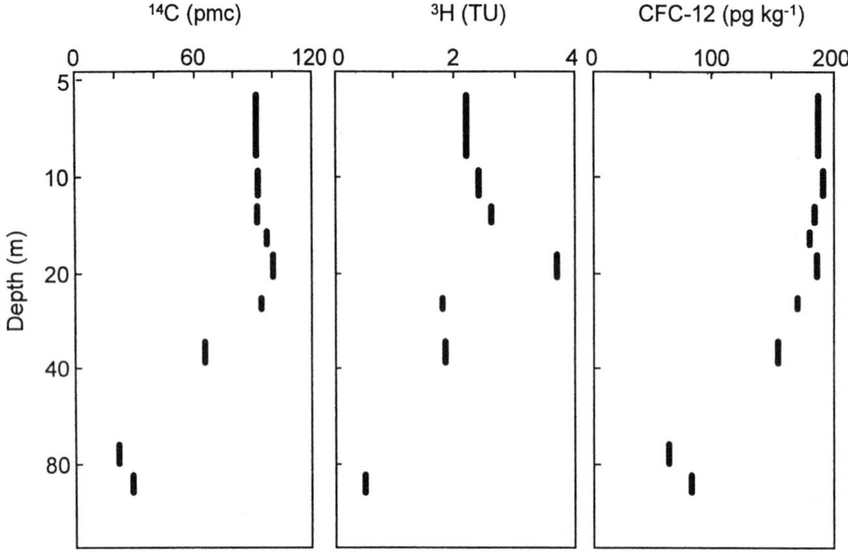

Figure 5. Vertical profiles of ^{14}C, ^{3}H, and CFC-12, obtained from the piezometer nest at Pearce Road, Clare Valley. Analytical uncertainties (one standard deviation) are between 0.9 and 4.8 pmc, 0.03 and 0.2 TU, and 4 and 6 pg kg^{-1} for ^{14}C, ^{3}H and CFC-12 respectively. Concentrations of ^{3}H and CFC-12 below 60 m depth appear to be above analytical detection limits, but are believed to represent background levels introduced during well construction.

with generally high hydraulic conductivity and young groundwater ages, and a lower flow system with older groundwater ages. The mean vertical temperature gradient above 37 m is significantly lower than that at depth, which is indicative of more rapid vertical (downward) water movement in the upper zone.

5.1. *The Upper Flow System*

The presence of ^{3}H and CFC-12 to an almost 40 m depth indicates relatively rapid vertical water movement. Above 25 m, CFC-12 concentrations range between 170 and 190 pg kg^{-1}, equivalent to apparent CFC-12 ages between 12 and 14 years. The apparent CFC-12 age of 12 years immediately below the water table represents the amount of time required for these gases to diffuse through the unsaturated zone [*Cook and Solomon*, 1995]. The decrease in apparent CFC-12 age between 5 and 25 m is approximately 2 years ($\delta t_a/\delta z = 0.1$ yr m^{-1}). However, this does not, in itself, constrain either the recharge rate or the vertical water velocity through the fractures. [The measured distribution of tracers in the upper 20 m of the aquifer (above a 25 m depth) could be achieved with a recharge rate as low as 1 mm yr^{-1} (e.g., $D = 10^{-4}$ m^2 yr^{-1}, $\theta = 0.02$, $b = 100$ μm, $B = 10$ m, $V_w = 100$ m yr^{-1}, in [3] and [7]), for example, or as high as 500 mm yr^{-1} ($D = 10^{-4}$ m^2 yr^{-1}, $\theta = 0.02$, $b = 100$ μm, $B = 0.1$ m, $V_w = 500$ m yr^{-1}). Similarly, it does not greatly constrain the vertical water velocity, V_w, which may be as low as 10 m yr^{-1}, or as high as several meters per day.)] Only by combining tracer data, outcrop measurements of fracture spacing, and hydraulic conductivity information is it possible to provide reasonable constraints on the major flow parameters.

A mean fracture spacing of $2B = 0.16$ m was measured on outcrops, and a hydraulic conductivity of $K = 0.1$ m d^{-1} was measured in single-well pumping tests (Figure 4). (Although single-well pumping tests are influenced by both the vertical and horizontal hydraulic conductivities, we have assumed that the vertical hydraulic conductivity is similar to the measured value. This is probably a reasonable assumption for a predominantly vertically fractured system.) Substituting this into (1) gives $b = 32$ μm. Using $D = 10^{-4}$ m^2 yr^{-1}, $\theta = 0.02$ and $\delta t_a/\delta z = 0.1$ yr m^{-1} gives $R = 102$ mm yr^{-1} and $V_w = 256$ m yr^{-1} using (3) and (7). Equation (2) then gives $\delta i/\delta z = 2.8 \times 10^{-3}$, slightly higher than that measured in the piezometers. Figure 6 depicts the sensitivity of the calculated parameters (b, R, V_w) to the measured or estimated parameters (B, K, θ, D, $\delta t_a/\delta z$). Importantly, the estimated recharge rate is particularly insensitive to the estimated hydraulic conductivity. In particular, an order-of-magnitude error in the hydraulic conductivity would result in a factor-of-two error in the estimated fracture aperture and minimum fracture velocity, but less than a 5% error in the estimated recharge rate. A hydraulic conductivity slightly larger than 0.1 m d^{-1}, and

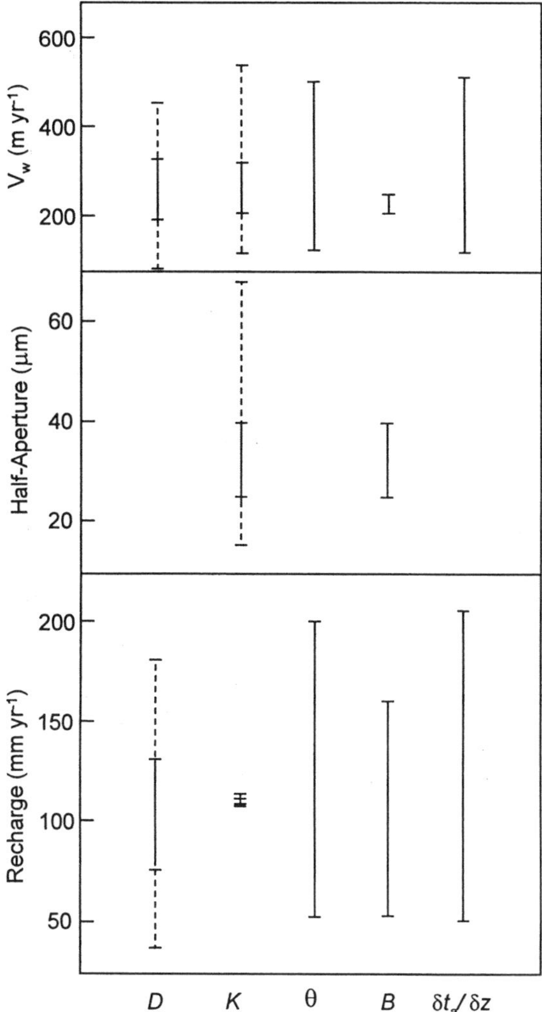

Figure 6. Sensitivity of calculated fracture parameters to measured or estimated parameters for the upper flow system at Pearce Road. Solid bars denote variability in calculated parameters arising from a factor-of-two variation in measured parameters. Broken lines show sensitivity to order-of-magnitude variation in hydraulic conductivity and diffusion coefficient, which is considered to be the more realistic uncertainty on these parameters. Note that the estimated fracture half-aperture is determined only by the estimated hydraulic conductivity and fracture spacing, and so is totally insensitive to the diffusion coefficient, matrix porosity, and vertical age gradient. Of particular note is the extreme insensitivity of the estimated recharge rate to the hydraulic conductivity.

hence fracture half-aperture greater than 32 µm, would result in a vertical head gradient more consistent with that measured at the site, while not affecting the estimate of recharge. Estimates of both recharge and vertical flow velocity are inversely proportional to the age gradient. The recharge rate is also sensitive to the fracture spacing, and there must be some doubt about this value, which was measured at an outcrop several kilometers from the borehole installations. At our site, the measured age gradient is close to the resolution of the dating technique (2 years in 20 meters), and so the error associated with this measurement is more than would otherwise be the case.

5.2. The Lower Flow System

Below 25 m, significant decreases in CFC-12 and ^{14}C concentrations are observed. The piezometer screened between 33 and 38.5 m coincides with the location of a large subhorizontal fracture, and a large increase in age would be expected immediately below such a feature, as vertical flow is intercepted and diverted laterally. This piezometer likely intercepts water from both the upper and lower flow systems, and the groundwater ages obtained are consistent with this water being a mixture of young and old water. Below 70 m, ^{14}C activities are between 22 and 29 pmc. The ^{3}H and CFC-12 concentrations appear to be above detection limits at this depth, but are believed to represent background levels introduced during well construction.

The measured hydraulic conductivity of the lower flow system is approximately 10^{-3} m d^{-1}. Using the above fracture spacing of $2B = 0.16$ m, (1) gives $2b = 14$ µm. The decrease in ^{14}C activity of 75 pmc occurs over a depth of approximately 50 m, giving $\delta t_d/\delta z = 230$ yr m^{-1}. Substituting into (3) and (7) using $D = 10^{-4}$ m^2 yr^{-1} and $\theta = 0.02$ gives $R = 0.05$ mm yr^{-1} and $V_w = 0.5$ m yr^{-1}. Equation (2) then gives $\delta i/\delta z = 1 \times 10^{-4}$, which is consistent with that measured in the piezometers (i.e., too low to be measurable). Of course, at these depths it may no longer be reasonable to assume that the fracture spacing is similar to that measured in near-surface exposures. If, instead, we assume $2B = 1.0$ m, (1) gives $2b = 25$ µm. Substituting into (3) and (7) then gives $R = 0.08$ mm yr^{-1} and $V_w = 3.1$ m yr^{-1}. Equation (2) then gives $\delta i/\delta z = 2 \times 10^{-4}$. In either case, the leakage rate below the subhorizontal fracture is extremely low.

6. DISCUSSION

In porous-media aquifers with simple geometries, the vertical flow velocity at the water table can be directly determined from a single measurement of groundwater age. If the aquifer porosity is known, then the recharge rate can be determined. In fractured media, where fracture spacings are large enough such that the system cannot be considered an equivalent porous media, a single groundwater age does not constrain either the vertical fracture velocity or the recharge rate without independent information on fracture properties. Furthermore, ages obtained with environmental tracers will usually not be equal to the hydraulic age of the water sampled, and different tracers may provide different apparent ages.

Using data from a field site in the Clare Valley, we have shown how flow parameters can be constrained by considering both environmental tracer data and hydraulic data simultaneously. The parameters are constrained much more tightly than would have been possible using either approach independently. In particular, water fluxes estimated using hydraulic approaches are usually highly dependent on measured values of hydraulic conductivity and hydraulic head gradients. Hydraulic conductivity is very difficult to measure in fractured media and usually highly spatially variable. Hydraulic head gradients may be temporally variable. In contrast, the recharge rate estimated using the combined approach is highly insensitive to the hydraulic conductivity, and the hydraulic head gradient was not directly used in the calculations.

The very low hydraulic conductivity of the rock matrix at our site has allowed us to assume that groundwater flow is exclusively through fractures and that groundwater samples obtained from the aquifer are representative of water within the fractures. In part, our approach may have been facilitated by the relatively high vertical hydraulic conductivity at our field site. We have analyzed subsurface distributions of environmental tracers using analytical solutions for groundwater flow through vertical fractures. At sites dominated by horizontal fractures, this approach may not be justified. However, our approach does not necessarily rely on flow being only vertical. If fractures are inclined, then the calculated flow velocity simply reflects the vertical component of the flow, not the actual flow rate within the fracture (which would be greater). However, we have assumed a constant fracture spacing, and where fractures are not parallel, then the interaction between tracer concentrations within the fractures and matrix becomes more complex. In effect, we are representing a highly complex network of fractures using average values for fracture and matrix properties. We use these average values to solve simple analytical equations and produce estimates of flow velocities. However, the effect of averaging spatially variable properties has not been properly examined. It is possible that the use of spatial averages causes some biasing of calculated parameters, whether using hydraulic or environmental tracer methods. This is one area needing further work.

7. CONCLUSIONS

In fractured rock aquifers, concentrations of environmental tracers are a function of fracture spacings and apertures, matrix porosities and diffusion coefficients, and vertical flow velocities. The large number of variables, some of which are very difficult to measure, makes interpretation of environmental tracer data difficult. Similarly, extreme spatial variability of hydraulic properties makes hydraulic estimates of flow velocities subject to error. Understanding of fractured rock systems can only be achieved by combining as much data as possible. At our field site, flow parameters can be best constrained by combining available tracer and hydraulic data. Despite the large vertical variations in hydraulic conductivity, profiles of ^{14}C, CFC-12, and ^{3}H are remarkably smooth. Based on the distribution of tracers, and also on the hydraulic properties of the system, two separate flow systems are identified. In the upper flow system, the vertical water velocity through the fractures is estimated to be approximately 250 m yr^{-1} and the recharge rate approximately 100 mm yr^{-1}. The leakage rate to the deeper flow system is estimated to be less than 0.1 mm yr^{-1}.

REFERENCES

Busenberg, E., and L. N. Plummer, Use of chlorofluorocarbons (CCl_3F and CCl_2F_2) as hydrologic tracers and age dating tools: the alluvium and terrace system of central Oklahoma, *Water Resour. Res.*, *28*(9), 2257–2283, 1992.

Cook, P. G., and D. K. Solomon, Transport of atmospheric trace gases to the water table: implications for groundwater dating with chlorofluorocarbons and krypton 85, *Water Resour. Res.*, *31*(2), 263–270, 1995.

Cook, P. G., D. K. Solomon, L. N. Plummer, E. Busenberg, and S. L. Schiff, Chlorofluorocarbons as tracers of groundwater transport processes in a shallow, silty sand aquifer, *Water Resour. Res.*, *31*(3), 425–434, 1995.

Cook, P. G., D. K. Solomon, W. E. Sanford, E. Busenberg, L. N. Plummer, and R. J. Poreda, Inferring shallow groundwater flow in saprolite and fractured rock using environmental tracers, *Water Resour. Res.*, *32*(6), 1501–1509, 1996.

Cunnold, D. M., P. J. Fraser, R. F. Wiess, R. G. Prinn, P. G. Simmonds, B. R. Miller, F. N. Alyea, and A. J. Crawford, Global trends and annual releases of CCl_3F and CCl_2F_2 estimated for ALE/GAGE and other measurements from July 1978 to June 1991, *J. Geophys. Res. Atmospheres*, *99*(D1), 1107–1126, 1994.

Grisak, G. E., and J. F. Pickens, Solute transport through fractured media. 1. The effect of matrix diffusion, *Water Resour. Res.*, *16*(4), 719–730, 1980.

Kalin, R. M., Radiocarbon dating of groundwater systems, in *Environmental Tracers in Subsurface Hydrology*, edited by P. G. Cook and A. L. Herczeg, pp. 111–144, Kluwer, Boston, 1999.

Khalil, M. A. K., and R. A. Rasmussen, The environmental history of probable future of fluorocarbon 11, *J. Geophys. Res.*, *98*(12D), 23091–23106, 1993.

Lever, D. A., and M. H. Bradbury, Rock-matrix diffusion and its implications for radionuclide migration, *Mineralogical Magazine*, *49*, 245–254, 1985.

Maloszewski, P., and A. Zuber, On the theory of tracer experiments in fissured rocks with a porous matrix, *J. Hydrol.*, *79*, 333–358, 1985.

Maloszewski, P., and A. Zuber, Influence of matrix diffusion and exchange reactions on radiocarbon ages in fissured carbonate aquifers, *Water Resour. Res.*, *27*(8), 1937–1945, 1991.

Morton, D., A. Love, D. Clarke, R. Martin, P. G. Cook, and K. McEwan, *Clare Valley Groundwater Resources. Progress Report 1. Hydrogeology, Drilling and Groundwater Monitoring*, Report Book 98/00015, Primary Industries and Resources, South Australia, 1998.

Neretnieks, I., Age dating of groundwater in fissured rock: influence of water volume in micropores, *Water Resour. Res.*, *17*(2), 421–422, 1981.

Robertson, W. D., and J. A. Cherry, Tritium as an indicator of recharge and dispersion in a groundwater system in Central Ontario, *Water Resour. Res.*, *25*(6), 1097–1109, 1989.

Sanford, W. E., Correcting for diffusion in carbon-14 dating of ground water, *Ground Water*, *35*(2), 357–361, 1997.

Sudicky, E. A., and E. O. Frind, Contaminant transport in fractured porous media: analytical solutions for a system of parallel fractures, *Water Resour. Res.*, *18*(6), 1634–1642, 1982.

Tang, D. H., E. O. Frind, and E. A. Sudicky, Contaminant transport in fractured porous media: analytical solution for a single fracture, *Water Resour. Res.*, *17*(3), 555–564, 1981.

Therrien, R., and E. A. Sudicky, Three-dimensional analysis of variably-saturated flow and solute transport in discretely-fractured porous media, *J. Contaminant Hydrol.*, *23*(1), 1–44, 1996.

Van der Kamp, G. ,Evaluating the effects of fractures on solute transport through fractured clayey aquitards, *Proceedings of the 1992 Conf. of the International Association of Hydrogeologists*, Canadian National Chapter, Hamilton, Ontario, 1992.

Vogel, J. C., Investigation of groundwater flow with radiocarbon, in *Isotopes in Hydrology*, IAEA, Vienna, pp. 355–369, 1967.

Peter G. Cook, CSIRO Land and Water, PMB 2, Glen Osmond, SA, 5064, Australia

Craig T. Simmons, Flinders University of South Australia, Bedford Park, Australia

Use of Chlorine-36 Data to Evaluate Fracture Flow and Transport Models at Yucca Mountain, Nevada

Andrew Wolfsberg, Katherine Campbell, and June Fabryka-Martin

Earth and Environmental Sciences Division, Los Alamos National Laboratory, Los Alamos, New Mexico

An extensive set of ^{36}Cl data has been collected in the Exploratory Studies Facility (ESF), an 8 km long tunnel at Yucca Mountain, Nevada, for the purpose of developing and testing conceptual models of flow and transport at this site. These data have been used in conjunction with a numerical model to establish upper and lower bounds on infiltration rates, estimate groundwater ages, evaluate hydrologic parameters for fractured volcanic tuff, and develop a conceptual model for the distribution of fast flow paths. At several locations, the measured signals are high enough to be unambiguous indicators of at least a small component of bomb-pulse ^{36}Cl fallout from atmospheric testing of nuclear devices in the 1950s and 1960s, implying that some fraction of the water traveled from the ground surface to the level of the ESF during the last 50 years. Characterization of the structural settings of these samples as well as predictive modeling generally support the conceptual model that a through-cutting fault in conjunction with sufficient infiltration would be required to transmit bomb-pulse ^{36}Cl to the sample depth in less than 50 years. Away from such fault zones, the ages of water samples at the ESF appear to be significantly controlled by the thickness of the nonwelded tuff between the ground surface and the ESF.

1. INTRODUCTION

The total system performance assessment of the potential high-level nuclear waste repository at Yucca Mountain, Nevada, addresses the effectiveness of multiple barriers blocking the migration of radioactive waste to the accessible environment. One of the barriers is the partially saturated volcanic tuff between this potential repository and the water table, referred to as the unsaturated zone or UZ (other barriers include the waste form itself, the waste canisters, the engineered barrier, and the saturated zone). Assessment of the unsaturated zone performance at Yucca Mountain requires understanding both of the percolation flux through the system as well as the solute transport pathways. Naturally occurring and anthropogenic environmental tracers are a valuable interpretive resource for evaluating flow and transport processes over the time and space scales that a defensible assessment must address.

Chlorine-36 is one of several cosmogenic and anthropogenic nuclides (others include carbon-14 and tritium) that can serve as an interpretive natural tracer. In this study, measured isotopic ratios of ^{36}Cl/Cl are used in conjunction with geologic interpretation and model simulations to evaluate flow and transport pathways, processes, and model parameters in the unsaturated zone at Yucca Mountain. Through synthesis of geochemical and geologic data, model results provide insight into the validity of alternative hydrologic parameter sets, flow and transport processes in and away from fault zones, and the

Dynamics of Fluids in Fractured Rock
Geophysical Monograph 122

Published 2000 by the American Geophysical Union

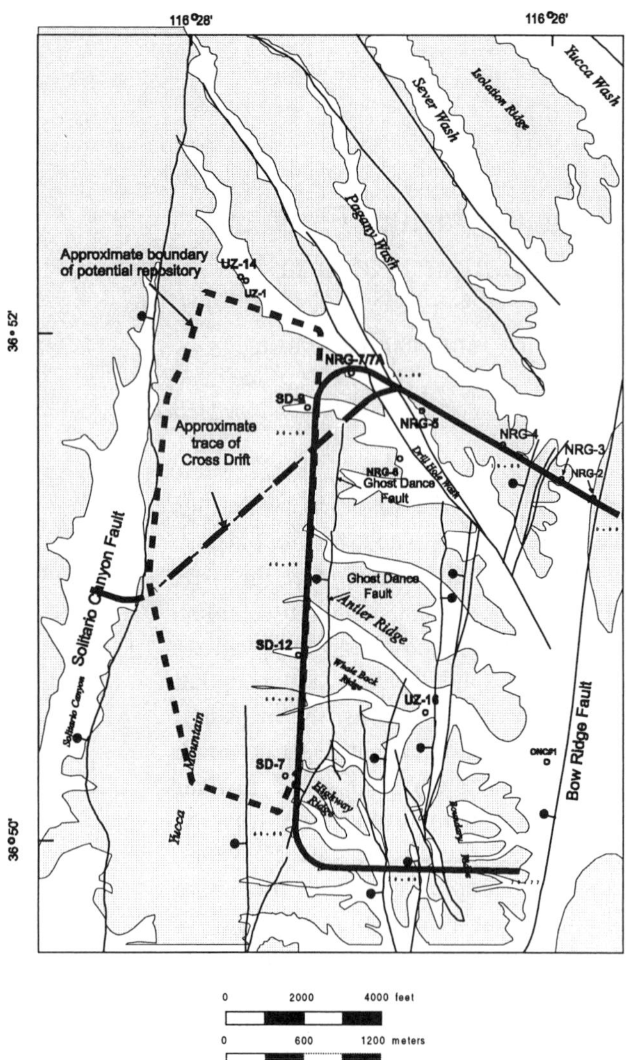

Figure 1. Plan view of ESF, Cross Drift, and proposed repository footprint.

applicability of $^{36}Cl/Cl$ ratios for evaluating alternative conceptual models. (Note: statements referring to alternative models do not imply that there is a prevalent or even accepted model. Rather, there are simply multiple alternatives under consideration.)

Measurements of $^{36}Cl/Cl$ ratios in the exploratory studies facility (ESF), an 8 km long tunnel that largely resides at the same depth and in the same formation as the proposed repository, offer a unique spatially distributed data set for assessing unsaturated zone flow and transport processes and testing flow and transport models. Figure 1 shows the location of the ESF with respect to the potential repository. The ESF sample locations are referred to by station number in such a way that the number represents a location. For instance, Station 59+50 is a location 5950 m into the tunnel as measured from the north entrance.

The $^{36}Cl/Cl$ ratios measured for samples from the potential repository horizon reflect flow and transport processes from ground surface to the potential repository, rather than from the potential repository to the accessible environment. Nonetheless, they still provide valuable insight into the assessment of Yucca Mountain UZ hydrology. First, they provide a data set against which predictive models of flow and transport in the UZ can be validated. Second, the hydrostratigraphy from the ground surface to the potential repository has characteristics that are similar to the hydrostratigraphy between the potential repository and the water table. Thus, understanding flow and transport processes in the upper portion of the UZ system provides valuable insight regarding the lower portion, which remains largely inaccessible for study beneath the repository block. For example, a key issue of concern relates to fast transport pathways. The ^{36}Cl data and associated modeling suggest strongly that faults may provide one of the necessary local conditions to sustain fast paths through a unit otherwise dominated by slow flow in the matrix. Finally, the percolation flux through the UZ is one of the most important parameters to quantify because it affects both waste canister performance and solute travel times to the accessible environment. Chlorine-36 measurements at the ESF can be used quite effectively to assess lower and upper bounds on the percolation flux through the UZ system.

2. CHLORINE-36 AS AN ENVIRONMENTAL TRACER

Chlorine-36 (half-life: 301 ka) is produced in the atmosphere as a result of high-energy cosmogenic nucleons (protons and neutrons) bombarding isotopes of argon and stable chlorine-35. Compared to isotopes such as carbon-14, for which the atmospheric reservoir is well mixed both temporally and spatially, chlorine-36 is removed rapidly from the atmosphere by precipitation. In very dry areas, chloride may remain at the surface in saline playas, from which it may be removed and redeposited downwind, but in general it is carried underground with percolating groundwater, for which it acts as a conservative (nonsorbing) tracer. Thus, the $^{36}Cl/Cl$ ratio over a period of tens of thousands of years reflects atmospheric concentrations at the time of precipitation. (The ratio is generally reported in order to eliminate the highly variable influence of evapotranspiration, which affects all isotopes of chlorine equally.)

Where the residence time of groundwater is of the same order of magnitude as the ^{36}Cl half-life, bounding limits can be established for the age of the groundwater based on the extent of decay of the meteoric ^{36}Cl. In the present application, the travel times of interest are generally much shorter than the ^{36}Cl half-life, although there is some evidence that a small fraction of the groundwater follows very long paths through the unsaturated zone. At the other extreme, high concentrations of ^{36}Cl were added to meteoric water during a period of global fallout from atmospheric testing of nuclear devices. At its peak in the late 1950s, the ^{36}Cl/Cl ratio associated with this "bomb-pulse signal" probably exceeded $200,000 \times 10^{-15}$. Atmospheric ratios had returned to background levels on the order of 500×10^{-15} by the mid-1980s. The bomb-pulse signal is sufficiently strong that it can be used to test for the presence of fast transport paths, even though the signal is considerably diluted as it percolates downward through the unsaturated zone.

Finally, data from fossil pack rat urine in middens from southern Nevada indicate that atmospheric ^{36}Cl/Cl ratios were higher at the close of the Pleistocene Epoch (before about 11 ka) than at present by a factor of two [*Plummer et al.*, 1997]. A decrease in cosmogenic production rates, beginning between about 14,000 and 9,000 yr before the present due to variations in geomagnetic field intensity, is also suggested by the ^{14}C record, although the decrease for ^{14}C is not as large of an effect as for ^{36}Cl. The effect for ^{36}Cl/Cl may have been enhanced locally by large-scale shifts in the average position of the jet stream, which controls the transfer of ^{36}Cl from the stratosphere into the troposphere, and hence the latitudinal distribution of ^{36}Cl deposition, and by changes in rainfall patterns between the Pleistocene and Holocene. Whatever the explanation, this stepwise shift in the background signal, as well as the bomb-pulse signal, can be seen in the Yucca Mountain data and may be used to constrain model parameters such as infiltration rates.

3. HYDROGEOLOGIC SETTING

The stratigraphic structure of the unsaturated zone of Yucca Mountain is comprised of alternating layers of welded and nonwelded volcanic tuffs, which are tilted, uplifted, fractured, and faulted. Characterization of these units and their subunits has advanced as more information from boreholes, the ESF, and other studies has become available. The classification of *Scott and Bonk* [1984] provided a framework around which subsequent hydrostratigraphic models were developed. *Buesch et al.* [1996] developed a detailed classification scheme for the subunits of concern in this study, namely those subunits from the surface down to the major stratigraphic unit of the proposed repository. Using the geologic interpretations from borehole data, tunnel data, and surface outcrops, *Clayton et al.* [1997] developed a three-dimensional hydrostratigraphic model of the Yucca Mountain vicinity. This model provides the framework for most recent Yucca Mountain site-scale flow and transport models [*Bodvarsson et al.*, 1997, 1999; *Wu et al.*, 1999; *Robinson et al.*, 1997, 1999; *Viswanathan et al.*, 1998].

Figure 2 shows a schematic of the hydrostratigraphy encountered from the surface, through the potential repository horizon, and down to the water table. For each of these hydrologic units and subunits, *Flint* [1998] has characterized the hydrologic properties of the rock matrix (permeability, porosity, etc.). Below any surface alluvium, which varies in thickness from 0 to 50 m, the first major hydrologic unit encountered is the Tiva Canyon welded tuff (TCw). The saturated matrix permeability of the TCw is relatively low (on the order of 10^{-18} m^2), but the high degree of fracturing leads to a much higher bulk permeability. Below the TCw is the Paintbrush nonwelded tuff (PTn). The matrix permeability of the PTn is generally several orders of magnitude greater than the TCw, but the degree of fracturing is substantially less. In situ air-permeability measurements suggest significantly lower bulk permeability in the PTn, as compared to the welded units above and below it [*LeCain, 1997*; *LeCain and Patterson*, 1997]. Most of the ESF samples have been collected from the Topopah Spring welded tuff (TSw), below the PTn, which is the horizon of the potential repository. The TSw is much thicker than the TCw and PTn together and is broken into subunits based on the lithophysal content and amount of devitrification that has occurred as a secondary process. Like the TCw, the matrix permeabilities in the TSw are low but the bulk permeability is substantially greater due to a high degree of fracturing. Finally, between the TSw and the water table are the Calico Hills, Prow Pass, and Bullfrog nonwelded tuffs, collectively referred to as the CHn. A unique characteristic of these lower bedded tuffs is the degree of alteration to zeolitic material that has occurred. The zeolites are of particular interest for radionuclide transport sensitivity studies because they have been shown to be effective in sorbing neptunium and other actinides. However, they will not be discussed in any further detail here because this paper focuses on flow and transport from ground surface to the potential repository.

Recent efforts to characterize the hydrologic properties of the various units have involved laboratory measurements of porosity, permeability, and characteristic curves [*Flint*, 1998]; air-permeability tests [*LeCain*, 1997, *LeCain and Patterson*, 1997]; and numerical modeling to match

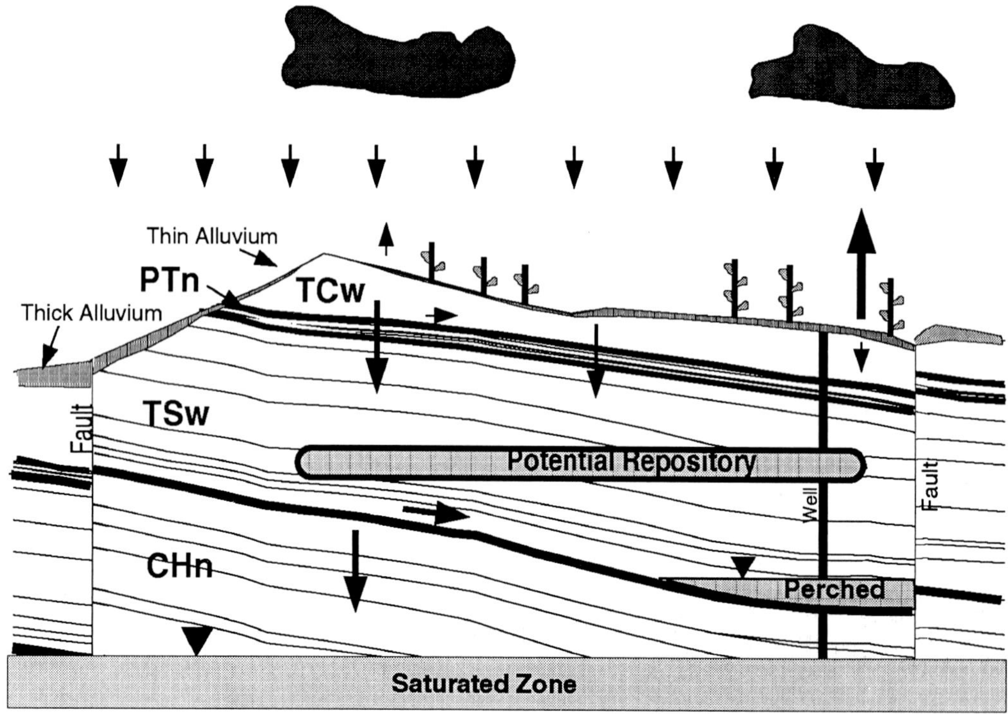

Figure 2. Stratigraphy and conceptual flow model of the unsaturated zone at Yucca Mountain.

measured saturations [*Bandurraga and Bodvarsson*, 1999, *Wu et al.*, 1999]. Although extremely beneficial in characterizing matrix material properties, these studies do not capture the bulk response of the system, particularly the combined influence of fracture-matrix interactions on transport processes. Therefore, the environmental tracer study described here is used to validate and enhance our conceptual and operational model of flow and transport in Yucca Mountain. Recently, the Busted Butte in situ transport test [*Bussod*, 1999] began providing results for solute migration in unsaturated tuff at spatial scales up to 10 m. However, environmental tracers such as ^{36}Cl provide a much longer time scale and larger sample volume for characterization, although under less controlled conditions.

4. CONCEPTUAL MODEL OF UNSATURATED ZONE FLOW AND CHLORINE-36 TRANSPORT AT YUCCA MOUNTAIN

A conceptual model of the spatial distribution of net infiltration developed by *Flint et al.* [1996] describes the effect of precipitation, runoff, evapotranspiration and redistribution of water in the shallow unsaturated zone. Field measurements of water content profiles indicate that fracture flow in the upper welded tuff unit (TCw) is initiated when the soil becomes saturated, or nearly saturated, at the soil/tuff interface. Thus, the amount and timing of precipitation, combined with soil thickness, soil properties, and bedrock properties, determine whether infiltrated precipitation can percolate below the zone dominated by evapotranspiration to become potential recharge to groundwater. Infiltration also has a significant temporal component; net infiltration occurs primarily in winter over a 3 to 4 week period, with 0 to 2 significant infiltration events per year over the central block under current climatic conditions [*Flint et al.*, 1996].

The complexity of the spatial and temporal patterns of infiltration makes it difficult to correlate measurements taken at depths of a few hundred meters with observable surface features. We are not concerned in this paper with modeling infiltration, which is treated as a boundary condition—defined as either a constant rate or a spatially variable rate, based on the work of *Flint et al.* [1996]. Rather, the purpose of this paper is to show how the combined analysis of the ^{36}Cl data with a UZ transport model provides estimates of upper and lower bounds for net infiltration over the past 10 to 30 thousand years and may provide insight into the spatial redistribution with depth of the surface infiltration rates of *Flint et al.* [1996].

Below the zone of evapotranspiration, water percolates through the unsaturated zone through a variety of pathways. What follows is a summary of our current conceptual model underlying numerical calculations for which model parameters and boundary conditions can be evaluated.

- Downward migration through the TCw occurs via rapid fracture flow to the top of the PTn. This aspect of the conceptual model is supported by neutron probe data obtained from shallow boreholes and by the extensive presence of bomb-pulse ^{36}Cl in almost all samples and cuttings from shallow boreholes [*Fabryka-Martin et al.*, 1997]. Intrinsic permeabilities in the TCw matrix are on the order of 1×10^{-18} m^2 [*Flint*, 1998; *Wu et al.*, 1999], but with 1 to 2 fractures per meter (f/m), there is little resistance to flow in this unit, leading to high fracture water velocities and short travel times through this unit. Thus, groundwater ages in the TCw are generally young, although the small component of the total flow that partitions into the low-permeability matrix may have very long residence times.
- A transition from fracture-dominated flow to matrix-dominated flow occurs when the flow enters the PTn, at least in areas away from discrete fracture flow paths expected to be associated with through-cutting faults. The PTn matrix material has relatively high intrinsic permeabilities (on the order of 1×10^{-13} m^2) and porosities (approximately 30%), low fracture densities (0.3 to 0.8 f/m), and observable heterogeneity of properties in vertical profiles. Low saturations in this unit tend to favor imbibition of water from fractures into the matrix. Given these properties, together with variability in PTn thickness and percolation flux, the ages of matrix water at the bottom of the PTn are expected to be between 500 and 20,000 yr. Older ages are expected in the north, where the PTn is thicker. However, in fault zones where fracture flow may be sustained throughout the entire PTn thickness, travel times through this unit may be on the order of years for a very small fraction of the total volumetric flux.
- In general, the transition from fracture to matrix flow in the PTn is expected to diminish seasonal and even decadal climate variability. However, fault zones that provide continuous fracture pathways through the PTn may not completely damp this climate variability. Two of the principal questions concerning the conceptual model are (1) what hydrogeologic features control isolated fast pathways through the PTn, and (2) to what extent does the PTn damp extreme episodic infiltration pulses to provide a uniform flux at the potential repository horizon. *Fabryka-Martin et al.* [1997] investigated the difference between a model that assumed a time-invariant infiltration rate and one that assumed episodic infiltration that varied daily. The simulations indicated that, away from fault zones, the PTn tends to damp episodic variability on a yearly basis. In fault zones, however, the daily variability in infiltration flux may be preserved at the potential repository horizon.
- A transition from matrix-dominated flow back to fracture-dominated flow occurs at the bottom of the PTn and the top of the TSw. In the central block, TSw permeability is on the order of 1×10^{-17} m^2 with fracture densities of 2 to 3 f/m, allowing for fast travel times from the base of the PTn through the level of the potential repository horizon to the top of the basal vitrophyre of the TSw.

5. CHLORINE-36 DISTRIBUTION IN THE UNSATURATED ZONE

Approximately 6000 m of the ESF are in the TSw unit, with the north and south ESF access drifts transecting the overlying TCw and PTn units. In the ESF, 285 rock samples have been analyzed, and 201 of these come from the TSw. The majority of the sampling locations target specific features that are visible on the tunnel walls. More than half of the samples were collected in fault zones (including those with exposures of large faults known to correlate with faults mapped at the surface as well as minor offsets of only local extent). Others sampled fractures, breccias, unit and subunit contacts, lithophysae, and zones of high moisture content. Approximately 20% of the samples were systematic, collected at predetermined intervals along the tunnel, but more than half of these samples also turn out to be associated with features such as steeply dipping joints. Fewer than 15% of the samples were collected away from such features, which are prevalent in the densely welded units.

Sample preparation involves leaching the rock samples (consisting of 1 to 5 kg of material) with an equal mass of deionized water to extract soluble salts. The samples are gently crushed, but no more than is necessary to reduce them to 1–2 cm fragments because the chlorine of interest is on the outer surfaces of particles or fractures and it is desirable to minimize the extraction of fluid inclusion salts, which contain much older chlorine with negligible ^{36}Cl. Another potential source of bias in the results is the drilling water used in the tunnel, which is traced with bromide. Measured ^{36}Cl/Cl ratios are corrected for these effects using measured Br/Cl ratios [*Fabryka-Martin et al.*, 1997].

Replicate analyses for samples of homogeneous material have been shown to agree within the estimated measurement error. However, for heterogeneous material with a bomb-pulse component, replicated results have

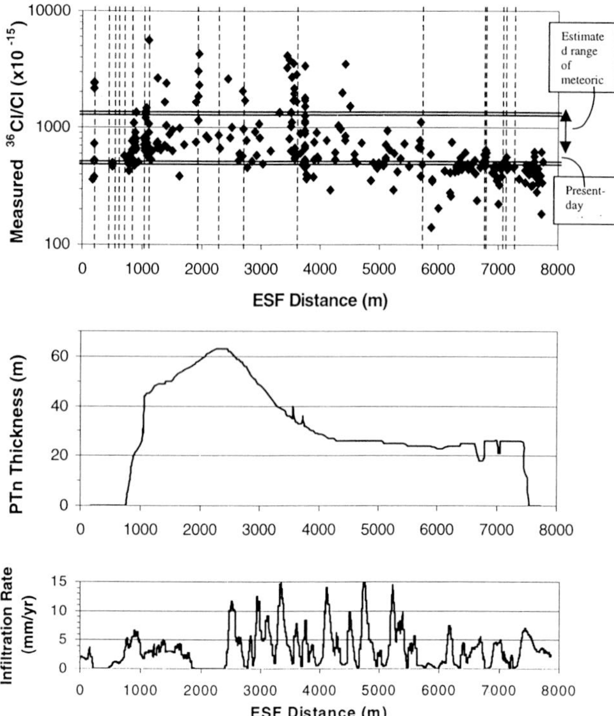

Figure 3. Measured $^{36}Cl/Cl$ ratios in the ESF and locations of faults mapped at the surface, overlying PTn thickness, and estimated infiltration rate [*Flint et al.*, 1996]. Vertical lines indicate locations of faults mapped at the surface.

differed by up to a factor of two. Even greater differences are found if one of the samples is crushed to a greater extent, confirming the role of fluid inclusions in the rock as a potentially significant, downward biasing factor.

6. DATA ANALYSIS

In an earlier work, *Murphy* [1998] used outlier identification techniques to differentiate ESF samples with a bomb-pulse component from all others. The main numerical result of this work was the selection of a cutoff for the $^{36}Cl/Cl$ ratio of about 1250×10^{-15}. However, a semi-log plot, (Figure 3) of the ratios as a function of distance along the ESF tunnel, as well as a histogram of these data (Figure 4) suggest that the simple two-component mixing model (background and bomb-pulse) described by *Murphy* [1998] cannot fully capture the structure of the background distribution. The complexities of the structure at background levels are only revealed when the ratios are plotted on a logarithmic scale because the high bomb-pulse results obscure the low-level variations on a linear scale. In particular, there is a trend to the part of the data from the TSw that correlates well with the thickness of the overlying PTn, extracted from the geologic model of *Clayton et al.*, [1997], which is also shown in the middle of Figure 3. Figure 4 suggests that the background below 1250×10^{-15} is weakly bimodal and needs to be represented by a mixture distribution.

A preliminary three-component statistical model was developed by *Fabryka-Martin et al.* [1998b]. Currently, a four component model is being completed and documented. In addition to the bomb-pulse component, with a mean ratio well above 1250×10^{-15}, the mixing of Holocene, Pleistocene, and so-called "Old" components are considered, where Old represents water old enough that decay of ^{36}Cl leads to ratios significantly less than the Holocene level of about 500×10^{-15}. The mean values for Holocene and late Pleistocene $^{36}Cl/Cl$ ratios in the packrat middens measured by *Plummer et al.* [1997] are 575×10^{-15} and 1050×10^{-15}, respectively. The shape of Figure 4 indicates that the data distribution may be fit with such components in addition to the bomb-pulse and Old components.

One of the most important questions associated with interpretation of this data set is whether measured ratios above the Holocene background are late Pleistocene (≥ 10 ka) or a mixture of bomb-pulse and Holocene water. Examination of Figure 3 provides significant insight into this issue. Neglecting for the moment the unambiguous bomb-pulse measurements, those greater than 1250×10^{-15}, there is a clear trend in the background data that follows the thickness of the overlying PTn. Where the PTn is thick, the ratios are generally higher than present-day background. As the PTn thins to the south, the ratios approach the Holocene background signal, with a few dropping well below 500×10^{-15}. Ignoring those very low values, the trend of decreasing values with decreasing PTn thickness is consistent with the conceptual model. Where the PTn is thicker, travel times to the ESF are longer, increasing the probability of a Pleistocene component. Where the PTn is thinner, travel times are shorter, providing predominantly Holocene water at the ESF. Those samples falling well below 500×10^{-15} may be indicative of very old water, in which the atmospheric ^{36}Cl component has decayed substantially. These may be representative of zones below which mixing due to lateral transport from higher infiltration zones occurs to a lesser degree than in the north. Certainly, *Flint et al.* [1996] estimate that there are zones with zero net infiltration along the trace of the southern ESF.

All of the unambiguous bomb-pulse measurements are made on samples from the northern ESF. The only possible bomb-pulse sample in the south occurs where the Ghost Dance fault crosses the ESF about 5700 m into the tunnel. Because the bomb-pulse signals occur in the north, where the PTn is thick, it is unlikely that the other elevated signals in the north indicate mixing of Holocene water with bomb-pulse water. If that were the case, then it would make sense that such signals would be also found in the south. An alternative proposed explanation is that all of the water containing bomb-pulse ^{36}Cl had already flushed through the ESF horizon in the south, such that the current signals represent water younger than about 20 yr because the atmospheric signal has returned to Holocene background. However, when testing this hypothesis, *Fabryka-Martin et al.* [1997] found ubiquitous bomb-pulse signatures in surface samples above the southern ESF, indicating that the bomb-pulse ^{36}Cl has, in fact, not been flushed out of the overlying rocks at this point.

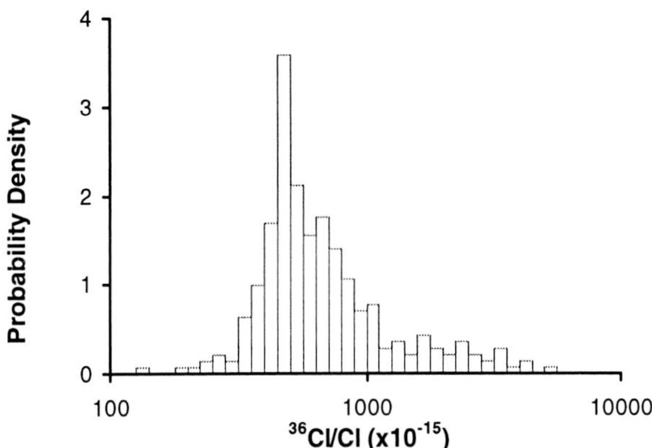

Figure 4. Histogram of ^{36}Cl/Cl ratios measured in ESF samples.

7. CORROBORATION WITH OTHER UZ DATA SETS

Our conceptual model suggests that the elevated, but not bomb-pulse, ^{36}Cl/Cl ratios in the north reflect Pleistocene water due to residence times of at least 10,000 yr in the PTn. Carbon-14 and stable hydrogen isotopes provide evidence for the presence of Pleistocene water in the subsurface at Yucca Mountain [*Yang et al.*, 1996, 1998]. Uncorrected ^{14}C ages of perched waters and pore waters from the TSw basal vitrophyre range up to 20,000 yr, although questions regarding corrections to the ages remain to be resolved. For example, the initial ^{14}C activity in the soil zone and the extent of isotopic exchange with gas-phase CO_2 or with calcite still need to be established. The lighter values measured for hydrogen isotopes in the perched waters suggest that much of the water at this depth originated during a time of colder climate, e.g., during the last period of glaciation, as compared to the origin of pore water in the PTn. While the presence of bomb-pulse ^{36}Cl below the PTn shows that some water moves quickly into the deep subsurface, the absence of clear bomb-pulse signals in perched water samples indicates that the flux of modern water (> 50 yr) from the surface downward through fractures to perched water bodies is probably small.

8. TESTING OF THE CONCEPTUAL MODEL WITH CHLORINE-36 DATA

The preceding conceptual model suggests ways to use the ^{36}Cl/Cl ratio data from the ESF. The Cl in any particular sample is assumed to represent a mixture of water parcels, each having different travel times as a result of traveling along different flow paths. As indicated above, different signals may be observed in these mixtures: the extremely high bomb-pulse level, the background throughout the Holocene, a higher background during the late Pleistocene, and older water in which the cosmogenic ^{36}Cl has measurably decayed. Of course, these signals are actually distributions along a continuum, not discrete or well-defined levels.

The first hypothesis concerns the bomb-pulse signal. The second hypothesis addresses the natural or background signal.

8.1 Hypothesis 1

8.1.1 Hypothesis 1a. The presence of bomb-pulse ^{36}Cl in TSw samples collected from the ESF records movement of waters in 50 yr or less through the entire PTn above those sampled locations, indicating that some flow followed structural pathways (e.g., faults), largely bypassing the PTn matrix. Based on our conceptual model, three conditions appear to be necessary for travel times of less than 50 yr to these sampled locations: (a) the residence time of water in the surface alluvium must be less than 50 yr; (b) a continuous fracture path must extend from the alluvium/bedrock contact to the sampled depth; and (c) the magnitude of surface infiltration must be sufficiently high to initiate and sustain at least a small component of fracture flow along such a connected pathway all the way through the PTn.

8.1.2 Hypothesis 1b. An alternative but related hypothesis requires only localized high saturation within the matrix to provide sufficient pore-water velocities for solute travel times of less than 50 yr through the PTn. Although such water contents have not been observed, they may be feasible if local conditions converge flow paths into an effectively saturated finger. The primary practical difference between 1a and 1b is that 1b does not require a structural feature for a fast flow path.

The major distinction between Hypotheses 1a and 1b is that if fast paths are restricted to fault zones, then waste could be placed so as to avoid such locations. If fast paths are randomly distributed, then structural mapping provides no design benefit for protecting canisters against episodic seepage or for isolating waste from potential fast paths in the nonwelded tuff below the TSw. Proponents of 1b argue that locations where fast paths can be supported may be random or highly uncertain. Figure 3 shows some correlation between the mapped faults and locations where bomb-pulse ^{36}Cl is found. However, for Hypothesis 1a, we would expect to actually find ESF bomb-pulse ^{36}Cl in a fault zone only if the fault were vertical. Otherwise, the TSw is amply fractured to provide multiple pathways once the bomb-pulse ^{36}Cl penetrates to the bottom of the PTn. Further statistical analysis would be required to quantify the probability of bomb-pulse measurements being correlated with fault zones. Recent observations at the Busted Butte Unsaturated Zone Transport Test [*Bussod,* 1999], which show that capillary forces dominate flow in bedded tuff similar to the PTn, may be instructive. Even with the application of extremely high infiltration rates, locally saturated channels have not been observed in the matrix at this field site. Finally, it is worth noting that fast paths do not indicate increased flux, they merely indicate travel times from the surface to the ESF. Conceivably, the flow rate into the potential repository could be the same or greater away from a fast pathway as it is within it, with the difference being how long ago water entered the system. However, fast paths from the potential repository to the water table are important for contaminant transport simulations and repository performance assessment, so understanding the physical controls on the system is critical.

8.2 Hypothesis 2

Away from structural pathways through the PTn, the travel time for most water encountered below the PTn unit is greater than 50 but less than 50,000 yr, so neither the bomb-pulse signal nor the decayed component contributes significantly to the result. The thickness of the overlying PTn unit is the most significant determinant of total travel times to these locations, although some of the other factors indicated by the conceptual model—thickness of alluvium and infiltration rates, in particular—may also be influential.

9. NUMERICAL SIMULATIONS

Simulations of flow and chloride (including ^{36}Cl) transport in the unsaturated zone at Yucca Mountain are performed using the finite element heat and mass transfer code, FEHM [*Zyvoloski et al.*, 1997]. FEHM has been applied to comprehensive site-scale transport models of the unsaturated zone at Yucca Mountain [*Robinson et al.* 1997] and numerous process level model studies, including reactive transport of radionuclides from the proposed repository to the water table [*Viswanathan et al.*, 1997; *Robinson et al.*, 1999]. In this study, all simulations utilize the dual-permeability formulation to represent flow and transport in coupled fracture and matrix continua. *Wu et al.* [1997, 1999] developed the hydrologic properties for both the matrix and fractures of each stratigraphic unit through inversion to saturation measured in boreholes. Using these properties, *Robinson et al.* [1997] matched the observed saturations at Well SD-9 with FEHM using the infiltration estimate of *Flint et al.* [1996] for that spatial location.

Dual permeability transport simulations that capture high fracture velocities as well as fracture-matrix interactions accurately and efficiently are achieved with the residence time transfer function (RTTF) particle tracking algorithm [*Zyvoloski et al.*, 1997; *Robinson et al.*, 1997]. Dual-permeability RTTF particle tracking simulations in this study involve releasing a swarm of particles instantaneously with the infiltrating water. Each particle then travels along a unique pathway governed by advection in fractures, imbibition into the matrix, diffusion into the matrix, and advection in the matrix. These pathways lead to a distribution of arrival times at all locations in the model, including the ESF. The distribution of particle arrival times at any location can then be interpreted directly as the distribution of ages represented by a sample collected there, hence the mixture of fluid ages in such a sample. This type of interpretation is used with the one-dimensional simulations in this study. The method differs significantly from traditional particle tracking methods, which map a particle's trajectory by interpolating velocities at locations between grid points. Because such interpolations are not required by the RTTF method, it is computationally very efficient, handling several million particles rapidly on a modern workstation. Further, because it is a cell-based method using mass flow rates directly from the flow solution, models based on unstructured grids, such as those used in this study, pose no additional complications.

A further benefit of the RTTF method for this study is the efficiency it lends to incorporating a time-variant solute input function. The age distributions are convolved with the reconstructed time history of ^{36}Cl/Cl ratios for infiltrating water and decay corrected to simulate the spatial distribution of ^{36}Cl/Cl ratios through out the domain. This method is used for the two-dimensional simulations in this

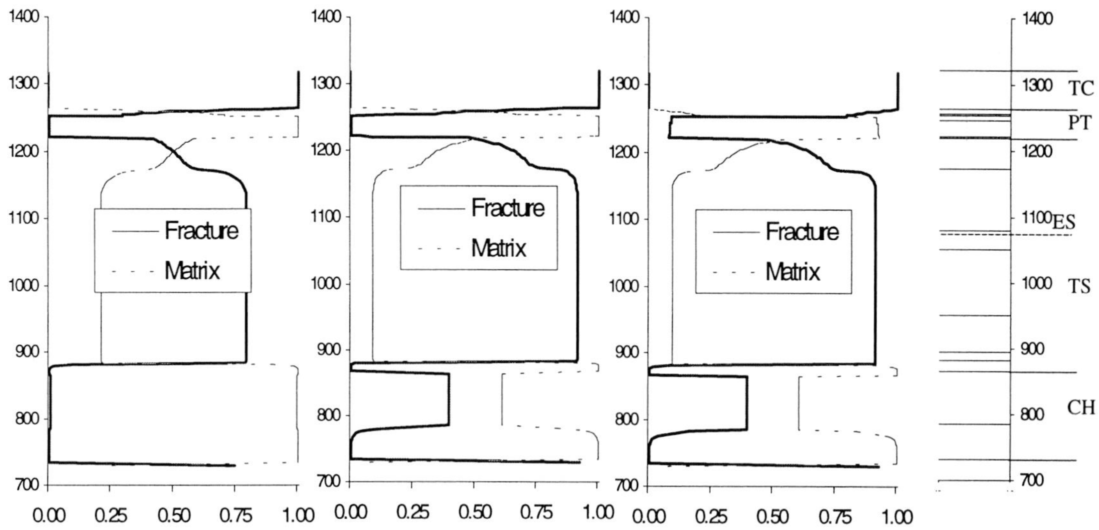

Figure 5. Station 36+00 stratigraphic profile and dual permeability flow simulations showing flux distribution between matrix and fracture continua for A) base case 1 mm/yr; B) base case, 5 mm/yr; and C) PTn fault properties, 5 mm/yr.

study. Here, a reconstructed time history of $^{36}Cl/Cl$ ratios for infiltrating water over the past 500,000 years, which is consistent with the data of Plummer et al. [1997], is used in the model input. *Fabryka-Martin et al.* [1997] describe the historical signal and its derivation in detail.

9.1 One-Dimensional Model: In and Away From Fault Zones

For the one-dimensional simulations described below, a vertical profile was extracted from the three-dimensional site model [*Clayton et al.*, 1997] at ESF Station 36+00. Although the site model does not contain a fault at this location, the Sundance fault is mapped at ground surface near this location and several ESF samples from here were found to contain bomb-pulse $^{36}Cl/Cl$ ratios (see Figure 3). The vertical, one-dimensional model was discretized to capture the stratigraphic units and their 20 subunits in the UZ profile at this location. This simple model was then used to evaluate the effects of infiltration rate and PTn properties in and away from fault zones on UZ flow and transport rates. Starting with the base-case properties of *Wu et al.* [1997], flow in this dual permeability system was simulated for infiltration rates ranging from 0.1 to 5.0 mm/yr. Then, in a second set of simulations, the PTn properties were modified from the base case to represent increased fracture permeability due to faulting; and the bulk fracture permeability in the PTn was increased by one order of magnitude, consistent with air-permeability testing in that unit [*LeCain*, 1997]. The properties of the TCw and TSw were not modified in these simulations because modeled travel times through them are already short using base-case properties.

Figure 5 shows the stratigraphy and simulated fluxes in the matrix and fracture continua for the dual permeability models. In simulations using base-case properties, infiltration rates of 1 to 5 mm/yr were not capable of sustaining fracture flow through the PTn. However, a small component of fracture flow was sustained throughout the PTn with the fault-zone properties for infiltration rates greater than about 1 mm/yr. In these simulations, infiltration is time averaged over the entire year. In reality, infiltration occurs as episodic events associated with rainstorms and occasional snow melt. The suitability of a time-averaged infiltration boundary condition for these simulations has been evaluated by *Fabryka-Martin et al.* [1997]. That study indicates that the PTn is highly effective at damping episodic infiltration events away from fault zones. In fault zones, however, subannual infiltration events are not perfectly damped in the simulations and some variation is preserved at the bottom of the PTn. Episodic infiltration in fault zones has little effect on travel time to the ESF. Rather, it indicates that the flux at the potential repository horizon may vary at a subannual time scale in fault zones.

We modeled water-age distributions at the ESF by combining the flow field simulations with the RTTF particle-tracking module in FEHM. Figure 6 shows the distribution of water ages at the ESF for base-case properties and 1 mm/yr infiltration. The wide distribution

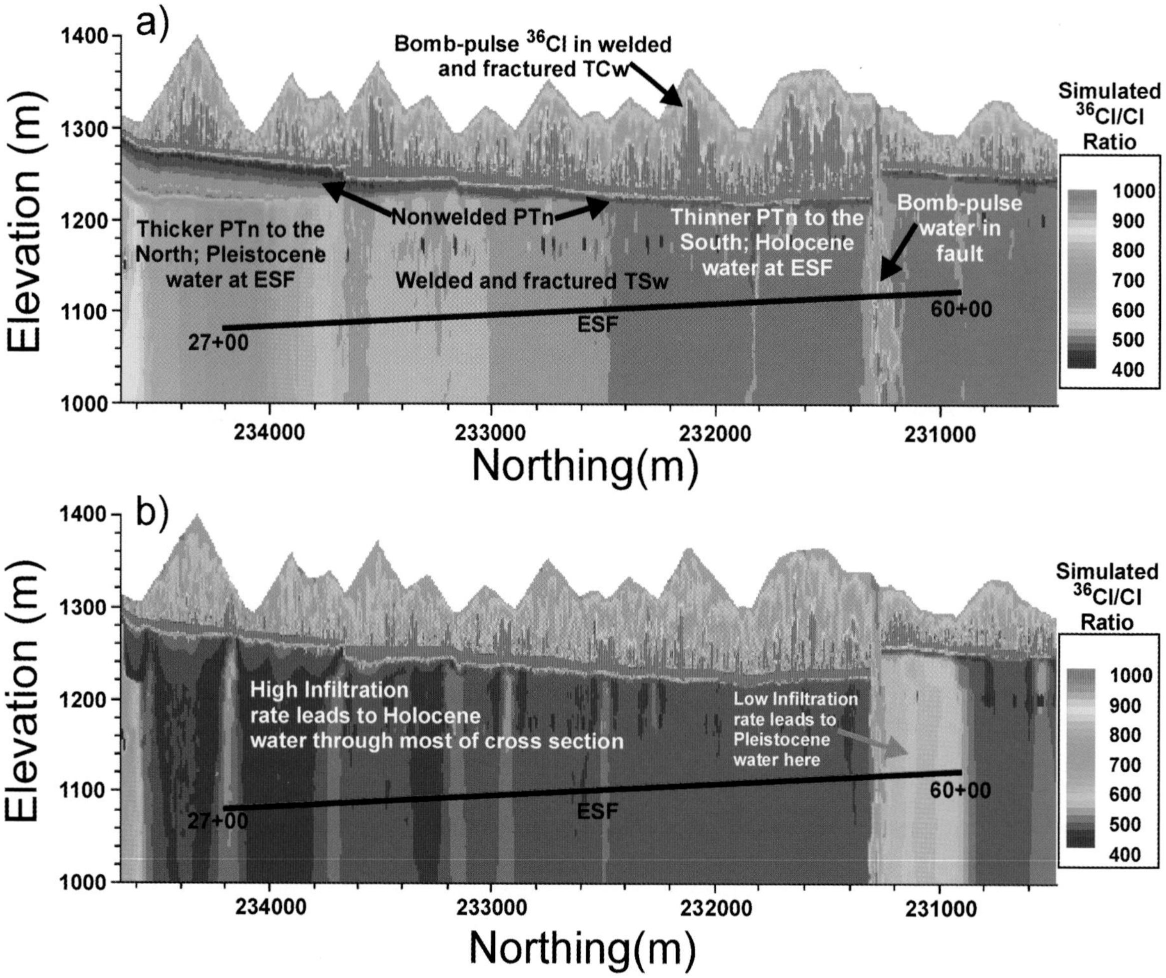

Plate 1. Simulated $^{36}Cl/Cl$ ratios on a north-south cross section along main drift of ESF. Infiltration rate is 1 mm/yr over the entire domain. Infiltration rate is from *Flint et al.* [1996] and varies spatially over the entire domain (Figure 3). Note that only for the fixed infiltration rate is the trend of Pleistocene water in the north and Holocene water in the south preserved.

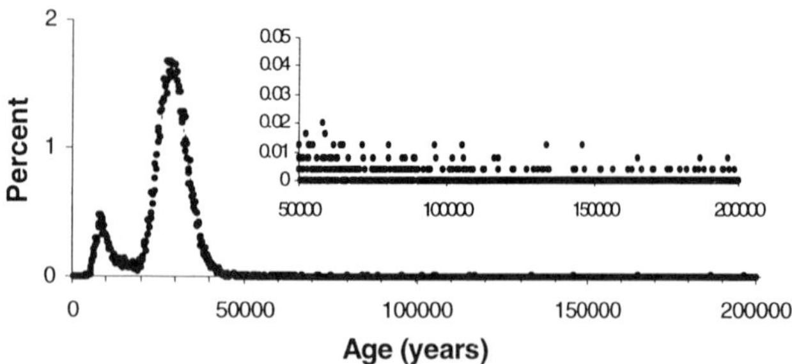

Figure 6. Simulated age distribution for a single sample collected at the ESF. Note the mixture of water from before and after the end of the Pleistocene, ten thousand years ago. The youngest age contribution in this simulated sample is 2800 yr. This one-dimensional simulation at ESF Station 36+00 uses base-case properties and 1 mm/yr infiltration rate.

of ages may seem surprising for a one-dimensional model. However, there are nine different hydrologic subunits between ground surface and the ESF, each of which has continuously changing fracture and matrix flow pathways. For this simulation, the younger ages are most representative of fracture flow in the welded tuffs and upper PTn while the older ages represent components of matrix flow in the welded tuffs as well. Integrating under age distribution curves such as Figure 6 provides cumulative age distribution curves (Figure 7). Although the two different types of curves contain the same information, the latter are more useful for comparing the results of multiple different simulations. Figure 7 shows simulated cumulative age distributions for a sample at the ESF using the one-dimensional column model and infiltration rates ranging from 0.1 to 5.0 mm/yr.

Although simple in design and implementation, this one-dimensional model is nonetheless a valuable tool for interpreting the data and assessing alternative conceptual models. Most importantly, these simulations demonstrate that the conceptual model for solute transport in and away from fault zones is consistent with the ESF ^{36}Cl data, as well as with other independent estimates of infiltration rates, such as those based on measurements of moisture content in shallow boreholes in conjunction with alluvial thickness and precipitation [*Flint et al.*, 1996], or on chloride concentration in ESF pore-water samples [*Fabryka-Martin et al.*, 1998a]. Namely, for base-case properties and infiltration rates between 1 and 5 mm/yr, the distribution of ages in a sample at the ESF includes some Pleistocene water, although more for 1 mm/yr than for 5 mm/yr infiltration rates. For this same range of infiltration rates, some fracture flow is sustained throughout the PTn when fault zone properties are used, providing a contribution of bomb-pulse ^{36}Cl to an ESF sample. Lower infiltration rates are inconsistent with the data because no travel times from the ground surface to the ESF of less than 50 years are simulated in fault zones. Furthermore, away from fault zones, infiltration rates less than 1 mm/yr lead to simulated ages that are mostly greater than 100,000 years. Such long travel times would result in a signal of decayed ^{36}Cl, inconsistent with elevated signals at the end of the Pleistocene and the elevated ratios measured in the ESF. On the other hand, infiltration rates greater than 5 mm/yr result in bomb-pulse ages in fault zones, but only in Holocene age samples away from fault zones. Thus, the data and the modeling serve to narrow the viable range of average annual infiltration rates to about 1 to 5 mm/yr at Station 36+00.

9.2 Two-Dimensional Model Along Main Drift of ESF

A two-dimensional transect along the main drift of the ESF is used to investigate effects of spatially varying properties, including the PTn thickness and the infiltration rate. Also, the simulated $^{36}Cl/Cl$ ratios achieved by convolving model age distributions with the time-varying source of *Fabryka-Martin et al.* [1997] are demonstrated with this example. Plate 1 contrasts the model results for a simulation using the spatially varying infiltration rate of *Flint et al.* [1996] (see bottom of Figure 3) with results obtained for a simulation with a fixed infiltration rate of 1 mm/yr. The fixed infiltration rate of 1 mm/yr preserves the trend of higher ratios in the north, where the PTn is thicker, and lower ratios in the south, where the PTn is thinner. However, when the spatially varying infiltration rate of *Flint et al.* [1996] is used, the only trend in the simulated ratios is an inverse relationship with the model infiltration rate. The simulation with the fixed infiltration rate matches the trend of the data significantly better than the simulation with the spatially varying infiltration rate. This result suggests that if the spatially varying infiltration estimates are realistic, then the site-scale flow and transport models need to simulate more mixing and lateral redistribution of infiltration in the PTn. If, on the other hand, vertical flow paths dominate through the PTn, then these results suggest that either the estimated infiltration rates of *Flint et al.* [1996] are too high or that the hydrologic properties of *Wu et al.* [1997] do not provide enough pore-water residence time in the PTn or, perhaps, in the matrix of the upper TSw. Thus, the combination of ESF ^{36}Cl data examined in conjunction with dual permeability flow and transport models provides additional calibration targets for site-scale models which ultimately consider flow into and radionuclide transport away from the proposed repository.

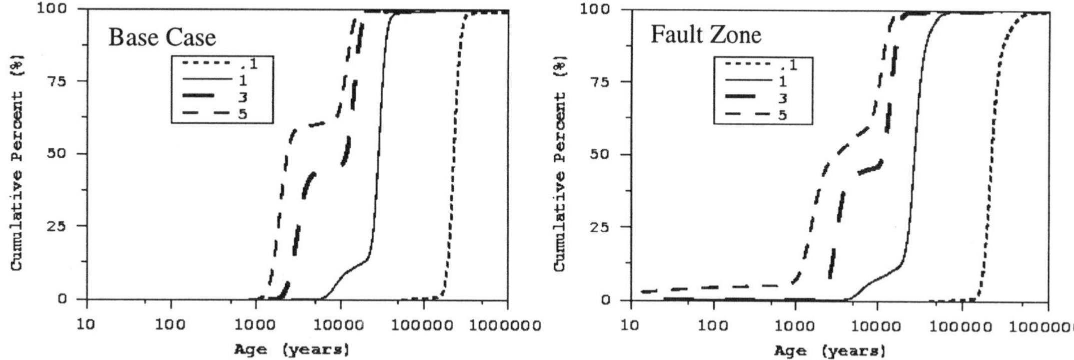

Figure 7. Simulated cumulative age distributions at ESF Station 36+00 with one-dimensional model and varying infiltration rate (specified as mm/yr in legend). Fault zone properties involve one order of magnitude increase in PTn bulk fracture permeability. Young ages simulated with fault-zone properties are consistent with bomb-pulse ^{36}Cl measurements near fault zones in the ESF.

10. SUMMARY

Analysis of measured ^{36}Cl/Cl ratios in ESF samples coupled with dual permeability flow and transport modeling provide an effective tool for evaluating conceptual models and parameter sets associated with UZ characterization for Yucca Mountain. The first part of the conceptual model addresses fast paths from ground surface to the ESF. The ^{36}Cl data confirm the existence of fast paths without ambiguity. Because travel times through the PTn matrix could not be less than 50 yr at observed field saturations, unique local conditions are required to sustain fast paths. There is debate as to whether fast paths are correlated with through-cutting faults or whether they could result from the convergence of flow paths leading to local, saturated fingers in the PTn matrix. Locally saturated fingers penetrating the entire PTn with velocities high enough to constitute fast paths have not been observed. Bomb-pulse ^{36}Cl has been found in the Ghost-Dance fault alcove in the ESF [*Fabryka-Martin et al.*, 1998b], and there is evidence that increased fracturing in the PTn occurs in fault zones. However, flowing water in fault-zone fractures has not been observed in the field. Preliminary inspection of the data seems to indicate a correlation between measurements of bomb-pulse ^{36}Cl and fault zones mapped at ground surface. However, statistical analysis is required to quantify the correlation. The hypothesis that fault zones are required for fast paths is strengthened with numerical transport modeling. We have not found that, using base-case hydrologic rock properties, model results predict travel times from ground surface to the ESF in less than 50 yr. However, slight property modifications to simulate increased PTn fracturing in fault zones, consistent with field measurement of air permeability, lead to prediction of bomb-pulse ^{36}Cl at the ESF.

The second part of the conceptual model addresses transport times to the ESF away from fast pathways. The measured ^{36}Cl/Cl ratios greater than present-day background either contain some bomb-pulse ^{36}Cl or they represent infiltration that occurred at the end of the Pleistocene, more than 10,000 yr ago. The correlation between elevated, non- bomb-pulse signals and greater PTn thickness is strong, supporting the hypothesis that those measurements in the northern ESF represent ages greater than 10,000 yr. Further, there are no data or conceptual models that support the notion that bomb-pulse signals should be pervasive in the north—where the PTn is thicker—but nonexistent in the south. Finally, the same model that predicts bomb-pulse ^{36}Cl at the ESF in fault zones also predicts a Pleistocene component at the ESF away from fault zones, where the PTn is thicker. The model provides upper and lower bounds on infiltration rates for simulations that consistently predict some Pleistocene Epoch water in the northern ESF, where the PTn is thick; mostly Holocene water in the south where the PTn is thin; and bomb-pulse signals in the ESF in fault zones where the infiltration rate is more than about 1 mm/yr but not greater than 5 mm/yr.

Acknowledgments. This work was supported by the Yucca Mountain Site Characterization Program Office as part of the Civilian Radioactive Waste Management Program. This project is managed by the U.S. Department of Energy, Yucca Mountain Site Characterization Project. The authors thank Bruce Robinson, Schön Levy, Don Sweetkind, Ed Kwicklis, and Alan Flint for invaluable discussions and assistance with this work. We are

grateful to Carl Steefel and John Apps for their external review and comments on the manuscript, which have led to its improvement.

REFERENCES

Bandurraga, T. M., and G. S. Bodvarsson, Calibrating hydrogeologic parameters for the 3-D site-scale unsaturated zone model of Yucca Mountain, Nevada, *J. Cont. Hydr.*, *38*(1–3), 25–46, 1999.

Bodvarsson, G. S., T. M. Bandurraga, and Y. S. Wu, *The Site-Scale Unsaturated Zone Model of Yucca Mountain, NV, for the Viability Assessment*. Technical Report LBNL-40378, Lawrence Berkeley National Laboratory, Berkeley, Calif., 1997.

Bodvarsson, G. S., W. Boyle, R. Patterson, and D. Williams, Overview of scientific investigations at Yucca Mountain—the potential repository for high-level waste, *J. Cont. Hydr.*, *38*(1–3), 2–24, 1999.

Buesch, D. C., R. W. Spengler, T. C. Moyer, and J. K. Geslin, *Revised Stratigraphic Nomenclature and Macroscopic Identification of Lithostratigraphic Units of the Paintbrush Group Exposed at Yucca Mountain, Nevada*, U.S. Geological Survey Open-File Report 94-496, U.S. Geological Survey, Denver, Colo., 1996.

Bussod, G. Y., H. J. Turin, and W. Lowry. *Busted Butte Unsaturated Zone Transport Test: Fiscal Year 1998 Status Report,* Technical Report LA-13670-SR, Los Alamos National Laboratory, Los Alamos, N.M., 1999.

Clayton, R. W., W. P. Zelinski and C. A. Rautman, *(CRWMS), ISM2. 0: A 3-D Geological Framework and Integrated Site Model of Yucca Mountain*, Doc ID B00000000-01717-5700-00004 Rev 0, MOL. 19970122. 0053, Civilian Radioactive Waste Management System Management and Operating Contractor, 1997.

Fabryka-Martin, J. T., A. L. Flint, D. S. Sweetkind, A. V. Wolfsberg, S. S. Levy, G. J. C. Roemer, J. L. Roach, L. E. Wolfsberg, and M. C. Duff, *Evaluation of Flow and Transport Models of Yucca Mountain, Based on Chlorine-36 Studies for FY97*, Yucca Mountain Project Milestone Report SP2224M3, Los Alamos National Laboratory, Los Alamos, N.M., 1997.

Fabryka-Martin, J. T, A. L. Flint, L. E. Flint, D. S. Sweetkind, D. Hudson, J. L. Roach, S. T. Winters, L. E. Wolfsberg, A. Bridges, Use of chloride to trace water movement in the unsaturated zone at Yucca Mountain, *Proceedings, 1998 8th Annual International High Level Radioactive Waste Management Conference*, pp. 264–268, American Nuclear Society, La Grange Park, Ill., 1998a.

Fabryka-Martin, J. T, A. V. Wolfsberg, S. S. Levy, K. Campbell, P. Tseng, J. L. Roach, and L. E. Wolfsberg, *Evaluation of Flow and Transport Models of Yucca Mountain, Based on Chlorine-36 and Chloride Studies for FY98*, Milestone Report SP33DDM4, Los Alamos National Laboratory, Los Alamos, N.M., 1998b.

Flint, L. E, *Characterization of Hydrogeologic Units Using Matrix Properties, Yucca Mountain, Nevada*, USGS Water Resour. Invest. Rep. 97-4243, U.S. Geological Survey, Denver, Colo., 1998.

Flint, A. L., J. A. Hevesi, J. A., and L. E. Flint, *Conceptual and Numerical Model of Infiltration for the Yucca Mountain Area, Nevada*, USGS Yucca Mountain Project Milestone Rep. 3GUI623M, U. S. Geol. Surv. Water Resour. Invest. Rep., U.S. Geological Survey, Denver, Colo., 1996.

LeCain, G. D, *Air Injection Testing in Vertical Boreholes in Welded and Nonwelded Tuff, Yucca Mountain, Nevada*, USGS Water Resour. Invest. Rep. 96-4262, U.S. Geological Survey, Denver, Colo., 1997.

LeCain, G. D., and G. L. Patterson, *Technical Analysis and Interpretation: ESF Air-Permeability and Hydrochemistry Data through January 31, 1997*, ACC: MOL. 19970415. 0387, USGS Yucca Mountain Project Milestone SPH35EM4, U.S. Geological Survey, Denver, Colo., 1997.

Murphy, W. M, Commentary on studies of 36Cl in the Exploratory Studies Facility at Yucca Mountain, Nevada, *Symposium Proceedings, Scientific Basis for Nuclear Waste Management XXI*, edited by I. G. Mckineley and C. McCombie, *506*, pp. 407–414, Material Research Society, 1998.

Plummer, M. A., F. M. Phillips, J. T. Fabryka-Martin, H. J. Turin, P. E. Wigand, and P. Sharma, Chlorine-36 in fossil rat urine: an archive of cosmogenic nuclide deposition during the past 40,000 years, *Science 277*, 538–541, 1997.

Robinson, B. A., A. V. Wolfsberg, H. S. Viswanathan, G. Y. Bussod, C. W. Gable, and A. Meijer, *The Site-Scale Unsaturated Zone Transport Model of Yucca Mountain*, Yucca Mountain Project Milestone SP25BM3, Los Alamos National Laboratory, Los Alamos, N.M., 1997.

Robinson, B. A., H. S. Viswanathan, and A. J. Valocchi, Efficient numerical techniques for modeling multicomponent groundwater transport based upon simultaneous solution of strongly coupled subsets of chemical components, *Adv. Water Res*, *23*(4) 307–324, 2000.

Scott, R. B., and R. Bonk, *Preliminary Geologic Map of Yucca Mountain, Nye County, Nevada, with Geologic Sections*, U.S. Geological Survey Report OFR-84-494, US Geological Survey, Denver, Colo., 1984.

Viswanathan, H. S., B. A. Robinson, A. J. Valocchi, and I. R. Triay, A reactive transport model of neptunium migration from the potential repository at Yucca Mountain, *J. Hydr.*, *209*, 251–290, 1998.

Wu, Y. S., A. C. Ritcey, C. F. Ahlers, A. K. Misra, J. J. Hinds, and G. S. Bodvarsson, Providing base-case flow fields for TSPA-VA: Evaluation of uncertainty of present day infiltration rates using DKM/base-case and DKM/weeps parameter sets, LBNL Yucca Mountain Project Milestone SLX01LB2, Lawrence Berkeley National Laboratory, Berkeley, Calif., 1997.

Wu, Y. S., A. C. Ritcey, and G. S. Bodvarsson, A site-scale model for fluid and heat flow in the unsaturated zone of Yucca Mountain, Nevada, *J. Cont. Hydr.*, *38*(1–3), 185–215, 1999.

Yang, I. C., G. W. Rattray, and K. M Scofield, Carbon and hydrogen isotopic compositions for pore water extracted from cores at Yucca Mountain, Nevada, *Proceedings, 1998 8th Annual International High Level Radioactive Waste Management Conference*, pp. 27–32, ACC: MOL. 19980330.0135, American Nuclear Society, La Grange Park, Ill., 1998.

Yang, I. C., G. W. Rattray, and P. Yu, *Interpretation of Chemical and Isotopic Data from Bore Holes in the Unsaturated Zone at Yucca Mountain, Nevada*, USGS Water-Resources Investigation Report 96-4058, MOL. 19970715.0408, TIC: 236260, U.S. Geological Survey, Denver, Colo., 1996.

Zyvoloski, G. A., B. A. Robinson, Z. V. Dash, and L. L. Trease, Summary of the Models and Methods for the FEHM Application—A Finite Element and Heat- and Mass-Transfer Code, Technical Report 13307-MS, Los Alamos National Laboratory, Los Alamos, N.M., 1997.

Andrew Wolfsberg, Katherine Campbell, and June Fabryka-Martin, Earth and Environmental Sciences Division, Los Alamos National Laboratory, Los Alamos, NM 87545

Pressure Interference Tests in a Fractured Geothermal Reservoir: Oguni Geothermal Field, Northern Kyushu, Japan

Sabodh K. Garg

Maxwell Technologies, Inc., San Diego, California

Shigetaka Nakanishi

Electric Power Development Co., Ltd., Tokyo, Japan

The subsurface stratigraphy in the Oguni Geothermal Field of the northwestern Hohi geothermal region, Japan, consists of a sequence of indurated sediments and volcanics overlying a granitic basement. The Hohi formation and the upper part of the Shishimuta formation constitute the principal geothermal aquifers. The feedzone pressures indicate that the northwestern Hohi region consists of two pressure zones: i.e., a high-pressure zone in the southern part of the Oguni Geothermal Field, and a low-pressure zone in the central and northern parts of the northwestern Hohi area. To delineate the permeability structure for the Oguni Geothermal Field, numerous pressure transient tests were performed. Analyses of pressure interference data indicate that the low-pressure zone has a transmissivity of 100–250 × $10^{-12}m^3$. In contrast to the high transmissivity of the low-pressure zone, the pressure interference data imply that the high-pressure zone has only a modest transmissivity of 8–15 × $10^{-12}m^3$. The pressure interference data are consistent with the presence of one or more no-flux boundaries between the low- and high-pressure zones. In addition, a significant degree of permeability heterogeneity is indicated within both the low- and high-pressure parts of the northwestern Hohi geothermal region, reflecting the fractured nature of the geothermal resource.

1. INTRODUCTION

The Oguni and the Sugawara Geothermal Fields together comprise the northwestern Hohi geothermal region, Kumamoto and Oita Prefectures, Kyushu, Japan (see Figure 1). An area of numerous hot springs, it is approximately 40 km southwest of the coastal resort of Beppu, and some 20 km north of Mt. Aso, an active caldera. The New Energy and Industrial Technology Development Organization (NEDO) carried out a 200-km^2 regional exploration program in the Hohi geothermal area during the years 1979–1985. The NEDO work resulted in the identification of a high-permeability geothermal prospect in the northwestern Hohi region. The Electric Power Development Co., Ltd. (EPDC), initiated a geothermal exploration program in the Oguni area in 1983. The Oguni field, delineated by the "GH" series boreholes, is located at the northeast end of Kumamoto Prefecture (Figure 1). The topography of the Oguni field is dominated by Mt. Waita, which rises to an elevation of about

Dynamics of Fluids in Fractured Rock
Geophysical Monograph 122

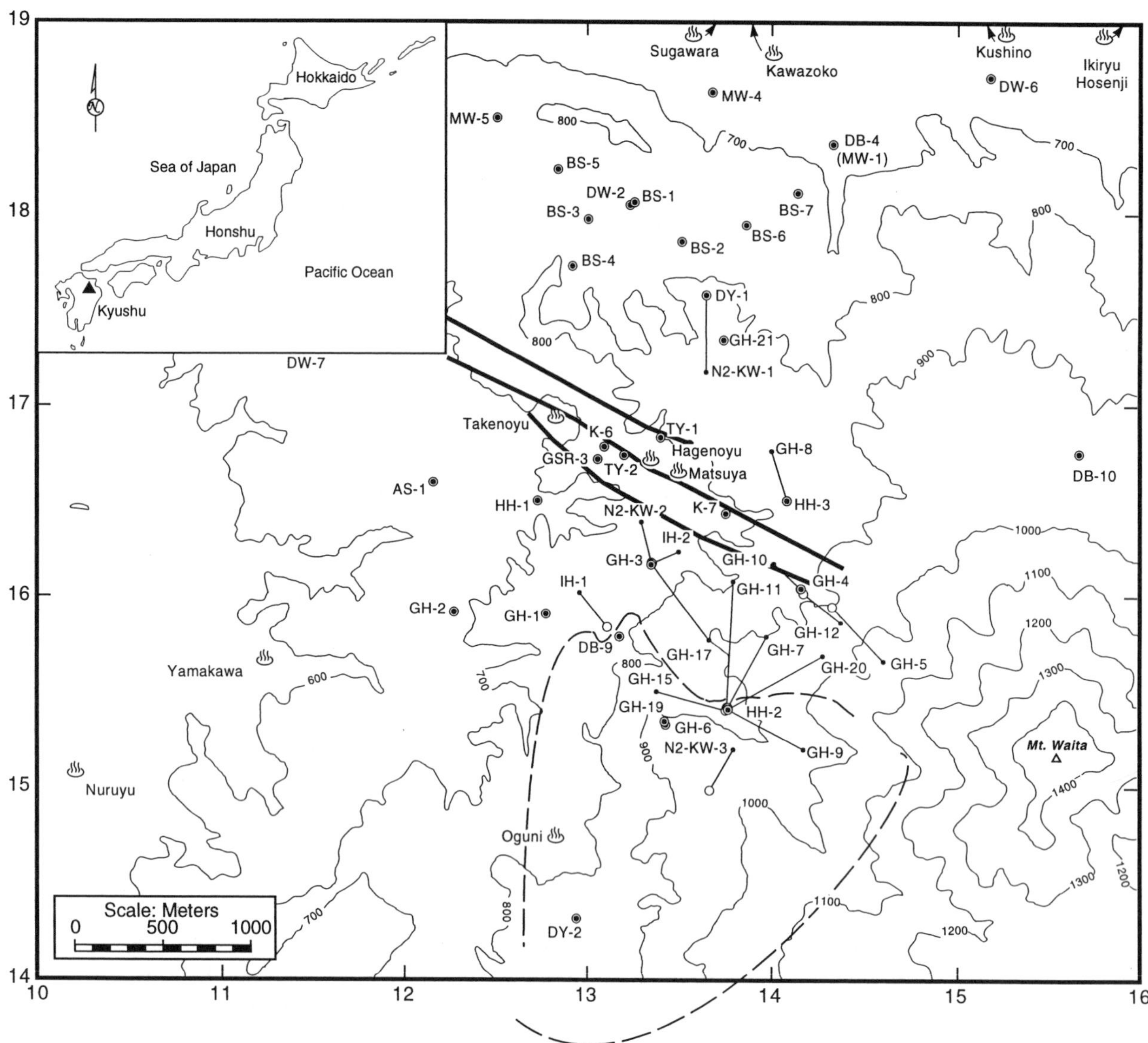

Figure 1. The Oguni and Sugawara geothermal fields, northwestern Hohi area, Kyushu, Japan. The Oguni Geothermal Field is delineated by the "GH" series boreholes; the "BS" series boreholes are drilled in the Sugawara Geothermal Field. The high-pressure zone boreholes are enclosed within the dashed boundary. The WNW–ESE faults are indicated by heavy solid lines. The inset map of Japan shows the location of the Hohi area (solid triangle). All horizontal distances are with respect to the Central Kyushu Co-ordinate System (CKCS). The origin of CKCS is located at 33°0' north latitude and 131°0' east longitude.

1500 m ASL (meters above sea level) to the southeast of the field. Many of the boreholes are located on the flanks of Mt. Waita. Striking WNW-ESE is the valley containing the hot spring areas and towns of Takenoyu and Hagenoyu. To the north of the valley is the Sugawara plateau where NEDO has drilled a number of boreholes (BS series, Figure 1) for a binary power plant. Although the northwestern Hohi region has been subdivided into two separate geothermal fields (Oguni, Sugawara), the area constitutes a single hydrological unit.

The subsurface stratigraphy in the northwestern Hohi region consists of a sequence of indurated sediments and volcanics overlying a granitic basement (Figure 2). The stratigraphic sequence above the basement is composed of the Pliocene Taio formation, the late Pliocene/early Pleistocene Shishimuta formation (pre-Kusu group), the

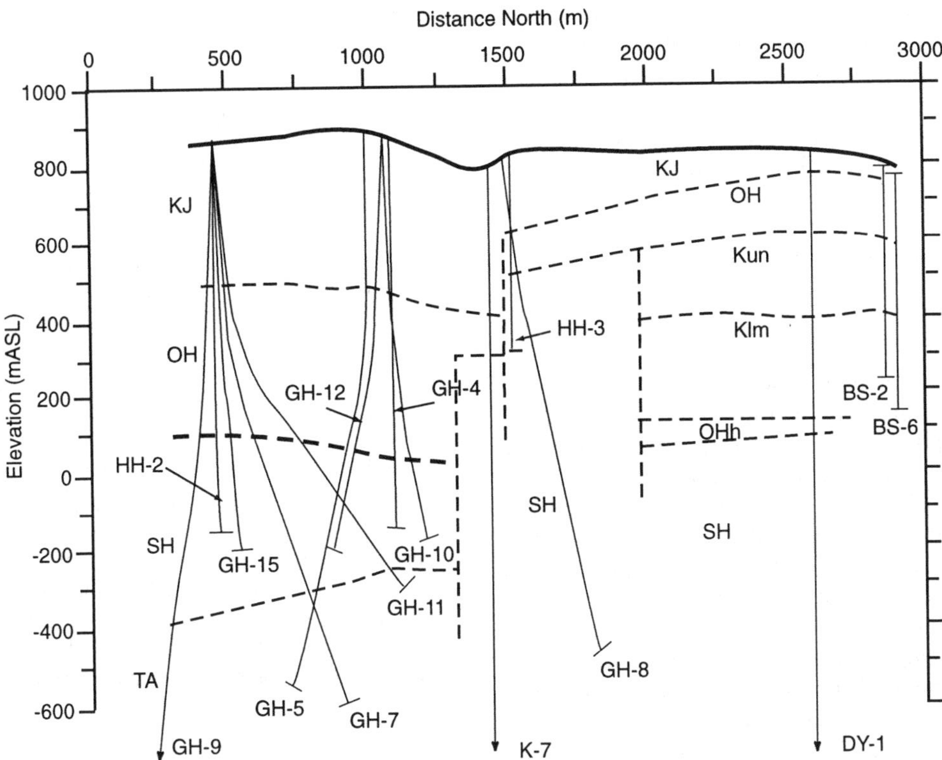

Figure 2. A schematic south-north stratigraphic cross section striking north along east-west CKCS coordinate 14 (see Figure 1). The ordinate is meters north from north-south CKCS coordinate 15. The following abbreviations are used for formation names: KJ (Kusu); Kun (Kusu/Nogami mudstone); Klm (Kusu/Machida lava); OH (Hohi); OHh (Hohi/Hatchobaru lava); SH (Shishimuta); TA (Taio).

lower-to-middle Pleistocene Hohi and Kusu formations, and the upper Pleistocene Kuju formation. The Nogami mudstones (part of the Kusu group) and the Kuju volcanics appear to function as a caprock for the geothermal system. Based upon feedpoint locations, it appears that the Hohi formation and the upper part of the Shishimuta formation constitute the principal geothermal aquifers. The Hohi formation consists of lavas and pyroxene andesitic rocks, whereas the Shishimuta formation is primarily dacitic pyroclastics. The elevation of the top of the Shishimuta formation varies between about 50 and 150 m ASL except in the Takenoyu-Hagenoyu area, where it rises to about 500 m ASL. This local rise in the formation is interpreted to be a small upthrown block. The Hohi and Kusu formations are interfingered in a complex manner. Both of these groups are truncated in the vicinity of the uplift of the Shishimuta rocks in the Takenoyu-Hagenoyu area.

The granitic basement (not shown in Figure 2) was encountered in two boreholes, DW-7 and DY-2, at about –960 m ASL in the Oguni area, and drops off steeply to the northeast in the Sugawara area. The faults that strike WNW-ESE across the Oguni Geothermal Field (Figure 1) may be of a type known as a "drape fault." This type of fault is associated with the cracking of a weak material overlying a stronger material in which there is an abrupt change in slope. The weak material "drapes," or hangs over the edge of the precipice where the slope changes, and gravity induces tensile cracks that splay from the precipice. At Oguni, the underlying material is the granitic basement, which begins to plunge beneath the Takenoyu fault. The weaker material consists of the volcanic flows. The mechanism for these types of faults has been confirmed by laboratory experiments [*Logan et al.*, 1978]. If this mechanism applies to Oguni, then the WNW-ESE trending faults in Figure 1 would extend down to the basement.

Garg et al. [1993] have analyzed drilling information and downhole pressure, temperature, and spinner surveys for 45 boreholes in the northwestern Hohi area in an effort to deduce feedzone depths and pressures. The feedzone pressures imply that the northwestern Hohi region consists of two pressure zones, a high one in the southern part of the Oguni field and a low one in the central and northern parts of the area (Figure 1). At present, the reasons for the existence of two pressure zones in close proximity (within at most a few hundred meters) are poorly understood. However, the steep dipping of the basement to the northeast and the existence of WNW-ESE trending drape faults may be the primary mechanism.

The stable feedzone pressures for the low-pressure zone boreholes can be described by the following correlation (Figure 3):

$$P = 56.0888 - 0.08531z^*$$
$$z^* = z + 7.619(x_N - 15)$$

where P is in bars (absolute), z^* and z are in m ASL, and x_N is the distance in kilometers from the origin of the Central Kyushu Coordinate System. However, the pressures for high-pressure zone boreholes do lie considerably above the straight line in Figure 3, as do those from shallow boreholes that do not penetrate the deep reservoir. The vertical pressure gradient in the low-pressure zone is 8.531 kPa/m and corresponds to a hydrostatic gradient at ~ 195°C. This implies fluid upflow in regions of the reservoir where temperatures exceed 195°C. The pressure correlation also implies that pressures decrease to the north. The lateral pressure gradient is ~ 0.65 bars/km. Thus, in the natural state a regional flow exists to the north in the northwestern Hohi area.

The average reservoir temperature at Oguni is about 225°C. The maximum stable preproduction subsurface temperature of 240°C was measured in the GH-10 and GH-20 wells at total vertical depths of 1027 and 1576 m, respectively. Temperatures decline rapidly to the east and west of the 1-km-wide subsurface zone defined by the Oguni boreholes GH-4, GH-10, GH-11, and GH-12 [*Garg et al.*, 1993]. To the south and north of these latter boreholes, the temperature decrease is much more gradual. The Oguni reservoir fluid is a relatively homogenous sodium-chloride brine of moderate salinity with an average chloride concentration of about 1100 mg/L. The reservoir fluid is a single-phase liquid. None of the geothermal boreholes provide any direct evidence of a two-phase zone at depths greater than 300 m. However, the presence of boiling above a 300-m depth is suggested by the occurrence of warm and boiling steam-heated sulfate and bicarbonate spring waters in the area.

To delineate the permeability structure for the Oguni Geothermal Field, EPDC performed numerous pressure transient tests. The available data set includes (1) cold fluid injection tests in single boreholes, (2) pressure drawdown and buildup tests in single boreholes, and (3) pressure interference tests involving multiple boreholes. This chapter focuses on analyses of the Oguni pressure interference test data from both slim holes and large-diameter wells.

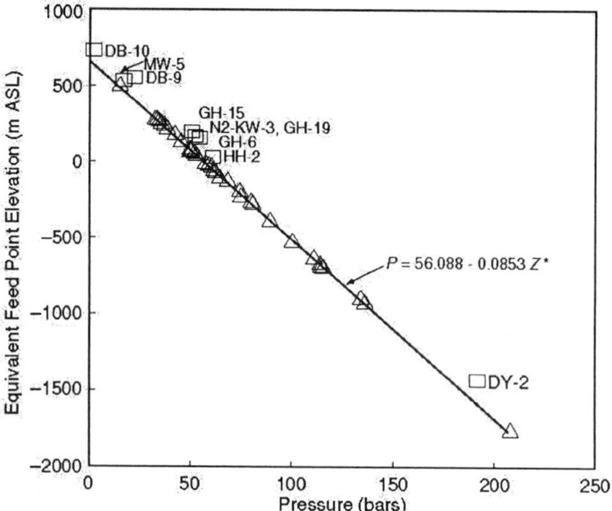

Figure 3. Correlation of pressure with equivalent feedpoint elevation of low-pressure zone boreholes (Δ). Also shown as □ are high-pressure zone boreholes GH-6, GH-15, GH-19, HH-2, DY-2, and N2-KW-3, and shallow slim holes DB-9, DB-10, and MW-5, which are not included in the pressure correlation.

2. PRESSURE INTERFERENCE TESTS

In parallel with EPDC activities, NEDO conducted a research project at the same location from 1990 to 1993 called "The Development of Geothermal Reservoir Evaluation Technology" [see, e.g., *Kawano et al.*, 1989]. As part of this project, EPDC—under contract to NEDO-drilled slim holes N2-KW-1, N2-KW-2, and N2-KW-3, and installed downhole gauges of the capillary tube type in slim holes GH-3, GH-4, GH-5, N2-KW-1, N2-KW-2, and N2-KW-3. These six slim holes were used as shut-in observation boreholes during four separate production/injection tests carried out by EPDC from April 15 to April 27, 1992 (Table 1). With the exception of slim hole N2-KW-3, all the other observation boreholes lie in the low-pressure zone. During Test 1, low-pressure well GH-20 was discharged for 12 days from April 15 April 27, 1991. The separated liquid water was injected into low-pressure well IH-1 and high-pressure well GH-15. Test 2 (June 5 to July 20, 1991) employed GH-11 as the production well and IH-1 and GH-15 as the injectors. For Test 3 (September 5 to October 20, 1991), low-pressure wells GH-12 (producer) and IH-2 (injector) were used as active boreholes. Test 4 (December 13, 1991, to April 27, 1992) involved simultaneous discharge from boreholes GH-10, GH-11, GH-12, and GH-20, and injection into boreholes GH-17, IH-1, IH-2, GH-15, and GH-19. The feedpoint locations and the feedpoint formations of all the

Table 1. Production, injection, and observation boreholes used during the 1991–1992 tests at the Oguni Geothermal Field.

Test	Dates	Injection wells	Production wells	Observation boreholes
1	April 15, 1991 – April 27, 1991	IH-1, GH-15[a]	GH-20	N2-KW-1, N2-KW-2, N2-KW-3,[a] GH-3, GH-4, GH-5
2	June 5, 1991 – July 20, 1991	IH-1, GH-15[a]	GH-11	
3	September 5, 1991 – October 20, 1991	IH-2	GH-12	
4	December 13, 1991 – April 27, 1992	GH-17, IH-1, IH-2, GH-15,[a] GH-19[a]	GH-10, GH-11, GH-12, GH-20	

[a]High-pressure zone borehole.

Table 2. Principal feedzone locations for all the production, injection, and observation boreholes involved in the Oguni pressure interference tests along with the feedpoint formation. All the horizontal distances (m North and m East) are with respect to the Central Kyushu Co-ordinate System.

Well	Feedpoint Depth (m ASL)	Feedpoint Coordinates m (North)	m (East)	Feedpoint Formation
N2-KW-1	–39	17225	13684	Shishimuta
N2-KW-2	–80	16323	13270	Shishimuta
N2-KW-3	158	15129	13734	Hohi
GH-3	–249	16201	13312	Shishimuta
GH-4	–40	16066	14181	Shishimuta
GH-5	–210.5	15794	14525	Shishimuta
GH-10	–146	16175	14029	Shishimuta
GH-11	–282	16090	13790	Taio
GH-12	114	15947	14294	Hohi
GH-15	178	15474	13529	Hohi
GH-17	25	16051	13432	Shishimuta
GH-19	154	15325	13424	Hohi
GH-20	–702	15701	14467	Taio
IH-1	442	15866	13111	Hohi
IH-2	235	16248	13463	Hohi

production, injection, and observation boreholes involved in these pressure interference tests are given in Table 2.

Because of the relatively simple borehole configuration (Figure 4) employed during the first three interference tests, pressure data from these tests are especially useful for defining the reservoir permeability structure. As can be seen from Figure 4, observation slim holes N2-KW-2 and GH-3 are much closer to injection well IH-1 than production well GH-20. The measured pressure response in slim holes N2-KW-2 and GH-3 shows that these slim holes do not respond to injection into IH-1. The casing in well IH-1 was perforated at a 313-m depth in June 1988. Apparently, the recompleted well IH-1 injects fluid into a shallow aquifer, and injection into well IH-1 plays no role in the pressure response recorded in the various observation boreholes. Flow histories for all the active production and injection wells, with the exception of IH-1, are illustrated in Figure 5. For analysis purposes, it is convenient to consider Tests 1 and 2 together. Analyses of pressure responses monitored in the low-pressure zone observation slim holes

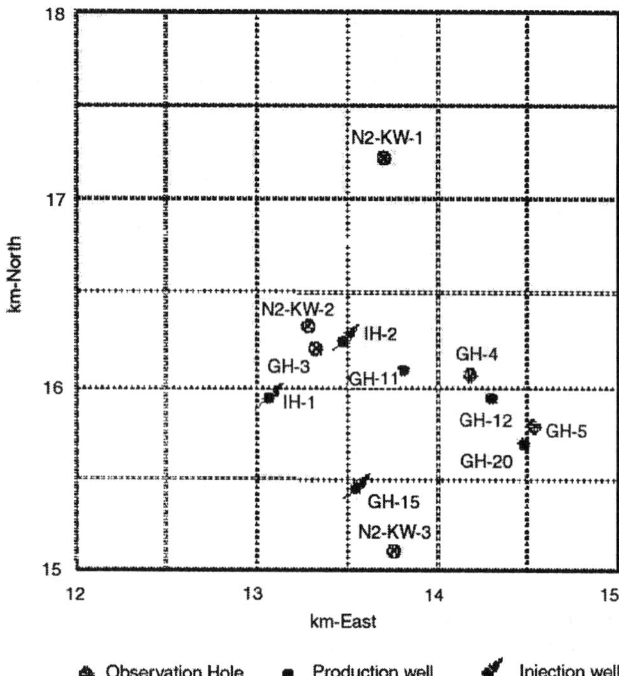

Figure 4. Horizontal feedzone locations of the various production, injection, and observation boreholes involved in Oguni Tests 1–3.

GH-3, GH-4, GH-5, N2-KW-1, and N2-KW-2 are given in the following paragraphs. The pressure response observed in high-pressure slim hole N2-KW-3 is considered in the next section.

An examination of the observed pressure responses shows that the pressures were declining slowly in slim holes GH-3, GH-4, N2-KW-1, and N2-KW-2 prior to the start of Test 1. In the case of slim holes N2-KW-1 and N2-KW-2, the pressure decline is almost certainly due to inadequate heatup time after cold water injection and interzonal flow. Secular drift of the pressure sensor may be responsible for the observed pressure decline in slim holes GH-3 and GH-4. Because of this declining trend in pressures, it was decided to represent well pressure $p(t)$ as follows:

$$\Delta p(t) = p_i - \alpha t - p(t)$$

where

p_i = pressure at gauge depth just prior to discharge from GH-20,

α = a parameter representing secular decline in pressure,

$\Delta p(t)$ = pressure decline/increase due to production/injection operations, and

t = time

Both the initial pressure p_i and decline parameter α were regarded as unknown parameters in the analyses of pressure interference data. The in situ density, viscosity, and temperature of water are assumed to be 840 kg/m^2, 1.2 × 10^{-4} Pa/s, and 220°C, respectively.

2.1. Observation Slim Holes GH-3 and N2-KW-2

Slim holes GH-3 and N2-KW-2 were drilled from the same pad. Both the vertical and horizontal distances between the feedpoints of these holes are relatively small. Not surprisingly, slim holes GH-3 and N2-KW-2 exhibit similar pressure responses. To analyze the observed pressure response during Tests 1 and 2, the line source solution [see, e.g., *Streltsova,* 1988] was used. It is assumed that boreholes GH-20, GH-11, GH-3, and N2-KW-2 fully penetrate an areally infinite reservoir. In the process of modeling the pressure interference response, it was found that the introduction of a no-flux boundary led to much better agreement between the measured and computed pressures. For the sake of simplicity, it was somewhat arbitrarily assumed that the no-flux boundary is located to the south of the observation boreholes. (The relative locations of the high- and low-pressure zones indicate that the no-flux boundary is probably oriented in a NW-SE direction). An automated nonlinear regression algorithm [*McLaughlin et al.,* 1995] was used to produce the best match between the computed and measured pressures (Figure 6) and to determine the unknown parameters in the mathematical model (p_i, α, permeability—thickness product kh, storage ϕch, and distance to no-flux boundary Y). The final model parameters are:

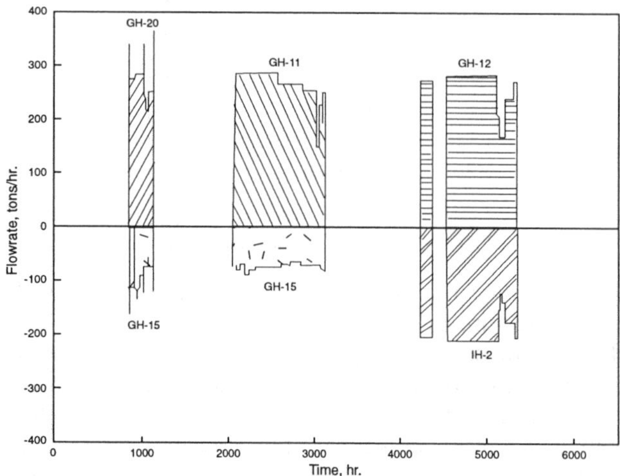

Figure 5. Flow histories of "active" production and injection wells involved in Tests 1–3. All times are in hours since 00:00 hours on March 11, 1991.

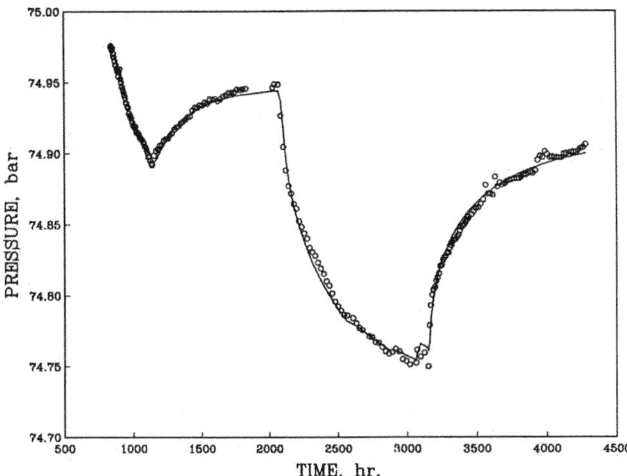

Figure 6. Comparison of measured pressures (o) in slim hole GH-3 with computed response (—) due to production from wells GH-20 and GH-11 (Tests 1 and 2).

GH-3.

$p_i = 74.97$ bars
$\alpha = 1.9 \times 10^{-4}$ Pa/s
$kh = 260 \times 10^{-12}$m^3
$\varphi ch = 7.0 \times 10^{-7}$ m/Pa
$Y = 1310$ m south of GH-3
(14,890 m north of CKCS)

N2-KW-2.

$p_i = 65.16$ bars
$\alpha = 6.1 \times 10^{-4}$ Pa/s
$kh = 220 \times 10^{-12}$m^3
$\varphi ch = 7.0 \times 10^{-7}$ m/Pa
$Y = 1560$ m south of N2-K2-2
(14,760 m north of CKCS)

The formation parameters (kh, φch, Y) inferred from separate fits to pressure data from slim holes GH-3 and N2-KW-2 are quite similar. The computed secular pressure decline (αt) during the roughly 3500 hours of pressure interference test for GH-3 (0.02 bars) is small relative to the total pressure signal (>0.20 bars). While the predicted secular pressure decline for N2-KW-2 (~ 0.08) is not small as a percentage of the total pressure signal, the close agreement between the formation parameters inferred from GH-3 and N2-KW-2 gives confidence in the model results. The inferred value for the storage parameter φch (~ 7×10^{-7} m/Pa) is fairly high. Since the permeable zone at Oguni is, at most, a few hundred meters in thickness and the formation porosity is about 0.1, the latter result implies that the reservoir rocks are very compressible. This is a reasonable conclusion since the feedzones for both GH-3 and N2-KW-2 are in the Shishimuta formation, which is primarily dacitic pyroclastics. Unpublished laboratory test data obtained on cores recovered from several Oguni slim holes indicate that dacitic pyroclastics (lapilli and coarse tuffs) have porosities exceeding 0.1 and compressional velocities of about 3 km/s. The rather low value for the compressional velocity implies that the Shishimuta formation is highly compressible. As already indicated above, the boundary orientation (i.e., an east-west direction south of the observation boreholes) was fixed arbitrarily. A different choice for boundary orientation will yield a different value for distance to the boundary. The essential point is that the pressure-interference data from GH-3 and N2-KW-2 (see also the following discussion for GH-4 and GH-5) imply the presence of a linear barrier.

In contrast with the pressure responses observed during the first two pressure interference tests, the pressure data recorded during Test 3 were found to be less useful for inferring formation properties. Slim holes GH-3 and N2-KW-2 are much closer to injection well IH-2 than they are to production well GH-12. Not surprisingly, the pressure response of GH-3 and N2-KW-2 is dominated by injection into well IH-2 (see e.g., Figure 7). For practical purposes, fluid production from GH-12 may be ignored in analyzing the pressure interference data. Apparently, GH-3 and N2-KW-2 are connected by a modest transmissivity (kh ~ 3×10^{-12}m^3) and low storativity (φch ~ 5×10^{-9} m/Pa) fracture zone to IH-2. Most likely, the interwell formation between GH-3/N2-KW-2 and IH-2 is intersected by a few permeable fractures. These permeable fractures presumably join the rest of the fracture network at some distance from GH-3, N2-KW-2, and IH-2.

2.2. Slim Hole GH-4

The pressure response of GH-4 (Figure 8) to discharge from wells GH-20 and GH-11 (Tests 1 and 2) was fitted in a manner similar to that for slim holes GH-3 and N2-KW-2. The computed model parameters are:

$p_i = 57.96$ bars
$\alpha = 2.7 \times 10^{-6}$ Pa/s
$kh = 150 \times 10^{-12}$m^3
$\varphi ch = 1.9 \times 10^{-6}$ m/Pa
$Y = 880$ m south of GH-4
(15,190 m north of CKCS)

The permeability-thickness value derived from the pressure response of GH-4 is considerably less than values obtained from pressure data for slim holes GH-3 and N2-KW-2. This indicates some permeability heterogeneity in the Oguni geothermal field. The large value for the storage parameter

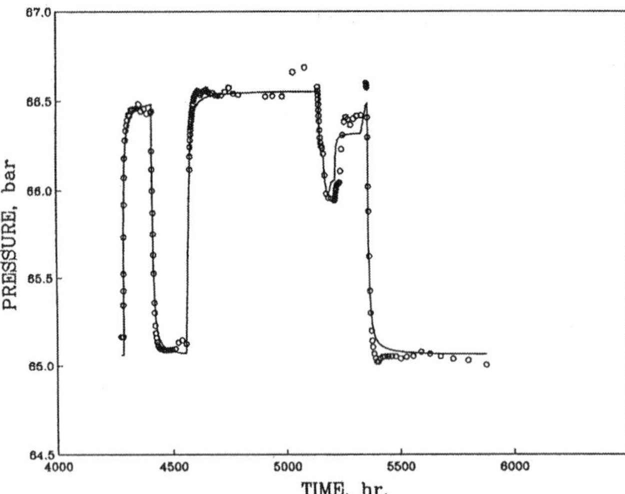

Figure 7. Comparison of measured (o) and computed (—) pressures in slim hole N2-KW-2 due to injection into well IH-2 (test 3).

φch implies that the reservoir rocks are very compressible. Similar to the feedzones for slim holes GH-3 and N2-KW-3, the feedzone for GH-4 is in the dacitic pyroclastics of the Shishimuta formation.

Slim hole GH-4 is much closer to well GH-12 than well IH-2. Not surprisingly, the pressure response of slim hole GH-4 during Test 3 was controlled by production from GH-12. For practical purposes, injection into IH-2 had negligible influence on the observed pressure response. Analysis of pressure-transient data indicates that boreholes GH-4 and GH-12 are connected by a modest transmissivity ($kh \sim 7 \times 10^{-12} m^3$) and a low storativity ($\varphi ch \sim 9 \times 10^{-9}$ m/Pa) fracture zone.

2.3. Slim Hole GH-5

The pressure record for slim hole GH-5 displayed several gaps during production from well GH-11 (Test 2). Because of these gaps in the pressure data, it was decided to consider only the early part of the GH-5 pressure record, which corresponds to production from GH-20 (Test 1). For analysis purposes, the initial pressure in slim hole GH-5 was kept fixed at 62.57 bars. Furthermore, it was assumed that any secular pressure drift is unimportant (computations with nonzero pressure drift did not lead to an improvement in results). The response of slim hole GH-5 to production from GH-20 was modeled using the line source model. Since the feedpoints of boreholes GH-5 and GH-20 have a vertical separation that is much greater than the horizontal separation, the actual distance between the feedzones was used in the analysis. Analysis of GH-5 pressure response during the field-wide pressure interference test based upon the anisotropic point-source model is described in a later section. Calculations were carried out for (1) the infinite (no boundary) and (2) the single barrier (barrier south of GH-5) cases. Inclusion of a barrier (see Figure 9) leads to much better agreement between the computed and measured pressures. The inferred formation properties are:

$$kh = 120 \times 10^{-12} m^3$$
$$\varphi ch = 1.9 \times 10^{-7} \text{ m/Pa}$$
$$Y = 1010 \text{ m south of GH-5}$$
(14,780 m north of CKCS)

The above formation properties are similar to those derived from analysis of the GH-4 pressure response.

The response of slim hole GH-5 to production from GH-12 and injection into IH-2 (Test 3) was analyzed using the line-source model. Since slim hole GH-5 is much closer to GH-12 than it is to IH-2, it is not surprising that production from GH-12 dominates the pressure response of GH-5. In an attempt to assess the effect of IH-2 on GH-5 response, model parameters were evaluated both with and without accounting for injection into IH-2. It appears that inclusion of IH-2 injection data leads to a somewhat better fit between the measured and computed pressures (see Figure 10). The model parameters for the best fit are:

$$i = 62.64 \text{ bars}$$
$$\alpha = 1.07 \times 10^{-3} \text{ Pa/s}$$
$$kh = 47 \times 10^{-12m3}$$
$$\varphi ch = 5.3 \times 10^{-7\,m}/\text{Pa}$$

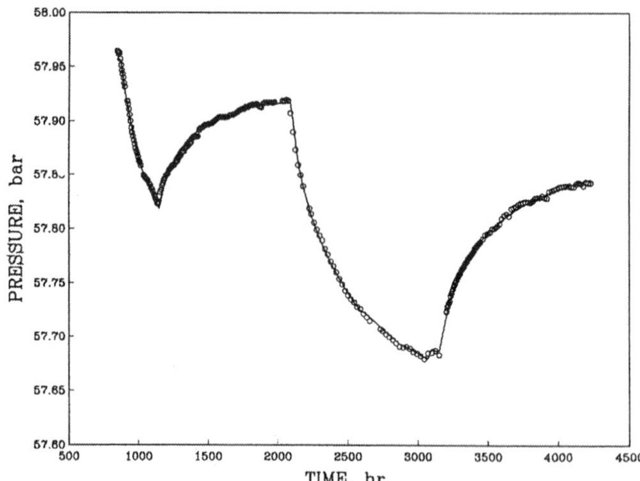

Figure 8. Comparison of measured pressures (o) in slim hole GH-4 with computed response (—) due to production from wells GH-20 and GH-11 (Tests 1 and 2).

The negative value of α indicates a linear increase in pressure, which may be due to continuing pressure buildup from earlier pressure interference tests (i.e., production from wells GH-20 and GH-11). The *kh* value inferred from the present test is substantially less than that obtained from Tests 1 and 2. The smaller kh value in this case is perhaps associated with the relatively small volume sampled by the pressure interference signal between GH-5 and GH-12.

2.4. Slim Hole N2-KW-1

Among the observation boreholes, slim hole N2-KW-1 is the farthest removed (1400 ± 300 m) from production wells GH-20 and GH-11 (see Figure 4). As already mentioned, pressures in N2-KW-1 were declining prior to discharge test of GH-20. Attempts to model the pressure response in N2-KW-1 due to production from GH-20 and GH-11 using the line-source model were largely unsuccessful. It was next decided to model the pressure response associated with production from these two wells separately. Again, the comparison between the measured and computed pressure response due to discharge from GH-20 was poor. This may be due to the secular drift in N2-KW-1 pressure and the relatively small pressure change associated with discharge from GH-20. The pressure response corresponding to GH-11 discharge can be adequately fitted (Figure 11) using the line-source model with the following parameter values:

$$p_i = 57.09 \text{ bars}$$
$$\alpha = 3.1 \times 10^{-4} \text{ Pa/s}$$
$$kh = 100 \times 10^{-12} \text{m}^3$$
$$\varphi ch = 9.4 \times 10^{-6} \text{ m/Pa}$$

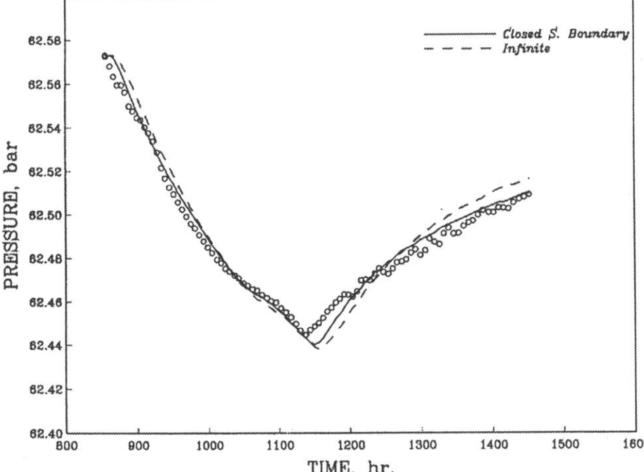

Figure 9. Comparison between measured (o) and computed (---- infinite reservoir, — linear barrier south of GH-5) pressure response of slim hole GH-5 due to production from well GH-20 (Test 1).

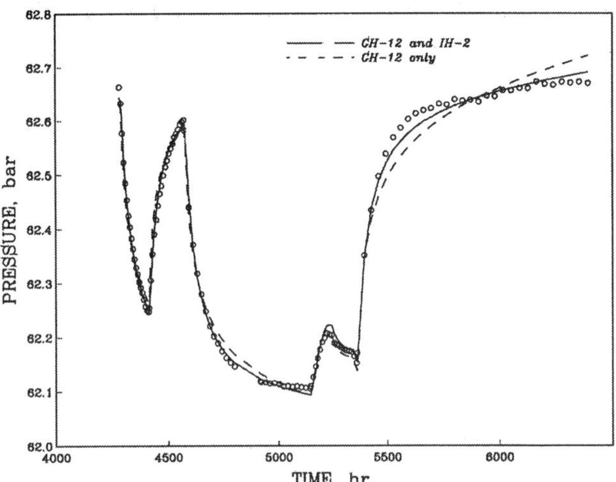

Figure 10. Comparison between measured (o) and computed (---- production from GH-12 only, — production from GH-12 and injection into IH-2) pressure response in slim hole GH-5 (Test 3).

Apparently, the pressure response of N2-KW-1 during Test 3 was affected by some unknown mechanism (e.g., sensor drift). Attempts to model the pressure response of N2-KW-1 using the production (GH-12) and injection (IH-2) well data were fruitless.

3. FIELD-WIDE DISCHARGE AND PRESSURE INTERFERENCE TEST

During the period from December 13, 1991, to April 27, 1992, EPDC conducted a field-wide production/injection test in the Oguni field. Low-pressure zone wells GH-10, GH-11, GH-12, and GH-20 were used as production wells. Separated liquid water from the production stream was injected into low-pressure zone wells GH-17, IH-1, and IH-2, and high-pressure zone wells GH-15 and GH-19. The measured flow-rate histories for the various wells are given in *Garg et al.* [1993]. Like the earlier pressure interference tests, slim holes GH-3, GH-4, GH-5, N2-NW-1, N2-KW-2, and N2-KW-3 were used as shut-in observation boreholes during the field-wide discharge test. Injection into well IH-2 was found to dominate the pressure response recorded in slim holes GH-3 and N2-KW-2. Also pressure changes in GH-4 were found to be related to flow rate changes in GH-12. Analysis of the pressure record of N2-KW-1 gave formation parameters similar to those obtained from Test 1. The remainder of this section analyzes the pressure responses of GH-5 and N2-KW-3.

3.1. Slim Hole GH-5

The horizontal distance between the major feedpoint for slim hole GH-5 and that for GH-20 is 110 m. The vertical

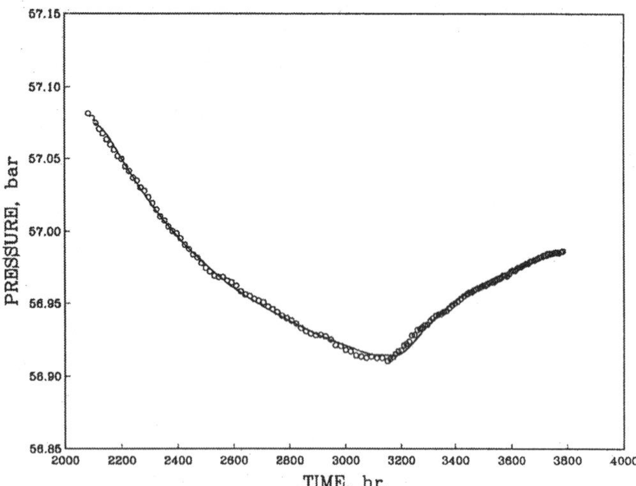

Figure 11. Comparison between measured pressures (o) in slim hole N2-KW-1 and computed response due to production from well GH-11 (Test 2).

separation between these feedzones is ~492 m. The horizontal separation (277 m) between the GH-5 and GH-12 feedzones is also less than the vertical separation (325 m). While the feedzone for GH-12 is shallower than that for GH-5, the feedzone for GH-20 is deeper than that for GH-5. The line-source model is inappropriate for modeling the pressure response of GH-5 to simultaneous production from GH-12 and GH-20. Accordingly, it was decided to use the anisotropic point-source (distinct horizontal k_x and vertical k_z permeabilities) model to fit the pressure response of GH-5. Initially, all the "low-pressure zone" production/injection wells (GH-10, GH-11, GH-12, GH-20, GH-17, IH-2) were included in the mathematical model. For analysis purposes, the initial pressure was kept fixed at 62.69 bars. The unknown parameters in the model (k_x, k_z, φc, α) were varied so as to minimize the deviations between the measurements and the computed pressure response. The fit between the computed and measured pressures was found to be rather poor at early times (i.e., at the start of Test 4). To improve the agreement between the measured and computed pressures, it was decided to model the pressure response of GH-5 by ignoring injection wells IH-2 and GH-17. Since the feedzones for IH-2 and GH-17 are located at quite a distance (both horizontally and vertically) from the feedzone for GH-5, it can be argued that the influence of these wells on the pressure response of GH-5 should be minimal. The best fit between the measured and computed response (Figure 12) was obtained using an anisotropic point-source model with a single impermeable boundary. The impermeable boundary is assumed to lie in a horizontal plane above the feedzone for GH-5. The inclusion of drift parameter (α) has a very minor influence (Figure 12) on the fit. The final model parameters are:

$$k_x = 91 \times 10^{-15} m^2$$
$$k_z = 36 \times 10^{-15} m^2$$
$$\varphi c = 4.1 \times 10^{-10} \text{ Pa}^{-1}$$

Y = 325 m above the feedzone for GH-5 (114.5 m ASL)

The vertical distance to the boundary implies that this boundary lies close to the feedzone for GH-12. Since GH-12 was known to influence the pressure response of GH-5, the impermeable boundary was constrained to lie above the feedzone for GH-12. Assuming a formation thickness of 1000 m, a k_xh value of $91 \times 10^{-12} m^3$ is obtained. The latter value for k_xh is comparable to the estimates obtained from pressure transient analyses presented earlier. Since the fluid compressibility is only $\sim 10^{-9} \text{Pa}^{-1}$, and porosity φ is about 0.1, the computed value for φc implies that the formation is extremely compressible. As noted above, the feedzone for GH-5 is in the dacitic pyroclastics of the Shishimuta formation, which should be extremely compressible in agreement with the calculations.

3.2. Slim Hole N2-KW-3

High-pressure-zone slim hole N2-KW-3 responds to injection into GH-15 (Tests 1/2 and 4) and GH-19 (Test 4). Preliminary analyses of these pressure interference tests

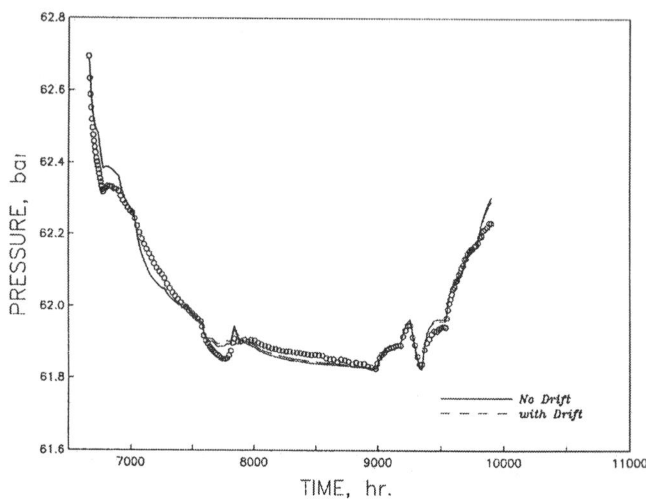

Figure 12. Comparison between measured (o) and computed (—, ----) pressures in slim hole GH-5 for the field-wide test. Only the low-pressure zone production wells (GH-10, GH-11, GH-12, and GH-20) were included in the mathematical fit.

gave inconsistent results for formation parameters. The only difference between the two tests (1/2 and 4) is that GH-19 was not involved in the earlier one.

To understand the reasons for the difference between the reservoir parameters inferred from the two tests, it was decided to restrict attention to the fall-off portion of the pressure records. The fall-off record for the field-wide test consists of two distinct parts, corresponding to shut-in of GH-15 and GH-19. The fall-off data show that a flow-rate change in either GH-15 or GH-19 produces an almost instantaneous pressure response in N2-KW-3. To obviate the uncertainties associated with the flow-rate amplitude just prior to shut-in and the shut-in time, it was decided to exclude pressure data around the shut-in time from the fall-off pressure analysis. Because the fall-off record from the field-wide test is only about 10 days in duration, the fall-off record for Test 2 was also truncated to about 10 days. The line-source solution was used to match the fall-off pressure response for both tests. In each case, the best fit was obtained using a single no-flux boundary. The boundary orientation was arbitrarily assumed to be to the east of N2-KW-3. Both the distance to the boundary and the boundary orientation are nonunique. The model parameters obtained by minimizing the deviations between the measured and computed pressures are (see Figures 13 and 14):

Tests 1 and 2 (GH-15/N2-KW-3).

$p_i = 57.7$ bars
$kh = 8.8 \times 10^{-12} m^3$
$\varphi ch = 1.2 \times 10^{-7}\ Pa^{-1}$
$Y = 640$ m east of N2-KW-3
(14,370 m east of CKCS)

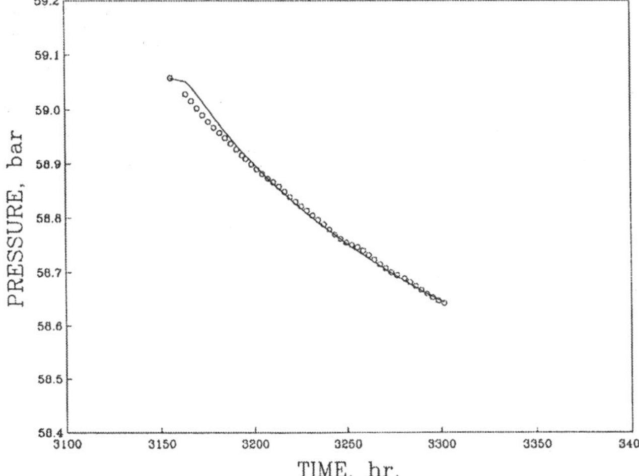

Figure 13. Comparison between measured (o) and computed (—) response in slim hole N2-KW-3 (Test 2) following shut-in of injection well GH-15 on July 20, 1991.

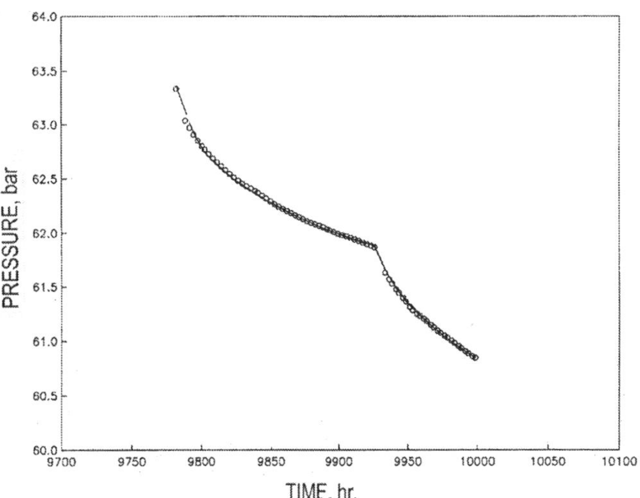

Figure 14. Comparison between measured (o) and computed (—) response in slim hole N2-KW-3 (Test 4) following shut-in of injection wells GH-15 (April 21, 1992) and GH-19 (April 27, 1992).

Test 4 (GH-15 and GH-19/N2-KW-3).

$p_i = 57.2$ bars
$kh = 14.4 \times 10^{-12}\ m^3$
$\varphi ch = 4.1 \times 10^{-8}\ Pa^{-}1$
$Y = 860$ m east of NW-KW-3
(14,590 m east of C*KCS)*

The inference of a linear no-flux boundary is in keeping with the notion of hydraulic separation between the low- and high-pressure reservoirs. The model parameters (*kh* and *φch*) obtained from the two fall-off tests are, however, very different. This difference is indicative of heterogeneity in the high-pressure reservoir. GH-15 exhibited rather poor injectivity during short-term injection tests. During production, the pressure drop in GH-15 is sufficient to cause in situ flashing. GH-19, on the other hand, has an extremely high injectivity (no production tests have been carried out on GH-19). Thus, it is likely that formation permeability in the vicinity of GH-15 is substantially lower than that near GH-19. Since the feedzones of the wells are in either the lavas or the pyroxene andesitic rocks of the Hohi formation, the difference in formation permeability is expected.

4. CONCLUSIONS

Pressure interference tests have been invaluable for characterizing the detailed permeability structure of the Oguni Geothermal Field. The low-pressure zone in the northwestern Hohi reservoir (Table 3) has a transmissivity

Table 3. Formation parameters inferred from Tests 1 to 4.

Observation borehole	Test	kh ($10^{-12}m^3$)	ϕch (m/Pa)	Location of no-flux vertical barrier[a]
GH-3	1 and 2	260	7.0×10^{-7}	14,890 m north
N2-KW-2	1 and 2	220	7.4×10^{-7}	14,760 m north
GH-4	1 and 2	150	1.9×10^{-6}	15,190 m north
GH-5	1	120	1.9×10^{-6}	14,780 m north
GH-5	3	47	5.3×10^{-7}	
N2-KW-1	2	100	9.4×10^{-6}	
N2-KW-3[b]	1 and 2	8. 8	1.2×10^{-7}	14,370 m east
	4	14.4	4.1×10^{-8}	15,990 m east

[a] Central Kyushu Co-ordinate System.
[b] High-pressure zone.

of 100–250 × $10^{-12}m^3$. The east-west transmissivity, as determined from the pressure response of slim holes GH-3 and N2-KW-2 to discharge from wells GH-20 and GH-11, is around 250 × $10^{-12}m^3$. The north-south transmissivity obtained from the pressure response observed in N2-KW-1 is about 100 × $10^{-12}m^3$. The latter transmissivity values (100–250 × $10^{-12}m^3$) apply only at the reservoir length scale (i.e., over distances of the order of 1 km or more). For shorter length scales (100-to-500-m separation between the production/injection and observation boreholes), the inferred transmissivity values (e.g., from the pressure response of GH-3 and N2-KW-2 to injection into well IH-2) are considerably smaller. In contrast to the high transmissivity of the low-pressure zone, the high-pressure zone has only a modest transmissivity of 8–15 × $10^{-12}m^3$ (Table 3). The pressure interference test data are consistent with the presence of one or more no-flux boundaries between the low- and high-pressure zones. The computed locations of the no-flux boundary (Table 3) do not, however, conform to the spatial distribution of high- and low-pressure zones. As an example, analysis of pressure data from observation boreholes GH-3, GH-4, GH-5, and N2-KW-2 implies that the impermeable barrier is located to the south of the no-flux boundary, which must be present between southernmost low-pressure well GH-20 and northernmost high-pressure well GH-15. This discrepancy between the computed and actual locations of the barrier is almost certainly due to the simplifying assumption that the no-flux boundary can be represented by a linear barrier oriented north-south or east-west. In reality, the barrier is likely to have a more complex configuration.

Analysis of Oguni pressure interference data presented in the preceding sections indicates a significant degree of permeability heterogeneity within both the low- and high-pressure reservoirs. Test data from several production/injection and observation boreholes (e.g., IH-2 and GH-3/N2-KW-2) imply that modest transmissivity/low storativity fracture zones connect these boreholes. The apparent dependence of transmissivity on the length scale is an essential characteristic of fractured reservoirs. The inferred transmissivity from a particular interference test is an indication of the permeable fractures within the formation volume connecting the production/injection and observation boreholes. If the latter formation volume is too small, then only a few permeable fractures will be encountered, and the transmissivity obtained from such a test will not be representative of the overall reservoir. A detailed knowledge of the permeability structure at both the interwell and reservoir scales is essential for designing the injection and production strategy for the geothermal field. Otherwise, incorrect placement of injection wells may result in premature cooling of the production zone.

Acknowledgments. We thank Jim Combs, Tsuneo Ishido, Malcolm Grant, and Grimur Bjornsson for constructive reviews of the manuscript. Dr. Larry Owusu performed some of the computations reported in the paper.

REFERENCES

Garg, S. K., J. W. Pritchett, T. G. Barker, L .A. Owusu, J. Haizlip, and A. H. Truesdell, *Reservoir Engineering Studies of the*

Oguni Geothermal Field (Phase 3), Report SSS-FR-92-13899, S-Cubed, La Jolla, Calif., March, 1993.

Kawano, Y., H. Maki, T. Ishido, and Y. Kubota, NEDO's project on geothermal reservoir engineering: A reservoir Engineering study of the Sumikawa Geothermal Field Japan, in *Proceedings: Fourteenth Workshop on Geothermal Reservoir Engineering,* pp. 55–59, Stanford, Calif., 1989.

Logan, J. M., M. Freidman, and M. T. Stearns, Experimental folding of rocks under confining pressure: Part VI, Further studies of faulted drape folds, *Geological Society of America Memoir*, *151*(79), 100, 1978.

McLaughlin, K. L., T. G. Baker, L. A. Owusu, and S. K. Garg, DIAGNS: An interactive workstation-based system for well test data diagnostics and inversion, in *Proceedings: World Geothermal Congress Florence, Italy,* pp. 29412944, 1995.

Streltsova, T. D., *Well Testing in Heterogeneous Formations,* John Wiley and Sons, New York, N.Y., 1988.

Sabodh K. Garg, Maxwell Technologies, Inc., 8888 Balboa Ave., San Diego, CA 92123

Shigetaka Nakanishi, Electric Power Development Co., Ltd., 15-1 Ginza 6-Chrome, Chuo-ku, Tokyo, 104 Japan

Evaluation of Geothermal Well Behavior Using Inverse Modeling

Stefan Finsterle,[1] Grimur Björnsson,[2] Karsten Pruess,[1] and Alfredo Battistelli[3]

Characterization of fractured geothermal reservoirs for numerical prediction of fluid and heat flow requires determination of a large number of hydrologic, thermal, and geometric properties. For use in a computer model that is based on a simplified conceptual model, these properties must be capable of reflecting the complex multiphase flow behavior in a fracture network, including fracture-matrix interaction. In this paper, we discuss the potential of inverse modeling techniques to provide model-related input parameters based on a joint inversion of field testing and actual production data from a geothermal reservoir. Using synthetically generated data, we demonstrate the need to simultaneously analyze multiple data sets in a joint inversion. The impact of parameter correlations on the estimated values and their uncertainties are also discussed. Inverse modeling techniques are then applied to data from the Krafla geothermal field in Iceland, in an attempt to estimate some critical reservoir parameters such as steam saturation after 20 years of production from that two-phase system. We conclude that inverse modeling is a powerful tool not only to provide input parameters to a numerical model, but also to improve the understanding of fractured geothermal systems. Its efficiency and the insight gained from the formalized error analysis allow an evaluation of alternative conceptual models, which remains the most crucial step in geothermal reservoir modeling.

1. INTRODUCTION

We have employed inverse modeling techniques to study and characterize multiphase flow in fractured geothermal reservoirs. Both field and synthetically generated data have been analyzed to investigate the theoretical possibilities and limitations of these techniques, as well as the usefulness of inverse modeling in actual geothermal-reservoir engineering problems.

Flow of water, steam, gas, and heat in fractured geothermal reservoirs is strongly influenced by the geometry and hydrological characteristics of the reservoir fracture network. The response to production and injection is affected by the coupling between fluid flow in the fractures and heat transfer from and to the adjacent matrix blocks. Extraction of hot fluids and injection of cold water lead, respectively, to vaporization and condensation effects near the production and injection wells. Furthermore, as a result of pressure and temperature declines during production of high-salinity geothermal fluids, precipitation of solids may occur, reducing fracture and matrix porosity and thus the overall permeability of the reservoir. On the other hand, injection of fresh water may dissolve solids. Changes in sodium chloride concentrations therefore contain information about fluid flow through the fracture

[1]Ernest Orlando Lawrence Berkeley National Laboratory, Earth Sciences Division, Berkeley, California

[2]Orkustofnun, Grensasavegur 9, 108 Reykjavik, Iceland

[3]Aquater SpA, San Lorenzo in Campo (PS), Italy

Dynamics of Fluids in Fractured Rock
Geophysical Monograph 122

network, indicating potential connections between injection and production wells. This connectivity is crucial for both pressure support resulting from injection and unwanted thermal interference. Temperature data obtained in production and observation wells are affected not only by the hydrologic characteristics but also by the thermal properties of the reservoir, which govern the conductive heat exchange between the matrix blocks and the fluids flowing in the fractures. The size and shape of the matrix blocks also determine the effectiveness with which thermal energy can be extracted from the reservoir rocks.

Numerical modeling is a useful tool for the design and optimization of injection operations, which are a means to sustain energy extraction from partially depleted geothermal reservoirs. The reliability of such model predictions depends on the accuracy with which the coupled processes described above are accounted for. In order to capture the salient features of the geothermal reservoir, an appropriate conceptual model must be developed, and accurate geometric, thermal, and hydrologic parameters must be determined.

Developing the conceptual model is the most important and challenging task in geothermal reservoir modeling. As discussed above, fractures play a dominant role in geothermal reservoirs, and their properties must be assessed over a wide range of scales. Coupled flow of fluid and heat is affected by the aperture distribution in individual fractures, the geometry and density of fractures on an intermediate scale, and the large-scale connectivity. Thus, the parameters and processes involved span many orders of magnitude of the spatial scale. It is currently impractical to simulate and characterize such a multiscale system in a single model. One approach is to estimate effective fracture properties on the scale of interest. In other words, the parameters and processes represent some average behavior on a specific scale. In our case, this scale is approximately the zone of influence of a production well or the distance between an injection and a production well. As will be discussed in Section 4 below, the estimated parameters must be interpreted according to the scale on which they are estimated.

Inverse modeling—automatic calibration of the numerical model against field data—is a means to obtain model-related parameters that can be considered optimal for a given conceptual model. However, the large number of parameters needed to fully describe coupled nonisothermal multiphase flow in fractured-porous media often leads to an ill-posed inverse problem, which is predisposed to yield nonunique and unstable solutions. It is therefore crucial to carefully identify and maximize the information content of the data used for calibration, and to assess and minimize correlations among the parameters to be estimated.

The iTOUGH2 code [*Finsterle*, 1999] provides inverse modeling capabilities for the TOUGH2 family of multiphase flow simulators [*Pruess*, 1991]. With iTOUGH2, any TOUGH2 input parameter can be estimated based on any type of data for which a corresponding TOUGH2 output is calculated. Parameter estimation is supplemented by extensive residual and error analyses. In this study, we make use of the EWASG module [*Battistelli et al.*, 1997], which describes three-phase (liquid, gas, and solid) mixtures of three components (water, sodium chloride, and a noncondensible gas). The dependence of brine density, enthalpy, viscosity, gas solubility, and vapor pressure on salinity is taken into account. Precipitation and dissolution of salt are also included, along with associated porosity and permeability changes. The method of Multiple Interacting Continua (MINC) [*Pruess and Narasimhan*, 1982] is used to resolve the pressure, temperature, and saturation gradients between the fractures and the matrix. The MINC concept is based on the notion that both the fractures and the matrix can be treated as interconnected continua and that changes in matrix conditions will be controlled by the distance from the fractures. Note that in the MINC formulation, fracture spacing is simply an input parameter to the mesh generator that produces the computational grid.

In the first part of this study, we perform synthetic inversions to investigate whether fracture properties can accurately be identified and to determine what data are required to constrain fracture property estimates. We then apply inverse modeling concepts to the analysis of field data from a single geothermal well. Theoretical as well as practical aspects of field-scale modeling are addressed. Parameter estimates and their correlations are provided for key reservoir characteristics such as permeability, porosity, and steam saturation after 20 years of production.

Section 2 is a short summary description of the inverse modeling concept. In Section 3, we apply iTOUGH2/EWASG to synthetically generated production data from a hypothetical fractured geothermal reservoir with high salinity and CO_2 as the noncondensible gas. Section 4 describes the analysis of pressure and enthalpy data from a new well in the Krafla high-temperature geothermal field in Iceland. A summary and concluding remarks are provided in Section 5.

2. INVERSE MODELING CONCEPT

In standard simulation practice, site-specific parameter values describing hydrogeological and thermophysical properties are entered into a numerical model along with appropriate initial and boundary conditions. The model then predicts the future state of the system (e.g., pressures, temperatures, concentrations). This is referred to as forward modeling. In inverse modeling, observations of the system at discrete points in space and time are used to estimate site-specific model parameters. Estimation occurs by automatic history matching of observed and computed data. The core of inverse modeling is an accurate, efficient, and robust simulation program that solves the so-called forward problem. It must be capable of simulating the flow and transport processes that govern the observed system response. The system under consideration requires a problem- and a site-specific conceptual model. The task of developing a representative conceptual model is the most important part of any simulation study. In inverse modeling in particular, an error in the conceptual model will lead to a bias in the estimated parameters, which is usually much larger than the uncertainty introduced by random measurement errors [*Finsterle and Najita*, 1998].

Next, an objective function has to be selected to obtain an aggregate measure of deviation between the observed and calculated system response. The choice of the objective function can be based on maximum likelihood considerations, which for normally distributed measurement errors leads to the standard weighted least-squares criterion [*Carrera and Neuman*, 1986]:

$$S=\mathbf{r}^{\mathrm{T}}\mathbf{C}_{zz}^{-1}\mathbf{r} \qquad (1)$$

Here, $\mathbf{r}$ is the residual vector with elements $r_i = z_i^* - z_i(\mathbf{p})$, where z_i^* is an observation (e.g., pressure, temperature, flow rate, etc.) at a given point in space and time, and z_i is the corresponding simulator prediction, which depends on the vector $\mathbf{p}$ of the unknown parameters to be estimated. The i-th diagonal element of the covariance matrix $\mathbf{C}_{zz}^{-1}$ is the variance representing the measurement error of observation z_i. This element is used to weigh data of different qualities and to scale data of different observation type, making the objective function dimensionless.

The objective function S has to be minimized in order to maximize the probability of reproducing the observed system state. Because of strong nonlinearities in the functions $z_i(\mathbf{p})$, an iterative procedure is required to minimize S. A number of minimization algorithms are available in iTOUGH2. They reduce the objective function by iteratively updating the parameter vector $\mathbf{p}$ based on the sensitivity of z_i with respect to p_j. Details about the minimization algorithms implemented in iTOUGH2 can be found in *Finsterle* [1999]. The Levenberg-Marquardt method [*Gill et al.*, 1981] yields good results for strongly nonlinear minimization problems.

Finally, under the assumption of normality and linearity, a detailed error analysis of the final residuals and the estimated parameters is conducted. These analyses provide valuable information about the estimation uncertainty, the adequacy of the model structure, the quality of the data, and the relative importance of individual data points and parameters. While details can be found in *Finsterle* [1999], we note here simply that the estimated error variance,

$$s_0^2=S/(m-n) \qquad (2)$$

can be used as an aggregate measure-of-fit, where m is the number of data used for calibration and n is the number of parameters being estimated. A linear approximation of the estimation covariance matrix is given by

$$\mathbf{C}_{pp}=s_0^2(\mathbf{J}^{\mathrm{T}}\mathbf{C}_{zz}^{-1}\mathbf{J})^{-1} \qquad (3)$$

where $\mathbf{J}$ is the Jacobian or sensitivity matrix with elements $J_{ij} = -\partial r_i/\partial p_j = \partial z_i/\partial p_j$. The covariance matrix $\mathbf{C}_{pp}$ not only shows potentially high estimation uncertainties resulting from insufficient sensitivity of the observed data, but also reveals correlations among the parameters that may prevent an independent determination of certain properties of interest. In addition to its efficiency, it is mainly the formalized sensitivity, residual, and error analyses that make inverse modeling preferable to the conventional trial-and-error model calibration.

3. IDENTIFICATION OF FRACTURE PROPERTIES

In this section, we perform synthetic inversions to analyze whether key properties of a fractured geothermal reservoir can theoretically be identified based on a combination of data sets produced by monitoring a production well.

iTOUGH2/EWASG [*Battistelli et al.*, 1997] was used to simulate production from a hypothetical single-layer geothermal reservoir with high salinity and CO_2 as the noncondensible gas. A MINC model was developed with nearly impermeable matrix blocks; fracture density was considered one of the unknown parameters to be estimated by inverse modeling. Because of salt precipitation near the production well and depletion of fluid reserves in the reservoir, the production of steam declines rapidly and almost ceases within a relatively short time. After five years of exploitation, freshwater is injected through three wells located a few hundred meters from the production well. Figure 1 illustrates the model domain, with saturation distribution in the matrix and fracture continua shown above and below the symmetry axis, respectively, five years after the beginning of liquid injection. Injection of cold water leads to a reduction of steam saturation in the immediate vicinity of the injection wells. Evaporation of injectate, however, increases the reservoir pressure, driving steam towards the production well, enhancing both the flow rate and the total heat produced. Furthermore, salt that

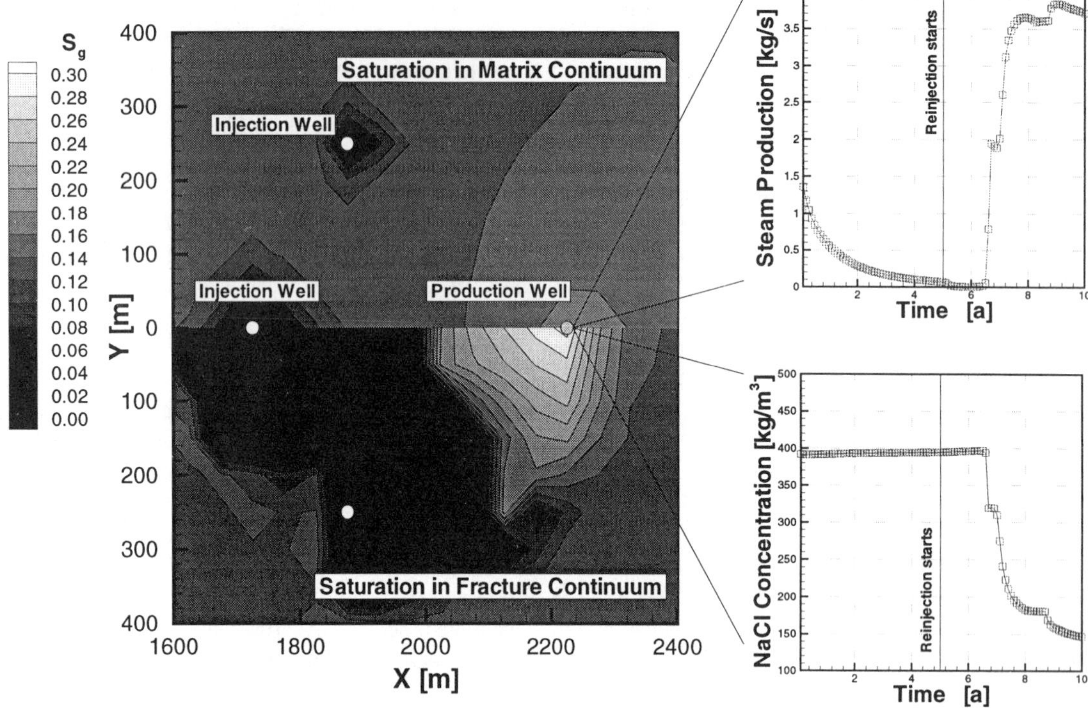

Figure 1. Simulated steam saturation distribution in matrix (upper left) and fracture (lower left) continua five years after injection of cold water into the geothermal reservoir. Steam production (upper right) and sodium chloride concentration in the production well (lower right) as a function of time. Simultaneous inversion of these data can be used to calibrate the numerical model.

precipitated during boiling is redissolved by the freshwater, potentially increasing the permeability.

Sensitivity coefficients are calculated to identify the potential contribution of each of the observation types to the inverse problem at hand. Moreover, the uncertainty of the estimated parameters is evaluated, along with the correlation coefficients, to detect dependencies among the parameters. Figure 2 shows contours of the objective function (1) in the parameter plane spanned by $\log(k_f)$ and ϕ_f Parameter combinations that lead to an equally good fit to the data—as measured by the objective function—lie on continuous surfaces in the n-dimensional parameter space. They can also be considered to have the same probability of being the true parameter set. The contour plot also reveals whether the solution to the inverse problem is unique and well posed or whether multiple minima exist or instabilities prevail. The topography of S near the minimum reveals estimation uncertainty and the correlation structure.

Figure 2a shows the objective function obtained when only steam production data are available for the inversion. While a unique global minimum can be identified, it is rather flat and elongated, indicating large estimation uncertainties and strong correlations between the two parameters. If all available data are inverted simultaneously, a well-constrained minimum results, which is accurately identified by the Levenberg-Marquardt minimization algorithm. The projection of the solution path is shown in Figure 2b. A more detailed error analysis confirms that the joint inversion of all data greatly improves the identifiability of key hydrologic and thermal properties. Adding concentration data to the inversion considerably reduces the correlation among some of the parameters, allowing for a more independent and more stable estimation of reservoir properties. Other characteristics such as fracture spacing remain difficult to determine because of their strong correlation with hydraulic and thermal parameters. For example, the amount of heat exchanged between the fluids in the fractures and the matrix can be increased by either increasing the heat conductivity or decreasing the fracture spacing. The two parameters are therefore strongly negatively correlated and cannot be determined independently.

In this section, we have discussed results from an inversion of synthetically generated production data from a geothermal field. We have demonstrated that a joint inversion of all available data may allow the identification

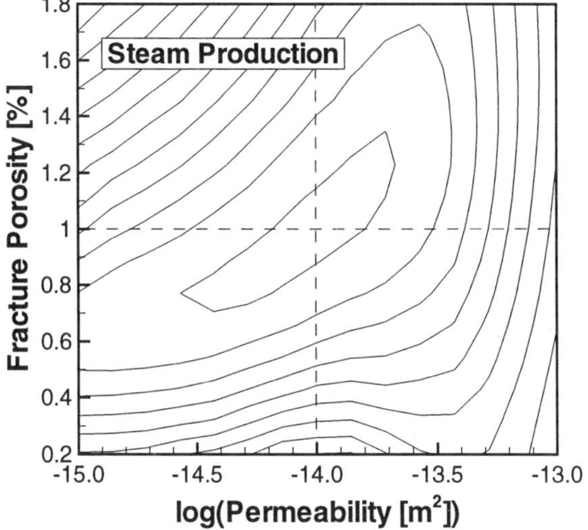

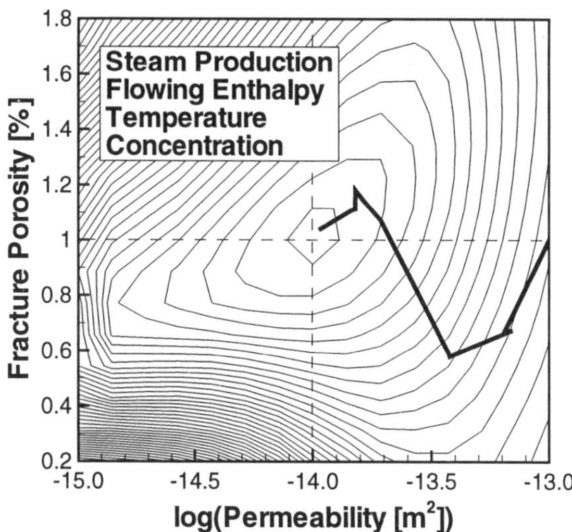

Figure 2. Contours of the objective function in the parameter plane $\log(k_f)$–ϕ_f using (a) steam production data only, and (b) all available data. The plane intersects the true parameter set indicated by dashed lines. The projection of the solution path taken by the Levenberg-Marquardt minimization algorithm is shown by the heavy line in (b).

of effective parameters describing coupled fluid and heat flow through a fractured reservoir. The true values may not be identified for parameters that are strongly correlated. Nevertheless, model predictions based on the estimated parameter set are expected to be reliable as long as the flow processes in the reservoir are not drastically changed.

As previously mentioned, the adequacy of the conceptual model is a key requirement for successful inverse modeling. Since the conceptual model is perfectly known in the synthetic example discussed above, the conclusions may be too optimistic. Changing the model structure—for example, using a single-porosity model to match data generated with a MINC model—would provide insight into the relative importance of the conceptual model and its parameters. An actual field example such as the one described in the following section also makes clear the importance of the conceptual model development, and thus reveals both the strengths and limitations of the inverse modeling approach.

4. FIELD EXAMPLE

4.1. Motivation

The Krafla geothermal field has been exploited for over 20 years [*Ármannsson et al.*, 1987]. As of now, 32 deep wells have been drilled, providing sufficient steam to operate a 60 MW_e power plant. Achieving this electrical generation rate turned out to be troublesome and time consuming. Volcanic activity occurred in the Krafla area during the period 1975–1984 [*Björnsson*, 1985]. Magmatic gases invaded part of the well field, resulting in severe scaling and corrosion problems in many wells. New areas were selected for additional drilling, but well flow rates turned out to be lower than expected. As a result of these difficulties, only one of the two 30 MW_e turbines was operated between 1978 and 1997. From 1996 to 1998, however, drilling of six new wells provided the additional steam necessary for operating both power units at full capacity.

The purpose of this field study is to jointly analyze well completion, warm-up period, and early production data in an attempt to characterize flow and saturation conditions in the vicinity of Well KJ-31, a new well in the Krafla field. Moreover, we would like to learn the capabilities and limitations of iTOUGH2 in matching data that result from propagating phase fronts in two-phase fracture-dominated systems. The estimated parameter values will be incorporated into a 3D Krafla model currently under development [*Björnsson et al.*, 1998].

4.2. The Krafla Geothermal Field

Figure 3 gives an overview of the present Krafla well field and some of the main geological features. The geology of the Krafla area is dominated by an active central volcano, consisting of a caldera and a N-S trending fissure swarm. Normal fault zones of the same direction serve as main permeability channels. Generally, fluid flow is from the NNE to the SSW. Some WNW-ESE striking faults and fractures have also been identified as possible flow paths.

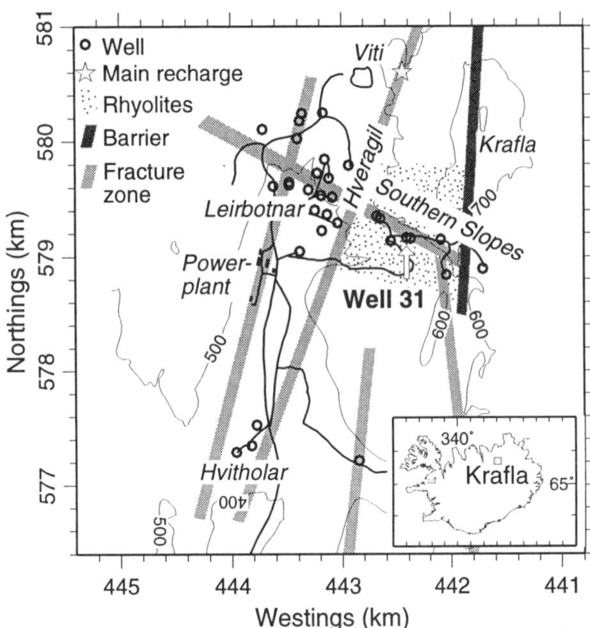

Figure 3. Well locations, subfields, and main geological features at Krafla.

The Krafla system can be divided into three subfields [*Ármannsson et al.*, 1987]. The Leirbotnar Reservoir is by far the largest one, divided into a shallow, liquid zone, 195–215°C and 200–1000 m in depth; and a deep two-phase zone below 1200 m, where temperatures and pressures follow the boiling curve (300–350°C). East of Leirbotnar are the Southern Slopes of Mt. Krafla. Here, near the Hveragil Gully the reservoir is characterized by boiling curve behavior from the surface to more than 2000 m in depth. Temperatures are, however, reversed and much cooler in the deeper and eastern part of the Southern Slopes (below 1200–1400 m). Horizontal rhyolite intrusions at the 900–1200 m depth are widespread. The main feed zones in the wells are frequently associated with this layer. The Southern Slopes are bounded to the east by a vertical low-permeability barrier, but share the Hveragil fault/upflow zone with the Leirbotnar field. Finally, there is the Hvíthólar field to the south, at the caldera rim. It acts as an outflow zone of the geothermal reservoir, with presumably the hot fluids coming from the north.

The successful 1996–1998 expansion of the Krafla power plant generated interest in additional development of the geothermal field. Simulations using the new 3D reservoir model are currently underway to characterize and predict the present and future system response to production [*Björnsson et al.*, 1998]. Inversion techniques have not been applied to this model because of the large computer time requirements. However, some subsets of the extensive Krafla database may be suitable for inverse modeling, as presented here.

4.3. Well KJ-31 Data Sources

Well KJ-31 was drilled into the Southern Slopes in October 1997 to a depth of 1450 m. Three feed zones were inferred from circulation losses during drilling: a major feature was identified at 1050 m depth, and two smaller feed zones were detected at depths of 850 and 1200 m. Transmissivity was estimated based on data from a step-rate injection test using a conventional single-phase, single-porosity isothermal reservoir simulator. Total circulation losses exceeding 50 L/s and an estimated transmissivity of $2 \times 10^{-12} m^3$ was taken as a preliminary indicator for a good producer. For comparison, well KJ-14, also drilled in the Southern Slopes, had a maximum circulation loss of 50 L/s and a step-rate injection-based transmissivity of 2.2×10^{-12} m^3 [*Bodvarsson et al.*, 1984]. This well has been a good producer since 1980, yielding a near constant flow of 10 kg/s of dry steam. While the enthalpy of well KJ-31 rose to that of dry steam after 10 days of discharge, the steam flux stabilized at 5–6 kg/s, or only half of the value expected from the well-completion data. Downhole data collected during the warm-up period showed that a reservoir pressure drawdown of 10 bars has taken place since 1981, which was when nearby well 14 was drilled.

4.4. Conceptual Models

The conceptual model developed for this study was made as simple as possible to avoid overparameterization of the associated inverse problem. A single, horizontal, 100-m-thick reservoir layer at a depth of approximately 1000 m was considered reasonable. This is because both circulation losses during drilling and the downhole temperature and pressure profiles observed during the warm-up period were dominated by this depth interval. A radial numerical grid was generated with a double-porosity inner zone and a single-porosity outer zone. This conceptual model accounts for both the fracture-dominated conditions near the well and the combined fracture-matrix character of the far-field Southern Slopes rhyolites. Linear relative permeability curves were specified, with initial estimates for steam and water residual saturations of 5% and 40%, respectively. Similar relative permeability curves have been employed in other simulation studies of Icelandic geothermal systems [*Bodvarsson et al.*, 1990; *Björnsson*, 1999]. An important feature in the warm-up data is a pressure pivot point of 68 bars observed at a 1000-m depth. By definition, the pivot point is a depth in a well where all pressure profiles collected during warm-up intersect (their slope varies with temperature). The pressure at the pivot point thus reflects a constant pressure boundary, assumed to be at the Southern Slopes rhyolites. It is also taken to be the initial reservoir pressure prior to drilling and well completion testing. To account for the cold plume created by drilling fluid invasion, a constant inflow of 20 kg/s of 40°C water is assumed for the drilling period between September 30, 1997, and the beginning of the step-rate injection test on October 8, 1997. Both estimates are based on observed circulation losses and well logs.

The data to be calibrated against are the downhole pressures observed during the cold water injection test, the three temperature data points monitored during the warm-up period, and the enthalpy measured during hot fluid discharge (Figures 4 and 5). The injection data are corrected for different sandface and wellhead flow rates. The discharge flow rate is given as a well boundary condition. Rock density, heat capacity, and thermal conductivity are fixed at 2600 kg m^{-3}, 1000 J kg^{-1} °C^{-1}, and 2.5 W m^{-1} °C^{-1}, respectively. Time zero is on September 30, 1997, at the time when the 1050-m feed zone was encountered when drilling well KJ-31.

Even this simple conceptual model requires a large number of hydrogeological, thermal, and geometrical input parameters. Moreover, initial conditions as well as a data shift representing the unknown depth of the steam-bearing rhyolites must be determined. In order to reduce the number of parameters to be estimated by inverse modeling, a set of preliminary inversions was performed using a single-porosity model. While the pressure data during the injection period were well matched, this oversimplified model was not able to reproduce the rapid increase in enthalpies during the first 10 days of production. It became obvious that storage of heat and its transfer from the matrix to the fluids in the fracture network are essential mechanisms that need to be accounted for in the model. Fracture spacing, fracture porosity, and permeability are key factors allowing concurrent matching of pressure, temperature, and enthalpy data.

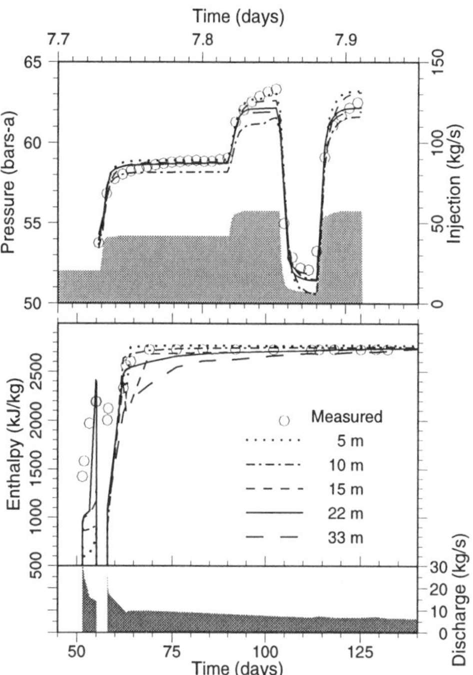

Figure 5. Well KJ-31. Measured and simulated pressures at 780-m depth and enthalpies matched with models of different inner zone radii. The shaded curves show the injection and production rates.

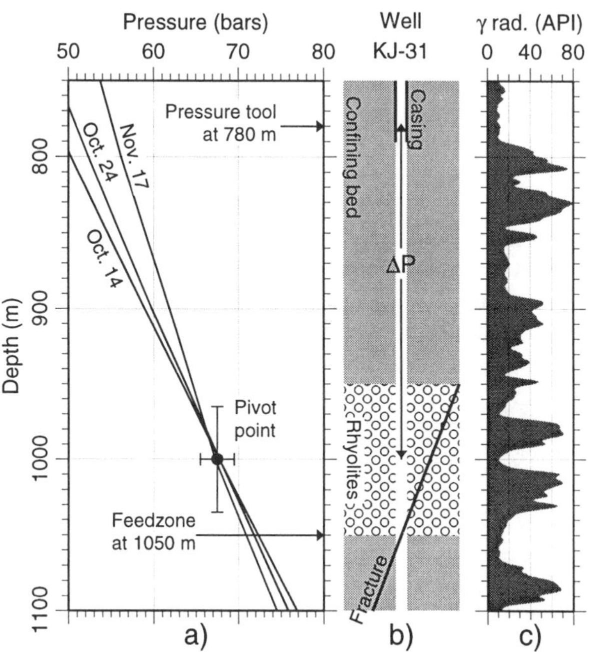

Figure 4. (a) Downhole pressure during warm-up, (b) the conceptual reservoir model and the pressure shift parameter ΔP, and (c) the natural gamma radiation log of Well KJ-31.

4.5. Inverse Modeling Results

We performed a number of inversions using different numerical grids representing inner zone radii, which varied between 5 and 33 m. The parameters estimated include the inner zone fracture permeability k_f, the fracture porosity ϕ_f, the mean fracture spacing a_f, the outer zone effective permeability k_{OZ}, porosity ϕ_{OZ}, the reservoir steam saturation prior to testing S_R, and the constant pressure difference ΔP during injection, representing the elevation difference between the pivot point in the Southern Slopes rhyolites and the pressure transducer placed at a 780-m depth. Figure 6 shows the pivot point pressure and its uncertainty. A slight calibration error in the pressure tool may shift the pivot point by several tens of meters, making

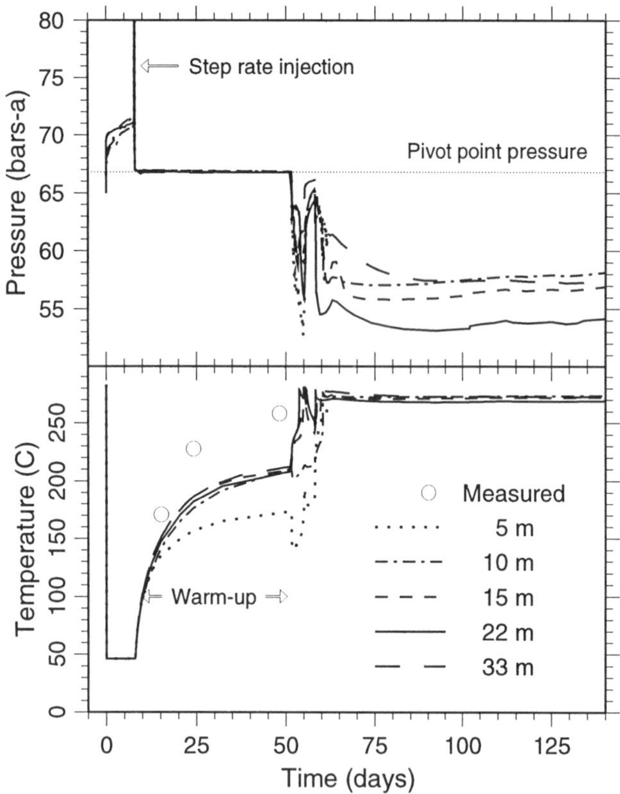

Figure 6. Well KJ-31. Simulated wellbore pressure and temperature histories at a 1000-m depth matched with models of different inner zone radii.

it necessary to treat the pressure difference ΔP between the tool depth and the elevation of the permeable reservoir as an additional parameter to be estimated. Also notice the approximately 50-m elevation difference between the pivot point and the feed zone at the 1050-m depth. This may indicate that a dipping fracture connects the overlying, horizontal rhyolites to the well. Figure 6 also shows the gamma radiation log, which indicates the SiO_2 content of the formation. Count rates exceeding 30 API units represent rhyolites and justify the 100-m reservoir thickness used in our conceptual model. The matrix porosity and permeability of the inner zone were determined to be insensitive and were thus held constant at 10% and 1 mD, respectively.

A comparison of the matches shown in Figure 4 indicates that a 22-m inner zone radius reproduces the observed data best, especially the enthalpies at very early times, where a sharp increase was predicted prior to the decline caused by a temporary shut-in of the well.

Figure 5 shows pressure and temperature histories in well KJ-31. The simulated wellbore temperature underpredicts the measured data. Note however, that the temperature data are unreliable because even minor internal wellbore flow can alter the temperature substantially. We account for this uncertainty by specifying a large measurement error to the temperature data, reducing their relative weight in the inversion. Only a mild pressure drawdown around the well is induced by production, which, assuming dry steam wellbore flow, should result in a wellhead pressure of 40 to 50 bars. This is considerably higher than the 10-bar wellhead pressure observed in the field. The discrepancy could be partly explained by non-Darcyan flow phenomena such as turbulence and sonic velocities at the sandface. However, the pressure loss could also be a result of fracture clogging near the well resulting from the calcite-rich circulation fluid used during drilling. The calcite may have precipitated during the warm-up period, reducing the permeability in the vicinity of the well and thus limiting the maximum discharge rate.

As stated in the introduction, one of the study objectives is to estimate the reservoir steam saturation S_R in the Southern Slopes rhyolites after 20 years of production and 10 bars of pressure drawdown. As liquid mobility greatly influences the enthalpy of the produced fluid, we added the residual liquid saturation S_{lr} as a parameter to be estimated. Several test inversions were performed, yielding different estimates for residual liquid saturation S_{lr} and reservoir steam saturation S_R, whereas all the other parameters converged to consistent values. Moreover, the sum of the estimates S_{lr} and S_R tended to be one in all inversions, i.e., reservoir water saturation is near irreducible saturation with the exception of the zone around the well (which is at higher saturation on account of drilling fluid invasion). This observation indicates that neither S_{lr} nor S_R can be estimated independently, as will be discussed in detail in the next section. The strong correlation between these two parameters results from the fact that shortly after the initial discharge of cooler fluids, well KJ-31 produced dry steam, consistent with all the other deep wells at Krafla. Liquid becomes immobile almost immediately after relatively small amounts of steam start to occupy the fracture network. As a result, the effective residual liquid saturation in a combined fracture-matrix system is estimated to be very high, and the sum of residual liquid saturation and reservoir steam saturation are near one. It should be noted, however, that liquid saturation in the interior of large matrix blocks could be above residual. Its small mobility does not affect the composition of fluids near and in the fractures.

Fixing the inner zone radius at 22 m, we performed an inversion with initial guesses for S_{lr} and S_R of 50% and 20%, respectively. Figure 7 presents the match obtained; Table 1 shows the initial parameter guesses, the best estimates, and their uncertainties. The permeability estimates are consistent with the results from the injection

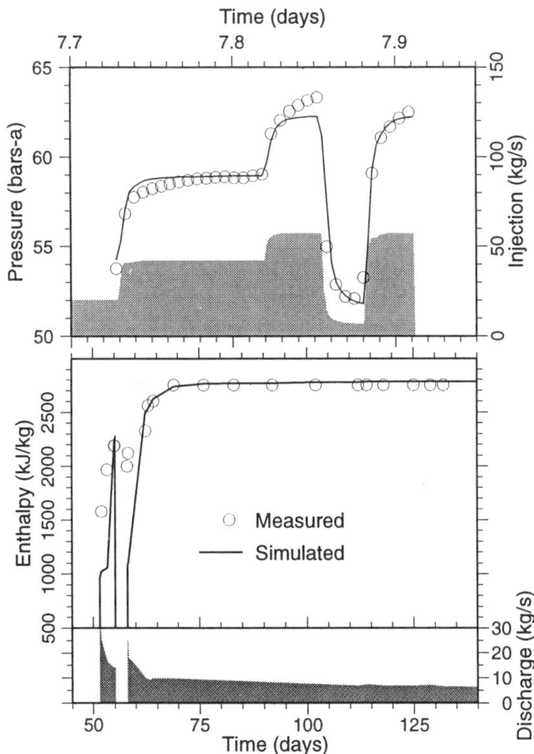

Figure 7. Well KJ-31. Measured and simulated pressures (top) and enthalpies (bottom) using the best-estimate parameter set. The shaded curves show the injection and production rates.

test. Of special interest is the estimated pressure shift of –16.8 bars, implying that the characteristic depth of the steam-bearing rhyolites is only about 170 m below the pressure gauge, at a depth of 950 m. This finding supports the conclusion drawn from the field data that a dipping fracture may be connecting the 1050-m feed zone of Well KJ-31 and the overlaying Southern Slopes rhyolites, which act here as a constant pressure boundary.

4.6. Sensitivity and Error Analysis

Here we discuss some aspects of the sensitivity and uncertainty analyses performed by iTOUGH2. The estimation covariance matrix was calculated using Equation (3). Generally strong correlations exist among the outer zone properties and the pressure shift, whereas the inner zone parameters can be estimated more independently. This difference in parameter identifiability is mainly a result of inner zone and fracture parameters being determined by both pressure and enthalpy data, whereas information about the outer zone can only be drawn from long-term enthalpy data. Data of different types have the potential to provide complementary information about different parameters, reducing correlations. If only one

Table 1. Initial Guess and Best Estimate Parameter Set

Property	Guess	Estimate	Uncertainty
$\log(k_f, \mathrm{m}^2)$	–11.5	–12.4	0.1
$\log(k_{OZ}, \mathrm{m}^2)$	–14.0	–13.6	0.3
ϕ_f, %	1.0	1.2	0.3
ϕ_{OZ}, %	10.0	6.4	4.4
S_R, %	20.0	22.2	8.9
a_f, m	10.0	12.1	4.2
ΔP, bars	–18.0	–16.8	2.2
S_{lr}, %	40.0	77.3	4.8

data type contributes to the estimation of a certain parameter, it is usually highly correlated with other parameters and exhibits a larger estimation uncertainty. The largest positive correlation occurs between the inner zone permeability and the pressure shift. It indicates that an increased permeability can be partly compensated for by applying a larger shift to the observed pressure data.

It is important to realize that the error analysis is performed at a single point in the parameter space, i.e., the best estimate parameter set. Because of the strong nonlinearities in the multiphase flow equations, the correlation structure changes if any of the input parameters is changed. Contouring the objective function throughout the parameter space can reveal a complete picture of parameter sensitivities and their correlations. As an example, we have evaluated the objective function in a two-dimensional parameter subspace spanned by the initial reservoir steam saturation S_R and the residual liquid saturation S_{lr}. The three panels of Figure 8 show the objective function when only enthalpy data are used (top), when only pressure data are used (middle), and when enthalpy, pressure, and temperature data are used simultaneously (bottom). The strong correlation between S_R and S_{lr} is evident. Parameter combinations along the diagonal and in the domain $S_R + S_{lr} > 1$ all lead to immobile liquid and thus the same good match to the dry steam enthalpy data. Since many parameter combinations lead to immobile liquid and thus an identical system behavior, the minimum of the objective function is nonunique. Similarly, a nonunique minimum is obtained when only pressure data are used. Residual liquid saturation is not sensitive at all, i.e., the pressures observed during cold water injection do not depend on the residual liquid saturation, whereas the enthalpy during production does. Independent information about the residual liquid saturation must be obtained to constrain the solution.

A high residual liquid saturation seems reasonable for a fractured system given that conductive heat exchange from the low-permeability matrix leads to a steam-saturated fracture network from which the fluid is discharged. As a result, a large fraction of the total water stored in the

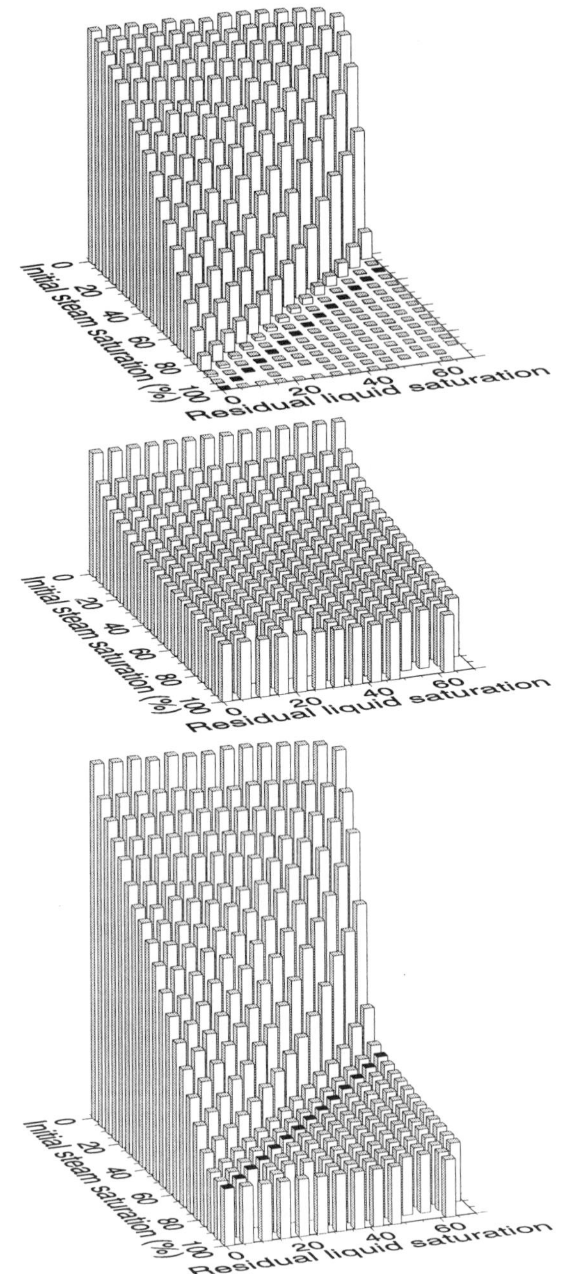

Figure 8. Histograms of the objective function in the parameter plane $S_R - S_{lr}$ using (top) enthalpy data, (middle) pressure data, and (bottom) enthalpy and pressure data. Black squares represent the diagonal $S_R + S_{lr} = 1$.

system becomes nearly immobile in a two-phase environment. This general behavior is confirmed by field observations. The Krafla wells drilled to date distinctively produce either dry steam or single-phase liquid from their feed zones, indicating that the saturation range with two-phase flow conditions is very narrow. The residual liquid saturation must be interpreted as an effective parameter describing fluid flow through the fractured rhyolites in the outer zone (Southern Slopes). As soon as steam is present in the fracture network, the flow paths for liquid water are effectively blocked, limiting liquid mobility to a small range of total saturation. The 3D model of the Krafla geothermal field currently under development supports this conclusion [*Björnsson et al.*, 1998].

5. SUMMARY AND CONCLUSIONS

We have demonstrated that inverse modeling techniques, which allow a joint analysis of multiple data sets, can provide model parameters with reduced estimation uncertainty. The key advantage of performing joint inversions lies in the fact that inherent nonuniqueness can be reduced, increasing the accuracy of the estimates and thus improving the reliability of subsequent model predictions. The synthetic inversions discussed in Section 3 show the benefit from jointly inverting complementary data sets. However, remaining correlations among certain hydrologic and thermal properties prevent a unique identification of their true values. Results from synthetic inversion may be too optimistic, for they do not reflect uncertainty introduced by errors in the conceptual model.

The theoretical study was complemented with an application of inverse modeling to actual field data in an attempt to better understand the production characteristics of well KJ-31 in the Krafla geothermal field in Iceland. A joint multiphase flow analysis of pressure and enthalpy data showed the need to accurately represent fracture flow as well as matrix-fracture heat exchange mechanisms using a double-porosity approach to a distance of 22 m from the well. The inversions confirmed the permeability estimates obtained from the analysis of the step-rate injection test. However, only a mild pressure drop was calculated during discharge, suggesting that the well output may be increased by declogging the fractures near the well, which might have been affected by calcite precipitation from drilling fluid invasion. The cold-water step-rate injection tests and their single-phase analysis are thus still appropriate as a first measure of future well output.

The study also showed that steam saturation cannot be estimated with a high degree of accuracy because it is strongly correlated to the residual liquid saturation. The inversion suggests, however, that the sum of the two estimated values is close to one (i.e., water saturation is close to irreducible), because calibration occurs against enthalpy data corresponding to production of dry steam during almost the entire discharge period. If a high residual liquid saturation is assumed (which can be justified given the fractured nature of the low-permeability reservoir), a correspondingly low initial steam saturation is determined by inverse modeling.

The estimates derived in this study represent effective parameters; they are related specifically to the scale and conceptual model used during the inversion. Simulating complex, multiphase flow phenomena in a fractured geothermal reservoir using a simplified conceptual model requires that reservoir parameters be newly defined or interpreted. As a result, the estimated fracture properties may significantly deviate from those locally measured in the field. Nevertheless, they best reflect the impact of fracture flow on the behavior of the geothermal reservoir on the scale of interest. This optimal relation of the parameters to the prediction model is a significant advantage of inverse modeling over conventional methods to determine reservoir parameters. However, it is also important to always be aware of this model dependence because it limits the applicability of the estimated parameter set to conditions that are similar to those encountered during the calibration of the model.

Estimating geothermal reservoir parameters by matching data obtained during past production and subsequently performing model predictions is a prime example of how inverse and predictive modeling are interrelated, yielding improved forecasts of reservoir performance.

Acknowledgments. This work was supported, in part, by the Assistant Secretary for Energy Efficiency and Renewable Energy, Office of Geothermal Technologies, of the U.S. Department of Energy, under Contract No. DE-AC03-76SF00098, and by the Research Division of Orkustofnun, Iceland. The authors wish to thank A. Gudmundsson at Orkustofnun for valuable information and discussions on Well KJ-31, and Landsvirkjun, the National Power Company, for its permission to publish the Krafla well data. Much information was drawn from project reports written in Icelandic; they are not cited in the reference list due to the limited audience familiar with the Icelandic language. We would like to thank M. O'Sullivan, M. Lippmann, C. Oldenburg, and G. S. Bodvarsson for careful reviews of this manuscript. Several of the illustrations were made using the public-domain GMT package [*Wessel and Smith*, 1995].

REFERENCES

Ármannsson H., Á. Gudmundsson, and B. Steingrímsson, Exploration and development of the Krafla geothermal area, *Jökull, 37*, 13–30, 1987.

Battistelli, A., C. Calore, and K. Pruess, The simulator TOUGH2/EWASG for modeling geothermal reservoirs with brines and a non-condensible gas, *Geothermics*, *26*(4), 437–464, 1997.

Björnsson, A., Dynamics of crustal rifting in NE Iceland, *Geophysics, 90*, 151–162, 1985.

Björnsson, G., Predicting future performance of a shallow steam zone in the Svartsengi geothermal field, Iceland, in *Proceedings, Twenty-Fourth Workshop on Geothermal Reservoir Engineering*, Stanford University, Stanford, Calif., 1999.

Björnsson, G., G. S. Bodvarsson, and O. Sigurdsson, *Status of the Development of the Numerical Model of the Krafla* Geothermal *System*, Orkustofnun Report GrB/GSB/OS-98-02, Reykjavik, Iceland, 1998.

Bodvarsson, G. S., S. M. Benson, O. Sigurdsson, V. Stefánsson and E. T. Elíasson, The Krafla geothermal field, Iceland, 1. Analysis of well test data, *Water Resour. Res.*, *20*(11), 1515–1530, 1984.

Bodvarsson, G. S., S. Björnsson, A. Gunnarsson, E. Gunnlaugsson, O. Sigurdsson, V. Stefansson, and B. Steingrimsson, The Nesjavellir geothermal field Iceland; 1. Field characteristics and development of a three-dimensional numerical model, *J. Geothermal Science and Technology*, *2*(3), 189–228, 1990.

Carrera, J., and S. P. Neuman, Estimation of aquifer parameters under transient and steady state conditions, 1, Maximum likelihood method incorporating prior information, *Water Resour. Res.*, *31*(4), 913–924, 1986.

Finsterle, S., *iTOUGH2 User's Guide*, LBNL-40040, Lawrence Berkeley National Laboratory, Berkeley, Calif., 1999.

Finsterle, S., and J. Najita, Robust estimation of hydrogeologic model parameters, *Water Resour. Res.*, *34*(11), 2939–2947, 1998.

Gill, P. E., W. Murray, and M. H. Wright, *Practical Optimization*, Academic Press, Inc., London, 1981.

Pruess, K., *TOUGH2—A General Purpose Numerical Simulator for Multiphase Fluid and Heat Flow*, LBL-29400, Lawrence Berkeley Laboratory, Berkeley, Calif., 1991.

Pruess, K., and T. N. Narasimhan, On fluid reserves and the production of superheated steam from fractured vapor-dominated geothermal reservoirs, *J. Geophys. Res.*, *87*(B11), 9329–9339, 1982.

Wessel, P., and W. H. F. Smith, New version of the Generic Mapping Tool released, *EOS Trans, 76*, 1995.

Stefan Finsterle, Ernest Orlando Lawrence Berkeley National Laboratory, Earth Sciences Division, Berkeley, Calif. 94720, U.S.A.; Grimur Björnsson, Okustofnun, Grensasvegur 9, 108 Reykjavik, Iceland; Karsten Pruess, Ernest Orlando Lawrence Berkeley National Laboratory, Earth Sciences Division, Berkeley, Calif. 94720, U.S.A.; and Alfredo Battistelli, Aquater SpA, San Lorenzo in Campo (PS), Italy.

Basic Research Strategies for Resolving Remediation Needs in Contaminated Fractured Media

P. M. Jardine,[1] S. C. Brooks,[1] G. V. Wilson,[2] and W. E. Sanford[3]

This paper describes a multidisciplinary endeavor to develop and employ basic research strategies for guiding remediation efforts in contaminated fractured subsurface media at the U.S. Department of Energy's Oak Ridge National Laboratory. It gives a brief overview of research activities spanning the past 15 years that have sought to provide both an improved understanding of and a predictive capability for contaminant transport processes in highly structured, heterogeneous subsurface environments that are complicated by fracture flow and matrix diffusion. Our approach involved a multiscale (cm-to-ha) experimental and numerical endeavor to investigate coupled time-dependent hydrological and geochemical processes that control contaminant migration in unsaturated and saturated soil and rock. Novel tracer techniques and experimental manipulation strategies were applied at laboratory, intermediate, and field scales to unravel the influence of multiple processes on contaminant fate and transport. The basic research strategies have significantly improved our conceptual understanding of time-dependent solute migration in fractured subsurface media. This information has proven useful in decision-making processes regarding selection of effective remedial actions and interpretation of monitoring results after remediation is complete.

1. INTRODUCTION

The disposal of low-level radioactive waste generated by the U.S. Department of Energy (DOE) during the cold war era has historically involved shallow land burial in unconfined pits and trenches. The lack of physical or chemical barriers to impede waste migration has resulted in the formation of secondary contaminant sources because radionuclides have moved into the surrounding soil and bedrock. At the Oak Ridge National Laboratory (ORNL), located in eastern Tennessee, U.S., the extent of the problem is massive. Thousands of underground disposal trenches have contributed to the spread of radioactive contaminants across tens of kilometers of landscape. Subsurface transport processes are driven by a large annual rainfall (~1400 mm/yr), where as much as 50% of the infiltrating precipitation results in groundwater and surface water recharge (10% and 40%, respectively). This condition promotes the formation of secondary contaminant sources since lateral storm flow and groundwater interception with the trenches enhance the migration of waste constituents into the surrounding subsurface environment. The subsurface medium at ORNL is comprised of fractured saprolite and shale bedrock, which is conducive to rapid preferential flow, and is coupled with

[1]Environmental Sciences Division, Oak Ridge National Laboratory, Oak Ridge, Tennessee

[2]Southern Nevada Science Center, Desert Research Institute, Las Vegas, Nevada

[3]Department of Earth Resources, Colorado State University, Fort Collins, Colorado

Dynamics of Fluids in Fractured Rock
Geophysical Monograph 122

significant matrix storage. Fractures are highly interconnected and surround low-permeability, high-porosity matrix blocks. When storm flow infiltrates the subsurface medium, large hydraulic and geochemical gradients result among the various flow regimes, causing nonequilibrium conditions during solute transport.

In these systems, the soil and bedrock matrices (secondary sources) have been exposed to migrating contaminants for many decades, and thus account for a significant inventory of the total waste. An important limitation on defining the remediation needs of the secondary sources is a lack of understanding of the transport processes that control contaminant migration. Without this knowledge base, it is impossible to assess the risk associated with the secondary source contribution to the total off-site migration of contaminants. The objectives of our research were to help resolve this dilemma by providing an improved understanding of contaminant transport processes in highly structured, heterogeneous subsurface environments that are complicated by fracture flow and matrix diffusion. Our approach involved multiscale experimental and numerical basic research to address coupled hydrological and geochemical processes controlling the fate and transport of contaminants in fractured saprolite and shale rock. Undisturbed soil columns (cm scale), in situ soil blocks (m scale), and well-instrumented field facilities (ha scale) were used in conjunction with novel tracer techniques and experimental manipulation strategies to determine the impact of coupled transport processes on the nature and extent of secondary contaminant sources.

2. LABORATORY-SCALE ASSESSMENT OF SOLUTE MIGRATION IN FRACTURED MEDIA

Undisturbed columns are used in tracer transport experiments to assess the interaction of hydrology, geochemistry, and microbiology on the fate and transport of nonreactive and reactive solutes. The primary purpose for the use of this research scale is to quantify transport mechanisms that are operative at the field scale but difficult to quantify at these larger scales. Undisturbed columns (typically 15 cm diameter × 40 cm length) are obtained from subsurface media identical to those used in the disposal of low-level radioactive waste on the Oak Ridge Reservation (ORR). This material is a weakly developed acidic inceptisol that has been weathered from interbedded shale-limestone sequences within the Conasauga formation. The limestone has weathered to massive clay lenses devoid of carbonate, and the more resistant shale has weathered to an extensively fractured saprolite. Fractures are highly interconnected, with densities in the range of 200 fractures/m [*Dreier et al.*, 1987]. Fractures surround low-permeability, high-porosity matrix blocks that have water contents ranging from 30% to 50%. Detailed information regarding the soil hydrodynamics, geochemistry, and mineralogy can be found in *Watson and Luxmoore* [1986], *Wilson and Luxmoore* [1988], *Wilson et al.* [1989, 1992, 1993, 1998], *Jardine et al.* [1988, 1993a,b, 1998], *Luxmoore et al.* [1990], *Kooner et al.* [1995], and *Reedy et al.* [1996].

At the column scale, several techniques are employed to assess nonequilibrium processes that result from the large difference in hydraulic conductivity of fractures vs. matrix blocks [*Jardine et al.*, 1998]. The techniques include (1) controlling flow path dynamics with manipulations of pore-water flux and soil-water tension; (2) isolating diffusion and slow geochemical processes with flow interruption; (3) using multiple tracers with different diffusion coefficients, and (4) using multiple tracers with grossly different sizes. For example, controlling flow-path dynamics by manipulation of the soil water content with pressure head variations is an excellent technique to assess nonequilibrium processes [*Seyfried and Rao*, 1987; *Jardine et al.*, 1993a,b].

The basic concept is to collect water and solutes from selected sets of pore classes in order to determine how each set contributes to the bulk flow and transport processes that are observed for the whole system. In heterogeneous systems, a decrease in pressure head (more negative) will cause larger pores, such as fractures, to drain and become nonconductive during solute transport. Since advective flow processes tend to dominate in large-pore regimes, a decrease in pressure head, which will restrict flow and transport to smaller pores, will limit the disparity of solute concentrations among pore groups. By minimizing the concentration gradient in the system, the extent of physical nonequilibrium is decreased. Figure 1 conveys this concept by showing the breakthrough curves of a nonreactive Br^- tracer at three different constant pressure heads in an undisturbed column of weathered fractured saprolite from the ORR. The increasing asymmetry of the breakthrough curves with increasing saturation (less negative pressure head) indicates enhanced preferential flow coupled with mass loss into the matrix. As the soil becomes increasingly unsaturated, breakthrough curve tailing becomes less significant because of a decrease in the participation of larger pores (fractures) involved in the transport process. These findings suggest that mass-transfer limitations (nonequilibrium conditions) become increasingly negligible for these unsaturated conditions because fracture flow has been eliminated.

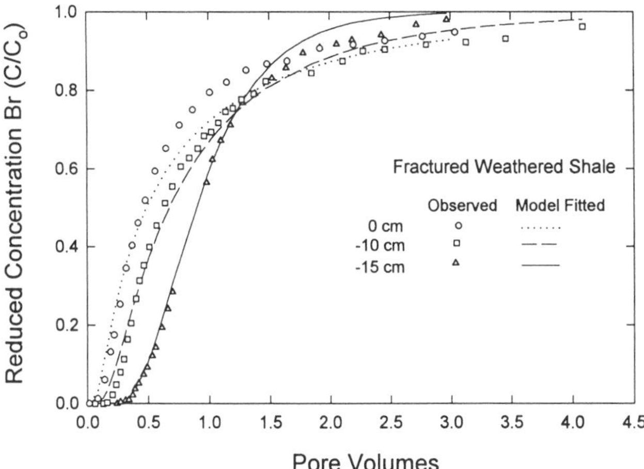

Figure 1. Breakthrough curves for a nonreactive Br^- tracer as a function of pressure head (h) in an undisturbed column of fractured weathered shale. For conditions where h = 0 cm, transport occurred under saturated flow and the entire fracture network was conductive. When h = –10 cm, the primary fracture network became nonconductive, and when h = –15 cm, the primary fractures and a portion of the secondary fractures became nonconductive. The model-fitted curves used the classical convective-dispersive model with optimization of the dispersion coefficient to the observed data [modified from *Jardine, P.M., et al.*, 1993a, with permission].

Another useful technique for isolating diffusion or slow time-dependent geochemical reactions involves flow interruption during a portion of a tracer displacement experiment [*Murali and Aylmore*, 1980; *Brusseau et al.*, 1989; *Ma and Selim*, 1994; *Hu and Brusseau*, 1995; *Reedy et al.*, 1996; *Mayes et al.*, in press]. The technique consists of inhibiting the flow process during an experiment for a designated period of time and allowing a new physical or chemical equilibrium state to be approached. When physical nonequilibrium processes are significant in a soil system, the flow-interruption method will cause an observed concentration perturbation for a conservative tracer when flow is resumed.

Interrupting flow during tracer injection will result in a decrease in tracer concentration when flow is resumed, whereas interrupting flow during tracer displacement (washout) will result in an increase in tracer concentration when flow is resumed. The concentration perturbations that are observed after flow interrupts are indicative of solute diffusion between pore regions of heterogeneous media. Conditions of preferential flow create concentration gradients between pore domains (physical nonequilibrium), resulting in diffusive mass transfer between the regions. Therefore, during injection, tracer concentrations within advection-dominated flow paths (i.e., fractures, macropores) are higher than those within the matrix. Upon flow interruption, the relative concentration decrease that is observed indicates that solute diffusion is occurring from larger, more conductive pores into the smaller pores. During tracer displacement or washout, the concentrations within the preferred flow-paths are lower than those within the matrix. Thus, solute diffusion is occurring from smaller pores into larger pores, and a concentration increase is observed when flow interruption has been imposed.

The utility of the flow interrupt method for confirming and quantifying physical nonequilibrium can be observed in Figure 2, which shows Br^- breakthrough curves at two fluxes in an undisturbed column of fractured weathered shale from the ORR. The observed concentration perturbations on the ascending and descending limbs of the breakthrough curves are the result of prolonged flow interruption and the system approaching a new state of physical equilibrium. The concentration perturbations that are induced by flow interruption are significantly more pronounced at larger fluxes. This is because the system is further removed from equilibrium at the larger fluxes since a greater concentration gradient exists between advection-dominated flow paths and the soil matrix.

These and other techniques for assessing nonequilibrium processes in structured media are discussed by *Jardine et al.* [1998]. When combined, the techniques not only improve our conceptual understanding of time-dependent contaminant migration in subsurface media, they also provide the necessary experimental constraints that are needed for the accurate numerical quantification of the nonequilibrium processes that control solute migration.

3. INTERMEDIATE-SCALE ASSESSMENT OF SOLUTE MIGRATION IN FRACTURED MEDIA

A logical progression from laboratory-scale undisturbed columns is the use of intermediate scale in situ pedons for assessing the interaction of coupled processes on the fate and transport of solutes in the fractured weathered shales (Figure 3). This research scale, unlike the column scale, encompasses more macroscopic structural features common to the field (e.g., dip of bedding planes, more continuous fracture network, convergent flow processes), yet allows for a certain degree of experimental control since the pedon can be hydrologically isolated from the surrounding environment.

Numerous undisturbed pedons have been constructed on the ORR for the purpose of monitoring contaminant fate and transport issues. The pedons are undisturbed blocks of soil 2 m × 2 m × 3 m deep, with three excavated sides

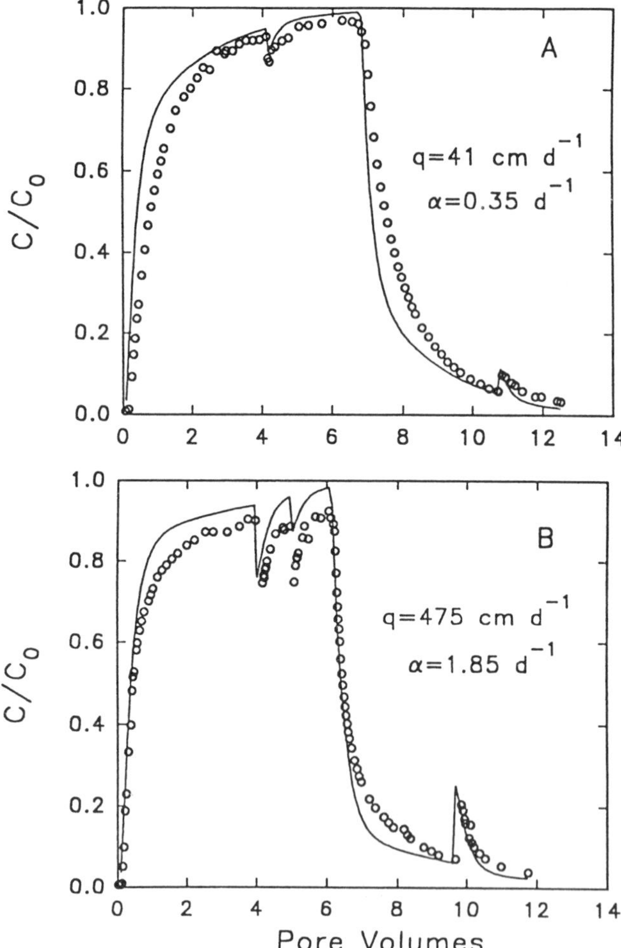

Figure 2. Breakthrough curves with flow interruption at two specific discharges for a nonreactive Br^- tracer in an undisturbed column of fractured weathered shale. Flow interruption was initiated for 7 days after (a) approximately 4 and 11 pore volumes of tracer were displaced at a flux of 41 cm d^{-1}; and (b) approximately 4, 5, and 10 pore volumes of tracer were displaced at a flux of 475 cm d^{-1}. The solid lines represent simulations using a two-region model with optimization of the mass transfer coefficient, α, that accounts for mass exchange between different pore regions [modified from *Reedy, O.C., et al.*, 1996, with permission].

refilled with compacted soil and a concrete wall with access ports placed in good contact with the front soil face. The pedon is instrumented with a variety of solution samplers designed to monitor water and solutes through various pore regimes as a function of depth. Fritted glass plate lysimeters of varying porosity and bubbling pressures are held under different tensions to derive solutions from large-, medium-, and small-pore regimes [e.g., *Jardine et al.*, 1990a]. Large-pore samplers have very coarse fritted glass plates that are tension free in order to capture rapid drainage through macropores and fractures. Sampling medium-sized pores, such as secondary fractures and mesopores, involves the use of coarse fritted glass plates maintained at tensions between 10 and 30 cm. Small-pore samples (i.e., soil matrix) have fine fritted glass plates that are maintained at a 250–500-cm tension and collect water and solutes from the micropore regime. The conductivity of the fine fritted glass is sufficiently low to minimize sampling contribution from medium and large pore regimes.

Numerous tracer experiments have been conducted using both nonreactive and reactive tracers with various steady-state infiltration rates or transient flow conditions driven by storm events [Jardine *et al.*, 1989, 1990a,b; G. V. Wilson, ORNL, personal communication,. The purpose of the experiments was to determine transport properties and mass transfer rates for the various pore regimes. An example of tracer mobility (Br^-) through the soil for two different infiltration rates can be seen in Figure 4. At an infiltration rate of 30 cm/d, Br^- is transported exclusively through medium and small pore regimes indicative of secondary fractures and the soil matrix, respectively. Flow through the large-pore regimes, indicative of primary fractures, is essentially excluded because the imposed infiltration velocity is not sufficient to accommodate their conductivity. Larger infiltration rates (e.g., 300 cm/d) did produce flow through the primary fractures, and thus three distinct breakthrough curves can be observed for tracer movement through primary and secondary fractures, as well as the soil matrix. Large infiltration rates; however, create local-scale perched water tables within the soil, allowing small-pore-regime samplers to extract a portion of the larger pore water. This is why the ascending limbs of the three breakthrough curves at 300 cm/h have diminutive differences. Nevertheless, the distinction of tracer migration through the different pore regions allows for quantitative estimates of distinct pore flow velocities, dispersion coefficients, and mass transfer rates between the pore classes at a scale one step closer to the realities of the field scale.

4. FIELD-SCALE ASSESSMENT OF SOLUTE MIGRATION IN FRACTURED MEDIA

4.1. Vadose Zone

Waste migration issues at the various DOE facilities throughout the U.S. are field-scale problems that are complicated by large-scale-media heterogeneities that cannot be replicated at the laboratory or pedon scale. For

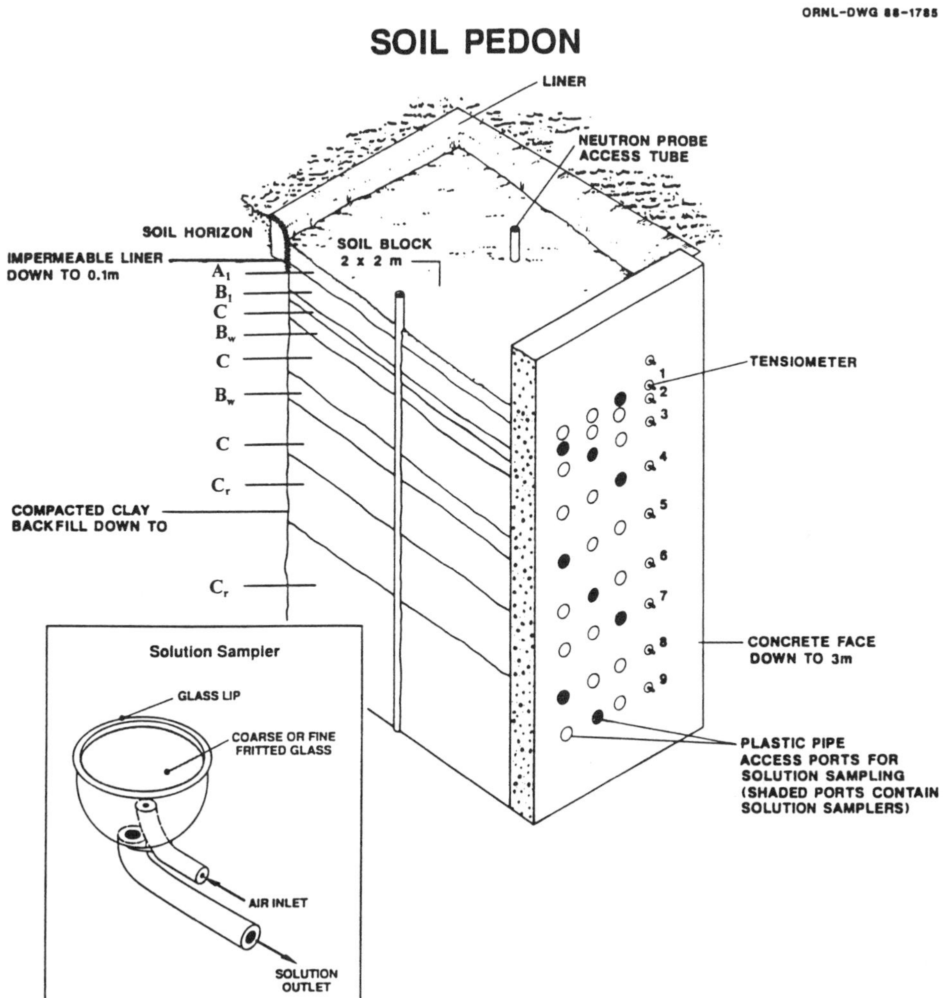

Figure 3. Cross-sectional diagram of the soil block (pedon) facility (2-m × 2-m × 3-m deep) that was used for tracer transport investigations in fractured weathered shale. The inset illustrates the fritted glass solution samplers that were installed laterally within the soil as a function of depth. Each depth interval contained coarse, medium, and fine frit samplers that were held at different tensions for the purpose of extracting pore water from primary fractures, secondary fractures, and the soil matrix, respectively [modified from *Jardine, P.M. et al.,* 1990a, with permission].

this reason, basic research efforts designed to assist with the remediation of contaminated sites must include solute fate and transport experiments at the field scale. At ORNL, field facilities have been constructed for understanding storm-driven contaminant mobility in the unsaturated zone [*Luxmoore and Abner*, 1987; *Wilson et al.*, 1993]. The facilities contain buried line sources for tracer release to simulate leakage of trench waste and are equipped with an elaborate array of water- and solute-monitoring devices.

The Melton Branch Watershed field facility is situated on the Conasauga formation and contains the same subsurface material that was used in the disposal of low-level radioactive waste at ORNL. In fact, this field facility houses the intermediate-scale pedon and is the location used for excavating undisturbed columns. The most unique feature of the facility is a set of subsurface weirs that collect lateral storm flow from a 2.5 m deep × 16 m long trench that was excavated across the outflow region of the subwatershed (Figure 5). Six massive stainless steel pans serve to intercept subsurface drainage from different portions of the landscape. The three upper pans intercept lateral drainage through the soil A- and B-horizons, and the three lower pans intercept lateral drainage that elutes through the soil C-horizon. Pan-intercepted water, as well

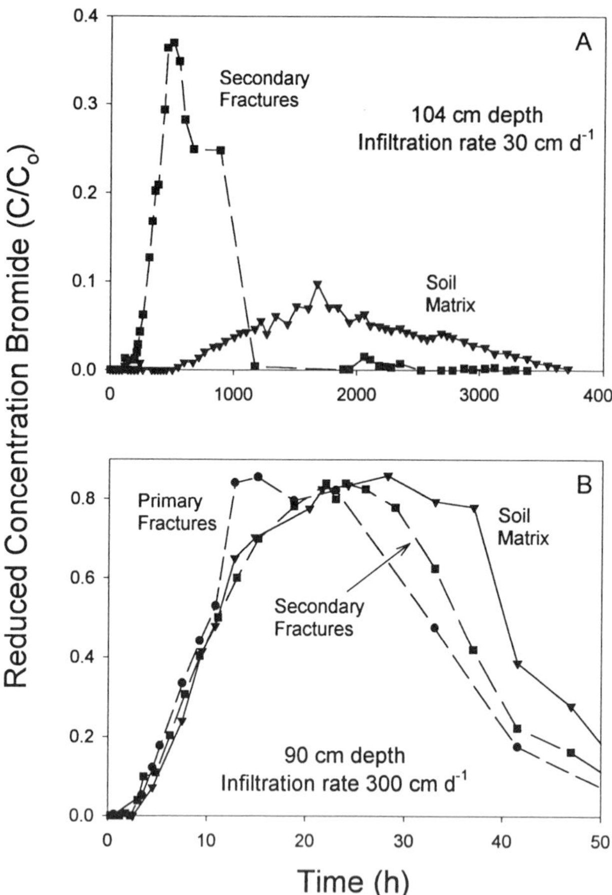

Figure 4. Breakthrough curves for a nonreactive Br^- tracer in discrete pore regimes during two infiltration experiments on the fractured weathered shale soil block. At an infiltration rate of 30 cm d^{-1} (a), only secondary fractures and the soil matrix are conductive, whereas at an infiltration rate of 300 cm d^{-1} (b), both primary and secondary fractures and the soil matrix are conductive.

as overland flow and drainage under the pans, is routed into tipping bucket rain gauges situated in two H-flumes. These flumes are equipped with ultrasonic sensors for measuring water levels. The tipping buckets and ultrasonic sensors have computer data acquisition equipment that allows for real-time monitoring of tracer fluxes during storm events. Besides the ability to capture subsurface drainage, the field facility is equipped with numerous multilevel solution samplers, tensiometers, and piezometer wells, with the latter used to assess perched water table dynamics during storm events.

Two long-term, storm-driven tracer studies were conducted at this field facility to address the rates and mechanisms of contaminant mass exchange between the various pore regions of the media. An example of how this facility was used to quantify multiregion processes under field-scale transient flow conditions can be found in *Wilson et al.* [1993], who released a Br^- tracer from the ridgetop-buried line source during a storm event and monitored its mobility through the subsurface for nearly 8 months. During the release they observed that a small portion of the total injected Br^- mass (~5%) migrated very rapidly through the hillslope via lateral storm flow with subsequent export through the weirs, which were located 70 m from the line source. Calculating the eluted Br^- mass through all portions of the weir was possible because the water total flux and Br^- concentration were measured over time. Subsequent storm events over the 8-month period resulted in the export of ~25% of the injected Br^- mass (e.g., Figure 6a).

While the actual tracer release revealed a rapid transport through the fracture network of the soil, the mass transfer into the low-permeability matrix was significant because >50% of the applied tracer was found to reside within matrix porosity of the soil, primarily at a depth of 1.0–1.5 m. This depth was also where lateral storm flow occurred through the hillslope (Figure 6b). Strong evidence indicating matrix-fracture interactions during transient storm flow can be inferred from tracer breakthrough patterns at the subsurface weirs. Storm events that followed the release of tracer resulted in delayed tracer breakthrough pulses relative to the subsurface flow hydrograph (Figure 6b). This was caused by the time-dependent mass exchange of soil matrix Br^- with newly infiltrating storm water flowing along fractures. These observations are consistent with stable isotope ($^{18}O/^{16}O$) and solute chemistry (e.g., Si) analyses, which revealed that subsurface flow was predominately new water at peak flow (i.e., minimal contribution from the soil matrix) and almost exclusively old water (i.e., significant contribution from the soil matrix) during the descending limb of the subsurface hydrograph.

These results suggest that the storm-driven export of Br^- through the weirs was the result of tracer mass transfer from the soil matrix into the fracture network with subsequent mobility through the hillslope. The field-scale endeavors provided an improved conceptual understanding of how transient hydrodynamics and media structure control the migration and storage of contaminants in the subsurface. The field-scale findings were consistent with the multiregion conceptual framework established with laboratory- and intermediate-scale observations of flow and transport through these heterogeneous systems.

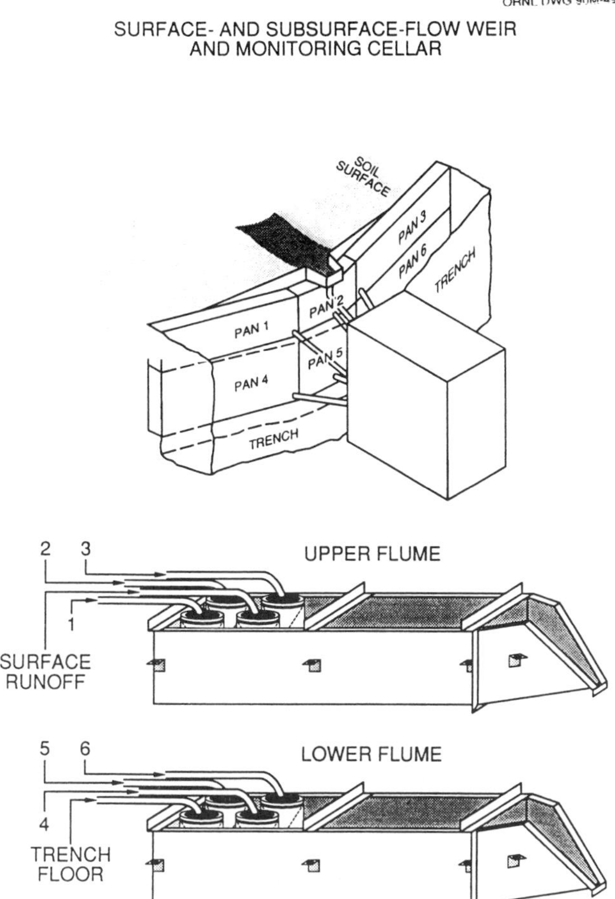

Figure 5. Schematic illustration of the subsurface weirs that intercept lateral storm flow from the subcatchment that is used for field-scale tracer injection experiments in the fractured weathered shale soil. The upper illustration depicts the six stainless steel pans pressed against the trench face (~40 m^2 area) for collecting free-lateral flow, and the lower illustration shows the two H-flumes that contain tipping bucket rain gages and ultrasonic sensors for continuous monitoring of subsurface flow [from *Wilson G.V, et al.*, 1993, with permission].

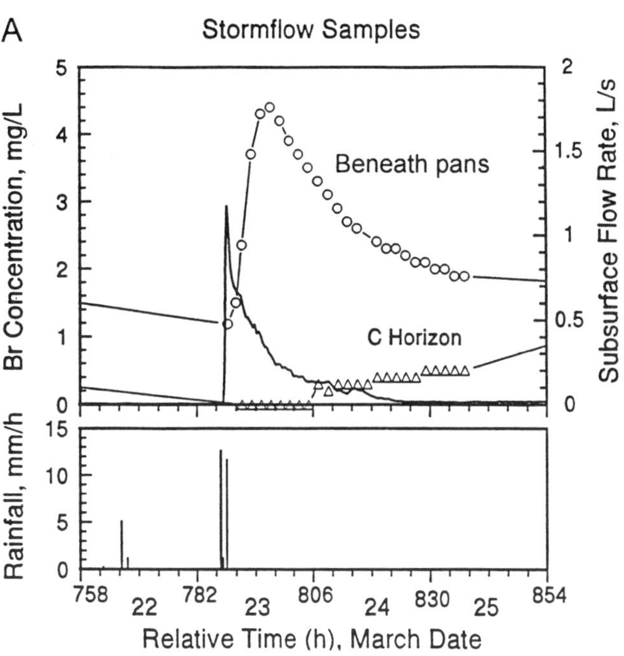

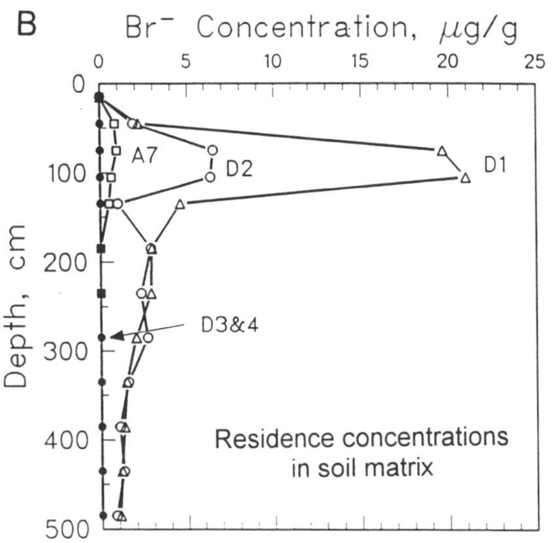

Figure 6. Storm flow and soil matrix tracer results following a release of Br$^-$ at the field-scale tracer injection facility in the fractured weathered shale soil. The upper graph (a) shows an example of a subsurface flow hydrograph (solid line) for the lower flume that resulted from a storm event, with the corresponding Br$^-$ concentrations from the C-horizon (lower pans) and from flow beneath the lower pans. The lower graph (b) shows Br$^-$ residence concentrations from the soil matrix as a function of depth for six sampling locations [see *Wilson et al.*, 1993, for site coordinates] downslope of the line source where tracer was released [modified from *Wilson, G.V., et al.*, 1993, with permission].

4.2. Saturated Zone

Lateral-moving storm flow is not the only process contributing to the migration of contaminants from primary waste trenches into the surrounding soils. At ORNL, waste trenches were often excavated to depths that approached the bedrock-saprolite interface. Seasonally fluctuating groundwater levels coupled with storm-derived perched water tables allow contaminants to easily access the underlying bedrock. Further, the very nature of the trench design allowed for rapid vertical infiltration of storm water and direct connection with the fracture network of the

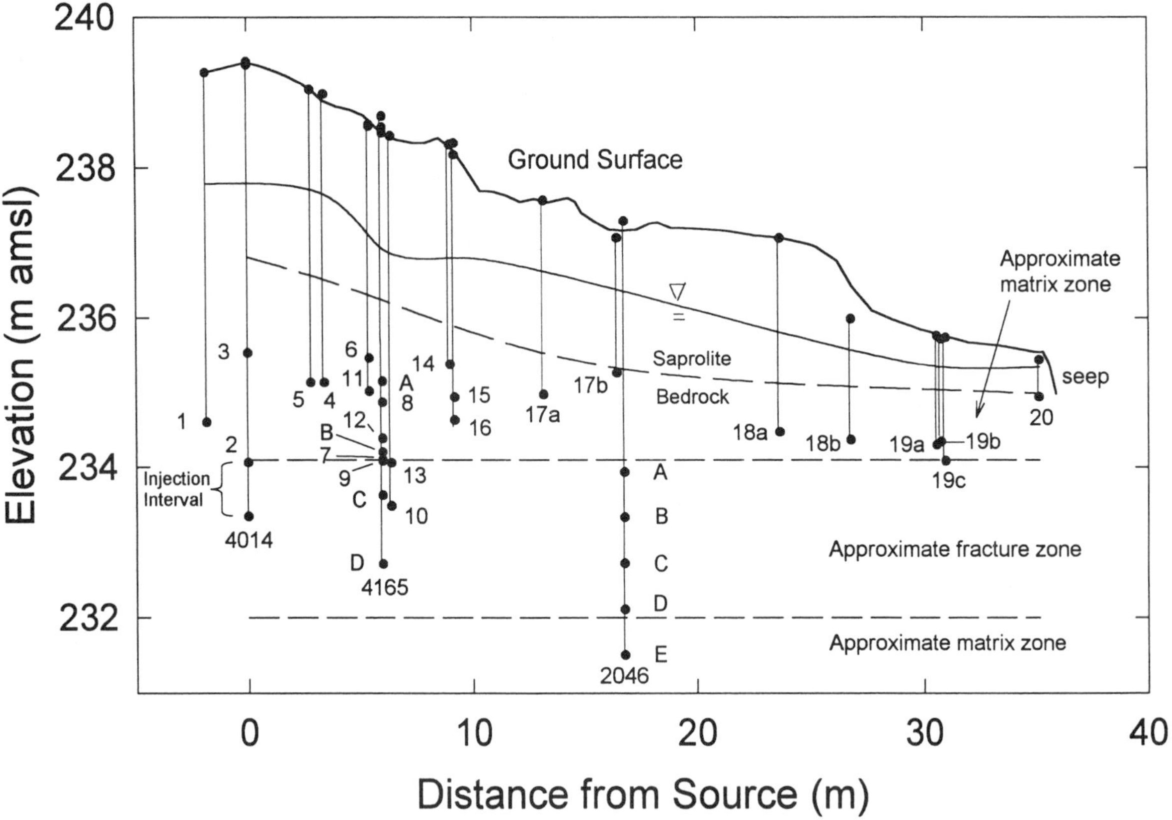

Figure 7. Cross section of the experimental field facility in the saturated zone of a contaminated fractured shale bedrock showing the location and sampling depth of all groundwater monitoring wells, with well 4014 serving as the tracer injection well. The wells form a strike parallel transect from the waste trenches to a seep that drains into a cross-cutting tributary. The approximate locations of the fracture and matrix regimes are also illustrated. Numbers designate specific wells and letters designate multilevel sampling ports within a given well [from *Jardine, P.M. et al.*, 1999, with permission].

underlying shale bedrock. In an effort to understand the importance of secondary source formation in the shale bedrock, a field facility was established within the saturated zone of a contaminated fractured shale bedrock in Waste Area Grouping 5 (WAG5) on the ORR [*Jardine et al.*, 1999]. A transect of multilevel groundwater monitoring wells was established along a geologic strike within a fast-flowing fracture regime and a slow-flowing matrix regime (Figure 7). A sophisticated computer-driven tracer injection system dispensed tracers into the fracture regime under natural gradient conditions. Two long-term steady-state natural gradient experiments, each with a duration of 1.5 to 2.0 yr, have been conducted using multiple nonreactive tracers (Br, He, Ne) and multiple reactive tracers [^{57}Co(II)EDTA, ^{51}Cr(III)EDTA, and ^{109}CdEDTA]. The multiple tracer technique takes advantage of the difference in the molecular diffusion coefficients as well as geochemical reactivity between the tracers. The experiments were designed to quantify the significance of fracture flow, matrix diffusion, and chemical reactivity on the formation and longevity of secondary contaminant sources in the shale bedrock.

Jardine et al. (1999) showed the utility of multiple dissolved gas and solute tracers for assessing physical nonequilibrium processes in the fractured shale bedrock. The natural gradient injection of Br^-, He, and Ne was initiated for 6 months, followed by a 12-month washout. Spatial and temporal monitoring of the tracers was performed in the matrix and fracture regimes of the bedrock using the multilevel sampling wells instrumented down gradient from the injection source. Tracers migrated preferentially along the strike, and their concentrations in the fracture regime quickly reached a consistent steady-state value. Thus, observed differences in tracer breakthrough into the matrix were a function of their molecular diffusion coefficients. The breakthrough of the three tracers 6.0 m from the source and 0.8 m into the matrix relative to the fracture is shown in Figure 8. The movement of He and Ne into and from the matrix was more rapid than Br^-, which is consistent with the larger molecular diffusion coefficients of the dissolved gases relative to Br^-. These results supported the notion that matrix diffusion contributed to the overall physical nonequilibrium process that controls solute transport in shale bedrock.

At greater distances from the source, the contribution of matrix interactions was still prominent and tracer breakthrough profiles remained suggestive of a diffusion mechanism, although at first glance this may not have been apparent (Figures 8b and 8c). At 13 m from the source and 0.6 m into the matrix, the three tracers broke though nearly simultaneously, with the concentration of the gas tracers eventually surpassing Br^- (Figure 8b). This was followed by tracer washout after the input pulse was terminated at 180 days. At 23.0 m from the source and 0.1 m into the matrix, the movement of Br^- into the matrix was actually more rapid than that of the noble gas tracers, which was exactly the opposite of what was observed 6.0 m from the source. This apparent paradox was caused by the preferential loss of gas tracers to the rock matrix closer to the source. Thus, Br^- remained within the advective flow field (fracture regime) for a longer time period, allowing it to be transported greater distances. Having been transported further down gradient, Br^- experienced the first opportunity to begin diffusing into the matrix at greater distances from the source. Eventually, He and Ne arrived at the same locations and also begin to diffuse into the matrix, lagging behind Br^- (Figures 8b, c). Because of the larger diffusion coefficients of the noble gases, the movement of He and Ne into the matrix was more rapid and, if given time, the He and Ne breakthrough curves would eventually cross over and surpass the Br^- breakthrough curves (see Figure 8b as an example).

The results of this study show that secondary contaminant sources form within the bedrock matrix, and that the importance of these sources increases with continued contaminant discharge through the bedrock fracture network. This is particularly important for reactive contaminant such as radionuclides, where matrix diffusion can enhance solute retardation by many orders of magnitude.

Geochemical processes also influence the behavior of radioactive contaminants in subsurface environments. Low-level radioactive waste generated at DOE facilities was typically composed of inorganic fission byproducts mixed with various chelating agents such as ethylenediaminetetraacetic acid (EDTA) [*Ayres*, 1971;

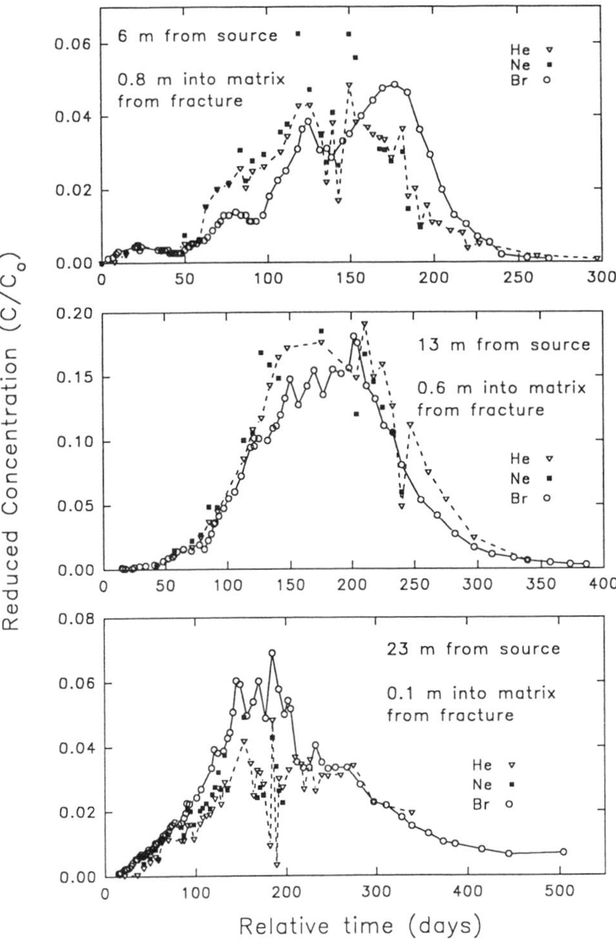

Figure 8. Observed breakthrough of three nonreactive tracers (He, Ne, and Br) within the matrix regime of the shale bedrock, at 6 m (well 8), 13 m (well 17a), and 23 m (well 18a) from the source (a, b, c, respectively). The three tracers have different free water molecular diffusion coefficients with He > Ne > Br [modified from *Jardine, P.M. et al.*, 1999, with permission].

Toste and Lechner-Fish, 1989; *Riley and Zachara*, 1992]. The presence of the complexing agent alters the geochemical behavior of the disposed contaminant in subsurface media.

Thus, a field experiment was conducted to assess the significance of fracture flow, matrix diffusion, and chemical reactivity on the migration of chelated radionuclides from waste trenches into the underlying bedrock. The natural gradient injection of ^{57}Co(II)EDTA, ^{51}Cr(III)EDTA, and ^{109}CdEDTA was initiated at the WAG5 field facility for 6 months, followed by a 12-month washout. The chelated radionuclides were significantly retarded relative to a nonreactive Br^- tracer (Figure 9), which is contrary to previous conceptual models that suggest chelated radionuclides move conservatively

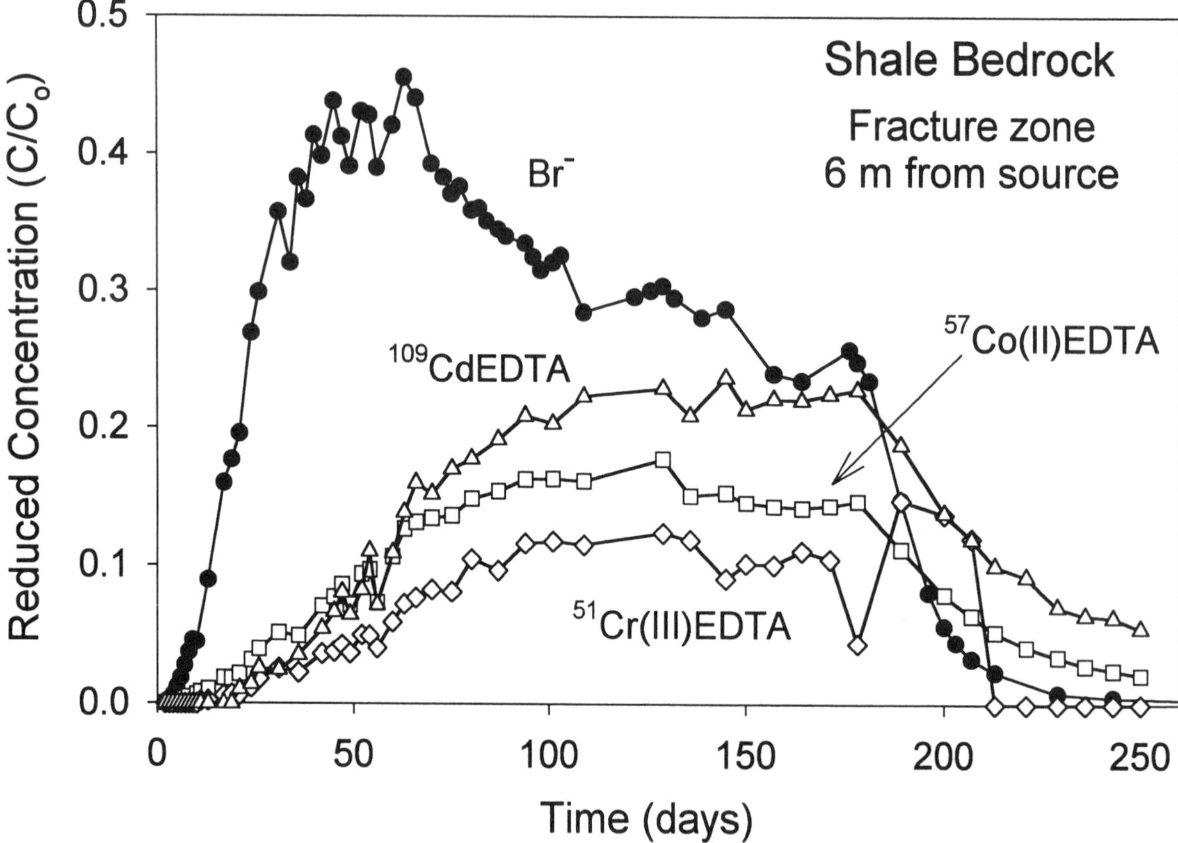

Figure 9. Observed breakthrough of three reactive chelated radionuclides [^{57}Co(II)EDTA, ^{109}CdEDTA, and ^{51}Cr(III)EDTA] and the nonreactive tracer Br^- within the fracture regime of the shale bedrock 6 m from the source.

through shale bedrock. Retardation mechanisms along fractures involved primarily geochemical processes, whereas retardation mechanisms within the matrix involved primarily physical processes (i.e., diffusion). Along fracture pathways, primary mineral coatings effectively dissociated the radionuclide-chelate complexes with the subsequent formation of Fe(III)EDTA$^-$. The dissociation reaction promoted the rapid stabilization of the free radionuclide as carbonate precipitates. By stabilizing the radionuclide in situ, the off-site migration of the contaminants was significantly reduced. In many instances, the increased retardation provided sufficient time for short-lived radionuclides such as ^{90}Sr and ^{60}Co (~29 and 5 yr, respectively) to decay, thereby eliminating the need for remediation.

REMEDIATION NEEDS AND BASIC RESEARCH

There are several ways in which our basic research strategies improve remedial options at contaminated sites.

- Basic research provides an improved conceptual understanding of the geochemical and hydrological processes controlling contaminant migration from secondary sources. This research can unravel complex, time-dependent coupled processes such as preferential flow, matrix diffusion, sorption, and redox transformations in heterogeneous media. Basic research strategies designed around novel tracer techniques and experimental manipulations not only improve our conceptual understanding of time-dependent contaminant migration in subsurface media, they also provide the necessary experimental constraints needed for the accurate numerical quantification of the coupled nonequilibrium processes.
- Basic research provides a direct measure of contaminant migration rates along fracture flow paths and into the soil and rock matrix. Such information is critical to contaminant fate and transport modeling and risk assessment modeling. Too often risk assessment models

treat soil and bedrock as inert media or assume that they are in equilibrium with migrating contaminants. Failure to consider the time-dependent significance of secondary contaminant sources will greatly overpredict the off-site contribution of contaminants from the primary trench sources and thus provide an inaccurate assessment of pending risk. Basic research furnishes the necessary information needed for more appropriate risk modeling strategies that can potentially translate into multimillion dollar savings by eliminating the need for certain remedial efforts.

- Basic research endeavors provide information that is necessary for improving our decision-making strategies regarding the selection of effective remedial actions and the interpretation of monitoring results after remediation is complete. Simple distinctions in how fracture networks operate to disseminate contaminants versus matrix storage mechanisms are basic research issues that can drastically influence the type and cost of remedial effort that is mandated. Basic research endeavors provide a more thorough understanding of coupled transport mechanisms so that remediation concepts are not limited to a purely empirical approach. This will allow engineers to develop remediation strategies targeted at specific problems and with a higher probability of success.

Acknowledgments. This research was supported by the Environmental Technology Partnership (ETP) program of the Office of Biological and Environmental Research, U.S. Department of Energy. The authors would like to thank Mr. Paul Bayer, contract officer for DOE's ETP program, for financially supporting this research. Oak Ridge National Laboratory is managed by Lockheed Martin Energy Research Corporation for the U.S. Department of Energy, under contract DE-AC05-96OR22464. Environmental Sciences Division, ORNL Publication No. 4951.

REFERENCES

Ayers, J. A., *Equipment Decontamination with Special Attention to Solid Waste Treatment*, Surv. Rep. BNWL-B-90, Battelle Northwest Laboratories, Richland, Wash., 1971.

Brusseau, M. L., P. S. C. Rao, R. E. Jessup, and J. M. Davidson, Flow interruption: A method for investigating sorption nonequilibrium, *J. Contamin. Hydrol.*, *4*, 223–240, 1989.

Dreier, R. B., D. K. Solomon, and C. M. Beaudoin, Fracture characterization in the unsaturated zone of a shallow land burial facility, in *Flow and Transport Through Unsaturated Rock, Geophysical Monograph 42*, edited by D. D. Evans and T. J. Nicholson, pp. 51–59, 1987.

Hu, Q., and M. L. Brusseau. Effect of solute size on transport in structured porous media, *Water Resourc. Res.*, *31*, 1637–1646, 1995.

Jardine, P. M., G. V. Wilson, and R. J. Luxmoore, Modeling the transport of inorganic ions through undisturbed soil columns from two contrasting watersheds, *Soil Sci. Soc. Am. J.*, *52*, 1252–1259, 1988.

Jardine, P. M., G. V. Wilson, R. J. Luxmoore, and J. F. McCarthy, Transport of inorganic and natural organic tracers through an isolated pedon in a forested watershed, *Soil Sci. Soc. Am. J.*, *53*, 317–323, 1989.

Jardine, P. M., G. V. Wilson, and R. J. Luxmoore, Unsaturated solute transport through a forest soil during rain storm events, *Geoderma*, *46*, 103–118, 1990a.

Jardine, P. M., G. V. Wilson, J. F. McCarthy, R. J. Luxmoore, D. L. Taylor, and L. W. Zelazny, Hydrogeochemical processes controlling the transport of dissolved organic carbon through a forested hillslope, *J. Contamin. Hydrol.*, *6*, 3–19, 1990b.

Jardine, P. M., G. K. Jacobs, and G. V. Wilson, Unsaturated transport processes in undisturbed heterogeneous porous media: I. Inorganic contaminants, *Soil Sci. Soc. Am. J.*, *57*, 945–953, 1993a.

Jardine, P. M., G. K. Jacobs, and J. D. O'Dell, Unsaturated transport processes in undisturbed heterogeneous porous media: II. Co-contaminants, *Soil Sci. Soc. Am. J.*, *57*, 954–962, 1993b.

Jardine, P. M., R. O'Brien, Wilson, G. V., and J. P. Gwo, Experimental techniques for confirming and quantifying physical nonequilibrium processes in soils, in *Physical Nonequilibrium in Soils: Modeling and Application*, edited by H. M. Selim and L. Ma, pp. 243–271, Ann Arbor Press, Inc., Chelsea, Mich., 1998.

Jardine, P. M., W. E. Sanford, J. P. Gwo, O. C. Reedy, D. S. Hicks, R. J. Riggs, and W. B. Bailey, Quantifying diffusive mass transfer in fractured shale bedrock, *Water Resour. Res.*, *35*, 2015–2030, 1999.

Kooner, Z. S., P. M. Jardine, S. Feldman, Competitive surface complexation reactions of SO_4^{2-} and natural organic carbon on soil, *J. Environ. Qual.*, *24*, 656–662, 1995.

Luxmoore, R. J., and C. H. Abner, *Field Facilities for Subsurface Transport Research*, DOE/ER-0329. U. S. Department of Energy, Washington, D. C., 1987.

Luxmoore, R. J., P. M. Jardine, G. V. Wilson, J. R. Jones, and L. W. Zelazny, Physical and chemical controls of preferred path flow through a forested hillslope, *Geoderma*, *46*, 139–154, 1990.

Ma, L., and H. M. Selim, Predicting the transport of atrazine in soils: Second-order and multireaction approaches, *Water Resourc. Res.*, *30*, 3489–3498, 1994.

Mayes, M. A., P. M. Jardine, I. L. Larsen, and S. E. Fendorf, Multispecies contaminant transport in undisturbed columns of weathered fractured shale, *J. Contamin. Hydrol.*, in press.

Murali, V., and L. A. G. Aylmore, No-flow equilibration and adsorption dynamics during ionic transport in soils. *Nature*. *283*, 467–469, 1980.

Reedy, O. C., P. M. Jardine, G. V. Wilson, and H. M. Selim, Quantifying the diffusive mass transfer of nonreactive solutes in columns of fractured saprolite using flow interruption, *Soil Sci. Soc. Am. J.*, *60*, 1376–1384, 1996.

Riley, R. G., and J. M. Zachara, *Chemical Contaminants on DOE Lands and Selection of Contaminant Mixtures for Subsurface Science Research*, DOE/ER-0547T, U. S. Gov. Printing Office, Washington, D.C., 1992.

Seyfried, M. S., and P. S. C. Rao, Solute transport in undisturbed columns of an aggregated tropical soil: Preferential flow effects, *Soil Sci. Soc. Am. J.*, *51*, 1434–1444, 1987.

Toste, A. P., and T. J. Lechner-Fish, Organic digenesis in commercial, low-level nuclear wastes. *Radioact. Waste Manag. Nucl. Fuel Cycle*, *12*, 291–301, 1989.

Watson, K. W., and R. J. Luxmoore, Estimating macroporosity in a forest watershed by use of a tension infiltrometer, *Soil Sci. Soc. Am. J.*, *50*, 578–582, 1986.

Wilson, G. V., and R. J. Luxmoore, Infiltration, macroporosity, and mesoporosity distributions on two forested watersheds, *Soil Sci. Soc. Am. J.*, *52*, 329–335, 1988.

Wilson, G. V., J. M. Alfonsi, and P. M. Jardine, Spatial variability of saturated hydraulic conductivity of the subsoil of two forested watersheds, *Soil Sci. Soc. Am. J.*, *53*, 679–685, 1989.

Wilson, G. V., P. M. Jardine, and J. P. Gwo, Modeling the hydraulic properties of a multiregion soil, *Soil Sci. Soc. Am. J.*, *56*, 1731–1737, 1992.

Wilson, G. V., P. M. Jardine, J. D. O'Dell, and M. Collineau, Field-scale transport from a buried line source in variable saturated soil. *J. Hydrol.*, *145*, 83–109, 1993.

Wilson, G. V., J. P. Gwo, P. M. Jardine, and R. J. Luxmoore, Hydraulic and physical nonequilibrium effects on multi-region flow and transport, in *Physical Nonequilibrium in Soils: Modeling and Application*, edited by H. M. Selim and L. Ma, pp. 37–61, Ann Arbor Press, Inc., Chelsea, Mich., 1998.

P. M. Jardine and S.C. Brooks, Environmental Sciences Division, Oak Ridge National Laboratory, P.O. Box 2008, MS-6038, Oak Ridge, TN 37831-6038

G.V. Wilson, Southern Nevada Science Center, Desert Research Institute, Las Vegas, NV

W.E. Sanford, Department of Earth Resources, Colorado State University, Fort Collins, CO